P9-DTX-563

THE CHEMICAL FORMULARY

THE CHEMICAL FORMULARY

A CONDENSED COLLECTION OF VALUABLE, TIMELY, PRACTICAL FORMULAE FOR MAKING THOUSANDS OF PRODUCTS IN ALL FIELDS OF INDUSTRY

VOLUME I

Editor-in-Chief
H. BENNETT

THE CHEMICAL FORMULARY CO.
950 THIRD AVE.
BROOKLYN, N. Y., U. S. A.
1933–1934

Printed in U. S. A.

PRESS OF
BRAUNWORTH & CO., INC.
BOOK MANUFACTURERS
BROOKLYN, NEW YORK

NRA
MEMBER
U.S
WE DO OUR PART

PREFACE

Chemistry as taught in our schools and colleges is confined principally to synthesis, analysis and engineering—and properly so. It is part of the proper foundation for the education of the chemist.

Many a chemist on entering an industry soon finds that the bulk of the products manufactured by his concern are not synthetic or definite chemical compounds but are mixtures, blends or highly complex compounds of which he knows little or nothing. The literature, in this field, if any, may be meagre, scattered or antiquated.

Even chemists, with years of experience in one or more industries, spend considerable time and effort in acquainting themselves on entering a new field. Consulting chemists, similarly, have problems brought to them from industries foreign to them. A definite need has existed for an up-to-date compilation of formulae for chemical compounding and treatment. Since the fields to be covered are many and varied an editorial board was formed, composed of chemists and engineers in many industries.

Many publications, laboratories, manufacturing companies and individuals have been drawn upon to obtain the latest and best information. It is felt that the formulae given in this volume will save chemists and allied workers much time and effort.

Manufacturers and sellers of chemicals will find in these formulae new uses for their products. Non-chemical executives, professional men and others, who may be interested, will gain from this volume a "speaking acquaintance" with products which they may be using, trying, or with which they are in contact.

It often happens that two individuals using the same ingredients in the same formula get different results. This may be the result of slight deviations or unfamiliarity with the intricacies of a new technique. Accordingly, repeated experiments may be necessary to get the best results. Although many of the formulae given are being used commercially many have been taken from patent specifications and the literature. Since these sources are often subject to various errors and omissions, due regard must be given to this factor. Wherever possible it is advisable to consult with other chemists or technical workers regarding commercial production. This will save time and money and avoid "head-aches."

It is seldom that any formula will give exactly the results which one requires. Formulae are useful as starting points from which to work out one's own ideas. Formulae very often give us ideas which may help us in our specific problems. In a compilation of this kind errors of omission, commission and printing may occur. We shall be glad of any constructive criticism in this, our first attempt.

To the layman, it is suggested that he arrange for the services of a chemist or technical worker familiar with the specific field in which he is interested. Although this involves an expense it will insure quicker and better formulation without wastage of time and materials.

H. Bennett
1933

CONTENTS

ADHESIVES

* Bakelite, Adhesive

Shellac	16
Pontianak Gum	8
Titanium Dioxide	2
Asbestine	22
Alcohol	22

Box Toe Adhesive

1. Rosin	1300 gm.
2. Shellac	200 gm.
3. Alcohol	1520 c.c.
4. Whiting	4000 gm.

Dissolve one and two in three and then work in four until uniform.

* Adhesive, Casein

Casein	50
Magnesium Oxide	3
Soda Ash	1
Water	500
Yeast	1
Sod. Borate	2

* Adhesive, Casein

Casein	75
Slaked Lime	15
Kieselguhr	5
Sodium Fluoride	7

Mix the above with water for use.

* Adhesive, Waterproof Casein

Soda Ash	15
Sod. Acetate	6
Sod. Fluoride	5
Slaked Lime	45
Casein	140
Basic Copper Carbonate	3

* Casein, Liquid Adhesive

Casein	100
Urea	90
Water	100

Mix together and allow to stand until dispersed and free from lumps; this may be hastened by heating to 140–160° F. with stirring. Addition of more water causes thickening or precipitation. This adhesive is fairly water-proof and not alkaline like most commercial casein adhesives.

A glue base which when mixed with water and alkalies produces a smooth glue (having a much longer "life" than a similar material made without casein and seed meal) is formed of dried blood albumin 90, dried milk casein 15–30, a seed meal high in protein material such as peanut, cotton-seed or soy-bean meal 30–45 and finely comminuted cellulose about 100 parts.

* Adhesive, Moisture-Proof Cellophane

Ethylene Glycolmonoethylether	20–80%
Lactic Acid	80–20%

The above is mixed with an equal volume of water.

Adhesive for Celluloid to Celluloid

Gum Camphor	1 part
Alcohol	4 parts

Dissolve the camphor in the alcohol and then add 1 part Shellac. Warm to dissolve. This cement is applied warm, and the parts united must not be disturbed until the cement is hard.

* Celluloid and Rubber, Adhesive for

Ethyl Crotonate is a solvent for both pyroxylin and rubber. Both surfaces are cleaned and each is wet with Ethyl Crotonate and pressed together.

Cellulose Ester Adhesives

1.

15 parts nitrocotton.
6 parts camphor.
79 parts acetone.
10 parts filler.

2.

20 parts scrap film.
60 parts ethyl acetate.
20 parts ethyl alcohol.
10 parts aluminium powder.

All formulae preceded by an asterisk (*) are covered by patents.

3.

16 parts nitrocotton.
10 parts ethyl acetanilide.
74 parts acetone.
15 parts starch.

4.

12 parts cellulose acetate.
8 parts tricresyl phosphate.
20 parts methyl alcohol.
30 parts ethyl acetate.
30 parts methyl acetate.
25 parts filler.

5.

12 parts nitrocotton.
4 parts ethyl acetanilide.
2 parts castor oil.
20 parts ethyl acetate.
20 parts methyl acetate.
17 parts methyl alcohol.
25 parts starch.

6.

14 parts scrap film.
2 parts ethyl acetanilide.
2 parts castor oil.
3 parts tricresyl phosphate.
13 parts ethyl acetate.
13 parts methyl acetate.
6 parts methyl alcohol.
21 parts acetone.
6 parts benzine.
20 parts starch.

7.

10 parts nitrocotton.
4 parts camphor.
2 parts tricresyl phosphate.
50 parts acetone.
20 parts butyl acetate.
14 parts filler.

Cellulose Ester, Adhesives for

SOLUTION I. 12.8 kg. alcohol-damp nitrocotton in 12.0 kg. methyl acetate.

SOLUTION II. 25.0 kg. first crepe latex dissolved in 72 kg. benzole,

or

SOLUTION I. 7.5 kg. celluloid in 7.5 kg. acetone, 7.5 kg. methyl acetate and 15 kg. ethyl acetate.

SOLUTION II. 17.5 kg. first crepe latex in 72 kg. benzole.

Solutions I and II are mixed and thinned to a suitable viscosity.

An even simpler method consists in dissolving celluloid in acetone or a similar solvent, the layer remaining after evaporation being highly adhesive, soft and elastic, and is not attacked by cold or warm water.

Resins may also be added to the straightforward celluloid solution, in which case a solvent must be selected which dissolves both celluloid and resin. Acetone is probably the most suitable in this connection. Cellulose acetate may be used in place of celluloid, and suitable resins are copal and rosin, the following mixture, for example, giving excellent results:

Celluloid	20 g.
Acetone	60 g.
Copal	5 g.
Rosin	5 g.
White lead	1 g.
Acetone	20 g.

In addition there may be added a small proportion of nitro-benzole, which improves the odour.

An adhesive layer of exceptional properties is obtained by using de-camphored celluloid and castor oil, which are thoroughly incorporated in ethyl acetate or acetone. This adhesive is stable for an unlimited period and may be made up on the following lines:

Castor oil	85 g.
Nitrocellulose	15 kg.

Solvents as required.

Pigments, fillers and odoriferous substances may also be incorporated.

---

* Cigarette Tip Adhesive

Nitrocellulose 1.5, rosin 13, tricresylphosphate 13.4, triacetin 1.6, ethylene glycol 2.5, glycol monoformate 5 and lithopone 45 kg.

---

* Decalcomania Adhesive

Glue	13.5
Water	28
Butanol	7.3
Toluol	9.7
Alcohol	26.8
Turkey Red Oil	14.7

---

* Glass to Cement Adhesive

Glass is coated on one side with a mixt. of Na silicate and a metal oxide, *e.g.*, ZnO, which readily forms a silicate. The glass is then heated gradually to 100°, preferably by heating it to 40°, maintaining that temp. for a few hrs., raising the temp. to 100°, and maintaining that temp. for 1–2 hrs. The solid coating thus obtained does not corrode the glass and adheres well to cement or gypsum.

Glass to Brass Adhesive

Caustic Soda	1
Rosin	3
Plaster of Paris	3
Water	5

Boil together until all lumps disappear and cool before using. This sets in about 20 min.

Quicksetting Insulating Adhesive

Modified Alkyd Resin	11–20
Pyroxylin Solution (35%)	64–73
Tricresyl Phosfate	4–8
Lacquer Thinner	11–21

This is useful on coils and radio parts.

---

* Latex Adhesives

Latex	100
Invert Sugar	2
Sod. Thiosulfate	3
Pot. Bichromate	2

Latex	100
Albumen	2
Carraghean Moss	5
Formaldehyde	3
Sod. Bichromate	3

Adhesive, Leather Shoe

Good leather adhesives for use by the shoe industry are based on nitrocellulose, rubber or casein. A nitrocellulose compn. contains nitrocellulose 200, AcOAm 15, AmOH 15, rosin 10, camphor 5, Venice turpentine 15 and linseed oil 20 parts. Soft leather is made to adhere especially well by the following compn.: gutta percha 85, rosin 25, asphalt 26, petroleum 130 and CS_2 300–350 parts.

---

* Adhesive, Mask

Beeswax	52
Lanolin	24
Venice Turpentine	15
Castor Oil	9

---

* Mica Adhesive

Gilsonite	2
Rubber	1
Benzol	3

Allow to swell and mix properly. This may be thinned down with benzol or naphtha.

---

* Adhesive, Heat Plastic

The following is used for special adhesive binding tapes.

Balata	10 lb.
Rosin	5 lb.
Mineral Oil	3 oz.

* Synthetic Resin Adhesive

For the prepn. of a transparent weatherproof resin to be used in the manuf. of reflectors for uniting glass particles to a support, a mixt. of PhOH 40, CH_2O soln. 100, and NaOH 1.2 parts is warmed to about 62° for about 2.5 hrs., treated with 3.3 parts of lactic acid, and warmed again to about 60° until the mixt. becomes sirupy.

---

* Adhesive for Silk or Rubber

Latex	5–15
Rubber	20–52
Rosin	1.5–5
Copal	3–10
Filler	6–25
Color	3–18
Gum Arabic	6–25

---

* "Masking" Adhesive Tape

For making a paper base or backing, the paper is first submitted so a preliminary treatment by a saturating solution involving a glue base. The saturating solution of the following materials in proportions:

36 pounds of dry glue
72 pounds of water

108 pounds (approx. 16° Twaddell) glue solution. Complete swelling is permitted, assisted by warming.

To this is then added 108 pounds of yellow glycerine.

108 pounds (approx. 16° Twaddell) glue solution
108 pounds pale yellow glycerine

216 pounds glue-glycerine water solution.

To this is added 216 pounds of water.

216 pounds glue-glycerine water solution
2 pounds Formaldehyde

434 pounds

The paper above described is preferably continuously submerged and passed through a bath of the saturating solution as above prepared and then passed through pressure rolls to squeeze off the excess and then dried by heating. It will be observed that just complete saturation is preferred as this step is closely related to the success or failure of the treatment.

The rubber resin compounds in their solvents may be spread upon the paper backing directly, utilizing a knife spreader to uniformly and equally distribute this material upon the base or backing. The solvent may thereafter be removed by evaporation, preferably without recovering the solvent and leaving the rubber mixture upon the paper backing.

RUBBER RESIN

2 lb. of plantation rubber.
5 lb. of Mexican or wild rubber, high in natural resin content.
1 lb. of zinc oxide pigment.

The ingredients above enumerated are compounded on a rubber mixing roll and then cut to the desired consistency in a rubber solvent, based upon the necessary viscosity for spreading this material. Ordinarily, the solvent is calculated by the number of pounds of solid compound in one gallon of solvent such as, for instance, 8 pounds of solid or compounded material and 1 gallon of benzol, which is commonly referred to as an 8 pound cut. The variations in proportions of solvent added will depend upon the desired thickness of adhesive coating required in the residuum.

It will be understood that the examples above given are for purposes of getting the requisite adhesiveness in temperate climates. An increase in resinous material or wild rubber may be made for material to be used in colder climates and in warmer climates the resin component may be reduced.

The resinous component may also be varied in its reactions to solvents by choice of the resinous material. Thus, for purposes of removal of the adhesive from some body to which it may be applied, it may be made soluble to various organic solvents, either benzol, gasoline, acetone or alcohol. Thus, where it is desirable to make a surgeon's tape, which is soluble in alcohol, an alcohol soluble resin is added in the examples above cited. Such resin may be Burgundy pitch. This will permit alcohol to be used in removing a piece of adhesive tape from any surface, such as from the skin of a patient, by merely soaking the backing of the tape in alcohol. The rubber, in any event, merely acts as a vehicle for the resin and the character of the adhesive in its reaction to solvents will be dependent upon the character of the resin incorporated with the rubber.

* Adhesive Tape

Plasticized Crepe Rubber	10
Cumarone Resin	2
Zinc Oxide	½

Compound to a plastic mass on a rubber mill and then ''cut'' to desired body with benzol or naphtha. Before applying to cloth or paper the latter should have the reverse treated with a flexible glue (formalized) to prevent soaking thru and sticking. Then apply above mixture with a knife spreader evenly and allow to dry.

Tape, Coating for Adhesive

Heat 10 parts Castor Oil to 270° C. and to it add slowly with stirring 6 parts shellac and 1 part rosin. The addition of glycerol or glycols produces more sticky products.

Tape, Masking

As above except that 9 parts of shellac is used.

Adhesive, Tin

1. Pot. Hydroxide	5
2. Water	56
3. Rosin	50
4. Rezinel No. 2	5

Heat one and two to boiling and while stirring vigorously run in three and four which have been melted together: stir until uniform and add

Water	50

---

* Adhesive, Vegetable

(a) Soya bean flour	30
Alum	1
Water	70
Caustic Soda 18%	13
{ Slaked Lime	4
{ Water	20
(b) Cottonseed flour	30
Alum	1
Water	70
Caustic Soda 18%	13
{ Slaked Lime	4
{ Water	20
(c) Low grade wheat flour	30
Portland Cement	10
Water	30
Caustic Soda 18%	30

Warm to 80° C. and add

Sod. Silicate	15

Adhesive Wax

Rosin	100
Paraffin Wax	10
Thin Mineral Oil	88

Sticky Wax

Rosin	100
Talc	16
Lanolin	60
Paraffin	8
Sapon. Wax	2

Melt together and while stirring rapidly add slowly a boiling caustic soda solution (10° Bé.) stir until uniform.

Adhesive for Wigs

Damar	20
Rosin	20
Beeswax	40
Venice Turpentine	20

Heat to 90° C. and stir until uniform; cast in sticks.

---

* Adhesive, Wood

Casein	23
Hydrated Lime	4
Pot. Chlorate	1.5
Sod. Fluoride	1.5
Soda Ash	1.9
Borax	4
Alum	1
Titanic Anhydride	1

This will not combine with tannins and oils present in wood.

---

* Adhesive, Wood Veneer

Pot. Dichromate	0.25–2.0%
Slaked Lime	1–1.5%
Tapioca or Cassava flour	balance

Mix with water for use.

Waterproof Adhesive for Wood

Light gasoline	0.5 gal.
Acetone	0.5 gal.
Soft cumarone	10.0 lb.
Pine oil	0.5 lb.
Tricresyl phosphate	0.25 lb.

Adhesive for Fixing Wood, Tin, etc. to Celluloid

Shellac	2 gm.
Spirits of Camphor	3 gm.
Alcohol	4 cc.

Warm together until dissolved.

---

* Adhesive, Water-Resistant

Peanut Meal	100 lb.
Hydrated Lime	16 lb.
Soda Ash	10 lb.
Sod. Silicate	30 lb.
Copper Sulfate	2 lb.
Water	400 lb.

The above is used in glueing wood.

Casein, "Dissolving"

3 to 4 parts of cold water by weight to each pound of dry Casein.

1 ounce 26° Ammonia to each pound of dry Casein.

If a heavy solution is required, use 3 to 1 proportion; if a thinner solution is desirable, use 4 to 1.

Pour water into a jacketed kettle, or a kettle heated by live steam, and add the Casein. Stir well to break down any lumps that may form and then add Ammonia. Stir the mixture after adding the Ammonia and immediately turn on the heat. Heat, while stirring, to about 160° F. Turn off the heat when this temperature is reached and continue to stir, preferably with a mechanical agitator, until the Casein is completely dissolved, which will take about half an hour.

If the temperature exceeds 160° during the heating, it is not serious, although it is advisable not to apply excessive heat, particularly when Ammonia is used, as there is a tendency to somewhat weaken the Casein and to darken it in color.

When the Casein is completely dissolved it may be diluted, if necessary, by the addition of warm water and used, as dissolved, either hot or cold, in the same manner as ordinary glue.

10 pounds Casein
1½ lb. Powdered Borax
40 to 60 pounds *cold* water

Stir cold for about 15 minutes or until the Casein commences to swell.

Then heat in a jacketed kettle for 40 to 60 minutes at a temperature not higher than 160° F. stirring constantly.

Ammonia 26° can be used in place of Borax.

To make a thin solution we suggest using equal parts of Ammonia 26° and Trisodium Phosphate or Borax and Trisodium Phosphate.

If a preservative is desired you can use about 2% of Benzoate of Soda or ¼ of 1% Carbolic Acid.

Note—do not dissolve Casein in a copper kettle as this tends to discolor the Casein particularly if the solvent is Ammonia.

Cork and Wood Flour, Binders for

A.	Rosin	100
	Dibutyl Phthallate	35
	Sod. Silicate	4
	Nitrocellulose	4
	Castor Oil	2

B.	Ester Gum	50
	Cumarone Resin	50
	Linseed Oil bodied	10
	Dibutyl tartrate	35
C.	Urea formaldehyde resin	50
	Cumarone Resin	25
	Rosin	25
	Tricresyl phosphate	20
	Dibutyl phthallate	20

---

* Binder, Oilproof and Waterproof

Lead Oxide	59.6
Iron Powder	2.0
Portland Cement	18.2
Slaked Lime	5.8
Glycerol	8.2
Water	6.2

This sets quickly and is resistant to shock.

---

* Cement, Acid Proof

SiO_2 powder ground from grains of good strength and of sufficient purity not to be attacked by acids is mixed with a hardening agent, *e.g.*, $NaBF_4$ or Na_2SiF_6, and a solution of Na silicate in which the SiO_2/Na_2O ratio is $\nless$ 3·5: 1. Graphite may be added as a lubricant.

Aquarium Cement

To 10 lbs. of glazier's putty add 1 lb. dry litharge, 1 lb. dry red lead, and 1 gill of asphaltum. Mix to a stiff consistency with boiled linseed oil and add sufficient lampblack to give a slate color.

Another well-known formula consists of 10 parts by bulk of plaster of Paris, 10 of fine sand, 10 of litharge, 1 part of powdered rosin, and sufficient boiled linseed oil to make a stiff putty. A third formula is as follows: Red lead 3 parts, litharge 7, fine sand 10, powdered rosin 1 part, and spar varnish sufficient to make a stiff cement.

In each case add the linseed oil or varnish little by little and mix the ingredients very thoroughly. If the putty should become too soft, merely add more of the dry materials as the exact proportions are not especially important.

Adhesive Cement (For Fine Furniture)

Casein (fine ground)	12 lb.
Lime (powdered, unslaked)	13 lb.
Mica (dry, ground)	15 lb.
Barium sulphate (barytes)	60 lb.

Mix all ingredients. Keep in dry container. To use, mix with water until pasty. Hardens in about 24 hours.

Bituminous Cement

A mixt. of asphalt 660, asbestos fiber 60, pulverized soapstone 100, infusorial earth 80 and sand 300 lb. is used with a softening agent formed from a mixt. of (*a*) asphalt 48.8 lb., "turpentine substitute" 2.9 gal. and coal oil 10.7 gal. and (*b*) paraffin wax 73.1, Al stearate 3.6 lb. and coal oil 9.7 gal. The product is suitable for sealing pipes and conduits.

---

* Cement, Dental

Zinc Oxyphosfate	3
Tin C. P.	1

---

* Glass Cement

Chlorinated Naphthalene	10 lb.
Ester Gum	10 lb.
Rubber Latex	1 lb.

Melt together and apply hot. This may also be used for uniting metals, wood, etc.

Cement, Safety Glass

Pyroxylin	12
Camphor	2
Ethyl Methyl Ketone	30
Alcohol	15
Gum Benzoin	2
Triacetin	5
Benzyl Alcohol	2.5

Waterproof Glass and Metal Cement

This cement will also stand fairly high temperatures.

Cement and litharge in equal parts are thoroughly mixed. Then glycerine in an amount equal in volume to half the volume of the mixed powder is added and the whole thoroughly mixed with a spatula. This cement will set under water.

To repair leaks in pipes, fill the hole with the cement and bind it in place with cheese cloth. Then daub a quantity of the cement on the cloth and wrap the whole tightly together with iron wire.

The powders may be mixed ready for use, but the glycerine must only be added as needed.

---

* Iron Cement

Ground birch charcoal	4
Am. Chloride	0.5
Rye flour	1
Soda Ash	1
Sod. Nitrate	0.25

Iron Cement (for castings)

Iron filings	128 lb.
Plaster of Paris	20 lb.
Whiting	8 lb.
Gum Arabic	8 lb.
Carbon Black	1 lb.
Portland Cement	4 lb.

Make into a paste with water directly before using.

Linoleum Cement

Clay	20
Red Oxide of Iron	20
Dextrin	60

The powders are thoroughly mixed and made into a paste of desired consistency with water.

* Cement, Linoleum and Tile

1. Sicapon	82
2. Paraffin	9
3. Glycerin	9

Heat 1 and 3 to 80° C. and add 2 which has been melted to it slowly with vigorous stirring until emulsified.

* Cement, Linoleum Backing

A satd. felt base is coated with an alkyd resin paint which may be made by heating together at 150–180° ethylene glycol 35, diethylene glycol 3.5–7.5, glycerol 8–13, phthalic anhydride 105 and drying oil acids 30 parts and dissolving the product in ethylene glycol monoethyl ether or similar low-boiling solvents.

* Cement, Oxychloride

Fused Calcium Chloride	111
Magnesium Sulfate	120
Calcined Magnesite	250
Casein	10
Water	204

* Cement, Pipe Thread

Graphite	55%
Sicapon	45%

To the above paste may be worked in amounts of oils or water to obtain a lubricating effect. This paste hardens under heat to seal joints effectively.

* Lute, Chlorine Resistant

Burnt Clay (finely ground)	65
Caustic Soda 40° Bé.	35

* Pipe Cement, Plastic

Asphalt	24–28
Tung Oil	4–8
Asbestos Fibres	40–48
Petroleum Naphtha	20–24

Rubber Cement (For Use on Leather Shoes)

Naphtha (62° Bé.)	9.8 pt.
Carbon Tetrachloride	5.4 lb.
Crepe Rubber	0.33 lb.

Makes 1 gal. cement on allowing to swell.

Raincoat Rubber Cement

Hevea Rubber	50
Litharge	20
Whiting	26.5
Rosin	2
Sulfur	1.5

Grind and mix thoroughly.

* Cement, Rubber to Metal

Crepe Rubber	68 lb.
Benzol	6800 lb.
Bromine	40–80 lb.

Allow to stand and shake slowly until uniform.

Cement, Rubber Tire

Crude Rubber	2 lb.
Rosin	2 lb.
Carbon Bisulfide	1 gal.

* Cement for Repairing Shoes

Portland Cement	10 lb.
Rubber	10 lb.
Rosin	1.5 lb.
Shellac	2 lb.
Sole Leather Scrap	6 oz.
Benzine	1 qt.

* Pipe Joint Compound

The following compound contains no poisonous materials and may be prepared in dry form which will keep indefinitely. It forms perfectly leak proof joints when applied as a paste by mixing with water.

Flour	66
Portland Cement	25
Talc	3
Lamp Black	3
Sea Sand	3

* Filler, Expansion Joint

Cottonseed Oil	16
Rosin	4
Diglycol Oleate	1

Melt the above and add

Sulfur	8
Silica Dust	4

Continue heating and stirring until thick.

Floor Crack Filler

Plaster of Paris	32 lb.
Silica	200 lb.
Dextrine Yellow	33 lb.

Make into a stiff dough with water before use.

Glue

Blood albumin (90 per cent solubility)	100 parts
Water	170 parts
Ammonium hydroxide (specific gravity 0.90)	4 parts
Hydrated lime	3 parts
Water	10 parts

Pour the larger amount of water over the blood albumin and allow the mixture to stand undisturbed for an hour or two. Stir the soaked albumin until it is in solution and then add the ammonia while the mixture is being stirred slowly. Slow stirring is necessary to prevent foamy glue. Combine the smaller amount of water and the hydrated lime to form milk of lime. Add the milk of lime, and continue to agitate the mixture for a few minutes. Care should be exercised in the use of the lime, inasmuch as a small excess will cause the mixture to thicken and become a jellylike mass. The glue should be of moderate consistency when mixed and should remain suitable for use for several hours. The exact proportions of albumin and water may be varied as required to produce a glue of greater or less consistency or to suit an albumin of different solubility from that specified.

Blood albumin (90 per cent solubility)	100	parts
Water	140–200	parts
Ammonium hydroxide (specific gravity, 0.90)	5½	parts
Paraformaldehyde	15	parts

The blood albumin is covered with the water and the mixture is allowed to stand for an hour or two, then stirred slowly. The ammonium hydroxide is next added with more stirring. Then the paraformaldehyde is sifted in, and the mixture is stirred constantly at a fairly high speed. Paraformaldehyde should not be poured in so rapidly as to form lumps nor so slowly that the mixture will thicken and coagulate before the required amount has been added.

The mixture thickens considerably and usually reaches a consistency where stirring is difficult or impossible. However, the thickened mass will become fluid again in a short time at ordinary temperatures and will return to a good working consistency in about an hour. It will remain in this condition for 6 or 8 hours, but when the liquid finally sets and dries, as in a glue joint, it forms a hard and insoluble film.

This glue may be used in either hot or cold presses. When cold pressed, however, it has only moderate strength, and for that reason is not to be depended upon in aircraft construction where maximum strength is required. If hot pressed, it is high in strength and very water resistant.

Flexible Bindery Glue

Glue No. 1	123 lb.
Glycerin	90 lb.
Water	123 lb.
Betanaphthol	½ lb.
Terpineol	½ lb.

Extra Flexible Bindery Glue

Glue No. 2	75 lb.
Glue No. 3	75 lb.
Glycerin	64 lb.
Water	144 lb.
Betanaphthol	½ lb.
Terpineol	½ lb.

Flexible Machine Bindery Glue

Glue No. 3	150 lb.
Glycerin	105 lb.
Water	135 lb.
Betanaphthol	½ lb.
Terpineol	½ lb.

Regular Bindery Glue

Glue No. 1	175 lb.
Glycerin	10 lb.
Water	175 lb.
Betanaphthol	½ lb.
Terpineol	½ lb.

Tablet Binding Glue

Glue No. 1	120 lb.
Glycerin	113 lb.
Water	113 lb.
Zinc Oxide	5 lb.
Betanaphthol	½ lb.
Terpineol	½ lb.

Glue for Cellophane

17½ parts gum arabic
52½ parts water
30 parts Glycerine
.05 part Formaldehyde

* Casein Glue, Water Resistant

Casein	39
Peanut Meal	39
Hydrated Lime	11
Trisodium Phosfate	4
Sodium Fluoride	7
Water	225–235

Add the solids slowly to the water while stirring with an efficient stirrer. Continue until smooth and free from lumps. Allow to stand 20–30 minutes and add a mixture of aldol ½, water 1, and 50% copper nitrate 2. Stir for 5 minutes when it is ready for use.

"Dissolving" Glue

In a 100 gal. steam jacketted kettle place 80 gal. water; to this add 100 lbs. glue and soak for one hour; turn on steam and cook glue until dissolved; do not heat above 110° F.

Cabinet Makers' Glue

Glue No. 2	87½ lb.
Glue No. 3	87½ lb.
Glycerin	10 lb.
Water	175 lb.
Betanaphthol	½ lb.
Terpineol	½ lb.

In the above formulae the glue is soaked in cold water over night and heated not over 150° F. and stirred until dissolved. The other ingredients are then dissolved in it and the liquid is then poured into molds where it sets on cooling.

Case Making Machine Glue

Glue No. 2	175 lb.
Glycerin	10 lb.
Water	175 lb.
Betanaphthol	½ lb.
Terpineol	½ lb.

Furniture Glue

Animal glue	10 lb.
Powd. white lead	2½ lb.
Powdered Chalk	5 oz.
Sodium salicylate	2 lb.
Wood alcohol	1¼ pt.
Water	19 lb.

Dissolve sodium salicylate in water. Dissolve animal glue in the same water. Mix lead and chalk; add to the sodium salicylate water and glue. Add wood alcohol to the batch.

Leather Sole Glue

Rosin	60
Crepe Rubber	40
Varnish	20

Digest on a water-bath and when dissolved cool and add

Naphtha	30

* Liquid Glue

Sod. Chlorate	3.5 lb.

is stirred into a hot solution of

Glue	10 lb.
Water	13 lb.

Liquid Glue

Borax	2
Water Boiling	4
Pot. Carbonate	1

Stir the above into

Glue	16
Water boiling	32

Masking Tape Glue

Glue (compatible with Calcium Chloride)	50
Water	35

Allow to swell for 3–4 hrs. Heat to 160° F. and then add while stirring

Glycol Bori-Borate	8
Glycerin	7

followed by

Calcium Chloride	0.35
Water	2

Care must be taken that temperature is kept below 170° C.

* Glue, Vegetable

Soya Bean Flour	100 lb.
Slaked Lime	10–20 lb.
Caustic Soda	5 lb.
Water	100 or more lb.

Mucilage

To 30 gallons water add 75 lbs. gum arabic, clean sorts. Mix at 160° F. until completely dissolved; add 6 lbs. carbolic acid, 1 lb. oil of cloves. Strain and fill.

Envelope Mucilage

Gum arabic	1 part
Starch	1 part
Sugar	4 parts

Water, sufficient to produce the desired consistency.

The gum arabic is first dissolved in water, the sugar added, then the starch, breaking up all lumps, after which the mixture is boiled for a few minutes in order to dissolve the starch, after which it is thinned down to the desired consistency with more water.

Mounting Paste

White dextrine	1	lb.
Gum arabic	1	oz.
Water	1½	pt.
Acetic acid	1	oz.
Oil of wintergreen	20	drops
Oil of cinnamon	20	drops
Salicylic acid	20	gr.

The dextrine and the gum, which should be pulverized, are dissolved in the water, and then the salicylic acid added and dissolved. This liquid is heated with the dextrine, and when the whole has become pasty, which should require a quarter of an hour, the acetic acid is added, stirring in slowly. The heating is continued, taking care not to boil the mass. The paste will soon become pearly, and should then be removed from the fire and the perfume oils added while it is cooling. It should be stirred thoroughly while the oils are being added.

Mucilage, Stick Form

Powdered white glue	10 parts
Powdered gum arabic	2 parts
Sugar	5 parts
Water	Sufficient

Mix the glue and gum, then stir in enough cold water to make the solution the consistency of thick syrup. Soak overnight to allow the glue and gum to absorb the water, then add enough water to again bring it to a thick syrup. Pour into a flat bottom pan that has been chilled and cut into sticks of desired size when almost solid. If poured into molds the molds should first be well greased and then chilled by setting upon cracked ice.

The addition of 0.1% of Moldex in the water used will prevent spoilage.

Decorators' Paste

	Pints by Weight
Rye meal	4
Fine whiting	2
Casein	1
Powdered alum	½

Mix the above ingredients together and rub to a fine powder. Use 2 lb. of the mixture to one quart of water either hot or cold.

Flour Paste

Wheat Flour	4 lb.
Cold Water	2 qt.
Boiling Water	3 gal.

Make smooth paste of flour and cold water and then pour into boiling water. Stir and boil for 5 minutes.

Library Paste

1.

Tragacanth (powdered)	20
White Dextrin	10
Wheat Flour	60
Glycerin	10
Cold Water	40
Salicylic Acid	3
Boiling Water	400

Mix the tragacanth with 160 parts of boiling water, stir well and set aside. Mix the dextrin and the flour with the cold water, stir well and add to the tragacanth mucilage. Pour into the resulting mixture the rest of the boiling water stirring constantly. Rub up the salicylic acid with the glycerin, add to the mucilage and boil for 5 to 6 minutes with constant stirring.

2.

White Dextrin	6	oz.
Diluted Acetic Acid	1	oz.
Oil of Clove	10	drops
Glycerin	1	oz.
Water to make	16	fl.oz.

Make a paste of the dextrin with 6 ounces of cold water, add 8 ounces of boiling water, boil 5 minutes with constant stirring, then add enough hot water to make 14 fluid ounces. Let cool then add the other ingredients.

Library Paste

Flour	16
Gum Acacia	12
Gum Tragacanth	3
Salicylic Acid	0.5
Clovel	0.6
Water	160

Use part of water to make a paste of flour. Heat another part of water with gums until dispersed. Mix these two well and other ingredients and bring to a boil while stirring.

Library Paste—Photo Mounting

White Potato Dextrine	15	lb.
Water	15	lb.
Glycerin	1	lb. 15 oz.
Formaldehyde	2½	oz.
Oil of Sassafras	2½	oz.

White Library Paste

To 30 gallons cold water, add 75 lbs. white potato dextrine. Break up all lumps then heat to 180° F. Add 6 lbs. carbolic acid and 1 lb. oil of wintergreen. Strain and fill into jars while hot. Allow to stand for three days.

Starch Paste

The strength of starch paste is increased by the addition of a small quantity of ammonium hydroxide. Paste may be rendered flexible by the addition of glycerine. The following formula produces satisfactory results:

100 grams Water
4 grams Ammonium Hydroxide
8 grams Paste Starch
1 gram Glycerine

Starch Paste

Corn or Tapioca Starch	4
Cold Water	8
Boiling Water	64

Make a paste of starch and cold water then pour into boiling water and stir until translucent.

Putty

Whiting	800
Corn Oil	20
Crude Cottonseed Oil	10
Thin Mineral Oil	69
Sod Oil	3

Elastic Putty

Turpentine	5
Rosin Oil	8
Linseed Oil and drier	5.5
Barytes	8.5
Whiting	73.0

Non-Shrinking Putty

White Lead	150 lb.
Raw Linseed Oil	16 gal.
Whiting	505 lb.
Silica	41 lb.
Flour Paste	41 lb.

Whiting Putty

800 lb. Whiting
23 gal. Raw Linseed Oil

980

White Lead-Whiting Putty

700 lb. Whiting
100 lb. White Lead
22 gal. Raw Linseed Oil

970

Metal Cap Seal

Rubber Factice	20
Gutta Percha	20
Asbestos Flour	60
Dark Red Iron Oxide	1.5

* Plastic Seal for Glass Jars

This composition withstands action of oils and fats.

Glue Edible	75
Casein	175
Talc	75
Titanium Dioxide	75
Diethylene Glycol	400
Paraformaldehyde	10
Am. Hydroxide	18
Water	900

Sealing Wax

Shellac (Button)	14
Rosin	24
Vermillion	1¼
Barytes	14
French White	4
Turpentine	1

Melt shellac and rosin; keep hot and work in pigment and finally the turpentine. Cast in sticks.

Sealing Wax

Shellac	84
Venice Turpentine	60
Rosin	21

Sealing Wax

Limed Rosin	3
Tallow	6
Turpentine	3
Precipitated Chalk	4
Red Lead	4

Sealing Wax—Red

Orange Shellac	39 lb.
Rosin	78 lb.
Turpentine	14 lb.
Whiting	56 lb.
Silex	35 lb.
Pale Vermillion	5¼ lb.

Sealing Wax—Brown—Cheap

Orange Shellac	26 lb.
Rosin—H grade	83 lb.
Turpentine	7¼ lb.
Whiting	32 lb.
Silex	31 lb.
Burnt Umber	4 lb.

Hard Wax Stopping for Filling Screw Holes in Wood

Carnauba wax	16 lb.
Paraffin wax	8 lb.
Rosin	8 lb.
Asphaltum	1 lb.

Melt the above together and apply hot.

Water	19 lb.

Cellophane Glue

Animal Glue	40%
Water	40%
Aqua Resin	20%

Use grade of glue common to paper box work; soak glue in cool water for around one hour, melt in water bath at 140° F. and stir in Aqua resin. Add sufficient water to produce the proper working consistency at 130–140° F.

Liquid Glue

Animal Glue	46.7%
Water	46.7%
Sodium Nitrate	6.6%

Dissolve sodium nitrate in cool water, stir glue into solution, allow to soak two hours, melt in water batch at temperature between 140–160° F. Heat a couple of hours or until mixture remains fluid at room temperature. Glue may be preserved by adding phenol or other common preservative.

Glue—Starch Paste

Starch (Cassava)	30%
Glue (Bone Glue)	10%
Water	60%

The starch and glue are put into solution separately and mixed hot. Any additional water necessary to produce the desired consistency is incorporated later.

Flexible Bookbinding Glue

Animal Glue	30%
Water	29%
Glycerine	30%
Preservative	1%

Soak the glue (medium grade hide) in the cool water for two hours, and melt at a temperature of 140° F. Stir the glycerine into the glue after the 140° F. temperature has been reached. In the event the glue is kept for a period of time, some effective preservative should be incorporated.

Flexible Paper Box Glue

Animal Glue	45%
Glycerine	15%
Water	39%
Preservative	1%

Soak the animal glue (bone glue suitable for paper box work) in cool water for approximately two hours and melt at 140° F. Stir the glycerine into the glue solution after the temperature has reached 140° F. If the glue is kept for a period of time, some effective preservative should be added.

Cement

Celluloid	32 oz.
Acetone	128 oz. or 1 gal.
Amyl Acetate	16 oz. or 1 pint
Methanol	16 oz. or 1 pint

Mix all the ingredients in a jar and allow to stand until dissolved—shaking from time to time.

Clean surface well before applying then apply a thin coating; first allow to dry then apply another coat and cement articles.

Adhesives for Hard Rubber

1. Carefully melt together 1 part gutta percha and 2 parts coal tar pitch. Immediately apply the fluid, homogeneous hot mass to the parts to be joined, these first having been degreased. Allow the repair to cool under pressure.

2. Broken hard rubber can be repaired by applying to the 2 surfaces to be joined, concentrated silicate of potassium and subjecting them to strong pressure.

3. Marine glue is made of 10 parts rubber dissolved in 120 parts benzol or turpentine. Add 20 parts asphalt or 18 parts gum lac and allow to digest until the mass is homogeneous. The solid glue, when it is to be used, is liquefied by careful heating; while the surfaces to be joined are first heated.

4. Melt together equal parts of pitch and gutta percha. Apply hot.

5. Dissolve 20 parts of rubber in 160 parts benzol or naphtha and mix with a solution of 20 parts gum lac and 50 parts mastic in the smallest possible amount of 90% alcohol.

When the surfaces to be adhered are smooth, it is always necessary to roughen them first by filing them lightly.

Oilproof Joint Cement

For use in connections of rubber and metal pipes carrying gasoline, oils, greases, etc.

A.	Aquaresin GM	25
	Lampblack	5–15

B.	Graphite	10
	Sicapon	20–40

Silicate Adhesive

Sod. Silicate	40
Water	10
Tescol	10

The water is mixed with the silicate and the Tescol is added a little at a time with good stirring. Do not add further quantities of Tescol until the previous portion is dissolved. This adhesive is less alkaline and not as brittle as most silicate adhesives. Further flexibility can be gotten by adding some glycerin to the Tescol.

Handling of Glue

Special precaution should be used in all cases to insure a soaking of the glue in the required amount of cold water for at least 4 hours. In order to effect solution of glue the temperature should be increased to about 160° F. Prolonged heating and excessive heating should be avoided, because this has been shown to result in extensive loss due to the hydrolyzing action of the water. In applying the heat, the most advantageous method is to apply heat (*e.g.*, steam or electricity) to a water jacket in which glue container is placed.

To employ glue such that the greatest benefit may be derived from its physical and chemical characteristics, the surface should be made so warm that the melted glue will not be chilled before it has time to effect a thorough adhesion.

For high class joint work only the better grades of hide glue should be employed.

For Veneer work the medium grades are indicated. In this case a high viscosity is desirable on account of the tendency of a thin liquid to penetrate the pores of the thin sheet of wood and show itself on the opposite surface.

Chipped Glass

Glue and Gelatine are allowed to rapidly dry out upon a plate of glass. As the glue loses moisture it contracts and adhesion of the gelatine is so great that it tears away the surface of the glass itself, chipping it into characteristic fern-like patterns. The general appearance of the design can be modified by varying the properties of the solution used, *i.e.*, addition of 6% alum and other salts. A brittle glue will give a different pattern than a tough glue. Sand sprinkled over film of gelatine is also employed to make certain patterns.

Sizing of Paper

Glue is used to serve for two distinct purposes in the manufacture of wall paper. It is employed as a binder for the clay, or other material with which the papers are grounded, and also as a sizing agent for the ground colors, especially for sun-fast wall paper.

The most general practice is to precipitate the color directly on an insoluble base as finely divided $BaSO_4$, draw off the precipitated mass after setting, wash, to free it of excess precipitant or reagent, and then separate from the excess of water by running it through a centrifugal hydroextractor. This heavy insoluble base (pulp color) is easily incorporated with glue solution in preparation of sized material.

In preparation of some pulp colors, a number of chemicals are employed in order that the exact shade of color desired may be produced. The viscosity of reagents employed and frequent failure to wash out completely the excess of precipitant or reagent has indicated the use of a good hide or bone glue.

In ordinary sized papers the glue is applied in one of two ways. The glue is either put into the beater with the paper pulp previous to making, or, the paper is run through a dilute bath of glue before drying. There is seldom anything used with the glue except at times a little alum to give paper a somewhat harder finish.

Coated paper is made by applying a mixture of high grade animal glue and various pigments or fillers, about the consistency of cream, to the paper after it has been finished. High gloss papers are of this type.

Sizing of Textiles

For this service hide glue finds extensive use because of absence of the most objectionable impurity SO_2 or sulphites. As the colors employed for dyeing fabrics are much more delicate than those used in paper and are usually soluble, the absence of traces of mineral acids or alkalies is also indicated.

Hide or extracted bone glue is used on cotton goods to stiffen and give body to the material. If solution of this glue is too thin it will penetrate the pores of cotton fibre to such a degree that the latter will be altogether too stiff to use, while if it is too viscous it will not be absorbed at all and will fail to dry out during passage through drying chamber. The desired results are obtained when a very dilute solution of this glue is treated with a solution of alum. The alum thickens the solution and is satisfactory because no precipitation will result.

Carpets, tapestries, burlap wall covering are all heavily sized with this grade of glue.

In the case of shade cloth where firmness with flexibility is desired—strong high grade glue is used.

All straws used in the manufacture of hats are sized. In this case a product that is more or less resistant to the action of water and also light in weight is desired. A final bleaching is given the material, by the use of oxalic acid, or lead acetate. Many manufacturers bleach their glue before sizing.

Adhesive Paste

Steep 4 oz. of ordinary gelatine in 16 oz. H_2O until it becomes soft, dissolve and while hot add 2 lb. of good flour paste and one part H_2O. Heat to boiling and when thickened remove from fire. While cooling add ¾ oz. silicate of soda and stir with wooden spatula.

Pastes for Paper and Fine Fancy Articles

Dissolve 100 parts glue in 200 water and add a solution 2 parts of bleached shellac in 10 of alcohol. Stir constantly while adding. Keep temperature below 50° C.

Paste for Fixing Labels (Machines)

Make 10% solution of glue and add to this 25% by weight of glue of dextrin. Mix while warm and add to every pound thereof ½ oz. each of boiled linseed oil and turpentine. This paste resists dampness and thus prevents printed labels from falling from metallic surfaces.

Paste for Joining Leather to Pasteboard

Dissolve 50 parts of glue with 50 parts water, add 1% Venice turpentine and next a thick paste made with 100 parts starch in water.

Cement for Attaching Metal Letters to Glass, Marble, Wood

Dissolve over a water bath 5 parts glue in a mixture of 15 parts copal varnish, 5 parts boiled linseed oil, 3 parts crude oil of turpentine and 2 parts of refined oil of turpentine and add 10 parts slaked lime to mixture.

Strong Paste

Glue	4 parts
Water	80 parts

in one pot

Starch	30 parts
Water	20 parts

so that a thin milky fluid without lumps is obtained. Mix two while hot and after cooling add 5–10 drops phenol.

Venetian Paste

Fish Glue	4 oz.
Cold Water	½ pt.
Venice Turpentine	2 fl. oz.
Rye Flour	1 lb.
Water	1 pt.
Boiling Water	2 qt.

Soak and dissolve glue and while hot stir in Venice turpentine. Make up rye flour and pour into boiling water. Stir and add glue solution. Will adhere to painted surface.

Label Paste

Soak glue in 15% Acetic Acid solution and heat to boiling and add flour.

Mucilage

Soak 5 parts of good glue in 20 parts of water and to liquid, add 9 parts glucose and three parts gum Arabic. Mixture may be brushed on paper while lukewarm. It does not stick together but adheres to bottles.

Glue for Cementing Glass
(To be exposed to boiling water)

Five parts hide glue, one part dissolved acid chromate of lime; the glue prepared, becomes, after exposed to light, insoluble in water in consequence of a partial reduction of chromic acid.

Leather to Metal Glue

Digest a quantity of nutgalls (approx. 1 part) reduced to powder in 8 parts distilled water for 6 hours and filter. If tannic acid is available use 5% solution instead. Dissolve 1 part by weight of glue in same quantity of water. Leather moistened with decoction of nutgalls or acid solution, and glue applied to metal previously roughened and heated. Dry under pressure.

Sausage Casing Glue

Glue for making sausage casings: Add to 1 quart of hide glue 20% solution, ¾ to 1 oz. bichromate of potash. Warm slightly when about to use it and before application moisten paper, latter must be dried rapidly and then exposed to light until yellow glue becomes brownish, boiled in sufficient quantity of water to which 2 to 3% alum added until chromate is dissolved out.

Wood Coating Glue

A sprayable coating composition suitable for use on wood, cloth, paper, etc., comprises a non-jellying stable solution of substantially 29 parts glue free from foreign substances of acid reaction in a solvent comprising alcohol about 33 parts and water about 35 parts and about 0.1 weight of the glue of a glue plasticizing substance such as glycerol or turkey red oil.

Glue for Hectograph

One part glue, 1 part glycerine and smallest amount of H_2O possible is used as hectograph mass for the transfer of matter, when with concentrated solution of aniline color.

Liquid Glue

Glue liquid is prepared by treating a hot solution of animal glue with a soluble perchlorate not having a tanning action. Sodium perchlorate 3.5 parts may be stirred into a hot solution of glue 10 parts in water 13 parts.

Glue liquid is prepared by treating animal glue with chloric acid, animal glue 10 parts, dissolved in water 15 parts, may be stirred with 20% chloric acid 3 parts.

Glue for Joints in Leather Driving Belts

Soak 1 part domestic isinglass and 25 parts glue in 75 parts water until thoroughly soft. Heat until solution has been effected. Add 0.2% Beta Naphthol and 0.1% Venice Turpentine C.P. Surfaces to be cemented should be free from grease, slightly roughened and glue applied at a temperature of 150° F.

Jeweler's Cement

Dissolve over the water bath 25 parts of fish glue in a small quantity of alcohol-water mixture 40%, add 2 parts of gum ammoniac. Separately dissolve 1 part of mastic gum in 5 parts alcohol-water solution. Mix the two solutions and keep in well stoppered bottles.

Stratena—Household Cement

Dissolve 12 parts of white glue in 16 acetic acid, and then add this solution to 2 parts gelatine in 16 of water. After mixing add 2 parts shellac varnish.

Banknote or Mouth Glue

Dissolve gelatine with about ¼ to ⅓ of its weight of brown sugar in as small a quantity of water as possible. When liquid cast mixture in thin cakes and when cold cut to size. When required for use moisten one end.

Paste for Cardboard

Dissolve 14 oz. of high grade glue in 26 oz. H_2O. Add 1 oz. of a solution composed of 1 part shellac in 7 parts alcohol and stir as long as solution is warm. Next dissolve ½ oz. of dextrine in 7 oz. of alcohol and 3½ oz. of H_2O, stir and place vessel in warm water until solution is complete. Mix two solutions and allow to cool. When wanted for use cut off a small piece and liquefy by warming.

Paste for Pads

Glue 4 parts, glycerine 2, linseed oil ½, sugar 4, dye to color. Dissolve glue and add glycerine with sugar and then add dye and stir in the oil. Use paste hot.

Waterproof Glue

Solution of glue by itself or mixed with pigments is used in painting walls in distemper. A waterproof coating is obtained as follows: Boil part of powdered gall-nuts and 12 parts H_2O until mass is reduced ⅔ of its bulk. Strain through cloth and apply solution to dry coat of distemper paint, the latter becoming thereby as solid and insoluble as oil paint. The tannin of gall-nuts acting only upon soft glue, the solution has to be applied so that the lower layer of the glue becomes thoroughly soaked through.

Waterproof Wrapping Paper

Dissolve 24 alum, 4 white soap in 32 water in one pot. In another 2 gum arabic, 6 glue in 32 parts water. Mix 2 solutions. Heat and immerse paper, dry.

Tungstic Glue
(Substitute for Hard India Rubber)

Mix thick solution of glue with tungstate of soda and HCl, by means of which a compound of tungstic acid and glue is precipitated which at a temperature of 86–104° F. is sufficiently elastic to admit of being drawn out into thin sheets. On cooling, this mass becomes solid and brittle and on heating is again soft and plastic. It can be used for all purposes to which hard rubber is adapted.

AGRICULTURAL SPECIALTIES

Apples, Removing Arsenic Spray Residue from

Removal of As to within tolerance limits is effected by washing with 0.33% HCl, provided no oil-spray has been used on the fruit. Accumulations of oil or wax may necessitate the use of 0.66–1.33% HCl. Apples were injured by 2% HCl. Oils having viscosity $>$65–75 or lighter oils applied very late in the season rendered As removal very difficult. Storage of apples at ordinary temp. prior to washing also increased the difficulty of cleaning, but cold storage had little effect. Kerosene emulsion, prepared with kaolin and used in conjunction with hot HCl, facilitated oil and wax removal. Heating the acid (35–40°) improved washing efficiency more than did increasing the concn. of HCl used.

Banana Plants, Combating "Panama Disease"

Best results were gotten by treating roots and surrounding soil of each plant with 1½ pints heavy gas oil (sp. gr. up to 0.8869).

Prevention Black Rot in Delphinium

Mercuric Chloride	1
Sod. Nitrate	1
Water	1280

Dissolve the above and saturate soil around roots.

---

* Disinfectant, Seed
Trichlorodinitrobenzene
Barium Dioxide
Talc

Fertilizers

Commercial fertilizers are compounded from various raw materials which contain one or more of the three necessary ingredients: Nitrogen, Phosphoric acid and Potash.

Different crops need different proportions of these chemicals and in general it is better to have the Nitrogen present in two or more forms such as Ammonium Sulphate, Sodium Nitrate, Organic (such as tankage, blood, cottonseed or other meals, etc.) The phosphoric acid is derived from super-phosphate or animal bone: the Potash from mineral salts such as Muriate, Sulphate or mixtures such as Kainit or Manure Salt, and in special cases, Carbonate. Typical formulae follow.

In a formula the first figure represents the percentage of Nitrogen, the second, Available Phosphoric acid and the third, Potash.

A simple formula 4–8–4

Ammonium sulfate (contains 20% Nitrogen)	400 lb.	equal	80 lb. N
Super-phosphate (contains 16% Available P_{205})	1000 lb.	"	160 lb. P_{205}
Muriate Potash (contains 50% K_{20})	160 lb.	"	80 lb. K_{20}
Earth (to make up one ton)	440 lb.		

4–8–7 Potato Fertilizer

Am. Sulfate (20% N)	100 lb.	contain	20 lb. N
Sodium Nitrate (16% N)	100 lb.	"	16 "

Blood (13% N)	340 lb.	"	44	"
Super-phosphate (16% P_{205})	1000 lb.	"	160	"
Muriate Potash (50% K_{20})	280 lb.	"	140	"
Earth	180 lb.			

Tobacco Fertilizer

	Pounds
Sulphate Ammonia (20.50% N)	293
Tankage (7% N)	286
Cottonseed Meal (5.50 N)	351
Superphosphate (18% P_{205})	778
Sulfate Potash (48% K_{20})	292
	2000

General Garden Fertilizer

	Pounds
Sulfate Ammonia (20.50% N)	293
Nitrate Soda (16% N)	125
Tankage (7% N)	286
Superphosphate (18% P_{205})	889
Muriate Potash (50% K_{20})	200
Filler	207
	2000

Grass Fertilizer

	Pounds
Sulfate Ammonia (20.50% N)	585
Castor Pomace (4.50% N)	440
Superphosphate (18% P_{205})	667
Muriate Potash (50% K_{20})	80
Filler	228
	2000

Corn Fertilizer

	Pounds
Sulfate Ammonia (20.50% N)	341
Tankage (6% N)	166
Superphosphate (18% P_{205})	1333
Muriate Potash (50% K_{20})	160
	2000

* Fodder, Preserving Green

Spraying with 6% Hydrochloric acid in the ratio of 5 lb. per 100 lb. of fodder prevents development of injurious organisms.

* Fungicide, Seed

The seed is dusted with

Copper Mercury Sulfocyanide	10
Talc	20

Grass Killer

Grass between the bricks or stones of a walk may be killed by adding a strong solution of calcium chloride in water.

Quack-Grass Killer

Sod. Chlorate	1 lb.
Water	1 gal.

Spraying two or three times yearly is efficaceous.

* Insecticide Against Lice

Aluminum Naphthenate 25 gm.
dissolve in
Turpentine 500 gm.
add
Acetone 375 gm.
Alcohol 125 gm.

To the above mixture add:

Sodium Salt of Benzyl Naphthalin Sulphonic Acid 20 gm.

Lettuce Bottom Rot, Control of

Ethyl Mercury Phosphate	1
Powdered Bentonite	2

Ornamental Bushes, Insecticide for

Kerosene	10 gal.
Soap Chips	5 lb.
Water	10 gal.
Nicotine Sulfate	1 oz.

* Peat Fertilizer

Peat which has been treated moist with HCl and then dried is stirred for 1–1½ hr. at 170–180° with a 1:1 mol. mixture of H_3PO_4 and KH_2PO_4; the product, after cooling, is mixed with H_2O, neutralised with NH_3, and dried.

Potato Blight Control

Dusting with following gives good results

Anhydrous Copper Sulfate	1
Slaked Lime	8

Potato Flake Fodder

Potato flakes contain all the solid constituents of the tubers and are an easily digested fodder material. The potatoes are washed, cooked or steamed under pressure, and then mashed to a pulp, which is dried as a film on steam-heated rollers, scraped off, broken up and stored. 400 kg. of potatoes contg. 18% starch yield 100 kg. of flakes contg. 12–15%

H_2O, 6–7% protein, 0.3–0.5% fat, 1.2–1.5% cellulose and 72–77% N-free exts.

Seed Potato Disinfectant

The dip is prepd. by adding to 25 gals. of water, a mixt. of 6 oz. of $HgCl_2$ dissolved in 1 qt. of com. HCl. Forty bu. of potatoes can be treated with 25 gals. of the dip. The soaking period is 5–40 min. according to the severity of *Rhizoctonia* and scab infection.

Seed Disinfectant

Hydrated Lime	95
Water	500

Stir well and add while agitating

Mercuric Chloride	5
Water	100

Filter and dry precipitate.

Sprout Killer

Sprouts or shoots of young trees can be killed by injecting into them a twenty per cent solution in water of sodium arsenite. Since this material is very poisonous it must be handled with the utmost care.

Sulphur Resin Spray

Stock spray made by mixing equal parts of potassium polysulphide solution (liver of sulphur) with potassium resin solution.

Potassium Polysulphide Solution

Flowers of Sulphur	4 lb.

dissolve in hot solution of caustic potash made by dissolving 5 lb. KOH in 10 lb. water.

Potassium Resin Solution

Made by heating.

Pine Resin	4 lb.
Potassium Hydroxide	2 lb.
Water	10 lb.

One gallon of stock solution to 50 gals. water gave combination fungicide and contact insecticide.

Tree-Bands, Insect

Rolls of corrugated paper are saturated with following and wrapped around trees

Mineral Oil	1½ lb.
Alpha Napthylamine	1 lb.
Paraffin Wax	4 oz.

---

* Weed-Killer

(Non-poisonous to cattle)

Calcium Chloride	20
Sodium Chlorate	30

Weed Killer for Seed Beds

Zinc Sulfate	8 gm.
Water	250 c.c.

Dissolve and apply above equally to every square foot of seed bed. Careless application will damage root tips. The second dose for a succeeding crop should be half of above strength.

ALLOYS

MAKING FUSIBLE ALLOYS

When making fusible alloys, melt the lead and bismuth together. When molten, add the tin with stirring. When the tin has been molten into the mix, adjust the temperature of the mix to about 300° C., and using the cadmium sticks in tongs as stirrers, work in the necessary cadmium. Cadmium burns easily in air, hence the temperature must be watched, and if it rise much above 300° C. this may happen.

Good metal can often be recovered from the dross formed in making fusible alloys by working the dross with the ladle or a stick against the side of the kettle.

Lipowitz Metal

Cadmium	3
Tin	4
Bismuth	15
Lead	8

Melt above together and add

Mercury	2

previously heated to 220° C.

Melting point of above is 143° F.

Rose Alloy

Bismuth	2
Lead	1
Tin	1

Melting pcint 200° F.

Electrical Fuse Alloy

Tin	94
Lead	344
Bismuth	500

Melting point 168° F.

---

* Alloy, Aluminum

An Al alloy not requiring hardening by heat treatment and suitable, *e.g.*, for internal-combustion engine pistons and piston rings, consists of Al 77.5–91, Cu 6–12, Ni 1–3, Cr 0.05–5 and Mg 0.5–2.5%.

Al alloys, particularly for internal-combustion engine pistons, contain Si 10–17, Ni 4–8 and either Cu 1–5 or Mg 0.5–2%.

* Aluminum Alloy

Copper	9–14%
Silicon	5–12%
Nickel	2–6%
Aluminum	Balance

This alloy is highly resistant to deterioration at elevated temperatures.

---

* Copper Alloy, Heat Treatment of

Wire composed of an alloy of Cu 20, Mn 30, and Ni 50% is annealed, preferably in vac. at 300–450° (350°), for 12–24 hr. The treatment increases the elastic limit and tensile strength.

---

* Alloy, Bearing

Tin	9–11
Antimony	9–11
Cadmium	1.4–1.8
Arsenic	0.9–1.7
Copper	1.2–1.6
Lead	Balance

---

* Alloy, Bearing

Aluminum	3–12.5
Copper	0.1–2
Magnesium	0.5–2.3
Zinc	Balance

---

* Alloy for Bearings and Knife Edges

Carbon	0.5–0.7%
Silicon	0.7–0.9%
Manganese	0.5–0.7%
Chromium	7.5–8.5%
Tungsten	7.5–8.5%

The remainder being iron, which may contain small amounts of impurities such as phosphorus, silicon and sulphur, and a surface hardened by treatment with ammonia at an elevated temperature.

---

* Alloy, Brake Drum

Nickel	0.5–30
Carbon	3–3.75
Silicon	1–2.5
Manganese	0.4–1

---

* Alloy, Copper

An alloy of high strength and electrical conductivity consists of

Copper	94
Beryllium	1
Chromium	5

---

* Alloy, Copper Bearing

Copper	62.3–46.2%
Tin	4–8%
Zinc	3–10%
Lead	30–35%
Calcium	0.2–0.5%

---

* Dental, Alloy
(Resilient and non-corrosive)

Gold	39.8
Copper	45
Nickel	14
Chromium	1
Platinum	0.2

---

* Alloy, Drill Bit

Tungsten Carbide	90–97
Molybdenum	0.5–5
Tantalum	2–9.5

---

* Alloy, Electrical Resistance

Chromium	85–95
Molybdenum or Tungsten	15–5

---

* Alloy, Electrical Contact Point

Silver	65
Copper	30
Nickel	5

---

* Hard Alloy

A process for obtaining alloys of high hardness consisting in forming an alloy of 5 to 25% of tin and the balance chiefly nickel and heating said alloy to a temperature lying between 900° C. and the melting point of the alloy, then rapidly cooling said alloy and subsequently annealing it at temperatures between 400 and 800° C.

A process for obtaining alloys of high hardness consisting in forming an alloy of 8 to 30% of molybdenum and the balance chiefly nickel and heating said alloy to a temperature lying between 900° C. and the melting point of the alloy, then rapidly cooling said alloy and subsequently annealing it at temperatures between 400 and 800° C.

---

* Alloy, Imitation Gold

Cobalt	1–5
Chromium	0.5–5
Tin	0.5–1
Zinc	0.1–0.5
Titanium	0.5–1
Silver	0.5–2

This is resistant to acid and heat and has a high power of elongation.

---

* Iron Alloy, Corrosion Resistant

Copper	0.2–0.5%
Tungsten	0.01–0.5%
Chromium	0.00–0.5%
Carbon	less than 0.05%
Manganese	less than 0.25%
Silicon	less than 0.02%
Phosphorus	less than 0.02%
Sulfur	less than 0.02%
Iron	Balance

---

* Lead Alloy

A tough, slightly hardened alloy suitable for storage battery plates consists of

Calcium	0.1–0.4%
Tin	0.5–2.0%
Lead	Balance

---

* Lead Coating Alloy

A lead alloy for coating wire is composed of

Antimony	0.6–1.4%
Bismuth	0.05–0.5%
Lead	Balance

---

* Alloy, Magnetic

An alloy of 70–30% Co and 30–70% Fe is melted with 0.5–4% V and, after rolling into sheet, is annealed at 900–1000° and allowed to cool slowly.

---

* Alloy, Permanent Magnet

Carbon	less than 0.2
Nickel	2
Silicon	5
Cobalt	15–55
Iron	40–80
Molybdenum	5–20

---

* Silver Alloy, Tarnish Resistant

Silver	80–95%
Nickel	0.1–2%
Cadmium	Balance

---

* Alloy, Silver Brazing

Alloys which are suitable for use in brazing contain Ag 48–52, Cu 12–16, Zn 14–18 and Cd 16–20%.

Alloys which are suitable for brazing purposes contain Ag 48–52, Zn 14–18, Cd 16–20, Cu 12–16 and P about 0.5–2%.

Non Tarnishing Silver Coating

Ag and its alloys are protected against atm. influences by dipping in a soln.

contg. CrO_3, Cu NH_4 chloride and (or) a persulfate, whereby an invisible coating is produced on the metal.

---

* Alloy, Stainless Silver

Silver	50–89.5
Zinc	0.5–20
Tin	10–40

---

* Alloy, Sulfur Resistant

Chromium	16–22%
Manganese	6–16%
Molybdenum	1–10%
Carbon	$>$ 0.3%
Iron	Balance

This has a high strength at 600° C.

---

* Alloy, Sulfur Resistant Steel

Chromium	6%
Silicon	0.75%
Tungsten	1%
Carbon	$>$ 0.5%
Iron	Balance

---

* Alloy, Thermocouple

Rhenium	3–15%
Platinum	Balance

---

* Alloy, Thermostatic Couple

(a)	Nickel	32–42
	Iron	Balance
(b)	Molybdenum	1–10
	Nickel	34–45
	Iron	Balance

---

* Tough Alloy

Nickel	2.5–18%
Copper	2.5–18%

Tungsten Carbide to make 100%. The above alloy is cast in forms.

---

* Alloy, Watch Spring

Nickel	30
Beryllium	0.1–0.5
Tungsten	8
Iron	Balance

---

* Brake-Shoes, Automobile

To prep. an alloy for brake shoes, 35–49% Cu and 1–2% Sb are melted in one crucible and 49–64% Pb in another. The melted Pb is gradually added to the melted alloy of Cu and Sb with const. agitation and heating. The product is poured into a mold provided with an iron gauze lining which serves as a skeleton and the whole is cooled until it solidifies.

Copper, Improving Electrical Conductivity of

The molten metal is deoxidised with 0.005–0.1% Li, the amount used being sufficient to leave 0.002–0.005% Li in the cast metal.

---

* Gold, Imitation

To 5 lb. 10 oz. of melted Cu are successively added 3 oz. fuller's earth, 7 oz. $Na_2B_4O_7$, 3 oz. ammoniated mercury, 12 oz. Sn, 3 oz. MgO and 1 oz. alc., and the mixt. is agitated and boiled. The resultant alloy simulates Au and is malleable, ductile, immune from tarnishing and suited for jewelry.

---

* White Gold, Untarnishable

A white Au (Au 50, Cu 30, Ni 11, Zn 9%) is rendered untarnishable by plating with Sn and afterwards heating to 240–250° to form a surface alloy.

Stainless "Invar"

Two alloys containing approx. 36.5% Fe, 54.5% Co, and 9% Cr have coeffs. of expansion $<10^7$ and -1.2×10^{-6}, respectively. Polished surfaces are unattacked by moist air, H_2O, sea-H_2O, etc. for many months.

Electrotype Metal

Tin	4%
Antimony	3%
Lead	Balance

Electrotype Backing Metal

Tin	4%
Antimony	3.5%
Lead	92.5%

---

* Resistance, Electrical Metal (Nichrome substitute)

Aluminum	5–10%
Manganese	0.5–5%
Carbon	0.05–1%
Iron	Balance

Linotype Metal

Tin	4–4.5%
Antimony	11.5%
Lead	Balance

Monotype Metal

Tin	7.3%
Antimony	16.8%
Lead	Balance

---

* Pewter or Brittania Metal

A soft white metal consists of Sn together with 1–15% of hardening metal selected from the Sb–Cu group and 0.005–0.1% of Al or Zn.

* Refining Type Metal

Used and partially oxidized type metal is refined by fusing with a reducing agent comprising, for example, the following components: rosin 200, basswood C 50, BaS 50, borax 50, NH_4Cl 17.5 and Na_2CO_3 15 parts, mixed at temps. sufficiently high to melt the rosin. When cool, 75 parts of $NaHCO_3$ are added. The mass is then powd.

Stereotype Metal

Tin	6.5–7%
Antimony	12.75–13%
Lead	Balance

* Steel, Armor Plate

Carbon	0.28–0.45%
Chromium	2–4%
Molybdenum	0.15–1%
Nickel	1–3.3%
Iron	Balance

* Steel, Non-Magnetic

Carbon	0.45–0.95%
Chromium	1.5–5%
Manganese	7–10%
Nickel	8–10%
Iron	Balance

* Steel, Non-Oxidizing
(For motor valves)

Carbon	0.45–0.65%
Manganese	0.3–0.6%
Silicon	1.3–2.5%
Chromium	11–14%
Cobalt	2–3.5%
Molybdenum	0.6–1.3%
Iron	Balance

* Steel, Razor Blade

Carbon	0.5–2.5%
Chromium	5–20%
Manganese	0.1–1.75%
Molybdenum	0.05–2%
Nickel	0.25–3.5%
Silicon	0.1–2%
Vanadium	0.05–1.5%
Iron	Balance

* Steel, Rustless

Chromium	14–20%
Carbon	$>$ 0.4%
Copper	0.5–6%
Molybdenum	0.4–3.5%
Iron	Balance

* Stainless Steel

A process for making a stainless metal composition which comprises intimately admixing finely divided particles of iron, nickel and chromium, which have clean surfaces and are of a size sufficiently small to pass through a 200 mesh screen, substantially in the proportion of iron 74%, nickel 8% and chromium 18%, subjecting the resulting admixture to a pressure of not less than 20,000 pounds per square inch, and heating the compressed mixture in a non-oxidizing atmosphere to a temperature above 900° C. but not substantially above 1200° C. to form a substantially homogeneous product.

* Steel, Stainless

Nickel	5–35%
Chromium	9–13%
Molybdenum	3–10%
Iron	Balance

* Stainless Steel, Bright Annealing

Articles such as sheets formed of nickel, stainless steel or Ni-Cr alloys with a bright surface are obtained by treating the metal with HNO_3 to render the bright surface passive and then heating to about 900° to 1100° for several hrs. in a reducing atm. to effect annealing without discoloration.

* Steel, Tool

A hard alloy for tools, implements and projectiles consists of W 38–98 and Be 2–9.5%, with or without up to 3% C and 57% Fe, the Be being at least 5% if the W is less than 82% and the Fe at least 5% if the C exceeds 2%. The W may be replaced wholly or in part by Mo and Cr, the Fe by Ni, Co, Mn or Ti and the C in part by Al, Mg, Si, B, Zr or Ce.

Steel, Tool

Iron	10–33
Carbon	1–4
Tungsten	64–46
Chromium	16–11.5
Cobalt	8–5.5

* Tool Steel

Steel (C=0.9–1%)	95–98.2
Cobalt	0.2–5

* Non-Tarnishing Acid Resistant Alloy

Thallium	10
Aluminum	10
Silver	80

White Gold

An alloy which possesses many of the physical properties of Pt including some degree of resistance to acids is prepared by alloying a primary alloy with a large proportion of Au. For a soft (hard, in parentheses) 18-carat white Au the primary alloy contains Au 37 (37.4), Ni 38.1 (44.5), Cu 16.4 (5.0), Zn 7.1 (11.1), and Mn 1.4 (2)%. This alloy is best prepared from granulated metals, and approx. 25% of the alloy is melted with 75% of Au in the second stage.

---

* Copper Refining Electrode Alloy

Thallium	10
Tin	20
Lead	70

The above is far more resistant than lead when used in electrolytic deposition of copper from acid solutions.

---

* Stainless Steel

Above a bath of molten Fe (500 kg.) with the desired C content is formed a slag of chromite (300 kg. containing 48% Cr_2O_3), NiO (30 kg.), CaO (150 kg.), CaF_2 (40 kg.), and bauxite (30 kg.), and to this is added a mixture of the same chromite (820 kg.), NiO (110 kg.), Al (316 kg.), and 75% ferrosilicon (82 kg.), whereby an exothermic reaction ensues with the direct production of stainless steel.

---

* Cold Drawing Wire Alloy

An alloy suitable for cold-drawn wire, etc., comprises Cu 91–99 (96.25), Sn 0.25–3.00 (1.75), Al 0.5–4.0 (1.0), Si 0.25–2.0 (1.0)%.

---

* Electrical Contact Alloy

An alloy of Au 30, Ag 70% is very suitable for contacts that are open for long periods.

---

* Low-Expansion Alloy

An alloy having a coeff. of expansion about 1×10^{-6} over a range of temp. depending on the amounts of the minor constituents comprises Cr 95–99, Fe 0.1–3.0, Si 0.1–2.0, C $\ngtr$ 1.0, and Mn+N+O $\ngtr$ 0.4%.

---

* Strong Malleable Cast Iron

White-Fe castings are packed with 4–15% of Fe_2O_3 into an annealing pot and heated first at 900–980° for 20–50 hr. to graphitise the free cementite, then at 730–650° for 10–50 hr. to graphitise the pearlitic cementite and decarburise the white Fe.

---

* Hardening Steel

Linseed oil is heated to the b.p., resin ⅛ lb. per gal.) is added, and the metal (Fe or steel) is immersed in the solution until it attains the same temp.; the metal is then removed, covered with powdered resin, and quenched in cold coal-oil.

---

* Rustproof Steel

Molten Fe or steel containing 3% Ti and 0.5–0.6% Mn is treated with 1–20% of a 50:50 Pb–As alloy, whereby the Pb separates to the bottom of the liquid mass and the As remains finely dispersed throughout the ferrite crystals and protects the resulting castings from rusting.

---

* Working Aluminum-Magnesium Alloys

In the working of aluminum-base alloys containing from about 5 to 15 per cent of magnesium, the steps comprising preheating the alloy at a temperature above about 550° F. but below the temperature of incipient fusion, cooling the alloy rapidly to a working range which is below about 600° F. and is also below the preheating temperature but is not lower than about 475° F. and working the alloy within said range.

---

* Blasting Resistance Wire

A bridge wire for blasting caps comprising gold about 58.4% and nickel about 41.6% alloyed together and drawn to the required size to have a desired electrical resistance per unit of length.

---

* Corrosion Proof Steel

A corrosion-proof steel free from graphite and Si consists of Cr 20–30, W 5–6, Ni 5–10, Cu 10–15, C 2.5–3 and the rest Fe.

---

* Sulfur Resistant Steel

An alloy suitable for high-pressure oil and steam fittings, etc., contains Fe together with Cr 6, Si 0.75, W 1 and C from a trace up to 0.5%.

ANIMAL PREPARATIONS

* Cattle Food

Dried Blood	75
Precipitated Chalk	5
Molasses	20

Lice and Mite Tablets (Poultry)

Calcium Sulfide	16.13
Silica Sand	7.52
Gypsum	6.48
Sugar	57.80
Starch	11.64

Poultry Louse Powder

Nicotine	0.28
Naphthalene	9.98
Sulfur	19.80
Sodium Fluoride	0.54

Veterinary Gall Salve

Tribromphenol	5.75 kg.
Petrolatum	67.15 kg.
Beeswax	9.2 kg.
Lard Compound (Paraffin added in summer)	29.9 kg.
Alum	13.8 kg.
Sulphur	27.6 kg.
Indigo	2.25 kg.

Melt the wax; add the other ingredients, and rub thoroughly through ointment mill.

Worm Expeller

Magnesium Sulfate	12.04
Calcium Sulfate	9.05
Calcium Silicate	6.85
Venetian Red	7.34
Sand	2.11
Nicotine	0.22

Mange Ointment

Mercurous Iodide Yellow	10 gr.
Salicylic Acid	½ oz.
Sulfur Sublimed	3 oz.
Coal Tar (neutral)	½ oz.
Pine Tar	3 oz.
Fish Oil	24 oz.
Diglycol Oleate	1 oz.

Shake well before using: apply at night and wash off next day.

Distemper Cure for Dogs

Fluid Extract of Buckthorn	1 oz.
Fluid Extract of Ginger	⅛ oz.
Syrup of Poppies	2 oz.
Simple Syrup	1 oz.
Cod Liver Oil	4 oz.

Shake well.

Dose—A tablespoonful is given twice daily.

Animal Condition Powder

Sulfur	5
Rosin	5
Fenugreek Seed	5
Flaxseed Meal	5
Magnesium Sulfate	5
Ginger African	4
Gentian Root	4
Copperas	4
Sod. Bicarbonate	4
Antimony	2
Salt	2
Pot. Nitrate	1

All of above materials should be powdered and then mixed thoroughly.

Mange Cure

Potassium Carbonate	8 gr.
Flowers of Sulphur	64 gr.
Oil of Picis	12 c.c.
Oil of Cade	12 c.c.
Linseed Oil	to make 11 liters

BEVERAGES AND FLAVORS

Almond Extract

Oil Bitter Almonds F.P.A.	1¼ fl. oz.
Alcohol	3 pt.
Water	5 pts.

Almond Flavor

1 Fluid Ounce Oil Bitter Almonds
40 Fluid Ounces Glycopon S
59 Fluid Ounces Water

Imitation Almond Flavor

Benzaldehyde (F.F.C.)	1.3
Glycopon XS	16
Glycerol	24
Water	128

Anise Flavor

3 Fluid Ounces Oil Anise
75 Fluid Ounces Glycopon S
22 Fluid Ounces Water

Caraway Flavor

2 Fluid Ounces Oil Caraway
70 Fluid Ounces Glycopon S
28 Fluid Ounces Water

Celery Flavor

4 Fluid Ounces Oil Celery
70 Fluid Ounces Glycopon S
26 Fluid Ounces Water

Thyme Flavor

3 Fluid Ounces Oil Thyme
70 Fluid Ounces Glycopon S
27 Fluid Ounces Water

Cinnamon Flavor

1 Fluid Ounce Oil Cinnamon
35 Fluid Ounces Glycopon S
14 Fluid Ounces Water

In making the above flavors the oil should be dissolved in the Glycopon by stirring at room temperature. The water is then added slowly with vigorous stirring. In some cases (where a clear flavor is desired) mix in a weight of magnesium carbonate equal to the weight of the oil used; stir and filter.

Coffee Aroma

Ethylmethylacetaldehyde 4, 2, 3-pentanedione 4, C_5H_5N 3, AcH 3, isovaleric acid 2, α-methylfurfurole 2, Ac_2 1, furfurole 1, PhOH 1, isoeugenol 1, methyl mercaptan 0.6, guaiacol 0.5, α-methylcyclopentenolone 0.5, thioguaiacol 0.4, furyl mercaptan 0.3, octyl alc. 0.2 parts.

* Coffee Extract

Ground Roast Coffee	40
Glycerol (Anhydrous)	160

Heat at 80–90° C. with stirring and filter.

Coffee Substitute

Coffee Bean Powdered	33
Sugar Powdered	5
Roasted Peanuts Powdered	62

Dry Ginger Ale Extract

8 oz. Solid Extract Jamaica Ginger
2 drams Oil Ginger
2 drams Oil Sweet Orange
1 dram Oil Limes, Distilled
¼ dram Oil Mace
¼ dram Oil Coriander
¼ dram Oil Lemenone

Grind the above in a mortar with 4 oz. powdered magnesium carbonate; then add 1 gallon Glycopon XS slowly while grinding in thoroughly; then add one gallon water slowly and stir thoroughly for 2 hours; add 2 oz. kieselguhr and filter through fine filter paper. The finished product should be aged to develop a finer aroma and taste.

4 oz. of this extract is used per gallon of syrup.

Ginger Ale

Jamaica Ginger, fine powder	8 lb.
Capsicum, fine powder	6 oz.
Alcohol	a sufficient quantity

Mix the powders intimately, moisten them with enough alcohol to make them distinctly damp but not wet, set aside for four hours, then pack in a cylindrical percolator and percolate with alcohol until ten pints have been collected; place the percolate in a bottle of at least 2-

gallon capacity and add 2 fluid drams of oleoresin ginger, shake and add 2½ pounds of finely powdered pumice stone and agitate frequently for twelve hours, then the next step is most important. Add 14 pints of water in one pint at a time, then shake briskly and add the next, after adding all the water set aside for twenty-four hours, agitating strongly every hour or so, then add:

Oil of Lemon	1½ fl. oz.
Oil of Rose Geranium	3 fl. dr.
Oil of Bergamot	2 fl. dr.
Oil of Cinnamon	3 fl. dr.
Magnesium Carbonate	3 oz.

First rub the magnesia with the oils in a mortar, add nine fl. oz. of the clear portion of the ginger mixture to which two ounces of alcohol have been added and continue trituration, rinsing the mortar out with the ginger mixture, pass the ginger mixture through a double filter and add the mixture of oils through the filter. Finally pass enough water through the filter to make three gallons of the finished extract which is to be used 4 fl. oz. to a gallon of syrup. Dilute the syrup, 1 fl. oz. with 6 fl. oz. of carbonated water; bottle.

Note: The ginger ale can be colored a darker color with caramel.

Soluble Ginger or Capsicum Flavor

12 Fluid Ounces Oleoresin Ginger or Capsicum
243 Fluid Ounces Glycopon AAA
6 Ounces Precipitated Magnesium Carbonate
189 Fluid Ounces Water

In making the above flavor, first mix the oleoresin thoroughly with Glycopon AAA and then add the magnesium carbonate, working it into an even paste. Add the water slowly with thorough stirring, then filter.

Havana Cigar Flavor

Coumarin, pure, cryst.	1 dr.
Methyl Benzoate	4 dr.
Essence Vanilla, Special	2 pt.
Oil Cascarilla	1 dr.
Oil Valeriana	½ dr.
Acetic Ether, Absolute	5 oz.
Glycopon XS	1 pt.

Kola Beverage

Fluidextract of Coca	4 fluidounces
Fluidextract of Kola	2 fluidounces
Spirit of Orange	1½ fluidounces
Lime Juice	1½ pints
Ginger Ale Extract	¾ fluidounce
Cologne Spirit	8 fluidounces
Sugar	6 pounds
Water	3 pints
Caramel	enough

Mix the fluidextracts, the Cologne spirit and the water, add the spirit of orange and set aside for two days shaking occasionally. Then filter, add the lime juice and the ginger ale extract and dissolve the sugar in the mixed liquids.

Pure Lemon Flavor

Dissolve 5 fluid ounces Lemon Oil in 95 fluid ounces Glycopon S; no heating is necessary.

The same proportions of oils of orange, limes, caraway, peppermint, wintergreen, etc., may be used as above to make 5% flavors.

Glycopon S will dissolve 10% of oils.

Glycopon XS will dissolve any quantity of oils, but should only be used in concentrated flavors because it has an ethereal odor.

Imitation Lemon Flavor

5 Fluid Ounces Citral
96 Fluid Ounces Glycopon AAA
189 Fluid Ounces Water

Imitation Lemon Flavor

½ oz. Citral
100 oz. Glycopon AAA
1 lb. Glucose 43° Baumé
60 oz. Water

Lemon Extract

Oil of Lemon, U.S.P.	6½ oz.
Alcohol, 190 proof	121½ oz.

Mix, let stand overnight, then filter.

Lemon Oil Emulsion

1. Gum Arabic	13 oz.
2. Terpeneless Oil of Lemon	20 oz.
3. Oil of Lemon	20 oz.
4. Glycerin	40 oz.
5. Water	to make 10 gal.

Mix one and four then mix in two and three to this add five slowly with good stirring. Beat intermittently until homogeneous. Then pass through an homogenizer.

Concentrated extract of lemon.—Shall be prepared from oil of lemon, or lemon peel, or both, and ethyl alcohol of proper strength, and shall contain not less than 20 per cent, by volume, of oil of lemon and not less than 0.8 per cent, by weight, of citral.

Extract of lemon.—Shall be prepared from oil of lemon or lemon peel, or both, and ethyl alcohol of proper strength. It shall contain not less than 80 per cent, by volume, of absolute ethyl alcohol, not less than 5 per cent, by volume, of oil of lemon and not less than 0.2 per cent, by weight, of citral derived solely from the oil of lemon or lemon peel used in its preparation.

Terpeneless extract of lemon.—Shall be prepared by shaking oil of lemon with dilute ethyl alcohol, or by dissolving terpeneless oil of lemon of proper strength in dilute ethyl alcohol, and shall contain not less than 0.2 per cent, by weight, of citral derived solely from oil of lemon.

Lemon flavor, nonalcoholic.—Shall be a mixture of 20 per cent, by volume, of oil of lemon (U.S.P. standard) and 80 per cent, by volume, of cottonseed oil. The cottonseed oil shall be thoroughly refined, winter pressed, sweet, neutral, and free from rancidity. The finished product shall be clear, free from sediment and rancidity.

Lemonade Powder for Soft Drinks

86 parts Cane Sugar
14 parts Dry Bordens Lemon Powder
$\frac{1}{10}$ Part Citric Acid
Color with a yellow certified food color.

The above powders are mixed and colored. Four ounces of above powder when mixed with pint of cold water will make delicious lemonade.

---

* Maté, Improving Taste and Odor

Mate	100
Acetaldehyde (1%)	100

Allow to stand for day and dry.

Imitation Maple Flavor

1 lb. Maple Base
1½ lb. Glycopon AAA
1 lb. Sugar Color

Balance water to make 1 gal.

Orange Extract

Oil Orange	6½ oz.
Alcohol	121½ oz.

Mix, let stand overnight, then filter.

Extract of orange.—Shall be prepared from oil of orange or orange peel, or both, and absolute ethyl alcohol of proper strength, and shall contain not less than 80 per cent, by volume, of ethyl alcohol, and not less than 5 per cent, by volume, of oil of orange.

Terpeneless extract of orange.—Shall be prepared by shaking oil of orange with dilute ethyl alcohol or by dissolving terpeneless oil of orange of proper strength in dilute ethyl alcohol, and shall correspond in flavoring strength to orange extract.

Orange flavor, nonalcoholic.—Shall be a mixture of 20 per cent, by volume, of oil of orange (U.S.P. standard) and 80 per cent, by volume, of cottonseed oil. The cottonseed oil shall be thoroughly refined, winter pressed, sweet, neutral, and free from rancidity. The finished product shall be clear, free from sediment and rancidity.

Orange Oil Emulsion

4 oz. gelatin
16 lb. water
24 lb. cane sugar
60 lb. invert sugar
20 oz. terpeneless oil orange
20 oz. oil orange

Dissolve the gelatin in the water, add the cane sugar and heat until dissolved. Then add the invert sugar and mix well; homogenize.

Orange Powder for Soft Drinks

80 Parts Cane Sugar
20 Parts Dry Bordens Orange Powder
⅕ Part Citric Acid

Color with an orange certified food color.

The above powders are mixed thoroughly. Four ounces of above powder when mixed with pint of cold water will make a delicious orange drink.

Peppermint Flavor

3 Fluid Ounces Oil Peppermint
70 Fluid Ounces Glycopon S
27 Fluid Ounces Water

---

* Tea Extract, Concentrated

A tea concentrate in paste form is made by the following method. It is noteworthy in that all bitter principles are eliminated without destroying any of the delicate flavoring principles. A quantity of tea leaves is submerged in two to four volumes of cold water in a sealed container. A quantity of powdered dry calcium hydrate, approximately 3 to 5% of the quantity of tea leaves is added to the water and the complete mixture is subjected to agitation through the manipulation of the container for about a half hour. After this period the extract is decanted or filtered off through a cloth or fine mesh and the complete residue is returned to the container for a second extracting which may be repeated as often as desired. The various extractives from the various extractive operations are mixed together and subjected to desiccation by a process known as spray drying. The concentrate is then mixed with 50% solution of glycerine and water to produce a relatively thick paste, packed in hermetically sealed containers, preferably collapsible tubes so that the paste may be positively sealed and measured quantities thereof readily dispensed.

Pure Vanilla Flavor

Oleoresin Vanilla	4 oz.
Glycopon AAA	2 pt.

Water to make 1 gallon.

Flavoring ingredients must be completely dissolved in Glycopon before any water is added. Filter clear after two or three days.

Vanilla Beans may be exhausted with Glycopon AAA diluted with water as completely as with alcohol. No solvent losses occur through evaporation.

Imitation Vanilla Flavors

1.	Vanillin	2 oz.
	Coumarin	½ oz.
	Glycopon AAA	32 fluid oz.

Water to 7 gallons.

2.	Vanillin	2 oz.
	Coumarin	1 oz.
	Glycopon AAA	28 fluid oz.
	Sugar	5 lb.

Water to 5 gallons.

3.	Vanillin	20 oz.
	Coumarin	4 oz.
	Glycopon AAA	1½ gal.
	Water	184 oz.

Take 1 lb. of above and add water to it slowly with stirring to make 2 gallons.

Imitation Vanilla Concentrate

4.	Vanillin	10 oz.
	Coumarin	3 oz.
	Glycopon AAA	128 oz.

5.	Vanillin	20 oz.
	Coumarin	10 oz.
	Glycopon AAA	1 gal.

Concentrated Vanilla Compound Flavor (Highest Quality)

For dilution with water up to 17 to 1.

60 oz. Glycopon AAA
6 oz. Vanillin
2 oz. Coumarin
4 oz. Oleoresin Vanilla
3½ lb. Glucose 43° Baumé
4 oz. Caramel Color
Balance water to make 1 gal.

The usual procedure on above formulae is to put the Vanillin and Coumarin in a container containing the required amount of Glycopon AAA; heat to 50° C. and stir until completely dissolved. Then allow to cool to room temperature and add to it *slowly* with *stirring* the required amount of water. If caramel color, prune juice, sugar or syrup is to be added, these should be dissolved first in the water.

Where a water-white Vanilla is desired, the solution of Vanillin in Glycopon AAA may be decolorized by the addition of a little tartaric or citric acid.

Non-Alcoholic Vanilla, Lemon and Almond Flavors

The following method for making a non-alcoholic flavor has been suggested:

Non-Alcoholic Vanilla Flavor

Vanillin	3.2 Gm.
Coumarin	0.19 Gm.
Glycerin	180.00 mils
Syrup	180.00 mils
Water	120.00 mils
Ether	120.00 mils
Color	sufficient

Dissolve the vanillin and the coumarin in the ether. Mix the glycerin, syrup and water, add to this ether solution of the vanillin and coumarin. Beat until the ether is entirely volatilized and then add the color.

The Paste type of flavors has been suggested for non-alcoholic lemon and almond. Soak 250 Gm. of gum tragacanth in 4 liters of distilled water for three or four days or until it is softened and has taken up as much water as it will hold. Now forcibly strain it through cheesecloth. Mix 120 mils of this mucilage with 360 mils of glycerin. This will serve as the vehicle for the flavor. For this quantity of paste add gradually and with constant trituration in a mortar 60 mils of oil of lemon.

For almond flavor use 120 mils of the paste and 360 mils of glycerin and to this add gradually and with constant trituration 15 mils of benzaldehyde which must be free from hydrocyanic acid and chlorine.

Compound Vanilla Extract

A.	Mexican Vanilla Beans	1 lb.
	Bourbon Vanilla Beans	1 lb.
	Water	2 gal.
	Alcohol	2 gal.
	Glycerin	26 oz.
	Rock Candy Syrup	2 pt.

Grind or cut the beans small and place in a porcelain jar or clean wooden keg; pour over them the water at a boiling temperature and macerate for twenty-four hours. Then add the alcohol and glycerin and macerate for forty-eight hours; lastly, add the rock candy syrup, stir well and macerate for not less than four weeks.

B.	Vanillin	2 oz.
	Alcohol	2 pt.

Mix and let stand for twenty-four hours; then add one pint rock candy syrup, and let stand for twenty-four hours longer; add one pint prune juice and let stand for twenty-four hours; then add five pints boiling water and let stand for two weeks. Filter.

To make the extract add one quart of solution (B) to one gallon of solution (A).

Vanilla Extract

Oleoresin Vanilla	4 oz.
Alcohol	4 pints
Simple Syrup	1¼ pints
Water	2¾ pints

Mix by stirring thoroughly. Simple syrup is prepared by dissolving 3½ lb. of sugar in one quart of water.

Pure vanilla extract.—Shall be prepared without added flavoring or coloring, from prime vanilla beans with or without sugar and/or glycerin; shall contain, in 100 cubic centimeters, the soluble matters from not less than 10 grams of vanilla beans; shall contain not less than 40 per cent, by volume, of absolute ethyl alcohol, and show a Wichman lead number not less than 0.70. The strength of the extract in respect to the vanillin and vanilla resins, which shall be derived solely from the beans used, shall be not less than 0.17 per cent vanillin and not less than 0.09 per cent vanilla resins.

Imitation vanilla, artificially flavored and colored.—Shall be a solution of vanillin and coumarin in dilute glycerol with 5 per cent, by volume, of true vanilla extract, colored with caramel. There shall be not less than 0.6 gram of vanillin, 0.1 gram of coumarin, and 35 centimeters of glycerol (U.S.P. standard), in 100 centimeters of the finished product.

Extra concentrated extract of vanilla.—Shall be prepared, without added flavoring or coloring, from prime vanilla beans, with or without glycerin; shall contain, in 100 cubic centimeters, the soluble matters from not less than 100 grams of vanilla beans, and shall contain not less than 30 per cent, by volume, of absolute ethyl alcohol, and when one part by volume, of the product is diluted with nine parts, by volume, of dilute alcohol (40 per cent, by volume) the resulting mixture shall comply with the requirements for vanilla extract except in regard to alcohol content. The label shall clearly indicate the strength of the product and if the product is not made directly from vanilla beans, the label should contain a statement to that effect.

4X strength, extract of vanilla.—Shall be prepared without added flavoring or coloring, from prime vanilla beans with or without sugar and/or glycerin; shall contain, in 100 cubic centimeters, the soluble matters from not less than 40 grams of vanilla beans; shall contain not less than 35 per cent, by volume, of absolute ethyl alcohol, and when one part, by volume, of the product is diluted with three parts, by volume, of dilute alcohol (40 per cent by volume) the resulting mixture shall comply with the requirements for vanilla extract, except in regard to alcohol content. The label shall clearly indicate the strength of the product and if the product is not made directly from vanilla beans, the label should contain a statement to that effect.

NON-ALCOHOLIC FLAVORS

Imitation Black Walnut Flavor

8 oz. Oil of Black Walnut Flavor
1½ lb. Glycopon AAA
1 lb. Glucose 43° Baumé
2 oz. Sugar Color
Balance water to make 1 gal.

Wintergreen Flavor

3 Fluid Ounces Methyl Salicylate
70 Fluid Ounces Glycopon AAA
27 Fluid Ounces Water

Chocolate Syrup

Heat 2 lb. chocolate.
Add 6 lb. 30° Bé. sugar syrup
Boil down to desired thickness
Add $\frac{1}{10}$ of 1% Sodium Benzoate

Fruit Syrup

One quart lemon, orange or other fruit pulp; 6½ lb. sugar; 5 pints water; ½–1 oz. citric acid, and 1 oz. Viscogum.

Directions:

Mix thoroughly 1 lb. of sugar with 1 oz. of Viscogum. Bring the 5 pints of water to a boil and add slowly while stirring the mixture of Viscogum and sugar. Then boil vigorously for one minute. If artificial color is desired, it may be added at this point. Now add the balance (5½ lb.) of sugar and cook until completely dissolved. Allow to cool to 180° F. and add the citric acid, previously dissolved in a little water. The fruit pulp is then added and slow stirring is continued until cool. If some additional flavor is desired it is added at this point. If a preservative is indicated then 3.6 grams of Benzoate of Soda is stirred in. The finished syrup is stirred slowly while bottling. It is advisable to shake each bottle the next day before packing for shipment. The pulp will now remain in suspension for long periods.

EMULSION FLAVORS

Formula (Cold Method)

A.	1. Lemon or Orange Oil	25 oz.
	2. Emulsone B	3–4 oz.
	3. Water q.s.	1 gal.
	4. Glycerin	10 oz.

Put (1) and (2) in a pot fitted with a beating stirrer. Start mixing to wet the gum thoroughly with the oil. Add (3) and (4) while beating vigorously. Continue beating until homogeneous. Continue beating intermittently for a few hours. If the above amount of essential oil is not desired, any part of it may be replaced with mineral or cottonseed oil.

Formula (Hot Method)

B.	1. Lemon or Orange Oil	25 oz.
	2. Emulsone B	2 oz.
	3. Water q.s.	1 gal.
	4. Sugar	16 oz.

Mix (2) and (4) intimately in dry pot. In a steam-heated kettle or double boiler, bring (3) to a boil; add the mixture of (2) and (4) very slowly while stirring. Cover the kettle and boil for two hours, while stirring. Allow to cool and add the oil slowly while beating vigorously. Continue beating until uniform. Continue beating intermittently for a few hours.

The above formulae can be used for making any emulsion flavor by substituting other oils or combinations of oils for lemon or orange oil. Thus oil of peppermint, wintergreen, cinnamon, clove, nutmeg or any combinations of these or other oils may be used to make emulsions of different flavors. The concentrations of the oils given in the above formulae may be varied to suit individual requirements. Emulsions made with 50% of some oils are so thick that they will scarcely flow. The viscosity of a weak oil emulsion may be increased by mixing some cotton or other edible oil with the flavor oil used, before emulsification.

The addition of 1% phosphoric or hydrochloric acid or a larger amount of a weaker acid increases the stability of these emulsions. The acid should be dissolved in the water used.

Essence Grape Aroma "Special"

Nerolin	20 gr.
Essence Cognac	10 mils
Sol. Methyl Anthranilate 1: 10	20 mils
Tinct. Cacao	20 mils
Fluid Ext. Valerian	2 mils
Sol. Benzoic Ether 1: 10	1 mil
Grape Juice	60 mils
Glycopon XS	200 mils

Pistache Essence

Oil Lemon, Handpressed	4 mils
Oil Bitter Almonds, F.F.P.A.	8 mils
Essence Strawberry Aroma	12 mils
Benzyl Acetate, pure	3 drops
Glycerine, pure	12 mils
Peach Flavor, pure	3 mils
Glycopon XS	120 mils
Green Color	½ gm.

Essence Prune Juice for Blending

Tinct. St. John's Bread	10 oz.
Extract Vanilla	5 oz.
Prune Juice	28 oz.
Prune Spirit	12 oz. 4 dr.
Essence Rum Kingston	2 oz. 4 dr.
Tinct. Lemosin Oak	30 oz.
Essence Raisin Wine	10 oz.
Essence Cognac Fine Champagne	5 oz.
Essence Figs	2 oz. 4 dr.
Essence Grape Aroma	2 oz.

Oil Blood Orange

Oil Sweet Orange, Handpressed	64 oz.
Oil Lemon, Handpressed	15 oz. 4 dr.

Oil Peach Blossom	2 dr.
Methyl Anthranilate, pure	1 dr.
Vanillin, pur. cryst.	16 oz.

Corn Ether

Glycopon XS	5000 gr.
Acetic Ether	1000 gr.
Fusel Oil	30 gr.
Coriander Oil	4 gr.
Oil Cognac	4 gr.

Oil Gin, Old Tom

Oil Coriander, pure	3 oz. 4 dr.
Oil Angelica Root	3 dr.
Oil Anise, Russian, Rectified	1 oz.
Oil Caraway, Dutch	4 dr.
Oil Juniper Berries, Rectified	7 oz. 4 dr.
Glycopon XS	1 pt. 8 oz.

Essence Gin, Old Tom

Essence Gin, Holland	1 gal.
Glycopon XS	1 pt.
Oil Coriander, pure	1 oz.
Oil Calamus	1 oz.

Essence Gin, London Dock

Oil Gin, Old Tom	6 oz.
Oil Gin, Holland	18 oz.
Oil Cassia, Rectified	4 dr.
Glycopon XS	64 oz.

Arrac Aroma Essence

Oil Birch	16 gr.
Oil Cognac	16 gr.
Oil Maraschino	25 gr.
Oil Celery	8 gr.
Rum Essence	250 gr.
Glycopon XS	250 gr.

Oil Gin Holland

Oil Lemon	1 dr.
Oil Anise	1 dr.
Oil Angelica Root	6 dr.
Oil Fusel	4 dr.
Oil Juniper Berries	20 oz.
Oil Rosemary Flavor	6 dr.
Oil Coriander	4 dr.
Glycopon XS	10 oz.

Essence Holland Gin

Oil Gin	1000 mils
Glycerine C.P.	200 mils
Glycopon XS	216 oz.

Essence Apple Aroma

Oil Apple Ethereal	750 mils
Oil Jasmine Flowers	3 mils
Amyl Valerianate, pure	20 mils
Vanillin	10 gr.
Tinct. Civet 4 oz. to 1 gal.	5 mils
Sol. Peach Aldehyde, pure 1:20	1 mil
Glycopon XS	2000 mils
Apple Cider	1500 mils
Water	750 mils

Oil Pear Ethereal

Benzyl Propionate	1 pt.
Amyl Acetate, pure	11 pt.
Butyric Ether, Absolute	4 pt.

Oil Neroli Artificial

Ambrettone	2 gr.
Oil Rose Geranium	5 gr.
Infusion Balsam Tolu	8 gr.
Glycopon XS	50 gr.
Phenyl Ethyl Acetate	20 gr.
Orange Oil	40 gr.
Rose Leaf Infusion	75 gr.
Oil Neroli Gen. Bigarde	100 gr.
Geranyl Acetate	100 gr.
Methyl Anthanilate	100 gr.
Inf. Orange Flowers	100 gr.
Linalol	100 gr.
Oil Petit Grain Algerian	150 gr.
Linalyl Acetate	150 gr.

Apricot Oil

Oil Neroli Art.	12 oz.
Oil Cognac White	14 oz.
Oenanthic Ether	14 oz.
Peach Aldehyde 100%	4 oz.
Vanillin	64 oz.
Oil Apple Ethereal	16 oz.
Acetic Ether	96 oz.
Valerian Ether Absolute	16 oz.
Glycopon XS	240 oz.

Essence Sweet Cherry

Heliotropin	60 gr.
Solution Jasmin, Concrete 1:10 in Glycopon XS	24 mils
Solution Peach Aldehyde, pure 1:20 in Glycopon XS	7½ mils
Cyclamic Aldehyde, pure	2 mils
Oil Bitter Almonds, F.F.P.A.	16 mils
Vanillin	84 gr.
Fluidextract Rhatany	35 mils
Oil Cloves	2¼ mils
Oil Cinnamon Ceylon	1¼ mils
Cherry Juice	800 mils
Glycopon XS	800 mils

Essence Whiskey Bourbon

Fusel Oil	1 gal.
Oil Bitter Almond	1½ oz.
Oil Rose Art.	48 min.
Vanilla Extract	32 oz.
Ess. Jamaica Rum	40 oz.
Pineapple Aroma	40 oz.
Acetic Ether	12 oz.

Essence of Jamaica Rum

Oil of Cassia 1 dr.
Oil of Birch Tar 25 drops
Oil of Ylang Ylang Natural 3 dr.
Oil of Orange Flower Natural 20 drops
Oil of Ceylon Cinnamon 15 drops
Rum Ether Pure 3 pt.
Acetic Ether 2½ oz.
Butyric Ether 1 oz. 1 dr.
Tincture of Saffron 1 lb. to a gal. 4 oz.
Extract of Vanilla Pure 3 oz.
Balsam Peru 2 dr.
Tincture Styrax U.S.P. 2 dr.
Coumarin 5 dr.

Essence Whiskey "Scotch"

Guaiacol, pure 4 dr.
Oil Cade, pure 1 oz.
Butyric Ether, pure 4 oz.
Essence Rye Whiskey 2 gal.

Essence Cognac Brandy

Essence Brandy 20 oz.
Extract Vanilla 4 oz.
Tinct. Orrisroot, Florentine (2 lb. to 1 gal.) 2 oz.
Oil Cognac, Genuine 1 oz.
Oil Bitter Almonds, Free from Prussic Acid 2 dr.
Essence Rum, New England 6 dr.
Acetic Ether, Absolute 2 oz. 2 dr.
Nitrous Ether, Absolute 2 oz.
Glycopon XS 10 oz.

Essence Slivovitz

Oil Bitter Almonds, F.F.P.A. 2 mils
Oil Neroli, Artificial 1 mil
Oil Cognac, Genuine, Green 2 mils
Vanillin 5 gm.
Essence Raspberry Aroma 300 mils
Essence Plum 300 mils
Essence Jamaica Rum 25 mils
Essence Raisin Wine 50 mils
Prune Spirit 100 mils
Glycopon XS 100 mils

Essence Nordhaeuser Korn

Carvol 10 oz.
Oil Caraway, Dutch 2 oz.
Oil Coriander, pure 30 drops
Acetic Ether, Absolute 4 dr.
Glycopon XS 60 oz.
Glycerine, Pure 18 oz.

Essence Nordhaeuser Korn

Rum Ether, Pure 2 gal.
Corn Fusel Oil 2 pt.
Oil Spice Gewuertz 2 dr.
Butyric Ether, Absolute 2 dr.
Tinct. Foenigraeci, Concentration 3 oz. 4 dr.

Essence Kartoffel Schnaps

Essence Rye Whiskey 8 oz.
Essence Nordhaeuser Korn 8 oz.

Oil Cherry Ethereal

Amyl Acetate, Pure 12 pt.
Amyl Butyrate, Pure 8 pt.
Benzaldehyde, free from Prussic Acid 12 pt.
Oil Lemon, Handpressed 16 oz.
Oil Sweet Orange, Handpressed 8 oz.
Oil Cloves, Pure 16 oz.
Oil Cassia, Leadfree 8 oz.
Vegetable Red Coloring.

Essence Rootbeer

Oil Sassafras, Pure 1 oz.
Oil Anise Russian, Rectified 1 oz.
Oil Lemon, Natural 1 oz.
Methyl Salicylate (Oil Wintergreen Art.) 18 oz.
Glycopon XS 6 oz.
Water 11 oz.
Bismarck Brown Color

Essence Rum New England

Oil Cinnamon, Ceylon 2 dr.
Oil Cloves, Pure 2 dr.
Oil Chamomile, Roman 4 dr.
Rum Ether, Pure 4 pt.
Butyric Ether, Absolute 3 oz.
Extract Vanilla 4 dr.
Acetic Ether, Absolute 3 oz.
Glycopon XS 8 oz.

Root Beer Oil

Methyl Salicylate 5 oz.
Safrol 8 oz.
Oil Orange 1 oz.
Oil Clove 2 drops
Oil Nutmeg 2 drops
Coumarin ½ oz.
Vanillin 1 oz.
Glycopon XS 64 oz.
Water q.s. 128 oz.
1 ounce of above flavors 2 gallons.

Oil Scotch

Oil Corn Fusel 6 oz.
Oil Bitter Almonds 4 dr.
Oil Coriander 4 dr.
Oil Cade 1 oz.
Guaiacol 2 dr.

Butyric Ether	4 oz.
Glycopon XS	4 oz.

New England Rum Essence

Nitrous Ether	250 gr.
Butyric Ether	250 gr.
Acetic Ether	250 gr.
Oil Lemon	3 gr.
Oil Cinnamon	3 gr.
Oil Neroli	1 gr.
Balsam of Peru	2 gr.
Rum Ess. No. 10	500 gr.

Tincture of Castorium

Castorium, Canadense	1 lb.
Glycopon XS	1 gal.

Tincture of Civet

Civet, Genuine	4 oz.
Glycopon XS	1 gal.

Tincture Foenugreek

Foenugreek, Powder	2 lb.
Glycopon XS	1 gal.

Essence Raisin Wine

Extract Vanillin	70 oz.
Essence Raspberry Aroma	2 oz.
Oenanthic Ether, Absolute	4 dr.
Geraniol Pure	2 oz. 2 dr.
Acetic Ether, Glacial	2 oz. 2 dr.
Glycopon XS	40 oz.
Methyl Anthranilate Pure	20 drops
Water	16 oz.

Oil Plum Ethereal

Oil Pineapple, Ethereal	4 pt.
Oil Jamaica Rum	4 pt.
Essence Slivovitz	4 pt.
Essence Peach Blossoms	4 pt.
Glycopon XS	6 pt.

Tincture of Foenugreek, Concentrate

Foenugreek, Powder	4 lb.
Glycopon XS	1 gal.

Tincture of Figs

Figs	4 lb.
Glycopon XS	4 pt.
Water	4 pt.

Tincture of Hickory

Hickory Bark, Powder	2 lb.
Glycopon XS	2 pt.
Water	4 pt.

Tincture of Lemosin Oak

Oak Bark, Powder	2 lb.
Glycopon XS	4 pt.
Water	4 pt.

Tincture of Maple Bark

Maple Bark, Powdered	2 lb.
Glycopon XS	4 pt.
Water	4 pt.

Tincture of Saffron

Saffron	1 lb.
Glycopon XS	1 gal.

Tincture of Sandalwood

Sandalwood, Powder	2 lb.
Glycopon XS	1 gal.

Tincture of St. Johns Bread

St. Johns Bread, Powder	2 lb.
Glycopon XS	4 pt.
Water	4 pt.

Tincture Orrisroot, Florentine

Orrisroot, Florentine, Powder	2 lb.
Glycopon XS	4 pt.
Water	4 pt.

Tincture of Almonds, Shells

Almonds, Shells	4 lb.
Glycopon XS	6 pt.
Water	2 pt.

Tincture of Arnica

Arnica Powder	1 lb.
Glycopon XS	1 gal.

Tincture Gum Benzoin, Siam

Gum Benzoin Siam, Powder	2 lb.
Glycopon XS	1 gal.

Tincture of Musk Tonquin, Grains

Musk Tonquin, Grains	4 oz.
Glycopon XS	1 gal.

Tincture Musk Artificial

Musk Artificial 100% Pure	4 oz.
Glycopon XS	1 gal.

Dissolve.

Tincture of Mastic

Gum Mastic Powder	1 lb.
Glycopon XS	5 pt.

Dissolve.

Essence Apple, Extra

Oil Apple, Ethereal	1500 mils
Peach Flavor	100 mils
Glycopon XS	5000 mils
Water	3500 mils
Vegetable Liquid Yellow Color	10 mils

Oil Absinthe, French

Oil Wormwood, American	10 oz.
Oil Star Anise, Leadfree	16 oz.
Oil Anise Russian, Rectified	12 oz.
Oil Fennel, Rectified	6 oz.
Oil Neroli, Artificial	½ dr.
Glycopon XS	3 oz.
Tinct. Gum Benzoin, Siam 2 lb. to 1 gal.	3 oz.

Oil Anisette

Oil Anise Russian, Rectified	465 mils
Oil Sweet Fennel, Rectified	20 mils
Oil Coriander, Pure	10 mils
Oil Star Anise, Leadfree	465 mils
Oil Angelica Root	30 mils
Oil Bitter Almonds, F.F.P.A.	8 mils
Oil Rose, Artificial	2 mils

Oil Alkermes, Cordial

Oil Cinnamon, Ceylon	100 gm.
Oil Cassia, Leadfree	200 gm.
Oil Cloves, Pure	200 gm.
Oil Mace, Distilled	450 gm.
Oil Rose, Genuine	1 gm.
Glycopon XS	50 gm.

Anisette Flavor

Oil Star Anise	100 gm.
Oil Anise	50 gm.
Oil Carvol	7 gm.
Oil Lemon	5 gm.
Oil Rose	½ gm.
Oil Neroli	2 gm.
Oil Cardamon	2 gm.

Essence Arac

Oil Neroli Petale, Extra	15 drops
Essence Jamaica Rum	42 oz.
Extract Vanilla	12 oz.
Essence Cognac Fine Champagne	2 oz. 4 dr.
Essence Raisin Wine	1 oz.

Essence Wild Cherry Aroma

Heliotropin	40 gm.
Solution Jasmine	24 mils
Peach Aldehyde	7½ mils
Oil Bitter Almond	23 mils
Vanillin	84 gm.
Fl. Extr. Phatany	35 mils
Oil Cloves	2½ mils
Oil Cinnamon	1¼ mils
Cherry Juice	800 mils
Glycopon XS	800 mils

Cognac Essence

Oil Bitter Almond	20 drops
Oil Cognac	50 gm.
Violet Flower Essence	25 gm.
Woodruff Essence	50 gm.
Oenanthic Ether	15 gm.
Acetic Ether	120 gm.

Oil Scotch Whisky Mix

Oil Fusel	6 oz.
Oil Bitter Almond	4 dr.
Oil Coriander	4 dr.
Oil Cade Pure	1 oz.
Guiacol Pure	2 dr.
Butyric Ether	4 oz.

1 oz. to 60 gal.

Oil Peach Blossom

Oil Neroli	16 oz.
Oil Cognac Genuine	14 oz.
Cenanthic Ether	14 oz.
Peach Aldehyde 100%	4 oz.
Oil Apple Ethereal	16 oz.
Acetic Ether Absolute	96 oz.
Valerianic Ether Absolute	16 oz.
Glycopon XS	240 oz.

Oil Bourbon 1–30

Oil Fusel	6 pt.
Butyric Ether	2 oz.
Oil Bitter Almonds	2 dr.
Oil Jam. Rum	16 oz.
Tinct. Castoreum	8 oz.
	122 oz.

Extract Bourbon 1–1

Oil Bourbon	6 oz.
Glycopon S	32 oz.
Sugar Color	20 oz.
Citric Acid S.	8 oz.
Tannic Acid Sol.	1 oz.
	67 oz.
Water	61 oz.
	128 oz.

Super Aroma Bourbon 1–5

Oil Fusel Rectified	240 oz.
Ess. Pineapple	½ oz.
Ess. Peach Blossom	½ oz.
Citric Acid Solution 50%	240 oz.
Solution Saccharin Saturated	¼ oz.
Oil Jam. Rum	13 oz.

Glycopon S	133 oz.
Tannic Acid Sol.	1 oz.
	626 oz.

Special Whisky Flavor 1–16

Super Aroma Bourbonette 1–5	100 oz.
Oil Bourbon Cyllo	4 oz.
	104 oz.

Bourbon 1 to 1

Oil Bourbon	40 oz.
Oil Combindlion	20 oz.
Glycopon XS	10 oz.
Tannic Acid Solution 1 lb. C.P. Tannic Acid Dissolved in 1 gal. Hot Water	10 oz.
Saccharin Solution 1 lb. Soluble Water Saccharin 5 gal. Boiling Water	½ oz.
Citric Acid Solution	10 oz.
Sugar Color 100%	200 oz.
Vanilla Ext. Imitation	2 oz.

Imit. Vanilla Ext. 1 oz. Vanillin. Dissolve in ½ gal. Glycopon S; ½ gal. Water.

Whisky Flavor 1–25

Oil Bourbon	100 oz.
Oil Fusel	200 oz.
Ess. of Peach Blossom	1½ oz.
Ess. of Pineapple Aroma	½ oz.
Tannic Acid C.P.	1 dr.
	300 oz.

Essence of Peach Blossom

Oil of Peach Blossom	1½ oz.
Peach Aldehyde 100%	2 dr.
Glycopon XS	6 pt.
Water	28 oz.

Gordon Gin Essence

Oil Juniper Berries	16 oz.
Oil Angelica Root	20 cc.
Oil Angelica Seed	20 cc.
Oil Coriander	40 cc.
Oil Lemon	60 cc.
Sweet Orange	20 cc.
Neroli	5 cc.
Geranium Rose	5 cc.

Glycopon XS to make 1 gal.
4 oz. of above to make 50 gal.

Essence Chartreuse

Oil Peppermint, Rectified	1½ dr.
Oil Lemon, Handpressed	2 dr.
Oil Cassia, Leadfree	1 dr.
Oil Cloves Pure	1 dr.
Oil Mace Distilled	1½ dr.
Oil Anise Seed, Russian, Rectified	1 dr.
Oil Angelica Root	40 dr.
Oil Bitter Almonds, F.F.P.A.	½ dr.
Oil Wormwood, American	20 dr.
Oil Neroli Bigrade, Petale, Extra	1 dr.
Oil Cognac, Genuine, White	15 dr.
Glycopon XS	20 oz.

Essence Concord Grape

Methyl Anthranilate, Pure	10 oz.
Glycopon XS	100 oz.
Glycerine, Pure	45 oz.
Vegetable Red Liquid	5 oz.

Essence Cognac Brandy

Essence Cognac Fine Champagne	5 oz.
Extract Vanilla	2 dr.
Tinct. St. Johns Bread	2 dr.
Glycopon XS	2 oz.
Glycerine, Pure	4 dr.

Essence Cognac Fine Champagne

Oil Cognac, Genuine, White	20 oz.
Oil Bitter Almonds, F.F.P.A.	1 oz. 2 dr.
Acetic Ether, Absolute	45 oz.
Glycopon XS	15 pt.
Essence Raisin Wine	10 oz.
Tinct. Lemosin Oak	5 oz.

Essence Creme de Menthe

Oil Peppermint, Twice Rectified	2 oz.
Menthol	2 dr.
Glycopon XS	35 oz. 4 dr.

Green Coloring.

Essence Whiskey "Rye"

Oil Fusel Potato	2 pt.
Oil Fusel Rye	18 pt.
Rum Ether, Pure	20 pt.
Oil Coriander, Pure	5 oz.
Oil Bitter Almonds, F.F.P.A.	2 oz. 4 dr.
Glycopon XS	50 pt.
Tinct. Catechu	1 pt.
Vanillin	2 dr.
Heliotropin	4 dr.
Tinct. Balsam, Peru, True	1 dr.

Essence Trester Brandy

Oil Cognac, Genuine	4 oz.
Oil Corn Fusel	5 oz.
Methyl Salicylate	3 oz.
Acetic Ether, Absolute	2 lb. 8 oz.

Glycopon XS	24 pt.
Water	3 pt. 12 oz.

Essence Tutti Frutti

Essence Benedictine	16 oz.
Essence Maraschino	16 oz.
Essence Curacao	16 oz.
Essence Violet Flowers	16 oz.
Oil Strawberry, Ethereal	32 oz.
Tinct. Vanilla 1 lb. to 1 gal.	32 oz.

Essence Rock and Rye Whiskey

Oil Corn Fusel	7 oz. 4 dr.
Oil Cognac, Genuine Green	4 dr.
Balsam Peru, True	4 dr.
Essence Jamaica Rum	4 dr.
Vanillin	2 dr.
Acetic Ether, Absolute	4 dr.
Coumarin	5 dr.
Essence Raisin Wine	12 oz.
Peach Flavor	4 dr.
Glycopon XS	35 oz.
Glycerine, Pure	16 oz.

Oil Benedictine

Oil Sweet Orange, Hand-pressed	72 oz.
Oil Angelica Root	6 oz.
Oil Calamus	3 oz.
Oil Cinnamon, Ceylon	3 oz.
Oil Mace, Distilled	3 oz.
Oil Celery	3 oz.
Glycopon XS	12 oz.

French Curacao

Oil Orange	10 oz.
Mace Oil	8 cc.
Cassia Oil	16 cc.
Cloves Oil	8 cc.
Lemon Oil	32 cc.
Rose Oil	1 cc.
Vanillin	1 dr.
Jam. Rum Ess.	2 oz.

Artificial Grape Oil

Benzyl Butyrate	10½ fl. oz.
Methyl Anthranilate	4½ fl. oz.
Methyl Salicylate	½ fl. oz.
Amyl Valerianate	½ fl. oz.
Fluid Extract Valerianate	3 fl. oz.
Port Wine	75 fl. oz.
Alcohol	150 fl. oz.
Grape Juice	50 fl. oz.
Glycerine	25 fl. oz.

Mix the first five with the alcohol, then add the other materials one at a time in the order given, stirring well after each addition. Let stand for 24 hours and filter.

Artificial Grape Syrup Form

Artificial Grape Oil	6 oz.
Tartaric Acid	2¾ lb.
Cream of Tartar	2 oz.
Tannic Acid	15 gm.
Grain Alcohol	3 pt.
Sugar Syrup	7 pt.

Color sufficiently to give the desired shade.

The syrup is made by dissolving 7 pounds granulated sugar in sufficient water to make one gallon.

Artificial Grape Flavor (Powder)

Tartaric Acid	2¾ lb.
Cream of Tartar	2 oz.
Tannic Acid	15 gr.
Granulated Sugar	10 lb.
Concentrated Grape Oil, Artificial	6 oz.

Mix the tannic acid with cream of tartar. (The tannic acid may be omitted if desired.) This should be mixed thoroughly, then mix this with about ½ pound of the acid (fine powdered.) Mix well, then work in the remaining acid in lots of ½ pound at a time, thorough mixing being essential. It is best done by sieving several times, mixing well after each sieving. Now work in the sugar the same way, so that the whole forms a perfectly even mixture. Now slowly work in the artificial grape oil, mixing thoroughly. Sufficient color is added to give the required shade when dissolved in water. Mix thoroughly and spread out until dry, then rub again through a sieve and put up in packages.

As the color will vary in strength, it will be necessary to experiment a little to get the exact quantity required to give the desired color when the product is made up into a finished drink.

In the strength given here, a teaspoonful will be sufficient to flavor strongly a quart of water.

Powdered Flavors

Put about 4 ounces of the powder into a mortar and spray or drop the mixed flavoring materials over it slowly, mixing well. When all have been added, gradually add the remainder of the acid, mixing well after each addition. The color should be dissolved in the flavoring mixture before adding the acid. When well mixed, place in a glass dish and stir often until it has dried out sufficiently to admit of packing. Best put up in glass bottles with closely fitting stoppers, but may be put up in cans. The quantity is sufficient for 45 gallons of liquid.

Raspberry

The base as above	1	lb.
Artificial Oil of Raspberry	1½	oz.
Bordeau S. Amaranth Color	2 to 5	gr.
Artificial Vanilla Flavor	1	dr.

Strawberry

The base as above	1	lb.
Ponceau 3 R Color	2 to 5	gr.
Artificial Oil of Strawberry	1½	oz.
Artificial Vanilla Flavor	1	dr.

Cherry

The base as above	1	lb.
Artificial Oil of Cherry	1½	oz.
Bordeau S. Amaranth Color	10	gr.

Pineapple

The base as above	1	lb.
Artificial Pineapple Oil	1½	oz.
Napthol Yellow, Color	10	gr.

Wild Cherry

Oil of Wild Cherry. (See formula below)	½	pt.
Distilled Water	½	gal.
Cologne Spirits	½	gal.
Red Color	¼	fl. oz.

Mix water and Cologne Spirits. Add the oil of Wild Cherry, mix and add the color. Mix well.

Use to:

1 gallon Simple Syrup.
1 ounce Extract.

Oil of Wild Cherry

Acetic Ether	10	fl. oz.
Benzoic Ether	5	fl. oz.
Oil of Bitter Almonds	5	fl. oz.
Amyl Valerianic Ether	2	fl. oz.
Benzoic Acid	2	fl. oz.
Glycerine	8	fl. oz.
Cologne Spirits	6	pt.

* Coffee Extract

Roast ground fresh coffee is percolated with hot water until exhausted. 5–20% Glycerin is then added. Excess water is driven off by heating in vacuo at temperatures up to 90° C.

Artificial Oil of Raspberry

Acetic Ether	5	oz.
Formic Ether	1	oz.
Methyl-Salicylic Ether	1	oz.
Nitrous Ether	1	oz.
Oenanthic Ether	1	oz.
Sebacylic Ether	1	oz.
Butyric Ether	1	oz.
Benzoic Ether	1	oz.
Amyl-Sutyric Ether	1	oz.
Succinic Acid	1	oz.
Saturated Solution Tartaric Acid in cold Alcohol	5	oz.
Glycerine	4	oz.
Tincture of Orris	100	oz.

Mix the succinic acid with the tincture, add the others and, lastly, the glycerine. One ounce of pure vanilla extract will improve this.

Artificial Oil of Pineapple

Amyl Butyrate	1	oz.
Butyric Ether	4	oz.
Sebacic Ether	1	oz.
Acetic Ether	4	dr.
Amyl Acetate	4	dr.
Pineapple Juice	4	dr.
Glycerine C. P.	4	oz.
Alcohol	50	oz.

Mix, adding glycerine last.

Artificial Oil of Peach

Ethyl Formate	5	oz.
Ethyl Butyrate	5	oz.
Ethyl Acetate	5	oz.
Ethyl Sebacate	1	oz.
Ethyl Valerianate	5	oz.
Oil of Bitter Almonds	5	oz.
Aldehyde	2	oz.
Glycerine	5	oz.
Amyl Alcohol	2	oz.

Alcohol enough to make up 100 ounces.

Artificial Oil of Cherry

Ethyl Benzoate	5	oz.
Ethyl Oenanthate	1	oz.
Ethyl Acetate	5	oz.
Benzoic Acid	1	oz.
Glycerine	3	oz.
Oil of Bitter Almonds	½	oz.

Alcohol enough to make up 100 ounces.

Artificial Vanilla Flavor

Vanillin	6	dr.
Cumarin	2	dr.
Alcohol	2	pt.

Water	5 pt.
White Sugar Syrup	1 pt.
Glycerine C. P.	1 pt.

Caramel color enough to give the desired shade.

Dissolve the vanillin and cumarin in the alcohol, then add the other materials and let stand for a few days before using. If not clear, filter. The syrup is made by dissolving 12 ounces of sugar in water enough to make a pint of syrup.

Artificial Oil Strawberry

Ethyl Butyrate	5 oz.
Ethyl Formate	1 oz.
Ethyl Salicylate	1 oz.
Ethyl Nitrate	1 oz.
Ethyl Acetate	5 oz.
Amyl Acetate	3 oz.
Glycerine C. P.	2 oz.

Alcohol enough to make up 100 ounces.

Beverage Colors
(Vegetable)

Yellow

Tincture of Turmeric

Turmeric (ground)	1 lb.
Dilute Alcohol	10 pt.

Exhaust by maceration and percolation. Keep in a dark place.

Saffron	1 lb.
Alcohol	5 pt.
Water	5 pt.

Mix alcohol and water and add saffron. Allow this mixture to stand in a warm place for several days, with occasional agitation, then filter. The tincture thus prepared has a deep orange color and when diluted, or used in small quantities, gives a beautiful yellow tint to syrups.

Orange

Solution of Annatto

It is prepared by dissolving pure annatto in alcohol, making it of any desired strength. Pure annatto only should be employed. Ordinary annatto used for dyeing may be purified by dissolving in a weak solution of sodium carbonate or other alkali by the aid of heat. Let cool, and add pure dilute sulphuric acid, drop by drop, stirring constantly, until the soda is neutralized. The pure annatto which precipitates must be washed thoroughly with water and dried.

This solution may be used for coloring ices and various other articles.

Red

Liquid Cochineal

Powdered Cochineal	1 oz.
Carbonate of Potassium	½ oz.
Alum	½ oz.
Bitartrate of Potassium	1 oz.
Alcohol	1 fl. oz.
Glycerine (C. P.)	6 fl. oz.
Water, enough to make	16 fl. oz.

Triturate the cochineal intimately with the carbonate of potassium and 8 fluid ounces of distilled water, then add the alum and bitartrate of potassium successively, put the mixture in a capacious vessel to boil, then set aside to cool, add alcohol and glycerine, filter the same and pass enough distilled water through the filter to make 16 fluid ounces.

Carmine Solution

Carmine, best	480 gr.
Ammonia Water	6 fl. oz.
Glycerine	6 fl. oz.
Water, to make	16 fl. oz.

Triturate the carmine to fine powder in a wedgwood mortar, gradually add the ammonia water, and afterwards the glycerine, under constant trituration. Transfer the mixture to a porcelain capsule, and heat on a water bath, stirring constantly, until the liquid is entirely free from ammoniacal odor. Then cool and add enough water to make 16 fluid ounces.

Carmine solution may also be prepared by triturating the carmine with just enough solution of potassa to dissolve it, then adding 2 fluid ounces of alcohol and enough water to make 16 ounces. Or, instead of the solution of potassa, use sufficient saturated solution of borax to dissolve the carmine, then add enough water to make 16 fluid ounces.

Carmine solution makes a brilliant color, and is largely employed, but it is not a satisfactory preparation to use at the soda fountain because the syrups are acid as a rule and will separate the carmine from its alkaline combination and cause its precipitation.

Brown Red

Compound Tincture of Cudbear

Cudbear, powder	120 gr.
Caramel	1½ av. oz.
Alcohol, of each	Sufficient
Water, of each	Sufficient

Macerate the cudbear with 12 fluid ounces of a mixture composed of 1 volume of alcohol and 2 of water for 12 hours, agitating frequently, then filter. Add the caramel, previously dissolved in 2 fluid ounces of water, and then pass through the filter enough of the beforementioned alcohol water mixture to make the whole liquid measure 16 fluid ounces.

This preparation may also be made by dissolving 1½ ounces of caramel in 2 fluid ounces of water, adding 4 fluid ounces of tincture of cudbear and then enough of a mixture composed of 1 volume of alcohol and 2 of water to make the whole measure 16 fluid ounces.

Chlorophyll

This may be employed in alcoholic solution for coloring preparations of a green tint. It may be purchased or it may be prepared as follows:

Digest leaves of grass, nettles, spinach, or other green herb, in warm water, until soft; pour off the water, and crush the herb to a pulp. Boil this for a short time with a ½ per cent solution of caustic soda, and afterwards precipitate the chlorophyll by means of dilute hydrochloric acid; wash the precipitate thoroughly with water, press and dry it, and use as much for the solution as may be necessary.

Tincture of Grass

Lawn Grass, fresh, cut fine	2 av. oz.
Alcohol	16 fl. oz.

Put the grass in a wide mouth bottle and pour the alcohol upon it. After standing a few days, agitating occasionally, pour off the liquid.

This is a useful preparation for giving a green color to essences, syrup of violets, etc. It can be used with alcohol or water.

Purple

Tincture of Litmus

Litmus, powder	2½ av. oz.
Water, boiling	16 fl. oz.
Alcohol	3 fl. oz.

Pour the water upon the litmus, stir well, allow to stand for about an hour, stirring occasionally, filter, and to the filtrate add the alcohol.

Root Beer Emulsion

Gum Arabic	17 lb.
Water	6½ gal.

Heat and stir until dissolved. Filter through cheese cloth.

Formula No. 1

(To make 10 gallons of concentrate.)

Oil of Wintergreen (synthetic)	1 pt.
Oil of Sassafras	1 pt.
Vanillin	4 oz.
Coumarin	1 oz.

Formula No. 2

(To make 10 gallons of concentrate.)

Oil of Wintergreen (synthetic)	2 qt.
Oil of Sassafras	1 qt.
Oil of Cloves	1 pt.
Oil of Cassia	1 pt.

Emulsification or absorption of the oils in the gum solution can be accomplished by simply gradually puring the oils into the gum solution, while the same is being vigorously agitated, it should be agitated for at least 10 minutes and as this process is going on you will note the tendency of the gum solution to thicken. If you have an emulsifier, so much the better; emulsifier, however, is not essential, as a good stirring with a wooden paddle or a large size cream whipper will do the work. While you are mixing this solution, about one gallon of sugar coloring (caramel coloring) should be added to give the concentrate a dark color. The resulting product is then made up with water to make exactly 10 gallons and then given another thorough stirring before being placed in container. One gallon of this product will flavor 20 barrels of root beer.

East India Lemon Sour Extract

Oil of Lemon	6 oz.
Oil of Limes	2 oz.
Alcohol, 95 per cent	½ gal.
Warm Water	½ gal.
Alum	½ dr.

Add the oils to the alcohol and shake well. Dissolve the alum in the water. Add the water gradually in small quantities, shaking well after each addition. Set aside to settle for 6 hours. A scum will form on top. Separate extract from this with rubber hose. Filter clear through magnesia.

Use to:

1 gallon Simple Syrup.
2½ ounces Lemon Sour Extract.
3 ounces Lemon Sour Acid.
½ ounce Yellow Color.

Ciderette Syrup

Sugar	46	lb.
Water	6½	gal.
Soluble Lemon Extract	4	fl. oz.
Butyric Ether	1	fl. oz.
Sugar Color	1½	fl. oz.
Citric Acid Solution	50	fl. oz.

Use to:

8-ounce soda bottle.
1 to 1¼ ounce Syrup.

Imitation Apple Flavor

Amyl Valerianate	6	oz.
Ether Acetic	3	oz.
Spirits of Nitrous Ether	3	oz.
Amyl Butyrate, Absolute	1	oz.
Aldehyde	½	oz.
Essence of Peach Blossom	½	oz.

Alcohol 95 per cent, enough to make 1 quart.

Cheap Apple Cider

Boiled Cider	2	gal.
Granulated Sugar	25	lb.
Tartaric Acid	¾	gal.
Water	30	gal.

Color to suit with sugar color. Thoroughly mix; let stand three days, then draw off and add one ounce of benzoate of soda to each ten gallons of cider. Keep in a cool place.

Sweet Artificial Drinking Cider

Boiled or Condensed Cider	8	gal.
Granulated Sugar	10	lb.
80 per cent Acetic Acid	9½	oz.

Water enough to make up to 50 gallons.

Note: A leading firm tells us that they have put up thousands and thousands of barrels of drinking cider using the above formula and it has given excellent results.

Strawberry Basic Ether Wild

Wintergreen Oil	6	dr.
Ceylon Cinnamon Oil	6	dr.
Vanillin	12	dr.
Coumarin	3½	oz.
Nerolin	5	oz.
Ethyl Benzoate	6	oz.
Methyl Salicylate	½	lb.
Ethyl Butyrate	½	lb.
Ethyl Acetate	1½	lb.
Benzyl Acetate	1½	lb.
Amyl Acetate	3	lb.
	8	lb.

Vanilla Basic Ether

Cardamom Oil	1½	oz.
Cinnamon Oil	2½	oz.
Clove Oil	4	oz.
Ethyl Oenanthate	8	oz.
Vanillin	8	oz.
Amyl Acetate	1½	lb.
Ethyl Acetate	3	lb.
	6	lb.

Cherry

Oil of Cherry Artificial

Amyl Acetate	6	fl. oz.
Amyl Butyrate	3	fl. oz.
Benzoic Ether	3	fl. oz.
Oil of Bitter Almonds (free from prussic acid)	8	fl. oz.
Oil of Lemon	2	fl. oz.
Oil of Orange	1	fl. oz.
Oil of Cloves	½	fl. oz.
Glycerine	10	fl. oz.
Oil of Cardamom	6	dr.
Cologne Spirits	30	fl. oz.

Cherry Compound

Dry Citric Tartaric Acid (½ Citric and ½ Tartaric)	1¼	lb.
Extract Cherry Concentrated	1	pt.
Vegetable Red Color in liquid form	8	oz.

Water, enough to make 1 gallon.

Brewed Ginger Ale

This gives a true flavored ginger ale.

Fifty barrels of hot water are run into the kettle and heated to boiling. Six hundred pounds of granulated sugar are now added, making sure that the same dissolves properly. This having been accomplished, seventy-five pounds of powdered ginger, twenty-one pounds of crystallized citric acid and eight ounces of powdered capsicum are introduced into the solution, which is permitted to boil for half-hour. Eighteen pounds of good quality hops are now added and the solution boiled for an additional three-quarters of an hour, whereupon it is made up to a volume of, at least, fifty-two barrels, cooled over the Baudelot cooler and run into a settling tub, where it is permitted to remain overnight.

The following morning the clear supernant liquid is withdrawn or, to work more economically, the whole solution may be filter-pressed and run into a clean vat or fermenter.

Having reached this stage, the beverage may be treated in one of two different ways. Either five barrels of this solution may be withdrawn, pitched with yeast and permitted to ferment completely and after completed fermentation freed of the yeast by filtration, returned to the main portion of the solution and stored for, at least, ten days. If preferable or more convenient, instead of withdrawing a portion of the solution to be completely fermented and subsequently returning the same, the entire solution can be carefully checked fermented by pitching with the customary amount of yeast and permitting the gravity to decrease no more than 0.8 of one per cent, after which the solution or beverage must be chilled almost to freezing, filtered and run into a clean and sterile vat, where it is to be stored for a period of ten days. The beverage is carbonated and filtered in the usual manner, as practiced in the manufacture of cereal beverages. It is advisable to carbonate twice, after which the beverage is ready for bottling.

The bottled ginger ale may be pasteurized if desired, although this is not necessary. If sold in bulk it is to be racked into freshly pitched packages and can be shipped without any danger of fermentation.

Soluble Ginger Ale Extract

(To be used in the proportion of 4 ounces of extract to 1 gallon of syrup.)

Jamaica Ginger, in fine powder	8 lb.
Capsicum, in fine powder	6 oz.
Alcohol, a sufficient quantity.	

Mix the powders intimately, moisten them with a sufficient quantity of alcohol and set aside for 4 hours. Pack in a cylindrical percolator and percolate with alcohol until 10 pints of percolate have resulted. Place the percolate in a bottle of the capacity of 16 pints, and add to it 2 fluid drams of oleoresin of ginger; shake, add 2½ pounds of finely powdered pumice stone, and agitate thoroughly at intervals of one-half hour for 12 hours. Then add 14 pints of water in quantities of 1 pint at each addition, shaking briskly meanwhile. This part of the operation is most important. Set the mixture aside for 24 hours, agitating it strongly every hour or so during that period. Then take

Oil of Lemon	1½ fl. oz.
Oil of Rose (or geranium)	3 fl. dr.
Oil of Bergamot	2 fl. dr.
Oil of Cinnamon	3 fl. dr.
Magnesium carbonate	3 fl. oz.

Rub the oils with the magnesia in a large mortar and add 9 ounces of the clear portion of the ginger mixture to which have been previously added 2 ounces of alcohol, and continue trituration, rinsing out the mortar with the ginger mixture. Pass the ginger mixture through a double filter and add through the filter the mixture of oils and magnesia; finally pass enough water through the filter to make the resulting product measure 24 pints, or 3 gallons. If the operator should desire an extract of more or less pungency he may obtain his desired effect by increasing or decreasing the quantity of powdered capsicum in the formula.

Lemon Extract (Terpeneless)

Oil of Lemon	30 lb.
Citral	8 oz.
Cologne Spirits	16 gal.

Put in a churn and work 2 hours. Of 11 gallons of water, add gradually about 5 gallons every hour and work for two hours more, then add 3 gallons water and work more. The whole process takes about 10 hours. After ten hours add 1½ gallons Cologne Spirits. Let stand for 48 hours and filter.

Use to:

1 gallon Simple Syrup.
1 ounce Extract.

A Root Beer

Oil of Sweet Birch or Methyl Salicylate	15 oz.
Oil of Cloves	¾ oz.
Oil of Sassafras	⅜ oz.
Oil of Lemon	½ oz.
Oil of Cassia	⅛ oz.
Mexican Vanilla Extract (best quality)	6 pt.
Cologne Spirits	15 pt.
Caramel	4 oz.
Oil of Nutmeg	2 dr.

Dissolve the oils in the alcohol. Mix the caramel with the vanilla extract. Pour the colored vanilla extract into the alcoholic solution. A brown precipitate will form, which acts as the clarifying agent, and may be filtered out after standing an hour to two hours.

Use to:

1 gallon Simple Syrup (10 pounds sugar to gallon water).
1 ounce Root Beer Extract.
¼ ounce Citric Acid Solution.
2 ounces Caramel.

B

Oil of Wintergreen	20	oz.
Oil of Sassafras	24	oz.
Oil of Anise	10	dr.
Oil of Cassia	1	oz.
Cologne Spirits	3½	gal.
Water	½	gal.

Use to:

1 gallon Simple Syrup.
2 ounces Extract.

C

Oil of Sassafras	2½	fl. oz.
Oil of Wintergreen	2½	fl. oz.
Oil of Sweet Orange	2	fl. oz.
Amyl Butyrate	2	fl. oz.
Oil of Spruce	½	fl. oz.
Oil of Cloves	2	dr.
Oil of Anise	2	dr.
Cologne Spirits	7	pt.
Water	2	pt.

Add a little at a time the oils to the Cologne Spirits, shake well, add the 2 pints of water and filter through pumice.

Use to:

1 gallon Simple Syrup.
1½ ounces Extract.

D

Oil of Sassafras	5	fl. oz.
Oil of Peppermint	½	fl. oz.
Oil of Tar	10	drops
Oil of Cinnamon	10	drops
Carbonate Magnesius	4	av. oz.
Cologne Spirits	½	gal.
Water	½	gal.

Use to:

1 gallon Simple Syrup.
½ fl. ounce Extract.
2 fl. ounces Sugar Color.

Sarsaparilla Extract

Oil of Wintergreen	4	oz.
Oil of Sassafras	4	oz.
Oil of Anise	1	oz.
Cologne Spirits	5	pt.
Powdered Pumice Stone	4	oz.
Granulated Sugar	8	oz.
Water	2½	pt.
Sugar Color	1	oz.

Dissolve the oils in two pints of the spirits. Each oil must be added separately and well shaken with the spirits before another oil is added. Now put the pumice stone and sugar in a Wedgewood mortar, add the mixture gradually and rub together to a paste. Mix the remainder of the spirits and water together, add the sugar color to these, and dissolve carefully. Mix the whole together gradually, stirring well until all combines, and filter through filter paper.

Use to:

1 gallon Simple Syrup.
1 ounce Extract.

Plain or Simple Syrup

Granulated Cane Sugar	30 lb.
Water (boiling)	7 qt.

Pour the sugar into the water gradually, stirring meanwhile, and when dissolved, strain through coarse cotton cloth. Do not cover container until thoroughly cooled. This will produce four gallons of syrup. The relative proportions of sugar and water are very important since, if a smaller amount of sugar is employed, fermentation sooner or later will ensue. If too much sugar is used, crystallization will surely follow, resulting in a liquid too thin to keep under ordinary temperature.

Beverage Acidulants

Citric Acid Crystals	4 lb.
Boiling Water	4 pt.

When dissolved, filter through filter paper using glass funnel. Keep in glass and avoid contact with metal.

Tartaric Acid Solution

Tartaric Acid Crystals	4 lb.
Boiling Water	4 pt.

Treat the same as above.

Mixed Acid Solution

Tartaric Acid Crystals	2 lb.
Boiling Water	4 pt.
Citric Acid Crystals	2 lb.

Treat the same as above.

Phosphoric Acid Solution

Phosphoric Acid 85%	4 lb.

Cold water to make one gallon.

Stand over night and filter through paper. Mix acid in stone jar and keep in glass bottle.

Cherry Acid Solution

Citric Acid	2½ av. lb.
Tartaric Acid	2½ av. lb.
Hot Water	1 gal.

Thoroughly dissolve and add Phosphoric Acid syrupy 2 fluid ounces.

Compound Cider Acid

Citric Acid Crystals	5 lb.
Tartaric Acid	5 lb.
Acetic Acid, pure 80%	1 pt.
Phosphoric Acid Syrupy	1 pt.

Place all the acid in a stone jar and add two or three gallons of boiling water, stirring until all is dissolved. Add water to make 6 gallons.

Foam Producers

Soap Bark Foams

Formula A—

Quillaja bark is used in the form of tincture and may be prepared as follows:

Quillaja, fine chips	5½ av. oz.
Alcohol	10 fl. oz.
Water	Sufficient

Mix the drug with 24 fluid ounces of water, boil for 15 minutes. Strain and add enough water through the strainer to make the volume equal to 22 fluid ounces. Mix the liquid when cool with the alcohol, let stand for 12 hours, filter, and to the filtrate add enough water to measure 32 fluid ounces.

If a cheaper preparation is desired, the alcohol may be replaced by water or by glycerine. If the former be used, the preparation must be preserved by the addition of a small amount of salicylic acid solution. Either of the latter is to be preferred to the alcoholic solution, as the alcohol has the tendency to cause premature expulsion of gas from the soda when served.

About one fluid ounce of this preparation is usually sufficient for one gallon of syrup.

Formula B—

Soap Bark (chips)	1 lb.
Boiling Water	10 pt.
Alcohol (95%)	1 pt.

Boil the soap bark in the water for 30 minutes. Allow to cool. Add the alcohol. Pack a small quantity of dry soap bark in a percolator to make a bed and percolate. One-half to 1 ounce of this is used per gallon of syrup.

Sapinone Foams

Formula A—

Sapinone	1 lb.
Glycerine	½ gal.
Water	½ gal.

Dissolve the sapinone in ½ gallon of clear water, then add glycerine. Use ½ dram to 1 gallon or 1 ounce to 15 gallons of syrup.

Formula B—

Sapinone	24 av. oz.
Water	1 gal.

Dissolve sapinone in water by agitation and when dissolved add

Formaldehyde	2 fl. dr.

Use 1 dram to 1 gallon or 1 ounce to 15 gallons of syrup.

Ginger Ale Extract

Oleo Resin Ginger	15 oz.
Oleo Resin Capsicum	2 oz.
Lemon Extract	5 pt.
Orange Extract	2½ pt.
Alcohol and Water	2 gal.

Use to:
1 gallon Simple Syrup.
3 ounces Extract.

Ginger Ale Extract (Belfast)

Oleo Resin Ginger	24 oz.
Oleo Resin Capsicum	5½ oz.
Oil of Lemon (Terpeneless)	36 oz.
Oil of Orange (Terpeneless)	12 oz.
Oil of Cassia	1½ dr.
Oil of Rose, Artificial	½ dr.
Oil of Cloves	1½ dr.
Cologne Spirits	5½ gal.
Water	3 gal.

Use to:
1 gallon Simple Syrup.
2 ounces Extract.

Ginger Ale Extract

Oil of Ginger	4 oz.
Oil of Capsicum	1 oz.
Lemon Extract	16 oz.
Orange Extract	8 oz.
Alcohol	3½ pt.
Water	3½ pt.

Ginger Champagne Syrup

Sugar	46 lb.
Water	6½ gal.

Soluble Extract of Ginger	8	fl. oz.
Soluble Extract of Orange	16	fl. oz.
Soluble Extract of Lemon	8	fl. oz.
Sulphurous Acid	4	fl. oz.
Vanilla Extract	2	fl. oz.
Sugar Color	2½	fl. oz.
Citric Acid Solution	32	fl. oz.

Use to:

⅛ ounce soda bottle.
1 to 1¼ ounces of above.

Cola Flavor:

Oil of Lemon	120 drops
Oil of Sweet Orange	80 drops
Oil of Nutmeg	40 drops
Oil of Cinnamon	40 drops
Oil of Coriander	20 drops
Oil of Neroli	40 drops
Alcohol, 95 per cent	1 qt.

Add in rotation, shaking well before adding next ingredient, and let stand 48 hours when it is ready to use.

Cola Syrup:

Sugar	60 lb.
Water	5 gal.

Dissolve sugar and bring to boil. Then, while boiling, stir in syrup:

Beet Sugar Color	40	fl. oz.

Let cool and add:

Phosphoric Acid Syrup	3	fl. oz.
Alkaloid of Caffeine (Dissolved in 8 ozs. boiling water.)	1½	av. oz.
Fluid Extract Cola Leaves	1½	fl. oz.
Fluid Extract Kola Nuts	2½	fl. oz.
Alcohol	1	pt.
Extract of Vanilla	5	fl. oz.
Cola Flavor	4	fl. oz.
Glycerine	4	fl. oz.
Lime Juice	16	fl. oz.

Let age for three days.

Note: The sugar color used in all Cola drinks must be the best.

Extract of Limes

Oil of Limes	6	fl. oz.
Alcohol, 95 per cent	½	gal.
Distilled Water	½	gal.

Orange

Orange Extract

Sweet Orange Oil	64	oz.
Oil of Bitter Orange	32	oz.
Grain Alcohol	6	gal.
Water	6	gal.

Use to:

1 gallon Simple Syrup.
1 ounce Extract.

Concrete Orange Extract

Gum Tragacanth	2	dr.
Glucose	4	oz.
Concentrated Oil of Orange	½	oz.
Oil of Bitter Orange	1	oz.
Concentrated Tincture of Orange	12	oz.
Citral	15	gr.

Use to:

1 gallon Simple Syrup.
1 ounce Extract.

Orange Champagne Syrup

Sugar	46	lb.
Water	6¼	gals.
Soluble Orange Extract	3	fl. oz.
Soluble Lemon Extract	20	fl. oz.
Soluble Lime Extract	5	fl. oz.
Sulphurous Acid	3	fl. oz.
Citric Acid Solution	3	fl. oz.
Orange color to suit.		

Use to:

Each pint bottle 2¼ ounces Extract.

Orange Extract

Alcohol (94 per cent)	2½	gal.
Terpeneless Oil of Orange	11	dr.
Neroli	5	drops
Water	2½	gal.

Orange Cider Compound

Acid Citric and Acid Tartaric (½ of each)	1¼	pt.
Orange Extract (above)	1	pt.
Sugar color	6	oz.
Vegetable Red Color (any red color will do)	2	oz.
Water, enough to make	1	gal.

Orangeade Substitute Powder

Terpeneless Oil of Orange	1	dr.
Orange Color No. 1	2 to 5	gr.
Citric Acid	12	oz.
Powdered Sugar	4	oz.

Mix the oil of orange with about 1 ounce of alcohol. Put the sugar in a mortar and gradually add the solution of the oil. The color should be dissolved in the oil solution before mixing with the powder. After the liquid has been mixed with the sugar, add the Citric Acid gradually, mixing well after each addition.

The same remarks concerning the use of the terpeneless oil apply to this as to the lemon powder. If ordinary oil of sweet orange is used, it will be necessary to employ at least one ounce to get as strong a product as made with the quantity of the terpeneless oil specified. Furthermore, a product made with the ordinary oil will not possess the same keeping qualities as one made with the terpeneless oils, as they will gradually acquire a turpentine odor and flavor by the oxidation of the terpenes contained in the oil.

Beer

3½ gallons of water with 11 pounds of fancy brewing malts and 1 pound of Soy Grits (Kreemko). The water in this malt mash is first raised to a temperature of 122 degrees F. and the malt together with the Body Grits is run in and the temperature again raised to 122 and maintained for 2 hours.

The cooker mash is made at the same time using 1 pound of malt and 4 pounds of rice and 3½ gallons of water. The water used in this cooker mash is raised to 122 degrees F. and the malt and the rice are run in and the mash brought again to 122 degrees F. and this temperature maintained for one hour after which temperature it is quickly raised to 170 degrees and maintained for ½ hour, after which it is raised to boiling and boiled for ½ hour. It is then emptied quickly into the malt mash from which the lauter has first been drawn, the lauter being nearly all of the liquid portion of the malt mash. This proportion of materials in the mashes and heated in the manner described will give a temperature of mixed mashes of 170 degrees Fahrenheit, which temperature is maintained for 20 minutes and then raised to 175 for complete saccharification. The lauter which was withdrawn from the malt mash is added back into the combined mashes immediately when the temperature of 170 degrees is reached.

This method is one in cereal chemistry producing a larger quantity of dextrin material, reducing the fermentable matter. After running off the first wort at about 18.5 balling the grains are sparged down to one yelding 11½ gallons of wort in the kettle. The Soy Grits (Kreemko) are employed in this manner in order to completely peptonize the protein content it contains. The material should not be incorporated in the cooker because Soy Body Grits contain but a trace of starch and need no cooking as the carbohydrate content amounting to about 35% is in the form of soluble dextrins and sugars.

Yogurt or Bulgarian Buttermilk

Propagate a small culture of the Bacillus Bulgaricus from day to day as indicated for the lactic culture for buttermilk. This culture may be obtained from various commercial laboratories. To prevent contamination by yeasts or gas-forming bacteria, it is necessary to carry this culture at a temperature of about 110° F. A small egg incubator may be used for this purpose.

Carry in a similar way a culture of the ordinary sour-milk organism, which may be obtained from many of the commercial laboratories.

Thoroughly pasteurize the milk to be fermented. If a small quantity—5 to 10 gallons, for instance—is to be made, it may be done by holding a can of milk in a tub or vat of water heated by a steam hose. If a larger quantity is made, one of the starter cans used in creameries will be found convenient. These are essentially cylindrical vats with mechanical stirrers and a jacket which can be filled with steam for heating or water for cooling. The milk should be held at a temperature of at least 180° F. for not less than 30 minutes.

Cool the milk to about 100° F. Draw off one-half and inoculate it with the culture obtained in the second operation. Inoculate the remaining half with Bulgaricus culture obtained in the first operation. The amount to be added will depend on the quantity of milk to be fermented, the time at which it is desired to have it curdled, and the temperature maintained during the fermentation. This can best be determined by experience. One pint should be sufficient for any amount between 10 and 20 gallons.

Buttermilk Lemonade

A refreshing and nutritious drink may be made by the addition of lemon juice and sugar to buttermilk, following the same procedure as in making ordinary lemonade. It will usually be found necessary to use more sugar and more lemon juice than in making lemonade with water. Buttermilk lemonade should be served very cold.

Kefir or Koumiss

Use buttermilk or freshly curdled sour milk. This should be thoroughly agitated to break the curd into fine particles. Buttermilk containing Bacillus Bulgaricus will give a flavor too acid for most tastes.

Add 1 per cent cane sugar (1½ oz. to the gallon). Add a small amount of yeast cake—one-fourth of a cake will be sufficient for 1 gallon of buttermilk. The yeast cake should be ground up in water so that it will be well distributed.

Bottle this preparation, leaving sufficient space to permit a thorough shaking of the contents. Strong round bottles of the type used for carbonated drinks should be used, as considerable pressure is developed by the fermentation. If the bottle is not provided with a sealing device the corks must be securely tied or wired in place.

Hold for 4 or 5 days at a temperature of 65 to 70° F., shaking every day to keep the curd well broken up. At the end of this time there should be considerable gas but not enough to blow the milk out of the bottle. It should have a pleasant acid taste with a slight bitterness. The fresh milk sometimes has a yeasty taste but this gradually disappears. If the milk is kept on ice it will remain in good condition for two weeks or more.

Carbonated Milk

The best results are secured when newly pasteurized milk or cleanly drawn fresh milk is treated with carbon dioxide in a tank, such as is used in bottling establishments in preparing carbonated drinks, and then placed in siphon bottles. When charged under pressures of from 70 to 175 pounds and kept at temperatures ranging from 35° to 60°, bottles of clean fresh milk or pasteurized milk kept from four to five months without perceptible increase in acidity.

Milk carbonated under a pressure of 70 pounds comes from the bottle as a foamy mass, more or less like kumiss that is two or three days old. It has a slightly acid, pleasant flavor, due to the carbon dioxide, and has a somewhat more salty taste than ordinary milk. In the case of carbonated milk pasteurized at 185° F., there is, of course, something of a "cooked" taste. Though the cream separates in the bottle, it is thoroughly remixed by a little shaking as the milk comes from the bottle and there is no appearance of separate particles of cream. All who have had occasion to test the quality of carbonated milk as a beverage agree in regarding it as a pleasant drink. In the case of milk bottled under a pressure of 150 pounds of carbon dioxide, the milk delivered from the siphon is about the consistency of whipped cream, but, on standing a short time, it changes into a readily drinkable condition. From the experience had, it would seem that carbonated milk might easily be made a fairly popular beverage.

Malted Milk Powders

50 parts Powdered Malt Extract
20 parts Powdered Skimmed Milk
30 parts Cane Sugar

Mix well. One teaspoonful when added to 8 ounces of a mixture of chocolate syrup, milk and ice cream and then mixed with the malted milk machine will make a delicious malted milk drink.

THE MANUFACTURE OF BUTTERMILK FROM SKIMMED MILK

The finest quality of buttermilk is probably that produced by churning clean-flavoured cream which has been properly ripened with the aid of a pure culture of lactic acid. Surplus skimmed-milk, may, however, in many cases, be profitably converted into an artificial buttermilk of practically the same composition and quality as the natural buttermilk.

In making artificial buttermilk the skimmed-milk may or may not be pasteurized. In either case about 10 per cent of clean flavoured lactic acid culture should be added to the skimmed-milk which is maintained at a temperature of 70° F. until coagulation takes place. If the time required to produce coagulation is too long the process should be hastened by increasing the percentage of culture used, rather than by raising the temperature. Raising the temperature above 70° F. will usually result in a product of inferior flavour.

As soon as coagulation has taken place the curdled milk is transferred to the churn which is revolved for thirty to forty minutes as in churning cream. If the skimmed-milk is allowed to stand long after coagulation takes place before being churned, the whey and curdy matter of the finished product will show a greater tendency to separate. The churning breaks the curd into fine particles producing a smooth velvety buttermilk which is difficult to distinguish from a good natural product. As soon as the artificial buttermilk is drawn from the churn it should be strained to remove any particles of curd which may not have been broken up in the churning process. The temperature of the product should at once be reduced to at least 50° F. to retard the development of acidity and of undesirable flavours.

Artificial buttermilk may also be satisfactorily produced in a small way in the home. A clean fruit jar of suitable size may be partially filled with clean fresh skimmed-milk which is allowed to sour naturally at a temperature of 70° F. to 75° F. When coagulated, the milk should be vigorously shaken for a few minutes in the closed jar. It may now be strained to remove any lumps of curd not finely broken up by the agitation after which it should be kept in a cool place. If a clean pleasant flavour is obtained by such natural souring and the artificial buttermilk is to be made frequently, it is advisable to add a few ounces of the first artificial buttermilk to the next quantity of skimmed-milk to be soured. Thus the desirable flavour may be reproduced from time to time in the same manner as yeast is propagated.

The composition of such artificial buttermilk is practically the same as that of natural buttermilk, the only difference being that the latter usually contains slightly more milk fat. The percentage of milk fat in the artificial buttermilk may be increased to approximately that of natural buttermilk by adding to each one hundred pounds of skimmed-milk before souring, two quarts of whole milk.

Butter Substitute

1. Water	120
2. Galagum C	1
3. Cottonseed Oil	40
4. Caustic Soda	0.02
5. Butter Flavor	to suit

Dissolve 4 in 1 and strew 2 on surface; bring to a boil while stirring; run 3 and 5 into it slowly with high speed intermittent stirring.

---

* Butter and Honey Cream

Liquid Honey	92
Butter	8

The above are warmed and passed thru an homogenizer.

Candy, Yeast

Glycerol	18
Citric Acid	4
Epsom Salts	2
"Yeast-Foam" (live-yeast)	100
Tapioca Starch	200

The above is mixed intimately and is ready for use in candy mixtures.

---

* "Non-Blooming" Chocolate Coatings

Chocolate liquor is heated to about 46° and not more than 3.5% of finely divided solid gelatin is added; water not more than 1% of the entire quantity is added as the mixing proceeds and sufficient fat such as cacao butter is added to bring the fat content to about 30%, the temp. is raised to about 60° and is maintained at this point for about 15 hrs.

Spiced Chocolate—I

2500 g. cacao
2500 g. sugar
36 g. powdered cinnamon
19 g. powdered cloves
8 g. powdered cardamom seed

Spiced Chocolate—II

4000 g. cacao
130 g. starch flour
70 g. powdered cloves
4000 g. sugar
125 g. powdered cinnamon
33 g. powdered cardamom seed
6 g. Peru balsam

Spiced Chocolate—III

2500 g. cacao
2500 g. sugar
65 g. powdered cinnamon
4 g. powdered coriander seed
44 g. powdered cloves
1 g. oil of lemon
8 g. powdered cardamom seed

Spiced Chocolate—IV

2500 g. cacao
2500 g. sugar
5 g. powdered cloves
110 g. powdered cinnamon
25 g. powdered cardamom seed
4 g. powdered nutmeg

Spiced Chocolate—V

2500 g. cacao
1800 g. sugar
2 g. powdered cardamom seed
50 g. powdered Ceylon cinnamon
50 g. vanilla
1 g. powdered nutmeg

Spiced Chocolate—VI (Leipzig)

2500 g. cacao
3000 g. sugar
30 g. powdered cardamom seed
200 g. powdered cinnamon
130 g. powdered cloves

Spiced Chocolate—VII (Vienna)

2500 g. cacao
2500 g. sugar
20 g. powdered cardamom seed
110 g. powdered cloves
210 g. powdered cinnamon
25 g. Peru balsam

Coffee Chocolate

2000 g. cacao
2000 g. sugar
500 g. ground coffee

* Candy Jellies

Moderately Firm Pectin Jellies for Cast or Slab Work

Ingredients

Water	2½ gal.
100 Grade Exchange Citrus Pectin	12 oz.
Acetate of Soda (U.S.P.)	1½ oz.
Citric Acid (crystals or powdered)	2¼ oz.
Glucose (43° Bé.)	20 lb.
Granulated Sugar	20 lb.
Color and Flavor	as desired

Directions

(1) Put 2½ gallons of water in a kettle and heat hot (170° F.). (Open fire or steam-jacketed kettle may be used.)

(2) *Thoroughly mix* 12 ounces of 100 Grade Exchange Citrus Pectin with about 6 pounds of granulated sugar.

(3) Add the Pectin-Sugar mixture to the warm water as it is being stirred with a paddle. Continue to stir and heat to boiling. Boil vigorously for a moment.

(4) Combine the acetate of soda and citric acid. Dissolve in a small portion of hot water.

(5) Add the acetate of soda-citric acid solution to the kettle and then the 20 pounds of glucose. Heat to boiling again.

(6) Add the remainder of the sugar (14 pounds) and cook to 222°–224° F., or to a good "sheet." (This temperature corresponds to 75–78% total soluble solids at sea level. It is sufficient to cook the batch to 10°–12° F. above the boiling point of water at your factory.)

(7) Add the color and flavor, then cast into starch at once. This formula will produce about 48 to 50 pounds of candy. The finished piece may be crystallized, sanded, iced, or coated with chocolate.

Note: Cooking the batch to 224° F. is recommended for slab work.

Refined Corn Sugar may be substituted

for all or a part of the cane or beet sugar given in the above formula.

Tart and Moderately Firm Pectin Jellies for Cast or Slab Work
(Especially for Fruit Flavors)

Ingredients

Water	2½	gal.
100 Grade Exchange Citrus Pectin	12	oz.
Acetate of Soda (U.S.P.)	3	oz.
Citric Acid (crystals or powdered	4	oz.
Glucose (43° Bé.)	20	lb.
Granulated Sugar	20	lb.
Color and Flavor	as desired	

Directions

(1) Put 2½ gallons of water in a kettle and heat hot (170° F.). (Open fire or steam-jacketed kettle may be used.)

(2) *Thoroughly mix* 12 ounces of 100 Grade Exchange Citrus Pectin with about 6 pounds of granulated sugar.

(3) Add the Pectin-Sugar mixture to the warm water as it is being stirred with a paddle. Continue to stir and heat to boiling. Boil vigorously for a moment.

(4) Combine the acetate of soda and citric acid. Dissolve in a small portion of hot water.

(5) Add the acetate of soda-citric acid solution to the kettle and then the 20 pounds of glucose. Heat to boiling again.

(6) Add the remainder of the sugar (14 pounds) and cook to 222°–224° F., or to a good "sheet." (This temperature corresponds to 75–78% total soluble solids at sea level. It is sufficient to cook the batch to 10°–12° F. above the boiling point of water at your factory.)

(7) Add the color and flavor, then cast into starch at once. This formula will produce about 48 to 50 pounds of candy. The finished piece may be crystallized, sanded, iced, or coated with chocolate.

Note: Cooking the batch to 224° F. is recommended for slab work.

Refined Corn Sugar may be substituted for all or a part of the cane or beet sugar given in the above formula.

Firm Pectin Jellies for Cast or Slab Work

Ingredients

Water	3	gal.
100 Grade Exchange Citrus Pectin	15	oz.
Acetate of Soda (U.S.P.)	1½	oz.
Citric Acid (crystals or powdered)	2	oz.
Glucose (43° Bé.)	20	lb.
Granulated Sugar	20	lb.
Color and Flavor	as desired	

Directions

(1) Put 3 gallons of water in a kettle and heat hot (170° F.). (Open fire or steam-jacketed kettle may be used.)

(2) *Thoroughly mix* 15 ounces of 100 Grade Exchange Citrus Pectin with about 8 pounds of granulated sugar.

(3) Add the Pectin-Sugar mixture to the warm water as it is being stirred with a paddle. Continue to stir and heat to boiling. Boil vigorously for a moment.

(4) Combine the acetate of soda and citric acid. Dissolve in a small portion of hot water.

(5) Add the acetate of soda-citric acid solution to the kettle and then the 20 pounds of glucose. Heat to boiling again.

(6) Add the remainder of the sugar (12 pounds) and cook to 222°–224° F., or to a good "sheet." (This temperature corresponds to 75–78% total soluble solids at sea level. It is sufficient to cook the batch to 10°–12° F. above the boiling point of water at your factory.)

(7) Add the color and flavor, then cast into starch at once. This formula will produce about 48 to 50 pounds of candy. The finished piece may be crystallized, sanded, iced, or coated with chocolate.

Note: Cooking the batch to 224° F. is recommended for slab work.

Refined Corn Sugar may be substituted for all or a part of the cane or beet sugar given in the above formula.

Tart and Firm Pectin Jellies for Cast or Slab Work
(Especially for Fruit Flavors)

Ingredients

Water	3	gal.
100 Grade Exchange Citrus Pectin	15	oz.
Acetate of Soda (U.S.P.)	2¾	oz.
Citric Acid (crystals or powdered)	3½	oz.
Glucose (43° Bé.)	20	lb.
Granulated Sugar	20	lb.
Color and Flavor	as desired	

Directions

(1) Put 3 gallons of water in a kettle and heat hot (170° F.). (Open fire or steam-jacketed kettle may be used.)

(2) *Thoroughly mix* 15 ounces of 100 Grade Exchange Citrus Pectin with about 8 pounds of granulated sugar.

(3) Add the Pectin-Sugar mixture to the warm water as it is being stirred with a paddle. Continue to stir and heat to boiling. Boil vigorously for a moment.

(4) Combine the acetate of soda and citric acid. Dissolve in a small portion of hot water.

(5) Add the acetate of soda-citric acid solution to the kettle and then the 20 pounds of glucose. Heat to boiling again.

(6) Add the remainder of the sugar (12 pounds) and cook to 222°–224° F., or to a good "sheet." (This temperature corresponds to 75–78% total soluble solids at sea level. It is sufficient to cook the batch to 10°–12° F. above the boiling point of water at your factory.)

(7) Add the color and flavor, then cast into starch at once. This formula will produce about 48 to 50 pounds of candy. The finished piece may be crystallized, sanded, iced, or coated with chocolate.

Note: Cooking the batch to 224° F. is recommended for slab work.

Refined Corn Sugar may be substituted for all or a part of the cane or beet sugar given in the above formula.

Tender Pectin Jellies for Cast Work

Ingredients

Water	2½ gal.
100 Grade Exchange Citrus Pectin	8 oz.
Acetate of Soda (U.S.P.)	1¼ oz.
Citric Acid (crystals or powdered)	2¼ oz.
Glucose (43° Bé.)	20 lb.
Granulated Sugar	20 lb.
Color and Flavor	as desired

Directions

(1) Put 2½ gallons of water in a kettle and heat hot (170° F.). (Open fire or steam-jacketed kettle may be used.)

(2) *Thoroughly mix* 8 ounces of 100 Grade Exchange Citrus Pectin with about 4 pounds of granulated sugar.

(3) Add the Pectin-Sugar mixture to the warm water as it is being stirred with a paddle. Continue to stir and heat to boiling. Boil vigorously for a moment.

(4) Combine the acetate of soda and citric acid. Dissolve in a small portion of hot water.

(5) Add the acetate of soda-citric acid solution to the kettle and then the 20 pounds of glucose. Heat to boiling again.

(6) Add the remainder of the sugar (16 pounds) and cook to 224° F., or to a good "sheet." (This temperature corresponds to about 78% total soluble solids at sea level. It is sufficient to cook the batch to 12° F. above the boiling point of water at your factory.)

(7) Add the color and flavor, then cast into starch at once. This formula will produce about 48 pounds of candy. The finished piece may be crystallized, sanded, iced, or coated with chocolate.

Note: Refined Corn Sugar may be substituted for all or a part of the cane or beet sugar given in the above formula.

Tart and Tender Pectin Jellies for Cast Work
(Especially for Fruit Flavors)

Ingredients

Water	2½ gal.
100 Grade Exchange Citrus Pectin	8 oz.
Acetate of Soda (U.S.P.)	2½ oz.
Citric Acid (crystals or powdered)	4¼ oz.
Glucose (43° Bé.)	20 lb.
Granulated Sugar	20 lb.
Color and Flavor	as desired

Directions

(1) Put 2½ gallons of water in a kettle and heat hot (170° F.). (Open fire or steam-jacketed kettle may be used.)

(2) *Thoroughly mix* 8 ounces of 100 Grade Exchange Citrus Pectin with about 4 pounds of granulated sugar.

(3) Add the Pectin-Sugar mixture to the warm water as it is being stirred with a paddle. Continue to stir and heat to boiling. Boil vigorously for a moment.

(4) Combine the acetate of soda and citric acid. Dissolve in a small portion of hot water.

(5) Add the acetate of soda-citric acid solution to the kettle and then the 20 pounds of glucose. Heat to boiling again.

(6) Add the remainder of the sugar (16 pounds) and cook to 224° F., or to a good "sheet." (This temperature corresponds to about 78% total soluble solids at sea level. It is sufficient to cook the batch to 12° F. above the boiling point of water at your factory.)

(7) Add the color and flavor, then cast into starch at once. This formula will produce about 48 pounds of candy. The finished piece may be crystallized, sanded, iced, or coated with chocolate.

Note: Refined Corn Sugar may be substituted for all or a part of the cane or beet sugar given in the above formula.

Firm Pectin Fruit Jellies for Slab Work

Ingredients

Water	3 gal.
100 Grade Exchange Citrus Pectin	15 oz.
Acetate of Soda (U.S.P.)	1 oz.
Citric Acid (crystals or powdered)	2 oz.
Glucose (43° Bé.)	20 lb.
Granulated Sugar	20 lb.
Fruit Pulp (2 No. 10 tins or)	13 lb.
Color and Flavor	as desired

Directions

(1) Put 3 gallons of water in a kettle and heat hot (170° F.). (Open fire or steam-jacketed kettle may be used.)

(2) *Thoroughly mix* 15 ounces of 100 Grade Exchange Citrus Pectin with about 8 pounds of granulated sugar.

(3) Add the Pectin-Sugar mixture to the warm water as it is being stirred with a paddle. Continue to stir and heat to boiling. Boil vigorously for a moment.

(4) Combine the acetate of soda and citric acid. Dissolve in a small portion of hot water.

(5) Add the acetate of soda-citric acid solution to the kettle and then the 20 pounds of glucose. Heat to boiling again.

(6) Add the remainder of the sugar (12 pounds), the fruit pulp (2 No. 10 tins), and cook to 224° F. or to a good "sheet." (This temperature corresponds to about 78% total soluble solids at sea level. It is sufficient to cook the batch to 12° F. above the boiling point of water at your factory.)

(7) If desired, color and flavor may be added, although flavor is seldom needed. The batch is poured at once into oiled or waxed paper-lined forms to the desired depth and allowed to stand until set. It is then cut to produce pieces of the desired size which may be crystallized, sanded, iced, or chocolate coated to produce extra fancy confections at low cost.

Note: Refined Corn Sugar may be substituted for all or a part of the cane or beet sugar given in the above formula.

Firm Pectin Honey Jellies for Slab Work

Ingredients

Water	3½ gal.
100 Grade Exchange Citrus Pectin	16 oz.
Acetate of Soda (U.S.P.)	1 oz
Citric Acid (crystals or powdered)	2 oz.
Honey (82–83% soluble solids)	20 lb.
Granulated Sugar	20 lb.
Color and Flavor	if desired

Directions

(1) Put 3½ gallons of water in a kettle and heat hot (170° F.). (Open fire or steam-jacketed kettle may be used.)

(2) *Thoroughly mix* 16 ounces of 100 Grade Exchange Citrus Pectin with about 8 pounds of granulated sugar to which has been added the 1 ounce of Acetate of Soda (U.S.P.).

(3) Add this mixture to the warm water as it is being stirred with a paddle. Continue to stir and heat to boiling. Boil vigorously for a moment.

(4) Add the remainder of the sugar (12 pounds) and cook to 219° F., or to a good "sheet." (This temperature corresponds to about 65% total soluble solids at sea level. It is sufficient to cook the batch to 7° F. above the boiling point of water at your factory.) Cool the batch to 170° F.

(5) Add the honey which should be at a temperature of about 170° F. Mix thoroughly with the batch, then add the acid solution. Pour on a slab at once. When the batch has set, the jellies are cut. They may be crystallized, sanded, iced, or coated with chocolate.

(6) This will produce about 50 pounds of candy.

Moderately Firm Pectin Coffee Jellies for Cast or Slab Work

Ingredients

Coffee Extract (see note)	2½ gal.
100 Grade Exchange Citrus Pectin	12 oz.
Acetate of Soda (U.S.P.)	2¼ oz.
Citric Acid (crystals or powdered)	4 oz.
Glucose (43° Bé.)	20 lb.
Granulated Sugar	20 lb.
Color and Flavor	if desired

Directions

(1) Put 2½ gallons of Coffee Extract in a kettle and heat hot (170° F.). (Open fire or steam-jacketed kettle may be used.)

(2) *Thoroughly mix* 12 ounces of 100 Grade Exchange Citrus Pectin with about 6 pounds of granulated sugar.

(3) Add the Pectin-Sugar mixture to the warm Coffee Extract as it is being stirred with a paddle. Continue to stir

and heat to boiling. Boil vigorously for a moment.

(4) Combine the acetate of soda and citric acid. Dissolve in a small portion of hot water.

(5) Add the acetate of soda-citric acid solution to the kettle and then the 20 pounds of glucose. Heat to boiling again.

(6) Add the remainder of the sugar (14 pounds) and cook to 222°–224° F., or to a good "sheet." (This temperature corresponds to 75–78% total soluble solids at sea level. It is sufficient to cook the batch to 10°–12° F. above the boiling point of water at your factory.)

(7) Add the color and flavor, if desired, then cast into starch at once. This formula will produce about 48 to 50 pounds of candy. The finished piece may be crystallized, sanded, iced, or coated with chocolate.

Note: Cooking the batch to 224° F. is recommended for slab work.

Moderately Firm Molasses Pectin Jellies For Cast or Slab Work

Ingredients

Water	2½	gal.
100 Grade Exchange Citrus Pectin	12	oz.
New Orleans Molasses	5	lb.
Glucose (43° Bé.)	15	lb.
Granulated Sugar	20	lb.
Citric Acid (crystals or powdered)	4	oz.
Color and Flavor	as desired	

Directions

(1) Put 2½ gallons of water in a kettle and heat hot (170° F.). (Open fire or steam-jacketed kettle may be used.)

(2) *Thoroughly mix* 12 ounces of 100 Grade Exchange Citrus Pectin with about 6 pounds of granulated sugar.

(3) Add the Pectin-Sugar Mixture to the warm water as it is being stirred with a paddle. Continue to stir and heat to boiling. Boil vigorously for a moment.

(4) Add the 5 pounds of New Orleans Molasses to the kettle and then the 15 pounds of glucose. Heat to boiling again.

(5) Add the remainder of the sugar (14 pounds) and cook to 222°–224° F., or to a good "sheet." (This temperature corresponds to 75–78% total soluble solids at sea level. It is sufficient to cook the batch to 10°–12° F. above the boiling point of water at your factory.) Add the citric acid dissolved in a small amount of hot water.

(6) Cast into starch at once. This formula will produce about 48 to 50 pounds of candy. The finished piece may be crystallized, sanded, iced, or coated with chocolate.

Note: Cooking the batch to 224° F. is recommended for slab work.

Chocolate Pudding Desert

23 parts corn starch
9 parts tapioca starch
18 parts cocoa powder
50 parts cane sugar
Vanilla Flavor to suit.

The above powders are very carefully mixed. Four ounces when carefully cooked up with a pint of milk will make a delicious pint of chocolate pudding.

Gelatin Dessert Powder

Gelatin Powder (best grade)	80
Sugar Powder	450
Tartaric Acid Powder	10

Thickening of Jams, Preserves and Other Fruit Pastes

For many specific uses, particularly in baking and for soda fountain use, true fruit as well as imitation fruit jams, preserves and pastes must be thickened. This thickening is necessary to prevent leakage in pies and pastries and too rapid flow when used as coatings and dressings. Here Galagum fills a long felt want with a resultant lowering of costs in addition.

The method for making 100 pounds of finished jam or preserves is as follows: Mix thoroughly 7 ozs. of Galagum with 35 ounces of cane sugar. The usual amount of sugar and fruit is boiled together in a steam-jacketed kettle. Start the stirring paddle when boiling begins and add VERY SLOWLY the above mentioned mixture of Galagum and sugar. Heat up to 221° F. and then turn off heat. Continue stirring until cool. If desired the jam may be worked on the cooling table, mixing it occasionally. The use of Galagum in this process increases the bulk or volume more than 5 per cent.

Imitation Jellies

The corn syrup imitation jelly is made as follows: The 8 pints of water is brought to a boil. Add slowly with stirring the 70 grams of Aacagum, which has been previously mixed with the 7 ozs. of Cerelose. Bring to a boil and cook for one minute. Now add the certified food color which has been dissolved in a little warm water. Then add the 7 lbs. of warm corn syrup. Stir until completely mixed and at no time need the temperature be higher than 200° F. Transfer the jelly to pail, allow to cool down about 150° F. Then add with stirring the 35 grams of phosphoric acid and fruit flavor. The jelly will set in several hours or allow to set all night.

The imitation cane sugar jelly is made exactly the same way as the corn syrup jelly with the exception that you mix the 70 grams of Aacagum with about 10 per cent of the weight of cane sugar. This mixture will aid the Aacagum considerably in going into solution when added to the hot water.

The phosphoric acid used in the above formulae was made by diluting 85 per cent phosphoric acid with an equal volume of water. The fruit flavors used were of the fruit oil type and were dissolved in Glycopon XS.

Jelly (Non Sweating)

Agar-Agar or Pectin	0.752–1%
Sod. Alginate	0.5–1%
Sugar	15–20%
Water	78–83%
Citric Acid	0.03–0.04%

Guava Jelly

Preparation of Juice:

Wash Guavas, and slice into small pieces with a sharp knife. For each pound of fruit add 2 pints of water and boil until soft (about 25 minutes), allow to stand until cold. Pour into cheese cloth bag and allow to drain pressing to extract all juice. This juice is then drained without pressing thru a clean flannel jelly bag.

Making the Jelly:

Bring the juice to a boil, and then add the sugar. Continue boiling until the jellying point has been reached, which is indicated by the flaking or sheeting from the spoon. The jellying point of the guava is 108° C. or 226½° F.

Kumquat Jelly

1 lb. Kumquats
1 lb. Sugar
1½ pints water

Wash kumquats, treated with soda, and then cut in halves. For each pound of fruit taken add 1½ pints water. Boil for 15 minutes then the kettle is covered and set aside for 15 hours. After again boiling for 5 minutes, remove from the stove, and allow to drain. Let this stand for one hour, then pour into a flannel jelly bag, press to obtain all possible juice, drip thru a bag to remove particles of fruit. The juice is then placed in a kettle and brought to a boil, at which time there is added 1 lb. sugar for each pound fruit taken. The jellying point is determined by dipping a spoon into the boiling solution, and then holding it above kettle allowing the syrup to drop. When it drops in flakes or sheets from the spoon pour immediately into clean, sterilized jelly glasses. When jelly is cold pour hot paraffin over it and store it away.

Fig Preserves

6 qts. figs
2 qts. sugar
3 qts. water

Add one cup soda to 6 qts. boiling water. Plunge figs into hot soda solution and allow to remain until white, milky fluid is extracted (about 15 minutes) or until water is cold enough to plunge hand into comfortably. Put figs thru two cold water baths to rinse well.

Cooking. Drain figs thoroughly and add gradually to the syrup you have made by boiling the sugar and water together 10 minutes and skimming. Cook rapidly until figs are clear and tender (about 2 hours).

Fig Jam

Select very ripe figs, wash and drain. To every gallon of peeled figs add 2 qts. sugar, mash and cook to the proper consistency. When nearing the finishing point be careful not to scorch. If using a thermometer, cook to 222° F. or 106° C.

Grapefruit Preserves

1 lb. grapefruit peel
¾ lb. sugar
1 pt. water
2 slices of lemon

Preparation: Select bright fruit with a thick peel, wash carefully. Cut peel into strips or shapes. To 1 lb. of fruit add 2 pints of water and the lemon. Boil for 15 minutes, change the water and boil again. Repeat the process as often as is necessary to remove as much of the bitter of the peel as is desired. Remove the peel and the lemon from the water and drop them into a boiling syrup made by adding ¾ sugar to 1 pint water for each pound of peel taken and boiling until the sugar is dissolved. After the peel is added boil until the peel is transparent and the syrup sufficiently heavy.

Peach Preserves

10 lb. peeled sliced cling stone peaches
7 lb. sugar
3 pints water
10 peach kernels

Bring sugar and water to a boil, add the peaches and kernels. Cook until the fruit is clear when lifted from the syrup. Pack in sterilized containers and seal.

Orange Marmalade

3 lb. oranges
3 lemons
1½ pint water
3 lb. sugar

Wash, remove the peel and seeds, cutting one half of the peel into very thin strips, and add it to the pulp and balance of the peel, which has first had the yellow portion grated off and has been passed through a food chopper with the pulp. Cover with water and let stand overnight. Boil for 10 minutes the next morning, allow to stand for 12 hours, add the sugar and again stand overnight. Cook it rapidly next morning until the jelly test can be obtained (about 222° F.). Cool to 176° F. pour into sterilized glasses, and seal with paraffine.

Green Tomato Mince-Meat

1 peck green tomatoes
2 lb. raisins
2½ lb. brown sugar
½ lb. suet or cocoanut
2 tsp. ground cinnamon
2 tsp. nutmeg
2 tsp. cloves
½ cup vinegar
2 tsp. salt.

Chop tomatoes fine and drain. Cover with cold water, heat thru and drain again. Add chopped raisins and other ingredients. Cook 30 minutes. Pack into sterilized jars and process 15 minutes.

Curry Powder (Spicing)

A.

Coriander Seed	16 oz.
White Pepper	1 oz.
Cayenne Pepper	½ oz.
Turmeric	1½ oz.
Ginger	1 oz.
Mace	½ oz.
Clove	½ oz.
Fennel	½ oz.
Celery Seed	½ oz.
Cardamom	½ oz.
Slippery Elm	4 oz.

B.

Indian Curry Powder

Coriander Seed	5 oz.
Turmeric	5 oz.
Cardamom	40 oz.
Cayenne Pepper	10 oz.
Fenugreek Seed	4 oz.

The above ingredients are mixed and allow to dry in a warm oven to drive off the moisture. It is then ground very fine and packed in tins.

* *Coffee Extract* is prepared by bringing 3 gallons of water to a boil and adding 1¼ pounds of Ground Coffee. The mixture is stirred well and set aside to draw for 10 minutes. The extract is then strained through a suitable cloth, or filtered, and will yield the needed 2½ gallons of extract.

Pineapple Icing

Pineapple (grated or crushed) 1 lb.

Thicken to proper consistency with icing sugar. Heat to 110° C. and apply while warm.

Lemon Icing

Hot Water	16 oz.
Sugar	120 oz.
Lemon Grating or Juice	2 oz.
Glucose Syrup	4 oz.

Orange Icing

Hot Water	16 oz.
Sugar	120 oz.
Orange Grating or Juice	2 oz.
Glucose Syrup	4 oz.

Maraschino Icing

Hot Water	16 oz.
Maraschino Juice	6 oz.
Chopped Cherries (to suit)	
Sugar	120 oz.
Glucose	4 oz.

Coffee Icing

Fresh Made Coffee	16 oz.
Sugar	96 oz.
Invert Sugar	8 oz.
Caramel Color	⅛ oz.

Vanilla Icing

Hot Water	16 oz.
Glucose	4 oz.
Sugar	112 oz.
Vanilla	½ oz.
Egg Whites	3 oz.

Chocolate Icing

Hot Water	16 oz.
Sugar	96 oz.
Melted Butter	4 oz.
Melted Chocolate	16 oz.
Inverted Sugar	8 oz.

Home Made Icing

Beat stiff:

Egg Whites	32 oz.
Salt	¼ oz.
Sugar	16 oz.
Vanilla (to suit)	

Boil together to 236–240° F.

Sugar	104 oz.
Glucose	8 oz.
Water	2 oz.

Add cooked syrup to beaten egg whites and beat until stiff. Add chopped fruits, nuts as desired.

Chocolate Fudge

Bring to a boil:

Chocolate	16 oz.
Butter	4 oz.
Sugar	16 oz.
Milk	16 oz.
Glucose	6 oz.

Cool to 120° F.
Then add and mix smooth

Vanilla	1 oz.
Sugar Icing	72 oz.
Egg Whites	2 oz.

Mix smooth.

Butterscotch Fudge

Cook to 235° F.:

Brown Sugar	64 oz.
Milk	32 oz.
Butter	8 oz.
Glucose	1½ oz.

Cool to 120° F.

Then add

Milk	16 oz.
Lemon Juice	1 oz.
Salt	⅛ oz.
Butter	8 oz.
Icing Sugar	128 oz.
Burnt Sugar	¼ oz.

Use Warm.

Light Meringue Icing

Beat until stiff:

Egg Whites	32 oz.
Salt	⅛ oz.
Vanilla	¼ oz.

Boil to 240° F.

Sugar	96 oz.
Glucose	8 oz.
Water	32 oz.

Add syrup to beaten whites, and beat up until desired consistency is reached.

Royal Icing

Beat light:

Egg White	16 oz.
Icing Sugar	96 oz.
Juice of Lemon	1 oz.
Cream of Tartar	⅛ oz.
Vanilla	¼ oz.

Fruit Cake Shrine

Bring to a boil:

Water	16 oz.
Glucose	12 oz.
Malt Extract Powdered	½ oz.

Add:

Gelatine (dissolved)	1 oz.
Flavor to Suit	

Apply while hot to baked cake.

Cocoa Icing

Beat together until smooth and glossy:

Plastic Cocoanut Butter	16 oz.
Invert Sugar	20 oz.
Water	12 oz.
Cocoa	20 oz.
Icing Sugar	88 oz.
Milk Powder	4¾ oz.
Salt	⅛ oz.
Vanilla	½ oz.

Marshmallow

Soak together:

Gelatine	3½ oz.
Cold Water	13 oz.

Then heat to 140° F. and add

Hot Water	24 oz.
Invert Sugar	16 oz.
Icing Sugar	104 oz.
Vanilla	1 oz.

Beat stiff and use while warm.

Marshmallow and Meringue Powders

Formula No. 1

25 lb. Dried Egg Albumen
25 lb. Galagum C
40 lb. Corn Starch
5 lb. Skimmed Milk Powder
5 lb. Powdered Alum
Vanillin to suit.

Mix the above well and run through a fine mesh sifter.

Formula No. 2

25 lb. Dried Egg Albumen
25 lb. Galagum C
10 lb. Tapioca Starch
35 lb. Cane Sugar (powdered)
5 lb. Skimmed Milk Powder
Vanillin to suit.

Mix the above well and run through a fine mesh sifter.

Formula No. 3

25 lb. Dried Egg Albumen
25 lb. Galagum C
25 lb. Corn Starch
20 lb. Corn Sugar (powdered)
5 lb. Skimmed Milk Powder

Mix the above well and run through a fine mesh sifter.

In any of the above formulae, where it is desired to reduce the amount of Galagum C, then the difference is made up with cane sugar.

The above meringue formulae are to be used as follows:

Take 5 oz. of meringue powder to 1 quart cold water and 3 lb. cane sugar. Put the cold water into a clean kettle, then add to it the sugar and meringue powder. Beat in the machine until the required stiffness is obtained. For marshmallow whip take 2 oz. of meringue powder, 1 quart cold water, 3 lb. of cane sugar and whip to the desired stiffness. Now dissolve thoroughly 2½ oz. of Gelatin in ½ pint hot water. Add this slowly to the beaten meringue, and continue to beat up until the desired consistency is attained.

* Flour, High Rising

To increase the vol. yield of bread, etc., made from dough free from egg yolk or egg-yolk substitute and substantially free from fat, a phosphatide, *e.g.*, lecithin from soy beans, to the extent of 0.05–1% is incorporated in the flour.

* Improvement of Flour

The addition of the following greatly improves baking properties of flour.

$(NH_4)_2S_2O_8$	2.5–5%
$CaH_4(PO_4)_2$	2.5–5%
Pot. Citrate	2.5–5%

* Flour Improver

The baking capacity of flour is increased by adding 0.1–0.5 gm. of following mixture to 1 kg. of flour:

Asparagine	3 lb.
Hydrogen Peroxide	10 lb.

Cool and mix the crystallized product with

Malt Diastase	3

* Baking Powder

Sod. Bicarbonate	300
Calcium Dihydrogen Phosfate	33
Sod. Hydrogen Pyrophosfate	405
Starch	262

Household Baking Powders

1.

Sodium Bicarbonate	28 parts
Mono Calcium Phosphate	35 parts
Corn Starch	27 parts

Mix the above powders thoroughly and store in airtight containers.

2.

Sodium Bicarbonate	28 parts
Calcium Acid Phosphate	29 parts
Sodium Aluminum Sulphate	19 parts
Starch Corn	24 parts

3.

Sodium Bicarbonate	28 parts
Mono Calcium Phosphate	12 parts
Sodium Aluminum Sulphate	21½ parts
Starch Corn	38½ parts

4.

Sodium Bicarbonate	28 parts
Sodium Aluminum Sulphate	28 parts
Corn Starch	44 parts

Bakers Baking Powder

5.

Sodium Bicarbonate	35 parts
Mono Calcium Phosphate	9 parts
Sodium Aluminum Sulphate	29 parts
Corn Starch	27 parts

6.

Sodium Bicarbonate	35 parts
Sodium Aluminum Sulphate	35 parts
Corn Starch	30 parts

7.

Sodium Bicarbonate	35 parts
Calcium Acid Phosphate	36 parts
Sodium Aluminum Sulphate	24 parts
Starch Corn	5 parts

8.

Sodium Bicarbonate	28 parts
Sodium Acid Pyrophosphate	20 parts
Mono Calcium Phosphate	22 parts
Corn Starch	30 parts

9.

Sodium Bicarbonate	27 parts
Cream of Tartar	60 parts
Corn Starch	13 parts

10.

Sodium Bicarbonate	27 parts
Cream of Tartar	45 parts
Tartaric Acid	6 parts
Corn Starch	22 parts

In all these formulas the powders are of course mixed thoroughly.

* Mold on Food, Preventing

The food is given a coating of glyceryl formate.

* Fruit, Prevention of Mold On

Citrus fruit is treated with a warm solution containing 2 ounces each of Borax and soda ash per gallon, preferably under high pressure.

* Eggs, Preservative For

Benzene	10,000
Crepe Rubber	500
Naphtha	10,000
Carbon Bisulfide	2,000
Sulfur	70
Paraffin Wax	500
Derris Root Extract	200

Ice Cream Powder

Dried Milk Powder	51
Sugar Powder	52
Sod. Carbonate	2
Cream of Tartar	4.4
Vanillin	0.06

One pound of above makes 10 lbs. ice cream.

Mayonnaise

Whole Eggs	4
Egg Yolks	16
Liquid Pectin	2½ oz.
Mustard Powder (yellow)	¼ oz.
Sugar	1½ oz.
Salt	1 oz.
Vegetable Oil	1 gal.
Mayonnaise Flavor	2 cc.
Tincture Capsicum (optional)	4 cc.
Lactic Acid	4 cc.
Vinegar	6½ oz.
Water	6½ oz.

Mayonnaise

Cottonseed Salad Oil	70.25
Egg Yolk	10.00
Vinegar (50 grain)	10.00
Water	3.90
Salt	1.45
Sugar	3.50
Mustard	0.80
White Pepper	0.10

This formula gives good resistance to freezing, keeps well and has good flavor and appearance.

Mayonnaise

Egg Yolk	8 oz.
Vinegar	8 oz.
Sugar	1¾ oz.
Oil	96 oz.
Salt	1½ oz.
Mustard	½ oz.
Water	10 oz.

Build up and run on colloid mill.

Milk and Cream, Increasing Viscosity of

To increase the viscosity and improve the consistency of milk or cream, the material is heated to 40—42° in 20—30 min., cooled to 2—3° in 20—30 min. and held at 2—3° for 1—2 days.

* Cream, Artificial

Butter Fat	19
Vegetable Fat	10
Milk Powder	7
Sugar	0.75
Gelatin	1
Borax	0.25
Water	62
Flavor	to suit

* Breast Milk, Artificial

A substitute for human milk is prepd. by adding to animal milk, or animal milk products, appropriate amts. of albumin, lactose, K_2CO_3, KCl, K_2HPO_4, Fe oleate and citrates of Na, Ca and Mg. Thus, to a heated mixt. of whey 100, 25% cream 180 and water 720 cc. there may be added albumin 3.6, lactose 52, K_2CO_3 0.267, KCl 0.3, K_2HPO_4 0.232, Na citrate 0.5, Ca citrate 0.54, Mg citrate 0.12 and Fe oleate 0.006 g.

* Milk Products, Preserving

A small proportion (suitably about 1.5%) of glycerol or other polyhydric alc. such as dimethylene and trimethylene glycol or propylene glycol is added to prevent development of rancidity in milk, dried milk, cream, butter, etc.

Manufacturing Cream Cheese (Hot Process)

The new method of manufacturing cream cheese involves a new principle; namely, the aggregation of the fat globules into large clusters by proper homogenization. This is accompanied by a partial coagulation of the casein in these fat clusters so that the entire mass sets to a permanent condition which is not materially affected by temperature.

Sweet cream of good flavor containing 40 to 42 per cent of milk fat is the basis for this cheese. From 3 to 5 per cent of soluble dry skimmilk is stirred into the cream. Then 0.5 to 0.7 per cent of finely ground agar free from objectionable flavor or odor should be added to this mixture while it is being constantly stirred.

The mixture should then be heated to 180 to 185° F. and held for 5 to 10 minutes for the agar to dissolve. It should then be cooled to 110° F. Add 0.75 per cent of common salt and 0.5 to 1 per cent of good commercial starter depending upon the rate at which acidity is desired in the cheese. The mixture should then be passed thru a coarse strainer and homogenized at 3,000 to 4,000 pounds pressure per square inch. The mixture should leave the homogenizer at the consistency of soft butter and slightly firmer than ice cream as it leaves the freezer.

The mixture should be placed immediately into the final molds before the temperature lowers to 100 or less because the finest body and texture is secured if the cheese is not mixed after the agar has set. The cheese can be chilled in the refrigerator to 70° and then placed in a 70° room for 10 or 15 hours for the acid flavor to develop.

The quantity of acid developed in the cheese can be varied not only by the percentage of starter but by the quantity of dry skimmilk. The more dry skimmilk the higher the acidity will be. Acid develops somewhat slowly in this cheese so that it may be necessary to increase the percentage of starter under special conditions.

When relish, olives, etc., are mixed with the cheese it is generally not necessary to use starter since the relish gives plenty of tartness and flavor to the cheese. The quantities used vary from 10 to 30 per cent. The cream can be homogenized at 120° thus making it possible to pack a much warmer cheese with less danger of the agar congealing before packing. It is desirable in such cheese to use fully 5 per cent of dry skimmilk to help prevent any whey drainage. If there is much juice from the relish it may be desirable to add it to the warm cheese before homogenization but such a procedure increases the acidity in the cream thereby causing excessive fat clumping. This may be offset by the use of lower homogenization.

---

* Nuts, Removing Rancidity of

In order to remove rancidity and discoloration, rancid nut meats are immersed in a NaOH soln. (suitably of about 5% strength) and subsequently treated with a HCl soln. (suitably of about 1% strength), washed and dried.

---

* Lard, Preventing Rancidity in

The addition of 0.05–0.1% Gum Guaic to lard or other fats prevents rancidity.

---

* Salt, Cooking and Fermentation

A salt mixt. contains cations of alkali and alk. earth metals in a physiol. equil. such that it does not modify the surface tension of an electroneg. lipoid soln. and that the optimum colloidal state for org. albumin is reached. Examples contain NaCl 86.81, KCl 5.54, $MgCl_2$ 3.53, $CaCl_2$ 4.1 g., and NaCl 74.55, K tartrate 7.47, Mg lactate 8.15, Ca lactate 9.78 g. The salt mixts. are used in *cooking, fermentation, tanning,* etc.

Sherbets

13.5 lb. Sugar
2¾ to 3 oz. Sheragum
Flavor, Water, Acid, Color

and mix to make 5 gallons of mixture.

1. Directions if not pasteurizing:

Mix well 3 oz. or slightly less of Sheragum with all of the sugar of the mix. Add this to the cold water in the vat, agitating all of the time. Add the flavor and mix thoroughly. If the flavor contains a high sugar content, cut down on the amount of sugar added. The amount of sugar given is satisfactory when orange or lemon sherbets are made. This mixture requires no aging, but if aged overnight will give a smoother product.

Freeze with cold refrigerant and when the mixture has started to thicken slightly add the acid (3–4 oz. of 50% citric acid). When the mix is a little stiffer, add 2 quarts of regular mix. Draw when frozen or when the overrun reaches 25 to 30 per cent.

The regular formula used by the plant may be used. The only things to watch are—that the gum is mixed well with a large quantity of sugar and added slowly to the cold water, or milk if milk is used. Do not add the acid until the mixture is being frozen. The mix may be added any time. We always add the mix at the freezer because if the mixture is very acid, it may curdle the mix.

2. Directions if product is pasteurized:

The same rule is followed, but that 2½ oz. of Sheragum will be sufficient in this case. Acid, color and flavor are not pasteurized. Since heating brings out a little flavor from the gum, the gum and enough sugar to carry it should be left out until the mixture is cooled.

---

* Sherbet (Water-Ice)

The following formula gives a smooth product of good stability.

(1) Prepare, in the form of a powder, the following ingredients, weighing and mixing same according to percentages given:

		Per cent
(a)	Corn Sugar	85.724
	Agar	2.857
	Pectin (160 grade)	4.286
	Citric Acid Powdered Crystals	5.713
	Gelatin	1.420
		100.

or

(b)	Corn Sugar	87.517
	Agar	2.872
	Pectin (160 grade)	4.877
	Citric Acid	4.734
		100.

(2) Prepare 100 pounds of ice mix, using 7 pounds of the above powder, 21 pounds sucrose (beet or cane sugar), 20 pounds of fruit, and 52 pounds of water.

In preparing the ice mix, the 7 pounds of powder should be dissolved in 25 to 30 pounds of water and heated to boiling temperature, boiling not to exceed approximately one minute, as excessive boiling in the presence of the acid will reduce the jellying strength of the pectin. The solution thus prepared is then added to the balance of the mix. In case concentrated fruits (such as concentrated orange juice) are used, additional amounts of water will be necessary for diluting the fruit concentrate to normal strength. In addition to the ingredients listed, additional flavoring and coloring may be desirable or necessary, depending upon the fruit used. It should be mentioned that a good ice, smooth, palatable, of desirable flavor, and possessing good keeping qualities, should contain a uniform amount of acid, preferably 0.60 to 0.65 per cent titratable acidity, calculated in terms of citric acid. Also, air incorporated to the extent of 30 to 40 per cent of the original liquid content is generally considered as sufficient for the production of the most desirable ice from a commercial standpoint.

In the two examples of powder given above, the first will give a greater yield than the second. In both instances the agar should be ground to pass approximately a 40 mesh screen. The pectin may vary in amount, depending upon the smoothness desired in the finished product and it will be understood that if a lower grade of pectin is used a corresponding increase in the amount will probably be required. In both examples the citric acid crystals should be ground to a powder and then mixed with the other ingredients.

Water Ices and Sherbets

The formulae give only the basis for the mixture and do not attempt to specify flavors and fruit juices to give the water ices or sherbets their characteristic flavors. The figures are given on the basis of 100 pounds of mix which is about 10½ gallons. The mix has a specific gravity of approximately 1.14 at 10° C. and weighs 9.5 pounds per gallon. The specific gravity varies greatly, depending largely upon the percentage of sugar and the temperature.

Water Ice

Cane Sugar	25.0 lb.
Corn Sugar	7.0 lb.
Agar	0.2 lb.
(3.2 ounces or 90.6 grams)	
Gum Tragacanth or High-grade India Gum	0.4 lb.
(6.4 ounces or 181.2 grams)	
Water, Fruit, Fruit Acid, Flavor, and Color	67.4 lb.

Overrun 20 to 25 per cent—Total yield 13 gallons.

Sherbet Using Milk

Cane Sugar	25.0 lb.
Corn Sugar	7.0 lb.
Agar	0.2 lb.
(3.2 ounces or 90.6 grams)	
Gum Tragacanth or High-grade India Gum	0.2 lb.
(3.2 ounces or 90.6 grams)	
Whole Milk	50.0 lb.
Water, Fruit, Fruit Acid, Flavor, and Color	17.6 lb.

Overrun 25 to 30 per cent—Total yield 13.5 gallons.

Sherbet Using Ice Cream Mix

Cane Sugar	25.0 lb.
Corn Sugar	7.0 lb.
Agar	0.2 lb.

(3.2 ounces or 90.6 grams)

Gum Tragacanth or High-grade India Gum	0.2 lb.
(3.2 ounces or 90.6 grams)	
Ice Cream Mix, without Sugar or Gelatin	10.0 lb.
Water, Fruit, Fruit Acid, Flavor, and Color	57.6 lb.

Overrun—25 to 30 per cent—Total yield 13.5 gallons.

The mixture should be prepared by first weighing most of the water or all of the milk, if any is used, leaving out enough water to dissolve the agar and to allow for fruit juices, etc. The sugars should be thoroughly mixed with the powdered gum tragacanth or high-grade india gum and slowly poured into the water while the water is being agitated rapidly. Powdered agar is preferable to granular or shreds because it can be more readily dissolved. The powdered agar should be poured into 50 times its weight of boiling water while the water is being agitated rapidly. The water with agar should continue to boil for about five minutes when the agar will be completely dissolved. The hot agar solution should be added to the mix as if it were a hot gelatin solution. The gelatinization strength of agar is reduced by boiling in acid solutions, but it is only slowly altered by boiling in water, so it is important that fruit acid should be added to the mix after the agar. All other ingredients used should be added to the mix at this time and the total weight brought up to the required amount with water, making allowance for the fruit and fruit acids or juices which are usually added at the freezer.

There is no necessity of aging water ices or sherbets made with agar and gum as stabilizers because the action of each takes place within a few minutes. Evidence of a weak gel formation should be readily observed at once if sufficient agar has been used, since agar solutions set at 40° to 42° C. and since the temperature of the cold mixes is much lower.

Vanilla Bean Flavoring Powder

25 Parts Ground Vanilla Bean
74 Parts Confectioners Powdered Sugar
1 Part Oil of Bitter Almond

Mix the above ingredients very thoroughly. Place in sifter top cans and use as powdered flavor over ice cream, cereals and baking.

Vanilla Sauce Powder

Corn Flour	100
Vanillin	0.5
Yellow Food Color	0.05

Procedure for Washing and Sterilizing Freezers Using Steam and Chlorine

At the conclusion of the freezing operation drain the ice cream from the freezer. Rince the strainer, hopper, and outside of the freezer, particularly at the head, with cold water. Fill the freezer two-thirds full of cold water. Run one-half minute and drain.

Fill the hopper full of water at 140° to 145° F. and add a half pound (1 cup full) of cleansing powder. Wash the strainer, hopper, and outside of the freezer with a brush. Drain the solution into the freezer, (the freezer should be at least two-thirds full) run one-half minute, and drain the freezer.

Remove the head, scrub with a brush, being certain to clean out the front bearing. Wash the bearing end of the dasher with a brush, remove from freezer, and wash. Replace dasher and head.

Fill the hopper full of water at 180° to 185° F. so that the screen is immersed. Let it stand 2 minutes to sterilize the hopper and screen. Drain into the freezer, (the freezer should be at least two-thirds full) run one-half minute, and drain.

Partially close the freezer gate so that it is about one-fourth open. Turn steam into the freezer, through a special removable pipe, with sufficient force to give a noticeable blowing of steam from the fruit hopper opening. Steam until the steam condensate dripping from the freezer is above 180° F. This will require 3 to 5 minutes for a 10-gallon freezer and 5 to 8 minutes for a 25-gallon freezer. Open the gate and let the freezer stand intact until ready for use.

Before using the freezer, fill the hopper with water at 100° to 110° F., making certain that the screen is covered. Add sufficient chlorine to give 100 p.p.m. and stir well. If desired, the chlorine solution can be pumped into the hopper from a special tank. Drain the chlorine solution into the freezer, operate the freezer for one-half minute, and drain. The freezer is then in excellent sanitary condition and ready for immediate use.

---

* Vitamin Concentrate

A vitamin concentrate in tablet form, consisting of a pulverized and compressed blend of the following raw vegetables containing vitamins A, B, C, D and E in substantially the following proportions:

	Per cent
Spinach	10 to 20
Green Cabbage	5 to 15
White Cabbage	25 to 40
Lettuce	3 to 7
Carrot	20 to 40

and dehydrated to less than 10% of their original moisture content.

For Fancy Cakes, Tea Cakes, Macaroons

Preparation of the Narobin solution:

Stir 10 grams Narobin powder (mixed with a little sugar to help solution) in one litre water, allowing about one hour for dissolving. (*Note:* the metric Kilo is equivalent to 2.2 pounds, and One Ounce is equivalent to about 28 grams).

Cakes:

12 Kilos Sugar
17½ Kilos Flour
5 Kilos Butter
5 Kilos Raisins
2 Litres Eggs
5 Litres Narobin Solution
200 Grams Baking Powder with Orange or Other Flavor

Sponge Cakes

12 Kilos Sugar
18 Kilos Flour
5 Kilos Margarine
500 Grams Powdered Milk
5 Litres Narobin Solution
2 Litres Eggs
250 Grams Baking Powder

Cake No. 2:

6 Kilos Sugar
7 Kilos Flour
5 Litres Eggs
3 Litres Narobin Solution with Vanilla Flavor
2 Kilos Melted Margarine

Cake No. 3:

6 Kilos Sugar
7 Kilos Flour
5 Litres Eggs
3 Litres Narobin Solution
1 Litre Egg White and ½ Litre Narobin Solution, beaten together
30 Grams Ammonia (baking powder)
2 Kilos Crisco (a vegetable substitute for butter)

Cake No. 4:

6 Kilos Sugar
7½ Kilos Flour
4 Litres Eggs
4 Litres Narobin Solution

Beat well together.

Macaroons:

6 Kilos Ground Nuts
15 Kilos Crystallized Sugar
5 Kilos Potato Flour
500 Grams Glucose
2 Kilos Glace Sugar
4 Kilos Rice Flour
2 Litres Narobin Solution

Uses of Narobin in Connection with Eggs, in General Baking, Pastry Making, Sponge Cake, and Other Cakes

Here, by the use of Narobin solutions, a saving from 10 to 25% of eggs, can be effected. Aside from economy, Narobin appears to make the egg whites rise, giving them body, and preventing lumping. Various formulae are given below, showing the use of Narobin solution to replace whole eggs, egg whites, etc., in various blends. For instance, formulas A and B are found to take the place of 20 quarts of whole eggs.

Narobin Solution is made up of 20 to 30 to 35 grams per litre of water,—which solution replaces one egg by 50 grams and each white or yolk by 25 grams, as a general basis for use.

Formula A and Variations:

12 Quarts Whole Eggs
5 Quarts Egg Whites
3 Quarts Narobin Solution (one ounce per quart water)

Mix the Narobin solution with the whites, then incorporate with the entire eggs (or yolks). Use in preparations the same as actual eggs. This formula takes the place of 20 quarts of whole eggs.

6 Litres Whole Eggs
2½ Litres Egg Whites
1½ Litres Narobin Solution (30 grams Narobin per litre water)

Can replace 10 litres of whole eggs.

It is equally possible, with very good results, to use 5 litres whole eggs, 2½ litres egg whites, 2½ litres Narobin solution.

Note: Egg whites may be replaced by powdered egg albumen. This is particularly recommended in winter.

Formula B and Variations

10 Quarts Whole Eggs
5 Quarts Egg Whites
5 Quarts Narobin Solution (same strength)

This is found to replace 20 quarts of whole eggs.

5 Litres Egg Yolks
2½ Litres Egg Whites
2½ Litres Narobin Solution (30 grams Narobin dissolved in one litre warm water)

This takes the place of 10 litres of egg yolks.

Formula B is recommended for spice cakes, etc., giving better results than by use of egg yolks alone. In summer it is better to reduce the proportion of Narobin, as well as the quantity of egg whites.

Formula C and Variations

10 Quarts Egg Yolks
5 Quarts Egg Whites
5 Quarts Narobin Solution (same strength)

Narobin solution is prepared by dissolving 30 to 35 grams Narobin powder in one litre warm water. Fifty grams of this solution replaces one whole egg, and 25 grams takes the place of one yolk or one white.

Formula D and Variations

Dissolve in one quart of water from 1 to 1¼ ounces Narobin; and it will be noted that one egg can be replaced by slightly less than two ounces of this solution; and one egg yolk, or one egg white can be replaced by about one ounce of this solution.

For 100 whole eggs, formula for replacement is:

75 Whole Eggs
45 Ounces of the Narobin Solution

(Narobin Solution—30–35 grams in one litre warm water.)

For 100 egg yolks, formula for replacement is:

75 Egg Yolks
22 Ounces of Narobin Solution

For 100 egg whites, the formula for replacement is:

75 Egg Whites
22 Ounces of Narobin Solution

Use of *less* Narobin (both in quantity and strength of solution) is recommended in summer than in winter. This is because the composition of the egg differs in the seasons—the fresh eggs in summer having more albumen than in winter; and correspondingly in winter they contain more yolk.

Formulas A, B, C and D should only serve as examples, and as definite starting points. Narobin gives better value in connection with storage eggs, giving them substantially more adhesive power, and facilitating their emulsion.

---

* Chocolate Margarine

Cocoanut Oil	1800 lb.
Cocoanut Oil (hydrogenated)	500 lb.
Cottonseed Oil	300 lb.

In preparing a mix of these ingredients, the hydrogenated and unhydrogenated cocoanut oils are preferably first mixed together by churning them at a temperature of approximately 90° F., at which temperature the oils are free-flowing, and the agitation continuing over a period of approximately three minutes.

The proportion of hydrogenated cocoanut oil used may vary considerably from the figure given above, keeping in mind the desired consistency and other characteristics of the final product.

After the cocoanut and hydrogenated cocoanut oil have been mixed as above, thirty gallons of water at a temperature of about 46° F. are placed in the churn with the fats and the whole mass is agitated for about five minutes. Then the 300 lbs. of cottonseed oil, which should be at a temperature of about 80° F., are mixed in. At this point the mass will be in a flowable state. The important feature during this step is to prevent the oils from graining. The temperatures above are selected with this in mind but are not critical.

The next step is to pour this flowable mass into cold water maintained at a temperature of about 34° F. to crystallize the fats. This is accomplished by flowing the fatty mass into a stream of cold water, whereupon the fatty mass is broken up and quickly chilled to produce small globules or granules of fat. The stream flows for such a distance that the fat and water will remain together for preferably less than two minutes. It will be understood that the vegetable fats may, however, be crystallized in any desired manner.

The fat mixture, after being crystallized, is mechanically removed from the water and dumped into a large mass of water at a temperature of 67° F. This latter temperature is important, and for best results should not be permitted to vary by more than 2° F. either way. The fat, being in a highly subdivided state before being charged into the water, is rapidly and uniformly brought to very nearly the temperature of the water and at that temperature coalesces again into a large unitary mass. The purpose of charging the fat base into water is to bring it quickly and uniformly to the desired temperature. This may be also accomplished by tempering in the air to the same temperature as specified but not so satisfactorily as by the use of water.

After leaving the water bath, the fat is placed in a continuous working device, preferably one having screw blades, and is worked until the water content is brought down to about 9%. When this point is reached, the mass is removed to a butter worker, comprising a revolving table and a wooden roll, and is worked on this device until the moisture content is reduced to somewhat below 9%, preferably to about 7%. The exact manner of working is not critical, neither is it necessary to use the specific types of machines mentioned above. Since the object of this working is to reduce the moisture content, it is evident that any device which will accomplish this object will be operative. However, the particular arrangement discussed above, or an equivalent thereof, has the virtue that all particles are properly worked so that the tendency for white particles to appear in the finished product is lessened.

After the moisture content has been reduced and the mass has been thoroughly worked to maintain all parts of the mass at the same temperature, so as to eliminate most, if not all, of the hard particles, the material is placed in a tempering room, where it is held at a temperature of about 64 to 68° F. for from twelve to eighteen hours.

It is desirable to work or mix the fatty material with the milk and aqueous chocolate syrup immediately after the moisture content has been reduced to the proper amount, but if this can not be done because of insufficient equipment, the fatty material must be kept under close temperature control in a tempering room.

After the tempering is finished, the material is mixed with milk, for example, about twenty gallons of milk for each 2600 pounds of fatty material.

Preferably, the milk should be slightly acid. This may be accomplished by adding a suitable amount of lactic acid culture and ripening the milk to about .85% acidity. At the time that the milk is added, a suitable amount, for example, about 3% by weight of an emulsifying agent should be added. These ingredients are then thoroughly worked together at room temperature.

About 1560 lbs. of a suitable chocolate syrup, also at ordinary room temperature, are then added and thoroughly worked into the mass, while the fats of the base are in a solidified state, that is, without the application of heat. At this time about ½% by weight of salt, based on the total weight of the mix, is added. After these ingredients have been worked together until the mass is smooth and of suitable texture, the mass is spread in thin layers to permit it to reach uniform temperatures in the shortest possible time, and is chilled to render it capable of being printed. The product may then be printed and packed for shipment and sale. If the product is to be packaged in glass or other similar containers, however, this chilling step is unnecessary.

The chocolate-bearing material used should be in the form of an aqueous syrup. A syrup of the following formula is satisfactory:

	Percent by weight
Sugar	62
Cocoa	12
Chocolate	7
Salt	0.2
Vanilla	0.01
Water balance, or about	18%

The emulsifying agent to use is one commonly sold under the name of "Emargol." This is a complex fatty mixture consisting of approximately 50 to 55% of moisture and 45 to 50% by weight of fatty matter. The active emulsifying agent in the fatty matter is monostearyl glycerine sodium sulphoacetate, which is present in the mixture to the extent of approximately 15 to 20% by weight.

The product of this process is a substantially permanent and homogeneous emulsion of an edible vegetable fat and an aqueous chocolate-bearing syrup, which is of smooth uniform texture and of semi-solid consistency and spreadable like butter. The flavor of the chocolate predominates over that of the other ingredients, thus making a new product entirely different from any spread for bread, cake, pastries or the like previously known.

---

* Cheese, Pasteurizing

The process of treating soft acid cheese having a pH of about 3.5 to 5 which comprises, adding an amount of an alkaline substance to the cheese to bring its pH to about 5.5 or 6.5, pasteurizing the mixture and then adding an amount of an acid substance to the cheese to bring its pH back to about 3.5 to 5.

Cultured Milk

Three different organisms are commonly used in the manufacture of cultured milk drinks in this country. The most common product is that made by souring milk under control conditions with pure cultures of S. lacticus. Some manufacturers prefer a heavy body and a sharper flavor which they secure by adding a small proportion of L. bulgaricus starter to that made with S. lacticus. For the acidophilus drink a third organism is used called L. acidophilus. All three of these starters can be secured from any commercial culture laboratory.

In some cases no butterfat is added, but a much more palatable product can be secured by the addition of sufficient cream to make a total fat content of 1–2 per cent.

Essential to Have Good Starters

Probably the most essential requirement for the successful manufacture of cultured milk is that the starter be kept pure. This means that proper facilities must be available for growing the cultures, and a competent person must be in charge. Even with the best of care, starters occasionally "go off" and need to be replaced with new stock.

Mother cultures should be grown in the laboratory. From these mother cultures the bulk cultures can be set. In no case should the attempt be made to carry starters by transferring from one vat or can to another. The transfer should be carefully made, using only sterile equipment, from the mother culture to what is to be the next mother. Since the preparation of the three starters varies somewhat each one will be considered separately.

Preparation of S. Lacticus Starter

A. Mother culture.

1. Use only high quality skim milk.
2. Place milk in glass container such as fruit jar and heat to 190° F. for 30 minutes.
3. Cool slowly to 72° F.
4. Using sterile spoon or pipette transfer about 10 cc. of the last mother culture to each quart of the sterilized milk. Cover bottle immediately.
5. Incubate at about 72° F. for about 18 hours or until curd is well set up.
6. Place in 40° F. room until used.

B. Bulk starter.

1. Use only high quality skim milk.
2. Heat to 180° F. for 30 minutes.
3. Cool to 72° F.
4. Add 1½–2 per cent of the mother culture and mix well.
5. Incubate at about 72° F. for 18 hours or until acidity of about .75 per cent is reached.
6. Break curd and cool immediately to at least 50° F. by pumping over surface cooler.

Preparation of L. Bulgaricus Starter

A. Mother culture.

1. Use only high quality skim milk.
2. Place milk in glass container such as fruit jar and heat to 190–200° F. for 30 minutes.
3. Cool slowly to 100° F.
4. Using sterile spoon or pipette transfer about 10 cc. of the last mother culture to each quart of the sterilized milk. Cover bottle immediately.
5. Incubate at 100° F. for about 18 hours or until firm curd is formed.
6. Place in 40° F. room until used.

B. Bulk starter.

In case only small quantities of bulgarlac are to be made it will not be necessary to prepare any bulk starter of the bulgaricus culture, as a sufficient amount of the mother culture can be prepared to supply the quantity needed to mix with the lactic starter. Otherwise proceed as follows:

1. Use only high quality skim milk.
2. Heat to 190° F. for 30 minutes.
3. Cool to 100–105° F.
4. Add 1½–2 per cent of mother culture.
5. Hold at 100° F. for 18 hours or until acidity of about 1.00 per cent is obtained.
6. Break curd and cool immediately to at least 50° F. by pumping over surface cooler.

Occasionally bulgaricus starter is sold for a cultured milk drink, but its flavor is so sharp and its body so viscous that it is better to mix it with the lactic culture. A desirable drink can be prepared by adding one part of the bulgaricus to nine parts of the lactic culture together with the amount of cream necessary to supply 2 per cent fat in the finished product.

This product has the advantage of a distinct acid flavor, a smooth and fairly heavy body with little tendency to whey off.

Preparation of L. Acidophilus Starter

The preparation of acidophilus cultures requires considerable care as slight contamination will ruin the culture.

A. Mother culture.

1. Sterilize selected milk in autoclave by heating to 240° F. for 15 minutes.
2. Cool to 100° F.
3. Add about 10 cc. of mother culture using sterile pipette. Cotton plug should be flamed before returning to flask.
4. Incubate at 100° F. for 18 hours.
5. Use immediately if possible; otherwise store at about 50° F.

Acidophilus cultures should be examined microscopically occasionally to make sure the culture is pure.

B. Bulk starter.

1. Use selected milk.
2. Heat to boiling or slightly higher for 30 minutes.
3. Cool to about 100° F., hold 30 minutes and again heat to boiling for 10 minutes.
4. Cool to 105° F.
5. Add 1–1½ per cent mother culture.
6. Incubate at 105° F. for 18–20 hours or until an acidity of about .70 per cent is reached.
7. Cool as rapidly as possible to 50° F.

Care must be taken to keep the temperature up to at least 100° F. during the incubation period. All possible sources of contamination should also be controlled as the culture must remain pure. These factors are so important that specially constructed vats are necessary for the successful manufacture of acidophilus milk on a commercial basis.

L. acidophilus cultures may be stored at 40° F. or lower for several days without affecting the number of living organisms.

Churned Buttermilk

Catering to the ideas of certain individuals who believe that the products and practices of our childhood are better than those of today, many dealers have placed on the market within recent years a type of fermented milk termed churned buttermilk. This product has been made in numerous ways, but in general there are three methods.

Probably the more common method is to ripen thoroughly and pasteurize a 2-per cent milk to about .75 per cent acidity. The ripened milk is then churned at a sufficiently high temperature to produce butter granules in the usual length of time or even shorter. The churning is stopped when the granules are about the size of small rice grains. The buttermilk is then pumped over a cooler and bottled. If butter coloring is added to the milk before churning a more distinct granule will be obtained.

The second method is to churn a good grade of highly colored cream until butter granules of desirable size are secured. The granules are then chilled in a 40° F. room until firm and are then added to starter that has been cooled to at least 50° F., in sufficient quantities to be visible in the bottle. A small quantity of cream added to the starter will improve the flavor. The main objection to this method is the fact that the finished product lacks the buttermilk flavor. Its main advantage is in the reduced volume of cream that must be churned.

Another method is to ripen 8–10 per cent pasteurized sweet cream to an acidity of about .35 per cent. Butter color is added and the mixture churned until granules of the proper size are secured. Enough cooled starter is then added to bring the fat content down to about 1 per cent. This gives a product of good flavor and fairly light body. The advantage of this method over the first is the greater ease of churning and the reduced volume of cream that must be handled in the churn.

Sour Cream

Commercial sour cream sometimes called Jewish cream, is the heavy bodied, smooth textured product of high acid flavor secured by processing and ripening sweet cream under control conditions. It is used as a spread for bread, as a dressing for vegetables, and in the making of sauces of various kinds.

There are several successful methods for preparing sour cream. Variations in plant equipment and plant conditions make it impossible to suggest a method applicable to all plants. Three general procedures will therefore be given.

A. Method for making sour cream without the use of a viscolizer or homogenizer.

1. Using enzyme
   a. Pasteurize the cream (18–20 per cent fat) by heating to 175° F. for 30 minutes.
   b. Cool to 85° F. and add 3 per cent starter and .5 cc. of rennet (diluted with 30 volumes of water) to each 100 pounds of milk.
   c. Pour cream into shotgun cans.
   d. Incubate at 85° F. until a firm curd is formed.
   e. Cool rapidly without stirring by placing can in ice water or 40° F. room.

This is a fast method for making a sour cream of good body.

2. Using cheese curd
   a. Pasteurize 32 per cent cream by heating to 145° F. for 30 minutes and cool to 72° F.
   b. Add 3 per cent starter and incubate at 72° F. for 18 hours.
   c. Mix 4 parts of soured cream with 1 part of cottage cheese curd and 1.5 parts of good starter which have been previously mixed and strained to remove curd particles.
3. Using skim milk powder
   a. Add 3 per cent skim milk powder to 20 per cent cream.
   b. Raise temperature gradually to 145° F. with constant stirring. Hold 30 minutes at 145° F.
   c. Cool to 72° F. and add 3 per cent starter and ⅓ cc. rennet (diluted with 3 volumes of water) to each 100 pounds of milk.
   d. Place in shotgun cans.
   e. Incubate 15 hours at 72° F.
   f. Cool without stirring by placing can in ice water or 40° F. room.

B. Method of making sour cream, using viscolizer or homogenizer.

1. Pasteurize 18–20 per cent cream at 180° F. for 30 minutes.

2. Homogenize at 180° F. using 3,000 pounds pressure on one valve. (Be sure homogenizer is thoroughly washed and sterilized previous to use.)

3. Cool to 72° F. and add 3 per cent starter.

4. When acidity of .6–7 per cent has been reached package and store at 40° F.

A slightly heavier body can be secured by adding 2 or 3 per cent of milk powder to the cream; or enough concentrated skim milk to increase the serum solids 2–3 per cent; or .25 per cent of high grade gelatin.

A better body can also be secured by ripening the cream in the final container if such a procedure can be made practical.

Brick Cheese

Perfectly sweet milk is set in a vat at 86° F. with sufficient rennet to coagulate it in 20 or 30 minutes. The curd is cut with Cheddar curd knives, is then heated to 110° or 120° F., and is stirred constantly. The cooking is continued until the curd has become so firm that a handful squeezed together will fall apart when released. The curd is then dipped into the mold, which is a heavy rectangular box without a bottom and with slits sawed in the sides to allow drainage. The mold is set on the draining table, a follower is put on the curd, and one or two bricks are used on each cheese for pressure. The cheeses are allowed to remain in the molds for 24 hours, when they are removed, the entire surface rubbed with salt, and the cheeses piled three deep. The salting is done each day for three days, after which the cheese is taken to the ripening cellar, which should be comparatively moist and have a temperature of from 60° to 65° F. Ripening requires two months.

Brie Cheese

This is a soft, rennet cheese made from cows' milk. The cheese varies in size and also in quality, depending on whether whole or partly skimmed milk is used. The method of manufacture resembles closely that of Camembert.

The milk used is usually perfectly fresh. It is not uncommon, however, to mix the evening's milk, when kept cool overnight, with the morning's milk. Some artificial coloring matter is added to the milk, which is then set with rennet at a temperature of 80° or 85° F. After standing undisturbed for about two hours, the curd is dipped into forms or hoops, of which there are three sizes in common use. The largest size is about 15 inches in diameter, the medium size about 12 inches in diameter, and the smallest size about 6 inches in diameter, all varying in height from 2 to 3 inches. After drainage for 24 hours without pressure being applied, the hoops are removed, and the surface of the cheese is sprinkled with salt. Charcoal is sometimes mixed with the salt used. The cheese is then transferred to the first curing room, which is kept dry and well ventilated. After remaining in this room for about eight days the cheese becomes covered with mold. It is then transferred to the second curing room or cellar, which is usually very dark, imperfectly ventilated and has a temperature of about 55° F. The cheese remains there for from two to four weeks, or until the consistency and odor indicate that it is sufficiently ripened. The red coloration which the surface of the cheese finally acquires has been attributed to an organism designated *Bacillus firmaticus*. The ripening is due to one or more species of molds which occur on the surface and produce enzymes, which in turn cause a gradual and progressive breaking down of the casein from the exterior toward the center. The interior of a ripened cheese varies in consistency from waxy to semiliquid and has a very pronounced odor and a sharp characteristic taste.

Brinza Cheese

This cheese from sheep's milk, or a mixture of sheep's and goats' milk.

The cheese is made in small lots, from 2 to 4 gallons of fresh milk being used at one time. This is put into a kettle and when the temperature of the milk is from 75° to 85° F. sufficient rennet is added to obtain coagulation in 15 minutes. The curd is broken up and the whey dipped, and the curd is placed in a linen sack and allowed to drain for 24 hours. It is then cut into pieces and placed on a board, where with frequent turnings it is allowed to remain until it commences to get smeary, which requires about eight days. The pieces are then laid one on top of another in a vessel holding from 40 to 60 pounds, where they remain for 24 hours, after which they are removed, the rind cut away, and the curd or partially cured cheese broken up in another vessel. After 10 hours salt is stirred in and the curd run through a mill, which cuts it very fine, when it is packed in a tub with beech shavings.

Camembert Cheese

This is a soft, rennet cheese made from cow's milk. A typical cheese is about 4¼ inches in diameter, three-quarters of an inch or 1 inch thick, and in the market in this country is usually found wrapped in paper and inclosed in a wooden box of the same shape. The cheese usually has a rind about one-eighth of an inch in thickness, which is composed of molds and dried cheese. The interior is yellowish in color and waxy, creamy, or almost fluid in consistency, depending largely upon the degree of ripeness.

Camembert cheese is made from whole milk or from milk slightly skimmed. It is not advisable to skim the milk unless it tests more than 3.5 per cent butterfat. The temperature of setting is from 78° to 87° F., and the quantity of rennet added for this purpose is sufficient to get the desired degree of firmness in from two to five hours. The curd is then transferred, usually with as little breaking up as possible, to perforated tin forms or hoops about 4¼ inches in diameter and the same in height. These rest upon rush mats, which permit it free drainage. The filling of the forms may be done at two or three different times, short intervals being necessary for the curd to settle. Each form holds the equivalent in curd of about 2 quarts of milk. After draining for about 18 hours, preferably in a room having a uniform temperature of 65° or 70° F., the cheese is turned. This is repeated frequently for about two days, when it is removed from the forms and salted on the outside. After 24 hours the cheese is carried to the curing rooms, which are maintained at temperatures of from 53° to 59° F. and with a high relative humidity. Curing the cheese is the most difficult part of the manufacturing process, for not only must there be a uniform and progressive development of the ripening agents, but the curd must be gradually desiccated at the same time. Proper conditions of humidity and temperature must be maintained and subject to regulation in order to favor the development of the needful mold, *Penicillium camemberti,* the bacteria, and yeasts. Although the growth of the mold is necessary in order to bring about a gradual breaking down of the casein, this growth should not be too vigorous and luxuriant; otherwise the product will be rendered unfit for commercial purposes. Following the growth of the mold, other organisms develop, giving the resultant cheese a reddish appearance instead of a white and blue, as is the case in the initial mold fermentation. From 15 to 20 days are required to bring about the proper balance between the various forms of life. At the end of that time the cheese is allowed to complete its ripening at the lower limits of the indicated temperatures and with a minimum of ventilation.

Cheddar Cheese

The milk, morning's and evening's mixed, is set at 85° F. with sufficient rennet to coagulate to the proper point in from 25 to 40 minutes. At the time of setting the milk should have an acidity of about 0.18 or 0.20 per cent. Color may or may not be used. The curd is cut when it breaks evenly before the finger. The cutting is done with curd knives made up of blades set about one-third of an inch apart in frames. In one frame the knives are set perpendicularly and in the other horizontally. When well cut the curd is in uniform cubes of about one-third of an inch.

After being cut, the curd is heated slowly and with continued stirring until it reaches a temperature of from 96° to 108° F. With the use of mechanical agitators, as is the common practice, the curd should be heated about 4° higher than when stirring is done by hand. After heating, the stirring is continued intermittently until the curd is sufficiently firm. This is determined by squeezing a handful, which should fall apart immediately on being released. The whey is then drawn. At the same time the acid should have reached about 0.20 per cent, or one-fourth of an inch, the latter of which is determined by measuring the length of strings when the curd is touched to a hot iron. The curd is then matted about 4 inches deep, sometimes in the bottom of the vat, sometimes on racks covered with a coarse linen cloth. After it has remained there long enough to stick together it is cut into rectangular pieces easy to handle, which are turned frequently and finally piled two to four deep; in the meanwhile the temperature of the curd is kept at about 90° F. When the curd has broken down until it has the smooth feeling of velvet, which requires from one to three hours, it is milled by means of a machine, which cuts it into pieces the size of a finger. It is then stirred on the bottom of the vat until whey ceases to run, which requires from one-half to one and one-half hours, when it is salted at the rate of 2 or 2½ pounds of salt to 100 pounds of milk. It is then ready to be put into the press. The curd is put into tinned-iron hoops of the proper size, which are lined with cheesecloth bandages. The hoops are put into presses and great pressure is applied by means of screws. The next morning the cheese is removed from the hoops and put on shelves in a curing room. Formerly it was kept in a curing room as long as six months, but at the present time it is covered with a coat of paraffin and put into cold storage when from 3 to 12 days old. There is a growing demand on the part of consumers for mild cheese, and consequently ripening must be carried on at a temperature below 50° F.

An important point in the process of manufacturing Cheddar cheese is the development of the desired quantity of acid, which is responsible for the proper breaking down of the curd before milling and salting. The maximum quantity of acid that can be developed in the whey without injuring the texture of the cheese should, therefore, be aimed at. It is very probable that too much weight has been placed on the desirability of a maximum development of acid, and that practically as good cheese can be produced without the high acid.

Some of the details in the manufacture of Cheddar cheese are varied to some extent, and other names may be used to designate the cheese so made. A stirred-curd cheese is one in which the curd particles are not allowed to mat together after the whey is drawn. The curd is stirred occasionally to prevent this matting process, but it differs from the sweet-curd cheese, as acid is allowed to develop before salting and pressing. Formerly a comparatively large quantity of stirred-curd cheese was made, but very little, if any, is made at the present time.

A washed-curd cheese varies from the regular Cheddar process in having the milled curd subjected to cold water for a short period. This process is evidently practiced to force the curd to take up a small percentage of the water and increase the yield. It results in a cheese which apparently breaks down or ripens much more rapidly than cheese made in the ordinary way. This ripening is very likely not due to the excess of moisture but to some other unexplained reason. Some States have prohibited the use of the State brand on washed-curd cheese.

Cheshire Cheese

This cheese is one of the oldest and most popular of the English varieties. It is a rennet cheese made from whole milk of cows, and is named for Chester County, England, where it is largely produced. It is made in cylindrical shape, from 14 to 16 inches in diameter, and weighs from 50 to 70 pounds. In making this cheese sufficient annatto is used to give the product a very high color. The process of manufacture varies in detail in different sections. Perfectly sweet milk, night's and morning's mixed, is set at a temperature of from 75° to 90° F. In one hour, the curd is cut usually with an instrument in which knives are set in a frame to cut cubes 1 or 1¼ inches square. This is pushed down through the curd and finally worked back and forth at an angle. This is continued for about an hour, or until the particles of curd are the size of peas. The curd is then allowed to settle and mat on the bottom of the vat for about an hour, when it is rolled up to one end, weighted down, and the whey drawn, after the desired degree of acidity has been obtained. The curd is cut in pieces of the right size to handle and is piled on racks. It is then run through a curd mill, salted at the rate of 3 pounds to 1,000 pounds of milk, and put into a hoop having a number of holes in the side, through which skewers can be thrust into the cheese to promote drainage. The cheese in the hoop is put into a heated wooden box called an oven, and sometimes light pressure is applied, the pressure increasing gradually until it reaches about 1 ton. The curing cellar or room is about 60° to 65° F. The time required for thorough ripening is from 8 to 10 months.

Cottage Cheese

Cottage cheese is sometimes made with a small amount of rennet, and the curd is heated to from 118° to 125° F. It may be made on a small or a factory scale. With this method the skim milk is pasteurized, cooled to 70° or 80°, and 1 to 5 per cent of a starter added. Rennet is then added at the rate of 1 c.c. per 1,000 pounds of milk. The curd is allowed to develop an acidity of about 0.55 in from 6 to 10 hours. The coagulum is then cut into ½-inch cubes. Water at a temperature of 115° is run over the curd in about an hour and the temperature of the wash water than gradually raised to 120°. The curd is then stirred until it will stand without breaking. It is then gradually cooked to a temperature of 118° to 126° in the course of one and one-half to three hours. When the curd may be squeezed in the hand and still retain its shape, the whey is withdrawn and the curd is washed two or three times in cold water. After the washing the water is withdrawn, and the curd ditched along the side of the vat or kettle, and drained for one hour. It is then placed in a cooler for 12 hours. To each 100 pounds of curd, 70 pounds of a mixture of milk and cream containing 10 per cent cream is added. The curd is then stirred for a few minutes. After creaming the cheese is placed in a cooler at 30° to 40° until ready to use or ship.

When the cheese is made on a factory scale a drier product is desired in order that it may be marketed successfully. For this reason the curd is generally cooked at a higher temperature than when made on a small scale. The main equipment necessary for making cottage cheese on a factory scale is a pasteurizing outfit and a channel-bottom Cheddar vat. Ordinarily from 5 to 10 per cent of a good lactic starter is added to skim milk, after which the milk is allowed to ripen at a temperature of 70° to 80° F. until curdled. The curd is then cut into cubes and gradually heated to from 115° to 125° in 30 to 45 minutes. When the whey has been removed, the curd is washed with cold water, drained, and piled along the sides of the vat. Ordinarily the cheese is salted at the rate of 3 or 4 ounces per 100 pounds of milk. Often the cheese is mixed with cream and then marketed in small, single service, paraffined paper containers, or in butter tubs.

With milk of a good quality a yield of 15 to 18 pounds of cheese per 100 pounds of skim milk is obtained. Cottage cheese should always be kept in a refrigerator or in a cooler until disposed of.

Cream Cheese

Genuine cream cheese is made from a rich cream thickened by souring or from sweet cream thickened with rennet. The cream for this cheese should always be pasteurized. This thickened cream is put into a cloth and allowed to drain, the cloth being changed several times during the draining, which requires about four days. It is then placed on a board covered with a cloth, sprinkled with salt, and turned occasionally. It is ready for consumption in from 5 to 10 days.

Another variety of cream cheese is made from cream with a low content of butterfat (6 or 8 per cent). A small quantity of a lactic-acid starter is added to the cream, and after the mixture is warmed to from 70° to 76° F. and thoroughly stirred, rennet is added at the rate of from 1 to 1½ ounces of commercial liquid rennet to 1,000 pounds of cream. Usually the cream is placed in shotgun cans holding about 30 pounds each. After setting for about 18 hours, the curd is poured, with as little breaking as possible, upon draining racks covered with cloths. After a few hours' drainage the cloths are drawn together, tied, placed upon cracked ice, and allowed to remain overnight. The curd is then pressed, salted, and worked to a paste by means of special machinery or by suitable substitutes. The cheese is then molded into pieces weighing from 3 to 4 ounces, wrapped in tin foil and, without curing, placed upon the market. The standard package of cream cheese is 3 inches by 2 inches by 1 inch. It is a mild rich cheese which is relished most when eaten a few days after it is made. Cream cheese is now quite extensively made in the larger factories of the United States, where the ever-increasing demand for it makes it one of the most popular varieties of soft cheese.

Edam Cheese

The perfectly fresh milk is set at 82° to 84° F.; color is added and sufficient rennet is used to coagulate the milk in 30 minutes. The curdled milk is divided evenly with a knife. After 20 minutes the whey is partly removed. The curd is further divided; after 10 minutes another portion of the whey is removed and stirring is resumed for 10 minutes. Then the temperature of the mixture is increased to 92°. The curd is now allowed to settle and the whey removed; then the layer of curd is cut into pieces, each part having the size of a cheese. These are left to settle in the molds, and they are then turned a few times; after being wrapped in cloth they are pressed two or three hours. After this they are salted, either by rubbing in salt and putting them in molds without lids, or by immersion in brine for three days. They are then stored for ripening and turned at intervals, which is the cause of their flattened shape. When they are a few weeks old they are marketed and the ripening process continues in the warehouses of the cheese merchants.

Emmenthaler (Domestic Swiss) Cheese

This is a hard, rennet cheese made from cows' milk, and has a mild, somewhat sweetish flavor. It is characterized by holes or eyes which develop to about the size of a cent in typical cheeses and are from 1 to 3 inches apart. Cheese of the same kind made in the United States is known as Domestic Swiss, and that made in the region of Lake Constance is called Algau Emmenthaler.

There is a slight difference in manipulation of the milk in making Emmenthaler cheese in this country as compared with Switzerland. In the latter country the evening's and morning's milk is mixed and made into cheese, while in the United States it is popularly believed that the evening's milk must be made into cheese immediately after milking, as is done with the morning's milk.

However, there is a growing tendency to make the cheese from milk delivered once a day or from milk that has been slightly ripened, as it is believed that the quality of the cheese is thereby improved.

Swiss cheese is made both with homemade rennet and with commercial rennet. When homemade rennet is employed usually no additional cultures are used. In some cases the homemade rennet is inoculated with a pure culture starter of lactobacillus bulgaricus. With modern methods it has been found desirable to use the following pure cultures: (1) The lactobacillus bulgaricus to check undesirable fermentation and to aid in controlling the ripening; (2) the use of an eye and flavor culture to aid in the development of eyes and flavor. These pure cultures are sent out by the Bureau of Dairy Industry of the United States Department of Agriculture or by State agencies.

It has been found that by clarifying the milk a much better quality of cheese can be produced, both in regard to eye formation and in improving the body of the cheese. Clarification tends to reduce the number and to increase the size of the eyes. It is estimated that fully two-thirds of the factories of Wisconsin now clarify their milk for the manufacture of wheel and block Swiss.

In making the cheese in Switzerland the evening's milk is skimmed; the morning's milk is heated to 108° or 110° F., and the cream from the evening's milk is added and both thoroughly mixed. The evening's milk cooled with a little saffron to color it, is then added, and the whole is mixed. The milk is then brought to a temperature of 90° in summer and 95° in winter, and sufficient rennet is added to coagulate the milk in 30 or 40 minutes. The whole process is carried through in a huge copper kettle holding 300 gallons. The rennet used is obtained by soaking the calf's stomach in whey for 24 hours. When the milk has thickened to almost the desired point for cutting, which is practically the same as for ordinary American or Cheddar Cheese, the thin surface layer is scooped off and turned wrong side up. This is supposed to aid in incorporating the layer of cream into the cheese. The curd is then cut very coarse by means of a so-called harp. The cheesemaker, with a wooden scoop in each hand, then draws the mass of curd toward him, that lying on the bottom of the kettle being brought to the surface. At this point the cheesemaker and an assistant commence stirring the curd with the harp, a breaker having first been fitted to the inside of the kettle to interrupt the current of the whey and curd. The harps are given a circular motion and cut the curd very fine—about the size of wheat kernels.

After this stage is reached heating is commenced. In Switzerland until recently all the heating was done over an open fire, the kettle being swung on a large crane; most of the factories have the same method at the present time. In this country the same method was followed in the early days of the industry, but at the present time inclosed fireplaces, into which the kettle can be swung and doors closed to retain the heat, are largely employed. This takes away much of the discomfort of the operation. In a few instances the kettle is set in cement and an iron car containing the fire is run under it. The most modern factories use steam, which appears to be the most satisfactory way. When the heating is begun the contents of the kettle are brought rapidly to the desired temperature, which may be from 126° to 140° F., the higher temperature often being necessary to get the curd sufficiently firm. In the meanwhile the stirring continues for about one hour, with slight interruptions near the end of the process, when the curd has become so firm that it will not mat together. The end of the cooking is determined by the firmness of the curd, which is judged by matting a small cake with pressure by the hands and noting the ease with which the cake breaks when heating the edge.

When the curd is sufficiently firm, the contents of the kettle are rotated rapidly and allowed to come to a standstill as the momentum is lost. This brings all the curd into a cone-shaped pile in the center of the kettle. One edge of a heavy linen cloth resembling burlap is wrapped around a piece of hoop iron, and by this means the cloth is slipped under the pile of curd. The mass of curd is then raised from the whey by means of a rope and pulley and lowered into a cheese hoop on the draining table. These hoops are from 4 to 6 inches deep and vary greatly in diameter. The cloth is folded over the cheese, a large follower is put on top, and the press is allowed to come down on the cheese. The press is usually a log swung at one end and operated by a double lever. Pressure is continued for the first time just long enough for the curd mass to retain its shape. The hoop is then removed, the cheese turned over, and a dry cloth substituted. The cheese is allowed to remain in the press about 24 hours, during which time it is turned and a dry cloth substituted six or more times.

At the end of the pressing, the curd should be a homogeneous mass without holes. The cheese is then removed to the salting board, covered with a layer of salt, and occasionally turned. In a day or two it is put into the salting tank in a brine strong enough to float an egg; it remains there at the discretion of the cheesemaker for from one to four days. Often no brine tank is used with Emmenthaler cheese.

The cheese is then taken to the curing cellar. In the best factories two or more cellars with different temperatures are available, and the cheeses are placed in them according to their development. If it appears that the cheese may develop too fast and have too many and too large eyes, it is placed in a cool cellar; if the reverse is true, a warm cellar is selected. The cellars vary in temperature from 55° to 65° F., though in extreme cases 70° or a little higher may be used. While the cheeses are in the ripening cellar, which in Switzerland may be from 6 to 10 months or longer, and in the United States three to six months, they should be turned and washed every other day for the first two or three months and less often subsequently. At the same time a little coarse salt is sprinkled on the surface. In a few hours this salt has dissolved, and the brine is spread over the surface with a long-handled brush.

The cheeses are very large, about 6 inches in thickness and sometimes as much as 4 feet in diameter, and weigh from 60 to 220 pounds. In shipping, a number of them are placed in a tub which may contain 1,000 pounds of cheese. Sometimes Emmenthaler cheese is made up in the form of blocks instead of in the shape of millstones. The blocks are about 28 inches long and 8 inches square in the other dimensions and weigh usually from 25 to 28 pounds.

Gorgonzola Cheese

This variety, known also as Stracchino di Gorgonzola, is a rennet, Italian cheese made from whole milk of cows. The interior of the cheese is mottled or veined with a penicillium much like Roquefort, and for that reason the cheese has been grouped with the Roquefort and Stilton varieties. As seen upon the markets in this country the surface of the cheese is covered with a thin coat resembling clay, said to be prepared by mixing barite or gypsum, lard or tallow, and coloring matter. The cheeses are cylindrical in shape, about 12 inches in diameter and 6 inches in height, and as marketed are wrapped in paper and packed with straw in wicker baskets.

The milk used in making this cheese is warmed to a temperature of about 75° F. and coagulated rapidly with rennet, the time required being usually from 15 to 20 minutes. The curd is then cut very fine, inclosed in a cloth and drained, after which it is put into hoops 12 inches in diameter and 10 inches high. It was formerly the custom to allow the curd from the evening's milk to drain overnight and to mix it with the fresh, warm curd from the morning's milk prepared in the same way. The curd from the evening's milk and that from the morning's milk, crumbled very fine, were put into hoops in layers with moldy bread crumbs interspersed among the layers. The cheese is turned frequently for four or five days, the cloths being changed occasionally, and is salted from the outside, the process requiring about two weeks. It is then transferred to the curing rooms, where a low temperature is usually maintained. At an early stage in the process of ripening, the cheese is usually punched with an instrument about 6 inches long, tapering from a sharp point to a diameter of about one-eighth inch at the base. About 150 holes are made in each cheese. This favors the development of the penicillium throughout the interior of the cheese. Well-made cheese may be kept for a year or longer. In the region where it is made, much of the cheese is consumed while in a fresh condition.

Limburg Cheese

This is a soft, rennet cheese made from cows' milk which may contain all the butterfat or may be partly or entirely skimmed. The best Limburg is undoubtedly made from the whole milk. This cheese has a very strong and characteristic odor and taste, weighs about 2 pounds, and is about 6 by 6 by 3 inches in size.

Limburg cheese originated in the Province of Lüttich, Belgium, in the neighborhood of Hervé, and was marketed in Limburg, Belgium. Its manufacture has spread to Germany and Austria, where it is very popular, and to the United States, where large quantities are made, mostly in New York and Wisconsin.

Sweet milk, without any coloring matter, is set at a temperature of from 91° to 96° F. with sufficient rennet to coagulate the milk in about 40 minutes. In foreign countries a kettle is used, but in the United States an ordinary rectangular cheese vat is found to be more satisfactory. The curd is cut or broken into cubes of about one-third of an inch and is stirred for a short time without additional heating. It is then dipped into rectangular forms 28 inches long, 5½ inches broad, and about 8 inches deep. These forms are kept on a draining board, where the whey drains out freely. When the cheese has been in the forms, with frequent turnings, for a sufficient length of time to retain its shape, it is removed to the salting table, where the surface is rubbed daily with salt. When the surface of the cheese commences to get slippery the cheese is put into a ripening cellar having a temperature of about 60° F. While in the cellar the surface of each cheese is frequently rubbed thoroughly. To ripen requires one or two months. When ripe the cheese is wrapped in paper, then in tin foil, and put into boxes, each containing about 50 cheeses.

Contrary to the popular belief, no Limburg is imported into this country at the present time. This type of cheese is made so cheaply and of such good quality in this country that the foreign make has been crowded out of the market.

Loaf or Process Cheese

It is defined as the clean, sound, heated product made by comminuting and blending, with the aid of heat and water and with or without the addition of salt, one or more lots of cheese into a homogeneous plastic mass.

At present it is estimated that one-half of all cheese made in this country is marketed as loaf or process cheese. American Cheddar, Swiss, Brick, Limburg, and even Camembert have been handled in this manner.

In the preparation of this product, cheese of different degrees of ripeness and of inferior quality with respect to flavor and texture may be used. Well-cured Canadian, well-cured Emmenthaler, or culture Swiss cheese is often used to impart a typical flavor. It is stated that as much as 20 per cent white American cheese is often blended with Swiss cheese in order to give the finished product the proper texture.

The method of manufacture consists in cleaning the surface of the cheese, grinding it, and then adding a small quantity of an emulsifier, such as sodium citrate, sodium phosphate, or rochelle salts, dissolved in water, and finally heating the mixture in jacketed containers with constant agitation until the cheese has reached the proper degree of consistency. It is then put into suitable containers either directly or by specially designed machinery. From 1 to 2 percent of emulsifiers are often used. Considerable skill is required in selecting the best kind of cheese to use as well as in regulating the manner and duration of the cooking. Ordinarily the cheese is gradually heated and stirred until a temperature of 140 to 160° F. is reached. The stirring is continued at this temperature for a longer or shorter period according to the nature and kind of cheese.

In the initial heating there is at first a slight separation of fat. This is followed by physical changes in the character of the curd so that the cheese becomes plastic and stringy. Upon further heating this plastic state is gradually broken down and a homogeneous mass with but slight plastic qualities is developed. When the cheese has reached this creamy condition and while still very hot, it is weighed and run into tin-foil-lined containers. Such packages render the cheese remarkably free from subsequent mold development.

Most of the process cheese manufactured in this country is made in a few large plants. At the present time there are no regulations as to the kind or quality of cheese that may be used in blending and no statement on the package as to whether or not emulsifiers are used.

Münster Cheese

Münster is a rennet cheese of the whole milk of cows, made in the vicinity of Münster, in the western part of Germany near the Vosges Mountains. Similar cheese made in the neighboring portion of France is called Géromé, and Münster cheese made near Colmar and Strassburg is sometimes given the names of those two cities.

The milk is set at about 90° F., with sufficient rennet to coagulate it in 30 minutes. The curd is then broken up and allowed to stand from 30 to 45 minutes without stirring, when it is dipped with a sieve, which gives slight pressure to the curd and holds back the small particles. After removing the whey the curd is scooped into forms or hoops, and caraway or anise seed is usually added. The hoops are made in two parts, the lower being 4 inches high and 7 inches in diameter, with holes in the bottom for draining, and the upper of the same dimensions. The whole resembles an ordinary cheese hoop with bandages. The hoop is lined with cheesecloth. After the curd has been in the hoop for 12 hours the upper part of the latter may be removed, the cheese turned, and the cloth removed. The cheese is now put into the upper portion of the hoop and turned frequently for from four to six days. In the meantime the temperature is held at 68° F. After salt has been rubbed on the surface daily for three days the cheese is taken to the cellar, which has a temperature of from 51° to 55° F., where it is allowed to ripen for two or three months.

Neufchâtel Cheese

This is a soft rennet cheese made extensively from either whole or skim milk of cows. Bondon, Malakoff, Petît Carré, and Petit Suisse are essentially the same as Neufchâtel but have slightly different shapes.

Neufchâtel cheese is made in the same manner as cream cheese, except that a little less rennet is used, perhaps 1 ounce of commercial liquid rennet to 1,000 pounds. Either whole milk or partly skimmed milk is used. Rennet is added to it at ordinary temperatures, and the curd when sufficiently firm is broken up, put into molds, and subjected to pressure. After being salted, the cheese is cured for from 8 to 15 days in a so-called drying room and then ripened in a cellar at a temperature of about 55° F. During the process of ripening the cheese becomes covered at first with a whitish mold and later with a blue mold in which red spots appear. After about one month it is ready for sale.

Parmesan Cheese

The milk, which has been skimmed to a greater or less extent, is heated in copper kettles to a temperature varying, according to the acidity of the milk, from 90° to 100° F. The kettle is then removed from the fire, rennet added, and the kettle covered and allowed to stand for 20 minutes to one hour, when the curd is cut very fine and cooked, with stirring, to 115° or 125° F. for from 15 to 45 minutes. The curd is removed from the kettle by means of a cloth, and after draining for a short time is put into hoops about 10 inches high and 18 inches or more in diameter, and lined with coarse cloth before filling. Pressure is then applied for 24 hours, the cheese being turned frequently and the cloths changed. The salting, which is begun in from one to three days after removing from the press, is continued for a considerable length of time, often 40 days. The cheeses are then transferred to a cool, well-ventilated room, where they may be stored for years, the surface being rubbed with oil from time to time. The exterior of the cheese is dark green or black, due to coloring matter rubbed on the surface. A greenish color in the interior has been attributed to the contamination with copper from the vessels in which the milk is allowed to stand before skimming.

Parmesan cheese when well made may be broken and grated easily and may be kept for an indefinite number of years. It is grated and used largely for soups and with macaroni. A considerable quantity of this cheese is imported into this country and sells for a very high price.

Roquefort Cheese

This is a soft, rennet cheese made from the milk of sheep. It is also stated from good authority that as much as 2.46 per cent of cows' milk and 0.18 per cent of goats' milk are mixed with the sheep's milk. There are, however, numerous imitations, such as Gex and Septmoncel, made from cows' milk, which resemble Roquefort. One of the most striking characteristics of this cheese is the mottled or marbled appearance of the interior, due to the development of a penicillium, which is the principal ripening agent.

Part of the milk is heated to 122° to 140° F. When this milk is mixed with the remainder the resulting temperature should be 76° to 82°, which is the setting temperature for the cheese. In from one to two hours after the addition of rennet the curd is cut until the particles are about the size of walnuts. The whey is dipped off, and the curd is put into hoops which are about 8½ inches in diameter and 3½ inches in height. The hoops usually are filled in three layers, a layer of moldy bread crumbs between each. The bread used for this purpose is prepared from wheat and barley flour, with the addition of whey and a little vinegar. It is thoroughly baked and kept in a moist place from four to six weeks, during which time it becomes permeated with a growth of the mold. The crust is removed, and the interior is crumbled dried, ground very fine, and sifted. The cheese is not subjected to pressure. It is turned usually one hour after putting into hoops and is not wrapped in cloths.

Formerly the manufacture of the cheese up to this stage was carried on by the shepherds themselves, but in recent years centralized factories have been established, and much of the milk is collected and there made into cheese. The cheese is then taken to the caves. These are for the most part natural caverns which exist in large numbers in the region of Roquefort. The temperature in these caves is 40° to 45° F., and the air circulates very freely through them. Recently artificial caves have been constructed and used. When the cheeses reach the caves they are salted, which serves to check the growth of the mold on the surface. One or two days later they are rubbed vigorously with a cloth and are afterward subjected to thorough scraping with knives, a process formerly done by hand, but now performed much more satisfactorily and economically by machinery. The salting, scraping, or brushing seems to check the development of mold on the surface. In order to favor the growth of mold in the interior, the cheese is pierced by machinery with from 20 to 60 small needles, which process permits the free access of air. The cheese may be sold after from 30 to 40 days or may remain in the caves as long as five months, depending upon the degree of ripening desired. During the process of ripening by scraping and evaporation the cheese loses from 16 to 20 per cent of the original weight. When ripened, it weighs 4½ or 5 pounds.

Stilton Cheese

This is a hard, rennet cheese, the best of which is made from cows' milk to which a portion of cream has been added. The cheese is about 7 inches in diameter, 9 inches high, and weighs 12 or 15 pounds. It has a very characteristic wrinkled or ridged skin or rind, which is probably caused by the drying of molds and bacteria on the surface. When cut it shows blue or green portions of mold which give its characteristic piquant flavor. The cheese belongs to the same group as the Roquefort of France and the Gorgonzola of Italy.

The morning's milk is put into a tin vat, the cream from the night's milk is added, and the whole is brought to a temperature of 80° F., when the rennet is added. It is claimed by some cheesemakers that the curd should be softer when broken up or cut than the curd for Cheddar cheese, whereas others believe that it should become very firm before it is disturbed, one or two hours being allowed for setting. When sufficiently firm, the curd is dipped into cloths which are placed in tin strainers. After draining for one hour, the cloths containing the curd are packed closely together in a large tub and allowed to remain for 12 hours, when they are again tightened and packed for 18 hours. The curd is ground up coarse, and salt is added, 1 pound to 60 pounds of curd. It is then put into tin hoops 8 inches in diameter and 10 inches deep. The cheeses remain in the hoops for six days, when they are bandaged for 12 days, or until they become firm, and are then placed in the curing room at 65° F. Ripened Stilton cheese of late is often ground up and put into jars holding from 1 to 2½ pounds.

Infants Milk

To make cow's milk more easily digestible by bottle-fed babies—one level tablespoon gelatine for each quart of milk is used. The gelatine is soaked for

10 minutes in ½ cup of cold milk taken from formula, then placed in boiling water and stirred until dissolution. Then add remainder of the milk.

Jelly Powders: In the manufacture of flavored gelatine, 10 parts gelatine is mixed with 85 parts sugar to which flavor, color and tartaric acid 2 parts are used to sharpen the flavor.

Gelatin in Ice Cream and other Food Products: ½ of 1% gelatine in ice cream prevents the formation of ice crystals by acting as an emulsifying agent improves the texture and body of the finished product.

* Non-Sweating Peanut Butter

1–5% of Diglycol Stearate or Glyceryl Monostearate or Cetamin is dissolved by warming and thorough mixing in the peanut butter mass.

* Protective Coating for Meats

The articles are dipped into a gelatin soln. contg. about 30% gelatin at a temp. of about 57–60° which has not been heated to a temp. over about 65° and which contains a hardening agent such as K alum and an emollient such as glycerol, and the coated article is then dipped into a gelatin soln. of about 27% at a temp. of about 43–46° which has not been heated to above about 65° and the coating formed is dried.

* Preserving Pepper Extracts

Extracts of red pepper are preserved by the addition of 0.01–0.05% thiosinamine.

* Sausage Casing

Cheese cloth or calendered muslin is coated with a viscous, gelatinous solution prepared by boiling down the extract from 25 lb. of fresh hog skins or hides with 8 gal. of H_2O to 15 lb. wt. and adding glycerin 2%, NaCl 10–20%, and KNO_3 1 oz. to 3 lb. of hide solution. The cloth is smoked for about 24 hr. to dry and harden the coating; alternatively, it may be treated with 2–3% CH_2O followed by hypochlorite.

CLEANERS, SOAPS

* Cleaning Compound

This product is claimed to be non-inflammable; for cleaning floors and oil paints.

Hydrogenated Naphthalin	35
Cyclohexanol	10
Sulphonated Oleates	10
Water	20
Turpentine	15
Ammonium Chloride	3
Isoamyl Acetate	2

Cleaning Compound, Bottle

Sodium Metasilicate	10
Soda Ash	20
Trisodium Phosphate	25

To Clean Bronze

Saturate a 5% acetic acid solution (or household vinegar) with ordinary table salt. This solution will clean bronze or brass; and if the metal is immediately polished and lacquered with clear lacquer, a reasonably permanent finish will result.

Cleaning Copper Coins

Sodium Cyanide	6–8 oz.
Water	1 gal.

Apply the above solution hot with a tampico brush, and when tarnish is removed, wash with clean cold water, then hot water and dry.

Caution.—This material is poisonous and care must be taken in handling.

Dry Cleaning Fluid

(Non-inflammable and quick acting)

Butyl Cellosolve	1
Diglycol Oleate	1
Water	1
Isopropyl Alcohol	10
Carbon Tetrachloride	14

Cleaning Fluid, Non-Inflammable

A. Carbon Tetrachloride 6¼ gal.
Deodorized Gasoline (68° Be) 3⅛ gal.
Chloroform 4 oz.

B. Carbon Tetrachloride 6 gal.
Deodorized Naphtha (57–59° Be) 3½ gal.
Benzol ⅜ gal.
Chloroform 4 oz.

* Cleaning Fluid

Methyl Acetone	2
Ethyl Acetate	1
Alcohol	1
Methanol	1

* Cleaner, Dairy Equipment

Trisodium Phosphate	30–50
Sod. Metasilicate	40–60
Soap	2–10
Soda Ash	8–10

Dry Cleaner

Oleic Acid	370 gm.
Stearic Acid	80 gm.
Potassium Carbonate	80 gm.
Water	70 gm.
Benzin	395 gm.
Stronger Ammonia Water	5 gm.

Melt the stearic acid and dissolve it in the warmed oleic acid. To this add the warm benzin and mix thoroughly. Dissolve the carbonate in the water and add this with constant stirring into the benzin mixture. Finally add the ammonia and beat into a homogeneous paste.

Gasoline Cleaning Cream

1. Cocoa Soap 5 gm.
Ammonia Water 8 cc.
Solution Potassa 4 cc.
Water, enough to make 30 cc.

Dissolve the soap, by the aid of heat, in 10 cc. of water, add the ammonia and solution of potassa, and sufficient water to make 30 cc. To this saponaceous cream carefully add, in small portions at a time, 5000 cc. of gasoline. This is stated to be an excellent cream for removing grease spots from clothing.

2. Spirit of Ammonia 20 gm.
Ether 50 gm.
Gasoline 150 gm.
Oil Lavender 5 gm.
Tincture Soapbark 225 gm.
Alcohol 500 gm.

3. Oleate Ammonia 2 oz.
Solution Ammonia 2 oz.
Ether 1 oz.
Benzine 5 oz.
Chloroform 1 oz.

Mix the solution and oleate; shake well and add the ether; shake, and add 5 ounces of benzine; agitate thoroughly; then add 1 ounce of chloroform and shake again. Allow to stand a few minutes and shake at intervals, when a mixture having the consistency of cream and showing but little tendency to separate will result.

* Deodorant Cleaner, Porcelain

Sod. Bisulfate	80
Pine Oil	4
Sodium Sulfate	16

Powdered Glove Cleaner

Cream of Tartar Powd.	480
Soap Bark	160
Whiting	96
Oil Birch Tar	12

* Hand Cleaner and Softener

Coarse Corn Meal	60–80 lb.
Glycerol	7–22 lb.
Soap	11–22 lb.

Color and perfume to suit.

Hand Wash, Mechanics Antiseptic

Chloride of Lime Powd.	175 gm.
Sod. Bicarbonate	359 gm.
Boric Acid	35 gm.
Water	30 oz.

For use on grimy hands to prevent dermatitis dilute with 10 times water and follow by thorough rinsing with mild soap and water.

Cleaning Paste for Mechanics

100 lb. Stearic Acid
54 lb. Caustic Soda Soln. 30° Be
10 lb. Soda Ash
836 Water

1000 lb.

Heat at 85° C. for about 10 minutes, stirring until uniform. Fine pumice stone may be incorporated as an abrasive if desired.

Kerosene Jelly Cleaner

1. Trihydroxyethylamine Stearate	5
2. Kerosene	16
3. Cresylic Acid	1
4. Water (Boiling)	45

Heat (1) and (2) until dissolved; add (4) slowly while stirring with high speed mixer then add (3).

The above makes an excellent antiseptic cleaner for woodwork, tile, porcelain, etc.

---

* Laundry Detergent

Soap	5.5 lb.
Water	29 lb.

Heat together until dissolved. Run into this slowly with rapid stirring:

Turpentine	11 oz.
Pot. Nitrate	4 oz.
Ammonium Hydroxide	12 oz.
Mineral Oil	17 oz.

Leather Cleaner

Castile Soap (Powd.)	6
Water	160

Boil until dissolved: cool and add

Ammonium Hydroxide	6
Glycerin	14
Ethylene Dichloride	7

Marble and Porcelain Cleaner

Diatomaceous Earth	3
Sulfuric Acid	9
Sodium Sulfate	88

---

* Marble and Porcelain Cleaner

Sodium Bisulfite	25
Sodium Sulfate	75

---

* Cleaner, Oil Painting

Tetralin	35
Hexalin	10
Sod. Sulforicinoleate	10
Turpentine	15
Water	20
Am. Chloride	3
Amyl Acetate	2

---

* Cleaner for Oil Paintings

A paste for cleaning oil paintings, delicate fabrics, precious wood, etc., is obtained by stirring a soln. of 3000 g. rice starch and 50 g. deodorant, *e.g.*, rose oil, almond oil, $PhNO_2$, in 9 l. H_2O into a mixt. of CCl_4 280, decahydronaphthalene 980, cyclohexanol 380, olive oil 340 and H_2O 240 g. and adding up to 1620 g. of 15° Bé. NaOH soln.

---

* Detergent and Paint Remover

Tallow 14 lb., coconut oil or the like 8.5 lb. and a soln. of NaOH 3.75 lb. in water 25 lb. are boiled together, water 75 lb. is added, with further boiling, and there are then also added silicate of Na or glycerol 3 lb., an aq. soln. of borax 0.5 lb., light mineral oil 6 lb., petroleum jelly 2.5 lb., pumice stone 20 lb., benzine 0.5 lb. and perfume 0.5 lb.

Printers Form Cleaner

Sod. Metasilicate	20 lb.
Water	50 gal.

Rifle Cleaner

Sperm Oil	10
Turpentine	10
Acetone	10
Kerosene	20
Lanolin	0.5

Rug Cleaner

Di-Glycol Oleate	44
Butyl Cellosolve	5
Ethylene Dichloride	12
Alcohol	15
Oleic Acid	11
Ammonium Hydroxide	11
Water	45

This may be made thinner by increasing the amount of water.

---

* Silk Stockings and Gloves, Detergent

Ammonium Hydroxide (0.880)	3
Gum Arabic	1
Oil Lavender Spike	½
Water	14

2 ounces of the above are used per gallon of wash water.

Cleaning Straw Hats

1. Hats made of natural (uncolored) straw, which have become soiled by wear, may be cleaned by thoroughly sponging with a weak solution of tartaric acid in water, followed by water alone. The hat after being so treated should be fastened by the rim to a board by means of pins, so that it will keep its shape on drying. Packets containing some of the acid in powdered form and wrapped in wax paper may be put up and sold for this purpose. Of course, printed directions for the use of the acid should accompany the packet.

2. Sponge the hat with a solution of:

Sodium Hyposulphite	10 parts
Glycerin	5 parts
Alcohol	10 parts
Water	75 parts

Lay aside in a damp place for 24 hours and then apply:

Citric Acid	2 parts
Alcohol	10 parts
Water	90 parts

Press with a moderately hot iron after stiffening with gum water if necessary.

3. If the hat has become much darkened in tint by wear the fumes of burning sulphur may be employed. The material should be first thoroughly cleaned by sponging with an aqueous solution of potassium carbonate, followed by a similar application of water, and it is then suspended over the sulphur fumes. These are generated by placing in a metal or earthen dish, so mounted as to keep the heat from setting fire to anything beneath, some brimstone, and sprinkling over it some live coals to start combustion. The operation is conducted in a deep box or barrel, the dish of burning sulphur being placed at the bottom, and the article to be bleached being suspended from a string stretched across the top. A cover not fitting so tightly as to exclude all air is placed over it, and the apparatus allowed to stand for a few hours. Hats so treated will require to be stiffened by the application of a little gum water, and pressed on a block with a hot iron to bring them back into shape.

Wall Paper Cleaner

Whiting	10 lb.
Magnesia Calcined	2 lb.
Fullers Earth	2 lb.
Pumice Powd.	12 oz.
Lemenone	4 oz.

Laundry Sours

Neutralizing scale for use in souring after a chlorine bleach on cotton, etc. 1 ounce of 56% acetic acid equals the following:

0.6 oz. Oxalic Acid
0.5 oz. Sulfuric Acid Conc.
1.4 oz. Nitre Cake (33%)
0.5 oz. Sodium Silico Flouride
0.6 oz. Sodium Acid Fluoride
1.0 oz. Muriatic Acid
1.0 oz. Sodium Bisulfite
2.0 oz. Lactic Acid (44%)

others that could be added are SO_2 gas, "hypo," formic acid, etc.

Laundry Blue

Ultramarine Blue	35
Aniline Blue Soluble	1
Soda Ash	30
Corn Syrup	7

Make into a paste with water and press in forms.

Liquid Laundry Blue

Prussian Blue	1
Distilled Water	32
Oxalic Acid	¼

---

* Soap

Cottonseed Fatty Acids	60
Hardwhite Stearin	20
Soda Ash	12
Caustic Potash	8

These are ground together to form a dry water soluble soap.

Soap, Castor Oil

To obtain a transparent, amber-colored castor-oil soap (*A*), mix 30 cc. KOH of 80% (wt./vol.) with 15 cc. industrial alc. and 99.4 g. castor oil. The resulting opaque jelly when put into a warm place will be clear after 10 min. To prep. from this a *compound soln. of cresol,* add further 142 g. cresol, shake, then add H_2O to make 300 cc. To prep. a more dil. soln. of *A,* add to the above quantity of *A* sufficient H_2O to make 225 cc. This soln., *liquid castor-oil soap* (*B*), is miscible with H_2O in all proportions, is permanent and may be used as a stock soln. for other prepns.

---

* Floating Soap

A substance capable of generating H is added to the soap or a constituent thereof before, during or after the saponification process. Thus, 20 g. of Al dust may be added to 100 kg. of hot liquid grained soap.

---

* Soap, Dry Cleaning

Oleic Acid	1
Cyclohexanol	1
Carbon Tetrachloride	1
Ammonia (26° Bé)	0.2
Water	0.5

Dry Cleaning Soap

Red Oil	1000
Pot. Hydroxide (50° Bé)	400
Hexalin	1000
Benzine or Carbon Tetrachloride	300
Water	300

The first two items are warmed to 70° C. and stirred until saponification is complete. Cool and stir in other ingredients.

Dry Cleaners Soap

50 to 55 parts good quality red oil (oleic acid).

12 to 14 parts caustic potash is added to the red oil and stirred until soap solution is reached.

34 to 36 parts denatured alcohol.

The red oil soap is added to the alcohol and the mix stirred for one hour.

Diglycol Oleate is used as a dry cleaning soap because of the following advantages:

1. Dissolves quickly and clearly in dry cleaning solvents.
2. Low surface tension increases penetration.
3. Possesses high detergent powers.
4. Does not build up pressure on filters.
5. Low cost.

One pint is usually used with 50 gallons of solvent.

Dry Cleaning Liquid Soap
(Non-Alkaline)

Diglycol Oleate	130
Tetralin	28
Naphtha	30

Drycleaners Soap

White Oleic Acid	6–10%
Triethanolamine	3– 4%
Carbontetrachloride	18–17%
Cleaners Naphtha	73–69%

Mix white oleic and triethanolmine and heat solution until hand warm. Then add carbontetrachloride and cleaners naphtha, stirring mixture slowly.

Dry-Cleaning Soaps

One of the major uses for Triethanolamine is in the preparation of dry-cleaning soaps. The first requisite of such soaps is that they be soluble in dry-cleaning solvents, a property which is a characteristic of Triethanolamine soaps. In practice a mixed Triethanolamine-potash soap can be used, the mixture being cheaper and at least as soluble as the Triethanolamine soap itself. A formula along these lines, which gives excellent results in dry-cleaning, has been worked out and thoroughly tested. It produces a soap which is soluble in naphtha in all proportions, and is therefore particularly adapted for use with filter systems. Being more completely saponified than ordinary soaps, it is more concentrated and hence less is required for use. The incorporation of Butyl Cellosolve in the formula gives a particularly effective coupling action, and allows the addition of water which is vitally necessary for good detergent action. It also assists in removing food-stains and other water-soluble spots and aids in brightening the colors of the cleaned garments.

This formula is composed of the following ingredients:

Naphtha Soluble Soap

Oleic Acid	107 lb.
Butyl Cellosolve	27 lb.
Cleaner's Naphtha	25 lb.
Triethanolamine	19.7 lb.
Potassium Hydroxide	8.3 lb.
Water	13.5 lb.

The oleic acid, Butyl Cellosolve and naphtha are thoroughly mixed and heated to 140° F. in the absence of flames. In a separate container the potassium hydroxide is dissolved in the water and mixed with the Triethanolamine. The water solution is then stirred into the oleic acid solution, and stirring is continued for about 30 minutes until a clear stable solution is produced.

Laundry Soap

Tallow Soap	75%
Steam-distilled Pine Oil	25%

The pine oil content of this laundry soap promotes excellent penetration and has been tested and proven to insure the removal of more dirt. Pine oil has no deleterious effect on any type of textile fibre. A laundry soap of this type works well at any temperature and will assist in the brightening of colors. It leaves a pleasant piney odor in the damp clothes, which disappears upon drying.

Liquid Soap

Eighty kg. palm-seed oil and 20 kg. sunflower seed oil are sapond. at 50° with 52 kg. 50 Bé. KOH. After the mixt. has stood, it is adjusted to the desired alky., and then the filling mass (consisting of 200 kg. cryst. sugar, 10 kg. K_2CO_3 and 10 kg. KCl dissolved in 1000 kg. water) is added.

---

* Liquid Soap, Non-Gelatinizing

Eight kilograms of coconut oil, 2 kilograms of tallow and 1.3 kilograms of olein (oleic acid) are saponified by the half-boiled process with 7.2 kilograms of caustic potash lye (40 degrees Be.) with the addition of 15 liters of water. Shortly after saponification is completed, 3.2 kilograms of a 50 per cent solution of potassium acetate are added. The soap is then allowed to cool. It is filtered to remove impurities.

Concentrated Liquid Soap for Silk Goods, Silk Stockings, Etc.

Water	55 parts
Solid Caustic Potash	5 parts
Diethylene Glycol	20 parts
Red Oil or Oleic Acid	20 parts
Yield	100 parts

Dissolve the caustic in the water, add the diethylene glycol, bring to a boil and add the red oil. Adjust either with red oil or alkali until the sample dissolved in alcohol is neutral to phenolphthalein.

Formula: Liquid Cleaning Soap

Rosin Soap (Anhydrous)	10%
Oleate Soap (Anhydrous)	10%
Steam-distilled Pine Oil	20%
Trisodium Phosphate	4%
Water	56%

This product makes a very efficient cleaner for use on all types of floors, woodwork, tile, porcelain, etc. The pine oil content insures penetration and a solvent action to assist the removal of greasy and oily films. This product has a pleasant piney odor that will act as a partial deodorant, and the pine oil content will also insure some disinfecting value.

Liquid Soap

The soap base may be made from one-third coconut oil and two-thirds soya bean oil. The proportions used in saponification are 10.75 parts by weight of soya bean oil, crude or bleached, 5.00 parts by weight of coconut oil and about 7.87 parts by weight of 50 degrees Bé potassium hydroxide. The soap obtained from this saponification is dissolved in 77 parts by weight of water to which a maximum of 0.5 part by weight of potash has been added.

Another soap is made from two-thirds coconut oil and one-third castor oil. The proportions used in saponification are 10.75 parts by weight of coconut oil, 5.0 parts by weight of pure castor oil and about 7.48 parts by weight of 50 degrees Bé potassium hydroxide solution. After saponification, the soap is dissolved in 76 parts by weight of water and as above a maximum of 0.5 part by weight of potash is added.

In making the soap from coconut oil and olein, the following proportions are used: 8.5 parts by weight of coconut oil, 5.0 parts by weight of best quality oleic acid and about 7.3 parts by weight of 50 degrees Bé potassium hydroxide solution. After saponification the soap is dissolved in 77 parts by weight of water and again up to a maximum of 0.5 part of potash is added.

It is very interesting to follow through the progress of saponification. At the beginning the temperature of the mixture rises slowly, since only a small part of the mixture is saponified under the initial conditions of the process. But the rise in temperature constantly becomes greater and the principal reaction of the saponification then takes place. Hence if the mixture has been agitated at a temperature of 65 to 70 degrees C., the temperature rises slowly to approximately 75 to 78 degrees C. Thereafter the rise is more rapid until approximately 85 degrees C. is attained. At this point the greater part of the contents of the kettle is saponified and the heat of reaction liberated becomes smaller and further increase of the temperature is slower. In most cases the temperature increases to approximately 94 to 96 degrees C. and remains constant at that point for some time. Then there comes a point at which the temperature in the kettle begins to fall. Saponification reaction may then be considered as finished and it only remains to saponify residual traces of unsaponified matter. Hence the mixture in the kettle must show at this point noticeable traces of caustic alkali, so that the saponification of the residual fat and oil may be affected when the mixture is well-agitated.

As the mass in the kettle is worked up, it first becomes thick and heavy, but then soon thinner and thereafter thicker and heavier again. When this happens, agitation is best stopped and the soap mass is allowed to remain quiescent for some minutes. Then the soap is fitted and tested. If sufficient alkali were present, technically complete saponification would be obtained. Thus, the results would be as good as those obtained by hot saponification of fats.

At this point the fitting of the soap begins. The soap must have a slight but clearly perceptible acrid taste. This test may be used when the complete saponification test is not made in the works laboratory. This test is, however, very simple and should be made. A small quantity of the soap is dissolved in distilled water. The solution must not be turbid, but absolutely clear. If there is a slight turbidity, this indicates the presence of unsaponified oils or fats. However, in this case, no traces of free caustic potash could be detected in the soap, since the correctly carried out half-boil process gives absolutely good results. If too little lye has been used in the saponification process, which may also happen when the potassium hydroxide solution employed is not 50 degree strength (this does not happen often), if the solution of potassium hydroxide is allowed to remain in storage tanks exposed to the air for too long a time so that considerable of the hydroxide is converted into the carbonate and the strength of the solution accordingly reduced, then the soap may be lacking in potash lye and in fitting the soap it then becomes necessary to add potassium hydroxide. In this case the potassium hydroxide solution is diluted with distilled or soft water to about 30 degrees Bé concentration, so that it can be mixed with the soap more readily and more uniformly. The fitting of the soap must be repeated in this case after a short time has elapsed and the same process is carried through until a definite excess of potassium hydroxide is detectable in the soap.

Alkali in Soap Base

If the excess of alkali is found to be too large when the soap base is tested, the taste of the soap being too sharp, then there must have been an error in measuring out the alkali for saponification of the fats and oils, on the assumption that there was nothing wrong with the latter and they were completely saponifiable. However, fats and oils, which are not completely saponifiable, and hence are not of first quality (technical grade), are not suitable raw materials for making liquid soaps. However, if the soap base contains too much alkali, then it is necessary to neutralize the same. This is accomplished by introducing a small quantity of coconut into the hot soap. Good results are also obtained with oleic acid. After the added fats or oils have been thoroughly mixed with the soap mass and saponified, the soap must be tested again after about ten to fifteen minutes and fitted.

As has been remarked above, if the soap base had a content of about 65 to 66 per cent of fatty acids, it need be dissolved only in three times its weight of distilled or soft water to give a liquid soap containing about fifteen to sixteen per cent of fatty acids. If the soap base contained only a slight quantity of alkalin excess and was used without further treatment, the liquid soap will be found to be practically neutral. On the other hand, if the proportion of excess potassium hydroxide in the soap base was quite large, then the liquid soap must be neutralized. An acid turkey red oil is used with best results for this purpose. This product dissolves rapidly and completely in the liquid soap to give a clear solution. Neutralization is therefore rapid and as complete as desired.

White Rose Soap

Soap Chips	100 kilos
Perfume:	
Geranium Algerian Oil	250 grms.
Rhodinol	250 grms.
Benzyl Acetate	250 grms.
Patchouli Oil	50 grms.
Clove Oil	100 grms.
Benzoin Siam Tincture	75 grms.
Musk Ambrette Residue	300 grms.
Aldehyde C14	5 grms.
No color.	

Violet Soap

Soap Chips	100 kilos
Orris Powder	100 kilos
Perfume:	
Orris Resinoid	100 grms.
Ylang Ylang Bourbon Oil	100 grms.
Bergamot Oil	250 grms.
Ionone Special for Soap	200 grms.
Musk Ambrette Residue	300 grms.
Benzyl Acetate	50 grms.
No color.	

Oriental Bouquet Soap

Soap Chips	100 kilos
Perfume:	
Lavender Oil	250 grms.
Patchouli	200 grms.
Vetivert Bourbon	200 grms.
Cananga Oil	200 grms.
Musk Ambrette Residue	150 grms.
Color	
Dark Green	100 grms.

Lilac Soap

Soap Chips	100 kilos
Perfume:	
Terpineol	400 grms.
Methyl Ionone	100 grms.
Phenylacetaldehyde	100 grms.
Hydroxicitronellal	200 grms.
Benzyl Acetate	100 grms.
Bromostyrol	50 grms.
Musk Artificial	50 grms.
Color	
Lavender Blue	75 grms.

Almond Blossom Soap

Soap Chips	100 kilos
White Almond Flour	10 kilos

Perfume:

Bergamot Oil	200 grms.
Iso-Eugenol	200 grms.
Nerolin	200 grms.
Bitter Almond Oil	100 grms.
Aubepine	100 grms.
Vanilla Tincture	75 grms.
Bromostyrol	15 grms.
Aldehyde C14	10 grms.

No color.

Eau de Cologne Soap Perfume

Low Priced Perfume

Soap Chips	100 kilos
Orris Powder	5 kilos
Bergamot	100 grms.
Lemon Oil	50 grms.
Rosemary	50 grms.
Nerolin	100 grms.
Cananga	50 grms.
Musk Tincture	10 grms.

No color.

Lavender Soap Perfume

Low Priced Perfume

Soap Chips	100 kilos
Lavender Oil	300 grms.
Rosemary	50 grms.
Nerolin	150 grms.
Civet Tincture	10 grms.

Colors

Light Green	100 grms.

Heliotrope Soap Perfume

Soap Chips	100 kilos

Perfume:

Heliotropin Crystal	500 grms.
Vanillin	100 grms.
Iso-Eugenol	100 grms.
Clove Oil	50 grms.
Bitter Almond Oil Artificial	100 grms.
Geranium Algerian Oil	100 grms.
Musk Artificial	30 grms.
Civet Tincture	20 grms.

Colors

Lavender Blue	75 grms.

Dissolved in water and put in the mixer with soap and oil.

New Mown Hay Soap

Orris Powder	5 kilos
Soap Chips	100 kilos

Perfume:

Bergamot Oil	250 grms.
Coumarin	250 grms.
Nerolin	200 grms.
Benzoin Siam Tincture	200 grms.
Musk Artificial	100 grms.

Color

Light Green	75 grms.

Red Rose Soap

Soap Chips	100 kilos

Perfume:

Geranium Algerian Oil	250 grms.
Phenylacetaldehyde	100 grms.
Rhodinol	100 grms.
Benzyl Acetate	100 grms.
Sandalwood Oil	250 grms.
Vertivert Bourbon	50 grms.
Benzoin Siam Tincture	100 grms.
Musk Artificial	50 grms.

Color

Light Cinnabar	150 grms.

Pine Oil Powder Scrubbing Soaps

The pine powder scrubbing soaps are specialty products since they are manufactured for specific use rather than for general use.

Manufacturers have found that cleaners may be recommended for many purposes; in addition, however, pine powder scrubbing soaps are invaluable to the public garage owner and filling station manager for dissolving grease and dirt from concrete flooring. Its light sudsing property is a great advantage in that it does not leave a slippery film. In addition its searching piney fragrance excellently dispenses many obnoxious odors.

The following is representative of the best grades:

Parts by Weight

50	Oleic Acid (Acid Number—195)
50	"I" Wood Rosin (Saponification Number—165)
13.3	Sodium Hydroxide (100%)
100	Yarmor Pine Oil
737	Soda Ash (58%)
4.7	Water

It is prepared in the following manner:

The oleic acid and "I" Wood Rosin are added to a vat and brought to a temperature of 80° C. The sodium hydroxide is dissolved in the specified amount of water. Temperature of the mass is then dropped to 60° C. and the sodium hydroxide solution is added by stirring in slowly. After complete saponification the Pine Oil is added by stirring in slowly. Add the soda ash to the previous mass and mix it in a mechanical stirring device similarly constructed to a cement mixer. The resultant product is free flowing.

The pine powder is sprinkled over the greasy floors and wet down with a hose. The usual scrubbing practice is followed. Or it may be dissolved in a bucket of hot water and applied in usual manner.

Pine Oil Liquid Hand Soaps

Liquid soaps usually are made with cocoanut oil-potash soaps, or a combination of palm-kernel oil and vegetable oil-potash soaps.

These soaps are diluted with water, depending upon the price the consumer wishes to pay for such a product. When high percentages of water are present large percentages of ethyl (or grain) alcohol, glycerol or sugar are added to lower the freezing point. Consequently, there is less chance for the soaps to solidify out of solution and cause a subsequent clouding of the finished product. A cloudy product causes sales resistance while a clear, transparent product does not.

Manufacturers of liquid soaps have found that the addition of pine oil increases the cleaning action of the soap. In addition, pine oil imparts a piney fragrance to the soap. The following formula was developed for use in a washroom dispenser:

Parts by Weight	
160.0	Cocoanut Oil (Saponification No. 257)
46.0	Potassium Hydroxide (89% Pure)
40.0	Yarmor Pine Oil
754.0	Water
1000.0	

It is prepared in the following manner:

Cocoanut oil of Ceylon Grade is added to a vat and heated to a temperature of 80°–85° C. The potassium hydroxide is then dissolved in a sufficient amount of the water to make a 15% to 20% solution. One-half the solution is then added to the cocoanut oil and stirred in slowly. The balance of water is then added followed by the balance of potassium hydroxide solution which is stirred in slowly. The temperature of the mix is then kept at 80°–85° C. for a period of from two to three hours with good agitation. After complete saponification, the solution is then cooled, chilled and filtered in this chilled state. The Pine Oil is then added by stirring in very slowly. A sufficient amount of water is then added to balance water loss during sustained heating to bring product to original weight.

Pine Oil Liquid Scrubbing Soaps

The scrubbing soaps on the market are either liquid or powder. The former are principally composed of soaps and solvents with lesser percentages of alkali, whereas, the latter are mostly alkali with slight traces of soap and solvent.

Pine Oil Liquid Scrubbing Soap is recommended for general use and is widely used in many institutions to preserve costly surfaces and for its deodorizing properties.

The following is a good formula for a liquid scrubbing soap:

Parts by Weight	
61.6	Oleic Acid (Acid Number—194)
61.6	"I" Wood Rosin (Acid Number—165)
16.3	Sodium Hydroxide (100%)
133.0	Yarmor Pine Oil
26.7	Tri-sodium Phosphate
700.8	Water
1000.0	

It is prepared in the following manner:

The Oleic Acid and "I" Wood Rosin are added to a vat and heated to a temperature of 80° C. The sodium hydroxide is then dissolved in a sufficient amount of the water to make a 15% to 20% solution. One-half of the alkali solution is then added to the mass and stirred in slowly. The remainder of the water together with the tri-sodium phosphate is then added by stirring in slowly. After temperature has been dropped to 60° C. the balance of the sodium hydroxide solution is added with vigorous agitation and continued for 15 minutes. After complete saponification the Pine Oil is added by stirring vigorously for several minutes.

The finished or completed product is light red to dark brown in color, dependent upon the type of rosin or oleic acid used.

Such a pine liquid scrub soap is especially adapted for fine tile, cork, rubber, linoleum, mastic, terrazzo and painted floors.

1. It is a powerful solvent.
2. It does not contain any injurious ingredients.
3. It is an efficient cleanser.
4. It removes grease and stains.
5. It deodorizes.

6. It repeats.
7. It is economical to manufacture.
8. It is a concentrated product and effects a great economy.
9. Use 4 oz. in a 10 quart pail of (preferably hot) water and then apply in usual manner.

Pine Oil Soap

Water	8.0 parts
Solid Caustic Soda	2.5 parts
Alcohol	10.0 parts
Pine Oil	18.0 parts
Red Oil (Oleic Acid)	17.5 parts
Water	44.0 parts
Yield	100.0 parts

Mix the ingredients while stirring in the order given at a temperature of about 40° C.; finally adjust with red oil or alkali until a sample dissolved in alcohol is neutral to phenolphthalein.

Pine Oil Scrubbing Soap

Potash Corn Oil Soap	96–97
Pine Oil	3– 4

Saddle Soap

Carnauba Wax	54
Soap Flakes	20
Tallow	26
Turpentine	21
Sperm Oil	6
Water	5

Soft Soap for Textile Purposes

83 parts Saponified Red Oil.
17 parts fair grade of animal grease.
3 parts 36° Baumé Caustic Soda Lye.
5 parts Carbonate Potash.
24 parts Caustic Potash.

Dissolve and mix the Carbonate of Potash and Caustic Potash with the Soda lye and add to the melted fat in a boiling kettle. Boiling should be accomplished with live steam. Add sufficient water to bring to the required soap content and continue boiling until the saponification is complete. Then, while still boiling, make the necessary correction by adding more fat or caustic as needed to bring about neutrality.

Saddle Soap

Beeswax	500
Caustic Potash	80
Water	800

Boil for 5 minutes while stirring. In another vessel heat

Castile Soap	160
Water	800

Mix the two with good stirring; remove from heat and add

Turpentine	1200

while stirring well.

"Waterless" Soap

A soap which may be used to clean hands without water consists of

Agar-Agar	2
Psyllium	3
Glycerol	50
Soda Ash	50
Soft Soap	50
Am. Hydroxide	25
Javelle Water	5
Water	815

Soap Paste)

Soap (66%)	70
Sod. Silicate	1.5
Soda Ash	3.5
Water	25.0

---

* Perborate Soap Powder

Mag. Sulfate	1 lb.
Water	10 lb.

Dissolve above and mix into

Sod. Silicate (75° Tw.)	10 lb.
Soda Ash	22.5 lb.
Soap (Melted)	50 lb.

When thoroughly mixed cool to 50° and work in

Sod. Perborate	9.5 lb.

This mixture is finally reduced to a powder.

---

* Soap Powder

Soap (Figured on Dry Basis)	10
Bentonite (Dry Basis)	2.5
Soda Ash	45

---

* Soap Powder, Non-Caking

Sod. Metasilicate	10
Neutral Soap	3.3
Soda Ash	20

---

* Soap Powder, Antiseptic

Soda Ash	75–85
Powdered Soap	14–18
Barium or Sodium Peroxide	1.6– 2
Trioxymethylene	0.1–0.35

Washing and Bleaching Powder

Sod. Perborate	8–10%
Sod. Persulfate	8–10%
Sod. Carbonate	65–70%
Sod. Tetraborate	15%

---

* Protective Cream

A cream for protecting hands from paint, lacquer grease, etc., consists of

Soap Flakes	19
Dextrin	4
Lanolin	2
Aquaresin	3
Water	72

---

* Soap Rancidity, Prevention of

0.05–1.0% of Dicyandiamide is added to the soap.

---

* Rancidity in Soap, Prevention of

The addition of 0.2% Sod. Sulfanilate is recommended.

---

* Soap Stabilizer

The addition of 0.2 to 0.4% triethanolamine oleate to soaps inhibits oxidation.

Rug Cleaning Soap

Oleic Acid	28 lb.
Butyl Cellosolve	5 lb.
Ethylene Dichloride	13 lb.
Triethanolamine	15 lb.
Water	125 lb.
Isopropanol	14 lb.

The oleic acid, ethylene dichloride and Butyl Cellosolve are mixed and then added to a solution made of the Triethanolamine and water. The mixture is well stirred and sufficient isopropanol is added to form a clear solution. The product emulsifies in water, and the emulsion made with an equal volume of water is recommended for cleaning rugs.

Paint and Tar Solvent

Xylene	140 lb.
Trichlorethylene	47 lb.
Ethylene Dichloride	61 lb.
Oleic Acid	40 lb.
Sulphonated Castor **Oil**	24 lb.
Isopropanol	33 lb.
Triethanolamine	16 lb.

This is made by mixing the xylene, trichlorethylene, ethylene dichloride, oleic acid and sulphonated oil, adding the isopropanol and triethanolamine and stirring to obtain an even, clear mixture. This solution is easily dispersed in water and makes a stable emulsion that is excellent for removing paint and tar from wool.

Powdered Scouring Compound

Rosin Soap	5%
Oleate Soap	5%
Steam-distilled Pine Oil	10%
Soda Ash	75%
Water	5%

This product makes a very efficient scouring compound for cleaning concrete floors, tile, marble, granite, etc. The pine oil content insures good penetration and is essential for the efficient removal of greasy and oily dirt. Yarmor Pine Oil is an excellent solvent for grease, oil, etc.

Sweeping Compounds

Although there are many sweeping compounds on the market made of sawdust, sand, ground feldspar, oil, wax emulsions, coloring matter, disinfectant, etc., it is believed that in many cases fine sawdust moistened with water at the time of use will prove satisfactory. Some prefer a compound containing sand, oil, etc.; for example, the Treasury Department at one time used a compound made up according to the following formula:

Sand	10 parts by weight
Fine Sawdust	3½ parts by weight
Salt	1½ parts by weight
Paraffin Oil	1 part by weight

Mix thoroughly.

Certain Government offices have advised us that a compound conforming to the following formula has been satisfactory in service:

Fine Sand	35%
Pine Sawdust	40%
Paraffin Oil	15%
Water (dye if coloring is desired)	10%

The Navy Department has used a compound consisting of a uniform mixture of clean, fine sand and finely ground sawdust properly impregnated with a refined heavy mineral oil and water. Such a compound must show on analysis: not more than 20 per cent of water, not more than 50 per cent of clean sand, not less than 5 per cent of refined heavy mineral oil, and the remainder finely ground sawdust. Some of the commercial compounds are colored with iron oxide or other pigment and contain naphthalene flakes.

Essential oils, such as oil of eucalyptus, oil of sassafras, etc., are frequently added to impart a pleasant odor to the compound or to mask any unpleasant odor that may be due to the ingredients used.

* Combined "Sour and Bluing"

The proportions in which to mix the compound is six (6) ounces of aniline dye to one hundred (100) pounds of boric acid, these proportions being best suited for souring and bluing under ordinary conditions, but the proportions of dye and boric acid can be increased or decreased as may be found necessary to completely neutralize the residual alkali in the cloth or clothes, and provide the proper degree or extent of acidity and bluing.

The invention provides a new product which may be packed for commercial and domestic use. The product being non-corrosive, free running, and harmless, is safely handled, can be easily weighed or measured, and overcomes the hazard of using strong acids and/or acid salts for the souring operation. The use of the product efficiently and completely neutralizes all the alkali contained in the cloth or clothes, provides acidity if desired or needed, thoroughly and evenly blues the cloth or clothes, cuts down the number of rinsing operations, and preserves the fabric.

Coloring Liquid Soaps

Pink

Rhodamine B Ex 1 lb. to 6000 gal.

Yellow

Pylam Yellow S–318 1 lb. to 1500 gal.

Blue

Alizarine Blue 1 lb. to 1500 gal.

Leaf Green

Naphthol Green 1 lb. to 1500 gal.

Olive Green

Chloro Green S–310 1 lb. to 1500 gal.

Amber

Bismarck Brown 1 lb. to 1500 gal.

Opal

Fluorescene 1 lb. to 3000 gal.

Coloring Milled Soaps

Average soap mill holds 200 pounds. For each batch use 197 pounds of No. 1 soap chips. Add 3 lb. of zinc oxide. Add proper perfume.

Pink—1/16 oz. of Rhodamine BX.
Green—1 oz. of Chloro Green S–310
Blue—1 oz. of Alizarine Blue A. S.
Yellow—1 oz. of Pylam Yellow S–318
Red—1 oz. Cloth Red
Amber—1 oz. Pylam Amber S–271
Rose—½ oz. Violamine 2R
Violet—1 oz. Pylam Violet S–333
Lemon—½ oz. Fluorescene

All the above dyes are dissolved in water before being added to the soap.

Dry Cleaning Soap on Ammonia Base

This soap is easily prepared cold by a simple mixing operation. A good soap for pressure filter systems, if good grade oleic acid is used.

Oleic Acid (preferably cold pressed)	32 gal.
Stoddard Solvent or Varnolene	15 gal.
Ammonia (0.920)	64 lb.

Mix these ingredients thoroughly; in cold weather the oleic acid should be warmed up.

Beer Pipe Cleaning Compound

Caustic Soda	12.5
Soda Ash	87.5

BLEACHING, COLORING, DYEING

Bleach for Animal Fats

Bleach for use with animal fats and oils is to use from 1½ lb. to 4 lb. Manganate of Soda or Permanganate Salts and from 2½ lb. to 6 lb. of Sulphuric Acid to each 100 lb. of fat.

Dissolve required quantity of Manganate of Soda or Permanganate Salts in from 20 to 25 times its own weight of boiling water. Dilute required quantity Sulphuric Acid with 10 times its own weight of water. Liquefy fat thoroughly at as low temperature as possible and then add slowly and with vigorous agitation the Manganate or Permanganate solution, continue agitation actively for 15 to 30 minutes, then add, also with vigorous agitation the dilute Sulphuric Acid and continue stirring for 15 minutes. Then steam is to be turned on and an active boil kept up until all brown stain disappears, which should be from 30 to 60 minutes from time boiling commences. Then settle and draw off spent solution and wash oil with water.

If using Manganate of Soda care must be taken not to add bottoms or undissolved portion. Permanganate Salts cost a little more but is more readily soluble.

* Bleaching Vegetable and Animal Oils

Fatty oils (etc.) are mixed with a dry $CaOCl_2$ product containing 50–60% of available Cl; in amount equiv. to 0.5–1.0% of available Cl on the oil, and heated at 70–90° until bleached; the separated oil is blown with superheated steam until free from available Cl.

Bleaching Angora Wool

A good method is to prepare a bath at 60° F., make alkaline with ammonia, add the required hydrogen peroxide, give the yarn (previously thoroughly wetted out) a few turns in the liquor and submerge and allow to stand over night. Remove from the bath the following morning and rinse in warm water.

* Cellulose Pulp, Bleaching

(A) Unbleached sulphite pulp is treated at room temp. as a flowable aq. suspension in a solution containing ½–1% of NaOH (on the wt. of air-dry pulp), washed, and bleached with an alkaline hypochlorite liquor. The NaOH steep reduces the resin content but does not affect the α-cellulose content.

(B) The above process is applied to pulp which is caused to flow as a continuous stream through a suitable system. The NaOH liquor is added to the raw pulp entering the system, and at a point reached by the pulp about 2 hr. later the bleach liquor is added and the temp. raised to 27°.

* Chlorine Free Bleaching Powder

Sodium Peroxide	12.5
Citric Acid	4.17
Soap (Powd.)	33.33
Sod. Carbonate	41.66
Sod. Silicate	8.34

* Chloride of Lime, Non-Hygroscopic

Chloride of Lime is ground intimately with 5–10% Calcium Sulfate.

Bleaching Cotton in Kier
(per 1000 lb. cotton)

Hydrogen Peroxide (100 Volume)	25 lb.
Sodium Silicate (Sp. gr. 1.14)	40 lb.
Sulfonated Corn Oil	4½ lb.

Heat to 185–195° F. for ½–1 hour. Rinse well and dry.

Bleaching Cotton

The goods to be bleached are impregnated with a solution of Turkey-Red Oil of from 5 to 10 per cent strength, according to the natural color of the cotton, wrung and centrifuged to get rid of the excess, and then dried. The goods are next boiled for six hours under pressure with from 1½ to 2 per cent of caustic soda, rinsed, slightly soured, rinsed again, passed through a very weak soap bath, again rinsed, and then dried. If the cotton is very pure and easily bleached the process may be simplified by putting the Turkey-Red Oil into the boiler with the lye. The process has special importance for bleaching makko-yarn, as that yarn, so largely used for finer counts, has been hitherto very difficult to bleach, requiring strong baths of chloride of lime.

Turkey-Red Oil may also be used to advantage in bleaching cotton by the usual chloride of lime method, as follows:

Goods may be treated with the oil before bleaching. Pad goods in a 5 per cent solution of the oil, and steam without pressure. The oil may also be added to the contents of the kier, whether this consists of lime, soda, or caustic soda. Two litres of Turkey-Red Oil per cubic meter of caustic soda at 3° Tw. are sufficient. The oil is added to the saturated liquor, which is afterwards introduced into the kier. There is no change required in the bleaching operation.

When lime is used, the oil is added to the lime after slaking, and then the necessary quantity of water is added. A milky liquid is thus obtained, which only settles very slowly, and which penetrates the goods perfectly, especially when tepid. The use of the oil in the lime boil gives better results than in the caustic soda boil.

Before the anti-chlorine bath it is advisable to wash well in soft water, in order to remove any undecomposed oil. Goods bleached with the aid of Turkey-Red Oil are much softer than those bleached without. The chemicking is easier and quicker, while at the same time less bleach may be used.

Bleachers, Chlorine

Hypochlorite Liquor Made with Liquid Chlorine

In 400 to 500 gallons water dissolve:
150 to 200 lb. Soda Ash
80 lb. Caustic Soda
100 lb. Chlorine

The Chlorine should be added to the alkaline solution slowly to prevent heating and loss.

Another method is to use a solution of Caustic Soda:

400 to 500 gallons water
125 lb. Caustic Soda
100 lb. Chlorine

Tanks or tubs of good depth should be used in making Hypochlorite solutions. If shallow solutions are used, the Chlorine will not absorb readily and the finished solution will not be stable.

Sodium Hypochlorite Bleach

To prepare Sodium Hypochlorite.

Dissolve 100 lb. of 33% Bleaching Powder in 40 gallons of water.

Dissolve 60 lb. of Soda Ash in 20 gallons of boiling water, afterwards diluting with 10 gallons of cold water.

The Soda solution is then to be mixed with the bleaching powder paste and well stirred for one-half hour and allowed to settle over night.

In the morning the clear solution is to be drawn off.

The residue should be washed with clear water, allowed to settle, and the top liquor added to the main solution.

The washing may be done for economy, several times, each time letting the solution settle and adding the top to the main solution.

Use only sufficient wash waters to bring the main solution to stand at 6° to 7° Tw.

Now add 1½ to 2 lb. Soda Ash. Dissolve and let stand over night, when all the lime will have been thrown out of solution.

It is then ready for use by simple dilution in water to the desired strength for bleaching.

(Sodium Hypochlorite) has advantages over the old-time Chloride of Lime solution. The goods come out softer. They rinse cleaner, and this insures better strength of the fibre and a more permanent white.

Bleach for Furs

Water	3 gal.
Hydrogen Peroxide	3 oz.
Pot. Persulfate	6 oz.
Sod. Pyrophosphate	6 oz.

Hypochlorite Bleach

Caustic Soda	120 lb.
Water	700–800 lb.

Stir until dissolved.

Put 100 lb. of above in carboy packed in ice and salt. Pass into it chlorine gas from a weighed cylinder on a scale. When 16 lb. chlorine has passed in and solution is still alkaline to phenolphthalein shut off chlorine. Keep temperature as low as possible. The resulting hypochlorite solution may be diluted as desired.

Javel Water

Bleaching Powder	20 lb.
Soda Ash	20 lb.
Water	60 gal.

Mix well until reaction is completed.

Allow to settle over night and siphon off the clear liquid.

Laundry Bleach

Soda Ash	23 lb.
Chlorine	7.6 lb.
Water	60 gal.

Laundry "Sour"

Oxalic Acid	3 lb.
Water	3 gal.

Heat with stirring until dissolved. Cool and add

Acetic Acid (56%)	8½ lb.

One pint of this sour is used per 200 lb. of goods.

* Bleaching Paper Pulp

The pulp is agitated at room temp. with 0.25–10% of a hydrosulphite ($Na_2S_2O_4$) in aq. solution and then, without subsequent washing, converted into paper.

Bleaching Rayon-Cotton Skeins

1. Treat for ½ hr. at 70° C. with 1% Sod. Sulfide.
2. Rinse until free from sulfide.
3. Treat with 0.1–0.25% sod. hypochlorite.
4. Treat with 0.25% Hydrochloric Acid.
5. Rinse acid free.
6. Repeat 3 and 4.
7. Rinse with soft water until free from acid and chlorine.
8. Rinse with 1% sulfonated oil or olive oil soap.
9. Extract excess solution and dry.

Bleaching Shellac for Water Solution

Dissolve 30 g. of orange shellac in 600 cc. of water containing 10 g. of anhydrous sodium carbonate, by warming on the steam bath. Let the solution stand over night for the wax to collect and the orpiment to settle out; then filter through a plaited paper into a 1-liter beaker. Sodium hyperbromite solution is prepared by dissolving 5.5 g. of caustic soda in 150 cc. of water and adding to this 3 cc. of bromine, drop by drop with vigorous shaking, and cooling. The bleaching solution is added to the filtered shellac solution and then the mixture allowed to stand for 15 minutes. Then acidify by adding 1:1 hydrochloric acid in small portions, with vigorous stirring. The beaker should stand in a vessel of cold water so that the shellac will be precipitated in granular form, and not in gummy masses. Filter off the shellac on a large Witt plate or Buchner funnel provided with a filter paper, and wash thoroughly with a large amount of cold water. Without drying or other treatment, the bleached shellac is dissolved by heating for a long time on the steam bath with 1,000 cc. of distilled water containing 7 g. of crystallized borax.

Bleaching Tussah Silk

Dilute 10 gallons hydrogen peroxide (12 vol. per cent) with 3 to 4 times the weight of water, and add waterglass until a feebly alkaline reaction sets in. After cleaning the Tussah silk well with boiling soap and a little soda, enter it at about 40° C. (105° F.) into this bath, to advantage charged with 4–8 oz. soap per 10 gallons, gradually raise the temperature to boiling heat, and leave for 6 to 8 hours or over night in this bath. When the bleaching is complete. rinse thoroughly, treat for several hours in a bisulphite bath and rinse well once more.

* Stripping Composition for Dyed Fabrics

1.	Sodium Hydrosulphite	90 gm.
	Petrolatum	10 gm.
	Sodium Caseinate	5–30 gm.

Instead of sodium caseinate, use isopropylnaphthalene sodium sulphonate, sodium ricinoleate, sulphonated oil, with or without soda ash, sodium bisulphite, or common salt.

2.	Sodium Hydrosulphite	90 gm.
	Oleic Acid	10 gm.

Soda ash sufficient to effect complete or partial saponification.

Instead of oleic acid, you can use stearic acid, sulpholeic acid, castor oil, corn oil or sulphonated castor oil. Instead of soda ash you may use borax or ammonium carbonate.

This gives a stable composition in cake or other solid form.

Water Soluble Colors

Dissolve the color in hot water. Filter to insure that you have no particles of undissolved color (these cause spots and blotches). Use from 2 to 3 ounces of color to a gallon of water. It is not necessary to make fresh color each time. It is important, however, to stir the color, if you have not used it in sometime. This is necessary, as some colors have a tendency to settle out of solution on long standing. A little stirring puts them back into solution again. Do not use a tin or iron container for your color solution. A chemical reaction will set up that will decrease the coloring power.

Alcohol Soluble Colors

Dissolve from 2 to 5 ounces of color per gallon of alcohol, depending on the shade. Filter and use as required. These colors are also soluble in acetone, ethyl acetate.

Oil Soluble Colors

These are soluble in perfume oils, oleic and stearic acid, as well as other fatty acids, vegetable and mineral waxes, vegetable and mineral oils; molten paradichlorbenzole. Also soluble in acetone, ethyl acetate and toluol.

When the colors are dissolved in oils, waxes or fatty acids, the solvents should be heated to insure full solution of the color. You will not get full money value or perfect solution if you dissolve the color in cold oils.

Milled Soaps

You can use water or alcohol soluble colors. Water colors preferred, as alcohol may cause blistering. Add the liquid color to the soap in an amalgamator if possible—preferably after the perfume and zinc oxide. If no amalgamator is used, distribute the color throughout the soap as much as possible, before milling. Spots and blotches are caused by undissolved color, so make sure that you have a clear color solution.

Cold, Half-Boiled and Boiled Soaps and Soap Bases

You can use water or oil soluble colors. If you use water soluble colors add the liquid color after saponification has started. Wherever possible, as in figged soaps, crutch in the color after saponification is completed. Do not add dry color to your mass and expect it to dissolve. You will have trouble. Some of the color will not dissolve and will spot your soap, and cause blotches when the soap is used. If you use an oil soluble color dissolve it in hot oil before you use it.

Liquid Soaps

Use water soluble colors only; first having dissolved them in hot water and filtered. Use as much of the solution as is necessary to give required shade. Do not over-color. Remember that 2 ounces of colored liquid soap looks much lighter than one gallon of the same colored soap. Make sure that the suds are not too deeply colored.

Bath Salts

Use water or alcohol colors.

When you use water soluble colors, it is best to make a solution as concentrated as possible. Color some of your salt very heavily and then mix this up with the rest of your salt. This will minimize the water used. Add the color before you add the perfume oils.

Light and Washing Fast Dyeing Process

A brown shade very fast to washing and light is obtained by printing fabric with a thickened paste (*A*) containing *m*-NH_2. C_6H_4· OH (I), HCl, and a substance capable of liberating CH_2O (*e.g.*, CH_2O, $NaHSO_3$), steaming for 4–8 min. in a Mather-Platt, and oxidizing in 25% aq. $Na_2Cr_2O_7$ at 60°, followed by soaping and washing. Mordant dyes, especially alizarin, may be added to *A*, and the resulting shade is deeper if $Cr(OAc)_3$ is also added. The brown pigment has an affinity for basic dyes, and these may be added to *A* or applied afterwards, whereby very deep shades are obtained. The HCl in *A* may be replaced by a mixture of HCO_2H or AcOH and NH_4Cl. An alternative printing process, whereby the same brown pigment is formed ultimately, consists in condensing CH_2O with (I) in the presence of an alkali and using the resulting transparent gelatinous product in the prep. of *A*.

* Rendering Liquid Hydrocarbons Fluorescent

Less than 0.05% of any of the following added to hydrocarbon oils or liquids imparts fluorescence.

Dehydrothio-toluidine or xylidine
Primuline Base
6-amino-2-phenyl benzthiazole
5-amino-2-phenyl benzoxazole
3: 9 di-benzoyl perylene

* Aluminum, Coloring

Alloys of Zn, Al and Cu are colored black by dipping them into a bath composed of equal vols. of (1) a 10% soln. of $CuSO_4$, and (2) a soln. of picric acid 1: 120 for about 6 sec. Various colors are obtained by using a bath contg. equal vols. of (1) a 12% soln. of Cu tartrate and (2) a 16% soln. of NaOH.

METAL COLORING

The coloring of metals depends to a great extent upon the skill of the operator as well as to the different chemicals and methods used. The brushing and relieving operations must be done by one familiar with these operations to produce uniform results. For the brushing operation fine crimped nickel silver or brass wire wheels are used and operated at 800 R.P.M., either wet or dry.

Tampico or muslin buff wheels are used for relieving operations. They are generally used with water and fine pumice and operated at 800 R.P.M.

The use of the sand blast is essential also in producing various shades of colors, as some very beautiful effects may be produced by the proper use of the sand blast machine, both before and after the coloring operation.

The colors produced by chemical means are oxides or sulphides, or a combination of both.

Black Finish for Aluminum

Water	1 gal.
Caustic Soda	1 lb.
Common Salt	4 oz.

Heat the water in an iron or earthenware vessel, and dissolve the caustic soda. Stir well, and add the salt. Keep at about 200° F. and place the aluminum article in for about fifteen minutes. Rinse thoroughly, and immerse in second bath made up as follows:

Hydrochloric Acid	1 gal.
Iron Sulphate	1 lb.
White Arsenic	1 lb.
Water	1 gal.

Dip the aluminum in this bath for a few seconds only. Rinse well in hot water.

Aluminum, Electrolytic Coloring of

Of 7 suitable electrolytes, H_3PO_4 (N) + NaOH (0.2 N) gives the best coating for coloring. The coating is formed at the anode by electrolyzing at 100 v. at 25°. The following dyes are suitable: Alizarin Sicc. (red); alizarin orange S W Pdr.; Azoflavine F F N (yellow); Union Green B; Water Blue; Alkali Violet R O O; Alizarin Black for silk Pdr.

---

* Silver Finish for Aluminum (Jirotka Process)

Immerse the aluminum in boiling bath of *one* of the following solutions.

A.	Water	2.5 lit.
	Silver Nitrate	25.0 gm.
	Potassium Carbonate	25.0 gm.
	Sodium Bicarbonate	25.0 gm.
	Potassium Bichromate	10.0 gm.
	or	
B.	Water	1 lit.
	Silver Nitrate	10 gm.
	Potassium Chromate	2.5 gm.
	Pot. Carbonate	100.0 gm.
	Sodium Bicarbonate	80.0 gm.

To obtain a bright surface immerse for not more than 10 to 15 minutes.

Oxidized Silver Effect on Aluminum

Dip the aluminum in a bath containing

Hydrochloric Acid	1 gal.
Arsenic	2 oz.
Iron Sulphate	1 oz.
Copper Sulphate	2 oz.

The aluminum must be absolutely clean and free from grease before dipping.

Silver Finishes

The silver finishes are sulphide finishes, and the chemicals used are either sodium, potassium, calcium, or ammonium sulphide. The potassium salt produces the hardest black and the ammonium salt the softest. Either salt is used in the proportion of ½ to 1 oz. per gallon of water, and used hot. To produce a black color the finish is obtained by either wet or dry scratch brushing, and the relief or gray finishes with the use of a rag or tampico wheel with fine pumice and water.

Coloring Copper

There are many formulae for the coloring of copper or copper plated work, and the color will depend upon the chemicals used, the temperature and the length of time the work is left in the coloring solution.

The work should be perfectly clean and free from any grease or finger marks.

Brown on Copper

1.	Potassium Chlorate	1 oz.
	Copper Sulfate	4 oz.
	Water	1 gal.

Use hot, scratch brush wet. If color is uneven, repeat coloring operation and scratch brush dry.

A darker or more red color is produced in this solution:

2.	Copper Sulfate	4 oz.
	Nickel Sulfate	2 oz.

Potassium Chlorate	1 oz.
Water	1 gal.

Finishing operations are the same as above.

Various shades of bronze from a chocolate color to a black can be produced in a solution made of:

3. Potassium Sulphide	½ to 1 oz.
Water	1 gal.

For the light shades use cold and a short time of immersion. For darker, use hot, with longer immersion.

Various colors are produced in any of the following solutions used either hot or cold.

4. Yellow Barium Sulphide	1 oz.
Water	1 gal.
5. Yellow Barium Sulphide	1 oz.
Calcium Sulphide	½ oz. (fl.)
Water	1 gal.
6. Golden Sulphurett Antimony	½ to 1 oz.
Caustic Soda	1 to 2 oz.
Water	1 gal.
7. Copper Sulfate	12 oz.
Acetic Acid	4 oz.
Caustic Soda	4 oz.
Water	1 gal.
8. Copper Sulfate	4 oz.
Copper Acetate	2 oz.
Potassium Chloride	6 oz.
Water	1 gal.
9. Copper Sulfate	8 oz.
Potassium Permanganate	1 oz.
Water	1 gal.

Royal Copper Finish

There are two methods of producing this finish, one with molten sodium nitrate and the other with the use of the blow torch. When any quantity of work is to be done, the nitrate method is recommended. The articles must be of either copper or have a heavy deposit of copper upon them. Best results are obtained by lead plating the copper before the heat treatment process.

To prepare the lead solution, dissolve 6 oz. of caustic soda in 2 quarts of water and add 2 oz. of litharge (lead).

Blue Color

Hyposulphite Soda	8 oz.
Lead Acetate	4 oz.
Water	1 gal.

Use at boiling temperature and immerse just long enough to produce blue color.

Green Color

Nitrate of Iron	2 oz.
Hyposulphite Soda	8 oz.
Water	1 gal.

Use boiling temperature.

Brown Color

Gold Sulphurett of Antimony	4 oz.
Caustic Soda	8 oz.
Water	1 gal.

Use at boiling temperature. Scratch brush dry and if color is not even and dark enough, repeat immersion and scratch brush operations.

Brown Color

Copper Sulfate	4 oz.
Potassium Chlorate	2 oz.
Water	1 gal.

The work is immersed in this solution for a minute or so, and without rinsing immerse in a sulphur solution made of liquid sulphur 1 ounce, water 1 gallon. The work is rinsed in cold water, and if color is not dark enough, repeat both dipping operations. Dry by using hot water and sawdust and scratch brush dry.

Brown Color

Liquid Sulphur	1 oz.
Water	1 gal.

The work is immersed in this solution for a minute or so, and then without rinsing immersed into a solution made of sulfuric acid 1 oz., nitric acid 1 oz., water 1 gallon. If color is not dark enough, repeat both dipping operations and scratch brush dry.

Verde Color

Copper Nitrate	16 oz.
Ammonium Chloride	4 oz.
Acetic Acid	1 qt.
Water	3 qt.

Immerse the work and let dry. If color is not uniform use a painter's sash brush which is moistened with the solution and stipple lightly.

Verde Antique Finish on Copper

Copper Nitrate	16 oz.
Acetic Acid	4 oz.
Water	1 gal.

Best applied hot and sparingly to previously moistened surface.

* Green Patina on Copper

The article is made the anode in a solution containing 10% $MgSO_4$, 2% $Mg(OH)_2$, and 2% $KBrO_3$, using a stainless steel or C cathode. The bath is operated at 95° with 4 amp./sq. dm. at 5 volts for 15 min.

Verde Antique Finish

Copper Nitrate	4 oz.
Ammonium Chloride	4 oz.
Calcium Chloride	4 oz.
Water	1 gal.

Green Finish on Brass

Brass articles are colored various shades of green by any of the following baths. When dry they should be lacquered to preserve the coating.

1. Hyposulfite of Soda	8 oz.
Acetate of Lead	2–6 oz.
or Nickel Sulfate	2–6 oz.
or Iron Nitrate	2–6 oz.
or Iron Chloride	2–6 oz.
Water	1 gal.

Use hot.

2. Sod. Bisulfite	4 oz.
Lead Acetate	1½ oz.
Water	1 gal.

Use hot and dip repeatedly.

3. Copper Sulfate	2 oz.
Iron Sulfate	2 oz.
Am. Carbonate	2 oz.
Water	1 gal.

Steel, Blue-Black Finish

A. Place object in molten Sodium Nitrate (700–800° F.) for 2–3 minutes. Remove and allow to cool somewhat; wash in hot water; dry and oil with mineral or linseed oil.

or

B. Place in following solution for 15 minutes.

Copper Sulfate	½ oz.
Iron Chloride	1 lb.
Hydrochloric Acid	4 oz.
Nitric Acid	½ oz.
Water	1 gal.

Then allow to dry for several hours; place in above solution again for 15 min.; remove and dry for 10 hr. Place in boiling water for ½ hr.; dry and scratch brush very lightly. Oil with mineral or linseed oil and wipe dry.

Coloring Brass Red

Electroplate in following solution at 110–120° F. at current density of 6 amp./sq. ft. using cast bronze or electrolytic copper anodes.

Copper Cyanide	3 oz.
Zinc Cyanide	½ oz.
Sod. Cyanide	4½ oz.
Sod. Carbonate	1 oz.
Rochelle Salts	2 oz.
Water	1 gal.

By adjustment of current and temp. any shade between copper and yellow brass may be produced. A sufficiently thick coating is needed so that it may stand an acid dip.

* Bronzing Iron and Steel

The bronzing bath consists of

Caustic Soda	126 lb.
Water	150 lb.
Pot. Cyanide	4 lb.
Litharge	39 lb.
Neutral Lead Chromate	1 lb.
Lead Peroxide	2 lb.
Chromium Oxide	2 lb.

Coloring Iron

Etching (*"browning," "bluing," etc.*).—Solutions of chemical reagents are applied to the steel with a cloth or sponge; the steel is allowed to oxidize for some hours while drying; the rust is then scraped off, leaving a thin adherent coat of oxide. The process is repeated a number of times, depending on the depth of color desired. The surface is then oiled. The following is a representative list of combinations of reagents that have been used for producing the respective colors:

Color, and Reagent for Producing

	Parts by Weight
Black:	
First formula—	
Bismuth chloride	20
Mercuric chloride	40
Copper chloride	20
Hydrochloric acid	120
Alcohol	100
Water	1000
Second formula—	
Copper-nitrate solution (10 per cent)	700
Alcohol	300
Third formula—	
Mercuric chloride	50
Ammonium chloride	50
Water	1000
Brown:	
First formula—	
Alcohol	45
Iron-chloride solution	45
Mercuric chloride	45
Sweet spirits of niter (ethyl nitrite + alcohol)	45
Copper sulphate	30
Nitric acid	22
Water	1000
Second formula—	
Nitric acid	70
Alcohol	140
Copper sulphate	280
Iron filings	10
Water	1000

Blue:

Iron chloride	400
Antimony chloride	400
Gallic acid	200
Water	1000

Bronze:

Manganese-nitrate solution (10 per cent)	700
Alcohol	300

Niter bath.—The cleaned steel is heated in fused sodium nitrate or potassium nitrate or a mixture of the two, often with the addition of manganese dioxide. The color acquired by the steel depends on the temperature of the bath, as well as its composition. Other fused oxidizing baths can probably be used also.

Temper colors.—The "temper colors" seen on steel when it is heated between 220° and 320° C. are due to a thin layer of oxide. Such a layer of oxide is often applied as a protecting coating, the blue color being the one usually used. The steel is heated in free air and the various colors will be produced at the following temperatures:

Temper Color	° F.
Pale yellow	418
Straw	446
Brown	491
Purple	536
Pale blue	572
Dark blue	599

The color depends somewhat on the duration of the heating and to a lesser extent on the nature of the steel.

Statuary Finish on Naval Bronze

To produce statuary finishes on naval bronze base the following solns. may be used: for light bronze, $KClO_3$ 1 oz. and $CuSO_4 \cdot 5H_2O$ 4 oz. per gal. water; for dark bronze $KClO_3$ 1 oz., $NiSO_4 \cdot 7H_2O$ 2 oz. and $CuSO_4 \cdot 5H_2O$ 4 oz. per gal. water; for dark to blue-black finish, K_2S or $(NH_4)_2S$ ¼–1 oz. per gal. water.

Black Finish for Tin

First clean tin thoroughly from grease by soaking in boiling caustic potash solution. Rinse and transfer immediately to bath made up of.

Hot Water	1 gal.
Antimony Chloride	6 oz.
Copper Chloride	12 oz.

Keep in until desired color is obtained, then rinse in hot water.

Coloring Artificial Flowers

(Made from Cotton, Muslin, Silk, Velvet)

Material is colored in two ways.

1. Before cutting to shape.
2. After cutting to shape.

Method (1). Material is put in frames and backed with a starch sizing to give body. Dye is then brushed on. Dye may also be added to the sizing. Dried and die cut to shape.

Method (2). After backing coat is put on, the material is die cut and then dipped into the dye solution.

Dye solutions prepared as follows:

Yellow

Auramine O	1 oz.
Denatured Alcohol	4 oz.
Water	4 oz.

Rose

Rhodamine B	1 oz.
Water	4 oz.
Denatured Alcohol	4 oz.

Purple

Pylam Purple	1 oz.
Water	4 oz.
Denatured Alcohol	4 oz.

Peacock Blue

Patent Blue	1 oz.
Water	2 oz.
Denatured Alcohol	2 oz.

Green

Pylam Brilliant Green	1 oz.
Water	4 oz.
Denatured Alcohol	4 oz.

Pink

Eosine	1 oz.
Water	2 oz.
Denatured Alcohol	2 oz.

Cerise

Rose Bengale	1 oz.
Water	2 oz.
Denatured Alcohol	2 oz.

* Sulfur Dyeing Process

The dull red-brown shade obtained by dyeing cotton with the acenaphthene S is rendered faster and changed to a clear red-orange shade by after-treatment at 100° for 20 minutes in a bath containing per liter, 4 cc. of NaOH (d 1.38), 1.5 g. of $Na_2S_2O_4$, and 10 g. of an alkylating or aralkylating agent.

Colors for Bath Salts

Yellow—Lissamine Fast Yellow	2 GS
Orange—Naphthalene Fast Yellow	2 GS

Pink—Rhodamine	BS
Green—Solway Green	GS

A 0.1% solution of dye is made in water. One pint of this solution is used to 100–150 lb. of bath salts.

Coloring Belt Edges

Brown

Bismarck Brown	1 oz.
Water	1 pt.
Borax Shellac Water Solution	1 pt.

Black

Nigrosine Crystals	1 oz.
Water	1 pt.
Borax Shellac Water Solution	1 pt.

Coloring Bone Buttons

Black

Pylam Ebony Black	1 oz.
Water	1 qt.

Heat to boil. Dye at 100° C.

Orange

Acid Orange	1 oz.
Water	1 qt.

Heat to boil. Dye at 100° C.

Red

Croceine Scarlet 3BX	1 oz.
Water	1 qt.

Heat to boil. Dye at 100° C.

* Coloring, Brandy

Sod. Acetate	1
Water	5
Corn Sugar	100

Heat until a dark brown color forms.

Coloring Concrete

Table of Colors to be Used in Concrete Floor Finish

Amounts of pigments given in table are approximate only. Test samples should be made up to determine exact quantities required for the desired color and shade.

Color Desired	Commercial Names of Colors for Use in Cement	Pounds of Color Required for Each Bag of Cement to Secure	
		Light Shade	Medium Shade
Grays, blue-black and black	Germantown Lampblack* or	½	1
	Carbon Black* or	½	1
	Black Oxide of Manganese* or	1	2
	Mineral black	1	2
Blue	Ultramarine blue	5	9
Brownish red to dull brick red	Red oxide of iron	5	9
Bright red to vermilion	Mineral turkey red	5	9
Red sandstone to purplish red	Indian red	5	9
Brown to reddish-brown	Metallic brown (oxide)	5	9
Buff, colonial tint and yellow	Yellow ochre or	5	9
	Yellow oxide	2	4
	Green chromium oxide or	5	9
	Greenish blue ultramarine	6	

* Only first quality lampblack should be used. Carbon black is of light weight and requires very thorough mixing. Black oxide or mineral black is probably most advantageous for general use. For black use 11 pounds of oxide for each bag of cement.

Silvering Dragees

Silvering operation should be carried out only in clean vessels. Gelatin solution is first prepared by softening 25 parts gelatin with little water and cooking softened mass and then passing liquid gelatin through filtering cloth. Gelatin is then mixed with 60 parts acetic acid in suitable flask. The smaller the original sugar-coated pills, the thinner the gelatin solution must be and the more acetic acid must be added. Silvering process should be carried out in room in which air is dry and as cold as possible, compatible with comfort of workers. Dragees are moistened with gelatin solution in ordinary kettle and operation is carried out by hand. Only smooth dragees should be used, because a fine, metallic luster can be produced only on smooth surface. Dragees must be perfectly dry before silver coating is applied. Silvering is accomplished by addition of silver powder to glass-lined kettle containing pills. This kettle is made so that it can be rotated and silvering takes place while kettle is in motion. Uniform speed of 80 to 100 R.P.M. is important. Pills must run out of kettle quite dry and then they are further dried on glass plate or dish. If silver coating does not possess required luster, then dragees are allowed to remain few days and are then run into glass-lined kettle again and moistened with little acetic acid. Hermetically sealed containers must be used for storing silvered dragees, because they lose luster on contact with air.

Another good method for silvering

dragees is to prepare a little gum solution or white syrup with which sugar-coated pills are moistened. They are then placed in box which is filled with few pieces of silver leaf. Box may be made of porcelain, glass, horn or wood. It is closed and rotated so that dragees roll around in it in continuous circle. Rotation continues as long as any metallic particles remain unattached to pills. More silver leaf is added as may be necessary and rotation of box is continued until perfectly silvered pills are obtained. If dragees contain medicaments, which react with silver, such as for example sulfur compounds, a collodion coating is applied before silvering or before sugar coating pill. Thus the pills may be placed in roomy dish and mixture of 2 parts collodion and one part ether is poured over them. Pills are rolled in solution until solid, uniformly lustrous spotless coating is obtained.

Fur Skin Dyeing

A typical acid dyeing process would be as follows: The dyestuff solution is sieved into the bath, 10 per cent Glauber's salt and 2 per cent acetic acid on the weight of the material are added. The goods are entered at 20 deg. C., raised to 40 deg. C., and a further 10 per cent Glauber's salts added. After dyeing for half an hour at this temperature, the bath is slowly raised to 65 deg. to 70 deg. C., and a further 2 per cent acetic acid added. If necessary, the bath may be cleared by the addition of from 1 to 2 per cent formic acid. It is advisable to allow the skins to cool down at least for half an hour in the baths, as this, while helping to exhaust the bath, also helps to increase the penetration.

After dyeing the skins are treated in a solution containing—

Olive Oil Soap	100 grm.
Olive Oil	20 grm.
Ammonia	10 grm.

per liter, for 15 minutes at 20 deg. C., then hydro-extracted without rinsing and dried. This process for acid dyestuffs gives very good results.

Fur Skin Dyeing

Chrome colors are applied in the same way as the acid dyestuffs, but they are dyed with the addition of potassium dichromate equal to half the weight of dyestuff. The skins are dyed for 1 to 2 hours at 70 deg. C., and it is advisable to replace the final addition of acetic acid by 1 per cent sulphuric acid. This has the effect of clearing the bath of dichromate and ensuring the action of the chrome. The chrome dyes are the fastest in general respects of all the soluble dyes, and although the process is expensive and laborious compared to the straight use of acid dyes, the results are well worth the extra trouble involved.

Leather, Applying Basic Dyes to

Before dyeing with basic dyes, tanned leather is treated for 30 min. with a liquor containing as much $CuSO_4$ as the dye to be afterwards applied, whereby the depth of shade obtained subsequently is 4–5 times that similarly obtained on non-treated leather, whilst exaggerated grain defects and a tendency for the dyed flesh side of the leather to be loose to rubbing (evident in leather not fixed after tanning) are avoided. The Cu treatment colors the tanned leather from a pale yellow to brown, but insufficiently to affect the shade obtained with the basic dye, and enables acid dyes to be satisfactorily replaced by basic dyes.

* Black Leather Dye

O-Dichlor Benzol	30	gm.
Spirit Soluble Nigrosine	7.5	gm.
Oleic Acid	5	gm.
Alcohol	48	gm.

Coloring Gasoline

Red—1 lb. Azo Oil Red
20,000 gallons gasoline.
Orange—1 lb. Azo Oil Orange
20,000 gallons gasoline.
Yellow—1 lb. Azo Oil Yellow
20,000 gallons gasoline.
* Green—1 lb. Anthraquinone Oil Green
30,000 gallons gasoline.
* Blue—1 lb. Anthraquinone Oil Blue
30,000 gallons gasoline.
* Violet—1 lb. Anthraquinone Oil Violet
30,000 gallons gasoline.

In commercial practice dye is first dissolved in benzol (1 lb. to 2 gal.).

The above dyes do not precipitate out of solution and have good light fastness.

To Whiten Yellow Gasoline

1. Determine Saybolt number of gasoline.
2. Dissolve Pyla-White in benzol (1% sol.).
3. Add Pyla-White in following proportion:

   1 lb. to 2500 bbl. Saybolt Color No. 16

1 lb. to 5000 bbl. Saybolt Color No. 18
1 lb. to 10000 bbl. Saybolt Color No. 20
1 lb. to 25000 bbl. Saybolt Color No. 22

4. Agitate until Pyla-White solution is thoroughly distributed.

Whitening is instantaneous.

Coloring Glycerin

Yellow—Auromine
Scarlet—Pylam Scarlet No. 1323
Green—Malachite Green
Blue—Methylene Blue
Orange—Chrysoidine
Violet—Methyl Violet
Black—Pylam Basic Black
Brown—Bismark Brown

Use from one to two ounces per gallon depending on depth desired.

Coloring Gelatine Solutions

1 oz. of color
1 pt. of water

This makes a stock solution.

Add as much of stock solution to the dissolved gelatine to give desired depth. The following shades are available:

Yellow—Tartrazine
Red—Pylam Brilliant Gelo Red
Blue—Patent Blue
Violet—Hastings Light Violet
Green—Mixture of Tartrazine and Patent Blue
Black—Acid Jet Black

Gelatine Backed Lantern Slides

Same as above.

Black Stain on Zinc

Nickel Chloride	4	oz.
Ammonium Chloride	6	oz.
Ammonium Sulphocyanide	2	oz.
Zinc Chloride	½	oz.
Water	1	gal.

The solution should be used at 100° F. Immerse the work until a black color of sufficient intensity is obtained.

Coloring Die Cast Zinc

Zinc weathers to a soft gray. To obtain other effects artificial coloring is necessary. This may be accomplished by electrodeposition or simple immersion (chemical coloring). Since the compounds of zinc are chiefly white, the process of coloring zinc necessitates the production on the zinc surface of a colored compound of some other metal. The compounds of copper are the most useful. By treating zinc with various copper solutions several colors may be obtained. All shades of black and brown produced by small changes in the procedure, such as time of dip, concentration, etc.

An adherent bright black can be readily produced by electrodeposition in the following bath:

Nickel Ammonium Sulphate (per gal.)	8 oz.
Zinc Sulphate	1 oz.
Sodium Sulpho-Cyanate	2 oz.

A fairly adherent black capable of being brushed to remove the coloring in the high lights results from a 5-second dip in the following solution:

Sodium Hydroxide (per gal.)	4 oz.
White Antimony Trioxide	½ oz.

Use at 158° to 167° F.

A similar result may be obtained by means of a 30-minute dip in the following solution:

Single Nickel Salts (per gal.)	10 oz.
Sodium Sulphate	15 oz.
Ammonium Chloride	1¾ oz.
Boric Acid	2 oz.

Black, brown, gray, gold, bronze, etc., may be produced in a large range of shades. Oiling with a light oil, or in some cases the use of a coat of clear lacquer will improve the luster and permanence of the deposit.

Colors produced by chemical means are reasonably permanent when used indoors. When exposed to outdoor atmospheres a relatively short life may be expected.

Coloring Zinc Die Castings

Formula No. 1

Copper Sulphate	125 grm.
Potassium Chlorate	60 grm.
Water	1 lit.*

* A full quart—to be exact 1.0567 quarts.

This solution should be heated to about 150° or 160° F., and the hot solution should be brushed on the castings.

Formula No. 2

Copper Sulphate	100 grm.
Nickel Ammonium Sulphate	100 grm.
Potassium Chlorate	100 grm.
Water	7 lit.

This solution is to be applied by immersion (dipping).

Formula No. 3

Antimony Chloride	90 grm.
Alcohol	800 grm.
Hydrochloric Acid	60 grm.

This solution is applied by immersion (dipping), pulled out and wiped with a dry cloth, then immersed again, withdrawn and wiped with linseed oil.

Solutions for producing a brown color are as follows:

Formula A

Copper Nitrate	200 grm.
Water	1 lit.

Use this at 65° F., and apply the liquid by immersion.

Formula B

Copper Sulphate	38 grm.
Sodium Carbonate	400 grm.
Ordinary Sugar	56 grm.
Water	1 lit.

Note: Sodium Carbonate comes in several grades, but the grade to use in this solution is what is designated as having ten molecules of water.

This solution is to be painted on and allowed to dry: then the castings are brushed with a dry brush to remove excess and non-adhering material. After this treatment warm the castings to about 130° F., or slightly higher.

* Paraffin Wax, Coloring

1. Dye	2
2. Trihydroxyethylamine Stearate	6
3. Paraffin Wax	400 or more

Melt (2) and dissolve (1) in it with stirring and then add to (3) which has been melted.

* Butter Coloring

Oil-soluble Yellow Food Color	2–3 grm.
Water	100 grm.
Gum Arabic	½ to 1 grm.

The color matter is preferably oil-free, even though of course it should be oil-soluble, so that as little foreign oil or fat as possible may enter into the finished butter or oleomargarine.

In order to avoid freezing of the aqueous compound, various additional ingredients may be added, especially during the colder seasons of the year, as for example glycerin, in sufficient amounts to accomplish the desired purpose.

Dyeing Cellulose Acetate

4 lb. of 4-nitro-2-methoxy-4′ dimethylaminoazobenzene (25% paste) are intimately mixed with 3 lb. of turpentine and 12 lb. of 50% Turkey red oil, sufficient b. H_2O being added to give a thin paste. The mixt. is heated to 80° and dild. to 10 gal. with b. H_2O. The clear soln. is poured into 300 gal. of soft H_2O contg. 2.5 lb. of olive oil soap. 100 lb. of cellulose acetate yarn is dyed with this soln. by treating for 1.5 hr. at 75°.

* Lubricating Oils, Stabilizing Color of

There is added to the oil 0.05–1% butyl diethanolamine.

* Coloring Paper

400 g. of rosin and 500 g. of aniline color are dissolved in 10 l. of alc. The soln. is applied to paper which is then dried.

Spotting Pencil

(For restoring color on fabrics, etc.)

Stearic Acid (D.P.)	50 parts
Japan Wax	50 parts

Required amount of oil dyes for shade.

Place material in a steam-jacketed vessel, preferably; melt slowly and agitate until thoroughly mixed. Pour into forms desired to cool.

Use

Stains or spots removed previously on fabrics and on last of original shade these spotting pencils can be used advantageously in restoring original shade.

Dyeing Straw Green

The light green which is so popular on straw hats at present is produced with basic colors in a bath made up of 5 per cent acetic acid and 5 per cent Malachite green crystals. The dyeing is continued at about 160 deg. F. for an hour or until the shade is acquired, after which the straw is removed, rinsed, hydro-extracted and dried at a low temperature.

Suede Brown, Dyeing

Sheepskins for suede are usually of a straight vegetable tannage, or vegetable-tanned and retanned in chrome. These should be given a good wash before coloring. They are then ready for the bottom. The selection of the mordant for bottoming depends largely on the shade of brown desired. Usually a bottom of sumac extract and fustic crystals will prove satisfactory. For a particularly dark shade a small amount of logwood crystals may be used with them. After drumming for fifteen to twenty minutes at 90° to 100° F., a striker such as titanium potassium oxalate or bichromate of potash is added, and drumming is continued for an additional ten or fifteen minutes. The drum is then drained and the skins given a slight rinse.

They are then ready for the first dyebath. This is usually a bath of Acid Colors. The skins are drummed in this bath for twenty minutes at 110° F. At the end of this time, if the color is not sufficiently exhausted, a small amount of formic acid is added and drumming continued for ten to fifteen minutes. Then the drum is drained.

The next step is the addition of the Basic Color. This may be made in one bath or in several, according to the shade desired. After obtaining the shade desired, drain and fat liquor in a fresh bath. The skins are then washed in the drum or in a tub and horsed up. After putting out, they are hung up to dry. When dry they are dampened back in the sawdust, then staked, and tacked on the boards. From the boards they are blocked and finally brushed.

For particularly dark shades on this stock, it is sometimes necessary to give a second coloring. After hanging up, the skins are wet back and then colored to the desired shade. They are then finished as previously stated.

The following formula is for Prado Brown, one of the popular brown shades. This is calculated for 1,000 square feet combination tanned sheepskins prepared for suede. After washing, bottom for fifteen to twenty minutes at 110° F. with

Fustic Crystals	3 lb.
Logwood	1 lb.

strike with:

Bichromate of potash	5 oz.

and run for ten minutes. Drain and rinse. Dye for twenty minutes at 110° F. with

National Resorcine Brown R	3 lb. 8 oz.
National Wool Orange A Conc.	1 lb. 12 oz.
National Buffalo Black NBR	10 oz.

then add:

Formic Acid	12 oz.

and run for ten minutes. Drain.

Top with:

National Bismarck Brown Y Extra	3¼ lb.
National Safranine A	8 oz.
National Methylene Blue 2B	10 oz.

Run at 110° F. for twenty minutes and add:

National Phosphine RN	20 oz.
National Safranine A	5 oz.

and run for 15 to 20 minutes. Then drain and the pack is ready for the fat liquor.

Fat liquor in a fresh bath for twenty-five minutes with:

Sulphonated Neatsfoot Oil	10 oz.

Then drain, wash and horse up. The skins are then hung up, dampened in sawdust, staked, tacked, blocked and brushed.

The selection of a good fat liquor is very important. This applies to chrome-tanned suede as well as vegetable-tanned suede. The use of too much fat liquor is to be avoided, as this will cause a sheen or a greasy appearance. One should also avoid the use of too much dye, particularly a Basic Color, as this will cause crocking. Washing the skins thoroughly and brushing after blocking will help to overcome this.

Chrome-tanned leather prepared for suede is colored in a similar manner to the process just given for combination tanned leather. However care should be taken to be sure the stock is thoroughly wet out before starting to color. This stock is much harder to wet out than the previous stock. The chrome-tanned leather also has a better affinity for the color, and it also may be colored at a slightly higher temperature. After wetting out, the leather is given a bottom of sumac extra, fustic crystals and logwood crystals if necessary.

This is drummed for fifteen to twenty minutes at 110° to 120° F. Then the mordant is struck with a suitable striker such as bichromate of potash and run for another ten minutes. The liquor is then drained off and the skins rinsed. The skins are then given a bath of an acid brown similar to National Para Brown PD, National Resorcine Brown R, or National Resorcine Brown RN, and run in this for twenty minutes at 110° to 120° F. By this time, if color has not sufficiently exhausted, add a small amount of formic acid and run for ten minutes. Then drain, and top with a basic brown. Run for twenty minutes at 110° F. Drum in a fresh bath with a small amount of Sulphonated Neatsfoot Oil and egg yolk. Wash in drum or tub and horse up. The skins are then hung up to dry. When dry, dampen in sawdust and stake them. Then dry well to bring up the nap. Tack on boards and then brush.

In horsing up suede, the skins should always be placed grain to grain. When placed in the dust, they should be put grain to grain, also.

Chrome-tanned suede may also be colored with Direct Colors. When used for this purpose, they should be applied directly to the leather.

Colored Waters (Non-Fading)

These are for filling bottles which are exposed to sunlight.

Amethyst

Sodium Salicylate	10 gm.
Tinc. Ferric Chloride	½ dr.
Distilled Water	2½ gal.

Blue

Copper Sulfate	4 oz.
Ammonia	sufficient to dissolve precipitate
Distilled Water	2½ gal.

Green

Nickel Sulfate	3 oz.
Sulfuric Acid	6 oz.
Distilled Water	2½ gal.

Garnet Red

Pot. Bichromate	16 oz.
Sulfuric Acid	16 oz.
Water	2½ gal.

Rose Red

Cudbear	2 oz.
Water	10 oz.

Macerate for two days and filter; dilute with water to the proper shade and add ½ oz. Ammonium Hydroxide to each gallon.

Orange

Pot. Bichromate	16 oz.
Nitric Acid	8 oz.
Distilled Water	2½ gal.

Water Stains

Red Mahogany

Azo Rubine	4 oz.
Pylam Red	4 oz.
Pylam Black	½ oz.
Acid Orange	3½ oz.

Dissolve in 3 gal. hot water.

Brown Mahogany

Azo Rubine	4 oz.
Pylam Red	4 oz.
Nigrosine Powder	2½ oz.
Acid Orange	5½ oz.

Dissolve in 4 gal. hot water.

Dark Walnut

Pylam Black	5 oz.
Acid Orange	1 oz.
Pylam Yellow	1 oz.

Dissolve in 2 gal. hot water.

Light Walnut

Pylam Black	2 oz.
Acid Orange	2 oz.

Dissolve in 1 gal. hot water.

Oak

Pylam Black	1 oz.
Metanil Yellow	7 oz.

Dissolve in 4 gal. hot water.

---

* Protecting Fruit against Mold

Borax	5–8 oz.
Casein	½–1 oz.
Glucose	½–1 oz.
Water	1 gal.

This is used for coating fruit and is allowed to dry on latter.

Spirit Stains

Red Mahogany

Pylam Spirit Black	½ oz.
Bismarck Brown	3 oz.
Basic Fuchsine	½ oz.

Dissolve in 1 gal. denatured alcohol.

Brown Mahogany

Pylam Spirit Black	4½ oz.
Pylam Spirit Orange	3 oz.
Basic Fuchsine	½ oz.

Dissolve in 2 gal. denatured alcohol.

Walnut

Bismarck Brown	3 oz.
Pylam Spirit Black	1 oz.

Dissolve in 1 gal. denatured alcohol.

Oak (Dark)

Pylam Orange	10 gm.
Bismarck Brown	3½ gm.
Malachite Green	2 gm.

Dissolve in 1 pint denatured alcohol.

Oak (Golden)

Pylam Orange	1 oz.
Auramine	1 oz.

Dissolve in 1 gal. denatured alcohol.

The above are soluble in alcoholic shellacs and lacquers containing alcohol.

* Moth Proofing

Am. Selenate or Selenious Acid	1–2
Water	1000

Allow material to soak in above for two hours; rinse with water and dry.

Coloring Wood

Water Stain

½ oz. of any Basic Color
1 quart of Water

This raises the grain. Gives best penetration.

Spirit Stain

½ oz. of any Basic Color
1 quart of Denatured Alcohol.

Good penetration. Raises the grain somewhat.

Oil Stain

½ oz. of Oil Soluble Color
1 quart of Benzol

Does not raise grain. Penetration—poor.

Varnish Stain

½ oz. of Oil Soluble Color
1 quart Varnish

Stir until thoroughly dispersed and allow to stand overnight.

Shellac Stain

Same as spirit stain. Substitute shellac solution for denatured alcohol.

COSMETICS

Violet Ammonia

Ammonia Water	12 pt.
Distilled Water	28 pt.
Perfume (see below)	1 oz.
Color	enough

Perfume for the Foregoing

Anisic Aldehyde	½ dr.
Benzyl Acetate	½ dr.
Ionone	1 dr.
Coumarin	1 gr.
Oil of Bergamot	15 min.
Oil of Neroli	10 min.
Tincture of Musk	4 oz.

Liquid Toilet Ammonia
(For Bath)

Ammonium Stearate (Paste)	8 oz.
Ammonia 28°	6 oz.
Water	50 oz.
Glycerine	2 oz.

Perfume to suit.

Borated Bathing Solution

Boric Acid	10 gm.
Alum. Powd.	2.5 gm.
Camphor	1.5 gm.
Alcohol	120.0 cc.
Water, enough to make	500.0 cc.

Pine Oil Bath Liquid

Turkey Red Oil	10 oz.
Fluorescein	$\frac{1}{10}$ oz.
Pine Oil	3 oz.
Water	3 oz.

Dissolve the fluorescein in the turkey red oil; add the pine oil and when well mixed add the water, stirring until a uniform liquid results. Strain if necessary.

Pine Needle Bath Tablets

A good formula for the production of pine needle extract bath tablets is as follows: 65 parts of common salt, 15 parts of borax, 17 parts of true pine needle extract, 3 parts of pine needle perfume oil, such as pine needle oil, bornyl acetate, oil of silver pine, oil of knee pine, rounded off with lavender oil, oil of sage, and strengthened with eucalyptus oil. About 10 to 15 parts of fluorescein are used for color.

A pine needle extract preparation which will give the bath a fine green color is made as follows: 25 parts of pulverized borax, 25 parts of common salt, 12 parts of calcined soda, 0.05 part of fluorescein and 1½ parts of oil of silver fir. Another formula calls for 5 parts of fluorescein, 10 parts of ammonia, 25 parts of oil of knee pine, 25 parts of oil of silver fir, 935 parts of 95% alcohol. Uranine may be used in the place of fluorescein with the result that a greener shade is obtained.

Pine Needle Concentrate (For Bath)

Many pine needle oil preparations now marketed, do not take into account that when they are put into water the oil floats on top and only makes contact with a very small portion of the body. By using the following formula the oil is emulsified and spreads uniformly through the bath, giving the entire body the benefit of the pine needle oil.

1. Pine Needle Oil	10 lb.
2. Sodium Sulforicinoleate	10 lb.
3. Water	5 lb.
4. Fluorescein	To Suit

Mix 1 and 2 until dissolved. Add 3 slowly with stirring. Add 4 and stir until dissolved.

The above formula when thrown into water disperses uniformly to give a milky green solution. Other oils may be substituted for Pine Needle Oil. If a lower cost is desired, part of the pine oil may be replaced by mineral, olive or cottonseed oil and a larger amount of water may be added.

Pine Needle Milk (For Bath)

Pine needle bath milk is prepared as follows: In one process the milky consistency and appearance is secured by emulsification with soap, gum tragacanth and the like. In a second process the same effect is secured with tincture of benzoin. Other directions call for lanolin as an aid in procuring the emulsified condition. The simplest formula calls for 2 parts of eucalyptus oil, 2 parts of lemon oil, 18 parts of oil of silver pine, 15 parts of knee pine oil, 400 parts of tincture of benzoin, 8,000 parts of alcohol and 3,000 parts of water. In another formula, 6 parts of soda soap are dissolved in 100 parts of alcohol; 10 parts of this mixture are triturated into a smooth paste with ½ part of gum tragacanth powder. Then there are added 4 parts of pine needle oil, 1 part of juniper oil and 12.5 parts of alcohol. As soon as this mass has been uniformly mixed, 15 parts of water are added and the emulsion is formed by vigorous shaking and agitation. At the end 50 to 60 parts of water are added.

Pine Needle Balsam

Pine needle balsam is prepared as follows: 3 parts of lavender oil are mixed with 20 parts of pine needle oil, 25 parts of knee pine oil, 1,000 parts of alcohol and enough chlorophyll to give desired green color. Following formula is for pine needle balsam with approximately 50% alcohol content: 100 parts of tincture of nutgalls, are mixed with 50 parts of aromatic tincture, 50 parts of sweet spirit of niter, 20 parts of ethyl acetate, 25 parts of pine needle oil, 50 parts of knee pine oil, 5,000 parts of 95% alcohol and 5,000 parts of distilled water. Sugar color or chlorophyll may be added to color the mixture.

A pine needle bath preparation may also be made as follows: 20 parts of bath chamomille, 40 parts of peppermint leaves, 100 parts of calamus root, 60 parts of woodruff herb and 80 parts of eucalyptus leaves, the entire mixture cut up into proper form, is treated with 4,800 parts of 96% alcohol and macerated for 14 days. Mixture is filtered and residue pressed. The filtrate is mixed with 120 parts of aromatic tincture, 50 parts of oil of Siberian fir needles free from terpenes, 20 parts of knee pine oil, 20 parts of juniper oil, 15 parts of eau de cologne and 275 parts of pure glycerin of 28° Bé. Residue after filtration may be digested with 4,000 parts of boiling water and filtered. The two extracts are united and colored green with chlorophyll.

Effervescing Bath Salts

Another important class of bath preparations contains oxygenated salts, which release oxygen gas during the bath. Preparations that develop carbon dioxide during the bathing process are closely allied to the former and the two may be grouped together in the class of effervescent bath salts. These are the preparations that have been recommended for attaining slimness of figure.

The simplest carbon dioxide releasing preparation contains sodium acid sulphate and sodium bicarbonate. While this preparation is effective, it is by no means so effective as the mixture which contains tartaric acid or potassium bitartrate. These chemicals increase the cost of the preparation, but they are well worth while adding. They are used in the place of the sodium acid sulphate. If 900 parts of sodium bicarbonate are used, then about 750 parts of pulverized tartaric acid or 1,200 parts of potassium acid tartrate are required. It is essential that this preparation should not react to produce carbon dioxide before it is actually used, and in order to prevent the reaction from taking place prematurely it is sufficient to add to it a water-absorbing salt, such as sodium sulphate, and about 200 parts are enough to give good results. Instead of the sodium sulphate, the same proportion of starch may be used. It is also useful to add a lather-producing agent so that the carbon dioxide is released in the bath in very fine bubbles. Such an agent is pulverized soap or dry crude quillaia bark extract or else a solution of casein in lye. These preparations may be used in connection with pine needle compositions as well.

A new formula for the preparation of bath salts that evolves carbon dioxide is the following: 90 parts of sodium carbonate, 75 parts of tartaric acid, 120 parts of starch, 15 parts of lemon oil and 5 drops of iönone. The oil and starch are mixed and other ingredients added and kneaded into a paste with ether. Approximately 1 part of gum benzoin is mixed with 30 parts of ether and used for the above purpose. Mixture can be pressed into tablets which are stable due to the starch contained in them.

An effervescent pine needle bath salt preparation is made as follows: 300 parts of sodium bicarbonate, 275 parts of pulverized sodium bisulphate, 12 parts silver fir oil. Uranine is added until color is yellow. Tablets may be pressed from this mixture.

Bath salts, which evolve oxygen, are generally made with the aid of sodium perborate. A catalyst must be used in making the preparation. Thus for 1,000 parts of sodium perborate, there are required 1.4 parts of manganese dioxide or 6.7 parts by weight of cobalt carbonate, or 40 parts of gypsum or 26.7 parts of magnesium fluoride.

An effective bath salt of this type contains 300 parts of sodium perborate and a catalyst composed of 6 parts of manganese sulphate and 9 parts of potassium bitartrate. Another new preparation of this type calls for 3 parts of sodium perborate, 4 parts of manganese sulphate, 11 parts of sodium tartrate. Pressed residues from sweet and bitter almonds can be used to good advantage as catalysts. These residues may be mixed with the dry oxygenated salts. They possess the additional property of creating a lather when the composition is dissolved in water.

Effervescent Bath Salts

Another preparation is made from 400 parts of pulverized sodium biborate, 200 parts of sodium sulphate, 300 parts of sodium bicarbonate, 225 parts of tartaric acid, 50 parts of lactose, 25 parts of talc and 15 parts of oleum pinus silvertris and oleum pinus pumilio. Ingredients are mixed 2 or 3 times and passed through a fine sieve, and then the coloring matter, for example fluorescein, is added. Addition of talc and milk sugar is necessary to be able to prepare tablets possessing a certain strength and stability.

It was mentioned earlier in this article that the use of herbs for the manufacture of bathing preparations gives excellent results. The herb extract may be made from a number of different botanicals, such as peppermint leaves, sage leaves, rosemary leaves, thyme and chamomille, which may be used in the proportion of 100 parts each. The botanicals must be used free from dust and are treated with 250 parts of 90% alcohol.

Production of this preparation is simpler and less troublesome, if a pine needle milk is prepared for direct use. The first step in the process is to prepare a 5% solution of 80% soda soap in 95% alcohol. Five parts of the finest pulverized white gum tragacanth are triturated with 100 parts of soap solution. Then 45 parts of pine needle oil and 5 parts of juniper oil dissolved in 125 parts of 95% alcohol are mixed with paste. Thereafter 550 parts of water at 30° C. are added and mixture is agitated for long time. A thick emulsion is formed, resembling a cod liver oil emulsion. This emulsion is ready for use and can be added directly to the bath. Astringent substances such as oak bark extract may be added to the emulsion, but this must be done during the manufacturing process.

Liquid Brilliantine

Light Mineral Oil	99%
Perfume (Flower Type)	1%

Mix and filter.

Jelly Brilliantine

Spermaceti	14 lb.
Beeswax	6 lb.
Mineral Oil	100 lb.
Perfume	1 lb.

Color to suit.

Melt the waxes in the mineral oil. Strain and allow to cool to about 115° F. Add perfume; stir until cold.

Liquid Brilliantine

Mineral Oil	100
Chlorophyll (Oil Soluble)	To Suit
Perfume	To Suit

Solid Brilliantine

Petrolatum	100 lb.
Chlorophyll	2 oz.
Perfume Oil	8 oz.

Face Clay

Clay	100 lb.
Water (Cold)	20 gal.
Tincture of Benzoin	3 pt.
Perfume	3 oz.

Add the water to the clay and grind till smooth. Evaporate until 150 lb. remain. Run through mill to smooth clumped particles; cool and mix in the benzoin and perfume. Fill in collapsible pure tin tubes.

* Corpse Tissue Filler

Zinc Oxide	50 lb.
Glucose	10 lb.
Borax	20–25 lb.
Plaster of Paris	3 lb.
Phenol	1 lb.
Alum	5 oz.

Mole and Blotch Covering

Collodion	1 gal.
Zinc Oxide	1 lb.
Geranium Lake	½ oz.
Yellow Ochre Lake	1½ oz.

Leg and Arm Blemish Covering

Stearic Acid	4 lb.
Diethylene Glycol	16 lb.

Heat to 180° F. and to this add while stirring the following solution heated to 140° F.

Caustic Potash	4 oz.
Water	16 pt.

When uniform work in following:

Zinc Oxide	15 lb.
Yellow Lake	12 oz.
Persian Lake	4 oz.
Perfume Oil	4 oz.

The colors may be varied to give more suitable shades.

Cuticle Remover

Pot. Hydroxide	2 oz.
Water	1 gal.
Phenyl Ethyl Alcohol	¼ oz.

Cholesterol-Lecithin Cream (Synthetic Hormone)

1. Lanolin, Anhydrous	20	gm.
Stearin	10	gm.
Cacao Butter	20	gm.
White Wax	20	gm.
Sweet Almond Oil, Preserved with Nipagin	200	gm.
Cholesterol	6	gm.
Lecithin	12	gm.
Water	80	gm.
Sodium Benzoate	1.5	gm.
Borax	15	gm.
Nipagin M.	0.8	gm.

Cholesterol and Lecithin Skin Creams

2. Lanolin, Anhydrous	30	gm.
White Wax	50	gm.
Spermaceti	10	gm.
Borax	2	gm.
Water	18	gm.
Cholesterol	1.5	gm.
Egg Lecithin	0.5	gm.

Lanolin Emulsion

Lanolin	80 lb.
Stearic Acid	15 lb.
Triethanolamine	5 lb.
Water	200 lb.

Preparation

Weigh out the Triethanolamine and stearic acid and add to the whole quantity of water. Heat the mixture in a kettle and, when the stearic acid is melted, stir to a creamy soap solution. Add the lanolin and continue heating without stirring until the lanolin is melted and the mixture is just below the boiling point.

At this point stir the mixture thoroughly until a thick creamy emulsion results. Continue stirring intermittently until the emulsion has cooled to room temperature.

Properties

This emulsion is a very smooth, lightly colored cream of excellent stability, and can be diluted to any desired consistency with water. Such a lanolin emulsion is essentially a water-soluble lanolin and can be used in place of the straight fat whenever washability is advantageous.

Variations

To overcome a slight rancid odor in lanolin it is suggested that one per cent terpineol by weight be added to the lanolin prior to emulsification. Moreover, only the purest anhydrous grade should be used for cosmetic and medicinal preparations. Lanolin, as a readily absorbed and beneficial oil, is recommended for use in many skin creams, and may readily be incorporated in vanishing creams, cold creams and shaving creams.

Uses

Sunburn creams, hand lotions, shaving creams.

Anti-Perspiration Cream

1. Lanolin Hydrous	1
2. Benzoinated Lard	90
3. Zinc Oxide	6.5
4. Salicylic Acid	1.2
5. Benzoic Acid	0.9
6. Perfume Oil	0.4

Dissolve (4) and (5) in small amount of alcohol; mix into (1) and then work into (2). Grind in (3) until smooth and then work in (6).

Almond Cream Liquid

Oil Sweet Almonds	1 lb.
Spermaceti	2 lb.
Beeswax	2 lb.
Castile Soap Powdered	3 lb.
Borax	2 lb.
Quince Jelly	1 lb.
Alcohol	1 pt.
Water	4 pt.

Melt the spermaceti and wax together. Dissolve the soap and borax in hot water. Mix these together and add balance of ingredients. Stir and filter through cloth.

Almond Cream for After Shaving

1. Potassium Carbonate	1 oz. 130 gr.
Distilled Water	15 oz.

Dissolve *Potassium Carbonate* in water, filter

2. Gum Tragacanth	175 gr.
Glycerin	10 oz.
Borax	1 oz.
Distilled Water	64 oz.

In 20 oz. hot water dissolve Borax then add Gum Tragacanth and Glycerin. Allow to stand 12 hours, stirring frequently. When gum has formed mucilage add the remaining 44 oz. of water while stirring and strain through muslin.

3. Stearic Acid triple pressed	5 oz. 260 gr.
Oil Sweet Almond	3 oz.
Ethyl Amino Benzoate	½ oz.

Melt acid and oil together and add Ethyl Amino Benzoate. Stir until dissolved and adjust temperature to 70° C.

Anti-Sunburn Cream

Stearic Acid	96
Trikalin	20
Glycerin	32
Water	400
Aesculin	10–25
Perfume	To Suit

Astringent Cream

1. Glycosterin	3 lb.
2. White Petrolatum	1 lb.
3. Astringent Powder No. 1	4 oz.
4. Water	15 lb.
5. Perfume	1 oz.

Heat (1) and (2) to 160° F. and add to it slowly (4) which has been heated to 200° C. Stir and work in (3) until uniform; add (5) just before pouring.

Absorption Base Cream

Absorption Base Creams are coming to the fore because of their beneficial effect on the skin because of their cholesterin and oxycholesterin content.

Parachol is a highly refined absorption base of the Eucerin type, which is used in producing high grade creams which are pure white—not yellow like most creams of this type and which are also free from the objectionable lanolin odor. Such creams do not dry out and will not corrode metal containers. The following formula may be used as a starting point. For special purposes, sulphur, bismuth subnitrate, mercury salts, titanium dioxide, salicylic and thymol or other products may be introduced.

1.	Parachol	10 lb.
	Parasterin	20 lb.
	Mineral Oil	10 lb.
2.	Water	25 lb.

Heat (1) in water, both, till melted, allow to cool to 45–47° C. Warm (2) to 45–47° C. and add in 7 or 8 different portions to (1), stirring vigorously, taking care not to add more water until previous portions are absorbed.

Bleach Cream

White Wax	1½ oz.
White Petrolatum	12½ oz.
Ammoniated Mercury	1¼ oz.

Bismuth Subnitrate	¾ oz.
Oil of Red Rose	40 drops

Melt the white wax in a double boiler. Add the petrolatum and stir until melted. Cool. Mix the ammoniated mercury and bismuth subnitrate. Add ¼ pound cold petrolatum mixture and mix in a paint mill. When smooth, add the balance of the petrolatum mixture and perfume.

Cleansing Cream

Stearic Acid	29 lb.
Lanolin (Anhydrous)	8 lb.
Mineral Oil (White)	50 lb.
Triethanolamine	3.6 lb.
Carbitol	10 lb.
Water	100 lb.

Preparation

Melt the stearic acid in the mineral oil, add the lanolin and bring the temperature of this oil solution to 70° C. Then add it to the solution of Triethanolamine and water which has been brought to the boiling point in a separate container. Stir vigorously to obtain a uniform emulsion and add the Carbitol solution of the perfume. Continue with even stirring until a smooth cream is obtained and then occasionally until cold. Too rapid stirring causes an undesirable aeration of the cream.

Properties

Cleansing creams contain a fairly high content of mineral oil and usually a wax base. The latter is not essential in a properly formulated cream although it is frequently used. The mineral oil content is normally quite high as it is this material which dissolves or suspends the dirt particles so that they may be readily removed by a cloth or absorbent paper. The higher percentage of Triethanolamine used in this type of cream than in a vanishing cream serves to completely emulsify the oil, aids in its penetration into the pores, and forms a cream which is readily removed with water. Carbitol exerts a soothing action on the skin and facilitates the cleansing action.

Variations

While various waxes and oils may be used in this type of cream, it is important that the correct proportion of Triethanolamine be used. A deficiency of the base is indicated by a thin emulsion, which is not readily washable, and a surplus by a granular cream which tends to separate on cooling. The water content can be increased or decreased slightly to change the consistency of the cream as desired.

Cleansing Cream

1. Mineral Oil (White)	54
2. Beeswax	18
3. Parachol	5.5
4. Borax	1
5. Water	21
6. Perfume	0.5

Melt together 1, 2 and 3. Dissolve 4 in 5 and heat to boiling. Add this to first mixture slowly with stirring; add perfume before solidification begins.

Cleansing Cream

1. Mineral Oil	80 lb.
2. Spermaceti	30 lb.
3. Glycosterin	24 lb.
4. Water	90 lb.
5. Glycerin	10 lb.
6. Perfume to suit.	

Heat 1, 2 and 3 to 140° F. and stir into it slowly 4 and 5 heated to same temperature. Add perfume, at 105° F. stir slowly until cold after allowing to stand for 5 minutes stir until smooth and pack.

Cleansing Cream

1.	Mineral Oil	78 lb.
	White Wax	5 lb.
	Spermaceti	28 lb.
	Trihydroxyethylamine Stearate (Special)	20 lb.
2.	Perfume	1 lb.
3.	Glycerin	4 lb.
	Water	92 lb.

Heat Nos. 1 and 3 separately to 200° F.; then add Nos. 1 to 2 slowly, stirring thoroughly. When the cream begins to set, the perfume is added and stirred in. Allow to stand over night. Stir thoroughly the next morning and package. This cream will not sweat oil during hot weather and will maintain its consistency.

Soluble Cleansing Cream
(Latherless Shaving Cream)

Creams of this type are made without heat. Merely beat together.

Ammonium Stearate (Paste)	250 oz.
Mineral Oil, White	25 oz.
Perfume to suit.	

Stir until most of the ammonia has evaporated.

This cream is particularly soothing to the skin and combines the properties of a vanishing and cold cream.

Cleansing Cream

A cream for removing dirt from the hands without the use of water contains casein 9, lime water 16, NH_3 0.5, soda 1, oxycellulose or hydrocellulose 9, perfume 0.5 and water 64 parts.

Liquid Cleansing Cream (Non-Greasy)

1. Beeswax	1.5
2. Spermaceti	6.5
3. Cherry Kernel Oil	6.0
4. Glycosterin	4.0
5. Water	122.0
6. Alcohol or Isohol	3.0
7. Galagum	1.0
8. Borax	3.0
9. Perfume	3.0
10. Glycerin	4.0

Melt together 1, 2 and 3. Heat while stirring 4, 5, 7 and 8 together until uniform. Mix these two solutions stirring until uniform. Stir in 6, 9 and 10 and mix until uniform.

Liquid Cleansing Cream

Stearic Acid	25 lb.
Lanolin (Anhydrous)	34 lb.
Mineral Oil (White)	57 lb.
Triethanolamine	9 lb.
Carbitol	75 lb.
Water	315 lb.
Quince Seed Mucilage	19 lb.
Terpineol	0.35 lb.

Preparation

Melt the stearic acid in the mineral oil, add the lanolin and terpineol and bring the temperature of this oil solution to 70° C. Add it to the solution of Triethanolamine and water which has been brought to the boiling point in a separate container. Stir vigorously until a good emulsion is formed and then add the quince seed mucilage, slowly, with continued stirring. Add the perfume to the Carbitol and stir this slowly into the cream. The stirring should be fast enough to keep the cream well mixed but not aerate it. If the stirring is not continued until the cream is cold, it thickens upon standing. The quince seed mucilage is made by adding 9½ ounces of quince seed to 20 pounds of water at 80° C., soaking 5 or 6 hours, and straining through a cloth. Some suitable material should be added to the quince seed mucilage to prevent its molding over a period of time.

Properties

The high percentage of Triethanolamine used in this cream serves to completely emulsify the oil and lanolin, aids their penetration into the pores and forms a cream which is readily removed with water, if desired. Carbitol exerts a soothing action on the skin and facilitates the cleansing action of the cream. Due to the high Carbitol and lanolin contents this cream is soothing and healing to the skin and can be used as a hand lotion as well as a cleansing cream.

---

* Procedure for Making Cold Creams

1. Dissolve borax in water, heating this to 150° F.

2. Melt in another pot beeswax, Glyco-Wax A and white mineral oil and keep at about 150° F.; add with stirring 3/7 parts Lily of the Valley (or other perfume).

Add 2 to 1 slowly with thorough stirring; continue stirring until cool enough to pour.

1.	Borax	2 parts
	Water	54 parts
2.	Glyco-Wax A	20 parts
	White Beeswax	26 parts
	White Mineral Oil	120 parts
3.	Perfume	1 part

Softer creams can be prepared by increasing the amounts of water in the above formulae.

If creams are packed when too warm the finished products will not look as well as if they are poured when cooler. The best time for packing is just before the cream begins to set.

Cold Cream

Stearic Acid	30 lb.
Lanolin (Anhydrous)	20 lb.
Beeswax (White)	16 lb.
Mineral Oil (White)	33 lb.
Triethanolamine	3.8 lb.
Carbitol	16 lb.
Water	95 lb.

Preparation

Melt the stearic acid, lanolin and beeswax in the mineral oil and heat to about 70° C. Prepare in a separate kettle a boiling solution of the Triethanolamine and water, and add to this the hot solution of waxes. Stir vigorously until a creamy emulsion is obtained and add the Carbitol to which the perfume has been added. Continue stirring until homogeneous and the product has reached the proper consistency. Pour into jars while still warm.

Properties

Cold creams are somewhat similar to cleansing creams in composition. They contain less oil and usually a mixture of fats and waxes of a type absorbed by the skin. Since cold creams usually remain in contact with the skin for several hours, they should contain the proper skin conditioners and the maximum absorbability of the fatty matter. The given cream is of good texture, is white and stable, and soothing in its action. It is also a washable cream.

Variations

The given formula should serve as a starting point for making up a cream to suit the individual preference and should not be considered as necessarily the best product obtainable. Great variation in the wax and oil constituents is allowable with little change in the basic ingredients. For example, vegetable and animal oils or fats may be substituted for all or a part of the mineral oil which is used only in the cheaper creams. Specific attention should be paid to the choice of perfumes, for some tend to discolor cosmetic creams after standing for a time. Neither Triethanolamine nor Carbitol, however, will have a deteriorating effect on perfumes properly chosen.

Cold Cream (Inexpensive)

Spermaceti	125
White Wax	120
Liquid Petrolatum	560
Borax	5
Distilled Water	190
Oil of Rose, Synthetic	q.s.

Melt the wax and spermaceti on the water bath and add the liquid petrolatum. Heat the distilled water and in it dissolve the borax. Add this warm solution to the melted mixture while both are warm and at about the same temperature. Beat rapidly; as soon as it begins to congeal add the oil of rose and beat until congealed. Dispense preferably in pure tin tubes.

* Cold Cream

Glyceryl Monostearate	18
Beeswax	1
White Petrolatum	6
Lard	4
Mineral Oil	7
Sweet Almond Oil	5
Glycerol	3
Water	55.5
Diethylaminoethyloleyl-phosfate	0.5

Cold Cream (Low Cost)

Glycosterin	20
Paraffin Wax	30
Petrolatum White	18
Mineral Oil	8
Water	200
Perfume	To Suit

Cold Cream (Cleansing Type)

White Wax	10 oz.
Paraffin	9 oz.
Ceresin	2 oz.
White Petrolatum	8 oz.
Liquid Petrolatum	3 lb.
Borax	1 oz.
Water, Distilled	1 pt., 4 fl. oz.

Cold Cream (Greaseless)

A very low priced light bodied but stable cream is made as follows:

1.	Glycosterin	22 lb.
	Petrolatum White	16 lb.
	Paraffin Wax	12 lb.
	Mineral Oil	32 lb.
2.	Water	128 lb.
	Borax	3 lb.
	Pot. Carbonate	2 lb.

Heat above separately to 80° C. and pour (2) into (1) slowly while stirring. Add perfume at 55° C. stir and pack. If cold packed a high gloss is given to surface by passing a flame lightly over surface in each jar.

Greaseless Quinosol Cream

180 grams stearin are melted in 6 to 7 liter vessel on water bath with 400 grams of water. Melted mass is allowed to remain on water bath and is mixed with boiling solution of 18 grams potassium carbonate in 400 grams water and stirred constantly with wood stirring rod, while carbonate solution is added in small portions. This is continued until uniform mass is obtained. Excess alkali in product must be neutralized with a little stearin. Then 300 grams C. P. glycerin, 40 grams lanolin and 10 grams beeswax are added and finally 1 to 2% (20 to 40 grams) perfume bouquet usually used in perfuming soap. When homogeneous product is obtained, vessel is removed from water bath and cooled to 55° C. while being constantly stirred. Then solution of 12 grams quinosol in 800 grams water, heated to same temperature, is added in portions. Mixture is agitated while being cooled to room temperature. It is permitted to stand for 1 to 2 days, then worked up again and finally filled into tubes or jars.

Cold Cream (Non-Greasy)

Glycosterin	22 lb.
Petrolatum (Vaseline)	16 lb.

Paraffin Wax	12 lb.
Mineral Oil	30 lb.
Water	100 lb.

Heat first four ingredients to 170° F. and stir together. Then slowly with stirring pour in the water which has been heated to the same temperature. Stir thoroughly and then allow to stand (hot) until air bubbles are gone. Add perfume and stir and pour at 110–130° F. Cover jars as soon as possible.

The above cold cream when made on a commercial scale costs less than 5 cents per lb. exclusive of perfume.

Liquid Cold Cream
(Water-soluble)

1. Mineral Oil	72 lb.
2. Trihydroxyethylamine Stearate (Special)	14½ lb.
3. Water (Warm)	160 lb.
4. Perfume	1½ lb.

Heat (1) and (2) until just melted together, and stir. Next add (3) slowly with thorough stirring and continue until the batch is homogeneous. Allow to stand one night and stir for 15 minutes before packing.

This cream washes off easily with cold water. The consistency can be changed by varying the amount of water in this formula.

Theatrical Cold Cream

Spermaceti	125 gm.
White Wax	120 gm.
Liquid Petrolatum	560 gm.
Borax	5 gm.
Water, Distilled	190 gm.

Cold Cream, for Sun and Wind Burn

Apricot Kernel Oil	54 oz.
White Beeswax	13 oz.
White Ceresin Wax	8½ oz.
Ethyl Amino Benzoate	½ oz.
Borax Powder	½ oz.
Distilled Water	25 oz.

Melt Apricot Kernel Oil, Beeswax and Ceresin Wax together and add Ethyl Amino Benzoate. Stir until dissolved. Adjust temperature to 65° C. Dissolve Borax in hot Distilled Water and filter. Adjust temperature to 65° C. Then add Borax solution slowly while stirring to the oil and wax mixture kept at the same temperature and stir until cold.

Vanishing Cream, for Sun and Wind Burn

Stearic Acid triple pressed	14 oz.
Apricot Kernel Oil	5 oz.
Ethyl Amino Benzoate	½ oz.
Potassium Carbonate	1 oz. 175 gr.
Borax	1 oz.
Distilled Water	70 oz.
Glycerin	9 oz.

Melt Stearic Acid and Apricot Kernel Oil together and add Ethyl Amino Benzoate. Stir until dissolved and strain through cloth. Dissolve Potassium Carbonate and Borax in Distilled Water and filter then add Glycerin. Adjust temperature of both the oil-stearic acid mixture and of the Borax, Potassium Carbonate solution to 75° C. then add slowly while stirring the melted stearic acid and apricot kernel oil mixture to the aqueous solution. Stir until completely emulsified and until temperature has dropped to about 40–45° C. Fill into jars or tubes.

Cold Cream

Mineral Oil	1 gal.
White Beeswax	2 lb.
Water (preferably distilled)	½ gal.
Powdered Borax (bolted)	2 oz.

Mix beeswax and oil in one container. Bring to 150° F. then reduce to 120° F. Dissolve borax in water. Bring to 120° F. Pour borax and water solution slowly into wax and oil solution stirring constantly but not rapidly. At 115° F., perfume and pour into containers.

Cold Cream

Beeswax	540 grams
Spermaceti	300 grams
Mineral Oil	1730 grams
Stearin	430 grams
Water	720 cc.
Borax	100 grams
Sodium Benzoate	10 grams
Perfume.	

The fat bases should be melted with mineral oil. The borax and benzoate of soda dissolved in water and brought to the boil and stirred while still hot into the molten fats. Allow to cool with slow agitation. Add perfume.

Greaseless Cream

Stearic Acid	4 oz.
Paraffine Wax	½ oz.
Glycerine	12 oz.
Add Ammonia 26°	½ oz.

When there is a perfect saponification, add 16 oz. warm distilled water in which must be dissolved 15 grams powdered borax.

Greaseless Cream

Stearic Acid	40 lb.
Water	22 gal.
Glycerine	3 gal. 1 pt.
Borax	3 lb. 12 oz.
Potassium Carbonate	18 oz.
Mineral Oil	1 pt.

Use 20 gal. water in kettle with Stearic Acid and melt. Stir well. Add potassium carbonate and borax dissolved in 2 gal. hot water. Beat until smooth. Stir constantly. Add mineral oil in about 15 minutes, gradually add glycerine. Heat all for ½ hour. Stir constantly until cool. Add perfume.

Greaseless Cream

Stearic Acid	14 oz.
Glycerine	12 oz.
Potash	4 oz.
Water	8 oz.
Borax	1 oz.
Perfume	To Suit

Greaseless Cream

Stearic Acid	30 oz.
Cocoa Butter	2½ oz.
Water	12 pt.
Add	
Borax	2½ oz.
Water	9 pt.
Add	
Sodium Carbonate	2 oz.
Water	4 oz.
Glycerine	15 oz.
Peroxide	15 oz.
Ammonia Water	10%
Perfume.	

Greaseless Cream

Stearic Acid	4 lb. 12 oz.
Glycerine	8 lb. 8 oz.
Water	14 pt.
Ammonia Water	4¼ oz.

Heat 2 lb. glycerine with 12 pints water into the ammonia. Then melt Stearic Acid. Add first mixture and balance of glycerine and water. Heat to 80° C.

Liquid Lanolin Cream

Liquid lanolin cream depends upon a suspension of lanolin by the aid of soap. The following is a satisfactory formula:

Hard Soap	1 dr.
Distilled Water	1 oz.

Dissolve and add

Hydrous Wool Fat	1 oz.
Glycerin	1 oz.

If a more liquid cream is desired the amount of soap may be increased to 1½ drachm, and the glycerin and hydrous wool fat reduced to ½ oz. each.

Lanolin Cream, Liquid

Liquid lanolin cream depends upon a suspension of lanolin by the aid of soap. The following is a satisfactory formula:

Hard Soap	1 dr.
Distilled Water	1 oz.

Dissolve and add

Hydrous Wool Fat	1 oz.
Glycerin	1 oz.

If a more liquid cream is desired the amount of soap may be increased to 1½ drachm, and the glycerin and hydrous wool fat reduced to ½ oz. each.

* Lemon Cream

The formulae given for cold creams can be modified to make a lemon cream by substituting Lemenone for the usual perfume to the extent of ½ of 1% and coloring yellow.

Cleansing Cream, Lemon

1. Lemon Juice	70
2. White Petrolatum	12
3. Parachol	17
4. Lemenone	1

Melt 2 and 3 and add 1 slowly with stirring. Then stir in 4 until uniform.

Liquefying Cream

Mineral Oil	7 lb.
Ceraflux	3 lb.
Petrolatum	2 lb.

Melt together at 220° F. and stir at room temperature until cold add perfume; pour into jars while liquid but at lowest possible temperature. This cream will not sweat oil during hot weather.

Creams, Massage

One formula suggests compounding 65 parts of mineral oil, 35 parts cetyl alcohol and 10 parts water. In another, 90 parts stearic acid, 9 parts potassium carbonate, 800 parts water are used to make soapy mixture by first melting stearic acid and then adding solution of carbonate in water and stirring until all carbon dioxide evolution has ceased. Then mass is cooled. It is mixed with 5 parts white beeswax, 20 parts anhydrous lanolin, 150 parts glycerin and perfumed with 6 parts oil of eucalyptus, 5 parts oil of pinus sylvestris and one part camphor. In another formula 65 parts mineral oil, 7.5 parts stearic acid, 7.5 parts white beeswax, 6 parts solid paraffin wax, 9 parts liquid paraffin, 0.5 part sodium carbonate, 0.5 part borax and 35 parts water are mixed together. Cream may be perfumed. Another cream contains 500 parts lanolin, 500 parts rose water, 500 parts lard, 200 parts glycerin, 15 parts cheiranthus, and 5 parts dianthus (clove pink).

Massage Cream

Glycerin	1 ounce
Borax	2 drachms
Boracic Acid	1 drachm
Oil Rose Geranium	30 drops
Oil Anise	15 drops
Oil of Bitter Almonds	15 drops
Milk	1 gallon

Heat the milk until it curdles and allow it to stand 12 hours. Strain it through cheese-cloth and allow it to stand again for 12 hours. Mix in the salts and the glycerin, and triturate in a mortar, finally adding the odors and the coloring. The curdled milk must be as free from water as possible in order to avoid separation.

Rolling Massage Cream

These creams are generally colored pink, with eosine. The general process for making these creams is as follows:

(1) To 128 parts of fresh milk add 2/10 of 1% of formaldehyde 40% solution or 1% sodium benzoate is added as an antiseptic, and enough of a 2% solution of eosine to give the proper shade. Mixture is warmed to about 50–55° C. on water bath while stirring gently, then strained if necessary.

(2) Prepare on the side, a 20% solution of alum or a concentrated solution of potassium sulphate in distilled water and bring it to the boiling point.

Bring mixture No. 1, (milk) to boiling point and pour while stirring slowly, the boiling mixture (No. 2). Stop heating, continue to stir gently, and let cool slowly at about 55° C.

When cool, and upper liquid is clear, strain on muslin previously wetted, allow precipitate to drain, wash with little cold water, drain again. Then pass through filter press if there is too large excess of water. Consistency of cream will depend upon quantity of water allowed to remain in casein. Then add to casein about 1% of perfume and 10 to 15% of glycerin or carbitol in order to prevent quick drying of casein, and put in tightly sealed jar. To obtain homogeneous product, it is recommended to pass the magma through an ointment mill before putting in jars. Addition of 1.5% sodium benzoate helps preservation.

Rolling Massage Cream

1.	Stearic Acid (Triple Pressed)	6.75 lb.
	Cocoa Butter	13.50 oz.
	Mineral Oil	2.25 lb.
2.	Corn Starch	12.00 lb.
	Boric Acid	2.40 lb.
	Water	5.60 gal.
	Moldex	1.50 gm.
3.	Glycerine	45 fl. oz.
	Ammonia 26 Baumé	12 fl. oz.
	Perfume (Rose)	4 oz.
	Color (Rose)	1 oz.

Mix the corn starch with cold water until smooth (no lumps). Add the boric acid. Heat until it forms a thick translucent paste, stirring continually, taking care to avoid overheating and burning the bottom of the pan. Take off the heat and add No. 3. Stir. Then add No. 1, which has previously been melted together at 200° F. Stir rapidly for about 1½ to 2 hours. Add color and perfume, and 2 oz. sodium benzoate dissolved in 4 oz. water. Pack cold.

Cream, Mosquito Repellent

White Mineral Oil	16 oz.
Beeswax U.S.P.	4 oz.
Spermaceti	1 oz.
Distilled Water	8 oz.
Borax	30 gr.
Butyl Salicylate	1.5 oz.

Mosquito Repellant Liquid

White Mineral Oil	95
Hexyl Salicylate	5

The above products are not malodorous or very volatile.

Nourishing Cream

1.	Beeswax	15 parts
	Mineral Oil	45 parts
	Lanolin (Anhydrous)	12 parts
	Glyco-Wax "A"	15 parts
2.	Water	25 parts
	Borax	1¼ parts
	Benzoate of Soda	½ part
3.	Perfume	½ part

Heat Nos. 1 and 2 separately to 200° F., then add 1 to 2 slowly with stirring in an emulsifier or beater. When the cream begins to set add the perfume. Allow to stand over-night; stir the next morning and package.

This cream possesses exceptional penetrating powers and is absorbed very readily by the skin.

Nourishing Cream
(Skin Food Type)

Glycosterin	12 lb.
Petrolatum (Vaseline)	4 lb.
Lanolin	6 lb.
Mineral Oil	12 lb.
Water	65 lb.

The procedure is the same as for the cold cream given above.

Nourishing Cream Cholesterol

White Wax	600 gm.
Spermaceti	100 gm.
Stearin	500 gm.
Lanolin, Anhydrous	600 gm.
Cacao Butter	400 gm.
Sweet Almond Oil (with preservative)	1,800 gm.
Cholesterol, Purest	120 gm.

After solution of the cholesterol has been effected, stir the following hot solution into the molten mass until pasty:

Sodium Benzoate	15 gm.
Borax	100 gm.
Water	1,700 gm.

Sun Burn Cream

Lanolin	2 lb.
White Petrolatum	8 lb.
Zinc Oxide	4 lb.
Glycerine	4 lb.

Mix the above thoroughly.

Tissue Cream

White Wax	5 oz.
Spermaceti	1 lb.
Petrolatum (Light Amber)	1 lb.
Mineral Oil	1½ pints
Lanolin (Hydrous)	2 lb.
Borax	⅜ oz.
Water	10 oz.
Benzyl Alcohol	1 drachm
Oil Bitter Almond	1 drachm
Oil Rose Geranium	1½ drachm
Oil Bergamot	2 drachm

Tissue Cream (Non-Alkaline)

1.	Spermaceti	10 lb.
	Lanolin	20 lb.
	Glycosterin	46 lb.
	Olive Oil	20 lb.
	Almond Oil	30 lb.
2.	Water	90 lb.
	Sodium Benzoate	¼ lb.
3.	Perfume	to suit

Heat (1) to 150° F. and run into it slowly with stirring (2) which has been heated to the same temperature. Add the perfume at about 105° F. and stir in. Pour at 95–100° F.

Tissue Cream

Lanolin	800 parts
Almond Oil	100 parts
Glycerine	100 parts
Benzoic Acid	2 parts
Perfume to suit.	

Melt lanolin on water bath, and add the oils and glycerine. Stir until of uniform consistensy. When cool, add perfume.

Tissue Cream with Cholesterin

Lanolin	325 grams
Cocoa Butter, odorless	200 grams
Beeswax, White	300 grams
Spermaceti	55 grams
Oleic Acid	50 grams
Stearic Acid	200 grams
Sesame Oil (with preservative)	800 grams
Cholesterin (Pure)	65 grams
Borax	50 grams
Water	800 grams
Sodium Benzoate	8 grams

Procedure: Melt the waxes, fats, and oil. Add the cholesterin. Make a hot solution of the borax, sodium benzoate and water and stir into the melted fats after the cholesterin has dissolved. Mix thoroughly and perfume to suit.

Tissue Cream with Lecithin and Cholesterin

Lanolin, Anhydrous	220 grams
Cocoa Butter, odorless	100 grams
Beeswax, White	200 grams
Stearic Acid T. P.	100 grams
Olive Oil (with preservative)	1000 grams
Lecithin	22 grams
Cholesterin	44 grams
Water	600 grams
Parahydroxybenzoic Acid	4 grams
Sodium Benzoate	10 grams

Procedure: Melt fats, waxes and oils, add cholesterin and lecithin. Stir in a solution (hot) of the water and sodium benzoate. Dissolve the parahydroxybenzoic acid in a small quantity of alcohol. Mix, perfume, and color.

Tissue Cream with Lecithin

Lanolin, Anhydrous	22 grams
Spermaceti	22 grams
Beeswax, White	40 grams
Cocoa Butter, odorless	28 grams

Almond Oil (with preservative)	390 grams
Lecithin	50 grams
Borax	5 grams
Sodium Benzoate	5 grams
Parahydroxybenzoic Acid	2 grams
Water	220 grams

Procedure as before.

Tissue Cream with Cholesterol, Lecithin and Turtle Oil

Beeswax, White	220 grams
Stearic Acid	100 grams
Cocoa Butter, odorless	200 grams
Lanolin	200 grams
Turtle Oil	1000 grams
Almond Oil (with preservative)	1000 grams
Cholesterin	58 grams
Lecithin	120 grams
Water	800 grams
Parahydroxybenzoic Acid	8 grams
Sodium Benzoate	12 grams
Borax	120 grams

Proceed as above.

Tissue Cream (Soft) with Cholesterin Base

Absorption Base	30 grams
Lanolin	5 grams
Water	55 grams
Beeswax, White	10 grams

Procedure: Melt the wax and lanolin, add the base and stir in the water (warm).

(Note: Consistency in the foregoing formulas can be adjusted by changing the wax content to suit.)

VANISHING CREAMS

Ordinary Type

Glyceryl Monostearate	10.0%
Glycerin	3.0%
Petrolatum	3.0%
Spermaceti	5.0%
Mineral Oil	2.0%
Stearic Acid	2.0%
Caustic Potash	0.1%
Titanium Oxide	1.0%
Water	73.9%

Pearly Type

Glyceryl Monostearate	2.5%
Stearic Acid	10.5%
Glycerin	4.5%
Ammonia (S. G. 91)	2.5%
Water	80.0%

Moderately Fatty Cream

Glyceryl Monostearate	12%
Petrolatum	6%
Lanolin	4%
Mineral Oil	6%
Almond Oil	6%
Glycerin	3%
Water	63%

Petrolatum Cream

Glyceryl Monostearate	10%
White Petrolatum	20%
Mineral Oil	10%
Water	60%

Vanishing Creams

Vanishing Creams made with Glycomine (a real forward step in cosmetics) enable anyone to produce perfect products, noteworthy because—

1. The use of caustic soda, potash and ammonia is eliminated.
2. No glycerin is necessary.
3. A most beautiful pearly finish results.
4. Closed jars will not dry or shrink.
5. It may be poured in jars when cold.
6. The batch is complete in 24 hours.

Formula

1.	Stearic Acid (C.G.)	20 lb.
2.	Glycomine	11 lb.
	Water	50 lb.
3.	Perfume	12 oz.

Heat No. 2 to 200° F. and add No. 1 (previously heated to 200° F.) to it slowly with stirring in an emulsifier or whipper. Continue stirring until mass is homogeneous. Allow to stand over-night. Add No. 3 and mix for 20 minutes. This cream is softer than the old-fashioned creams but typifies the highest grade modern vanishing cream. The pearliness in this cream increases with age and is helped by stirring cold the next day.

A softer cream con be produced by increasing the amount of water.

A harder cream is made by pouring hot or by increasing the amount of stearic acid; and also if stirring is very slow.

Astringent Cream

An astringent cream of the highest type is made by adding one ounce of our Astringent Powder No. 1 to the above vanishing cream just before it begins to thicken.

Vanishing or Foundation Cream

A.	Stearic Acid	4 lb.
	Lanolin	1 lb.
B.	Water	2 gal.
	Glycerin	2 lb.
	Pot. Carbonate	2 oz.
C.	Perfume Oil	2 oz.

In separate aluminum or enamel pots heat A and B to 180° F. Add B to A slowly with stirring until uniform. Stir in C at 110° F.

The above makes an excellent sunburn cream with or without the addition of 1% Quinine Ricinoleate.

Vanishing Cream

Stearic Acid	50 lb.
Lanolin (Anhydrous)	9 lb.
Triethanolamine	2.5 lb.
Carbitol	18 lb.
Water	120 lb.

Preparation

In one container melt the stearic acid carefully and add the lanolin. Heat the Triethanolamine and water separately to boiling and then add the melted fatty acid to it with constant stirring. When a smooth mixture is obtained, stir in the Carbitol to which has been added the perfume. Continue with even stirring while cooling until a heavy, smooth cream is obtained, and then stir occasionally until cold. The cream will become thinner as it cools and the acid crystallizes.

Properties

A vanishing cream should be completely absorbed without leaving a greasy residue. It should have no tendency to flake or roll and should impart a feeling of softness and smoothness to the skin. It should afford some protection against wind and sun and also act as a powder base. The given product gives these desired properties to the fullest extent, and is free from irritating effect.

Variations

An excellent suntan or sunburn cream can be made with the above formula using 40 lb. stearic acid and 20 lb. lanolin.

Stearic acid is the essential ingredient of a vanishing cream since it produces the desired "dryness" and pearliness. It should be a very pure product if no rancidity or discoloration is to develop. The grade of acid has some effect upon the consistency of a vanishing cream, and if it is very hard and waxy, more water will have to be added to give the proper body. As a rule, by variations in the amount of this ingredient, any desired consistency can be obtained. The speed of stirring also has an effect upon the body of the cream. During the cooling, as soon as a stiff smooth emulsion is obtained, stirring should be reduced until just sufficient to prevent crusting on top. Rapid stirring after this point has been reached will usually cause aeration and yield a thin cream.

Vanishing or Foundation Cream

Stearic Acid	24 lb.
Triethanolamine	1 lb.
Water	8 gal.
Glycopon S	12 lb.
Water	8 gal.
Perfume	8 oz.

In separate vessels heat stearic acid and all other ingredients except perfume to 180° F. Add one to the other and stir until uniform. Mix in perfume at about 105° F.

Vanishing Cream

Stearic Acid	35 lb.
Witch Hazel	6 gal.
Distilled Water	10 gal.
Glycerine	50 lb.
Castor Oil	8 oz.
Sodium Borate	8 oz.
Ammonia 28%	56½ oz.
Perfume.	

Melt stearic acid and castor oil in one container and in another heat Witch Hazel and Water in which has been dissolved the Sodium Borate. When at about 20° under b. p. of water, add ammonia to water solution and instantly introduce into this solution the stearic acid. Agitate cream for 12 hours until every trace of ammonia gas has passed off. Agitate again the next day for two hours. Add perfume.

Vanishing Cream

Stearic Acid	18 lb.
Glycerine	6 pints
Ammonia Water 26° Baumé	1 pint 2 oz.
Water	11 gal.
Perfume.	

Melt stearic acid at low heat. Mix glycerine with ammonia and 11 gal. of water. Add to stearic acid in several portions, heating and stirring until smooth and liquid. When all water has been added remove from fire. Add perfume. Stir occasionally until mass is cold. Strain cold through cheese cloth.

Vanishing Cream

Stearic Acid	16 lb.
Water	74 lb.
Glycerine	10 lb.
Borax	1½ lb.
Potassium Carbonate	½ lb.

When finished add

Glycerine	5 lb.
Perfume.	

Melt stearic acid and glycerine on water bath, keeping at 70° C. Dissolve Potassium Carbonate and Borax in water at 70° C. Add this solution very slowly constantly stirring to stearic acid and glycerine having turned off the heat. After all water is added, keep on stirring until cream forms. Then turn on the heat again and stir until whole mass is practically liquid. Turn off heat and stir till cold. Shortly before getting cool, add 5 lbs. glycerine.

Zinc Stearate Creams

Zinc stearate cream may be prepared with 150 grams glycerin, 100 grams water, 80 grams zinc stearate. Stearate is first triturated with glycerin and water gradually added. Cream is very soft, white and absolutely homogeneous. Sometimes ingredients of cream separate after long standing. This can be corrected by addition of about 5% of medicinal pulverized soap which ensures permanent cohesion of various ingredients in uniform mixture.

Five parts zinc stearate may be easily mixed with 50 parts petrolatum and is useful for many purposes, particularly in healing cuts.

Lanolin salve is made with 325 parts lanolin, 35 parts ceresin wax, 150 parts mineral oil and 150 parts water. Ceresin wax is melted in heated mineral oil and then lanolin is added and mixture allowed to cool. Mass is triturated into soft salve and water and perfume are worked in gradually. Five to 10% of zinc stearate is added to obtain preparation suitable for dry skin.

Liquid Body Deodorant

A.

Aluminum Aceto Tartrate	1 lb.
Rose Perfume (water soluble)	1 oz.
Water	5 gal.

B.

Aluminum Chloride (crystalline)	8 lb.
Hydrochloric Acid	4 oz.
Phenyl Ethyl Alcohol	4 oz.
Water	5 gal.
Color	to suit

Perspiration Deodorants

A. Liquid Type

Salicylic Acid	2 gm.
Aluminum Chloride	4 gm.
Cologne Spirit	30 mil.
Rose Water	54 mil.
Glycerin	10 mil.
Rose Colour	a trace

Dissolve the salicylic acid in the Cologne spirit, and the aluminium chloride in the rose water. Mix and add the glycerine. A more delicate perfume may be used.

B. Paste Type

Salicylic Acid	10 gm.
Levigated Zinc Oxide	60 gm.
Greaseless Cold Cream	480 gm.
Perfume to Suit.	

Deodorant Pencil

	Gr.	(White product) Gr.
Zinc Phenolsulfonate	5	10
Zinc Oleate	10	10
Aluminum Palmitate	7.50	7.5
Parachol	20.00	30
Glyco Wax	40.00	30
Titanium Dioxide	——	15

Rub first three ingredients to fine powder and add to liquified wax the Parachol mixture. Stir until just before solidification and pour into molds.

---

* Deodorant Pencil

White Kaolin	40%
Glyco Wax	20%
Mineral Oil	20%
Aluminum Chloride	20%

Melt wax in water bath and add mineral oil; keep at 90° C. and add the intimately mixed aluminum chloride crystals and kaolin. Stir with pressure until smooth paste is formed. Pour at once into molds and cool slowly.

Perspiration Deodorizing Cream

Beeswax	8 oz.
Liquid Petrolatum	24 oz.
Sodium Borate	100 gr.
Benzoic Acid	20 gr.
Salicylic Acid	400 gr.
Hot Water	16 oz.

Melt the wax and oil and heat to about 160 degrees F. Dissolve the other materials in the water, heat to the same temperatures as the wax solution, and pour it into the latter, beating briskly until the cream is formed. Here a comparatively high temperature of the solutions, plus a small amount of stirring, results in a glossy cream.

Perspiration Deoderant

Sod. Perborate	10
Sod. Bicarbonate	2

Glycerin	1
Rose Water	98

Tint pink with eosin.

Deodorant Powder

Methyl Salicylate	1.5 parts
Oil of Eucalyptus	2.0 parts
Thymol	12.0 parts
Menthol	0.5 parts
Boric Acid	39.0 parts
Acetanilid	43.0 parts
Starch	2.0 parts

Deodorant Powder

Zinc Peroxide	0.5 gm.
Betanaphthol Benzoate	0.1 gm.
Talcum	99.4 gm.

Depilatory

Strontium Sulfide	50
Zinc Oxide	50
Rice Flour	60
Perfume	to suit

Solid Eau de Cologne

1.3 parts of sodium hydroxide are dissolved in 40 parts water; and 8.5 parts of stearic acid are dissolved in 50 parts of 90% alcohol. Then the two solutions are thoroughly mixed and heated slowly until the liquid turns clear. The essence of Eau de Cologne is then added and the liquid cooled to avoid evaporation of the oils, but not enough to allow it to congeal. After the oil has become thoroughly mixed with the base, the solution is then poured into moulds and allowed to cool.

Eau de Quinine

The following is a formula for an inexpensive eau de quinine:

Tincture of Cantharidin	1 dr.
Quinine Hydrochloride	10 gr.
Tincture of Capsicum	20 min.
Glycerin	30 min.
Bay Rum, Prepared with Industrial Spirit	to 20 oz.
Tincture of Cudbear	sufficient to color

Eye Shadow

Mineral Oil	5 lb.
Lanolin	2 lb.
Petrolatum	1 lb.
Beeswax	1 lb.
Paraffin	2 lb.
Perfume Oil	4 oz.

Color with any of following combinations:

Blue

Ultramarine Blue	2 lb.
Zinc Oxide	2 lb.

Green, Light

Zinc Oxide	3 lb.
Green Lake	1 lb.

Gray

Ultramarine Blue	1 lb.
Carbon Black	1 lb.
Zinc Oxide	2 lb.

Brown

Burnt Umber	3 lb.
Zinc Oxide	1 lb.

Green, Dark

Green Lake	3 lb.
Zinc Oxide	1 lb.

Violet

Violet Lake	1 lb.
Zinc Oxide	3 lb.

Heat colors and wax mixture and grind in ointment mill; pack by pouring hot.

Eyebrow Sticks

Paraffin Wax	300 gm.
Cocoa Butter	300 gm.
Beeswax	100 gm.
Petrolatum	100 gm.
Carbon Black	sufficient

Mix thoroughly and run into molds to form sticks.

Non-lathering Hair Cleanser

Ammonium Stearate (Paste)	30 oz.
Water	2 oz.
Perfume to Suit	

This is made cold by simple mixing until homogeneous and until most of ammonia has evaporated.

Dandruff Mixture

Chloral Hydrate	1 dr.
Glycerin	4 dr.
Bay Rum	8 oz.

Mix.

Dandruff Remedy

Ammonium Carbonate	5
Alcohol	30
Glycerin	20
Rose Water	200

Dandruff Treatment

This complaint requires for its treatment and cure external medications in the form of ointments, shampoos and hair tonics, and these should contain antiseptics, parasiticides and stimulants. The following formulas indicate the type of preparation:

Scalp Tonic

Resorcin	10 gm.
Chloral Hydrate	5 gm.
Camphor	0.2 gm.

Tincture of Cantharides	10	gm.
Alcohol	50	gm.
Oil of Geranium,		
Oil of Bergamot,		
Oil of Lavender,		
Oil of Bitter Almond of each	0.25	gm.
Glycerin	2	gm.
Distilled Water to make	1,000	gm.

Color with trace of aniline dye. Filter perfectly clear and bright.

Apply to scalp three or four times a week and rub in thoroughly.

Ointment for Dandruff

Salicylic Acid	10	gr.
Precipitated Sulphur	15	gr.
White Petrolatum	1	oz.
Oil of Geranium,		
Oil of Bergamot of each	2	min.

Apply once or twice a week. Follow with shampoo the next morning.

Dandruff Ointment

Precipitated Sulphur	8	lb.
Oxyquinoline Sulphate	1	lb.
Lanolin	10	lb.
Petrolatum	61	lb.
Castor Oil	15	lb.
Tincture Fish Berries	1	lb.
Balsam Peru	2	lb.
Carbolic Acid 85%	2	lb.

Mix the sulphur with the castor oil rubbing thoroughly until lumps have disappeared. Mix the oxyquinoline sulphate with ten pounds of petrolatum, run through an ointment mill three times, add the sulphur castor oil mixture, mix thoroughly and run through mill again. Melt the lanolin and the rest of the petrolatum, add the remainder of the castor oil, mix thoroughly and then mix in the oxy-sulphur mass. Mix thoroughly, add the balsam Peru, continue mixing for thirty minutes.

Dressing for "Kinky" Hair

Beefsuet	16	oz.
Yellow Beeswax	2	oz.
Castor Oil	2	oz.
Benzoic Acid	10	gr.
Oil of Lemon	1	dr.
Oil of Cassia	15	drops

Melt the suet and wax, add the castor oil, and acid, allow to cool and add other oils.

---

* Hair Dyes

Mixture may consist of 0.5 part of haematoxylin and 0.3 part of orthoaminophenol. Striking blonde shade is obtained thereby. Mixture which contains 0.5 part of haematoxylin and 0.3 part of para-aminodiphenylamine, gives deep black shade which does not turn greenish black. Mixture consisting of 0.5 part of haematoxylin and 0.5 part of para-aminophenol gives fine chestnut brown shade, while mixture of 0.5 part of haematoxylin and 0.5 part of metaphenylenediamine gives beautiful platinum blond shade.

Hair-Dye, Non-Toxic

Colors such as 5: 5′ dichlorothioindigo or 5: 5′ dichloro-6: 6′ dimethylthioindigo which are bluish red and blend therewith suitable proportions of brominated-beta-napthalene indigo which is yellowish green with or without indigo to secure dark neutral shades. The amount of each color will depend upon the shades desired. The coloring material is dissolved in hot water to which a small amount of sodium hydrosulphite and ammonia are added and is ready for application to the hair.

In coloring the hair, it is first washed, if necessary, after which the solution is applied uniformly with cotton or a small brush. The solution is permitted to remain on the hair until a sufficient amount thereof is absorbed. The time required is variable, depending upon the shade to be produced. The hair is then again washed and dried. The exposure of the hair to the atmosphere after washing and while the hair is drying results in oxidation of color base to produce the desired color. The hair is then shampooed and dried. As a result of the operation, the hair is permanently colored and may be washed repeatedly without removing the color therefrom.

The solution as described affords all of the necessary material for the treatment, it being unnecessary, as is usual in many hair dyeing operations, to apply hydrogen peroxide or similar chemical agents. It is possible, therefore, to supply coloring material in a single solution of the leuco base adapted to produce the desired color or shade when the material is applied in the manner described to the hair. A typical example of such a solution consists of:

Color	10	gm.
Sodium Hydrosulphite	1	gm.
Aqua Ammonia	50	cc.
Water	1	liter

Such a solution is adapted to afford a deep color. In light shades the proportion of color would be considerably less. The hair after treatment as described is soft and free from harshness. There is no substantial loss of strength and the hair takes a permanent wave readily when treated by the usual waving methods. The colors are fast and do not change when the hair is exposed to strong light or becomes moist with perspiration.

Hair-Fixative

Water	20 gal.
Gum Tragacanth	1 lb.
Boric Acid	1 lb.
Moldex	2 oz.

Allow to stand over night and stir until uniform; then stir in

Perfume Oil	4 oz.
Color	to suit

Hair Fixers or Straighteners (Waxy)

A

Ceraflux	40 lb.
Glyco Wax A	10 lb.
Petrolatum, White or Yellow	100 lb.
Rosin	40 lb.

Melt together until clear and stir until uniform. Pour into jars while melted.

B

Beeswax	10 lb.
Petrolatum, Yellow or White	100 lb.
Paraflux	40 lb.
Flexoresin B1	40 lb.

Method as given for formula A. Formula A will give a very light colored product if white petrolatum, and FF rosin is used. Harder or softer product may be gotten by slight variation of the above.

Lemon Rinse

1. Lemonone	3 oz.
2. Isohol	14 lb.
3. Citric Acid	3½ lb.
4. Tartaric Acid	4½ lb.
5. Water	16 lb.

Dissolve 1 in 2 and add to it slowly with stirring 3 and 4 which have been dissolved in 5.

---

* Hair "Restorer"

Cholesterol	5
Ethyl Acetate	120

Allow to stand 24 hrs. and stir till dissolved. Add

Alcohol	800

After standing another 24 hrs. add

Balsam Peru	50

---

* "Hair-Restorer"

Vaseline	48
Beeswax	1
Olive Oil	3
Oil of Green Elder	3
Oil of Eucalyptus	3

Hair Tonic

Alcohol	10 gal.
Castor Oil	7 gal.
Quinine Ricinoleate	1 lb.
Perfume Oil	1 lb.

Hair Shampoos

The absence of alkalinity in Triethanolamine soaps and their harmless effect upon the skin has brought about their use not only in emulsified cosmetic creams but also in special cosmetic soaps. A very good hair shampoo, for example, is composed of a neutral cocoanut oil soap of Triethanolamine. For a variation, Carbitol may be added as a cleanser and stimulant for the scalp according to the following formula:

Shampoo

Oleic Acid	55 lb.
Cocoanut Fatty Acids	40 lb.
Triethanolamine	50 lb.
Carbitol	55 lb.

The product prepared in this way is a liquid soap of a clear red color, which can be diluted with water to any desired consistency or concentration.

Olive Oil Shampoo

Olive Oil	4 lb.
Oleic Acid	8 lb.
Cocoanut Oil	8 lb.
Caustic Potash	5 lb.
Alcohol	3 pt.
Water to Make	10 gal.

Dissolve the caustic potash in water. Mix and heat the oils to 120° F. Pour in the alkali solution and stir until saponified. Add two pints of the alcohol and heat to 180° F. Meanwhile prepare the following mixture and add foregoing

Glycerine	16 oz.
Borax	16 oz.
Potassium Carbonate	8 oz.
Oleic Acid	1 oz.

Dissolve the oleic acid in one pint of alcohol. Dissolve borax and potassium carbonate in glycerine with heat, mix thoroughly and add oleic solution. Add this mixture to soap base while still quite hot. Transfer to a refrigerating tank the day after soap has been finished, refrigerate to 40° F., filter and fill at once.

Lime Juice and Glycerin (for hair)

White Wax 500 gm.
Oil of Sweet Almonds 2 kilos, 500 c.c.

are melted together in a water-bath and added to:

Glycerin	300 gm.
Citric Acid	30 gm.

dissolved in a litre of rose water. Finally, there are added with stirring in an automatic mixer:

Alcohol (95 per cent.)	150 gm.
Oil of Lemon	75 gm.
Oil of Bitter Almonds	10 gm.

Dry Shampoo Powder

Cocoanut Oil Soap Powder	30%
Sodium Carbonate Monohydrated	45%
Borax	25%
Henna Leaves Powder	trace
Aniline Yellow	trace
Perfume	to suit

Mix together and sift. Keep in closed containers.

Soapless Shampoo

Sulfonated Olive Oil, concentrated	40 parts
Sulfonated Castor Oil, concentrated	10 parts
White Mineral Oil	15 parts
Water	35 parts
25% Solution of Caustic Soda to Clear	

Mix all the ingredients with the exception of the caustic soda, warm to 45–50° C. and add enough of the caustic soda solution (1 or 2%) until the mixture turns bright. Perfume as desired.

Soapless Shampoo

1. Sulfo Turk "A"	10 lb.
2. Mineral Oil	10 lb.
3. White Oleic Acid	10 lb.
4. Alcohol	2–10 lb.

Mix the above materials in the order given. If desired, the cast can be reduced further by adding an additional amount of water. The water should be added carefully with stirring. The addition of water should be stopped just before a cloudiness appears.

These shampoos are used by pouring a little into the hand and rubbing to a creamy consistency with water and then applying to the hair which must be wet.

Milky Hair Wash
(Kerosene)

1. Trihydroxyethylamine Stearate Special	10 lb.
2. Kerosene	150 lb.
3. Pine Oil	6 lb.
4. Water	250 lb.

Heat Nos. 1 and 2 to 140° F. and stir until dissolved; then stir in No. 3. Now allow No. 4 to run in slowly while stirring. If the pine oil is objectionable, however, any other oil may be substituted for it. It may be colored beautifully by means of any water-soluble dye free from salt.

Soapless Shampoo

Sapinone	10
Water	900
Alcohol	100
Perfume	15

Eau de Quinine Hair Tonic

Quinine Hydrochloride	30	gm.
Salicylic Acid	.25	oz.
Glycerine	4	oz.
Resorcin	4	oz.
Alcohol	52	oz.
Perfume and Color		
Water Q. S.	1	gal.

Hair Tonic, Honey and Flower

Oil of Orange	2	oz.
Oil of Lemon	1	oz.
Oil of Bergamot	½	oz.
Castor Oil	10	oz.
Honey	1	oz.
Oil of Cloves	1	dr.
Lavender	2	dr.
Geraniol	2	dr.
Coumarin	1	dr.
Synthetic Musk	½	dr.
Mineral Oil	1	gal.
Industrial Methylated Spirit	2	gal.

Scalp Tonic

Tannic Acid U.S.P.	0.5
Salicylic Acid U.S.P.	1.0
Castor Oil U.S.P.	24.5
Resorcinol Monoacetate	5.0
Alcohol	69.0
Perfume	sufficient

Hair Tonic—Cholesterol

Alcohol	75%
Glycerine	5%
Cholesterol	1%

Lecithin	1%
Distilled Water	12%
Perfume	1%
Chloroform	5%

Dissolve lecithin in chloroform add cholesterol and one gallon of alcohol. Mix the perfume with the alcohol, add the glycerine, add the lecithin-cholesterol mixture, agitate for one hour, add the water and agitate for two hours. Allow to stand over night and filter.

Hair Tonic—Dry Scalp

Castor Oil	1 gal.
Crude Carbolic 30%	8 oz.
Cresol U.S.P.	3 oz.
Lignol	1 gal.
Soya Bean Oil	2 gal.
Precipitated Sulphur	2 oz.

Mix the soya bean oil, the castor oil, heat to 100° F. and add the lignol. Take a small quantity of this mixture and rub up precipitated sulphur into a smooth paste. Mix with rest of oils. Add carbolic and creosol.

Hair Tonic—Oily Scalp

Water	15 gal.
Glycerine	2 gal.
Alcohol	30 gal.
Menthol	7 lb.
Resorcinmonoacetate	8 oz.
Perfume	sufficient

Dissolve menthol and perfume in alcohol, mixing rapidly. Add glycerine and 10 gallons of water. Dissolve resorcinmonoacetate in rest of water, add to the above and mix for three hours. Allow to stand overnight and filter.

Hair Setting Fluid

(Dries quickly and leaves no visible residue.)

1.	Glycomel	5 lb.
	Isohol	20 lb.
	Karaya Gum White	5 lb.
	Formaldehyde	1 lb.
	Lilac Oil	3 lb.
2.	Water	454 lb.
3.	Water	454 lb.

Mix together ingredients in (1). This is then poured slowly into (2) while stirring thoroughly until all particles are dispersed. This gives a concentrate. To make a finished product for use on the hair, this mixture is stirred into (3).

If a colored product is desired a little spirit soluble aniline green is dissolved in (1).

Permanent Waving Fluid

Permosalt	75 lb.
Ammonia 28 degree	72 lb.
Glycerine	7 lb.
Water	800 lb.

Stir the above until dissolved and filter the next day.

Hair Wave Concentrate

Gum Karaya	25 lb.
Alcohol	10 gal.
Liquid Glycol Bori-Borate	¼ gal.
Perfume Oil	8 oz.
Color	to suit

Shake and stir into water for use.

Finger Wave Dryer

1.

Potassium Carbonate	40 gm.
Borax	10 gm.
Mucilage of Tragacanth	100 cc.
Coumarin	5 gm.
Methyl Acetophenone	1 cc.
Alcohol	100 cc.
Rose Water	to make 1000 cc.

2.

Borax	600 gm.
Acacia	80 gm.
Boiling Water	18 liters
When cold add:	
Spirit of Camphor	75 cc.
Heliotropin	enough for perfume

Hair Wave Fluid

1. Trogeen	4 lb.
2. Glycopon S	16 lb.
3. Isohol	16 lb.
4. Water	128–256 lb.

Wet 1 thoroughly with 2 and 3 and allow to stand (overnight if possible). Stir 4 in slowly a little at a time. The viscosity of thickness of this fluid decreases with the use of more than a certain amount of water. This dries rapidly and does not leave a white deposit on the hair. It requires no preservative and will not spot.

Hair Wave Jelly

Gum Tragacanth	12 oz.
Alcohol	½ gal.
Water	3 gal.
Borax	8 gr.
Benzoic Acid	8 dr.
Perfume	3 dr.

Put the tragacanth into a vessel, add the water and borax and allow to stand until dissolved, a period which will depend upon whether the tragacanth is powdered in ribbons or lumps. Add alcohol to which perfume and benzoic has been added and mix thoroughly. Squeeze through muslin bag.

Hair Wave Liquid

Quince Seed	30 oz.
Water	10 gal.
Borax Powdered	20 oz.
Perfume	4 oz.
Benzoic Acid	3 oz.
Alcohol	10 oz.

Boil the water, add the quince seed and allow to stand overnight stirring occasionally. Add the borax solution (made with part of the water). Filter. Add perfume and benzoic acid solution and mix thoroughly.

Hair Curling Powder

Sodium Carbonate	15%
Sodium Bicarbonate	85%

Mix powders thoroughly.

Hair Wave Powder

Gum Karaya	100 lb.
Sod. Benzoate	2 lb.
Perfume Oil	1 lb.
Color	to suit

To use put in water to swell and stir till uniform.

Permanent Waving Solution

Permosalt	1 lb.
Water	5 gal.

Allow to stand overnight and filter. To this add

Sulfoturk C	13 oz.
Ammonium Hydroxide	125 oz.

A milky stable mixture results.

Permanent Wave Solutions

A. Hydrazine Hydrochloride	4
Water	96
B. Borax	3.75
Sod. Bicarbonate	3.50
Linseed Oil	0.17
Starch	0.40
Water	99.00

Hair Setting Preparations

Decoctions of quince seeds and of psyllium seeds are among those employed. For example, a decoction of 0.2 part of psyllium seed in 100 parts of distilled water, prepared by boiling for five minutes, and straining, mixed with an equal bulk of spirit, may be employed.

Hair Setting Lotion

Emulsone B, in Powder	0.1
Isopropyl Alcohol	10.0
Terpineol	0.25
Water, Distilled, or Rose	to 100.0

Thoroughly mix the emulsone B with 0.2 of isopropyl alcohol in a perfectly dry, capacious bottle. Add 8 of water all at once, and shake violently. Dilute with water, adding the rest of the isopropyl alcohol in which the terpineol has been previously dissolved, towards the end. After standing, it is desirable to filter the lotion, or to decant it from the sediment, if a perfectly clear product is required, and perfectly clear lotions make a much stronger appeal than cloudy ones. As is well known, terpineol has a lilac-like odor, and, especially if made with rose water, this lotion smells quite nice. The terpineol, however, may be replaced by any water-soluble perfume, a number of which, already compounded, are now on the market. A bare trace of carmoisine gives the lotion a pretty tint, or any other innocuous water-soluble dye can be employed.

Hair Whitener

Aniline Blue	2 oz.
Distilled Water	15 gal.

Dissolve blue in one half the water by allowing it to stand over night. Mix thoroughly, add the rest of the water and filter. It is undesirable to run this preparation through a mechanical filter because the stain is almost impossible to remove. It is better to filter in five gallon bottles reserved for this purpose.

Hand Cleaning Preparations

The following formulas make preparations for cleaning the hands by just using it and wiping off with a towel:

Liquid

Castor Oil	25
Sol. Caustic Potash (1–1)	10
Alcohol	60
Petrol	10
Water	20

Neutralize with oleic acid.

Solid

Oleic Acid	4 oz.
Turpentine Substitute	1 oz.
Alcohol	2 oz.
Castor Oil	1 oz.

Neutralize with solution of caustic potash (1–1). Add water 2 oz. to form a paste, incorporate 15 per cent borax powder.

Cleaning Artificial Dentures

The following formula has been found to be satisfactory in every way:

Precipitated Chalk	4 oz.
Heavy Magnesium Carbonate	1 oz.
Light Magnesium Carbonate	½ oz.
Powdered Soap	2 dr.

The dental plate brush should be slightly damp when using this powder.

Hand Cleanser and Conditioner

1. Mineral Oil	70 lb.
2. Olive Oil	8 lb.
3. Trihydroxyethylamine Stearate (Special)	14 lb.
4. Water	70 lb.
5. Perfume	2 lb.

Heat Nos. 1, 2 and 3 together to 140° F. and stir until homogeneous. Add No. 4 slowly while stirring and then stir in the perfume. Continue stirring until cool. By varying the amount of water a thicker or thinner preparation will be formed. The thicker preparations are put up in tubes and are now carried by men and women, especially motorists, who, when water is not available, merely put a little of this cleaner on their hands, rub it in and then wipe off with it the grease, oil, paint or dirt present. Not only is this an excellent detergent but it leaves the skin smooth, and produces a cooling sensation and prevents chapping during cold weather.

Lip Sticks

Vaseline	15 oz.
Beeswax	10 oz.
Spermaceti	400 gr.
Carmine	6 dr.
Perfume to suit.	

Melt and stir. Allow to cool some before adding perfume. Pour into molds.

Lipstick, Indelible

Stearoricinol	28 lb.
Mineral Oil	4 lb.
Lanolin (Anhydrous)	2 lb.
Petrolatum	2 lb.
Paraffin Wax	8 lb.
Beeswax	8 lb.
Bromo "Acid"	1 lb.
Lake Colors	5 lb.
Perfume Oil	1 lb.

By varying the colors correspondingly different shades may be gotten.

Orange Changeable Lipstick

Cocoa Butter	20 lb.
Castor Oil	12 lb.
Ceresine	15 lb.
Beeswax	5 lb.
Bromo "Acid"	4 oz.
Perfume Oil	1 lb.

Lipstick

1. Stearoricinol	4 to 6 oz.
2. Paraffin Wax	1 oz.
3. Beeswax	1 oz.
4. Bromo Acid	½ oz.
5. Geranium Lake	½ oz.
6. Perfume to suit.	

Melt and grind above in heated ointment mill 160° F. and mold.

No alcohol or other solvent is necessary as 1 is a powerful solvent.

The above formula gives an indelible stick which goes on evenly to form a coating free from objectionable gloss. After it penetrates it does not come off easily.

In hot weather the above formula should be modified by increasing the amount of Beeswax.

Lip Pomade

Mineral Oil	1 gal.
Petrolatum White	2 lb.
Ozokerite White	5 lb.
Beeswax White	2 lb.
Perfume	1 oz.
Color	to suit

Lotion Formulae

A.	Lanolin	12 lb.
	Mineral Oil	20 lb.
	Trihydroxyethylamine Stearate (Special)	4¼ lb.
	Glycosterin	2 lb.
B.	Glycerin	8 lb.
	Water	200 lb.
	Benzoate of Soda	¼ lb.
C.	Perfume to suit.	

Heat A and B separately to 180° F. and run B into A slowly while stirring. When temperature has dropped to 100° F. add perfume. Continue stirring until COLD.

The low cost and high quality of these lotions make them of great interest. This eliminates the use of spermaceti, almond oil and gums which are prone to spoilage and the technique is very simple.

These formulae can be made thinner by increasing the amount of Glycerin or thicker by decreasing the amount of Glycerin. They have excellent smoothing and nourishing properties for the skin because of their Lanolin and Glycerin content.

1. Lanolin	1 lb.
2. Tincture of Benzoin	20 oz.

3. Glycosterin 10 lb.
4. Witch Hazel 250 lb.

Melt 1, 2 and 3 together and run into this slowly with stirring 4 heated to 140° F.

After Shave Lotion

Menthol	1	dr.
Boric Acid	2½	oz.
Glycerine	5	oz.
Alcohol	5	qt.
Water, to make	5	gal.
Perfume		

Dissolve menthol in alcohol. Add Boric Acid, perfume, and glycerine. Stir thoroughly until everything is dissolved. Add water. Filter. This preparation may be colored by adding enough color to give shade desired.

Sun Burn or After-Shave Lotion

1. Emulsone B	50	gm.
2. Boric Acid	50	gm.
3. Isohol	100	gm.
4. Phenol	1	dram
5. Menthol	1	dram
6. Oil of Rose	1	dram
7. Glycopon AAA	400	gm.
8. Water	7	pt.
9. Titanium Dioxide	2	oz.

Rub No. 1 and No. 2 together with No. 3, add and mix in thoroughly Nos. 4, 5, 6 and 7. Mix Nos. 8 and 9 and stir into previous mixture rapidly for 4 minutes only. Strain through cheesecloth and bottle. This gives a thick soothing cream which is very popular.

Milky Powder Base or Lotion

1. Glycosterin	10 lb.
2. Water	300 lb.
3. Perfume	to suit

Heat 1 and 2 until melted. Stir while cooling, adding perfume at 105° F. By decreasing the amount of water more viscous products are obtained. By reducing the water to 100 lb. a paste cream is formed. The addition of Titanium Dioxide to the above forms a liquid powder or "night-white."

Almond Lotion

1. Mineral Oil	35 lb.
2. White Wax	2 lb.
3. Trihydroxyethylamine Stearate (Special)	8 lb.
4. Perfume (Almond)	1 lb.
5. Water	50 lb.

Heat Nos. 1, 2 and 3 together to 140° F. and stir until homogeneous. Heat No. 5 to 140° F. and run in slowly to the above mixture, stirring thoroughly. When the temperature has dropped to 105° F. add the perfume drop by drop, stirring until completely absorbed. Continue stirring until cool and package.

Honey and Almond Type Lotion

1. Glycosterin	8 lb.
2. Glycopon S	15 lb.
3. Glycerin	36 lb.
4. Honey	4 lb.
5. Water	240 lb.
6. Almond Perfume to suit.	

Heat 1, 2 and 3 to 140° F. and then add slowly with stirring 4 and 5 heated to same temperature. Finally add 6 and stir until cold.

Anesthetic Shaving Lotion

Boric Acid	160 gr.
Menthol	8 gr.
Benzocaine	6 gr.
Alcohol	6 oz.
Water	to 1 pt.

Dissolve the menthol and benzocaine in the alcohol and add gradually to the water in which the acid has been dissolved.

Lotion, Anti-Sunburn

Quinine acid sulphate is used in proportion of 4 parts, dissolved in 64 parts of water which also contains 1 part of citric acid and 12 parts of 95% alcohol. This solution is added to mixture of 4 parts of finest, pulverized gum tragacanth and 5 parts of glycerin. Solution is added to gum mixture in small portions with constant agitation. Preparation is easily made and is highly effective. It can be perfumed to taste.

Astringent Lotion

Water	24	oz.
Glycerine	½	oz.
Alum	1	oz.
Isohol	4	oz.
Lavender Oil	1	dram
Zinc Phenol Sulfonate	¼	oz.

Dissolve the Lavender Oil in the Isohol and stir into the water containing the other ingredients.

Artificial Sun Burn Liquids

A. Powd. Cudbear	20 lb.
Powd. Henna	4 lb.
Peanut or Almond Oil	32 lb.

Macerate at 120° F. for 3 hours and filter.

B.	Quinine Sulfate	2 lb.
	Witch Hazel	5 lb.
	Lanolin	10 lb.
	Peanut Oil	92 lb.
C.	Peanut Oil	60 lb.
	Olive Oil	35 lb.
	Bergamot Oil	1 lb.
	Laurel Berry Oil	3 lb.
	Chlorophyll	1 lb.

Formulae B and C above require exposure of skin to sun.

Astringent Lotion (Mild)

Alcohol	3½ gal.
Glycerin	4 pt.
Orange Flower Water	20 gal.
Zinc Phenol Sulfonate	1 lb.
Color	to suit
Perfume	to suit

Astringent Lotion

Witch Hazel Extract	5 gal.
Zinc Phenol Sulfonate	8 oz.
Color and Perfume	to suit

Astringent Lotion Cleanser

Alcohol	5 gal.
Glycopon S	4 lb.
Water	5 gal.
Phenol	2 oz.
Perfume	5 oz.

Color to suit.

Astringent Lotion

Alum	1 oz.
Pot. Carbonate	0.25 oz.
Glycerin	0.50 oz.
Rose Water	10.00 oz.

Water to make 1½ pints. Some of this water can be replaced by witch hazel.

Face Lotion

Triethanolamine	2.0 gm.
Calcium Carbonate	1.0 gm.
Aq. Hamamelis	65.0 gm.
Aq. Rosae	26.0 gm.
Triethanolamine	4.0 gm.
Borax	2.0 gm.
Petroleum Jelly	4.0 gm.
Alcohol	25.0 gm.
Distilled Water	65.0 gm.

Acne Lotion

Triethanolamine	10.0 gm.
Stearin	22.0 gm.
Petroleum Jelly	3.0 gm.

Emollient Cosmetic Wash

Triethanolamine	10.0 gm.
Stearin	15.0 gm.
Paraffin Oil	10.0 gm.
Distilled Water	65.0 gm.

Face Lotion

Triethanolamine	0.5 cc.
Glycerine 28° Bé.	4.0 cc.
Alcohol 95%	33.0 cc.
Water	62.0 cc.
Perfume	0.5 cc.

Hand Lotion

Macerate 3 oz. of Quince Seed in 2 quarts of cold water for 24 hours. Strain through linen cloth with force and add 1 quart of water to the strained mucilage. Mix: Bay Rum, 16 oz; Glycerin, 8 oz.; Orange Flower Water, 12 oz.; Alcohol, 26 oz. and add to the mucilage, followed by sufficient water to make 1 gal. of finished product.

Hand Lotion

Boric Acid	1 dram
Glycerine	6 drams

Dissolve by heat and mix with

Lanolin	6 drams
Petrolatum	1 oz.

The borated glycerine should be cooled before mixing. Add any perfume desired.

* Insect and Poison Plant Lotions

A.	Cyclo Hexylamine	25
	Alcohol	75
B.	Linoleic Acid	2
	Triethanolamine	1
C.	Stearic Acid	1
	Triethanolamine	2
	Zinc Oxide	1
	Water or Alcohol	to suit

Lemon Juice Lotion

Pectin	2.5
Lemon Juice	9.5
Water	88
Moldex	0.15

Skin Lotion

Zinc Phenolsulfonate	30 gr.
Alcohol	4 dr.
Glycerine	2 dr.
Tinct. of Cochineal	1 dr.
Orange Flower Water	1½ oz.
Rose Water to make	6 oz.

Lotion for Oily Skins

Boric Acid	1 dr.
Alcohol	0.5 oz.
Rose Water	5.5 oz.

Liquid White (for Skin)

Lotion for hand and arms contains 2,500 parts witch hazel extract, 5,000 parts rose water, 1,000 parts alcohol, 1,800 parts glycerin, 100 parts tallow, 100 parts magnesium carbonate, 50 parts magnesium stearate and 1,000 parts antipyrine. First, antipyrine is dissolved in witch hazel extract and rose water. Then glycerin is added. Perfume used is allowed to be absorbed by magnesium carbonate, magnesium stearate and tallow. Then alcohol is added. This suspension is strongly shaken for two days. Milk is filtered through coarse filter paper. The two preparations are united with vigorous stirring and decanted. This preparation is applied with cotton. Skin is rubbed and preparation is allowed to dry. Skin remains white the entire evening. Advantage of this preparation over ordinary liquid powder is that a dull white effect is obtained, lasting 4 to 6 hours.

Smooth-Skin Balm

The formula given above for Sun Burn with the exception that the Phenol is replaced by 1 dram Bismuth Oxychloride.

Skin Milks

Milky preparations for use on skin can be made with lanolin, cucumber milk and almond milk. In first case 50 parts lanolin are mixed with 3 parts medicinal soap, 20 parts glycerin, 300 parts rose water, 5 parts tincture of benzoin, 10 parts perfume bouquet and 612 parts water. In second case 30 parts lanolin are melted on water bath and 200 parts warm rose water, containing 10 parts pure potash soap and 20 parts glycerin in solution, are gradually added. Then mixture of 10 parts perfume composition and 30 parts tincture of benzoin are added, and mixture is removed from water bath and mixed with 700 parts warmed, freshly percolated cucumber juice. Mixture is agitated until it cools off. In third case 70 parts shelled almonds are crushed with addition of sufficient rose water to give stiff paste. Then 20 parts tincture of benzoin, 2 parts benzaldehyde and one part rose oil are mixed and added to paste along with 7 parts borax and 50 parts glycerin in sufficient rose water to give total of 1,000 parts. Mixture is allowed to stand several days and then filtered through hair sieve.

Sunburn Preparations

1.	Subnitrate of Bismuth	1½ dr.
	Powdered French Chalk	30 dr.
	Glycerine	2 dr.
	Rose Water	1½ oz.

Mix the powders, and rub down carefully with the glycerine; then add the rose water. Shake the bottle before use.

2.	Glycerine Cream	2 dr.
	Jordan Almonds	4 dr.
	Rose	5 oz.
	Essential Oil of Almonds	3 drops

Blanch the almonds, and then dry and beat them up into a perfectly smooth paste; then mix in the glycerine cream and essential oil. Gradually add the rose water, stirring well after each addition; then strain through muslin.

Swedish Face Tonic (After Shave Lotion)

1. Zinc Phenolsulfonate	½ oz.
2. Witch Hazel	15 oz.
3. Isohol	10 oz.
4. Glycerine	1 oz.
5. Balsam Peru	¼ oz.
6. Lavender Oil	10 gm.

Dissolve Nos. 1 and 2 and then dissolve Nos. 4, 5 and 6 in No. 3. Mix both solutions and stir thoroughly. Allow to stand overnight and filter.

Sunburn Liniment

Formula:

Water White Steam-distilled Pine Oil	75%
Medicinal Olive Oil	25%

The finished product will be almost water white and is an effective treatment for sunburn. The product is applied by rubbing directly on the sunburned surface of the skin.

Mascara

Trihydroxyethylamine Stearate Special	40 lb.
Carnauba Wax	10 lb.
Carbon Black	30–40 lb.

Melt with stirring and cast or extrude in sticks.

Liquid Mascara

Tinc. Benzoin (25%)
Black Dye (Oil Soluble)

Nail Preparations

Nail bleach consists of 3% borax, 7% glycerin (28° Bé.), 90% perfume water, 2.4% preservative. Bleach of greater potency is made with 65% hydrogen peroxide (3%), 34% distilled water, 1% alcoholic solution of ammonia, 0.5% terpeneless pineneedle oil. Liquor for removing nicotine stains contains 90% hydrogen peroxide (3%), 10% ammonia solution (density 0.96), or bisulfite liquor or sulfur dioxide may be used. Polishing powder contains 40% pumice powder, 15% talc and 45% stannous oxide, or 65% titanium dioxide, 10% talc and 25% pulverized pumice. Nail enamel consists of 7% white carnauba wax, 7% Japan wax, 2.5% spermaceti, 80.5% white petrolatum, 0.25% turpentine, 0.5% acetic acid (80° Bé.), one per cent ethyl alcohol (96 to 98%), 0.25% alcanin and one per cent perfume. Nail paste contains 99% white petrolatum and 0.5 to one per cent of non-poisonous, fat-soluble, scarlet red, or 15% white beeswax, 10% white ceresin, 30% sweet oil of almonds, 35% tartaric acid, 4% citric acid and 6% alum. Liquid cream for after-treating nails contains one per cent white beeswax, 4% glyceryl monostearate, 10% sweet oil of almonds or apricot kernel oil, 5% white petrolatum, 80% distilled water and one per cent preservative.

Nail White

Zinc White Sifted	5 grm.
Chloroform	20 grm.
Paraffin	2 grm.
Oil of Neroli	15 drops

Dissolve the paraffin in the chloroform and add the other ingredients with constant agitation.

---

* Nicotine Stain, Bleach for

A compn. especially suited for removing nicotine and other stains from the hands or delicate fabrics consists of an aq. soln. contg. alkali hypochlorite or hypobromite, the available Cl or Br being 0.5–15%, free alkali less than 8% and the former being 1–3 times the latter.

Nose Shine Preventer

Corn Starch	1 lb.
Glycerin	2½ lb.

Rub together.

Water	2 pt.
Turkey Red Oil	1 pt.
Eosin (0.1% solution)	7 oz.

Heat to 85° C. and add to above.

Zinc Oxide	2½ lb.
Zinc Stearate	1 lb.
Clay (Colloidal)	1½ lb.
Sienna (Raw)	1 oz.

Rub together at 30° C. and mix in.

Oil Red Rose	¼ oz.
Oil Lilac Blossoms	¼ oz.

Muscle Oil

Castor Oil Odorless	10 gal.
Alcohol	5 gal.
Perfume Oil	5 oz.

Solidified Perfume (Oils)

Trihydroxyethyl Amine Linoleate	1
Orange or other oil	1
Water	1

Add in above order stirring well.

Sun Tan Oil

The basis of all such bronzing preparations is generally a vegetable oil, preferably arachis oil (peanut oil), olive oil, or sesame oil. Arachis oil in particular is said to have a bronzing effect, but in nearly all cases it is accompanied by a special dye, such as the one indicated below.

The following formula may be used as a basis for experiments, and is said to have a bronzing effect as a result of direct application:

Arachis Oil	60 gm.
Bergamot Oil	1 gm.
Olive Oil	38 gm.
Waxoline Brown (Dye)	1 gm.

Sun Tan Oil

Cherry Kernel Oil	100
Green Color (Oil Soluble)	to suit
Moldex	0.1

Sun Tan Oil

Peanut Oil	98
Quinine Oleate	2

Perfume and color to suit.

Sun Tan Oil

Mineral or Olive Oil	95–98
Quinine Ricinoleate	5– 2
Oil Soluble Red or Orange	to suit

Beauty Pack

Tragacanth	25
Alcohol	40
Glycopon S	40
Calamine	80
Zinc Oxide	30

Zinc Stearate	50
Glycerin	20
Lime Water	1000

Dissolve the tragacanth in the alcohol and carbitol. Then add to the lime water. Rub up zinc stearate, zinc oxide and calamine with glycerin. Add tragacanth, alcohol, glycopon S, lime water mixture to calamine, zinc oxide, zinc stearate and glycerin mixture.

Soap Perfume, Tuberose

Cananga Oil	200 grm.
Phenylpropyl Alcohol	200 grm.
Benzyl Acetate	100 grm.
Amyl Salicylate	100 grm.
Phenyl Ethyl Alcohol	100 grm.
Petitgrain Oil (Paraguay)	60 grm.
Linalol	40 grm.
Ionine Beta	50 grm.
Heliotropine	50 grm.
Musk Xylol	40 grm.
Benzoin Resin	60 grm.
	1,000

---

* Water Soluble Perfumes (Jellified Perfumes)

1. Glycopon 297	1 lb.
2. Perfume	1 lb.
3. Water	½–1 lb.

Mix Nos. 1 and 2 together until uniform. Add No. 3 slowly with stirring until a jelly is formed. The water must be added VERY slowly, stirring thoroughly, but as soon as a turbidity appears no more water can be added. These jelly perfumes disperse in water to give a milky solution when concentrated and a slightly turbid solution when highly diluted. By incorporating medicinal ingredients, ointments, salves, etc., are made which are not sticky and wash off readily with water.

Extract of Cyclamen

Cyclamen Aldehyde	5 gm.
Hydroxycitronellal, very pure	25 gm.
Benzyl Ethyl Carbinol	10 gm.
Terpineol, very pure, middle distillate	5 gm.
Methyl Ionone	5 gm.
Citronellol, purified	10 gm.
Benzyl Acetate	2 gm.
Citral, Water-white, very pure	0.50 gm.
Alpha Ionone, Water-white, extra fine	10 gm.
Phenyl Ethyl Alcohol	10 gm.
Rhodinol Ex Geranium	10 gm.
Bergamot Oil	5 gm.
Cinnamic Alcohol	5 gm.
Jasmin Liquid Absolute	2 gm.
Grasse Rose Oil	0.50 gm.
Heliotropin, Crystallized	2 gm.
Infusion of Florentine Orris, 20 per cent	100 gm.
90 Per Cent Alcohol to produce 1 litre.	

---

* Plastic Cosmetic

Gelatin	100 lb.
Water	350 lb.

Allow to swell and stir in

Ethylene Glycol	52 lb.
Zinc Oxide	85 lb.
Castor Oil	50 lb.

to make a smooth plastic mass.

Face Powders

Base I—Medium Weight

Talc	50
Chalk Pptd.	15
Kaolin Bolted	20
Zinc Oxide	15
Zinc Stearate	5
Perfume Oil	12 oz.

Base II—Rice

Talc	45
Rice Starch	20
Zinc Oxide	15
Kaolin	10
Zinc Stearate	10
Perfume Oil	8 oz.

Base III—Light

Talc	60
Chalk Pptd. Light	15
Zinc Oxide	10
Zinc Stearate	10
Kaolin	5
Perfume Oil	10 oz.

Base IV—Heavy

Talc	45
Kaolin	30
Zinc Oxide	10
Titanium Oxide	10
Zinc Stearate	5
Perfume Oil	10 oz.

Coloring.

The raw colors as bought are mixed with talc in the ratio

1 Color
9 Talc

and are either ball milled or screened through fifty mesh wire screen and then bolted through a 120 mesh silk screen. The talc used is figured as part of the formula. These colors are then known as bases.

Geranium Lake Base
Burnt Sienna Base
Persian Orange Base

Yellow Ochre Base
Burnt Amber Base
Purple Lake Base or Violet Lake Base.

Approximate coloring for powders 100 lb. Base.

Rachel or Cream

Yellow Ochre Base	5 lb.

Peach

Persian Orange Base	5 lb.

Brunette

Burnt Sienna Base	4 lb.
Yellow Ochre Base	4 lb.

Flesh

Yellow Base	2 lb.
Geranium Base	1 lb.

Dark Rachel

Yellow Ochre	7 lb.
Burnt Sienna Base	3 lb.
Geranium Base	1 lb.

Suntan

Burnt Sienna Base	20 lb.

Violet or Lavender shades are secured with a Violet Lake Base. Greens with a Green Lake Base. Dullness in shades is secured with Burnt Amber Base.

Procedure: All materials are brushed through a thirty mesh screen into mixer and color added: Mixed for an hour or until a good distribution is effected. The perfume is rubbed into 2 pounds of Magnesium Carbonate and screened to break particles. The perfume and Magnesium Carbonate is then added to the balance of the ingredients, mixed again and all sifted through a 100 to 150 mesh silk screen.

Neroli Perfume Base

Neroli Petale	25
French Pettigrain	35
Nerol	10
Rhodinol	5
Linalool	5
Linalyl Acetate	3
Orange Flower Absolute	5
Methyl Anthranilate	5
Aldehyde ClO (10%)	2
Phenyl Ethyl Alcohol	5

* Carnation Perfume Base

* *Note.*—Do not use in creams or lipsticks as it is apt to irritate.

Isoeugenol	30
Eugenol	30
Rhodinol	10
Phenyl Ethyl Alcohol	10
Vanillin	3
Alpha Ionone	5
Synthetic Rose	7
Benzyl Salicylate	5

Medicated Perfume

Lavender Oil (42% Ester)	30
Camphor	10
Menthol	5
Thymol	5
Rosemary Oil	25
Methyl Salicylate	15
Benzaldehyde	5
Oil Bay Terpeneless	5

Dandruff Remover

Mineral Oil	5 gal.
Turkey Brown Oil	5 gal.
Medicated Perfume	1 lb.

Sweet Pea Perfume Base

Phenyl Ethyl Phenyl Acetate	5
Dimethyl Acetophenone	3
Ethyl Vanillin	1
Benzyl Acetate	5
Musk Ketone	5
Ylang Manila	5
Benzyl Salicylate	10
Synthetic Rose	2
Cinnamyl Alcohol	20
Hydroxycitronellal	20
Linalool	10
Hydrotropic Aldehyde	1
Neroli Petale	5
Terpineol	8

Face Powder

Osmo Kaolin	45
Zinc Oxide	10
Rice Starch	15
Magnesium Carbonate	7
Talc	18
Magnesium Stearate	5
Perfume (Compound)	2
Heliotropine	1

Sift through 120 mesh.

Face Powder (Heavy for Night Wear)

Osmo Kaolin	30
Titanium Oxide	30
Talc	23
Magnesium Carbonate	10
Magnesium Stearate	7

Perfume	3
Heliotropine	2

Sift through 120 mesh.

Bath Powder

Powdered Borax	1 lb.
Ammonia Muriat	2 oz.
Synthetic Violet	2 dr.
Synthetic Heliotrope	2 dr.

Liquid Powder

Zinc Oxide	3 lb.
Precipitated Chalk	3 lb.
Glycerine	1 pt.
Alcohol	4 pt.
Perfume	4 oz.
Water	4 gal.

Color
(See Face Powder)

Rachel—1 oz. Yellow Ochre Base
Tan—1 oz. Burnt Sienna Base
Flesh—1 oz. Geranium Base
Peach—½ oz. Persian Orange Base

Bath Powder

Powdered Borax	1 lb.
Ammonia Chloride	2 oz.
Synthetic Violet	2 dr.
Synthetic Heliotrope	2 dr.

Talcum Powder

Venetian Talcum Powder	700 gm.
Osmo-kaolin or Colloidal Clay	200 gm.
Magnesium Stearate	100 gm.
Benzyl Ethyl Carbinol	3 gm.
Alpha Ionone	2 gm.
Cyclamen Aldehyde or Cyclosal	1 gm.
Ethyl Vanillin Crystallized	0.5 gm.
Heliotropin Crystallized	5 gm.
Titanium or Zinc Oxide	25 gm.

Toilet Powder

Talcum	8 parts by weight
Boric Acid	1 part by weight
Starch	1 part by weight

Facial and Body Reducer

Camphor	5 oz.
Epsom Salt Powdered	10 oz.
Isohol	85 oz.
Tincture Iodine	1 c.c.
Water	5 oz.
Perfume	2 oz.

Stir quickly while bottling as this preparation separates quickly. Bottles should be labeled "Shake before using."

Rouge Compacts

Carmine	1 oz.
Talc	21 oz.
Gum Acacia	1¾ oz.
Ammonia	a few drops

Mix first three items in a mortar, add a few drops of ammonia and some water. Pound into a fine mass adding more water in small portions to form a stiff paste. Fill into molds immediately. The amount of carmine can be increased to obtain different shades.

Brushless Shaving Creams
Soapless Type

Glyceryl Monostearate	6.5%
Stearic Acid	6.5%
Mineral Oil	4.0%
Peanut Oil	4.0%
Glycerin	10.0%
Water	69.0%

Alkaline Type (Pearly Appearance)

A.	Cocoanut Oil	20.0 parts
	Suet	15.0 parts
B.	Caustic Potash (90%)	31.0 parts
	Caustic Soda (90%)	4.0 parts
	Borax	2.5 parts
	Water	142.5 parts
C.	Water	140.0 parts
D.	Stearic Acid	145.0 parts
E.	Glyceryl Monostearate	40.0 parts
	Stearic Acid	80.0 parts
	Water	380.0 parts

In the case of the alkaline type, "A" must be saponified with "B". To this add "C" and then "D," which has already been melted. The whole mass should be stirred for a few minutes at a temperature of about 80° C., so as to be sure that no lumps will form. It should then be allowed to cool without stirring. After one to two days, the mass will take on a pearly appearance. Then an emulsion made with "E" should be added while both are cold. In order to make the emulsion "E" smooth, it is advisable to take ten parts of the combined mass resulting from "A," "B," "C" and "D," and add this to "E" while the latter is still hot. The pearly appearance will temporarily vanish but after two days will again appear.

Shaving Cream, Brushless

Stearic Acid	12
White Mineral Oil	12
Paraffin Wax	5
Soap Flakes	3
Water	72

Brushless Shaving Cream

Liquid Creams

Stearic Acid	200 g.
Triethanolamine	10 g.
Water	800 g.

Thicker Creams

Stearic Acid	200 g.
Triethanolamine	10 g.
Anhydrous Sodium Carbonate	10 g.
Water	800 g.

Brushless Shaving Cream

Stearic Acid Triple	75 lb.
Sesame Oil	70 lb.
Spermaceti	10 lb.
Strong Ammonia Solution	10 lb.
Hot Water	315 lb.
Glycerin	30 lb.
Perfume q.s.	

Procedure.—Melt waxes and fats. Boil water, add ammonia, and pour into melted fats with constant agitation. When completely saponified stir slowly until quite cold. Add perfume.

Brushless Shaving Cream

Stearic Acid	50 gm.
Cocoa Butter	9 gm.
Sodium Carbonate Monohydrated	10 gm.
Borax	20 gm.
Glycerin	40 c.c.
Alcohol	32 c.c.
Water	400 c.c.
Perfume q.s.	

Procedure.—Dissolve the sodium carbonate, borax, and glycerin in hot water. Melt the fats and waxes and add the alkali solution. Stir briskly until effervescence ceases and a smooth white soap is formed. Stir slowly until cold; then add the perfume mixed with alcohol.

Liquid Rouge

Erythrosine	0.25 gm.
Eosin-bluish	0.40 gm.
Glycerin	80.00 cc.
Alcohol (grain, 190 proof)	560.00 cc.
Simple Syrup	100.00 cc.
Heliotrope Bouquet q.s.	
Distilled Water q.s.	1000.00 cc.

Dissolve dyes in glycerin-alcohol mixture. Add simple syrup and heliotrope; then add water.

Paste Rouge

By decreasing the amount of waxes in lipstick formula, an excellent paste rouge is made.

Paste Rouge

Beeswax	8 lb.
Stearoricinol	28 lb.
Mineral Oil	4 lb.
Lanolin Anhydrous	2 lb.
Petrolatum	2 lb.
Bromo Acid	1 lb.
Lake Colors	5 lb.
Perfume Oil	1 lb.

Perfumed Artificial Sea Salt

Potassium Chloride	1 oz.
Magnesium Chloride	6 oz.
Calcium Sulphate	1 oz.
Sodium Chloride	2 dr.
Coumarin	1 dr.
Alcohol	6 dr.

* Lathering Shaving Cream

1.	Mineral Oil	2 oz.
	Tallow Edible	4½ oz.
	Stearic Acid	10 oz.
	Coch. Cocoanut Oil	5 oz.
	Glyco Wax A	½ oz.
2.	Caustic Potash Lye 36° Bé.	17 oz.
	Caustic Soda Lye 36° Bé.	1½ oz.
3.	Water	23 oz.
	Boric Acid	1¼ oz.
	Glycopon AAA	2 oz.
4.	Stearic Acid (C.G.)	10 oz.
5.	Perfume	⅓ oz.

The above formula gives a profuse lathering cream equal to the best creams on the market. It gives a thick, rich, non-drying lather of the small bubble type, which softens the beard quickly and contains no uncombined alkali, making it non-irritating to the skin. This cream is pearly and the pearliness increases with age.

Heat (1) until melted and keep melted. Heat (3) until dissolved; then cool. Now add (2) to (3) and stir; then add this to (1) slowly with good stirring, keeping batch hot on a steam-bath; continue stirring until homogeneous. Keep hot and allow to stand covered for 30 minutes. Stir for 5 minutes. Melt (4) in a separate pot and run it into the above batch with good stirring; allow to stand covered for 30 minutes; take off steam-bath and stir until thick; add (5) when almost cold; stir thoroughly. Allow to stand covered for week or ten days, stirring each day for five minutes.

Shaving Cream (Lathering)

Melted mutton tallow (250 g.) and 50 g. ox tallow are saponified with 178 cc. 50° Bé. potassium hydroxide solution and boiled to sticky mass. Cool and mix with boiled solution of 150 g. stearin, 40 g. anhydrous lanolin, 50 g. potassium carbonate and 1200 g. water. Make up to 3000 g. with water.

Lather Shaving Cream

Cocoanut Oil	18 lb.
Stearic Acid	73 lb.
Caustic Potash Lye 39° Bé.	54 lb.
Glycerine	33 lb.
Water	27 lb.

Put oil and glycerine in kettle and heat to 120° F. and stir thoroughly. Add slowly 35 lb. lye and continue to stir until it thickens. Add balance of lye mixed with the water slowly with constant stirring until smooth. Allow to stand in kettle 24 hours, then add perfume. Fill into tubes.

Lathering Shaving Cream

1. Stearic Acid	30.0%
2. Cocoanut Oil	3.3%
3. Caustic Potash, 50° Bé.	18.8%
4. Caustic Soda, 20° Tw.	1.6%
5. Glycerin	5.0%
6. Water	41.3%
Perfume	to suit

Latherless Shaving Creams

Cream No. 1

Stearic Acid	50 lb.
Lanolin (anhydrous)	9 lb.
Carbitol	3 lb.
Triethanolamine	1.5 lb.
Borax	1.7 lb.
Water	135 lb.

Cream No. 2

Stearic Acid	40 lb.
Lanolin (anhydrous)	7 lb.
Mineral Oil (white)	18 lb.
Carbitol	3 lb.
Triethanolamine	3.3 lb.
Borax	3.7 lb.
Water	125 lb.

Preparation

Melt the stearic acid, which should be the purest grade obtainable, either alone or with the mineral oil depending upon which formula is followed. Add the lanolin and bring the temperature to about 70° C. Heat the water, Triethanolamine and borax in a separate container and when at the boiling point, add the acid solution. Stir vigorously until a smooth emulsion is obtained and then add the perfume dissolved in the Carbitol. During the further cooling of the cream, stir gently but continuously taking care to avoid rapid stirring, as this tends to aerate the cream.

Properties

Cream No. 1 is a white, pearly product somewhat like a vanishing cream and is preferable for oily skins. Cream No. 2 is a smooth white cream of greater body than the other, and is preferred for use on dry skins. Both creams are readily applied to give a smooth coating on the face, have a soothing after-effect and are readily washable. The consistency of these creams can be varied by altering the proportion of water, and other changes can be made along the lines indicated by the difference in the two formulae. A cream of good consistency can be made by combining the two formulae given above.

* Latherless Shaving Cream

Latherless creams of a highly pearly appearance are made by using the formula given above for vanishing cream. A little menthol may be incorporated to produce a cooling effect on the skin.

These shaving creams are particularly interesting because they do not contain caustic soda, potash or ammonia and, therefore, will not cause the most tender skin to smart or redden. They penetrate hairs and soften the skin, producing a remarkably clean and smooth shave. Since they are really vanishing creams, they not only clean the skin but do away with the necessity of after-shaving lotions and creams. An antiseptic shaving cream of this type is made by dissolving a small amount of any non-acid, non-irritating antiseptic in the batch.

* Latherless (Brushless) Shaving Cream
(Non-Irritating)

1. Mineral Oil	10 lb.
2. Glycosterin	10 lb.
3 Water	50 lb.

Procedure

Heat (1) and (2) to 150° F. and stir (3) into it heated to 150° F. slowly. A little perfume and menthol (if desired) is stirred in at 105° F. and stirring is continued until cold.

Shaving Cream, Latherless

Glycosterin	10 lb.
Ethylene Glycol	10 lb.
Mineral Oil White	8 lb.
Lanoline	2 lb.
Stearic Acid	34 lb.
Glycerin	2 lb.

Water	134 lb.
Menthol	0.2 lb.

Shaving Sticks

Stearic Acid	40
Cocoanut-oil	10
Caustic Potash 38° Bé.	23
Caustic Soda 38° Bé.	6
Glycosterin	4

Fats must be saponified at 70° Celsius. The reaction is rather strong, therefore the lye must be added more quickly than usual; to the saponified mass add Glycosterin and leave to the self-induced heating process for three hours, but stir through hourly. Put into forms or pass through a drying machine. A soap put into forms takes very long to harden. Good drying is necessary. The freshly machined sticks are too soft for cutting and must be left to harden several hours. After cutting wrap in tinfoil for preserving their soft and pliant quality.

Shaving Cream for Tubes

Stearic Acid	15
Peanut Oil	5
Cocoanut-oil (Cochin)	7
Caustic Potash Lye 40° Bé.	14
Water	16
Glycosterin	2

Stir as usual, add to the melted fats at 70° Celsius the mixed potash lye and water till sufficiently thick, leave till fully saponified and cooled. The melted Glycosterin and perfume is then stirred into the soft mass.

Shaving Cream

Lard	100
Olive-oil Sesame-oil	80
Cocoanut-oil (Cochin)	70
Glycosterin	5
Caustic Potash 40° Bé.	125
Solution of Potash 20° Bé.	15

Melt fats and Glycosterin, saponify with caustic potash lye; add the potash solution, perfume and pass through a 3-roll-mill. By addition of a little alcohol during the rolling the cream will get a silky shine.

After-Shaving Lotion

The following is a formula for a menthol after-shaving lotion:

Tragacanth (pdr.)	8 oz. (5 oz.)
Formalin	2 dr. (4 (dr.)
Menthol	2 oz. (1 oz.)
Cologne Oil	2½ oz.
Red Coloring	a sufficiency
Industrial Spirit	3 pt. (2 pt.)
Water	5 gal.

The alternative quantities are for a cheaper preparation

For Mosquito Bites

The following application is suggested as a means of preventing insect bites:

Cedar Oil	2 dr.
Citronella Oil	4 dr.
Spirits of Camphor	ad 1 oz.

This should be smeared on the skin of the exposed parts as often as is necessary. Cod-liver oil used in the same way has been highly recommended, and in combination with quinine it makes an effective "sunburn and midge cream," a formula being as follows:

Quinine Acid Hydrochloride	5 parts
Cod-liver Oil	20 parts
Anhydrous Wool Fat	75 parts
Oil of Lavender (or geranium)	a sufficiency

The irritation of a mosquito or fly bite may be allayed by gently rubbing the puncture with a moist cake of soap, or by applying a 1 per cent alcoholic solution of menthol, or 1–20 aqueous carbolic lotion. Hydrogen peroxide or weak ammonia solution dabbed on is also useful. If the bite shows signs of sepsis, constantly renewed hot boric fomentations should be applied, or if a limb is implicated, hot saline arm or leg baths.

Styptic

An excellent styptic powder results from the mixture of 50% powdered talc and 50% phthalyl peroxide. The latter often contains up to 40% of its weight as phthalic acid; this is beneficial and acts as a stabilizer. The mixture is antiseptic.

Styptic Pencils

The following are the methods adopted for the manufacture of alum pencils: White: Liquefy 100 gm. of potassium alum crystals by the aid of heat. Remove any scum and avoid overheating, particularly of the sides of the vessel in which liquefaction is being carried out. The molten liquid should be perfectly clear. Triturate a mixture of French chalk in fine powder, 5 gm., glycerin 5 gm. to a paste, incorporate with the liquefied alum and pour into suitable molds. A white appearance can be imparted to the resulting pencils by the addition of more

French chalk. Clear: Carefully liquefy potassium alum crystals so as to avoid loss of water of crystallization, adding a small amount of glycerin and water (about 5 per cent) until a clear liquid is obtained. This is poured, whilst hot, into suitable moulds, previously smeared with fat. The solidified pencils are rendered smooth by rubbing them with a moistened piece of cloth.

Styptic Pencils

Liquefy 100 grams of potassium alum crystals by the aid of heat. Remove any scum and avoid overheating particularly of the sides of the vessel in which liquefaction is being carried out. The molten liquid should be perfectly clear. Triturate a mixture of French chalk in fine powder, 5 grams, glycerin 5 grams, to a paste, incorporate with the liquefied alum and pour into suitable moulds. A whiter appearance can be imparted to the resulting pencils by the addition of more French chalk.

Witch Hazel Jelly

Boric Acid	1 oz.
Tragacanth	2 oz.
Witch Hazel	1 gal.

Wrinkle Remover

Distilled Extract of Witch Hazel	500 parts
Boric Acid	20 parts
Menthol	1 part
Glycerin	50 parts
Perfume (with a spirit basis)	100 parts
Elderflower water	329 parts
	1,000 parts

Dissolve the menthol in the perfume and add to the mixed liquids. Make up to volume as directed.

Skin Whitener
(Night White or Powder Base)

1. Glycosterin	10 lb.
2. Water	60 lb.
3. Titanium Dioxide	3 lb.

Heat 1 and 2 to 150° F. and stir until cold. Allow to stand overnight (very important). Stir the next morning and make sure that it is COLD. Then stir in Titanium Dioxide until uniform. In place of titanium, talc or zinc stearate may be used.

Removing Tattoo Marks

I.

Pepsin and papain have been proposed as applications to remove the epidermis. A glycerol solution of either is tattooed into the skin over the disfigured part; and it is said that the operation has proved successful. Papain, 5; water, 25; glycerol, 75; diluted hydrochloric acid, 1. Rub the papain with the water and hydrochloric acid, allow the mixture to stand for an hour, add the glycerin, let it stand for three hours and filter.

II.

Apply a highly concentrated tannin solution to the tattooed places and treat them with a tattooing needle as the tattooer does. Next vigorously rub the places with a lunar caustic stick and allow the silver nitrate to act for some time until the tattooed portions have turned entirely black. Then take off by dabbing. At first a silver tannate forms on the upper layers of the skin, which dyes the tattooing black; with slight symptoms of inflammation a scurf ensues, which comes off after fourteen or sixteen days leaving behind a reddish scar. The latter assumes the natural color of the skin after some time. The process is said to have good results.

Obviously such treatments are heroic and carry along with them the risk of permanent scarring. It is therefore a job for a trained dermatologist rather than for a layman.

Chypre Perfume Base for Face Powder

Coumarin	10
Santylyl Acetate	5
Musk Ketone	5
Musk Ambrette	2
Vetivertol Acetate	5
Patchouli	2
Isoeugenol	5
Methyl Ionone	5
Bergamot	25
Ylang Ylang Manila	10
Tolu Resin	5
Vanillin	2
Linalool	3
Mousse de Chene	7.5
Cinnamyl Alcohol	5.0
Labdanum Resin	3.5

Gardenia Perfume Base

Lilac Synthetic	20
Rose Synthetic	10
Lily Synthetic	30
Jasmin Synthetic	25

Phenyl Acetaldehyde (50%)	2
Methyl Naphthyl Ketone	6
Isoeugenol	2
Vanillin	2
Styralyl Acetate	3

Honeysuckle Perfume Base

Hydroxycitronellal	25
Alpha Ionone	10
Terpineol	5
Phenyl Ethyl Alcohol	6
Cinnamyl Alcohol	10
Vanillin	3
Jasmin Absolute	2
Mimosa Absolute	5
Neroli Absolute	1
Musk Ketone	2
Methyl Naphthyl Ketone	5
Linalool	5
Benzyl Acetate	5
Rhodinol	5
Cinnamyl Acetate	5
Heliotropin	5
Phenyl Acetaldehyde (50%)	1

Jasmine Perfume Base

Benzyl Acetate	50
Hydroxycitronellal	15
Cinnamyl Alcohol	10
Linalool	7
Ylang Ylang Manila	7
Para Cresyl Caprylate	2
Methyl Ionone	3
Benzyl Formate	1
Benzyl Propionate	3
Amyl Cinnamic Aldehyde	2

Lavender Perfume

French Lavender Oil	500
Spike Lavender Oil	100
Bergamot Oil	200
Geraniol	100
Sandalwood Oil	60
Rosemary Oil	80
Thyme Oil	20
Coumarin	30
Dimethyl-hydroquinone	10
Artificial Musk	3
Tincture of Civet	10
Mousse de Chêne	3
Labdanum Resin	3
Styrax Resin	3

Lilac Perfume Base

Terpineol	30
Hydroxycitronellal	15
Cinnamyl Alcohol	10
Rhodinol	10
Heliotropin	7
Rose Absolute	2
Jasmin Absolute	5
Phenyl Ethyl Alcohol	5
Anisic Aldehyde	7
Phenyl Acetaldehyde (50%)	5
Musk Xylene	3
Sandalwood Oil	1

Lily Perfume Base

Hydroxycitronellal	30
Terpineol	20
Methyl Ionone	5
Ylang Ylang	5
Rose Absolute	3
Jasmin Absolute	2
Heliotropine	5
Cyclamen Aldehyde	3
Phenyl Ethyl Alcohol	10
Vanillin	0.5
Methyl Phenyl Acetate	0.5
Nerol	6
Rhodinol	5
Linalool	5

Peach Blossom Odor
(for toilet creams)

Pure Peach Lactone	840 gm.
Amyl Acetate	25 gm.
Benzoic Aldehyde	10 gm.
Vanillin	90 gm.
Ethyl Valerianate	20 gm.
Ethyl Butyrate	25 gm.

Rose Perfume Base (Water Soluble)

Phenyl Ethyl Alcohol	70
Rhodinol	15
Phenyl Acetaldehyde	5
Methyl Phenyl Acetate	1
Vetivert Bourbon	2
Geranium Bourbon	2
Methyl Ionone	3
Aldehyde C10 (10%)	2

Sandalwood Perfume—I.

Sandalwood Oil	200
Cedarwood Oil	150
Patchouli Oil	15
Bergamot Oil	30
Eugenol	10
Vetiver Oil	20
Artificial Musk	5
Geranium Oil	30
Cassia Oil	5
Cananga Oil	5
Ext. of Mousse de Chêne	10
Styrax Resin	5
Coumarin	5
Dimethyl-hydroquinone	3
Tincture of Civet	20

Sandalwood Perfume—II

Sandalwood Oil	100
Cedarwood Oil	120

Geraniol	20
Terpineol	50
Hydroxy-citronellol	10
Artificial Musk	3
Styrax Resin	3

Violet Perfume

Ionone	400
Concrete Orris Oil	20
Cananga Oil	40
Methyl Heptin Carbonate	8
Sandalwood Oil	15
Benzyl Acetate	40
Artificial Otto of Rose	20
Bergamot Oil	20
Phenyl-ethyl Alcohol	10
Heliotropin	35
Cassie Extract	20
Styrax Resin	15
Artificial Musk	2
Ext. of Mousse de Chêne	5

Perfume for Windsor Soap (White)

Low Priced Perfume

Soap Chips	100 kilos
Caraway Oil	250 gm.
Anise Oil	100 gm.
Bergamot Oil	150 gm.

Perfume for Windsor Soap (Yellow)

Low Priced Perfume

Soap Chips	100 kilos
Caraway Oil	250 gm.
Cassia Oil	200 gm.
Clove Oil	50 gm.

Perfume for Almond Soap

Low Priced Perfume

Soap Chips	100 kilos
Bergamot Oil	150 gm.
Palmarosa Oil	75 gm.
Bitter Almond Oil	100 gm.
Mirbane Oil	75 gm.

Pompas Bouquet

Low Priced Perfume

Soap Chips	100 kilos
Cassia Oil	200 gm.
Clove Oil	100 gm.
Thyme Oil	100 gm.
Balsam Peru Tincture	100 gm.

Violet Perfume Bases, Synthetic

Constituents.	*Parma*		*Boise de Nice.*			*Classic.*	*Ordinary.*
	I.	*II.*	*III.*	*IV.*	*V.*	*VI.*	*VII.*
Ionone Alpha	260	400	500	350	350	300	150
Ionone Beta	140	—	—	—	—	—	250
Methylionone	200	—	—	250	250	—	—
Orris Concrete	—	50	—	25	—	—	—
Orris Resinoid	—	150	—	65	—	100	—
Cassie, Natural	—	20	—	—	10	—	—
Jasmin, Natural	—	15	—	25	—	20	—
Rose, Natural	—	10	—	—	—	10	—
Benzyl Acetate	50	25	100	40	100	30	100
Geraniol	—	—	100	25	—	—	—
Vetiverol	20	35	—	—	—	—	—
Musk Xylol	—	—	—	—	—	—	40
Musk Ketone	—	40	35	50	—	—	—
Methyl Heptin Carbonate	—	—	5	—	10	7.5	10
Methyl Octin Carbonate	5	5	—	—	—	—	—
Coumarin	—	35	—	—	30	—	—
Heliotropin	70	—	100	45	100	100	100
Vanillin	30	10	—	—	—	—	—
Phenylethyl Alcohol	100	60	—	75	—	140	150
Bergamot	—	50	—	—	—	125	50
Hydroxycitronellal	—	50	—	—	—	—	—
Violet Leaf Absolute	—	—	—	10	—	—	—
Methylnonyl Aldehyde	—	—	—	0.5	—	—	—
Linalol	75	—	—	25	—	—	150
Terpineol	—	—	—	—	85	—	—
Linalyl Acetate	50	—	40	—	50	—	—
Geranyl Acetate	—	—	—	—	20	—	—
Aldehyde C_{12}	—	—	—	—	5	—	—
Anisic aldehyde *ex* anethol	—	—	60	—	—	—	—

Toilet Soap Base

The following represent five standard and workable compositions of the stock used in making the soap base. The first mixture contains eighty per cent of fresh beef tallow, and twenty per cent of good grade coconut oil; the second, sixty-five per cent of beef tallow, fifteen per cent of lard and twenty per cent of coconut oil; the third, seventy per cent of bleached palm oil, fifteen per cent of sulphonated olive oil and fifteen per cent of coconut oil; the fourth, sixty-five to seventy per cent of beef tallow, ten to fifteen per cent of castor oil and twenty per cent of coconut oil; the fifth, sixty per cent of bleached palm oil, twenty per cent of beef tallow and twenty per cent of coconut oil.

In making soap bases of second quality good grades of fat refuse are used in large quantities and also palm kernel oil in the place of coconut oil. These raw materials can be converted into well-saponified soaps and of good keeping quality, but only when great care is paid to the details of the process. However, the soap base that is made in this manner cannot be perfumed satisfactorily.

The oldest and mostly used process for the manufacture of excellent soap bases is first to saponify the tallow, lard, palm oil, castor oil and the like and to salt-out the same once or several times. Then the coconut oil is added and the saponification continued and the soap salted out until a niger is obtained. This process has been improved by beginning the saponification of each batch of stock in a different kettle and after the batch has been completely saponified, the salted-out curd soap from a previous saponification is added. It is claimed that this method makes for technically complete saponification of the stock in a more easily and safely attained manner.

A third method of boiling the soap does not involve the addition of any salt. It has been used in various toilet soap works and has been found satisfactory over a period of years. The salting out of the curd from the previous boil as well as of the soap from the boil to which the curd soap has been added is accomplished with concentrated sodium hydroxide lye. The graining of the finished soap is also accomplished with dilute sodium hydroxide solution and not with salt water. The curd soap that is obtained after standing for thirty-six hours in the kettle is quite alkaline. However, the alkalinity of the soap disappears as the latter is dried. The result is that a product is finally obtained which can be readily milled into a perfectly neutral and stable toilet soap. This process has demonstrated its usefulness as it has been employed in practical operations for quite some years.

Half-Boil Process

A fourth process for the manufacture of soap base consists in complete saponification of the fatty mixture (neutral fats) only by the half-boil process. The soap is then comminuted to chips and these are dried in the usual manner as in all the soap making processes and thereafter milled. Toilet soaps that are manufactured by this process contain in excess of eight per cent glycerin. Hence it is evident that the soap is sufficiently plastic and easily millable. A long series of experiments has also proven that the soap is absolutely stable. Naturally a most important prerequisite of this soap making process is that the raw materials used must be absolutely pure and free from any odor as well as free from albumens. If the raw materials received into the plant are not of this quality, they must be purified by suitable means before being used in the kettles. Only when the temperature varies very markedly and when the humidity of the air is very high, close to 100 per cent, do soaps made in this manner become wet. On the other hand, soaps made by other processes of saponification as well as after-treatment become wet much more readily under considerably less severe conditions.

Some toilet soap manufacturers convert the soap base into toilet soap by the following process. The raw materials, consisting of tallow, lard and the like, are completely saponified in a large tank, provided with an agitating apparatus and situated close to the kettle. Saponification is carried out according to the emulsification-saponification process by the half-boil method using a small excess of lye. Then immediately after saponification the mass is added to the curd soap which has been subjected to several changes, the soap obtained from a previous boil. When the entire mixture has been saponified, then the soap is salted out, salt being used in making two changes. The soap is then finished in the usual manner. The emulsification and saponification of the stock, which is carried out in a single operation, gives a soap which is completely saponified. This process is therefore of considerable advantage.

Pearl Nail Enamel

High Viscosity Nitrocellulose	20 oz.
Low Viscosity Nitrocellulose	10 oz.
Cellosolve Acetate	1/4 pt.
Pale Dammer Gum	10 oz.
Butyl Acetate	1 qt.
Toluol	3 gal.
Ethyl Acetate	2 gal.
Pearl Essence	18 oz.
Dibutyl Phthalate	1 pt.

Lotion for Dry Dandruff

Tannic Acid	10 oz.
Chloral Hydrate	16 oz.
Witch Hazel	200 oz.
Castor Oil	5 oz.
Soya Bean Oil	50 oz.
Alcohol	800 oz.
Perfume to suit.	

Procedure: Dissolve the tannic and the chloral in the alcohol, add the witch hazel and the oils and mix thoroughly.

Lotion for Oily Dandruff

Zinc Sulphate	2 oz.
Phenol	1 oz.
Menthol	2 oz.
Glycerin	50 oz.
Water	120 oz.
Formalin	2 oz.
Alcohol	40 oz.
Perfume to suit.	

Procedure: Dissolve the zinc sulphate in some of the water. Dissolve the phenol and the menthol in the alcohol, add the glycerin, the formalin and the remainder of the water. Mix thoroughly and filter.

Other chemicals used in the manufacture of dandruff preparations include: crude oil, precipitated sulphur, oil of tar rectified, oil of camphor white, turkey red oil, oil of thyme, soya bean oil, thuja, cresol, lignol, sulphonated bitumen, lanolin, betanaphthol, croton oil, bismuth subcarbonate, mercuric salicylate, arsenic iodide.

Preparations for Scabies

Ointment

Potassium Sulphide	50 oz.
Water	250 oz.
Petrolatum	250 oz.
Lanolin	250 oz.
Titanium Dioxide	5 oz.
Mineral Oil	200 oz.
Perfume to suit.	

Procedure: Dissolve the potassium in the water. Take part of the petrolatum and mill in the titanium. Melt the rest of the petrolatum, the lanolin and the mineral oil and add the potassium solution. Then add the titanium mass. Mix thoroughly and mill again.

Lotion

Castor Oil	6 oz.
Oil Tar Rectified	10 oz.
Phenol	1 oz.
Formalin	1 oz.
Sesame Oil	160 oz.
Soft Soap	10 oz.
Alcohol	30 oz.
Perfume to suit.	

Procedure: Dissolve the soap in part of the alcohol using slight heat. Dissolve the formalin and the phenol in the rest of the alcohol. Mix the sesame, castor and tar oils, add the soap and then the formalin-phenol.

Other materials utilized in the preparation of ointments and lotions of this kind are: storax, creosote, ammoniated mercury, sulphonated bitumen, procaine hydrochloride, copper oleate, sublimed sulphur, balsam of Peru, titanium oxide, silver lactate, alcohol, olive oil, sesame oil, benzoated lard and a number of absorption bases.

Eczema Preparations

Ointments

Lanolin	200 oz.
Petrolatum	200 oz.
Beeswax	50 oz.
Phenol	5 oz.
Camphor	10 oz.
Oil Eucalyptus	50 oz.
Salicylic Acid	10 oz.
Perfume to suit.	
	525 oz.

Curling Liquid

Quince Seed	30 oz.
Water	10 gal.
Borax Powdered	20 oz.
Perfume Compound	4 oz.
Benzoic Acid	3 oz.
Alcohol	10 oz.

Procedure: Boil the water, add the quince seed and allow to stand overnight, stirring occasionally. Add the borax solution (made with part of the water). Filter. Add perfume and benzoic acid solution and mix thoroughly.

Extracting the quince seed hot increases the turbidity of the extract. If margin of profit is great enough it is better to extract the mucilage cold. As an additional precaution the quince seed should be cleaned by blowing. This wastes a little of the mucilage but it also removes clay and sand which the seed is apt to contain.

Curling Jelly

Gum Tragacanth	12 oz.
Alcohol	½ gal.
Water	3 gal.
Borax	8 gr.
Benzoic Acid	8 dr.
Perfume	3 dr.

Procedure: Put the tragacanth into a vessel, add the water and borax and allow to stand until dissolved, a period which will depend upon whether the tragacanth is powdered, in ribbons or lumps. Add alcohol to which perfume and benzoic has been added and mix thoroughly. Squeeze through muslin bag.

Hair Whitener

Aniline Blue	2 oz.
Distilled Water	15 gal.

Procedure: Dissolve blue in one half the water by allowing it to stand over night. Mix thoroughly add the rest of the water and filter. It is undesirable to run this preparation through a mechanical filter because the stain is almost impossible to remove. It is better to filter in five gallon bottles reserved for this purpose.

Liquid Brilliantine

Light Mineral Oil	99%
Perfume (Usually Flower Type)	1%

Procedure: Mix and filter.

Brilliantines are favorite sellers, the liquid being the best seller of the two. Although some chemists insist that brilliantines should be made from vegetable oils, the danger of rancidity in cases where the hair is not shampooed frequently is great and it seems advisable therefore to adhere to light mineral oil. The purpose of a brilliantine is to brighten the hair, to help hold it in place and to perfume it.

Jelly Brilliantine

Spermaceti	14 lb.
Beeswax	6 lb.
Mineral Oil	100 lb.
Perfume	1 lb.
Color to suit.	

Procedure: Melt the waxes in the mineral oil. Strain and allow to cool to about 115° F. Add perfume; stir until cold.

In addition to the hair tonics for the two primary scalp conditions, dry and oily, there are a multitude of others for which various claims are made. This group is so various that it would be impossible to give an adequate outline. We shall, therefore, content ourselves with giving two typical formulas together with one containing cholesterol. Much attention is being given to hair tonics containing lanolin derivatives, lecithin, etc.

Hair Tonic—Dry Scalp

Castor Oil	1 gal.
Crude Carbolic 30%	8 oz.
Cresol U. S. P.	3 oz.
Lignol	1 gal.
Soya Bean Oil	2 gal.
Precipitated Sulphur	2 oz.

Procedure: Mix the soya bean oil, the castor oil heat to 100° F. and add the lignol. Take a small quantity of this mixture and rub up precipitated sulphur into a smooth paste. Mix with rest of oils. Add carbolic and cresol.

Dry scalp is often a diseased condition, accompanied by dandruff. Often it is caused by poor circulation of blood. Above preparation should be rubbed into scalp at night, and, because odor is obnoxious, shampooed out in morning. Label should contain a statement to the effect that the longer the preparation is left on the better will results be.

Hair Tonic—Oily Scalp

Water	15 gal.
Glycerine	2 gal.
Alcohol	30 gal.
Menthol	7 lb.
Resorcinmonoacetate	8 oz.
Perfume q. s.	

Procedure: Dissolve menthol and perfume in alcohol, mixing rapidly. Add glycerine and 10 gallons of water. Dissolve resorcinmonoacetate in rest of water, add to the above and mix for three hours. Allow to stand over night and filter.

Hair Tonic—Containing Cholesterol

Alcohol	75%
Glycerine	5%
Cholosterol	1%
Lecithin	1%
Distilled Water	12%
Perfume	1%
Chloroform	5%

Procedure: Dissolve lecithin in chloroform, add cholosterol and one gallon of alcohol. Mix the perfume with the alcohol, add the glycerine, add the lecithin-cholosterol mixture, agitate for one hour add the water and agitate for two hours. Allow to stand over night and filter.

Dandruff Ointment

Dandruff ointment is usually a powerfully antiseptic salve, the following formula being typical of the class:

Precipitated Sulphur	8 lb.
Oxyquinoline Sulphate	1 lb.
Lanoline	10 lb.
Petrolatum	61 lb.
Castor Oil	15 lb.
Tincture Fish Berries	1 lb.
Balsam Peru	2 lb.
Carbolic Acid 85%	2 lb.

Procedure: Mix the sulphur with the castor oil rubbing thoroughly until lumps have disappeared. Mix the oxyquinoline sulphate with ten pounds of petrolatum, run through an ointment mill or milling rolls three times, add the sulphur castor oil mixture, mix thoroughly and run through the mill again. Melt the lanoline, and the rest of the petrolatum, add the remainder of the castor oil, mix thoroughly and then mix in the oxy-sulphur mass. Mix thoroughly, add the balsam of Peru, continue mixing for thirty minutes, add the tincture fish berries and the carbolic acid and mix again for twenty or thirty minutes. The machine best suited for this ointment is a pony mixer.

---

* Lip Stick

26.7% talc, 13.3% kaolin, 10.9% ponceau 3R 6.3% amaranth, 17.1% yellow ochre, 5.7% zinc oxide, 3.6% paraffin, 5.9% beeswax, 2.4% carnauba wax, 4.7% sulfonated oil and 3.4% petrolatum. Body materials, that is talc and kaolin, are mixed, then dry coloring matter is added including ponceau, amaranth and yellow ochre; then zinc oxide and finally binder which is first fused so that mixing with binder takes place above melting point of same. Mass is mixed well until plastic and poured into sticks. Some other ingredients mentioned include eosine Y, tartarazine, borax and bentonite. Sulfonated oil in lipstick aids dispersion of color on skin.

Lemon Juice Cleansing Cream

Pure Lemon Juice	70%
White Petrolatum	12%
Parachol	17%
Acid proof Lemon (to perfume)	1%
	100%

Mix the absorption base with the petrolatum with heat and mix until homogeneous. Allow to cool slightly and then slowly add the lemon juice while mixing rapidly. Add the acid proof lemon.

Translucent Jelly Cream

Stearic Acid	6%
Spermaceti	15%
White Petrolatum	30%
Mineral Oil	49%
Perfume Oil to Suit.	
	100%

Melt the stearic acid and the spermaceti, add the petrolatum and when melted stir in the mineral oil which has first been heated. When almost set stir in perfume.

Greasy Type Cleansing Cream

Spermaceti	23%
Petrolatum White	20%
Mineral Oil	57%
Perfume to Suit.	
	100%

Make as above.

Cold Cream

Mineral Oil	54 %
White Wax	18 %
Absorption Base Parachol	5.5%
Borax	1 %
Water	21 %
Perfume	.5%
	100.0%

Melt the white wax, add the mineral oil. Dissolve borax in part of water with heat. Add to melted fats. Heat rest of water, stir in absorption base until smooth and mix with fats. Agitate thoroughly and when just above solidifying point, add perfume.

Lecithin Lotion

Milky lotions (emulsions) are produced by dissolving lecithin in oil and agitating or churning the oil solution with neutral soap solution containing water or glycerine. In this way there form emulsions that are not too stable. Far more stable is the following emulsion: Two parts of monostearin glycerine ester, 1 part stearin alcohol, 5 parts stearin, 2 parts lanolin, 5 parts mineral oil (according to the particular fattiness desired 10–15 parts) and 2 parts lecithin are melted and 1 part potash in 5 parts glycerine and 40 parts hot water is stirred during heating into the fused mass. It is further heated until the mass no longer rises thick. Then it is stirred cold. It is then thinned after cooling with more water until the particular thin liquid state desired is attained. Instead of or in conjunction with the first two constituent parts a glycol stearate may be used.

Lecithin Nourishing Cream

Lanolin	15 gr.
Beeswax	15 gr.
Spermaceti	10 gr.
Petrolatum	35 gr.
Borax	1 gr.
Water	22 gr.
Cholesterin	1 gr.
Lecithin	1 gr.
Perfume	as required

Massage Cream

Spermaceti	10 gr.
Solid Paraffin	15 gr.
Mineral Oil	45 gr.
Lecithin	1.5 gr.
Cholesterin	0.5 gr.
Borax	1 gr.
Water	30 gr.
Perfume	as required

The solution of lecithin and cholesterin is accomplished best in the liquid or melted fats and waxy constituents. The melted mass is permitted to be cooled at say 40° C. and the hot solution of borax in water is poured first in small portions and then in larger portions into the fused mass while stirring thoroughly. Then it is stirred cold.

In the case of vanishing cream, it is somewhat more difficult to work in the lecithin. The simplest way is to dissolve the lecithin in the melted stearic acid (overheating should be prevented) and to mix the potash solution into it by stirring in the usual way. On the other hand saponification and emulsification might be affected by the lecithin. If any oil is permitted in the vanishing cream, lecithin is ground fine with warm mineral oil (1 part of lecithin to say ½–1 part of mineral oil), so that a mass is produced that can be distributed. As soon as the cream has been mixed and while it is still warm, the warm lecithin oil is stirred thoroughly into it. The whole of it is stirred cold.

Skin Smoothener

Boric Acid	3 drams
Tragacanth	8 grams
Glycerine	3 drams
Distilled Water	16 oz.

Boil—stir until a clear jelly is obtained.

Rolling Massage Creams

Creams of this type are made from freshly precipitated casein from milk. They at first, seem to disappear when rubbed on the skin, then on further rubbing, roll into small particles carrying with them the dust and dirt collected in the pores of the skin. They have the disadvantage of not keeping very well as the water contained in the casein evaporates rather quickly, especially if jars are not kept tightly closed, or are opened too frequently.

These creams are generally colored pink, with eosine. The general process for making these creams is as follows:

(1) To 128 parts of fresh milk add 2/10 of 1% of formaldehyde 40% solution or 1% sodium benzoate is added as an antiseptic, and enough of a 2% solution of eosine to give the proper shade. Mixture is warmed to about 50–55° C. on water bath while stirring gently, then strained if necessary.

(2) Prepare on the side, a 20% solution of alum or a concentrated solution of potassium sulphate in distilled water and bring it to the boiling point.

Bring mixture No. 1, (milk) to boiling point and pour while stirring slowly, the boiling mixture (No. 2). Stop heating, continue to stir gently, and let cool slowly at about 55° C.

When cool, and upper liquid is clear, strain on muslin previously wetted, allow precipitate to drain, wash with little cold water, drain again. Then pass through filter press if there is too large excess of water. Consistency of cream will depend upon quantity of water allowed to remain in casein. Then add to casein about 1% of perfume and 10 to 15% of glycerin or carbitol in order to prevent quick drying of casein, and put in tightly sealed jar. To obtain homogeneous product, it is recommended to pass the magma through an ointment mill before putting in jars. Addition of 1.5% sodium benzoate helps preservation.

Depilatory Cream

A formula for a depilatory cream is one part gum tragacanth, 10 parts water, 10 parts glycerin and six parts starch, together hot, and intimately mixed with

35 parts strontium sulphide, 3 parts sodium sulphide, 30 parts zinc oxide, 10 parts lanolin, 15 parts water and 0.2 part menthol. Formula for depilatory powder is 30 parts strontium sulphide, 20 parts calcium sulphide, 30 parts starch, 16 parts talc, 3 parts aluminum acetate and one part menthol.

Soothing Cream

Used to relieve skin irritation, especially after a depilatory has been used. A zinc oxide paste, containing 28 parts almond oil, 60 parts zinc oxide, 15 parts talc and 60 parts cold cream is useful; also a mixture of 30 parts lanolin and 90 parts soap-camphor liniment perfumed with oil of lavender.

Depilatory Perfumes

The essential oils, which have been found suitable for perfuming depilatories include oil of wintergreen, vetivert oil, patchouly oil, oil of thyme, lavender oil; also the aromatics, amyl salicylate, terpineol, benzyl acetate, menthol. About 2% is usually added. Lavender oil, particularly terpeneless, is much liked for this purpose, as it alleviates skin irritation.

Hair Lotions

One part cholesterin, 0.3 part lecithin in 200 parts of 96% alcohol and mixed with 3 parts castor oil. Another solution of 0.5 part oxyquiniline sulfate and 0.2 part salicylic acid in 75 parts 96% alcohol is added and mixture made up to 300 parts by weight.

Vanishing Cream

Five parts of cocoa butter are melted with 25 to 30 parts of pure stearin on water bath at not above 100° C. Warm solution, 60° C., of 100 parts water, seven parts potash, eight parts borax, 16 parts glycerin, 12 parts alcohol and 3 parts ammonia, is added to 30 parts of molten mass. Much carbon dioxide gas is liberated, which necessitates large kettle for operation. Vigorous agitation is required. After most of carbon dioxide has escaped, hot-filtered solution of 0.5 part agar-agar in 20 parts water is added and mixture stirred until cool. Perfume is added shortly before mass congeals. Cream is filled into containers after standing 1 to 2 days.

Mosquito Cream

Good results can be secured from composition containing 5 parts powdered wheat starch, 10 parts water, 45 parts glycerin 28° Bé., 30 parts lanolin and 5 to 10 parts oil of clove. Starch is rubbed into smooth paste with water; glycerin is mixed in and mass converted into jelly-like consistency by heating and agitating; it is then allowed to cool.

Nail Polish (Paste)

A good formula for a nail polish in paste form contains 100 parts of light colored rosin, 60 parts of stearin, 60 parts of yellow beeswax and 200 parts of ceresin wax. These ingredients are melted together on water bath and then 300 parts of white petrolatum are mixed in. Then a well mixed mixture of 200 parts of washed kieselguhr, 140 parts of zinc oxide and 100 parts of tin oxide is mixed with the waxy base. Before mixture is removed from water bath, coloring matter is added, for example alkanna pink, as well as 15 to 20 parts of perfume. These ingredients must be added shortly before mass becomes solid and is poured into containers.

Preparations for Baldness

Ointment

Pilocarpine Hydrochloride	20 oz.
Precipitated Sulphur	120 oz.
Parachol	60 oz.
Balsam of Peru	60 oz.
Resorcinol Monoacetate	30 oz.
Petrolatum	900 oz.
Water	60 oz.
Perfume to suit.	

Procedure: Dissolve the pilocarpine in water and mix with absorption base. Mill the sulphur and the monoacetate with part of the petrolatum. Melt the rest and stir in the absorption base and add finally the sulphur mass. Mix thoroughly.

Lotion

Mercuric Chloride	1 oz.
Salicylic Acid	5 oz.
Chloral Hydrate	5 oz.
Glycerin	25 oz.
Acetone	10 oz.
Alcohol	200 oz.
Water	825 oz.
Perfume to suit.	

Procedure: Take part of the petrolatum, add the salicylic, the phenol and the camphor and mill thoroughly. Melt the lanolin, the rest of the petrolatum and the beeswax, stir in the milled base and add the oil of eucalyptus.

Lotion

Oxyquinoline Sulphate	1 oz.
Tincture of Fish Berries	10 oz.
Glycerin	30 oz.
Tincture Benzoin	8 oz.
Witch Hazel	150 oz.
Water	10 oz.
Perfume to suit.	

Procedure: Dissolve the sulphate in water. Mix the fish berries with the glycerin, add the benzoin and the witch hazel. Then add the sulphate solution. Other chemicals used in the manufacture of eczema preparations are: calomel, iodoform, oil of wormwood, silver protein, sodium iodide, potassium iodide, pine tar, bismuth resorcinate, mercuric salicylate, bismuth subnitrate, red mercuric iodide, basic aluminum acetate, benzocaine, bismuth oxyquinolate, and various absorption bases.

Psoriasis Preparations

Ointment

Chrysarobin	3 oz.
Salicylic Acid	1 oz.
Rectified Oil of Pine Tar	10 oz.
Soft Soap	15 oz.
Petrolatum	28 oz.
Absorption Base	5 oz.
Perfume to suit.	

Procedure: Mill the salicylic and the chrysarobin with a part of the petrolatum. Melt the rest of the petrolatum and the absorption base, add the soap, the pine tar and the chrysarobin-salicylic mass and mix thoroughly.

Lotion

Oil of Mace	10 oz.
Olive Oil	10 oz.
Liquid Ammonia	15 oz.
Essence of Rosemary	5 oz.
Rose Water	50 oz.
Lecithin	½ oz.
Chloroform	3 oz.
Perfume to suit.	

Procedure: Add the olive oil to the mace and mix thoroughly, add the ammonia water and keep stirring until a saponaceous mass is evolved. Dissolve the lecithin in the chloroform. Mix the rosemary with the rose water and add the lecithin solution. Then add this to the first mixture very slowly; keep up a very slow mixing for about an hour afterward.

Other chemicals used in the making of psoriasis products are: lanolin, sesame oil, peanut oil, benzoic acid, bismuth subgallate, linseed oil, birch tar, chaulmoogra oil, neats foot, croton, cod-liver, and soya bean oils.

Procedure: Dissolve mercuric in part of water, chloral hydrate in the rest and mix the two solutions together. Dissolve the salicylic in alcohol, add glycerin, acetone, mix perfume and filter.

Other chemicals used in making alopecia preparations include: quinine hydrochloride, spirit of rosemary, ammonium chloride, tincture of nux vomica, cantharides, capsicum, calomel, lead acetate, quinine sulphate, glacial acetic acid, neats foot oil, lignol, soya bean oil, castor oil, colloidal silver, birch tar oil, etc.

Dandruff Preparations

Ointment

Lanolin	12 oz.
Water	15 oz.
Silver Lactate	3 oz.
Tincture Fish Berries	5 oz.
Sulphur Iodide	3 oz.
Balsam of Peru	15 oz.
Cocoa Butter	20 oz.
Petrolatum	60 oz.
Glycerin	10 oz.
Perfume to suit.	

Procedure: Dissolve the silver lactate in water and the sulphur iodide in glycerin. Melt the petrolatum, the lanolin and the cocoa butter, stir in the silver lactate solution, add the sulphur iodide solution and finally the balsam of Peru and the fish berries.

Lipstick

White Beeswax	20	grm.
Paraffin	5	grm.
Spermaceti	8	grm.
Cocoa Butter	10	grm.
Benzoated Lard	25	grm.
Parachol	20	grm.
Bromo Acid	3	grm.
Color Mixture for Shade	10	grm.
Para Oxybenzoic Acid Ester	.05	grm.
Perfume (with flavor character)	1	grm.

Procedure: Mix the colors first with the bromo acid. Melt the parachol and the lard, add the color mixture and grind through a paint mill three or four times. Meanwhile melt and mix the rest of the waxes, and, when the colors are ready, add the melted waxes and mix thoroughly. Heat should not be raised above the melting point of the waxes. As soon as the batch is finished it should be molded, keeping it so far as possible at a constant temperature.

Anti-Perspiration Liquid

Oxyquinoline Sulfate	1
Rose Water	500

Anti-Perspiration Powder

Oxyquinoline Sulfate	1
Talc	10

Freckle "Removers"

Two grams of zinc sulphophenylate, 30 grams of distilled water, 2 grams of ichthyol, 30 grams each of anhydrous lanolin and petroleum jelly and 2 grams of lemon oil or other suitable perfume, will give good results.

Preparations with a bleaching action are made containing 1500 grams of wool grease, 530 grams of almond oil, 110 grams of beeswax, 150 grams of borax, 150 grams of hydrogen peroxide (100% by volume) and 10 grams of yellow petrolatum.

Freckles Treatment

Alcohol	4 oz.
Stronger Rose Water	2 oz.
Tincture of Benzoin	15 dr.

Apply every night after scrubbing.

Perfume Sticks

Most suitable base for these perfumed crayons is acetanilide. It is used in proportion of 87.5 parts by weight. It is melted on water bath or over flame, provided it is carefully stirred while being heated. Temperature must not rise above 80° C. When it is molten, 10.5 parts of pulverized magnesium carbonate are mixed in until it dissolves entirely. Then there are added 35 parts of xylene musk, 17.5 parts of heliotropin, and 3.5 parts of Japan wax. When all ingredients have been melted, 8.4 parts of perfume dissolved in 4.2 parts of benzyl alcohol are added.

Mass will solidify rapidly and can be formed into shape while still warm. Amount of heliotropine added is maximum allowable limit, for more of this substance will make mass soft. Perfume must not be added in excess of that prescribed above, for the excess will simply ooze out of mass. When these perfumed crayons are properly packed in air-tight containers, they will last for years. When acetanilide and magnesium carbonate are used alone, then about 15% menthol or menthol and camphor should be added.

Wrinkle Cream

First requirements of skin creams for removing wrinkles is that they must be greaseless. Cream is naturally used as massage cream, for process of removing wrinkles involves massaging. Suitable formula for such cream is 1600 parts of rose water and 350 parts of glycerin. This mixture is brought up to boiling and 40 parts of potash soap added. Solution is boiled again and 18 parts of purified calcined potash added. In another vessel 180 parts of white stearin are melted. First solution is filtered through cloth to remove impurities. Then it is brought to boiling and molten stearin allowed to flow into vessel in thin stream while solution is vigorously agitated. Large vessel must be used for carrying out operation, for mass must not be allowed to boil over due to evolution of large quantities of carbon dioxide. If contents of kettle boil over, result is insufficient saponification of contents and poor product. This is noticed by formation of small lumps in cream. These lumps cannot be properly rubbed into skin and spoil entire action of cream. This cream is really a soft soap. Mass is cooled after being boiled long enough and is agitated thoroughly and perfumed with 15 parts of rose oil and one part of vanillin. Small amount of alcohol may be added either after or during addition of stearin. This is effective in preventing formation of lumps.

Concentrated Hair Wave

Gum Karaya White	5 lb.
Aquaresin G. M.	5–10 lb.

Rub together thoroughly and stir in

Isopropyl Alcohol (99%)	20 lb.

Perfume and color to suit.

This concentrate when thrown into water and stirred gives a uniform product whose thickness depends on amount of water used. This product differs from similar preparations in that it gives the hair lustre and does not flake off.

Eau De Cologne and Toilet Waters

Base A

Italian Lemon Oil	20 grm.
Bergamot	20 grm.
Neroli or Neroli Synthetic	35 grm.

Italian Sweet Orange Oil	10 grm.
Lavender 40–42% Ester	10 grm.
Orris Root Tincture	2 grm.
Ambreine or Ambrethene	3 grm.

Use 100 grams to 1 gallon 70% alcohol. Allow to stand for one week. Chill and filter while cold.

Perfume Bases

Floral Eau De Colognes (Acacia Type)

Base A (above)	100
Methyl Naphthyl Ketone	2
Anisic Aldehyde	1
Benzyl Acetate	1

Chypre Cologne

Base A	100
Oak Moss Absolute	3
Vetiverol Acetate	5
Patchouli	3
Coumarin	5
Santalol Acetate	4

Gardenia Cologne

Base A	100
Styralyl Acetate	2
Hydrotropic Aldehyde	0.5

Jasmin Cologne

Base A	100
Benzyl Acetate	5
Amyl Cinnamic Aldehyde	2
Hydroxycitronellal	3

Lilac Cologne

Base A	100
Benzyl Acetate	5
Terpineol	5
Anisic Aldehyde	1
Phenyl Acetic Aldehyde	1
Hydroxycitronellal	5

Orchidee or Treflé Cologne

Base A	100
Isobutyl Salicylate	10
Musk Ambrette 20% in Benzyl Benzoate	5

Carnation Cologne

Base A	100
Isoeugenol	5
Eugenol	5
Vanillin	2
Methyl Ionone	3
Phenyl Ethyl Alcohol	5

Rose Cologne

Base A	100
Rose Geranium	2.5
Rhodinol	5
Phenyl Ethyl Alcohol	7.5
Citronellal Acetate	2.5

Fancy Cologne

Terpeneless Lemon	3
Terpeneless Bergamot	15
Neroli Petale	25
Nerol	15
Terpeneless Bergamot	7
Phenyl Ethyl Alcohol	10
Hydroxycitronellal	15
Cinnamyl Acetate	5
Ambreine or Ambrethene	5

Jasmin Synthetic

Benzyl Acetate	400
Hydroxycitronellal	100
Linolool	50
Heliotropin	50
Amyl Cinnamic Aldehyde	50
Para Cresyl Caprylate	50
Ylang Ylang Oil	50
Jasmin Absolute	250

Rose Synthetic

Rose Otto	150
Rose Absolute	50
Rhodinol	200
Phenyl Ethyl Alcohol	300
Phenyl Ethyl Propionate	100
Alpha Ionone	50
Vetiverol Acetate	25
Rhodinol Acetate	25
Citronellol Butyrate	25
Phenyl Acetic Aldehyde 50%	50
Aldehyde C9 (10%)	15
Alcohol C10 (25%)	10

Carnation

(Do not use in Creams or Lipsticks)

Phenyl Ethyl Alcohol	100
Isoeugenol	250
Eugenol	300
Rose Otto	25
Rhodinol	100
Ethyl Vanillin	10
Musk Ketone	50
Benzyl Isoeugenol	50
Methyl Ionone	50
Oppoponax Resin	2
Tolu Resin	8

Oregon

Carnation Synthetic	250
Methyl Ionone	200
Peru Balsam	10
Tolu Balsam	10
Benzoin	50
Ylang Manilla	60

Jasmin Synthetic	50
Cinnamyl Alcohol	150
Rose Synthetic	50
Oppoponax Resin	5
Castoreum Absolute	5
Ambreine or Ambrethene	150

Jacinthe Synthetic

Phenyl Acetic Aldehyde 50%	200
Phenyl Acetic Aldehyde Dimethyl Acetal	50
Hydrotropic Aldehyde	50
Brom Styrol	10
Methyl Octrine Carbonate 10%	15
Clary Sage Oil	20
Ylang Manilla Oil	50
Methyl Ionone	50
Phenyl Ethyl Alcohol	100
Cinnamyl Alcohol	200
Rose Synthetic	50
Phenyl Ethyl Propinate	50
Phenyl Propyl Acetate	50
Terpineol	55
Vanillin	20
Musk Ketone	30

Tuberose Synthetic

Tuberose Natural	100
Cinnamyl Alcohol	50
Phenyl Propyl Alcohol	100
Ylang Manilla Oil	300
Benzyl Salicylate	100
Benzoin Resin	50
Tolu Resin	50
Styrax Resin	50
Methyl Ionone	50
Heliotropin	50
Methyl Salicylate	25
Aldehyde C12 (10%)	50
Alcohol C12 (25%)	25

Neroli Synthetic

Neroli Petale	250
French Pettigrain	300
Phenyl Ethyl Alcohol	100
Linalyl Anthranilate	100
Linalool	50
Nerol	100
Rhodinol	50
Phenyl Acetic Acid	5
Sweet Italian Orange Oil	45

Narcisse Synthetic

Ylang Bourbon Oil	150
Benzyl Acetate	100
Hydroxycitronellal	200
Terpineol	100
Cinnamyl Alcohol	100
Rose Synthetic	75
Coumarin	50
Jasmin Synthetic	50
Para Cresyl Phenyl Acetate	25
Para Cresyl Acetate	10
Methyl Para Cresol	10

Honeysuckle

Phenyl Ethyl Alcohol	100
Cinnamyl Alcohol	100
Heliotropin	50
Alpha Ionone	100
Mimosa Synthetic	50
Jasmin Synthetic	100
Rose Synthetic	50
Terpinol	50
Phenyl Acetic Acid	10
Musk Ketone	25
Musk Ambrette	25
Methyl Naphthyl Ketone	50
Para Cresyl Phenyl Acetate	10
Hydrotropic Aldehyde	10
Neroli Synthetic	50
Phenyl Ethyl Phenyl Acetate	50
Linalool	50
Nerol	50
Hydroxycitronellol	170

Treflé

Isobutyl Salicylate	250
Benzyl Salicylate	150
Ylang Bourbon Oil	150
Methyl Ionone	100
Isoeugenol	30
Eugenol	30
Bergamot Oil	100
Linalyl Acetate	50
Citronellol Acetate	65
Coumarin	50
Para Cresyl Phenyl Acetate	25

Violet Synthetic

Alpha Ionone	200
Beta Ionone	50
Methyl Ionone	150
Orris Resin	100
Cassie Synthetic	50
Jasmin Synthetic	50
Vetiverol Acetate	50
Coumarin	25
Vanillin	25
Bergamot	50
Hydroxycitronellal	50
Isobutyl Phenyl Acetate	50
Musk Ketone	50
Violet Natural	100

Ambre (Fixative)

Musk Ketone	30
Musk Ambrette	30
Labdanum Bleached	100
Orris Absolute	10
Methyl Ionone	50
Vanillin	50

Vetiverol Acetate	50
Coumarin	50
Clary Sage Oil	25
Bergamot Oil	125
Heliotropin	100
Benzyl Cinnamate	100
Resin Peru	50
Resin Tolu	50
Santalool Acetate	80
Resin Benzoin	50
Ambreine or Ambrethene	100

Mimosa Synthetic

Mimosa Absolute	100
Dimethyl Acetophenone	100
Isobutyl Salicylate	100
Phenyl Acetic Acid	25
Phenyl Acetic Aldehyde (50%)	25
Linalool	75
Benzyl Acetate	50
Coumarin	50
Cinnamyl Alcohol	200
Cinnamyl Acetate	75
Hydroxycitronellal	150

Cold Cream, Modern

Paraffin Wax	1	lb.
Cetamin	2	lb.
Petrolatum, White	1½	lb.
Mineral Oil, White	3	pt.

Heat to 180° F. and to it add with stirring

Water (Boiling)	1	gal.

When at 150° F., while mixing, add 1 dram perfume and mix till thick. Pack cold.

Lemon Cream

Follow above cold cream formula, using a little Tartrazine in the water and Citral in place of perfume.

Cucumber Cream

As above except using water soluble green color and cucumber perfume.

Strawberry Cream

As above except using water soluble pink color and strawberry perfume.

Lavender Cream

As above except using water soluble lavender color and lavender perfume.

Violet Cream

Follow cold cream formula using water soluble violet color and violet perfume.

Tangerine Cream

Follow cold cream formula using water soluble orange color and tangerine perfume.

Mint Cream

Follow cold cream formula using water soluble green color and peppermint perfume.

Wild Cherry Cream

Follow cold cream formula using water soluble cherry cold and wild cherry perfume.

June Type Cream

The most recent advance in an all purpose cream, sold in tubes, is exemplified by the following formula which gives a waxless cleansing, nourishing, stimulating and softening cream which also acts as a powder base.

A.	Glycosterin	16	lb.
	Mineral Oil, White	3	gal.
	Petrolatum, White	6	lb.
	Parachol	2	lb.
B.	Water	7½	gal.
	Glycopon AAA	4	lb.

In separate vessels heat A and B to 160° F. Add B to A slowly while stirring vigorously. A jelly like mass results. Add 4 oz. perfume and continue stirring. As temperature drops to 110° F. a transformation takes place—a beautiful white cream results; stirring is continued until cold when it is packed into tubes or jars. It may be packed warm by heating, with stirring, to 105–110° F.

This cream wipes off the skin without leaving a greasy film. It, nevertheless, penetrates and is readily absorbed by the skin.

To give a cooling effect on the skin, 1–2 oz. of menthol may be added with the perfume.

Modified forms of this cream may be made by the addition of water soluble colors and appropriate perfumes, oils or other materials to produce

Lemon Cream
Strawberry Cream
Cucumber Cream
Turtle Cream
Viosterol Cream
Lecithin Cream
Hormone Cream
Olive Oil Cream
Almond Oil Cream

Tissue Builder Cream

Paraffin Wax	1 lb.
Cetamin	2 lb.
Lanolin Anhydrous	1 lb.
Petrolatum, Amber	1 lb.
Mineral Oil	3 pt.

Heat above to 180° F. and while mixing add slowly

Water (Boiling)	1 gal.

Continue stirring and at 150° F. add 1½ drams perfume. This cream is poured into jars at 130–135° C.

Hair Milk

1. Mineral Oil, White	144 lb.
2. Trihydroxyethylamine Stearate	29 lb.
3. Water, Warm	320 lb.
4. Perfume	3 lb.

While stirring heat (1) and (2) until melted together. Add (3) slowly with stirring until uniform. Allow to stand overnight, stir moderately and package.

This preparation corrects dry scalp and hair and imparts a gloss to the latter and keeps it in place. It replaces old fashioned greasy hair oils and brilliantines.

Cold Cream

Mineral Oil, White	1 gal.
Beeswax, White	1 lb.
Ozokerite, White	1 lb.
Ceraflux	2 lb.

Heat to 170° F. and add to above, while mixing

Water	½ gal.
Borax	1½ oz.

previously heated to 170° F. When temperature is 140° F. add 1 oz. perfume and pour into jars at 130–135° C.

Tissue Cream

To the above mixture of waxes add

Lanolin Anhydrous	1 lb.

and replace the Beeswax, White by Yellow Beeswax.

Turtle Oil Cream

Same as Tissue Cream (above) with the addition of Turtle Oil ½ lb. and ½ oz. Moldex, dissolved in the water.

Cucumber Cream

Same as Cold Cream (above) except that a little water soluble green color is dissolved in the water and 1 oz. of cucumber perfume is used.

Lemon Cream

Same as Cold Cream (above) except that a little Tartrazine is dissolved in the water and as perfume either 1 oz. of Terpeneless Oil of Lemon or Citral is used.

Brushless Shaving Cream

Deramin	4 lb.
Water	5 gal.

Heat to 180° F. and pour into

Stearic Acid	15 lb.
Lanolin	1 lb.

previously heated to 180° F. while mixing moderately.

Add perfume 4 oz. when thick and mix until cold. If a cooling effect on the skin is desired 1 oz. Menthol may be added with the perfume.

Liquefying Cleansing Cream
Soft Type

Petrolatum, White	3 lb.
Ceraflux	2 lb.
Petrolatum, Liquid	1 gal.

Melt together and add 1 dram perfume; pour at lowest possible temperature.

Medium Type

Spermaceti	5 lb.
Petrolatum, White	8 lb.
Ceraflux	4 lb.
Petrolatum, Liquid	1½ gal.

Melt together and add 1½ drams perfume; pour at lowest possible temperature.

Hard Type (for Hot Climates)

Spermaceti	5 lb.
Petrolatum, White	8 lb.
Ozokerite	5 lb.
Petrolatum, Liquid	1½ gal.

Proceed as in Medium Type above.

Cold Cream

White Beeswax	150 gm.
White Mineral Oil	600 gm.
Water	240 cc.
Borax	10 gm.

Melt beeswax in mineral oil. Dissolve borax in water. Add two with vigorous stirring until cool. Perfume to suit.

Vanishing Cream

To make a quart.

Stearic Acid (Triple Pressed)	1920 gm.
Glycerin	960 gm.
Soda Ash	60 gm.
Borax	840 gm.

Distilled water to make 32 oz.

Melt stearic in glycerine and one-half the water. Dissolve soda ash and borax in other half. Mix two with stirring until cream is cooled sufficiently. Perfume to suit.

Pearly Vanishing Cream

This cream is non-beading as if it is free from glycerine.

Deramin	4 lb.
Water	5 gal.

Heat to 180° F. and pour into

Stearic Acid	16 lb.

previously heated to 180° F. while stirring, not too quickly. Add 4 oz. perfume when cream thickens and stir until cold. Allow to stand overnight and pack. The pearly finish becomes more pronounced with age.

This cream is noteworthy because it is free from ammonia, soda, potash and glycerin and therefore will not affect tender skins.

Astringent Cream

Add 4 oz. Astringent Powder to the above when cold. Or, preferably, grind the Astringent Powder into 1 lb. of the cream and then mix into the entire batch.

DECOLORIZING, DEODORIZING, DISINFECTING

* Ammonia Gas Mask Absorbent

First make a solution of

Sod. Silicate	43 kg.
Water	51 gal.

Sulfuric Acid	3500 c.c.
Water	13 liters

Add 10 liters of this acid solution slowly with stirring to the silicate solution. The balance of the acid solution is then poured in quickly while stirring vigorously. Stir until neutral or slightly acid. On standing for about ½ hour a glassy blue white jelly forms. This is transferred to a number of canvas bags and placed in a press. Pressure is applied to squeeze out as much water as possible. A rigid white gel is left. This is pressed thru a No. 4 screen and placed in a suitable mixer. To this is added the following solution

Malic Acid (Powd.)	20 lb.
Water	4 liters

heated slowly to 65–70° C. with good stirring. Add slowly another liter of water while stirring for 20 minutes. Transfer to a mill and grind to uniform size. Pour into shallow pans to depth of one inch and heat to 160° F.

---

* Refrigerator Deodorant

Take sour cherry charcoal 45%, coconut charcoal 25%, boxwood charcoal 20% and trioxymethylene 10%, all in granular form. This mixture is placed in a rotating cylinder or barrel, provided with agitating apparatus, and stirred for ten minutes or until a homogeneous product of uniform composition is obtained. The granules are then sieved to remove any pulverant material, and the mixture is then poured into a cylindrical container of perforated sheet metal, the perforations being of such size as to freely admit air, gases and vapors but too small to permit egress of the granular particles of carbon and trioxymethylene.

In the foregoing there is used sour cherry, coconut and boxwood carbon or charcoal and trioxymethylene in granular form, however, pulverant materials may be used with equally good results, or a central core of molten trioxymethylene may be used around which is disposed the gas absorbing materials, all of which may then be enclosed in a pervious container, a perforated cylinder of sheet aluminum approximately 3 inches tall and 2½ inches in diameter being one form in which I prefer to manufacture this apparatus. Such a receptacle has a capacity of approximately 105 grams and will contain:

	Grams
Sour Cherry Wood Charcoal	47.25
Cocoanut Shell Charcoal	26.25
Boxwood Charcoal	21.00
Trioxymethylene	10.00

In the foregoing, use pulverant trioxymethylene which is packed in a thimble or capsule of unsized paper. This capsule forms a central core about which is packed the granular absorbent material. The shape and size mentioned is one form suitable for use in a refrigerator having a capacity of approximately 3 cubic feet.

Another form product may be manufactured as follows:

	Per cent
Sour Cherry Wood Charcoal	45
Cocoanut Shell Charcoal	25
Boxwood Charcoal	20
Trioxymethylene	10

These materials are mixed to a stiff paste with ox blood, diluted with 10 volumes of water, and the mass is charred in suitable molds at 600° C. so as to form cakes or blocks which are available for use without the necessity of a containing receptacle.

---

* Garlic, Deodorizing

Garlic is chopped very finely and heated with water in a pressure cooker. The odorous materials are then removed by blowing air or steam thru it.

Deodorant Spray
For theatres, lavatories, etc.

Pine-needle Oil	
Formalin	of each 2 oz.
Acetone	6 oz.
Isopropyl Alcohol	to 20 oz.

For use as a spray 1 oz. is mixed with a pint of water.

Pine Deodorizing Spray

Pine spray—Pine oil 250, geranium oil 5, bergamot oil 5, lavender oil 15, rosemary oil 10, bornylacetate 15 and iso-PrOH 700 parts.

Spray for Movie Theatre

The following is a formula for a preparation suitable for spraying in theatres:

Pine-needle Oil	
Formalin	of each 2 oz.
Acetone	6 oz.
Isopropyl Alcohol	to 20 oz.

For use as a spray 1 oz. is mixed with a pint of water.

Theatre Sprays

(1)

Oil Lavender	60 c.c.
Oil Bergamot	30 c.c.
Oil Peppermint	5 dr.
Oil Cloves	30 dr.
Acid Benzoic	1.8 gm.
Alcohol sufficient to make	300 c.c.

The benzoic acid is dissolved in the alcohol and the volatile oils added.

(2)

Pine Needle Oil	2 oz.
Formalin	2 oz.
Acetone	6 oz.
Isopropyl Alcohol	to make 20 oz.

For use, mix 1 oz. of above with a pint of water for spraying.

---

* Disinfectant

A disinfectant having a coefficient of 2 plus in accordance with Rideal-Walker (1921 modification) may be prepared by mixing the following ingredients in the proportions given:

	Per cent by weight
Tar Acid Oil (25% tar acids)	68
Rosoap	30
Castor Oil Soap	2

A disinfectant having a coefficient of 3 plus may be prepared by mixing the following ingredients together in the proportions given:

	Per cent by weight
Insecticide Oil	56.0
High Boiling Tar Acids	4.0
Water Gas Tar Distillate	8.0
Rosoap	30.0
Castor Oil Soap	2.0

In the above examples, rather than adding the liquid resinate to the oily portions, it is preferable to add the oily portions such as the tar acids, insecticide oil, and water gas tar distillate, to the liquid resinate while stirring and heating. The remaining ingredients may be added in any order.

Water is added to the above compositions to produce a disinfectant in the form of an emulsion. Any desired proportion of water may be used. The emulsion formed is of an exceptionally stable character.

Any animal or vegetable oil soap may be substituted for castor oil soap to aid in stabilizing emulsions. For instance soya bean oil soap or stearic acid soap may be used. Sulphonated oils may also be used.

---

* Disinfectant, Deodorizing

Lime	85–93
Sod. Tetrasilicate	15–7
Alum	5

---

* Disinfectant Bleach

Toluene sulfodichloramide	24
Caustic Soda	10
Sod. Sulfate	110

When dissolved in water it is a strong bleach and disinfectant.

"Lysol" (Cresol Disinfectant)

Dissolve 25.5 gms. Caustic Soda in 140 c.c. water, warm this and add to a warmed mixture of 500 c.c. Cresylic Acid and 180 c.c. Rozolin; stir thoroughly and add water to make 1000 c.c.

Cresol Disinfectant

A.	Cresol	35
	Creosote	45
	Castor Oil Soap	20

B.	Rosin	17
	Washed Cresote	71
	Cresol	10
	Caustic Soda	2
	Water	2

White Cresol, Disinfectant

Cresol	50
Cresote	7.5
Glue	2.5
Castor Oil Soap	0.5
Caustic Soda	0.1
Water	60

Disinfecting Laundry

Soak for 1 hour in any of following and rinse.

Formalin	1
Water (at 140° F.)	200
Emulsified Cresol (Cresylic Acid)	1
Water	100

Use cold.

Lysol-Type Disinfectant
(Phenol Coefficient about 2.5)

Straw Colored Cresylic Acid (Phenol Coefficient about 5.0)	50 parts
Sulfonated Castor Oil, Con.	25 parts
25% Caustic Potash Solution	15 parts

Add the caustic potash while stirring to a mixture of the other two, and adjust either with alkali or red oil (oleic acid) until a sample dissolved in alcohol is neutral to phenolphthalein.

Pine Oil Disinfectant

A low priced disinfectant and deodorizer for spraying (when diluted with water) or general cleaning purposes.

Rosoap	10 lb.
Pine Oil	60 lb.

The pine oil is worked into the Rosoap, gradually, to avoid lumping. Part of the pine oil may be replaced by kerosene to lower costs. The above when stirred into water gives a milky white emulsin.

Pine Oil Disinfectants

Pine Oil Disinfectants are commonly made according to the Hygienic Laboratory Formula:

	Parts by Weight
Pine Oil	1000
"I" Wood Rosin (Acid Number—165)	400
Sodium Hydroxide (25% Solution)	200
	1600

It is prepared in the following manner: The Pine Oil and "I" Wood Rosin are heated together at a temperature of 80° C. in a jacketed steam kettle, the degree of heat is maintained until the rosin is thoroughly dissolved in the Pine Oil. The temperature is then dropped to 60° C. at which point the Sodium Hydroxide (25% solution) is added by stirring in very slowly. Saponification should be complete in thirty (30) minutes. This product has a predicted phenol coefficient of 3.5 to 4 determined by the Food and Drug Act Method against B-Typhosus.

The following formula was developed using a vegetable oil soap base:

Vegetable Oil Soap Base	20%
Yarmor Pine Oil	80%

Yarmor Pine Oil is added to the vegetable oil soap and stirred in slowly. No heating is required for this blend. This product has a predicted phenol coefficient of 5.2 determined by the Food and Drug Act Method against B-Typhosus.

The following label has been approved for disinfectants by the Government:

Pine Oil Disinfectants
Active Ingredients

Pine Oil	1
Soap or Base	2

Inert Ingredients

Moisture	3

(Moisture not to exceed 10% of total.)

Food and Drug Act Test—Phenol Coefficient (4). (Fill in blanks (1)—(2)—(3)—(4) to correspond with the disinfectant manufactured.)

Directions

In the bathroom.—To wash the bathtub, basin and toilet, apply in a 1 to 40 dilution in water.

In public places.—Schools, Hotels, Theatres, Stores, Office Buildings, Colleges, etc. Spray freely one part to forty parts of water.

In garbage receptacles.—To check the development of putrefactive action and breeding of flies. Spray the receptacle with a 1 to 40 dilution in water.

In the stable.—To help promote sanitation and destroy stable odors. Spray a 1 to 40 dilution in water.

In kennels, chicken houses, etc.—To kill lice, spray a 1 to 40 dilution with water on roosts and dropboards; to kill fleas, wash dogs in a 1 to 40 dilution in soapy water.

The Government has strict regulations to prevent labeling a product as a disinfectant if an adulterant is present.

Manufacturers should have a representative sample of their disinfectant tested for determination of phenol coefficiency.

The above procedures, if followed, insure the manufacturer of having a disinfectant labeled within the Government regulations.

A Steam-distilled Pine Oil Disinfectant made according to the prescribed rules and regulations insures the following:

1. Has a clear sparkling amber color.
2. Produces a snowy white emulsion in water.
3. Does not burn body tissues.
4. Is non-corrosive and non-toxic to humans.
5. Does not stain when in diluted form.
6. Leaves a clean piney odor whereever applied.
7. Kills typhoid, scarlet fever, diphtheria and cholera germs, etc.
8. Is free from suspended matter. This denotes uniformity.
9. May be used as an antiseptic for minor cuts and bruises as a wet dressing.

EMULSIONS

Emulsions

Theory

Since the theory and practice of emulsions is still in a highly disorganized state the theoretical side will be touched on but lightly.

An emulsion may be considered as a homogeneous suspension of tiny droplets of oil in water or water in oil. The oil in water type may be represented by the usual furniture polish (milky) and the water in oil type by butter. The term "oil" includes oils (mineral, vegetable, animal or essential), fats, greases, waxes, hydrocarbons (benzol, naphtha, turpentine, etc.), synthetics (thylene dichloride, nitrobenzol, etc.)—that is, something which does not mix with water.

Emulsification formulae and methods have been evolved chiefly through practice—by actually making innumerable emulsions. Because of the vagaries and eccentricities of emulsions practical workers have made greater technical advances in this field than the pure research chemists. Too often the trained chemist does not achieve as good emulsions as the lay worker—because the former rebels instinctly against empirical formulae and does not follow instructions as implicitly as the man "who knows he doesn't know." Moreover each new emulsion represents a new problem having numerous variable factors. These should not be underestimated if a good stable emulsion is desired. The technique and preparation of any particular formula should first be mastered before any variations are attempted.

Methods

Just as one man's food may be another's poison—so one method, which will give a perfect emulsion in one case. may produce a perfect failure in another. Thus no one method or emulsifying agent will serve universally. Specific technique will be given later in the case of the different emulsifying agents recommended.

When an emulsion of a solid melting above 100° C. is desired, it should first be melted with sufficient solvent or oil to reduce the combined melting point below 100° C. For example naphthalene with naphtha or other hydrocarbons; synthetic resins with hydrocarbons or vegetable oils.

Uses

Technical emulsions are used in numerous ways in many fields. The following are but a few of a large number of uses. Polishes, beauty creams, lotions, water-proofing, agricultural sprays, mayonnaise, cleaning compounds, lubricants etc. Many new specialty emulsions are likewise being created.

Summary

It must be borne in mind, however, that perfect results cannot be gotten until a few experimental emulsions are made in order to become familiar with working conditions. That is why experience shows that one of the given formulae should be mastered before attempting any variations.

Variations in raw materials, procedure, errors in proportions, etc., produce poor results. The formulae given have been repeated many times and will work if they are strictly adhered to.

Of course these formulae cannot fill every individual requirement. Variations are therefore necessary. In order to work out successful formulae, patience is essential. That which is worth while getting is worth while striving for. It is suggested that only one ingredient or proportion be varied at a time. This enables one to know exactly what produces the change in the finished product.

Emulsifying Agent

Ammonium Linoleate Paste

A cream colored paste; ammoniacal odor.

This is an excellent agent for emulsifying vegetable and fish oils, waxes, fat, resins, hydrocarbons and many other water insoluble products. When emulsifying a water insoluble product having a melting point of over 100° C., the latter should be first dissolved in naphtha, ethylene dichloride, turpentine or similar solvent. Alcohol as a rule should not be used as it breaks down most emulsions. Similarly acids, esters and salts must be avoided.

Procedure

Using proportions given in the following table, first dissolve the indicated amount of water in the Ammonium Linoleate Paste. This is done by covering the Ammonium Linoleate with the required amount of water and allowing it to soak over night. Work in slowly the next day until dissolved completely. Do not attempt to dissolve in any other way or lumps will result. To this add slowly with vigorous agitation the indicated amount of oil and continue stirring until homogeneous.

When a wax is to be emulsified the wax is melted and considered as an oil. In this case the water must be heated above the melting point of the wax. Most trouble is encountered in making wax emulsions because the solution of Ammonium Linoleate in water and the melted wax are not heated sufficiently. To play safe keep each of these solutions between 95 and 100° C., not allowing the temperature to drop below the melting point of the wax while adding one to the other. These formulae have been repeated numerous times with uniformly good results. If your emulsion is grainy or forms a film of wax on the surface, then the fault is in manipulation and not in the Emulsifier. Good wax emulsions cannot be made by hand or with a slow moving paddle. The vigorous agitation of a fast electric stirrer is essential.

Emulsions of the various inflammable hydrocarbons produce products of high cleansing powers and of a much higher flash-point.

In many synthetic reactions where better contact is desired between an aqueous and a water insoluble liquid recourse is had to emulsions. Similarly a water soluble solid may be dissolved in water and then emulsified with the water insoluble liquid.

Formulae
(All parts by Weight)

No.	Material Emulsified	Parts	Parts of Water	Parts Ammonium Linoleate Paste
1.	Kerosene.........	90	90	8
2.	Naphtha.........	90	100	7
3.	Benzol...........	90	100	7
4.	Gasoline..........	90	100	7
5.	Pine Oil..........	90	90	10
6.	Carnauba Wax....	90	620	12
7.	Beeswax..........	90	500	12
8.	Ozokerite.........	90	400	14
9.	Turpentine.......	90	100	8
10.	Nitrobenzol.......	90	100	8
11.	Orthodichlorbenzol	90	100	8
12.	Methyl Salicylate..	90	100	8

The above formulae can be lessened in cost by reducing the amount of emulsifier used. The minimum can be determined by experiment. Increasing the amount of water will give thin emulsions. Certain oil emulsions are improved by the addition of 1% or so of ammonia dissolved in water when making the emulsion.

Oil Emulsions
Using Triethanolamine Oleate

The procedure is to stir the triethanolamine oleate with the oleic acid until dissolved and then, while beating vigorously to run the oil and water into it in successive alternate portions. Emulsification takes place immediately and beating can be discontinued in a few minutes. These emulsions are very stable. As they are diluted, however, the degree of stability decreases. Salts, acids or other electrolytes disrupt these emulsions. The addition of small amounts of cresylic acid, alcohols or pine oil thicken them considerably.

Almond Oil Emulsion

Almond Oil	81
Triethanolamine Oleate	6
Oleic Acid	6
Water	81

Castor Oil Emulsion

Castor Oil	82
Triethanolamine Oleate	6
Oleic Acid	12
Water	82

Chinawood Oil Emulsion

Chinawood Oil	86
Oleic Acid	10
Triethanolamine Oleate	6
Water	78

Coconut Oil Emulsion

Coconut Oil	81
Triethanolamine Oleate	6
Oleic Acid	12
Water	82

Corn Oil Emulsion

Corn Oil	86
Triethanolamine Oleate	6
Oleic Acid	6
Water	82

Cod Liver Oil Emulsion

Cod Liver Oil	82
Triethanolamine Oleate	6
Oleic Acid	6
Water	80

Cotton Seed Oil Emulsion

Cotton Seed Oil	86
Triethanolamine Oleate	6
Oleic Acid	6
Water	80

Emulsifying Agent
Trihydroxyethylamine Stearate
(T. S. for short)

A light brown wax. Faint fatty odor.

In the formulae given below proceed as follows:

Melt the T. S. with the oil and add this to the water (some prefer to use warm water) slowly while stirring vigorously with an electric mixer. Warm water and very rapid stirring produce uniformly stable emulsions.

Formulae

	Material Emulsified	Parts	Parts Water	Trihydroxyethylamine Stearate
A.	Mineral Oil......	75	185	15
B.	Pine Oil	75	85	14
C.	Turpentine......	75	85	14
D.	Paraffin Wax....	85	200	10
E.	Eucalyptus Oil...	75	85	14
F.	Balsam Copaiba..	75	85	14
G.	Gasoline.........	75	85	14

Fuel or Lubricating Oil Emulsion

Fuel or Lubricating Oil	88
Triethanolamine Oleate	6
Oleic Acid	5½
Water	90

Lard Oil Emulsion

Lard Oil	88
Triethanolamine Oleate	9
Oleic Acid	4
Water	76

Linseed Oil Emulsion

Linseed Oil	86
Triethanolamine Oleate	6
Oleic Acid	6
Water	78

Menhaden Oil Emulsion

Menhaden Oil	86
Triethanolamine Oleate	6
Oleic Acid	6
Water	80

Neatsfoot Oil Emulsion

Neatsfoot Oil	86
Triethanolamine Oleate	6
Oleic Acid	6
Water	78

Olive Oil Emulsion

Olive Oil	86
Triethanolamine Oleate	6
Oleic Acid	6
Water	78

Palm Oil Emulsion

Palm Oil	86
Triethanolamine Oleate	6
Oleic Acid	6
Water	80

Paraffin Oil Emulsion

Paraffin Oil	80
Triethanolamine Oleate	12
Oleic Acid	6
Water	80

Peanut Oil Emulsion

Peanut Oil	86
Triethanolamine Oleate	6
Oleic Acid	6
Water	80

Perilla Oil Emulsion

Perilla Oil	84
Triethanolamine Oleate	6
Oleic Acid	6
Water	82

Rapeseed Oil Emulsion

Rapeseed Oil	85
Triethanolamine Oleate	6
Oleic Acid	6
Water	85

Rosin Oil Emulsion

Rosin Oil	80
Triethanolamine Oleate	6
Oleic Acid	4
Water	82

Rubberseed Oil Emulsion

Rubberseed Oil	84
Triethanolamine Oleate	6
Oleic Acid	6
Water	84

Sesame Oil Emulsion

Sesame Oil	86
Triethanolamine Oleate	6
Oleic Acid	6
Water	85

Soya Bean Oil Emulsion

Soya Bean Oil	86
Triethanolamine Oleate	6
Oleic Acid	6
Water	85

Sperm Oil Emulsion

Sperm Oil	82
Triethanolamine Oleate	6
Oleic Acid	6
Water	82

Oil in Water Emulsions, Coloring

Water soluble dyes are recommended.

The dyes are best dissolved in the water to be used in the emulsion.

If the emulsion is to be colored after completion, dissolve the color in as little water as possible and add the concentrated dye solution to the emulsion and stir vigorously. If impractical to add this small quantity of water, the dry dye should be added in small amounts to the emulsion and stirred vigorously.

Maroon Color
Pylam Water Maroon 1 lb. to 400 gal.

Yellow
Tartrazine 1 lb. to 1200 gal.

Black
Nigrosine Crystals 1 lb. to 1200 gal.

Violet
Alizarine Violet 1 lb. to 1200 gal.

Green
Pylam Chloro Green S-310 1 lb. to 1200 gal.

Rose
Violamine 2R, DuPont 1 lb. to 1200 gal.

Brown
Bismarck Brown 1 lb. to 1200 gal.

Amber
Pylam Amber S-271 1 lb. to 400 gal.

Coloring Washing Powders

Dishwashing and cleaning compounds made from T. S. P., modified soda, soda ash or combinations of same are colored as follows:

Favorite color is peach, which gives a greenish fluorescence when dissolved in water.

Color: Soluble Fluorescene,
1 lb. to 1250 lbs. compound

Coloring Mineral Oil

Brilliantines and light mineral oils for same are colored as follows:

Green
Pylakrome Green LX-799 1 lb. to 1600 gal.

Yellow
Azo Yellow 1 lb. to 1600 gal.

Blue
Alizarine Oil Blue 1 lb. to 2000 gal.

Violet
Grasol Violet 1 lb. to 2000 gal.

Red
Pylakrome Red No. 420 1 lb. to 1600 gal.

Water in Oil Emulsions, Coloring

Oil soluble colors are recommended. The colors are dissolved in the oil before emulsification.

Green—Pylakrome Oil Green 1119	8 oz. to 100 gal.
Azo Yellow	8 oz. to 100 gal.
Alizarine Oil Blue	8 oz. to 125 gal.
Alizarine Oil Violet	8 oz. to 125 gal.
Azo Oil Red	8 oz. to 100 gal.
Oil Black	8 oz. to 25 gal.

Black Wax Emulsions

To color any non-edible wax emulsion black, stir into it, with a high speed mixer about 10 parts Paris Paste per every 100 parts of wax present in the emulsion.

Rosin Emulsions

Rosin	700 gr.
Water	2100 cc.
Glue	150 gr.

Melt glue in water and while boiling hot, slowly add melted rosin, agitating violently. Continue agitation until perfectly smooth.

Rosin	700 gr.
Water	2100 cc.
Gelatine	150 gr.

Melt Gelatine in water and while boiling hot, add melted rosin slowly, agitating violently. Continue agitation until perfectly smooth.

Rosin	700 gr.
Water	2100 cc.
Stearic Acid	63 gr.
Triethanolamine	21 gr.

Melt Rosin and Stearic Acid together. Add Triethanolamine to water. Heat water to boiling point and stir in melted rosin. Stir until smooth.

Rosin	14 gr.
Water	84 cc.

Heat to boiling; turn off heat and add while stirring vigorously

Ammonium Hydroxide	5 cc.
Water	34 cc.

Stir until all lumps disappear (reheating may be necessary).

Asphalt Emulsions

Asphalt	500 gr.
Water	500 cc.
Bentonite	30 gr.
Quebracho	30 gr.
Soda Ash	10 gr.

Combine bentonite, Quebracho, soda ash and water and heat to 200° F. While stirring, add asphalt which has been heated to approximately 200° F. Continue stirring until asphalt is dispersed.

Asphalt	2800 gr.
Water	2800 gr.
Rosin Soap (50%)	118 gr.
Pine Oil	40 cc.

Add rosin soap to water and heat to 200° F. Heat asphalt to 200° and add pine oil. While agitating, slowly pour asphalt into water and continue agitating until a smooth emulsion is formed.

---

* Asphalt Emulsion

A very stable 1:1 bitumen–H_2O emulsion is formed by adding part of the melted asphalt, while stirring, to hot dil. NaOH; when a scum begins to form, 0.5% of oleic acid is added and the rest of the asphalt together with $< 2\%$ of clay.

---

* Pitch Emulsion

Bitumen or pitch is dispersed in an aq. paste of starch the p_H of which has been adjusted to 4.0–5.0 by the addition of tannic acid or material containing tannin. The product may contain 60% of bitumen, 1.5–2% of starch, 0.15–0.2% of tannic acid, and the remainder H_2O.

Soluble Oil Emulsions

The soluble oil method is particularly applicable for medium viscosity mineral oils and is not successfully applied to other oils or solvents. With such mineral oils, however, the method yields excellent emulsions which are quite stable. These oils usually require from 3.5 to 4.0 per cent Triethanolamine, depending upon the stability desired in the emulsion. The amount of oleic acid lies between 8 and 11 per cent, the amount varying especially with the type of oil. The more refined oils are the most difficult to emulsify as will be seen from the following table:

Soluble Oils

Type of Oil	Color	Oil	Oleic Acid	Triethanolamine
Cutting Oil...	Yellow	88 lb.	8.0 lb.	3.7 lb.
Textile Oil...	Bloom	87 lb.	8.8 lb.	3.5 lb.
Medicinal Oil.	White	86 lb.	10.0 lb.	4.0 lb.
Rayon Oil....	White	85 lb.	10.4 lb.	4.0 lb.

Formulation by this method requires great exactness, and it is always necessary to derive formulae for the specific oil to be emulsified because of the great variation in commercial petroleum products. Given an unknown oil, take 88 grams, add 8.0 grams of oleic acid and stir to a clear solution. Now measure carefully 4.0 grams of Triethanolamine into this solution and stir thoroughly. On holding this mixture up to the light, it will usually be cloudy or show minute suspended droplets. Now add oleic acid drop by drop, stirring thoroughly after each addition until the mixture becomes clear. It will now emulsify in water, but a few drops further of acid will give a slightly superior soluble oil. The total oleic acid can now be calculated and the whole formula reduced to the basis of 100 pounds.

Olive Oil Emulsions

Olive Oil	88 lb.
Oleic Acid	10 lb.
Triethanolamine	2 lb.
Water	80 lb.

Preparation

Working at ordinary temperatures add the Triethanolamine, oleic acid and 30 lbs. of the olive oil to the agitator. As soon as these three ingredients have been added, but not before, stir vigorously until the mixture is fairly homogeneous. Then slowly add with constant stirring 33 lbs. of water, obtaining a thick smooth emulsion.

Continuing with the same stirring rate, first add the remainder of the oil in small portions, and finally the remaining water in a similar manner. Emulsification is complete when the oil and water are evenly distributed.

Properties

The olive oil emulsion so prepared is pure white and creamy, and will be stable in the concentrated form in which it is made. If greater dilution is desired, water may be stirred into this emulsion in any proportion. When diluted to five times its volume, giving a 9% concentration of oil in water, no separation should occur within 24 hours.

Variations

When storage for an indefinite period of time is unnecessary, a technically satisfactory product can be similarly prepared with only 7 per cent oleic acid and 1½ per cent Triethanolamine.

Uses

Textile lubrication, shampoos, hand lotions.

Pine Oil Emulsion

Pine Oil	91 lb.
Oleic Acid	6 lb.
Triethanolamine	3 lb.
Water	100 lb.

Preparation

Add the oleic acid, Triethanolamine and 30 lbs. of the pine oil to the mixer and stir until the product is clear. Then add very slowly an equal volume of water stirring vigorously meanwhile.

When this mixture has become a smooth uniform emulsion, the remainder of the oil is gradually added with constant agitation. The rest of the water is next similarly added until emulsification is complete.

Properties

The pine oil emulsion so prepared is a creamy, white product which is indefinitely stable when concentrated. It can be further diluted as desired, the best results being obtained when the water is stirred into the product. At very high dilutions, such as is obtained with 1% oil in water, good dispersion and fairly high stability still characterize the emulsion.

Variations

To make this product as concentrated as possible and still maintain ready dilution with water, preparation is suggested as follows. Dissolve 3 lbs. of Triethanolamine in 40 lbs of water and add slowly, with high-speed stirring, a solution of 6 lbs. of oleic acid in 91 lbs. of pine oil.

Uses

Disinfectants and deodorants, textile wetting-out and scouring agents.

Light Mineral Oil Emulsion

Mineral Oil	88 lb.
Oleic Acid	8.0 lb.
Triethanolamine	3.7 lb
Water	

Formulation

The above formula was derived for a particular low viscosity lubricating oil and is typical of the formulation for a cutting oil.

Preparation

Weigh out the oleic acid and 8 pounds of the mineral oil and stir together to obtain a uniform solution. Then add the exact amount of Triethanolamine and stir until the solution is clear. Some warming will occur during the reaction of the acid and amine.

This soluble oil base is dilutable with the remainder of the oil at any time. Simply stir the remaining 80 pounds of the oil into the base, or four parts by weight of the oil to one part of the base.

Properties

Both the soluble oil base and the resulting soluble oil are stable indefinitely and will not separate on standing when made up in the proper proportions. The product emulsifies spontaneously when poured into water. The best method of emulsifying, however, is accomplished by stirring the oil with an equal volume of water until a smooth creamy mass is obtained, and this can be diluted further with water as desired.

Uses

Cutting oils, soluble greases.

Refined Mineral Oil Emulsion

Mineral Oil	87 lb.
Oleic Acid	8.8 lb.
Triethanolamine	3.5 lb.
Water	

Formulation

Typical of the partially refined mineral oils is the above formula which was derived for an oil suitable for an orchard spray. Similar formulae will be found for other oils of this type.

Preparation

Into a container equipped with a simple stirring device, pour 7 pounds of the mineral oil together with all of the oleic acid. Stirring for a few minutes produces a homogeneous solution to which should now be added the exact quantity of Triethanolamine. Mix this into the liquid until a clear solution results.

To the above product add the remainder of the oil and stir sufficiently long to obtain a uniform solution. In making shipments it will be sufficient to add one part of the oil base to four parts of the untreated oil without stirring.

Properties

The soluble oil so prepared will not deteriorate or separate on storage. It will emulsify spontaneously when added slowly to water and will form a stable white emulsion. Dilution, however, is best performed by first stirring well with an equal volume of water and then diluting to the extent desired.

Uses

Orchard spray, hand cleaner, shampoo.

White Paraffin Oil Emulsion

Paraffin Oil	85 lb.
Oleic Acid	10.4 lb.
Triethanolamine	4.0 lb.
Water	

Formulation

The refined white paraffin oils require somewhat more Triethanolamine and oleic acid to become readily soluble. The above formula is an example of the proportion of ingredients for a particular oil of this type. Preparation of the emulsion will be considered on the basis of this particular oil.

Preparation

In weighing out the ingredients, all measurements except those of the oil should follow the formula very exactly. Add the oleic acid to 5 pounds of the oil and stir until homogeneous. Then add the Triethanolamine and mix thoroughly until a clear viscous solution is obtained. The mass will heat up on account of the saponifying action of the amine upon the oleic acid.

To this soluble oil base, the remaining 80 pounds of paraffin oil may be added when desired, or simply one part by weight of this base to four parts of the oil. Stirring sufficient to mix the two materials only is necessary. A perfectly clear liquid will result if the proportion of oleic acid is correct.

Properties

This soluble oil is readily emulsified into any quantity of water, although it is preferable to agitate it first to a thick creamy emulsion with an equal volume of water before further dilution. At a concentration of 5 per cent oil in water, the emulsion should be perfectly stable for 24 hours.

Uses

Rayon lubrication, cosmetic creams.

Neat's Foot Oil Emulsion

Neat's Foot Oil	88 lb.
Oleic Acid	10 lb.
Triethanolamine	2 lb.
Water	80 lb.

Preparation

Add together the oleic acid, Triethanolamine, and 30 lb. of the Neat's foot oil at ordinary temperatures. Mix thoroughly in the agitator and gradually add 33 lb. of water, stirring vigorously meanwhile. A thick, uniform emulsion will result.

Continuing with vigorous stirring, add slowly the remainder of the oil and then the rest of the water. Discontinue stirring when an even mixture is obtained.

Properties

The Neat's foot oil emulsion prepared as above is a uniform white and stable product. The stability decreases somewhat on dilution, although in a 10% concentration of oil in water, no separation may be expected to occur within 24 hours. Dilution down to 1% is possible, provided the water is carefully stirred into the original thick emulsion.

Variations

When the emulsion is to be used shortly after preparation, the percentages of Triethanolamine and oleic acid can be considerably reduced. This is best brought about by stirring further oil and water alternately into the original emulsion by the procedure given above. On the other hand, the Triethanolamine method of emulsification can be used.

Uses

Leather soaking, silk soaking, leather dressing.

Linseed Oil Emulsion

Linseed Oil	88 lb.
Oleic Acid	10 lb.
Triethanolamine	2 lb.
Water	80 lb.

Preparation

Working at ordinary temperatures, thoroughly mix the oleic acid, Triethanolamine and 30 lb. of the linseed oil. Add 33 lb. of water to this mixture slowly with constant, vigorous stirring. This procedure yields a thick, smooth emulsion.

The remainder of the oil is then added in small portions, maintaining the same stirring rate, and the rest of the water is added similarly. Stirring is discontinued as soon as the last of the water has been evenly dispersed.

Properties

This emulsion is of the oil-in-water type, and is a stable, creamy product which can be further diluted with water. The emulsion should be stored in an airtight container as oxidation of the oil decreases the stability of the emulsion.

Variations

In the case of linseed oil, it is often desirable for the emulsion to be of the water-in-oil type. If the procedure given above is followed, using 1% Triethanolamine and 3% free fatty acid instead of the indicated proportions, such an emulsion will result.

The given formulae have been tested on a boiled linseed oil and will require some alteration for raw linseed oils. These different oils have a variable free fatty acid content which affects chiefly the amount of oleic acid to be added in emulsification.

Uses

Emulsion paints, linoleum coatings.

This is the most general of the emulsification methods and can be successfully used to emulsify most of the products. In the same way that previous methods are particularly valuable for certain products, however, the Triethanolamine method is invaluable in specific cases. It is recommended for mineral solvents, such as gasoline, naphtha, kerosene and benzol, and for many of the emulsion mixtures, such as polishes and cosmetic creams.

A number of polish and cosmetic formulae are given later to explain the use of this method in the latter connection. The following are typical of the formulae for emulsions of the mineral solvents, the ingredients being given as usual on a weight basis.

Triethanolamine Method

Type of Solvent	Solvent	Oleic Acid	Triethanolamine	Water
Kerosene..	89	8	3	100
Naphtha...	82	14	4	100

In using this method, two solutions are made up, to be called the "oil solution" and the "water solution" respectively. The solvent and oleic acid are mixed and being mutually soluble, form a clear oil solution. In a separate container, the water and Triethanolamine are similarly dissolved together at ordinary temperatures to form a clear water solution.

The oil solution is then added in its entirety to the water solution, and the mixture at once violently agitated. A white emulsion results instantly. To obtain good stability it is important that stirring be as rapid as possible at the start, and then be continued intermittently a few times after the emulsion has formed.

Kerosene Emulsion

Kerosene	89 lb.
Oleic Acid	8 lb.
Triethanolamine	3 lb.
Water	100 lb.

Preparation

The preparation of this emulsion is typical of the procedure used for any liquid. In one container weigh out the above quantities of kerosene and oleic acid and mix these two liquids thoroughly. In a separate container stir together the water and Triethanolamine until a homogeneous solution is obtained.

The oil solution is now poured into the water solution, and the resulting mixture is stirred or agitated vigorously. After the emulsion is well formed, it should be stirred occasionally, a few minutes at a time.

Properties

This method produces a pure white emulsion of kerosene which possesses excellent stability. It is of the oil-in-water type and can be diluted to any extent desired by the addition of water.

Variations

With the given formula the amount of water in the kerosene emulsion may be reduced to 50 pounds, the emulsion remaining of the oil-in-water type. If the proportion of water is further lowered, and particularly if the emulsion is prepared by gradually adding the water solution to the oil solution, the resulting product will usually be of the water-in-oil type. The latter type is also favored by an increase in the percentage of oleic acid.

Uses

Polishes, cleaning compounds, insecticides.

The water method of emulsification has been developed for use particularly with waxes and other semi-solid materials, such as greases and asphalts, and for the preparation of the wax polishes. It gives very good results, however, in other emulsion problems and is a convenient method whenever stearic acid is preferable to liquid fatty acids.

The following tabulation presents suggestions for the formulae to be adopted for waxes. The proportions of the ingredients are given on a weight basis.

Water Method

Type of Wax	Wax	Stearic Acid	Triethanolamine	Water
Beeswax...	88	9	3	300
Carnauba..	87	9	4	400
Paraffin...	88	9	3	300
Lanolin....	80	15	5	200
Japan.....	85	12	3	400

In this method, the wax or oil is emulsified by means of a water solution of the soap which is made from the Triethanolamine and stearic acid. The water is measured out into a container or kettle which can be heated. The Triethanolamine is then stirred into this and then the stearic acid is added. On heating, the acid gradually melts and can be stirred into the water to give a smooth soap solution, and the temperature is raised to just below the boiling point. The wax is now melted in a separate container and its temperature brought to 85–95° C. This is then added to the water solution and the whole at once stirred vigorously to obtain a good emulsion. Stirring is then continued gently until the product has cooled.

Carnauba Wax Emulsion

Carnauba Wax	87 lb.
Stearic Acid	9 lb.

Triethanolamine	4 lb.
Water	400 lb.

Preparation

Weigh out the stearic acid, water and and Triethanolamine, and heat the mixture in a kettle to 100° C. After the acid has melted completely and the solution is boiling gently, stir carefully until the acid has been dissolved and a smooth soap solution is obtained.

In a separate steam-heated container melt the carnauba wax until a temperature of 85–90° C. is reached. Do not allow the temperature to rise above 95° C., or the wax will be darkened in color. Now add the molten wax to the boiling soap solution and stir vigorously until an even dispersion of the wax results. Stir gently, but continuously, until the emulsion has cooled to room temperature.

Properties

The carnauba wax emulsion, when prepared as described, is a very smooth, cream-colored product. It is rather viscous when cold, but of thinner consistency when warm, and is a very stable emulsion. It may be diluted with water if desired.

Variations

The substitution of oleic for stearic acid in the above formula produces an emulsion somewhat less stable but much less viscous. It therefore permits a considerably higher concentration of wax to be used. When other ingredients are to be added they are best included prior to emulsification by methods given in the polish formulae.

Uses

Leather dressings, auto polishes.

Paraffin Wax Emulsion

Paraffin Wax	88 lb.
Stearic Acid	9 lb.
Triethanolamine	3 lb.
Water	300 lb.

Preparation

Mix the water, Triethanolamine and stearic acid and heat to 100° C., allowing the mixture to boil gently. Then stir carefully so that a smooth soap solution is obtained with a minimum of foam. In a separate container melt the paraffin wax and bring its temperature to 90° C. Add the hot wax immediately to the boiling soap solution and stir vigorously until the wax is evenly dispersed. Continue to stir the emulsion slowly while cooling.

Properties

The paraffin wax emulsion so prepared is a creamy, white product, the consistency and stability of which are but little affected by temperature changes. In its concentrated form, no separation will occur over a period of months. To make a more dilute product, water may be stirred into this emulsion.

Variations

The wax and stearic acid are melted together over a steam bath until the temperature is 90° C. and thoroughly mixed. The mixture is then poured quickly into a boiling solution of the Triethanolamine and water, and is at once violently agitated. The emulsion is then stirred gently until it has cooled to room temperature. The same procedure is to be followed if oils or solvents are to be mixed with the wax, the only change being a substitution of a portion of the wax in the given formula.

Uses

Sizing and waterproofing, cosmetic creams, polishes.

Triethanolamine Emulsions

The soap method of emulsification has proved to be the most practical method of treatment for the majority of animal and vegetable oils, as well as for certain types of organic solvents.

Typical of the oils to be emulsified by this method, as well as the formulae to be developed, is the following table which gives the proportions of ingredients by weight:

Soap Method

Type of Oil	Oil	Oleic Acid	Tri-ethanol-amine	Water
Castor Oil	82	16	2	80
Cottonseed Oil	88	10	2	80
Lard Oil	87	10	3	80
Linseed Oil	88	10	2	80
Olive Oil	88	10	2	80
Neat's Foot Oil	88	10	2	80
Dichlorethyl Ether	83	12	5	100
Ethylene Dichloride	86	10	4	100
Lubricating Oil	89	9	2	100
Pine Oil	91	6	3	100
White Paraffin Oil	82	14	4	100

The procedure by this method consists in stirring the oil and water alternately into the soap made from the Triethanolamine and oleic acid. To one-third of the oil is added the total oleic acid and Triethanolamine and these are stirred together until homogeneous. Stirring vigorously, a volume of water equal to the oil present is now added slowly, a thick creamy emulsion resulting. The remainder of the oil can next be added with continuous stirring, and finally the rest of the water in a similar manner. The following pages detail some emulsions prepared by this method.

Ethylene Dichloride Emulsion

Ethylene Dichloride	86 lb.
Oleic Acid	10 lb.
Triethanolamine	4 lb.
Water	100 lb.

Preparation

Mix together the oleic acid, Triethanolamine and 30 pounds of the ethylene dichloride until a clear solution is obtained. This will become somewhat warm due to the saponification of the fatty acid by the Triethanolamine. Now add slowly, with thorough stirring, 50 pounds of the water, finally obtaining a thick creamy emulsion.

Then with continued stirring of this emulsion, add first the remainder of the oil in small portions and finally all of the water likewise.

Properties

The emulsion resulting above is pure white and stable. It may be further diluted with water to any extent desired, the emulsion containing 20% of solvent showing no separation in 24 hours. Emulsions of chlorinated hydrocarbons can not be stored indefinitely because of a gradual hydrolysis in the presence of water. It is, therefore, recommended for use within a few weeks of its manufacture.

Variations

Ethylene dichloride is frequently used in textile scouring agents in emulsified form. For such uses a high proportion of soap to solvent is desirable, and in this case a soluble ethylene dichloride can be made. A clear solution results when 50 pounds of this solvent are stirred with 34 pounds of oleic acid and 16 pounds of Triethanolamine, and this mixture emulsifies instantaneously when added to water.

Uses

Scouring and wetting agents, polishes, insecticides.

Emulsifying Agent

Di-Glycol Stearate

A light colored wax. Practically odorless (m.p. 58–60° C.). This is absolutely free from alkalies or amines.

One part of Di-Glycol Stearate when melted in 10–30 parts of boiling water produces, on stirring, while cooling, a uniform milky dispersion of the wax in water which is very stable. The consistency varies with the amount of water used. They may be also used as lubricants to be squirted between spring-leaves or other inaccessible places. On evaporation of the water a film of non-flowing wax remains behind as a lubricant. These make excellent suspending media for titanium dioxide, carbon black, graphite, silica and other abrasives.

Formulae

A	Di-Glycol Stearate	10	Pine Oil.....40	Water 40
B		10	Mineral Oil...50	Water 500
C		10	Paraffin Wax 40	Water 250
D		10	....	Water 50
E		10	....	Water 300

Procedure

The oil or wax is melted with the Di-Glycol Stearate. The water is heated to a temperature above the melting point of the wax and added slowly while stirring vigorously. Continue stirring until cool. By varying the amounts of water, emulsions of varying consistency are obtained. They are very white in color and stable. Other oils and waxes may be emulsified in a similar way.

Formulae (A), (B), (C), (D) all useful as polishes.

Formula (A) serves as a liniment, disinfectant or deodorant. The pine oil may be replaced by turpentine, citronella oil or perfume compounds.

Formula (B) with a little perfume dissolved in the oil makes an excellent lotion or liquid cleansing cream.

Formula (D) with a little perfume is used as a lotion or powder base.

Formula (E) serves as a greaseless ointment in paste rouge base (with the addition of Glyco Wax B).

Emulsifying Agent

Miscibol (Pot. Oleo-Abietate)

A viscous paste; resinous odor. Alkaline reaction.

Used in place of Turkey Red or Sulfonated oils where an acid product is undesirable. For making "soluble" oils.

The following formulae gives clear solutions without heating. When these solutions are thrown into water they diffuse rapidly to give milky emulsions.

A.	Pine Oil	6 lb.
	Miscibol	1 lb.

B. Pine Oil 5 lb.
Kerosene 1 lb.
Miscibol 1 lb.
Water 1 lb.

Asphalt Emulsion

A hot dil. aq. soln. of alkali such as a soln. of NaOH of about 0.5% strength is prepd. and about an equal wt. of asphalt is melted; part of the melted asphalt is stirred into the hot soln. slowly until scum begins to form on the surface, then a small quantity (suitably about 0.5% of the final product) of oleic acid is added, followed by addn. of the rest of the asphalt while the temp. and agitation are maintained and a small proportion of clay is added to give desired stability and adhesiveness.

Asphalt Emulsion

A sodium oleate solution is made up to a concentration of 20 per cent by the addition of oleic acid and caustic soda to water at 90° C. This is then diluted with 9 times its volume of water heated to 90° C. The 2 per cent soap solution is run through the colloid mill with an equal amount of asphalt heated to not exceeding 100° C. The resultant emulsion contains equal parts of asphalt and water, with 1 per cent by weight of soap.

Carnauba Wax, Kerosene Emulsion

Carnauba Wax	16.0 gr.
Kerosene	20.0 cc.
Ammonium Linoleate	2.4 gr.
Water	200.0 cc.

The ammonium linoleate was placed in a vessel and covered with the water (cold) and allowed to stand overnight. The following day it was warmed and stirred until completely dispersed in the water, taking care that no lumps were left. This was taken to 90° C. and stirred by means of a high speed mixer. The wax was melted, taken to 100° C., and the kerosene added and stirred until the wax was dissolved in it. This was then added to the hot ammonium linoleate dispersion and the agitation continued until the emulsion was cool. This gave a fluid emulsion.

Carnauba Wax, Mineral Oil Emulsion

Mineral Oil (Spindle)	19 cc.
Carnauba Wax	18 gr.
Ammonium linoleate	2.4 gr.
Water	102 cc.

The ammonium linoleate and water were allowed to stand overnight as above. Then heated to 90° C. and stirred by means of high-speed mixer. The wax and oil were heated together until the wax dissolved in the oil, and taken to 100° C. This solution was then added to the ammonium linoleate dispersion in water, and stirred rapidly. This gave a paste emulsion.

* Colloidal Lecithin

Lecithin	1 lb.
Glycerol	1 lb.
Gelatin	2 lb.
Water	100 lb.

Warm and stir until dispersed.

Cumar Emulsion

Cumar	50
Naphtha	50

Allow to stand overnight and stir until dissolved. Add

Blendene 20

while stirring with a high-speed agitator; add slowly

Water 100

Stir vigorously for 5–10 minutes.

Halowax Emulsions

Formula No. 1

Water	3 lb.
Halowax	1 lb.
Stearic Acid	53 gm.
Triethanolamine	27 gm.

Formula No. 2

Water	3 lb.
Halowax	3 lb.
Stearic Acid	108 gm.
Triethanolamine	54 gm.

Formula No. 3

Water	3 lb.
Halowax	13 oz.
Halowax Oil No. 1000	3 oz.
Stearic Acid	108 gm.
Triethanolamine	54 gm.

Melt the wax and stearic acid together and stir. Heat the water and triethanolamine until they start to boil. Add the wax to the water and stir with an electric stirrer and then run through the colloid mill. A good emulsion is obtained if it is cooled quickly after coming from the colloid mill.

Lanolin Emulsion (Fluid)

Diglycol Oleate (Light)	10 gm.
Lanolin (Anhydrous)	30 gm.

Warmed till dissolved. Added to the above with rapid agitation

Water	60 cc.

made slightly alkaline with Caustic soda (¼%). Stir five to ten minutes.

Paradichlorbenzene Emulsion

Paradichlorbenzene	12 gm.
Glycol Stearate	3 gm.
Water	150 cc.

Melt the glycol stearate in the water (about 90° C.). Stir rapidly (high-speed mixer). Melt the paradichlorbenzene, preferably on water bath and add slowly to the stearate dispersion in water. Continue stirring until cool.

Rosin, Turpentine Emulsion

Rosin	11.0 gm.
Turpentine	2.5 gm.
Ammonium Linoleate	2.0 gm.
Water	50.0 cc.
Ammonia	15.0 cc.

The ammonium linoleate and water were taken up in the usual way (see above). Heated, and mechanically agitated (high-speed mixer). The rosin and turpentine were heated together and added to the ammonium linoleate dispersion in water to which had previously been added the 15 cc. of ammonia. Stirring was continued until cool.

This gave a paste emulsion.

---

* Rubber Emulsion

A mixt. of rubber 100, benzene 10–100, glue 1 and casein 1 part is masticated while slowly adding an aq. soln. of K oleate until the rubber constitutes the dispersed phase of the batch. The product is suitable for use as a cementing medium.

Raw Tallow Emulsion (50%)

Raw Beef Tallow (Good Quality)	80–100 lb.
Trihydroxyethylamine Stearate	9 lb.
Water	90–100 lb.

(6–8 ounces of Trisodium phosphate added to water may prove advantageous if water used is of a high degree of hardness.)

This is a substitute on an equal basis for commercial 50% Sulfonated Tallow in sizing preparations.

---

*Trichlorethylene Emulsion

Trichlorethylene	67 lb.
Turkey Red Oil	4.5 lb.
Bentonite	2 lb.
Water	26 lb.
Amonnia (26° Bé.)	0.45 lb.

---

*Sulfonated Mineral Oil

This is useful as an emulsifying agent and "spreader"—for various emulsions—particularly agricultural sprays.

Two volumes of lubricating stock such as brown neutral oil are mixed under continuous and rapid agitation with one volume of acid, ordinary 95 to 97 per cent sulfuric acid, for thirty minutes at a temperature approximately 35° to 40° C. The reactivity of the oil in commercial batches tends to cause excessive heating and the material should be suitably precooled or else the mixture intimately cooled to avoid the rise of temperature above 50° C.

One feature of this process is the control of temperature during the reaction and terminating the reaction in the minimum time so that the useful product acids produced will not be destroyed. The temperature of the reacting mass of sulfuric acid and mineral oil must not be allowed to rise materially above 50° C. and that the lowest temperatures compatible with a satisfactory reaction produce the best grade of water soluble product acid. With proper temperature control either ordinary concentrated sulfuric acid or 20% fuming acid may be employed.

The production of this useful product acid is also possible by the use of a proportional quantity of fuming sulfuric but it is then preferable to add the fuming acid gradually to the oil while the agitator is running and also to cool the reacting mass as with a water bath surrounding the agitating vessel and preferably cooled agitators, etc.

At the end of the reaction time one volume of water is added to this mass, and mixed by the same rapid agitation until uniformly distributed. The mass is then allowed to stratify into three layers. The upper layer consists of oil and oil soluble sulfonic acids. The middle layer consists of some oil, sulfonic acids, sulfonic tars, organic sulfur bodies, sulfuric acid, water, some sulfurous acid and the bodies in which I am interested. The bottom layer consists essentially of sulfuric acid and water.

The oil layer (upper) and the acid layer (lower) are then removed from the middle layer which is a thick greenish black mass and may even show a purple color in thin layers.

The separated middle layer is then dissolved in three volumes of alcohol (methyl or ethyl) and dry finely powdered soda ash (sodium carbonate) added under agitation until the strong free acids are neutralized and precipitated as salts insoluble in alcohol of this concentration.

The aqueous alcohol is then filtered to remove the precipitated salts. The resulting alcoholic filtrate is then further diluted with an equal volume of water and either exactly neutralized with a solution of sodium hydrate, or may be made alkaline with ammonia.

The filtrate is then placed in a still to remove and recover the excess of alcohol. When the alcoholic content of the filtrate has been reduced to approximately ten per cent by volume, the filtrate is removed from the still and placed in a closed agitating vessel where it is heated to 50° C. and mixed with approximately one-third its volume of benzol (benzene) to purify it. After sufficient agitation the mixture is allowed to stratify and the lower layer containing product in solution is drawn off from below. This benzol washing is repeated three or four times or until the sulfonic tars have been practically all removed.

After the solution has been washed with benzol it is returned to the still where the dissolved benzol and remaining alcohol are removed by further distillation.

The resulting purified product may be reduced to any desired consistency by evaporation. When reduced so as to contain 30 to 40 per cent solids product is a brown colored syrupy liquid completely soluble in water and in alcohol and contains practically no oil soluble matter. The color is variable from light brown to nearly black depending on the time and temperature of the original reaction and on the oil stock used. In general the lighter colored acids are of superior quality.

ETCHING, ENGRAVING, LITHOGRAPHING

Brass, Black Pickling of

Copper Carbonate	750
Ammonia Hydroxide	150

Immersion from 3 to 8 minutes is indicated.

Etching Glass

A. Sod. Fluoride	9 dr.
Pot Sulfate	108 gr.
Water	1 pt.
B. Hydrochloric Acid (conc.)	10 fl. dr.
Zinc Chloride	216 gr.
Water	1 pt.

Mix solutions A and B in equal amounts before use.

Etching Steel

The following solution is used.

Nitric Acid	32 oz.
Hydrochloric Acid	3 oz.
Denatured Alcohol	16 oz.
Water	96 oz.

---

* Stainless Steel, Etching

Iron Chloride	100
Water	50–75

Use at 25–37.5° C.

Etch Resist

In etching glass it is necessary at times to block off portions which one desires to keep unetched. A solution for this purpose is composed of the following:

Asphaltum	12.5%
Bees Wax	4.5%
Ceresine Wax	58 %
Stearic Acid	25 %

ETCH SOLUTIONS FOR LITHOGRAPHIC PLATES

Etches for Zinc Plates

Ammonium Nitrate	3 oz.
Ammonium Phosphate	3½ oz.
Calcium Chloride	¼ oz.
Hydrofluoric Acid	½ oz.
Gum Arabic Soln. (Saturated)	80 oz.

Phosphoric Acid	1 part
Gallic Acid	2 parts
Gum Arabic Soln.	8 parts
Water	14 parts

Gum Arabic Solution	32 oz.
Ammonia Water (16%)	3 oz.
Phosphoric Acid	1 oz.
Hydrofluoric Acid	5 or 6 dr.

Pour each of the above ingredients into gum separately and stir continuously. Keep 24 hours before using.

Etches for Either Zinc or Aluminum

Sod. Phosphate	1 part
Pot. Phosphate	2 parts
Sod. Nitrate	1 part
Pot. Nitrate	2 parts

Dissolve in 1 gal. of hot water and add 2 ozs. Phosphoric Acid.

Ammonium Nitrate	2 parts
Ammonium Phosphate	2 parts
Gum Arabic Soln.	20 parts
Water	75 parts

Ammonium Nitrate	1 part
Ammonium Biphosphate	1 part
Gum Arabic Solution	10 parts
Water	38 parts

Etches for Aluminum Plates

(a) Dissolve 2 ozs. of Pulverized Amm. Bichromate in 16 ozs. water.

(b) Mix 1 oz. of the soln. resulting from "A" with the folowing.

(1) (20%) Phosphoric Acid	1 oz.
(2) Gum Arabic Soln.	8 oz.
(3) Water	8 oz.

(a) Sod. Phosphate	½ oz.
(b) Sod. Nitrate	½ oz.

Dissolve (a) and (b) in ½ gal. of hot water and add 1 oz. (80%) Phosphoric Acid. Use this etch without gum, spreading it evenly over the Plate, by means of a soft sponge or a brush made of camels or badgers hair.

(a) Phosphoric Acid (85%)	1 oz.
(b) Gum Arabic Soln.	32 to 40 oz.

(a) 1 gal. of chemically pure HNO_3 with 7 Gals. of H_2O

(b) Dissolve zinc to the point of saturation in this HNO_3 solution.

(c) Take 1 oz. of resulting soln. and ½ oz. of gum arabic soln. and mix with a gallon of water.

---

2 oz. Bichromate of Ammonia
1 pt. Gum arabic Solution
1 tps. of the following:
2½ oz. Phosphoric Acid (85%
into
84 oz. Gum Solution

Gum Solution is water saturated with gum arabic and filtered.

Etches for Stone

HNO_3 added to gum solution until action of acid is plainly visible when it is applied to the stone.

Counter Etches

A. For Zinc Plates.

Alum	4 oz.
HNO_3	1 oz.
H_2O	1 gal.

Mix and cover plate thoroughly, then dry plates as quickly as possible.

For Aluminum Plates

1. Potash Aluminum	869 dr.
HNO_3	185 c.c.
H_2O	12 gal.
2. H_2O	1 gal.
HNO_3	½ oz.
Acetic Acid	2 oz.
HCL	1 oz.

(Mix thoroughly.)

3. Acetic Acid 99.5%	2 oz.
Potash of Aluminum	½ oz.
Water	60 oz.
Acetic Acid (99.5%)	2½ oz.
Nitric Acid Sp. Gr. 1.403	1½ oz.
H. F. Acid (Tech.)	1 oz.
Water	2 gal.

---

* Paste Acids

(for etching, cleaning and soldering)

1. Suspendite	6 lb.
2. Water	70 lb.
3. Muriatic Acid	28 lb.

Allow 1 and 2 to stand over-night and then mix until smooth. If necessary use warm water. When cold add 3 slowly and stir until uniform.

Antique Gold Finish

Gold Cyanide	½ oz.
Silver Cyanide	¼ dwt.
Sodium Cyanide	6 oz.
Sodium Carbonate	2 oz.
Water	1 gal.

A very small quantity of lead dissolved in caustic soda is added to this solution. In preparing the lead solution dissolve 1 ounce of lead carbonate and 4 ounces of caustic soda in 1 quart of water, and add 20 to 30 drops to each gallon of solution.

Operate solution at 110° F., with 4 to 5 volts. Use 18 karat green gold anodes. Agitation of the work is essential to produce the antique finish. After the smut is produced relieve on a small rag wheel, using bicarbonate of soda moistened with water. The work is lacquered to protect the finish.

Gun-metal Finish

After the work has been polished and cleaned, it is placed in the following solution for ten to fifteen minutes:

Ferric Chloride	2 oz.
Mercury Nitrate	2 oz.
Muriatic Acid	2 oz.
Alcohol	8 oz.
Water	8 oz.

After immersing the work in this solution it should be hung up to dry for 10 to 12 hours. Repeat the immersion and drying operation, then brush lightly with a fine crimped steel wire wheel. Finally, oil with paraffin or linseed oil, and remove excess oil with a soft cloth.

Photo Engravers Collodion

Nitrocellulose (15–20 sec.)	3
Ether	48.5
Alcohol	48.5

Filter and bottle.

Etching Filler

A filler for etched lines in metal to make them more distinctive has the following formula:

White Bees Wax	10 gr.
French Chalk	5 gr.

Melt together.

Etching Aluminum Reflectors

Water at 45° C.	950 c.c.
Hydrofluoric Acid (48%)	50 c.c.

Rotate reflector every 30 seconds.

Pour off and wash with running water.

Introduce 50–50 Nitric acid to remove black film.

Pour off and rinse with water.

Swab gently wtih soft cloth or cotton to remove last thin film of deposit.

* Desensitizing Lithographic Plates

1. Preparation of the Coating Solution.

Formula—A

Dissolve

¼ oz. of dry ammonium bichromate in 4 oz. of water

Add this solution to

15 oz. of fresh, strained gum arabic solution having a density of 14° on the Baumé hydrometer, or a specific gravity of 1.115 at 60–70° F.

Formula—B

This solution can also be made without the use of a hydrometer by completely dissolving

4⅜ oz. of air dry gum arabic in 10 oz. of water,

straining through at least four thicknesses of cheese cloth, and making up to 15 ounces, with water.

To this is added a solution made by dissolving

¼ oz. of dry ammonium bichromate in 4 oz. of water

In either case the solution, if correctly made, will measure approximately 12½° on the Baumé hydrometer, or 1.094 specific gravity at 60–70° F.

Precaution:—This solution will give the best results if made up fresh from sweet gum arabic solution, each day. It should be kept in a cool place and in a brown bottle to protect it from light action and should under no circumstances be used if it has been standing longer than two days.

The proportion of gum arabic to ammonium bichromate need not be limited to the exact figures given above, but may be varied with experience to as low as 12 ounces of 14° Baumé gum solution, or 3½ ounces of dry gum arabic, to ¼ ounce of ammonium bichromate. This proportion produces a harder film but one that under certain conditions may show a tendency to take a light tint. The proportion given in the formulae A and B (above) has worked well in practice and should be followed at least until experience has been gained.

The viscosity of the coating solution may be reduced, if desired, by adding to it a little water. The diluted solution will produce a somewhat thinner coating but the characteristics of the surface will not be appreciably affected.

2. Preparation of the Image.

Zinc and aluminum plates and lithographic stones on which the design has been produced by any of the methods commonly used, can be treated. The preparation of the design should be carried to the point at which it is ready for the final etch. The final etch may be applied in the usual manner before the application of the coating, but this is not necessary since the bichromated gum film alone produces complete desensitization.

Precaution:—The surface must be "clean" The image or design must be sufficiently well developed and greasy to stand the usual etching and gumming. The coating should not be applied to a photo-litho plate or stone until the image has been strengthened by rolling up sufficiently to stand etching and gumming. If the ink is too spare, or if the image is under developing ink alone, there may be difficulty in washing out the design after coating and exposing. Surfaces on which there is crayon or tusche work should be etched, gummed, washed out and rolled up well before application of the coating solution.

3. Coating the Plate or Stone.

Dampen the surface and if it has been under gum, wash it off with a clean sponge. Take off the excess moisture with a sponge or clean rag. Pour a little of the bichromated gum solution from the bottle onto the surface and rub it down evenly with a clean dry rag. With another clean dry rag or piece of cheese cloth rub until the coating is thin and dry exactly in the same manner as in gumming up.

Precaution:—Just as in gumming, the coating must be thin so that the work is not covered. After the coating is hardened, streaks are difficult to wash out and should be avoided. If the work has been too sparsely inked it will be difficult to prevent streaks.

4. Exposure.

After coating, the entire surface is exposed to the rays of an arc lamp to harden the film. Any light source of sufficiently high intensity can be used. Since arm lamps are in general use for printing down, they are recommended for this purpose also. A variety of these is available, having varying intensities, and we give the following suggestions for approximately correct exposure:

a. Lengths of exposures to be given, using a 25 ampere single open arc printing lamp, using ½″ photographic white flame carbons.

Distance from arc	Time of exposure (minutes) Aluminum	Zinc
48 in.	8½	10
60 in.	13½	16
78 in.	23	28

b. Lengths of exposure to be given, using a 30 ampere double open arc printing lamp, using ½″ photographic white flame carbons

Distance from arc	Time of exposure (minutes) Aluminum	Zinc
48 in.	3¼	4
60 in.	5	6
78 in.	8½	10

Due to the fact that arc printing lamps of various makes vary somewhat in current and voltage characteristics and in light intensity, the above tables are only to be taken as a guide. It is generally safe to assume that the same exposure which will give a satisfactory albumen print on a metal plate or stone, through the average dry plate negative in a printing frame, is the proper exposure to give the bichromated gum desensitizing film. The distance from the light source to the plate or stone during exposure should be at least as great as the length of the diagonal of the plate, in order to secure uniform light action.

5. Washing.

After exposure, the plate or stone is immediately washed for not less than two minutes in running water to remove the unchanged ammonium bichromate, the excess of water is wiped off with a sponge or rag and the plate is gummed up with ordinary gum arabic solution in the usual way and dried. It is then ready to be washed out and put under asphaltum.

Precaution:—If the excess ammonium bichromate is not thoroughly removed from the film by washing, the hardened gum film will gradually become grease-receptive unless the printing operation is begun within a short time. If an improperly washed plate stands in the dry condition longer than 2 or 3 days it will scum.

6. Printing.

Plates or stones prepared in this way should be handled in the usual manner except that in most cases they can be run without etch or dope in the fountain water. Should a tendency to fill up be encountered, the work should be gummed out and the surface gummed up and dried. A little weak etch may be added to the fountain if thickening persists. Chromic etches should be avoided. A satisfactory formula is as follows:

½ oz. 85% phosphoric acid
16 oz. 14° Baumé gum arabic solution.

From ¼ to ½ ounce of this mixture added to a gallon of dampening water is usually all that is needed to keep the work open.

Because of the nature of the coating of hardened gum, the plates appear "slick" to the pressman and therefore seem to carry more water than they actually do. The first reaction of the pressman is to cut down his water, and this usually results in a scum caused by the plate being too dry. A little experience is necessary to enable him to learn how to control the water. Since the valleys of the grain are filled with the water absorbent coating, printing is actually done with less surface moisture than on ordinary plates, but of course some water must be carried. Should a scum appear, do not etch the plate until you have tried increasing the water a little. If the scum disappears then, the trouble is due to too little water.

Electrotyping

The first step in the production of an electrotype consists in the preparation of an impression or "mold" in wax of the form to be reproduced. The molding wax usually consists of ozokerite to which various substances have been added to produce the desired physical properties. The molten wax is poured upon one side of a metallic plate, consisting of lead, copper, or aluminum. The wax-coated metal is termed a "case." After taking the impression of the form by the use of suitable pressure at a slightly elevated temperature, usually by means of a hydraulic press, the resultant "mold" is "trimmed" and "built up" to produce the desired degree of relief in the finished plate.

The mold is then coated with graphite, applied by a wet or a dry process, or both. After washing out the excess of graphite, the form is either introduced directly into the depositing bath, or, in some cases, is given a preliminary treatment (so-called "oxidizing") with copper sulphate solution and fine iron filings, whereby a thin film of copper is deposited by "immersion" upon the graphite. The baths are usually contained in lead-lined wooden tanks, with copper cross bars, from which the anodes and cathodes are suspended. Electrical connection to the graphited cathode surface is made by means of the suspending hook by either of two methods. In the one known as the "case connection," the hook is in direct contact with the metallic plate of the case, portions of the wax being removed in order to bring the metal and the graphite surface into electrical contact, while the back of the case is insulated with wax. In the method now more generally employed, and known as the "face connection," the hook is in contact with a small copper plate imbedded in the wax near the top of the form and in contact with the graphite surface. In the latter method the metallic plate itself is not in the circuit, and there is less tendency for copper to deposit upon any accidentally exposed portions of it.

After the copper is deposited to the desired thickness (usually 0.006 to 0.010 inch (0.15 to 0.25 mm.)) the case is taken from the bath, and the copper "shell" is loosened by means of hot water. After trimming the edges, the back of the shell is treated with soldering fluid (usually an acidified solution of zinc chloride) and coated with "tin foil" containing about 35 per cent of tin and 65 per cent of lead, after which it is laid face downward upon a heated pan. After the tin foil is melted upon the back of the shells, molten electrotype metal (usually containing from 3 to 4 per cent each of tin and antimony and from 92 to 94 per cent of lead) is poured over them to the desired depth. The electrotypes thus produced are cleaned, cut, and trimmed to the desired size, "finished" to a plane surface and shaved to the proper thickness. They may be subsequently curved if desired.

In many cases, for the most perfect reproduction of halftone or other work in low relief, molding in thin sheet lead at high pressures is practiced. The lead mold thus produced is cleaned with alcohol to remove grease, and is then treated with a dilute solution of chromic acid or a chromate. This forms a thin film of lead chromate, which prevents the deposited metal shell from adhering too tenaciously. The subsequent steps are similar to those involved when wax molds are used.

For the better classes of work, especially color process halftones, or for plates requiring very severe service, nickel electrotypes (commonly called "steel" or "nickel steel") are frequently employed. In their preparation, a thin layer of nickel (usually about 0.001 inch or 0.025 mm.) is first deposited upon the wax or lead mold, copper is then deposited back of the nickel, and the resultant nickel-copper shell is treated as above. The true "nickel electrotype" thus made should not be confused with a nickel-plated electrotype in which nickel is deposited upon the surface of a finished copper electrotype.

During recent years a thin coating of chromium, usually about 0.0002 inch (0.005 mm.) has been often applied to nickel or copper electrotypes that are to be used for very long runs, for example in the printing of cartons and labels.

EXPLOSIVES, PYROTECHNICS, AND MATCHES

* Blasting Composition

Am. Chlorate	54
Barium Nitrate	29.5
Aluminum Powder	1.5
Aluminum Granules	9
Rosin	3

* Explosive

An explosive consists of $NaClO_3$ 5–50, $NaNO_3$ 5–50, $o\text{-}O_2NC_6H_4Me$ 5–15, $(O_2N)_2C_6H_3Me$ 5–15, and sawdust 5–15 parts.

* Explosive, Blasting

Sod. Chlorate	77.1
Dinitrotoluol	17.1
Castor Oil	5.05
Paraffin Wax	0.75

or

Pot. Chlorate	75
Dinitrotoluol	18.4
Mononitro Naphthalene	1.06
Castor Oil	4.8
Paraffin	0.74

* Fuse, Blasting

A mixture such as Pb thiocyanate 30–50, $KClO_3$ 10–30 and ground smokeless powder 30–50% or one of similar character which may contain Cu thiocyanate, a nitrate or a perchlorate is used as ignition material around the ignition wire of an electric fuse, and a detonating charge may be used comprising tetranitromethylaniline associated with a mixture of Hg fulminate and $KClO_3$ in the proportions of 90 and 10%, respectively.

* Powder, Ballistic

Nitrocellulose (13.15% N)	85
Dinitrotoluol	10
Dibutyl Phthallate	5
Diphenylamine	1
Pot. Sulfate	1

* Igniter, Blasting Cap

Pot. Ferricyanide	20–40
Pot. Chlorate	10–30
Nitrocellulose	70–30

* Ammunition Primer

Mercury Fulminate	10–40
Antimony Sulfide	20–45
Barium Nitrate	10–60
Lead Trinitroresorcinolate	5–35

* Percussion Cap Primer

Lead Azide	5–15
Cu Silicide	10–15
Barium Nitrate	10–15
Tetrazene	10–15

* Electrical Detonator

Lead Sulfocyanide	30–50
Pot. Chlorate	10–30
Smokeless Powder	30–50

* Cartridge Primer, Explosive

Mercury Fulminate	30
Lead Trinitro Resorcinate	10
Barium Nitrate	29
Lead Sulfocyanide	10
Abrasive	20
Binder	1

* Primer, Explosive

Stable to shock and friction.

Pot. Chlorate	28
Antimony	55
Zinc Dust	17
35% Gelatin Solution	sufficient to act as a binder

* Primer, Explosive

Mercury Fulminate	37
Barium Nitrate	32
Antimony Sulfide	28
Ground Glass	3
Trinitro Toluol	4–8

---

* "Tracer" Bullet Composition

An improved light emitting composition is a mixture of two parts of magnesium powder and three parts of bismuth oxide, which when pressed under a load of ten cwts. into tracer pellets for shot gun cartridges, gives excellent results, having the desired properties of certainty of ignition, brightness of trace, and freedom from danger of possible toxic effects. Similar results are obtained with a mixture of one part magnesium powder and one to two parts of sodium bismuthate.

A composition containing three parts of bismuth oxide, two parts of magnesium powder or other suitable metallic powder, and half a part of strontium peroxide. The addition of the strontium peroxide changes the white colour of the flame emitted by the tracing composition to a reddish colour and in brilliant sunlight the trace is much more discernible than a completely white light.

---

* Explosive Primer

Zirconium (Powd.)	10
Mercury Fulminate	35
Barium Nitrate	40
Antimony Trisulfide	15

Pyrotechnics
"Red Fire"

Strontium Nitrate	66 parts
Potassium Chlorate	25 parts
Powdered Orange Shellac	9 parts

Strontium Carbonate	16 parts
Potassium Chlorate	72 parts
Orange Shellac Powdered	12 parts

Potassium Chlorate	37 parts
Strontium Nitrate	50 parts
Shellac Powd.	13 parts

Strontium Nitrate	8 oz.
Sugar	4 oz.
Potassium Chlorate	1 oz.
Potassium Perchlorate	15 oz.
Strontium Nitrate	80 oz.
Flowers of Sulphur	20 oz.
Wood Charcoal (powdered)	1 oz.
Gum Kauri (red gum)	2 oz.
Vaseline-sawdust Mixture	10 oz.

The sawdust and vaseline mixture is made by rubbing 8 oz. of sawdust with 6 oz. of melted vaseline.

Potassium Perchlorate	4½ oz.
Strontium Nitrate	20 oz.
Sulphur	5½ oz.
Rosin	½ oz.
Sugar	½ oz.
Antimony, Powdered	¼ oz.
Vaseline-sawdust Mixture	10 oz.

Perchlorate Potash	12½ parts
Nitrate Strontia Powdered	50 parts
Powdered Charcoal	1 part
Powdered Sugar	4 parts
Red Gum	15 parts

Potassium Chlorate	6 parts
Strontium Nitrate	2 parts
Strontium Carbonate	1½ parts
Gum Kauri (red gum)	2½ parts

Green Fire Composition

Barium Chlorate	90 gm.
Powdered Orange Shellac	10 gm.

This mixture is made by mixing the above two ingredients together.

Barium Chlorate	23 parts
Barium Nitrate	59 parts
Potassium Chlorate	6 parts
Orange Shellac	11 parts
Stearic Acid Powd.	1 part

Barium Chlorate	55 parts
Barium Nitrate	33 parts
Shellac	12 parts

Barium Nitrate	6 parts
Potassium Nitrate	3 parts
Sulphur	2 parts

Barium Nitrate	18 parts
Shellac	4 parts
Mercurous Chloride	4 parts
Potassium Chlorate	2 parts

Barium Nitrate	3 parts
Potassium Chlorate	4 parts
Gum Kauri (red gum)	1¼ parts

Blue Fire Composition

Potassium Chlorate	6 parts
Ammonio-sulphate of Copper	8 parts
Shellac	1 part
Willow Charcoal	2 parts

Potassium Chlorate	40 parts
Copper Sulphate	8 parts
Rosin	6 parts

White Fire Compositions

Potassium Nitrate	24 parts
Sulphur	7 parts
Charcoal (wood)	1 part

Potassium Nitrate	7 parts
Sulphur	2 parts
Powdered Antimony	1 part

Potassium Perchlorate	3½ oz.
Barium Nitrate	17 oz.
Powdered Sulphur	3½ oz.
Finely Powdered Aluminum	5 oz.

Potassium Perchlorate	7 oz.
Barium Nitrate	34 oz.
Flowers of Sulphur	7 oz.
Aluminum Bronze (dust)	2 oz.
Aluminum Flakes	7 oz.

Pyrotechnic

A nonhygroscopic successively exploding composition consists of

Pot. Chlorate	35 lb.
Magnesium Oxide	35 lb.
Phosforus Trisulfide	12 lb.
Gum Arabic	1 lb.
Pot. Dichromate	5 lb.
Clay and Sand	8 lb.

---

* Pyrotechnic "Snakes"

Naphthol Pitch	300
Tetryl	100
Nitrocellulose	20
Nitric Acid	250
Linseed Oil	60
Stearic Acid	0.5
Graphite	0.5

---

* Pyrotechnic Starter

Calcium Silicide	10
Lead Dioxide	15
Fused Silica	30
Copper Oxide	30

Showers of Fire

Potassium Nitrate	18 parts
Sulphur	8 parts
Lampblack	5 parts

This composition burns with a yellowish color, throwing out streamers of golden sparks, due to the lampblack which is used. The mixture burns slowly and is suitable for filling paper tubes.

Potassium Nitrate	10 parts
Sulphur	2 parts
Charcoal	2 parts
Iron Filings (fine)	7 parts

For loading into ordinary paper cases.

Potassium Nitrate	36 parts
Sulphur	2 parts
Charcoal (wood)	10 parts

For loading into paper cases.

Light Sticks

Fill thin paper tubes of about ⅜″ outside diameter and 1′ long with the colored fire compositions, alternating. One end of the tube should be closed tightly to a depth of 3″ with clay or sand. Fill with powder of the desired color and close end by pasting a piece of tissue paper around it, after inserting a fuse.

Boil a handful of sawdust or wood shavings in a cup of water containing a teaspoonful of potassium nitrate. When dry, it will burn with a whitish yellow flame, sizzling as it burns. Add ½ teaspoon of strontium nitrate to the water before boiling the sawdust in it. When the sawdust is then immersed and dried it will burn with a red flame. Barium nitrate will make the flame green; copper sulphate, blue.

Homemade Sparklers

White Potassium Chlorate	10 oz.
Granulated Aluminum	2 oz.
Charcoal	1/16 oz.

Mix to consistency of thick cream with a solution of 2 oz. of dextrine in a pint of water and coat upon wires or slender wooden sticks.

For red sparkler add 1½ oz. powdered strontium nitrate.

For green sparkler add 2 oz. powdered barium nitrate.

Smoke Composition

White: Powdered Potassium Nitrate	4 oz.
Powdered Soft Coal	5 oz.
Sulphur	10 oz.
Fine Sawdust	3 oz.
Red: Potassium Chlorate	15 parts
Paranitraniline Red	65 parts
Lactose (powdered)	20 parts
Green: Synthetic Indigo	26 parts
Auramine Yellow O	15 parts
Potassium Chlorate	33 parts
Lactose (powdered)	26 parts
Yellow: Precipitated Red Arsenic Sulphide	55 parts

Powdered Sulphur	15 parts
Potassium Nitrate	30 parts

Smoke, Composition for Producing

Tetrachlorethane or Chloronapthalenes	40–50
Zinc Filings	55–25
Pot. Nitrate Sod. Nitrate Calcium Silicide Pitch	5–15%

* Gas Bomb, Combined Tear and Smoke

Chloroacetophenone	12.5–30
Hexachlorethane	27.3–52.5
Ammonium Chlorate	5
Zinc Dust	2
Zinc Oxide	2

* Yellow Smoke Composition

One substance well adapted to produce yellow smoke is: potassium bichromate 66 parts, bismuth tetroxide 20 parts, magnesium 14 parts; that is, the ingredients are substantially in the proportion 13: 4: 2, respectively.

Another substance that gives the same result is: potassium bichromate 65 parts, bismuth subnitrate 20 parts, and magnesium 15 parts, the proportions having approximately the same relation as above given.

The compositions specified gives off puffs of yellow smoke that are particularly adapted for use in daylight fireworks and various daylight signaling devices.

* Match, Repeatably Igniting

These matches are ignited by friction; extinguished; and may be used again and again.

Pot. Chlorate	35
Calcium Plumbate	3.7
Sulfur	0.9
Benzoyl Peroxide	3.7
Powdered Glass	14.2
Hexamethylene Tetramine	42.5
Glue Solution	sufficient to bind
Formaldehyde	0.5

Mold into rods and coat with thin sod. silicate.

* Match, Repeatably Igniting

From the following is molded a match which ignites on rubbing and may be blown out and used repeatedly.

Pyroxylin	50
Pot. Chlorate	20
Powd. Glass	10
Camphor	8
Pyridine	4
Am. Oxalate	2

* Blasting Fuse

Fuses for "touching-off" by electricity are made from an explosive mixt. giving no gas and a hot flame. Reduced Fe and $K_2Mn_2O_8$, or of Sb and $K_2Mn_2O_8$, or of CaO_2 and C-Mg, all in the proportions 1: 1.

* Black Powder

Pot. Nitrate	72
Sulfur	6.5
Charcoal	21
Turkey Brown Oil	0.5–2

The dry materials must be in fine powder. The Turkey Brown Oil is used to slow up the rate of combustion.

FIRE PROOFING, FIRE EXTINGUISHING

Dry Fire Extinguisher

Ammonium Sulphate	30 lb.
Sod. Bicarbonate	18 lb.
Ammonium Phosphate	2 lb.
Red Ochre	4 lb.
Silex	46 lb.

Fire Extinguishing Fluid

Carbon Tetrachloride	94–95
Solvent Naphtha	5
Ammonia Gas	0.5–1

The above minimizes production of toxic fumes when extinguishing fires.

---

* Fire Extinguishing Liquid (Non-Freezing)

Pot. Carbonate	34–42
Ethylene Glycol	5–6%
Pot. Chromate	2–3%
Water	Balance

---

* Fire Extinguishing Foam

Large quantities of roasted peanut shells are ordinarily available as a waste product at little or no cost and this fact as well as the facility with which they may be digested, the high percentage of yield, and the excellence, color and non-staining characteristics of the product make roasted peanut shells preferable to the husks of other legumes, which may, however, be used.

In the preparation of stabilizer, the roasted peanut shells or hulls, together with any discarded peanut shells or hulls, together with any discarded peanut kernels or peanut vines which may be mixed with the shells, may be shredded and charged into a vertical extraction cylinder and mixed therein with approximately 10% of commercial caustic soda. Warm water, or the wash water from a previously treated batch of shells, is then introduced into the cylinder and dissolves the soda. The cylinder is then closed and water at a temperature of approximately 290° F. is forced into the cylinder at a pressure of approximately 60 pounds per square inch, the air in the cylinder being vented therefrom. Approximately 800 gallons of solution is used for each thousand pounds of shells and the mixture is cooked for approximately an hour and a half with occasional agitation by the injection of steam, which also serves to keep up the temperature.

When the charge has been sufficiently cooked, the liquor is withdrawn and boiled down from about 5% solid contents to about 40% solid contents. If desired, the stabilizer may be fortified by the addition to the boiled extract of minute quantities of borax, sodium resinate, benzoate of soda or para-formaldehyde.

The stabilizer may be used in liquid form but is preferably dried to a cake in a suitable vacuum drier and the cake ground to a crystalline powder of approximately 40 mesh.

The proportions of the stabilizer to be used will vary with the foaming ingredients used therewith and the kind of foam desired. By using approximately 3 ounces of the powdered extract, 22 ounces of bicarbonate of soda and 30 ounces of aluminum sulphate and combining these ingredients with a suitable volume of solvent, preferably approximately one pound of powder to one gallon of water, a stiff, tenacious foam will be produced which has great mobility, may be conveyed through conduits with but little deterioration, does not stain or discolor materials with which it comes in contact, and which owing to its light color, is readily visible so that it can be determined whether it is being projected to the right spot.

The dry powdered extract is preferably combined with the bicarbonate of soda and this mixture is preferably fed separately from the aluminum sulphate into a stream of water flowing through a hose or pipe. The powder may be introduced into the water by means of an ejector or ejectors creating sufficient suction to draw the powders into the running stream in proper proportions or by mechanical feeders. The powders are dissolved by contact with the water in the hose or pipe to form foam which is ejected from the nozzle.

All formulae preceded by an asterisk (*) are covered by patents.

Fireproofing

Periodically the question of fireproofing woodwork, curtains, and drapings crops up, especially in regard to trade displays and exhibitions. The use of sodium acetate for fireproofing wood has been known for a long time, and a solution of 15 per cent. strength has been found the most suitable concentration. Better results are obtained if the sodium acetate is reinforced with a small quantity of disodium phosphate. For flame proofing planks a solution containing 228 grams sodium acetate crystals and 33 grams disodium phosphate crystals per litre should be used. The planks are given three coatings with this solution, time being left between each application to allow the liquor to soak in. For efficient working the application of about 70 grams anhydrous sodium acetate per square metre of wood surface is necessary. The depth of penetration depends on the thickness and nature of the wood. In the case of air dried pine boards of 17 mm. thickness a total penetration of 15 mm. was found, the boards being coated on both sides. If the wood has been well dried out it is advisable to give a preliminary treatment with water.

For coating curtains, paper, etc., the L.C.C. recommends 1 lb. of ammonium phosphate and 2 lb. of ammonium chloride to 1½ gallons of water, or alternatively 10 oz. borax and 8 oz. boracic acid per gallon of water. The second formula is stated to be better for delicate articles. The fabrics should be dried without rinsing, and in all cases a small piece of the cloth should be treated first, in order to find the effect on colour and texture.

* Fireproofing Solution

Fibrous materials are immersed or treated under pressure with following:

Am. Chloride or Carbonate	12
Boric Acid	8
Ammonium Hydroxide	8
Water	70

Fireproofing Canvas

Am. Phosphate	1 lb.
Am. Chloride	2 lb.
Water	1½ gal.

Impregnate with above; squeeze out excess and dry.

Fireproofing Light Fabrics

Borax	10 oz.
Boric Acid	8 oz.
Water	1 gal.

Impregnate; squeeze and dry.

* Fireproof Coating

Wood covered with following is resistant to fire and heat.

Asbestos	40
Magnesite	30
Magnesium Chloride (30% solution)	to make paste

* Metal, Fireproofing

A metal article is protected and rendered fire-resisting by coating it with a layer of high-melting asphalt, then with a layer of asbestos paper or felt satd. with a mixt. of asphalt 60 and chlorinated polyphenyl 40% and then with a layer of the asphalt.

* Paint, Fireproof

A fireproof paint made exclusively of inorg. materials contains 20% silicate, 15% KOH, 20% dil. H_3PO_4, 15% finely divided asbestos, 15% ZnO and 15% kaolin, intimately mixed.

* Fireproofing Paper

Craftboard or paper is satd. with a soln. contg. $Al_2(SO_4)_3$ 8 oz., Ti sulfate ½ oz. and water 1 gal. heated to 140–180° F. The craftboard or paper is removed and partially dried and then dipped in a soln. contg. Na_2SO_4 10 lb., Na_3BO_3 5 lb., Na silicate soln. (39° Bé.) 14 gal. and water 28 gal. heated to 140–180° F. The craftboard or paper is partially dried and pressed in desired shape.

* Fireproofing

Paper or wood is impregnated with

Cryolite or Sodium Fluosilicate	1–3
Aluminum Sulfate	1–2
Sod. Silicate	3–5
Water	12–24

* Rayon, Fireproofing

Rayon cloth is immersed in 10% Phosphoric Acid at 40° C. for 15 min.

Fireproofing Textiles

The cloth is impregnated with

Borax	70
Boric Acid	30
Water	600

and dried.

Ammonium Chloride	20 kg.
Zinc Chloride 30 per cent.	300 l.
Ammonia 28 per cent.	350 l.
Water	100 l.

The Paris Municipal Laboratory, recommended the following process: Prepare a 2 per cent. solution of aluminum sulphate and a 5 per cent. solution of silicate of soda. Mix and enter the cloth. After squeezing and drying the aluminum silicate formed is insoluble.

Another method consists in padding the fabric in a solution of ammonium phosphates, then steeping in an ammoniacal solution of magnesium chloride. The compound formed on the fiber is insoluble in water. The fabric is rinsed to remove the excess of magnesium chloride and dried.

Tungstate of zinc resists washing, and this makes it preferred at times to tungstate of alumina. The most usual method consists in padding in a solution of stannate of soda at 14 deg. B., and then drying. The goods are then entered into a bath of the following composition:

Tungstate of Soda, 35 deg. B.	4 parts
Acetic Acid, 9 deg. B.	1 part
Ammonium Hydrochloride, 4 deg. B.	3 parts
Acetate of Zinc, 17 deg. B.	2 parts

After centrifuging and drying the fabric is hot-calendered to evaporate the acetic acid.

Perkin recommends the following method: Pad with a solution of stannate of soda at 26 deg. B., and dry, then treat with a solution of ammonium sulphate at 10 deg. B., squeeze, dry and wash in water to remove the excess of ammonium sulphate. This last step is not indispensable, as the sulphate has flameproofing properties. The stannate of soda combines intimately with the fiber and the ammonium sulphate precipitates the oxide which combines also with the fiber.

Asbestos Dope

Asbestos.—The cloth is painted with a dope containing asbestos which hardly interferes with suppleness. An interesting composition is:

Asbestos	350 gr.
Silicate of Soda, 36 deg. B.	350 gr.
Water	1,000 gr.

The particles crumble and shrink. Continue heating for about 12 hours. Crush and screen to uniform sizes; replace in pans and reheat at 185–195° F. until proper state of dryness is reached (about 8 hours). The dried material is of a granular glassy light yellow color. This material is air-cooled and sifted thru No. 6 and No. 8 screens.

Fireproofing Wood

Wood can be effectively fireproofed by immersion in a 15 per cent aqueous solution of anhydrous sodium acetate with the addition of about 3–4 per cent of crystalline sodium phosphate ($NA_2HPO_4 \cdot 12H_2O$).

---

* Fireresisting Fiberboard

Wood fiber is satd. with 10% NaOH soln. and the treated fibers together with mineral wool fibers are introduced into a body of a carrier liquid such as water and agitated to bring the fibers into suspension (the mineral wool fiber comprising 50–90% of the total fiber in the suspension) and a product such as fiberboard or the like is then formed from the suspension.

For Chemical Advisors, Special Raw Materials, Equipment, Containers, etc., consult Supply Section at end of book.

FLUXES AND SOLDERS; WELDING

Soldering Solutions

Zinc Chloride made by completely neutralizing hydrochloric acid with zinc is most universally used. In addition to this rosin, ammonium chloride and a mixture of 15% zinc chloride, 25% glycerine and 60% water are satisfactory for copper, brass, steel, terne plate, tinned steel, monel metal, etc. Hydrochloric acid is necessary on galvanized steel.

A well-made soft-soldered joint will develop 5000 to 6000 lb. per sq. inch in shear.

Silver solders consist of silver 20% to 70%, copper 50% to 18%, zinc balance, Borax or Boric acid mixture used for fluxes. Melting points of silver solder vary according to composition usually 200 to 300 degrees F. below those of the usual brazing—brasses and about 1100 to 1200 degrees F. above ordinary soft solder.

Aluminum solder is a 12% silicon and 88% aluminum melting at about 580 degrees C. (1076 degrees F.).

Soldering Solution for Stainless Steels

Zinc Chloride, Commercial	37 gm.
Glacial Acetic Acid 99.9%	23 gm.
Hydrochloride Acid, Com. 34.5% Hcl.	40 gm.

Soldering Solution for Rustless Irons

Hydrochloric Acid, specific gravity 1.18	60 gm.
Ferric Chloride, Lump Form, Pulverized	33 gm.
Nitric Acid, Specific Gravity 1.42	2 gm.

Add in order named.

Tinning Flux—Zinc chloride stick from saturated solution in water.

Non-Corrosive Soldering Flux

Rosin	1 oz.
Denatured Alcohol	4 oz.

Solder

Tin	1
Lead Phosphide	0.1
Lead	98.9

---

* Soldering Fluid

Zinc Chloride	10–50
Glycerol	0.5–50
Alcohol	5–50
Water	1–50

---

* Liquid "Solder"

Heavy Clear Lacquer	57.5
Benzol	23
Aluminum Powder	19.5

This has good heat resistance and is non-corrosive.

---

* Solder, Aluminum

Zinc	40–60
Tin	40–60
Cadmium	1–10
Iron	0.5– 5
Rosin	2
Tallow	2– 5

---

* Solder, Aluminum

Lead	25
Zinc	40
Tin	20
Aluminum	5

Solder, Aluminum

Tin	60
Zinc	40

---

* Solder, Aluminum

Zinc	50
Tin	20
Lead	15
Magnesium	10
Calcium	5

Solder, Aluminum

Aluminum	30
Zinc	20
Tin	15
Copper	5
Bismuth	10
Silver	10

All formulae preceded by an asterisk (*) are covered by patents.

* Aluminum Solder

Zinc	10–30
Aluminum	7–15
Copper	1– 5
Bismuth	1– 8
Tin	Balance

* Solder, Aluminum

Aluminum	30
Zinc	20
Tin	15
Copper	5
Bismuth	10
Silver	10–20

* Solder, Brazing

Phosforus	2.5–10
Zinc	5–50
Copper	Balance

Solder, Brazing

Copper	40–55
Zinc	60–45

* Chain Solder

A solder composition which may be applied to greasy machine-made chain in the hank, rubbed into the joints, and excess rubbed off so that after heat-treatment none will remain on the surface of the chain, is composed of powdered Sn 2 pts., powdered Cu 1 pt., red P 1 pt.

* Solder, Copper and Brass

Iron Chloride	8
Zinc Chloride	8
Lard	26
Rosin	2
Glycerol	6
Tin	12½
Lead	12½

Flux, Soldering

Zinc Chloride	71
Am. Chloride	29

* Soldering Flux, Anti-Fermentive

Salicylic Acid (Powd.)	20
Rosin (Powd.)	20

Fuse together

Calcium Fluoride	25
Borax	75

* Solder Flux

Fuse together

Zinc Chloride	8–20
Stearic Acid	88

Zinc Solder Flux

Cadmium Chloride	40
Lead Chloride	40
Ammonium Bromide	16
Sodium Fluoride	4

Pewter, Soldering

The surfaces are cleaned thoroughly. As a flux there is used a mixture of rosin and olive oil. A good solder consists of

Bismuth	50
Tin	25
Lead	25

Solder, Silver

Silver	20
Copper	45
Zinc	30
Cadmium	5

* Solder, High Speed Steel

Powd. Soft Steel	85
Fused Powd. Boric Acid	8
Borax	2
Powd. High Speed Steel	5

Solder, Stainless Steel

Tin	66
Lead	34

Solder, "Stainless Steel"

Manganese	20
Copper	25
Nickel	5
Silver	49
Gold	1

Soldering Paste

Water	10 parts
Zinc Chloride	25 parts
Ammonium Chloride	2 parts
Dark Petrolatum	65 parts

Dissolve the salts in the water and stir into the petrolatum.

* Welding Flux

Calcium Fluoride	1
Borax	3

Melt together and cast into sticks.

* Flux, Welding

Pot. Carbonate	3
Pot. Chloride	3.7
Lithium Chloride	6.9
Pot. Sulfate	7.2
Borax	20
Boric Acid	21
Soda Ash	38.2

Welding Rod Composition

Tungsten	1 –12%
Chromium	1 –10%
Nickel	0.1 – 5%
Aluminum	0.1 – 8%
Vanadium	0.1 – 2%
Carbon	1.75– 4%
Manganese	0.5 – 5%
Silicon	0.2 – 3%
Molybdenum	0.1 – 6%
Iron	Balance

Welding Rod Composition

Carbon	0.60– 0.85%
Manganese	11 –13.5 %
Nickel	2.5 – 3.5 %
Silicon	$<$ 0.60%
Iron	Balance

* Welding Electrode Coating

Cotton cloth impregnated with follow-lowing mixture is used:

Talc	10
Feldspar	30
Ferromanganese (low C)	10
Sod. Silicate	24

* Welding Rod

Copper	80 –96
Tin	1 –10
Phosphorus Copper	0.2– 1

* Welding Rod

Nickel	20–30
Copper	10–20
Iron	Balance

* Welding Rod, Bronze

A bronze rod is coated with the following composition and used with a blow-pipe flame:

Boric Acid	49
Borax	9
Iron or Copper Oxide	30
Sod. Silicate	12

* Welding Rod for Bronze to Iron

Copper	80 –96
Tin	1 –10
Phosfor Copper	0.2– 1

* Welding Wire, Nickel

Magnesium	0.2–0.02%
Silicon	4 –0.05%
Titanium	2 –0.05%
Nickel	Balance

Solder (Powder Form)

Iron Filings	100 parts
Ammonium Chloride	50 parts
Sulphur in Powder Form	25 parts

Mix well.

* Aluminum Solders

A.	Tin	66–69%
	Zinc	27.5–28.5%
	Aluminum	2.5–6.5%

B.	Tin	47.5–49%
	Zinc	47.5–49%
	Aluminum	2.5–5%

C.	Tin	37–45%
	Lead	37–45%
	Zinc	9–21%
	Aluminum	1– 5%

* White Metal Welding Composition

Copper	5
Antimony	5
Zinc	90

FOAMING, DE-FOAMING, FLOTATION

* Foam Prevention Agent

To reduce foaming of glycol antifreeze mixtures from 0.01–0.10% Calcium Acetate is added.

FUELS

Solidified Alcohol

Alcohol	1000.0 cc.
Stearic Acid	60.0 gm.
Caustic Soda	13.5 gm.

Dissolve the stearic acid in 500 cc. of the alcohol, and the caustic soda in the remaining alcohol. Warm to 60° C., mix, and allow to solidify.

Solidified Alcohol

Denatured Alcohol	1000 cc.
Soap Chips (Well Dried)	28–30 gm.
Gum Lac	2 gm.

Heat alcohol to 140° F., add soap and lac, mix till completely dissolved, allow to cool.

* "Anti-Knock" Fuel

Mercuric Cyanide dissolved in a little glycerol is added to gasoline to extent of 0.01–0.1%.

* Fuel Briquettes

A non-caking or poorly caking fuel, *e.g.*, anthracite or semi-coke, is mixed intimately with 10–25% of a finely-ground caking coal and with a small amount, *e.g.*, 1–2%, of a binder such as pitch, the mixture is briquetted, and the briquettes are embedded in a neutral refractory material, *e.g.*, small coke, and carbonized. For household fuel carbonization is effected at 600–650°, whilst briquettes for industrial purposes are produced by carbonization at 900–1000°.

* Briquets, Fuel

In making fuel briquets with an anthracite base, culm 85–90, asphalt 5–10 and pulverized bituminous coal about 5% are used together. The asphalt is rendered freely fluent by heating, the culm is heated to about the same temp. and mixed with the asphalt and the bituminous coal is then added and intermixed.

* Jelly, Benzine

Soap	20 gm.
Alcohol	20 gm.

Boil together; cool; run in slowly with stirring

Benzine	500 gm.
Water	2 gm.

* "Canned Heat"

A solidified fuel which gives an intense smokeless flame and which will not explode or evaporate is made as follows:

Ceylon Cocoanut Oil	50 lb.
Crude Palm Oil	12½ lb.
Pale Rosin	37½ lb.
Caustic Soda Lye 38° Bé.	50 lb.
Water	2½ lb.
96 per cent Alcohol	8 oz. (about)

In operation the 50 pounds of Ceylon cocoanut oil is placed in a suitable vessel and the remaining ingredients mixed therewith according to the following method:

The cocoanut oil (Ceylon cocoanut oil) and rosin are melted over a moderate fire. The palm oil is then added and also melted. The melted rosin fat is strained and when it shows a temperature of about 176 degrees Fahr. it is stirred in the lye in a fine jet. When combination has been effected, the water is added to the thick colloid mass, which is thereby rendered somewhat more liquid by the addition. The alcohol is now crutched in and the mass is permitted to rest for about an hour; the pan or receptacle being well covered. A more intimate union is thereby produced. The somewhat thick, transparent colloid is then brought into the frame, again drawn through with the crutch and allowed to stand without being covered. To this emulsification agent is added 75 per cent of denatured alcohol, and the whole heated to a boiling point after which it is allowed to cool; the colloid thus formed being then ready for use.

* Coal and Coke, Improving Appearance of

The following method is useful in restoring the lustre of weather beaten or discolored coal; to allay dust; to pre-

All formulae preceded by an asterisk (*) are covered by patents.

vent freezing; preventing corrosion of metal contacted; to improve efficiency of combustion.

1. Sicapon or Lignin Liquor	100
2. Fuel Oil	4
3. Water	280

Run (2) into (1) slowly with rapid stirring then follow up with (3).

Coal, Coloring

The coal is immersed in the following solution; the time of immersion influences the shade of coloring.

Water	100	gal.
Iron Chloride	2½	lb.
Pot. Ferricyanide	3	lb.

---

* Prevention of Dusting of Coal or Coke

A light mineral oil of about "32 gravity" and having a flash point of about 175° is atomized onto agitated coal or coke so as to deposit a film on the pieces amounting to about 1 gal. of oil per ton of material, which serves to suppress dust. U. S. 1,886,633 relates to a similar product.

Fire Kindler

1. Cork Dust	50
2. Sawdust	50
3. Paraffin	80
4. Pot. Chlorate	10
5. Sugar	10

Dissolve (4) and (5) in a minimum amount of water and mix thoroughly with (1) and (2). Place in heated dough mixer and pour in melted (3); mix until uniform and cast in blocks.

Fire Starters

Rosin or Pitch	10
Sawdust	10 or more

Melt and mix and cast in forms.

Fire Kindlers

Paraffin Crude	30
Rosin Pitch	10
Wood Flour	60

Compress strongly into bricks.

Rosin Dark	30
Petroleum Oil Thin	5
Sawdust	65

Mix and compress strongly into bricks.

Distillery Waste	20
Paraffin Crude	10

Mix in a heated dough mixer. Mix in

Sawdust	60
Charcoal or Coal Dust	10

Compress strongly into bricks.

---

* Gasoline Gum Formation Inhibitor

0.001–0.1% of cresol is added to the gasoline.

0.01–0.15% lecithin is added to the gasoline.

Gasoline Fuel, Modified

The following composition gives satisfactory service for buses, trucks and tractors.

Light Creosote Oil	90
Solvent Naphtha	10
Gasoline	50

---

* Gasoline, Solidified

Thirty-five grams of stearic acid are dissolved in five hundred cubic centimeters of ethyl alcohol by warming, and then seven cubic centimeters of a thirty per cent solution of sodium hydroxide (30% sodium hydroxide and 70% water) is added and the heating is continued until the reaction is complete. Forty five hundred cubic centimeters of gasoline is now slowly added and the resulting mass is then set aside to cool and gelatinize.

Coconut Oil	32 parts
Sodium Hydroxide (30%)	9 parts
Water	60 parts
Ethyl Alcohol	3 parts
Gasoline	6000 parts

The preferred manner of preparing this form of the composition is as follows:

A mixture of thirty-two cubic centimeters of coconut oil, seven cubic centimeters of a thirty per cent solution of sodium hydroxide, and sixty cubic centimeters of water are heated on a steam bath until the coconut oil is melted. Three cubic centimeters of ethyl alcohol is then added and the mass is boiled until the reaction is complete, whereupon six thousand cubic centimeters of gasoline are slowly added while stirring and the resulting mass is then set aside to cool and gelatinize.

The resulting gelatinous composition is a glutinous solid that is readily handled and which is well adapted for use as a solid fuel in lieu of dangerous and highly inflammable liquid fuels such as alcohol or gasoline. Also this composition is well adapted for other uses such as removing spots and stains from cloth ing and other apparel, and with the addition of antiseptic agents provides a desirable germicide.

Solidified Gasoline

Gasoline	0.5 gal.
White Soap (Fine Shaved)	12 oz.
Water	1.0 pt.
Household Ammonia	5 oz.

Heat the water, add soap, mix and when cool add the ammonia. Then work in slowly the gasoline to form semi-solid mass.

---

* Special Gasoline

The addition 0.2–1% oxidized paraffin wax to gasoline serves to act as a lubricant in automobile cylinders.

---

* Gasoline, Stabilizing

Decolorization and stabilization against development of undesired color odor or gum deposition is effected by adding a tri- or other poly-hydroxybenzene (suitably 1 lb. of pyrogallol to 75,000 lb. of oil).

---

* Internal Combustion Fuel

Gasoline	60–90
Tertiary Butyl Alcohol	40–10

---

* Kerosene, Solidified

Kerosene	96.5
Albumen	1.5

Heat the above to 40° C. cool add to this

Acetone	2

Remove precipitated albumen and solidify by heating to 60° C.

Fuel Oil

Fuel Oil	460 cc.
Degras	5 gm.

Dissolve by vigorous stirring; run in slowly following solution

Pot. Nitrate	6½ gm.
Borax	2½ gm.
Water	38 cc.

Finally pass through colloid mill.

The above mixture ensures perfect, rapid and complete combustion.

---

* Motor Fuel

Gasoline	70
Benzol	20
Methyl Formate	10

---

* Motor-Fuel, ''Anti-Knock''

The following is added to gasoline to prevent ''knocking.''

Aniline	1.5
Acetone	0.2
Alcohol	0.8

Special Fuel

Aluminum Powder	95
Sulfur Powder	5

* Engine Carbon Removers

A. Sulfur	0.5
Phosforus	0.5
Naphtha	99

B. Aniline	2
Benzol	2
Alcohol	2

C. Furfuryl Alcohol	10
Xylol	10

GLASS, CERAMICS, ENAMELS

* Casting Slip, Ceramic

Am. Hydroxide	0.10–0.62
Caustic Soda	0.01–0.14
Rochelle Salt	0.01–0.10
Oxalic Acid	0.01–0.10
Litharge	0.000003–.006

Crucibles, Refractories

Flake Graphite	21
Crushed Silicon Carbide	45
Flint	11
Borax	5
Tar	18

* Crucibles, Non-Porous

$Al_2O_3xH_2O$ is heated for 1 hr. at 1000°, mixed with kaolin (5:2) and $2N$-HCl, and ground in a ball mill. The paste is heated until viscous, dried in the air for 1 day, and heated at 900° for 4 hr. and finally at 1650° for 0.75 hr. The product is non-porous and temp.-resistant. The linear shrinkage undergone is about 27%.

*Refractory Lining

A metallic pot is lined with a mixture of

Slacked Lime	100
Borax	6
Vitreous Enamel	26
Glue Solution	12

It is dried and baked at 500–700° C.

Marking Glass

40° Bé. Sodium Silicate can be used as a marking ink on glass. It adheres well after drying. After a few weeks, the dried silicate is washed off, the glass will be found etched. If desired, colored pigments may be added to the silicate to make it show up better.

*Safety Glass

Laminated or safety glass which ordinarily consists of two sheets of glass cemented by a suitable binding material to the opposite sides of a sheet of tough reinforcing material, such as cellulose ester plastic of which celluloid is a common example. Among the cements or binders heretofore used are the so-called glyptal resins, such resins comprising the condensation products of a polyhydric alcohol with a polybasic acid. Unmodified resins of this type when used as cements, have in general certain favorable characteristics, one of which is that the sheets will not separate due to the absorption of moisture by the cement, but have certain undesirable properties when used in the manufacture of safety glass, one of the objections to the resins being their tendency to polymerize and become brittle and lose their holding power in the course of time. They also have the property of acting as plasticizers or solvents of the cellulose plastic and are themselves not soluble in non-solvents of cellulose ester plastic. I have found that resins of this kind can be improved for the desired purpose; made soluble in non-solvents of cellulose ester plastic solvents; and made non-solvents of cellulose plastic by the use of fat or fatty acid or oil in certain proportion during the polymerization of the resins.

The following formulae set forth in five examples of proportions of ingredients required to form modified polyhydric alcohol polybasic acid condensation resins:

Example No. 1

	Parts
Glycerol	94
Phthalic Anhydride	148
Fatty Acids (obtained from Soya Bean Oil)	120

Example No. 2

	Parts
Glycerol	94
Phthalic Anhydride	148
Fatty Acids from Castor Oil	40
Corn Oil	100

Example No. 3

	Parts
Glycerol	94

All formulae preceded by an asterisk (*) are covered by patents.

Phthalic Anhydride	123
Fatty Acids (obtained from Linseed Oil)	210

Example No. 4

	Parts
Glycerine	92
Phthalic Anhydride	185
Oleic Acid	141
China Wood Oil	20

Example No. 5

	Parts
Glycerol	92
Phthalic Anhydride	175
Butyric Acid	44
China Wood Oil Acids	50
Soya Bean Oil	50

The amounts of phthalic anhydride and of fatty acid are interchangeable according to their acid equivalents: one mol. of phthalic being equivalent to 2 mols, of fatty acid, or ⅔ mol. of oil, where it is used as the source of fatty acid. The proportions used need not necessarily be molecular quantities, since certain advantageous effects either in the preparation or in the final product may be obtained by using an excess of one or two ingredients.

Other monobasic acids, such as benzoic, propionic, butyric, lactic, salicylic, their analogues or substitution products, can be used in place of the fatty acids from oil or polybasic acid to esterify part of the hydroxyl groups of the polyhydric alcohol. Likewise polyhydric ethers, such as the polyglycerols and diethylene glycol, or the ether derivatives of a polyhydric alcohol, such as mono-ethyl-ether of glycerol, may be used in place of part or all of the polyhydric alcohol. Examples of other polyhydric alcohols are glycol and mannitol, and of other polybasic acids, succinic, sebacic, tartaric, citric, malic, maleic and lactic.

Among the non-solvents of cellulose ester plastic which may be used as solvents of the above resins are the following: heavy coal tar naphtha, toluol, benzol, xylol, carbon tetrachloride, cumene and ethyl benzene. The proportion of resin to solvent ranges from two to twenty parts in one hundred. In applying the cement, the resin is dissolved in the solvent and sprayed onto the faces of the glass sheets in a thin film or coat. This film is allowed to dry out in part or in whole after which the sheets are assembled and subject to heat and pressure following the usual practice in laminating safety glass, the temperature preferably being between 200 and 250 degrees F. and the pressure being about 150 pounds per square inch. If desired, the solvent used may be made up of a mixture of several solvents.

"Horak" Glass

"Horak" glass, made in Czechoslovakia, is said to possess great elasticity, and to be resistant to sudden changes of temperature. The composition is:

	Per cent.
Sand	60–70
Boric Acid	15–30
Potassium Carbonate	1–2
Sodium Carbonate	3–6
Zirconia	1–3
Titanium Dioxide	1–3

---

* Golden Luster on Glass

Cl_2 is passed into an aq. soln. of $FeCl_3$, and the soln. is mixed with H_2O_2. The soln. is sprayed on a glass surface heated to 700° to produce a golden luster.

---

* Refractory Glass

The glass is made of SiO_2 65–75, B_2O_3 10–15, Al_2O_3 2–5, alkali oxide 4–10 and ZnO 3–10%. The linear expansion coeff. is 4×10^{-6}. It is resistant to acids and alkalies.

Glass, Resistant

Silicon Dioxide	70
Boron Oxide	16–20
Litharge	10
Iron Oxide	5

This glass is resistant to high temperatures, quick temperature changes and is easily worked.

Glass, Ruby

The following is added to the basic glass batch

Selenium	2 %
Cadmium Sulfide	1 %
Arsenic Trioxide	1 %
Carbon	0.5%

---

* Glass, Safety

There are provided between glass sheets alkyd intermediate layers, and there is applied a relatively low pressure, e. g., 50 lb. per sq. in., at about 110° for 5 min.; subsequently the temp. is lowered to 70–75° while the pressure is slowly increased to about 200 lb. The finished product is removed from the press after cooling to room temp.

* Glass, Substitute

The following when applied to wire or cloth net forms a transparent material which may be used in place of glass. It is non-breakable and transmits ultra-violet light and is used for poultry houses, playrooms, etc.

Cellulose Acetate	100
Triphenyl Phosfate	12–17
Dibutyl Tartrate	12–17
Acetone 80–90	sufficient to get
Alcohol 10–20	thickness desired

* Glass, Ultra-Violet Transmitting

Silica Sand	53–57%
Feldspar	23–27%
Calcined Pot. Carbonate	8%
Bone Ash	12%

The above is used for ultra-violet light incandescent lamp bulbs.

Thermal Glass

A material, as elastic as fused SiO_2, more workable and stable in the flame, and less liable to surface-cracking, is prepared by fusing (at $>$ 1700°) a mixture of 90–99 pts. of SiO_2 and 10–1 pts. of beryl.

* Glass Ultra-Violet Transparent

A batch for making glass especially transparent to ultra-violet radiation of wave lengths below 3200 A. U. consists of silica 560, borax 527 and powd. metallic Zn 8 parts.

* Glass, Ultra-Violet Ray Transmitting

A colorless glass having high ultra-violet transmission is formed by melting a F-contg. batch in a carboniferous container. The batch may comprise SiO_2 35.5–40, H_3BO_3 16.5–32.5, Al_2O_3 21–27 and CaO 11–17% together with CaF_2.

Bohemian Plate Glass—I

50.0 Kg Quartz
20.0 Kg Potash
8.5 Kg Calcite
100.0 g Arsenic

Bohemian Plate Glass—II

50 Kg White Sand
12 Kg Potash
9 Kg Calcite
60 g Pyrolusite

Belgian Plate Glass

50 Kg White Sand
17 Kg Sulphate
40 Kg Calcite
1 Kg Coal Dust
230 g Arsenious Acid

German Plate Glass

50 Kg White Sand
17 Kg Sulphate
3 Kg Soda
18 Kg Calcite
1 Kg Coal Dust
500 g Arsenious Acid

English Plate Glass

50 Kg White Sand
14 Kg Sulphate
18 Kg Calcite
520 g Coal Dust
500 g Arsenious Acid

French Plate Glass

50 Kg White Sand
17 Kg Chalk
19 Kg Sulphate
500 g Coal Dust
510 g Arsenious Acid

Glaze, Acid Resisting

Lead Oxide	0.8
Sodium Oxide	0.1
Iron Oxide	0.1
Silicon Oxide	1.5
Boron Oxide	0.4

Glazes, Alkali-free Lime

Satisfactory bright glazes having a maturing temp. of cones 11 to 13 were produced. A good cone 13 bright glaze was produced with 100 limestone, 26 kaolin, 245 calcined kaolin and 396 sand. With mat glazes it was found necessary to use at least 3 mols. of SiO_2 to prevent crazing. A good cone 11 mat was produced with limestone 100, kaolin 26, calcined kaolin 112 and sand 96. These glazes are especially resistant to abrasion and chem. action and therefore are recommended for chem. porcelain, cooking utensils, insulators and tech. stoneware. A good magnesia-lime, alkali-free glaze was produced with calcined magnesite 19, limestone 78, kaolin 26, calcined kaolin 45 and sand 144.

* Glaze for Copper Tankards

Silicon Dioxide	45.5
Pot. Oxide	8
Sod. Oxide	14
Boron Trioxide	19.5
Aluminum Oxide	6.5
Calcium Fluoride	6.5

* Enameling Copper

Cu and Cu alloys are given an intensive treatment with acid and then coated with an enamel free from products capable of tarnishing, *e.g.*, one contg. SiO_2 45.5, K_2O 8, Na_2O 14, Ba_2O_3 19.5, Al_2O_3 6.5 and CaF_2 6.5. The enamel is applied directly by pulverization and the objects are baked at a temp. (800°) at which all the constituents of the enamel melt to form a limpid covering.

Vitreous Enamel

240 grams borax, 410 grams potash feldspar, 30 grams saltpetre, 120 grams sodium carbonate, 30 grams calcium spar, and 170 grams quartz are fused together to produce 1,000 grams of lump enamel. This is crushed, ground with 60 grams of tinting substance and about 20 grams of zirconia opacifier. The latter should contain about 1 gram of salt of unstable acid, for example, sodium nitrate of formate.

* Enamel, Vitreous

Sod. Silicate (23% Na–74% Si)	68
Aluminum Hydrate	5
Borax	13
Sod. Antimonate	12
Cryolite	6
Barium Carbonate	8

Vitreous Enamel, Acid Proof

An acid proof enamel suitable for use in chemical apparatus consists of

Sand	527
Kaolin	65
Borax	57
Calcium Carbonate	85
Soda Ash	230
Sodium Silico Fluoride	42

* Enameling Iron

Fe articles, particularly sheet Fe, are provided with a colorless ground-enamel coating, free from CoO or NiO, by applying a suitable mixt. of readily fusible and difficultly fusible substances in the form of a moist pulp and then heating the article for a short time, so that the difficultly fusible substances do not completely dissolve in the readily fusible substances, with the result that a coarse-grained coating is produced. A suitable mixt. contains borax 36.3, feldspar 36.3, quartz 32, soda 6, $NaNO_3$ 6 CaF_2 1.8 and kaolin 10 parts.

* Opacifier, Enamel

An opacifier contg. NaZr silicate and Zr silicate is prepd. by heating a mixt. of Zr silicate about 78 and Na_2CO_3 about 22% to about 900–950° and cooling and disintegrating the product.

Vitreous Enamel Opacifier

Sod. Antimonate	3
Tin Oxide	1

Removing Vitreous Enamel

Place article in a boiling 30% Caustic Soda solution and enamel will dissolve.

Enamel for Gold Dental Crowns

Silicon	6.5
Borax	2
Soda Ash	1.65
Sodium Nitrate	0.3
Cryolite	1.2
Tin Oxide	0.5

* Vitreous Enamel, Translucent

Fe which has a particularly low fusibility has its constituents limited in the following manner, boric anhydride 10–23, Ba or Sr 5–25, K 3.5–19, SiO_2 4–13, Zn 0–23, Ca 0–10, Al_2O_3 0–3 and F 0–3%. The enamel is applied by heating the article to redness and powdering the enamel thereon so that it immediately melts.

* Porcelain Composition

Pyrophylite is used to replace all the silica and some of the feldspar in a porcelain compn., to obtain a product of higher dielectric and mech. strength and of lower porosity. A mixt. may be used formed of china clay 34, ball clay 13–19, feldspar 37–22 and pyrophylite 16–25%.

* Porcelain, Insulating

China Clay	34 lb.
Ball Clay	13 lb.
Feldspar	37 lb.
Pyrophylite	16 lb.

Electrical Porcelain

Kaolin	40–55
Quartz	25–32
Feldspar	20–28

Grind very finely; mix well and "fire" at 1400° C.

---

* Refractory Composition

(For crucibles and furnace linings)

Barium Oxide	31–51
Aluminum Oxide	17–37
Silicon Dioxide	22–42

The above is not corroded by aluminum or its alloys.

Enameling Steel

The preparation of the steel for enameling consists in giving it such treatment as is necessary to leave a clean surface, free from any foreign matter that will injure the enamel when applied and burned. The treatment required depends upon the nature and size of the piece of ware and the kind of foreign matter that is to be removed.

The sand blast is used in cleaning large ware and such as can not be easily cleaned by pickling. When the sand blast is used, no other treatment is required, since grease, rust, and any other foreign matter is readily removed by it. This is the most effective method of cleaning steel and one that gives an excellent surface for enameling. For small pieces it is much more expensive than pickling, and it is economical only in making large pieces or special shapes of comparatively high value.

Treatment Preliminary to Pickling

Nearly all light steel ware is cleaned by the pickling process. The preliminary treatment before the ware is placed in the pickling acid varies. Grease and carbonaceous matter must be removed from the ware before placing in the pickling solution, and three general methods are in use for doing this; scaling, washing in caustic alkali solutions, or the use of proprietary cleaning compounds.

Scaling.—Scaling or heating the ware to redness is the method most generally employed. During the process of shaping the ware from the sheet of steel it invariably collects grease from machinery and workmen's hands, and one method of removing such carbonaceous matter is to burn it off. Especially is this the case when handling large numbers of small pieces. To do this, the ware should be carefully stacked on grates in such a manner as to admit free access of air to all parts of every piece of ware. Care must be taken to prevent flat surfaces from coming into contact with each other, and space must be provided between the different pieces of ware to admit sufficient air to completely oxidize all carbonaceous matter present. It must be remembered that the heat treatment forms an iron scale which must subsequently be removed by acid, and consequently the time and temperature should not be carried beyond that necessary to burn off the oil.

Removing Grease with Caustic Soda.—Caustic soda or potash may be used for removing fatty materials, especially if they are present in small amounts. In this process the steel article is immersed in a boiling solution of caustic soda or potash and allowed to remain for a few minutes. It is then taken out and washed free from alkali in clear water. This precaution is necessary because the adhering alkali solution would rapidly neutralize the pickling acid into which the steel is next placed for the removal of rust and other deleterious impurities.

Pickling

After the oil and carbonaceous matter have been removed from the surface, it is necessary to remove all rust and oxide of iron. The pickling solution used is one of either sulphuric or hydrochloric acid.

1. Mixing the Raw Materials

General practice in mixing the raw materials consists in weighing the batch, which generally approximates 500 pounds, into a box and then turning the mixture over a few times with a hoe or shovel. In the case of colored enamels it is considered mixed when the coloring oxide is uniformly distributed, imparting a uniform gray color to the batch. In white enamels the practice is to turn the mixture a certain number of times, which is considered to be sufficient. Here is one of the places where enamelers can improve their practice and raise the standard of their ware by doing away with slipshod methods and resorting to more thorough, exact, and economical methods. Rotating drums and other forms of mixing machines give much more satisfactory results.

Every enameler, and even the uneducated laborer who has worked around the smelter, has observed that the enamel smelts more quickly when most thoroughly mixed. This is simply the practical application of the well-known scientific principle that the speed of chemical reactions is directly proportional to the area of surface of contact between the reacting substances. If a fire brick were crushed to a powder and mixed into the batch it would go into solution in the melt and disappear with the other ingredients of the batch, while that same brick when laid in the wall of the smelter will stand for months without being eaten away. This same principle applies to all the refractory ingredients of the batch. A large piece of flint stone will go through a melt and come out with only the sharp edges eaten off. The length of time required for smelting the enamel depends directly upon the fineness of the raw material, especially flint and feldspar, and upon the thoroughness with which they are mixed. It follows, then, that better mixing of the raw materials means less labor, less fuel, less time of smelting, and less wear and tear on the smelter.

It is not only from an economic standpoint that thorough mixing is advisable. The quality of the white enamels is inversely proportional to the length of time spent in producing a thorough melt. Long smelting results in a considerable reduction in opacity. Fine grinding and thorough mixing insures a uniform fusion product in the shortest possible time and hence minimum solution of opacifying agents and minimum reduction in opacity.

2. Melting

In the smelter the enamel mixture is melted and fined until no lumps of unfused or undissolved material can be detected in a string of the glass drawn from the melt. The melting process begins with the fusion of the least refractory ingredients or fluxes—borax, soda ash, etc.—at relatively low temperatures. The liquid attacks the more refractory substances both by solution and by chemical reaction. The formation of eutectics between the raw materials and the compounds resulting from chemical reaction facilitates the melting process.

If the smelting process is continued for a sufficient length of time a perfectly homogeneous glass in which all constituents would be in equilibrium would result. Such a condition is not obtainable, especially in white enamels. The melting should proceed only to the point where a stable borosilicate glass is formed, in which the opacifying agents, fluorides, tin oxide, and antimony compounds are carried in suspension. Longer smelting results in a considerable solution of these materials, as well as decomposition of the fluorides and consequent reduction in the opacity of the enamel. No opacity is obtained from tin or antimony oxides after they are once taken into solution. Quick smelting is therefore to be desired, and this again calls attention to the value of fine grinding and thorough mixture of the raw materials.

3. Tempering Enamel Slips

In preparing enamel slips for application to the ware the frit is ground wet and contains 5 to 10 per cent (by weight) of plastic clay. To increase the viscosity of the slip and aid in holding the enamel in suspension, a flocculating agent is added. In white or cover enamels magnesium sulphate is generally used for this purpose. In ground coats borax is almost universally employed, since nearly all other salts which have a similar effect on the slip are likely to cause rusting of the steel during the drying of the ware.

1. Fine grinding makes the frit more easy to float, but enamelers dare not grind too finely, because of difficulty in getting a uniform coating on the ware. Ground coat enamels especially must be coarse, not finer than 100 mesh, and, better, 80 mesh.

2. Lead enamels would, of course, be more difficult to float than lighter ones, but lead is seldom used in enamels for sheet iron. However, all frits are relatively high in specific gravity as compared with clays and therefore settle more readily.

3. Settling is easily prevented by making the slip thick, approaching a paste, but in order to apply them by dipping or spraying, slips must be sufficiently fluid to flow. With such a consistency heavier substances will settle unless a floating agent is used.

4. Viscosity has been described as the friction between two liquids flowing in contact with each other, or between a liquid and a solid moving in it; in other words, resistance to flow. The efficiency of a floating medium in preventing the settling of heavier particles, therefore, depends upon its viscosity or resistance to the motion of particles passing through it. The floating medium in the case of enamels is not to be considered as the water, but as the clay substance in suspension in water.

High viscosity is also required in enamel slips to prevent them from flowing down the sides and into the corners of the ware after dipping. A steel body, being nonabsorbent, offers a different problem from that of a porous body dipped in a glaze slip. The absorption by the porous body prevents the flowing of the glaze, but the enamel slip must stay in place by virtue of its viscosity, although it is possible that surface tension also plays an important rôle here.

5. It is evident that a sufficient amount of the floating medium to prevent settling can readily be added, but other considerations limit the amount of clay which can be used with any glaze or enamel, about 10 per cent being the maximum permissible in the latter. The efficiency of the clay as a floating agent is therefore highly important, especially in enamels where the frit is of higher specific gravity and more coarsely ground than in glazes or engobes, and where the amount of clay used is necessarily small.

1. Application of the Enamel

There is no more vitally important operation in the entire process of enameling than the application of the first coat of enamel. A piece of ware which has passed through the operations of forming and cleaning has acquired considerable value to the manufacturer on account of the labor expended upon it. In the application of the ground coat it is possible to enhance this value or to destroy it, or, still worse, to so treat it that it will pass through the succeeding operations and still be worthless as a finished piece of ware. Given a good ground coat, properly applied and burned, the finishing of the ware is simple. The very best ground coat improperly applied or burned can give only a poor piece of ware, regardless of what its previous cost or future treatment may be. Every possible precaution should therefore be taken to insure a suitable coating on the steel.

Four different methods are used for applying the enamel to the steel—slushing, draining, spraying, and dusting. The choice of method depends upon the size and shape of the ware and the nature of the enamel. The chief factor to be considered in the application of the enamel is to obtain a coating of uniform and sufficient thickness on the surface of the ware. If a thin and uniform coating is not obtained, the enamel will burn off the portions where it is too thin and will not be sufficiently burned where it is thick. Either of these defects will cause the finished ware to be defective. The method best suited to produce this result, with due consideration to the cost of the operation, is the one generally used.

Slushing.—By far the greatest proportion of enameled ware is slushed, especially in the case of all light wares and such as can be easily shaken to distribute the enamel uniformly. The operation consists in dipping the piece of ware into the enamel slip, removing it and shaking it in such a way as to leave a thin and uniform coating over the entire surface of the metal. There are two factors of vital importance in securing proper results by this method—the consistency of the enamel slip and the skill of the operator. The consistency of slip for slushing is such as is termed "short"; that is, it has a high viscosity and will not run down or drain off from vertical surfaces after dipping.

To the novice it would seem a simple matter to dip a piece of steel into a tub of slip, shake off a little, and obtain a nicely coated piece of ware. As a matter of fact, considerable practice is required to acquire skill sufficient to slush even simple shapes uniformly, while extensive training and a very high degree of skill is required in the handling of complicated shapes.

Draining.—This method is frequently applied to perfectly flat ware, such as signs, and to simple shapes. The piece of ware is dipped in the slip and is then set on edge to allow the excess to run off at the bottom. The consistency of the slip, which is very different from that used in slushing, is the principal factor in the success of this operation. In this case the viscosity is much lower, so that the slip will flow down the vertical surface, but at the same time its consistency must be such that it will form a good coating and adhere to the ware after the excess drains off. It must also be sufficiently viscous to keep the enamel in suspension and not allow it to settle onto the bottom of the tank.

Spraying.—For applying enamel to complex shapes and heavy ware, spraying is frequently resorted to. It is too expensive to use on the ordinary grades of ware, but for special shapes with many corners and sharp angles, or any piece of ware which can not be slushed uniformly, spraying is the best method of coating. It is wasteful of material and requires skill to obtain good results, but if proper care is used any piece of ware can be very uniformly coated by spraying. The piece may be placed on a whirling rack and turned while the spray is being applied.

The consistency of the enamel is highly important again in this case. The enamel must be ground sufficiently fine to prevent stopping the nozzle of the sprayer, but for best results it must not be too finely ground. Its viscosity must be high to prevent flowing. Since the distribution of the slip over the surface is accomplished in this case by the movement of the spray and not by shaking the piece, it is possible to work with a higher viscosity than in slushing.

Dusting.—This method of application is very common in cast-iron work, but in steelwork it is used only on heavy wares, such as condensers for chemical works, etc. It has a decided advantage in the production of acid-resisting wares, because no raw materials are added to the frit, whereas when any of the other methods of application are used, it is necessary to add some raw clay and soluble salts to the frit in order to get a slip of the proper consistency. These raw materials are invariably decidedly injurious to the enamel, especially where resistance to chemical corrosion is desired. While an enamel is a glassy coating, it is far from being a solid glass; and the more raw material added in grinding the frit the further is the finished enamel removed from this condition, since these raw materials are only to a very slight extent combined with the frit during the brief burning operation. Because of this fact the dusting method is decidedly the best to use for making enameled ware to resist chemical corrosion.

In carrying out this process, the ground coat, as well as cover coat, is frequently dusted on. The metal is wiped with a wet sponge or cloth, and the powder dusted on while the metal is still wet. Sometimes an adhesive agent is added to prevent the enamel from falling off when dry.

The methods used for cover enamels are the same as those used for ground coats. The quality of workmanship in applying cover coats is far less important than in applying ground coats. If a piece of ware is perfectly coated with the ground coat, the cover coat may be quite imperfectly applied and still give good results. Of course there are limits to this, and the more uniformly the enamel is applied the better it will be. It should be said, further, that best results are always obtained with thin enamels. Barring the properties of whiteness and opacity, the excellence of enamels is inversely proportional to their thickness. This is especially true of the ability of the ware to withstand bending and abrasion. In view of these facts the aim should always be to keep the enamel as thin as possible, while at the same time obtaining the desired opacity and color.

2. Drying

Ground-coat enamels should be dried as rapidly as possible to prevent rusting of the steel. This will be controlled to some extent by the flocculating agents used in the slip, but rapid drying is the best practice in any case. If an alkaline flocculating agent is used for tempering the ground coat, it can be dried in the open air without serious rusting; but if chlorides or sulphates are used, rusting is almost sure to result even with rapid drying. This rust may or may not be visible after the ware is dry, but it is quite sure to make its appearance, when the ground coat is burned, in the form of spots where the iron oxide has reacted with the enamel to such an extent as to form a spot-like iron scale. When these spots are formed, it is practically impossible to cover them with cover enamel. They will show in the finished ware either as dark spots or as pits in the surface. While proper drying of the ground coat can not entirely prevent this trouble in an improperly tempered enamel, it will always reduce the trouble, and when the ware is not dried rapidly the trouble is likely to come even in the best tempered enamel.

The rate of drying of cover enamels is of less importance than in drying ground coats. However, rapid drying is here again desirable. One of the chief reasons for this, especially in white enamel, is the fact that dirt in the form of factory dust sticks to the ware while wet, and therefore rapid drying of the white enamel makes for pure white ware. Another point in favor of rapid drying of finished ware is the need of space for storing the ware. After the enamel is dry the ware can be handled and stored in much less space than when wet, and in making some classes of wares, such as cooking utensils, the problem of finding room for storing sufficient ware to keep the furnaces going is sometimes troublesome. There are two common defects caused by improper drying. Water streaking, caused by moisture from drying ware condensing on the cold surface of wet ware and running down vertical surfaces in streaks, can be avoided by proper circulation of air in the dryer. When ware is dried too rapidly the enamel will crawl. This is caused by the formation of shrinkage cracks due to driving off the moisture from the clay too rapidly. These cracks do not show in the dry ware, but when it is burned the enamel crawls and collects in beads. This defect will be caused when a piece of wet ware is set on a hot piece of metal or when the drying is very sudden. The same defect may result from rough handling of the dry ware, a sudden sharp blow breaking the bond between the dry enamel and steel, which results in crawling.

3. Burning Enamels

General Description.—Muffle furnaces are almost invariably used for burning light wares and especially white ware. For burning heavy steel wares open furnaces are used.

The ware is set on pointed projections from iron grates, which should be kept sharp so that the least possible part of the grates comes in contact with the enamel. Only pieces of approximately the same size and weight should be burned together, since only a few minutes are allowed for burning a fork of light steel ware, and if there is much difference in the size of the ware it will heat up to the temperature of the furnace at different rates. As a result of this the lighter ware will be sufficiently fired before larger pieces have acquired the desired temperature, and some of the ware will be sure to be imperfectly fired.

In setting the ware on the grates preparatory to firing, care should be taken to see that ample space is left between all surfaces. Heavy parts like handles on dishpans and ears on kettles should be removed as far as possible from all other surfaces. The reason for this is not only to permit these heavy parts to heat up as rapidly as possible but also to prevent them from absorbing radiated heat from parts near them, thereby retarding the rate at which these parts are heated.

It frequently happens that there will be a small area on a piece of ware underburned while the piece as a whole is properly burned. Investigation of the cause of this will reveal the fact that this underburned spot was in close proximity to some heavy piece of metal or other surface which absorbed the heat while the main body of the piece of ware was free to heat up rapidly. A good burner will strike the happy medium and leave his ware in the furnace long enough to fire the heavy parts properly but not long enough to burn off the light parts. The nature of the enamel influences very materially the burner's ability to properly burn light and heavy parts, but he can greatly facilitate matters by using proper care in setting his ware on the grates.

The temperatures used for burning enamels differ widely, depending upon the enamel and the ware. General practice is to burn the ground coat at much higher temperature than the finishing coats. This is not due to the fact that the ground coat necessarily has a higher softening temperature than the finishing coats, but rather to the fact that it has been found that the general excellence of the ware is improved by this procedure.

Ground-Coat Frit

Borax	90
Soda Ash	23
Potash Feldspar	110
Quartz	70
Manganese Dioxide	18
Saltpeter	18
Cobalt Oxide	1.5
Fluorspar	27

Mill Addition

	Per cent
Clay	8
Water	50
Magnesia	.25
Borax in Boiling Solution	2.0

White-Coat Frit

Borax	100
Soda Ash	54
Potash Feldspar	110
Flint	110
Saltpeter	23
Fluorspar	13
Barium Carbonate	25
Antimony Oxide	20
Zinc Oxide	25
Cryolite	25

Mill Addition

	Per cent
Tin Oxide	6
Clay	6
Magnesium Sulphate	.5

The key to the burning temperatures used, with their index numbers and the time required in each case to produce the best results in the enamel, are as follows:

Ground Coat

Index Number	Temperature, °F.	Time, Minutes
1	1700	4
2	1800	3
3	1900	2

First White

Index Number	Temperature, °F.	Time, Minutes
1	1600	2
2	1700	1½
3	1800	1

Second White

Index Number	Temperature, °F.	Time, Minutes
1	1500	3
2	1600	2
3	1700	1½

Ultra Violet Glass

A glass of the compn. B_2O_3 82, Li_2O 13.6, BeO 4.4% may be prepd. by fusing below 950° in a Pt or Al_2O_3 crucible. This glass and the K_2O and Na_2O analogs should be valuable substitutes for quartz in optical work. They transmit light after long exposure to air and ultra-violet light, and can be fused to ordinary glass.

* Gilding Glass

Glass contg. SiO_2 74.6, B_2O_3 8.8, Al_2O_3 4.3, alkali 3.7 and bivalent oxide 4.6%, is coated with an ethereal oil soln. of Au resinate contg. 3–30% Au and fired.

* Safety Glass

One surface of each of 2 sheets of glass is provided with a skin coating of a compn. formed from gelatin 1, nitrocellulose 1, a mutual solvent such as HOAc 70, a gelatin solvent such as water 14 and a nitrocellulose solvent such as acetone 14% and the coated surfaces are united with an intervening sheet such as a pyroxylin compn.

* Refractory Brick

A compn. for making refractory articles such as furnace bricks or crucibles contains plastic infusible clay 100, powd. glass 10, borax 5 and NaCl 15 kg. The proportions may be varied.

* Brick, Sound Proofing

Bricks which have good sound-deadening properties are formed from slag 40, slate $Ca(OH)_2$ 20, $CaSO_4$ 7, K_2SO_4 2, Ca silicofluoride 1 part and water.

Acid Resistant Enamel

	I	II	III
Quartz	47.3	52.5	55.8
Felspar	22.4	19.1	17.5
Soda	29.8	32.0	33.0
Marble	16.8	13.7	13.4
Boric Acid	6.2	3.5	...
	122.5	120.8	119.7

Percentage of Bases

	I	II	III
SiO_2	62.0	65.1	67.2
B_2O_3	3.5	2.0	...
Al_2O_3	4.1	3.5	3.2
K_2O	3.1	2.6	2.4
Na_2O	17.9	19.1	19.7
CaO	9.4	7.7	7.5
	100.0	100.0	100.0

Enamel I has a cubical coefficient of expansion of 322.8×10^{-7}, II of 321.4×10^{-7}, and III of 342×10^{-7}.

The preliminary surface treatment of the iron before enamelling is most important. Not only the gross irregularities, but all surface impurities should as far as possible be removed. The usual method employed today is the sand-blast, using a mixture of relatively coarse sand and steel scrap, blown at a pressure of four to six atmospheres. It is necessary that this cleaning process be applied as soon as the casting has cooled, and it is a frequent practice for the castings to be heated to a dull red after the sand-blasting, this heating particularly favoring the decomposition of the iron carbide.

The technology of the application of the acid-resisting enamels differs from that of the ordinary enamelling process only in regard to the final coat; in both cases the application of the ground coat is the same. This ground enamel, the so-called frit, is chosen so as to have a wide temperature softening interval, and consists usually of two parts of ground flint and one part of borax, with small quantities of felspar and fluorspar. This ground mass is applied to the cast iron by the wet process, and is burned on at a temperature of about 1000° C. It is white, and makes a very firm bond with the metal. The wet covering enamel, finely ground, is sprayed on to this frit, and it is advantageous to incorporate a little clay in the grinding mill. On to the layer of wet coating enamel there is sieved a fine powder, closely similar in composition to the enamel itself, after which the coating is thoroughly dried. This application process is generally repeated several times.

Burning and Cooling Operations

The temperature of burning depends upon the fusibility of the enamel, and is usually about 1000° C. Burning is usually effected in muffle furnaces, but in the case of very large pieces, in non-muffle furnaces of special construction. According to German Patent 478,632, burning is effected in an electric furnace under vacuum; by this means the formation of bubbles is stated to be completely avoided. Numerous highly resistant enamels give a surface of comparatively dull lustre, and it is sometimes the practice to give a final coat of highly lustrous enamel. This latter usually is not very resistant to acid, and is dissolved off when the vessel is put into use. The enamelled pieces should be cooled very slowly. If the cooling is too rapid, owing to the comparatively poor heat conductivity of the enamel, stresses are set up which lead to the formation of surface cracks. Really well-controlled cooling improves the acid resistance of the mass, for it is a well-known physical-chemical principle that has been confirmed in practice, that badly cooled glasses are less resistant to the leaching action of liquids than are well-cooled glasses.

INK, CARBON PAPER, DUPLICATORS, CRAYONS, ETC.

Black Carbon Paper

75% of these materials in proportions suitable for grade desired.

Candelilla Wax
Beeswax
Crude Montan Wax
Mineral Oil

Toners (Oil Soluble)	10%
Peerless Carbon Black	15%

This is ground hot. It is a base formula which may be modified to suit conditions.

* Carbon Paper

Glassine paper is coated with hectograph ink. This may be dusted lightly with talc and is ready for use.

Flexible Printing Roller

Casein Glue Solution	10
Glycerin	5
Molasses	5
Clovel	1

Mix until uniform and pour into forms.

* Stencil Sheet

Coat paper with the following material

Nitrocellulose	12.6
Acetone	225.0
Alcohol	135.0
Resin	2.7
Glycerine	45.0

* Carbon Paper

A suitable paper is coated with

Gutta Percha	30
Lamp Black	17
Carnauba Wax	30
Petrolatum	40

Carbon Paper

Crystal Violet Base or Methyl Violet Base 300 parts

are dissolved in

Red Oil 600 parts

This is introduced into approximately

Sesame Oil 3500–4000 parts

and added to

Carnauba Wax 3500 parts

melted at 105–110° C.

* Chemical Printing on Wall Board

The material, *e.g.*, plaster board, is printed with 5–10% aq. $KMnO_4$ and

All formulae preceded by an asterisk (*) are covered by patents.

heated to form MnO_2. Such printing is readily eradicated when desired.

Animal Marking Crayon

Tallow	180
Rosin	5
Rozolin	2

Melt together and add while stirring a mineral pigment such as Prussian Blue, Red Iron Oxide, etc. Cast in glass or metal tubes.

Blackboard Crayon

Calcium Carbonate (precipitated)	60 lb.
Kaolin Clay	40 lb.
Saponified Oleic Acid	5 lb.
Caustic Soda	¾ lb.

The Oleic Acid and Caustic Soda are mixed, warm, in a separate kettle and added to the clay mix along with enough water to bring to about the consistency of putty. The mixing is done in a standard type dough mixer or other clay mixing equipment.

* Cloth Marking Crayon

In making the crayon, pure chalk in the proportion of about 500 parts is thoroughly permeated with one to four parts of dye of a suitable character in alcoholic solution, a binder such as dextrin, in the proportion of about ten parts, being used to assist in the molding of the chalk into sticks.

The chalk particles should have a porous structure, giving a very large adsorption surface for the dye, which preferably is an alcoholic solution of aniline dye of the desired color.

While the preferable method of making the crayons is to permeate the chalk in a powdered condition, the dye may be added after the molding of the chalk and binder into sticks. In this event, substances of a suitable nature to assist penetration of the dye may be used, as for instance, butanol. These substances not only assist in the penetration of the dye, but they retard the drying, and make it more uniform. The butanol when used, assists penetration, but substantially the same results may be obtained without butanol, by extending the time of soaking. The chalk is dried in air, or in ovens, as may be desired.

Crayons so obtained are used in the following manner. The fabric, such as silk, rayon or the like, is moistened over the part which is to receive the design. The design is then drawn upon the cloth with the crayon, it being understood that the design may be in several colors, if desired. A portion of the chalk is abraded from the stick by the rubbing upon the fabric, and these abraded portions, of infinitesimal size, yield up their dye to the moistened fabric. After drying, the chalk particles may be brushed away, if desired, leaving the design in the form of a permanent impression upon the fabric. Dye applied with the improved crayons is less likely to ''bleed,'' than if applied in any other manner, and it is not removed by washing.

It will be understood that the fabric is stretched prior to the application of the design. It may be moistened before or after stretching, and is moistened over substantially the area to be occupied by the design.

While it is stated that the fabric is stretched prior to the application of the design, this is not essential in all cases, but is a matter of choice with the designer. It is apparent that the design might be drawn by mechanism suitable for the purpose instead of manually.

The particles of chalk also absorb moisture from the cloth, thus making a definite flow of moisture into the mark or design, removing any likelihood of the color running or bleeding.

Drawing Crayons

Black

Kaolin	24	lb.
Carbon Black	22	lb.
Garnet Shellac	12	lb.
Denatured Alcohol	1	gal.
Turpentine	½	gal.

Dissolve shellac in alcohol; add turpentine and then mix in solids and grind to smooth paste. Mould and dry slowly.

Blue

Soapstone	34	lb.
Chinese Blue	14	lb.
Garnet Shellac	12	lb.
Denatured Alcohol	1	gal.
Turpentine	½	gal.

Method—as under Black.

Wax Drawing Pastels

Black

Hard Soap	80
Beeswax Crude	60
Spermaceti Crude	28
Carbon Black	14
Burnt Umber	5
Prussian Blue	4

Melt waxes and soap, mix in pigments and grind until smooth; pour hot in molds; and plunge into cold water to "set."

Red

Hard Soap	28
Saponified Japan Wax	28
Spermaceti	16
Carnauba Wax	2
Beeswax Crude	8
Orange Chrome Yellow	12

Method—as under Black.

Lithographic Crayon

Sod. Stearate	7
Beeswax	6
Carbon Black	1

Lithographic Crayon

Beeswax	30
Tallow	25
Soap	20
Shellac	15
Lamp Black	6

Heat in enamelled pot to melt together. Then heat strongly until vapor ignites. Allow to burn for a while and smother flame with cover of pot. Take out a sample and test for elasticity. If not satisfactory ignite again in same way.

Marking Crayons

Ceresin	40
Carnauba Wax	35
Paraffin Wax	20
Beeswax	5
Talc	50
Chrome Green or Other Pigment	15

Crayon, Tailors' Marking

Carnauba Wax	11
Stearic Acid	2
Ceraflux	76
Ozokerite	6
Terra Alba	5

Tailors' Chalk

Yellow

Chalk (Powd.)	28
Soapstone	18
Pipe Clay	10
Yellow Ochre	7
Lemon Chrome Yellow	1½

Make into a paste with water and mold.

White

French Chalk	20
Pipe Clay	20
White Curd Soap	6

Make into a stiff paste with water and dry.

Black

Soapstone	56
Bone Black	8
Yellow Soap	6
Gum Arabic	2
Glycerin	1

Dissolve gum in water, add glycerin, mix in pigments; grind to a smooth paste with water and mold.

Warehouse Chalk

Gypsum	40
Soapstone	55
Carbon Black	6
Petrolatum	1

Mix to a uniform paste with a thin glue solution and mold.

Wax Crayons

The manufacture of wax crayons follows very closely that of the moulded candle, both in procedure and materials and an attempt to go into details would be endless and rather futile. A finely divided dry color is usually more suitable as the coloring medium and usually more dependable. The dry color is added to the wax combination after the wax is melted in a steam jacketed aluminum kettle. Mechanical agitation is continued until the kettle has been emptied in order to prevent any tendency of the color to settle to the bottom. The wax should be maintained as nearly to the melting point as practicable and rapid cooling is perhaps more important here than in candles. A good starting point on the wax combination would be as follows:

Double Pressed Saponified Stearic Acid	40 lb.
Paraffine	45 lb.
Beeswax	10 lb.
Carnauba Wax	5 lb.

Dry color to suit.

The above proportions may be changed to create a harder or softer crayon and Candelilla Wax may be added or substituted for the Beeswax. Care should be taken not to make the crayon too hard as a tendency of the points to crack or flake will be noted.

Mimeograph Moistening Compound

Powdered Soap	8 oz.
Castile Soap	5 oz.
Glycerin	4 oz.
Water to make	1 gal.

Non-Offset Compound

No. 1 Lithographic Varnish	35
Soft Cup Grease	35
Paraffin Wax	10
Beeswax	20

Melt together; cool and run in mill.

Ink: Copying and Record

All the ingredients in the standard ink must be of the quality prescribed in the current edition of the United States Pharmacopoeia.

Tannic Acid	23.4 gm.
Gallic Acid Crystals	7.7 gm.
Ferrous Sulphate	30.0 gm.
Hydrochloric Acid, Dilute	25.0 gm.
Phenol (Carbolic Acid)	1.0 gm.
Soluble Blue	3.5 gm.

Water to make 1 liter at 20°C. (68°F.)

Here as in all other formulae, ''water'' means distilled water, if it can be had. Rain water is second choice.

Dilute hydrochloric acid, U.S.P., is of 10 per cent strength. Concentrated hydrochloric acid as commonly sold is a water solution containing about 36 per cent by weight of hydrochloric acid gas, so as to make the 10 per cent acid, 100 parts by weight of concentrated acid must be diluted with 260 parts by weight of water.

Soluble blue is one of the comparatively few dyes that are not precipitated by the other ingredients of the ink. When buying a supply of it, be careful to say that it is to be used for making ink.

To make the ink, dissolve the tannic and gallic acids in about 400 milliliters of water at a temperature of about 50° C. (122° F.). Dissolve the ferrous sulphate in about 200 milliliters of warm water to which has been added the required amount of hydrochloric acid. In another 200 milliliters of warm water dissolve the dye. Mix the three solutions and add the phenol. Rinse each of the vessels in which the solutions were made with a small quantity of water, and use the rinsings to make the volume of ink up to 1 liter at room temperature. Be sure the ink is well mixed before it is bottled. If sealed hermetically in a glass bulb, the ink will keep for years with practically no formation of sediment. So when bottling the ink, have good tight corks and fill the bottles almost to the corks.

This ink is primarily for records, and is not like most copying inks. However it will make one good press copy when the writing is fresh, and this will generally suffice.

Ink: Writing

Except for the phenol and dye, this ink is half as concentrated as the record and copying ink. It is similar to some of the commercial writing fluids and fountain pen inks. The standard is made in the same way as the preceding ink, and from materials of the same quality. If made with slightly more hydrochloric acid than the formula calls for it will keep longer without depositing sediment, but it will be more corrosive to steel pens.

The standard formula is:

Tannic Aid	11.7 gm.
Gallic Acid Crystals	3.8 gm.
Ferrous Sulphate	15.0 gm.
Hydrochloric Acid, Dilute	12.5 gm.
Phenol (Carbolic Acid)	1.0 gm.
Soluble Blue	3.5 gm.

Water to make 1 liter at 20° C. (68° F.).

Writing Ink
(8 times concentrated)

The ingredients are best dissolved as follows:

	Dissolved in
2 ounces Ferrous Sulphate	1⅔ oz. of dil. Hydrochloric 3 oz. of Water
0.47 oz. of Soluble Blue 0.13 oz. of Phenol	3 oz. of Water
1.55 oz. of Tannic Acid 0.50 oz. of Gallic Acid	6 oz. of Water
For washing, etc.	2⅓ oz. of Water

Dissolve first the Dye and Phenol; pour into this mixture the acid solution of Iron and then the Tannic-Gallic Acid solution. All solutions should be heated to about 180° F. and the final mixture stirred well for some time and then allowed to cool. Let stand quietly for 2 or 3 days and decant.

Writing Ink

1. Nutgalls Powd.	8
2. Logwood Chips	8
3. Iron Sulfate	4
4. Gum Acacia	4
5. Aniline Black	1
6. Water	167

Dissolve (4) in ½ gal. water and (5) in 3 gal. water; filter and mix these two solutions. Boil (1), (2) and (3) in remaining water for 2½ hours and strain. Mix this liquid with previous solution.

Writing Ink—Red

Eosine	1 oz.
Gum Arabic	1 oz.
Phenol	½ oz.
Water	1 gal.

Writing Ink—Blue Black

Naphthol Blue Black	1 oz.
Gum Arabic	½ oz.
Phenol	¼ oz.
Water	1 gal.

Red Writing Ink

Water, Warm	250 gal.
Crocein Scarlet	15 lb.
Carbolic Acid	1½ lb.

Blue Writing Ink

Water, Warm	250 gal.
Methylene Blue	15 lb.
Carbolic Acid	1½ lb.

Jet Black Writing Ink

Water, Warm	250 gal.
Nigrosene	15 lb.
Carbolic Acid	1½ lb.

Directions

Dissolve all color in 25 gallons of hot water (about 160° F.), add balance of warm water while mixing. Allow to stand several days then decant without stirring up any sediment.

Concentrated Ink, Powder and Tablets

Concentrated ink that meets all the requirements of the specification can be made by cutting down the amount of water to a minimum, so as to make a pasty mass or a thick fluid with the solids only partly dissolved. Instead of hydrochloric acid, which is volatile, an equivalent quantity of sulphuric acid is used; that is, 1.77 grams of the usual concentrated acid of 95 per cent strength (66 deg. Baumé).

Ink: Red

The standard ink is made by dissolving 5.5 grams of crocein scarlet 3B in 1 liter of water

Hectograph Ink

Years before some of the modern duplicating devices had been invented, the hectograph was used for printing small editions of circular letters, etc., and it is still in rather wide use. The original is written with a special ink that contains a large proportion of a dye that has good tinting strength. The letter is then pressed face-downward upon a gelatin-glycerin or a clay-glycerin pad, which absorbs a considerable amount of the ink. From this pad it is possible to print a number of increasingly paler copies upon other sheets of paper. The name, hectograph, ''hundred writing,'' exaggerates somewhat, unless copies so pale as to be barely legible are counted. In experimenting with quite a number of dyes, it was found that the following would give at least 30 copies with unbroken line, and numerous other copies that were easily legible, though there were breaks in the strokes of the pen. Methyl violet gave the most copies, the best red dye was rhodamine B, and emeral green and Victoria blue were the best of their colors.

The ink used in making these tests was prepared according to the formula:

Acetone	8
Glycerin	20
Acetic Acid, Coml. 30%	10
Water	50
Dextrin	2
Dye	10

Stamp-Pad Ink

A solution of dye in water could be used on a stamp pad, but it would soon dry out. A mixture of equal volumes of glycerin and water remains moist under all atmospheric humidities, though the water content of the mixture fluctuates. In each 100 milliliters of the mixture of glycerin and water dissolve 5 grams of dye. The following are used for making the standards of different colors in the specification: water-soluble nigrosine (black), soluble blue, light green, magenta (red), and acid violet.

Recording Inks

For outdoor recording instruments the Weather Bureau uses inks made by dissolving about 10 grams of dye in 1 liter of a mixture of equal volumes of glycerin and water. As this mixture will freeze in some parts of the country, it is sometimes necessary to add a certain proportion of alcohol to the ink.

For recording instruments in the laboratory, the ink needs to contain only enough glycerin to prevent its drying at the tip of the pen. A mixture of 1 volume of glycerin and 3 volumes of water has been found satisfactory.

Almost any water-soluble dye might be used were it not that some of them rather unaccountably make blurred lines on the usual card and paper charts. Dyes that have been found to work well are crocein scarlet, fast crimson, brilliant yellow, emerald green, soluble blue, methylene blue, methyl violet, Bismarck brown, and water-soluble nigrosine.

Indelible Marking Ink

Dissolve 5 grams of silver nitrate in its own weight of water, and add ammonia water (not household ammonia) until the precipitate that first forms just dissolves. Separately dissolve 5 grams of gum arabic in 10 milliliters of warm water, and 3 grams of anhydrous sodium carbonate (or 3.5 grams of the monohydrate) in 15 milliliters of warm water. Mix the three solutions and warm until the mixture starts to darken. This ink should be used with a gold or a quill pen if possible, but if not, with a clean steel pen. The writing should be exposed to direct sunlight or pressed with a hot iron to develop the color. The ink must be kept in the dark.

Aniline black inks are made in one or in two solutions, the argument for the latter being that the chemical reaction that produces the color must take place largely in the fibers where the mark is wanted. There is no chance for the color to be formed in the bottle before the ink is applied to the fabric, and to make a sediment that can not penetrate into the fibers. However, excellent one-solution inks can be bought.

For a two-solution ink the following has been recommended:

Solution A.

Copper (Cupric Chloride)	85
Sodium Chlorate	106
Ammonium Chloride	53
Water	600

Solution B.

Gum Arabic	67
Water	335
Aniline Hydrochloride	200

Keep in separate bottles. Immediately before use mix 1 volume of A and 4 volumes of B.

Blue-Print Ink

For writing on blue prints use the following which bleaches white:

Soda Ash	10 gm.
Water	50

Ink for Brass

Copper Acetate	1
Water	15

Add sufficient ammonium hydroxide to dissolve the blue precipitate formed.

* Printing Ink for Cellulose Acetate Film

Cellulose Acetate	3
Ethylene Glycol Monomethyl Ether	50
Ethylene Glycol Monomethyl Ether Acetate	50
Color	to suit

* Ink, Concentrated Writing

A paste ink suitable for writing on diln. with water comprises water 2 oz., white potato dextrin 1 oz., gallic acid 336 grains, tannic acid 120 grains, granulated Fe_2SO_4 252 grains, HCl 130 minims, carbolic acid 1.5 drams, glycerol 2 drams, blue aniline A 217 grains, indigotin 68 grains and HOAc.

* Ink, Gold Bronze

Cresylic Acid	8
Sulfuric Acid	4
Borax	15
Flour	60
Chrome Yellow	3
Gold Bronze Powder	10
Varnish	10
Water	90

Heat to a boil while stirring and make thicker or thinner by altering amount of water.

When the finished ink is too heavy it may be reduced by petrolatum or varnish.

Writing and Copying Ink

	Fountain Pen Ink	Copying Ink
Tannic Acid	1.55 oz.	3.10 oz.
Gallic Acid	0.50 oz.	1.00 oz.
Ferrous Sulphate	2.00 oz.	4.00 oz.
Hydrochloric Acid (dilute)	1.67 oz.	3.34 oz.
Phenol	0.13 oz.	0.13 oz.
Soluble Blue	0.47 oz.	0.47 oz.

Dissolve the Tannin and Gallic Acid in about 3 pints of warm water (of about 130° F.) and add to it the Dilute Hydrochloric Acid (of about 7° Bé.) and then the solution of Ferrous Sulphate and Phenol in about 2 pints of water. Bring up to 1 gallon, mix well and let stand quietly for 4 days. Then decant without stirring up any sediment formed.

Ink for Glass and Porcelain

Shellac	4
Borax	1
Water	150

Warm and stir until dissolved; cool and filter. Add

Dye	1

Ink for Writing on Glass

Pale Shellac	2 oz.
Venice Turpentine	1 oz.
Sandarac	¼ oz.
Oil of Turpentine	3 fl. oz.

Dissolve by gently heating and then add one of the following pigments.

Black—Lamp Black	½ oz.
Blue—Ultramarine	½ oz.
Green—Brunswick Green	½ oz.
Red—Vermilion	½ oz.

Waterproof Ink for Glass

Shellac Bleached	10
Venice Turpentine	4
Rosin Oil	1
Turpentine	15
Indigo Powder	5

---

* Ink, Graining

Gum Arabic	2.5
Ethylene Glycol	60
Water	7.5
Pigment	30

Hectograph Ink

Acetone	8
Glycerin	20
Acetic Acid (28%)	10
Water	50
Dextrin	2
Dye	10

Dissolve dextrin in hot water with stirring; cool and add other liquids and dye.

Hectograph Ink

Fuchsine	1 oz.
Alcohol	1 oz.
Glycerin	¼ oz.
Phenol	½ oz.

Hectograph Mass

Good Grade Powdered Glue	2 parts
Water	1 lb.
Glycerine	4 lb.

Proceed as in printers' rollers composition.

---

* Ink, Indelible

A mixt. of castor oil 15, glycerol 15, aniline oil 3, Ph_2NH 5 and a small amt. of methyl violet is one example and a mixt. of soybean protein 15, aniline oil 5, β-naphthol 1, Ph_2NH 5 and varnish 5 is another.

---

* Intaglio Ink

Heat 3½ lb. Gilsonite under pressure with 1 gal. solvent naphtha until dissolved; cool and beat in a 20% water solution of dextrin.

Printing Inks

Printing Inks may be divided into three classes—typographic, lithographic and rotographic. They consist principally of a pigment, vehicle and drier.

Typographic Inks

Typographic inks are printed from a raised surface. They dry principally by oxidation and penetration. Magazine and book inks dry largely by oxidation. Representative formulae would be as follows:

Black

Carbon Black	20 lb.
No. O Lithographic Varnish	30 lb.
Rosin Oil	30 lb.
Cobalt Drier	10 lb.
Stearine Pitch	5 lb.

Yellow

Chrome Yellow	75 lb.
No. O Lithographic Varnish	25 lb.
Lead-Manganese Drier	2 lb.

Red

Lithol Red	45 lb.
No. O Lithographic Varnish	50 lb.
Drier	5 lb.

Besides these pigments, formulae contain many other colors, depending upon their use and desired shade. News inks, which come under the typographic class, dry principally by penetration, assisted in some cases by oxidation. The following would be representative formulae:

News Inks

Black

Carbon Black	12 lb.
Mineral Oil	85 lb.
Methyl Violet	1 lb.
Stearine Pitch	2 lb.

Blue

Peacock Blue	15 lb.
White Extender	7 lb.
No. 2 Lithographic Varnish	20 lb.
Mineral Oil	58 lb.

Red

Lithol Red	12 lb.
White Extender	10 lb.
Mineral Oil	25 lb.
No. O Lithographic Varnish	25 lb.
Rosin Oil	27 lb.

Lithographic Inks

The lithographic process depends upon the fact that oil or greasy substances and water will not mix. Most present day lithographic printing is done from grained zinc or aluminum plates. The original designs or characters are made onto the plates by the artist actually drawing or painting the original onto the grained plate or by transferring the designs from another print by transfer ink or by a photo litho process, whereby the design or negative is developed on the metal plate after it has been sensitized with an albumen coating.

This coating which has no affinity for water, allows the ink to transfer from a rubber roll to the plate and then to the paper. Lithographic inks, in composition, are very similar to typographic inks. Generally a heavier lithographic varnish is used as a vehicle. The only essential difference in pigments is that they must not bleed in water or weak acids to any great degree.

Vehicles.—The vehicles in printing inks are, as already mentioned. Lithographic varnish is nothing more than a heat bodied linseed oil. It may range in viscosity anywhere from 2 poises to 500 poises. Rosin oils and mineral oils may be either of high or low viscosity. Although the above oils are most commonly used in typographic inks, china wood oil, perilla oil and fish oil are also used.

Other ingredients may be found in inks such as waxes, resins and sometimes solvents.

Driers.—Driers are made from lead, manganese and cobalt compounds. These are dispersed in various oils and varnishes. Generally lead and manganese driers are used in light colors while cobalt is used in the darker colors. The kind of driers used are also dependent on the application.

Rotographic Inks

Rotographic inks are printed from an etched surface. They dry almost completely by evaporation. Generally solvents such as Toluene, Xylene and High Flash Naphtha are used to dissolve the resins which, together, make up the vehicles. Practically any resin soluble in the above mentioned solvents may be used. A formula would contain approximately

Pigment	33⅓ lb.
Resin	33⅓ lb.
Solvent	33⅓ lb.

Until recently only black and brown pigments were used, but at present rotographic inks may be made in other colors.

Printing inks are made by wetting and dispersing solid pigment colors in a suitable liquid medium. The vehicle used is usually a combination of oils and varnishes together with small amounts of driers, wax and grease compounds. The ink is manufactured by first mixing the ingredients in a change can or kneading mixer and then ground on steel roller mills.

In formulating a printing ink, only those pigments should be used that will meet the requirements of the printed matter, such as permanency to light, alkali proof, etc., and the method of printing used (either typographic, planographic or intaglio). The skillful blending of these pigments in a formula produces practically any desired color in the chromatic scale.

The specific gravity and oil absorption of the pigments will govern the ratio of pigment to vehicle. The type of vehicle will vary according to the body, tack, penetration, hardness of printed films, and drying properties that is desired to give to the ink. These in turn are governed by the method of printing used, type of press, size of the form, and nature of the stock the ink is printed on.

The final test of the suitability of a printing ink is its ability to work well on the printing press, print perfectly and to adhere properly to the printing surface.

The commercially available pigments, the properties of each and typical formulae containing these pigments are listed in the following:

YELLOW PIGMENTS

Chrome Yellows

These are Lead Chromates made from soluble lead salts and bi-chromate of soda. They range in shade from a light Primrose to a deep Orange.

Properties	*Light*	*Medium*	*Orange*
Resistance to Light	Good	Excellent	Excellent
Resistance to Varnish Bleed	Excellent	Excellent	Excellent
Resistance to Water Bleed	Excellent	Excellent	Excellent
Resistance to Paraffin Bleed	Excellent	Excellent	Excellent
Resistance to Alcohol	Excellent	Excellent	Excellent
Resistance to Alkali	Poor	Fair	Good
Resistance to Acid	Good	Fair	Poor
Hiding Power	Excellent	Excellent	Excellent
Baking Temperature	230° F.	320° F.	320° F.

The Chrome Yellows can only be used in making opaque colors and therefore only for the first color in three or four color process work.

Process Yellow

Primrose Yellow Dry	4 lb.
Lemon Yellow Dry	1 lb. 8 oz.
Magnesium Carbonate Dry	1 lb.
No. 1 Lithographic Varnish	1 lb.
No. 2 Lithographic Varnish	3 lb.
No. 5 Lithographic Varnish	3 oz.
No. 6 Lithographic Varnish	½ oz.
Lead Manganese Paste Drier	½ oz.

Process Yellow

Chrome Yellow Medium Dry	8 lb.
Magnesium Carbonate Dry	1 lb. 8 oz.
No. 1 Lithographic Varnish	1 lb.
No. 2 Lithographic Varnish	5 lb.
No. 5 Lithographic Varnish	4 oz.
No. 6 Lithographic Varnish	2 oz.
Lead Manganese Paste Drier	¾ oz.

The above formulae may be modified to be made stronger and more opaque by omitting the magnesium carbonate and using more of the chrome yellow pigments.

Offset Tin Printing Yellow

No. 1 Transparent Lithographic Varnish	20 lb.
No. 00 Transparent Lithographic Varnish	2 lb.
No. 2 Transparent Lithographic Varnish	4 lb.
No. 3 Transparent Lithographic Varnish	2 lb.
C. P. Medium Chrome Yellow Dry	55 lb.
Gloss White Dry	15 lb.
Offset Ink Wax Compound	1 lb.

on last pass over mill and add

No. 7 Lithographic Varnish	1 lb.

Yellow Lakes

These give transparent inks which are used for process colors. They are made from the auramine, quinoline, tartrazine and naphthol yellow S dyes. They all have very poor permanency to light. They are destroyed by alkalies and acids but have good resistance to lithographic varnish and paraffin. They are slightly soluble in water and alcohol.

Process Transparent Yellow

Tartrazine Yellow Lake Dry	4 lb. 12 oz.
No. 0 Lithographic Varnish	2 lb. 8 oz.
Cobalt Linoleate Liquid Drier	3 oz.
No. 00 Lithographic Varnish	1 lb.
Lead Manganese Paste Drier	6 oz
Paraffin Wax	2 oz.
Kerosene Oil	6 oz.
Amber Petrolatum	3 oz.

Cadmium Yellows

These are precipitated cadmium sulfides and are produced in a light and dark shade. They are very permanent to light and very resistant to alkali, water, alcohol, oils and paraffin, but are affected by acids.

Cadmium Yellow

Cadmium Yellow Light Dry	15 lb.
No. 1 Lithographic Varnish	4 lb.
No. 3 Lithographic Varnish	8 oz.
Lead Manganese Drier	4 oz.
Wax Compound	4 oz.
Aluminum Hydrate Dry	1 lb.
No. 0 Lithographic Varnish	1 lb.

ORANGE PIGMENTS

Orange Mineral

Orange mineral is also known as red lead. It is a very opaque and bright orange pigment of very good permanency. In formulating inks with orange mineral it is necessary to use a lighter pigment and a heavy varnish with it, due to its high specific gravity and low oil absorption, otherwise it will work very poorly on the press. It has a strong drying action on drying oils and therefore does not require the addition of any driers.

Opaque Orange Ink

Orange Mineral Powder, Dry	30 lb.
No. 0 Lithographic Varnish	6 lb.
No. 1 Lithographic Varnish	12 lb.
No. 3 Lithographic Varnish	3 lb.
Persian Orange, Dry	8 lb.
Alumina Hydrate, Dry	13 lb.
No. 6 Lithographic Varnish	1 lb. 8 oz.

Persian Orange

Persian Orange is made by precipitating the azo dyestuff orange II with either barium chloride or stannous chloride on freshly prepared alumina hydrate base.

Properties

Resistance to:

Light	Poor
Varnish Bleed	Good
Water Bleed	Good
Paraffin Bleed	Good
Alcohol	Poor
Alkali	Fair
Acid	Poor
Hiding Power	Very transparent
Baking Temperature	270° F.

Transparent Orange

Persian Orange Dry	7 lb. 8 oz.
No. 0 Lithographic Varnish	6 lb.
Woolgrease	12 oz.
Cobalt Linoleate Liquid Drier	8 oz.
Lead Manganese Paste Drier	4 oz.

Permanent Orange

Permanent Orange is made by coupling a diazotized solution of dinitroaniline with beta-naphthol.

Properties

Resistance to:

Light	Excellent
Varnish Bleed	Slightly soluble
Paraffin Bleed	Good
Water Bleed	Good
Alcohol	Good
Alkali	Good
Acid	Good
Hiding Power	Good
Baking Temperature	270° F.

For stock ink grind 1 part color with 1 part No. 0 Lithographic Varnish.

RED PIGMENTS

Lithol Toners

The colors are produced by coupling Tobias Acid and Beta Naphthol. The various shades ranging from Orange to Maroon are produced by forming different metallic salts. The sodium salt is the Orange shade. The barium and calcium salts are bluer.

Properties	*Sodium Lithol Tones*	*Barium Lithol Tones*	*Calcium Lithol Tones*
Resistance to Light	Fair	Fair	Fair
Resistance to Varnish Bleed	Excellent	Excellent	Excellent
Resistance to Water Bleed	Good	Excellent	Excellent
Resistance to Paraffin Bleed	Excellent	Excellent	Excellent
Resistance to Alcohol	Fair	Good	Good
Resistance to Alkali	Good	Good	Good
Resistance to Acid	Good	Excellent	Excellent
Hiding Power	Good	Fair	Good
Baking Temperature	....	165° F.	230° F.

Gloss Die Stamping Red

Gloss Stamping Varnish	33 lb.
No. 1 Burnt Plate Oil	2 lb.
Plate Paste, Drier	6 lb.
Blanc Fixe Dry	27 lb.
Paris White (Whiting) Dry	28 lb.
Calcium Lithol Toner Red Dry	4 lb.

Cylinder Press Red

Sodium Lithol Toner Dry	8 lb.
Barium Lithol Toner Dry	5 lb.
Gloss White, Dry	5 lb.
Magnesium Carbonate, Dry	5 lb.
No. 0 Lithographic Varnish	20 lb.
Boiled Linseed Oil	2 lb.
Lead Manganese Paste Drier	2 lb. 8 oz.
Cobalt Linoleate Liquid Drier	1 lb.

Red For Lake C

This color is made by coupling chlor toluidine Sulfonic Acid with Beta Naphthol and as in the case of Lithol Reds, various shades are produced by forming different metallic salts.

The sodium salt is the orange shade also called bronze orange. The barium salt is the red shade.

Red for Lake C is an excellent color for printing inks, possessing excellent working qualities, brilliance and transparency. The shades as a rule run much yellower in undertone than lithol reds.

Properties	*Sodium Salt*	*Barium Salt*
Resistance to Light	Fair	Fair
Resistance to Varnish Bleed	Excellent	Excellent
Resistance to Water Bleed	Good	Excellent
Resistance to Paraffin Bleed	Excellent	Excellent
Resistance to Alcohol	Fair	Fair
Resistance to Alkali	Fair	Good
Resistance to Acid	Fair	Good
Hiding Power	Good	Very poor
Baking Temperature	....	230° F.

Job Press Bright Red

No. 1 Lithographic Varnish	4 lb.
No. 0 Lithographic Varnish	5 lb.
Lead-Manganese Paste Drier	1 lb. 8 oz.
Barium Red for Lake C Dry	4 lb.
Gloss White Dry	7 lb.
Aluminum Hydrate Dry	3 lb.
Cobalt Drier	4 oz.

Cylinder Press Red Ink

No. 1 Lithographic Varnish	35 lb.
No. 00 Lithographic Varnish	12 lb.
Neutral Wool Grease	4 lb.
Paste Drier (Manganese Resinate Lead Acetate)	4 lb.
Gloss White, Dry	22 lb.
Barium Lake for Red C, Dry	23 lb.

Offset Process Red

No. 1 Lithographic Varnish	32 lb.
No. 3 Lithographic Varnish	4 lb.
Barium Red for Lake C, Dry	44 lb.
Aluminum Hydrate, Dry	8 lb.
Blanc Fixe Dry	8 lb.
Offset Ink Wax Compound	2 lb.
Paste Drier (Lead Acetate Manganese Borate)	2 lb.

Scarlet Ink

The pigment of Scarlet Ink is the lead lake of the scarlet dye formed by the combination of Xylidine and R salt. It is ground directly into the ink from the pulp and is marketed in this form only because the color cannot be dried without losing practically all of its strength. Scarlet Ink is very transparent and has a good finish, or gloss.

Properties

Resistance to:	
Light	Fair
Varnish Bleed	Fair
Water Bleed	Fair
Paraffin Bleed	Good
Alcohol	Good
Alkali	Fair
Acid	Fair
Hiding Power	Very poor
Baking Temperature	165° F.

Para Reds

These colors are produced by coupling Para Nitraniline with Beta Naphthol. Two distinct types; namely, Light and Dark Para Reds are used.

Properties	*Para Red Light*	*Para Red Dark*
Resistance to Light	Good	Good
Resistance to Varnish Bleed	Poor	Poor
Resistance to Water Bleed	Fair	Poor
Resistance to Paraffin Bleed	Poor	Poor
Resistance to Alcohol	Poor	Poor
Resistance to Alkali	Fair	Fair
Resistance to Acid	Good	Poor
Hiding Power	Excellent	Excellent
Baking Temperature	230° F.	230° F.

Para Reds are well suited for label and carton work due to their good permanence to light and excellent covering.

Label Red

No. 0 Lithographic Varnish	5 lb.
No. 1 Lithographic Varnish	8 lb.
Medium Bodied Rosin and Mineral Oil Varnish	6 lb.
Para Red Dark Dry	6 lb.
Para Red Light Dry	2 lb.
Aluminum Hydrate Dry	8 lb.
Wool Grease	1 lb. 8 oz.
Cobalt Linoleate Liquid Drier	2 lb.
Wax Compound	8 oz.
Barium Sulfate Dry	10 lb.

Toluidine Red

This color is made by coupling Meta Nitro Paratoluidine with Beta Naphthol. It is lighter and brighter than Light

Para Red and is noted for its excellent light-fastness and opacity.

Properties

Resistance to:

Light	Excellent
Varnish Bleed	Good
Water Bleed	Fair
Paraffin Bleed	Poor
Alcohol	Poor
Alkali	Excellent
Acid	Good
Hiding Power	Excellent
Baking Temperature	320° F.

For stock ink grind 1 part color with 1 part No. 0 Lithographic Varnish.

Madder Lake

This is made from the synthetic alizanine dyestuff dihydroxyanthraquinone.

Properties

Resistance to:

Light	Excellent
Varnish Bleed	Good
Water Bleed	Good
Paraffin Bleed	Good
Oils and Grease	Excellent
Alcohol	Good
Alkali	Changes to blue-violet
Acid	Poor
Hiding Power	Poor

The madder lakes are used mainly for oil and grease proof inks such as on butter-wrappers.

Madder Lake Ink

Madder Lake, Dry	5 lb. 8 oz.
No. 0 Lithographic Varnish	3 lb.
No. 2 Lithographic Varnish	5 lb.
Lead-Manganese Paste Drier	8 oz.
Cobalt Linoleate Drier	12 oz.

Pigment Scarlet
(Bluish Shade of Red)

This color is manufactured by precipitating the soluble acid azo dyestuff (made by coupling diazotized anthranilic acid with R salt) with barium chloride. The precipitation is made in the presence of freshly prepared alumina hydrate base thus forming an insoluble lake pigment.

Properties

Resistance to:

Light	Excellent
Varnish Bleed	Good
Water Bleed	Slt. soluble
Paraffin Bleed	Good
Alcohol	Good
Alkali	Poor
Acid	Poor
Baking Temperature	270° F.

For stock ink grind 1 part color with 1 part No. 0 Lithographic Varnish.

Eosine and Phloxine
(Bluish Shade Red)

These are lake pigments made from triphenyl methane dyestuffs. The dyestuff is precipitated with basic lead acetate on an alumina hydrate base. These colors are very clean, brilliant and transparent and are used in the manufacture of process reds, where permanency is not required.

Properties

Resistance to:

Light	Very poor
Varnish Bleed	Good
Water Bleed	Poor
Paraffin Bleed	Poor
Alcohol	Poor
Alkali	Very poor
Acid	Very poor
Heat	Very poor

Process Red

Phloxine Toner Red, Dry	12 lb.
Alumina Hydrate, Dry	10 lb.
No. 0 Lithographic Varnish	7 lb.
No. 1 Lithographic Varnish	14 lb.
No. 6 Lithographic Varnish	1 lb.
Wax Compound	3 lb.

BLUE PIGMENTS

Iron Blues

This class of colors is made from sodium or potassium ferrocyanides and ferrous sulfate. Three distinct types are supplied for printing inks namely, Milori, Bronze and Prussian blues.

Properties

Resistance to:

Light	Excellent
Varnish Bleed	Excellent
Water Bleed	Excellent
Paraffin Bleed	Excellent
Alcohol	Excellent
Alkali	Poor
Acid	Excellent
Baking Temperatures	320° F.

Job Press Blue

Bronze Blue, Dry	9 lb.
Permanent Violet, Dry	1 lb. 8 oz.

No. 0 Lithographic Varnish	9 lb. 8 oz.
Lead Manganese Paste Drier	4 oz.
No. 6 Lithographic Varnish	4 oz.
No. 1 Lithographic Varnish	2 lb. 8 oz.
Barium Sulfate, Dry	6 lb.
Petrolatum	4 oz.

Label Blue

Bronze Blue, Dry	8 lb.
No. 0 Lithographic Varnish	2 lb. 4 oz.
No. 1 Lithographic Varnish	1 lb.
Mineral Ink Oil	4 lb. 8 oz.
Barium Sulfate, Dry	3 lb. 8 oz.
Aluminum Hydrate, Dry	6 oz.
Permanent Violet, Dry	4 oz.
Wool Grease	6 oz.
Cobalt Linoleate Liquid Drier	4 oz.
Lead Manganese Paste Drier	12 oz.

Lichtdruck or Photogelatin Blue

No. 1 Lithographic Varnish	44 lb.
No. 3 Lithographic Varnish	3 lb.
Milori Blue, Dry	50 lb.
Multon Tallow	1 lb.

Steel Plate Blue

Bronze Blue, Dry	52 lb.
Barytes, Dry	14 lb.
No. 0½ Plate Oil	27 lb.
No. 1 Plate Oil	7 lb.

Peacock Blue

Peacock blue is a lake color produced by precipitating the acid dyestuff, erioglaucine or patent blue, on alumina hydrate base with barium chloride. This color is greener than the iron blues and of much greater cleanliness and transparency. It is chiefly used in the manufacture of process inks.

Properties

Resistance to:

Light	Very poor
Varnish Bleed	Good
Water Bleed	Poor
Paraffin Bleed	Good
Alcohol	Poor
Alkali	Poor
Acid	Poor

Process Blue

Peacock Blue, Dry	8 lb. 8 oz.
No. 0 Lithographic Varnish	4 lb.
No. 1 Lithographic Varnish	3 lb.
Cobalt Linoleate Liquid Drier	1 lb.
Wax Compound	8 oz.

Alkali Blue Inks

(Reflex Blue Toner)

These inks are made from C. P. Alkali Blue, and organic dyestuff and lithographic varnish. The alkali blue color in dry pigment form is unsatisfactory for grinding, so that it is necessary to add the varnish to the wet precipitated pulp and then to displace and drive off the water present.

Properties

Resistance to:

Light	Fair
Varnish Bleed	Excellent
Water Bleed	Excellent
Paraffin Bleed	Excellent
Alcohol	Poor
Alkali	Good
Acids	Excellent

The Reflex on Alkali Blue toners are very strong, clean, bronzy colors and are used for toning up the iron blues and also black inks.

Glassine and Cellophane Blue

Spec. Hard Grip Varnish (No. 1 Lithographic Varnish and Amberol)	25 lb.
Cobalt Linoleate Drier	8 lb.
Beeswax (Melted into Drier)	2 lb.
Red Shade Reflex Alkali Blue Ink	60 lb.
No. 00000 Lithographic Varnish	5 lb.

Permanent Blue Toner

This is a very clean, strong, reddish blue made from the basic dyestuff, Victoria blue and phosphotungstic acid. It is produced in both the dry state and ink form.

Properties

Resistance to:

Light	Good
Varnish Bleed	Excellent
Water Bleed	Excellent
Paraffin Bleed	Excellent
Alcohol	Poor
Alkali	Fair
Acids	Fair
Baking Temperature	270° F.

Blue Lake Ink

Aluminum Hydrate, Dry	3 lb.
Magnesium Carbonate, Dry	3 lb.
Permanent Blue Toner, Dry	2 lb. 8 oz.
No. 0 Lithographic Varnish	9 lb.

Cobalt Linoleate Liquid Drier	1 lb.
Lead Manganese Paste Drier	12 oz.
Wax Compound	8 oz.

Ultramarine Blue

Is an artificially prepared pigment made by heating together finely divided soda ash or sodium sulfate, china clay, sulfur and some form of carbon, without contact with the air. The color of ultramarine blue is reddish and very clean but it has the disadvantages of very weak tinctorial value, poor working qualities on the press and wearing action on copper plates due to its sulfur content.

Properties

Resistance to:

Light	Excellent
Varnish Bleed	Excellent
Water Bleed	Excellent
Paraffin Bleed	Excellent
Alcohol	Excellent
Alkali	Excellent
Acid	Poor

Ultramarine Blue Ink

Ultramarine Blue, Dry	15 lb.
Aluminum Hydrate, Dry	4 lb.
No. 1 Lithographic Varnish	8 lb.
No. 2 Lithographic Varnish	2 lb.
No. 3 Lithographic Varnish	8 oz.
Cobalt Linoleate Liquid Drier	2 oz.

PURPLE PIGMENTS

Methyl Violet

Methyl Violet is made from the basic dyestuff methyl violet by precipitation with tannic acid and tartar emetic. It is ground in lithographic varnish from the wet pulp and sold in ink form.

Properties

Resistance to:

Light	Very poor
Varnish Bleed	Good
Water Bleed	Good
Paraffin Bleed	Good
Alcohol	Poor
Alkali	Poor
Acids	Fair

It has very poor drying qualities when used in inks and should be used with sufficient cobalt drier.

Permanent Purple

Permanent purple is also derived from the methyl violet dyestuff but is precipitated with phosphotungstic acid instead of tannic acid.

Properties

Resistance to:

Light	Good
Varnish Bleed	Excellent
Water Bleed	Excellent
Paraffin Bleed	Excellent
Alcohol	Poor
Alkali	Fair
Acid	Fair

For stock ink grind 1 part color with 1 part No. 0 Lithographic Varnish.

GREEN PIGMENTS

Chrome Green (or Milori Green)

These are produced in various shades by precipitating greenish chrome yellow upon freshly precipitated greenish iron blue. These pigments are opaque and have the same properties as the chrome yellows and iron blues.

Properties

Resistance to:

Light	Good
Varnish Bleed	Excellent
Water Bleed	Excellent
Paraffin Bleed	Excellent
Alcohol	Excellent
Alkali	Poor
Acid	Good
Hiding Power	Excellent

Job Green

Milori Green, Dry	8 lb. 8 oz.
No. 0 Lithographic Varnish	1 lb.
No. 1 Lithographic Varnish	4 lb.
No. 2 Lithographic Varnish	6 lb.
Copal Green Varnish	4 lb.
Primrose Yellow, Dry	22 lb.
Aluminum Hydrate, Dry	1 lb.

Milori Green Ink

Milori Green, Dry	12 lb.
No. 1 Lithographic Varnish	3 lb.
No. 0 Lithographic Varnish	2 lb.
Copal Green Varnish	1 lb.
No. 00 Lithographic Varnish	8 oz.

Green Lakes

The green lake pigments are all transparent lake colors made from various organic dyestuffs, such as acid green, malachite green and Victoria green. All of the green lakes are very fugitive except those that are precipitated with phospho-tungstic or phospho-molybdic acid. The use of green lake pigments in printing inks enable one to obtain much cleaner and brighter greens than can be made from milori green or the chrome yellows and iron blues.

Various shades and color strengths are produced by the color manufacturers and the properties of each should be determined before used for any particular purpose.

Light Green Lake

Green Lake Light, Dry	7 lb. 8 oz.
No. 1 Lithographic Varnish	8 lb.
No. 2 Lithographic Varnish	1 lb.
Quinoline Yellow Lake, Dry	3 lb.
No. 6 Lithographic Varnish	12 oz.
Cobalt Linoleate Liquid Drier	1 lb. 4 oz.
Lead Manganese Paste Drier	1 lb. 8 oz.
Wool Grease	12 oz.

Black Pigments

The most important black pigments are lampblack and carbon black or gas black. The former is produced by the burning of oils and fats with incomplete combustion. Carbon black is produced by the burning of gas with insufficient air for complete combustion. In both cases the black carbon soot is deposited and collected by various methods. Carbon black is used more extensively today than lampblack for the manufacturing of printing inks as it is much stronger, blacker and gives glossier inks. Lampblack produces duller inks and is used for that purpose. The black pigments have a great retarding action on the drying of oils and varnishes so that it is necessary to use larger amounts of driers in formulating black inks. The iron blues and alkali blue toners are usually added to the black pigments to give the effect of greater depth and blackness to black inks.

Heavy Job Black

Carbon Black	8 lb.
Bronze Blue, Dry	5 lb.
Alkali Blue Toner	3 lb.
No. 1 Lithographic Varnish	5 lb.
No. 3 Lithographic Varnish	10 lb.
No. 5 Lithographic Varnish	4 lb.
Gloss Varnish	3 lb.
Cobalt Drier	2 lb. 8 oz.
Lead Manganese Paste Drier	3 lb.

Bond Ledger or Job Black

No. 3 Lithographic Varnish	16 lb.
Gloss Varnish (Lithographic Varnish and Amberol)	19 lb.
Amber Petrolatum	3 lb.
Highgrade Carbon Black, Dry	22 lb.
Blue for Black in Ink Form	14 lb.
Paste Drier (Manganese Resinate Lead Acetate)	12 lb.
Cobalt Linoleate, Liquid Drier	14 lb.

Halftone Black for Coated Stock

Boiled Linseed Oil	16 lb.
No. 00 Lithographic Varnish	12 lb.
No. 3 Lithographic Varnish	12 lb.
Concentrated Cobalt Linoleate Drier	8 lb.
Soft Wax Non-offset Compound (see below)	12 lb.
Red Shade Reflex Alkali Blue, Ink	10 lb.
High Grade Carbon Black, Dry	18 lb.
Blue for Black in Ink Form	12 lb.

Web Press Black for Newsprint

Heavy Body Mineral Ink Oil	33 lb.
Second Run Rosin Oil	22 lb.
Rosin Varnish (60 parts Mineral Oil and 40 parts Rosin)	34 lb.
News-grade Carbon Black, Dry	10 lb.
Blue Toner (10% Methylene Blue in Oleic Acid)	1 lb.

Lithographic Black

No. 3 Lithographic Varnish	24 lb.
No. 1 Lithographic Varnish	24 lb.
No. 7 Lithographic Varnish	1 lb.
Red Shade Reflex Alkali Blue Ink	9 lb.
Finest Grade Carbon Black, Dry	32 lb.
Concentrated Cobalt Drier	10 lb.

Copper Plate Black

No. 1 Burnt Plate Oil	26 lb.
No. 2 Burnt Plate Oil	4 lb.
Hard Black (Bone Black) Dry	37 lb.
Soft Black (Bone Black) Dry	16 lb.
Plate Paste Drier	10 lb.
Prussian Blue, Dry	7 lb.

Bookbinder's Black

No. 0 Lithographic Varnish	15 lb.
Gloss Copal or Kauri Varnish	25 lb.
Concentrated Cobalt Linoleate Drier	10 lb.
High Grade Carbon Black Dry	25 lb.
Bronze Blue in Ink Form	15 lb.
Reflex Alkali Blue, Red Shade, Ink	10 lb.

White Pigments

These may be divided into two classes, transparent and opaque.

1. Transparent White Pigments

Aluminum Hydrate.—Is made by precipitation from alumina sulfate and sodium carbonate.

It is the most transparent and best working pigment available. Its very low specific gravity and high oil absorption makes it an excellent inert extender to be used with other pigments, especially those of high specific gravity and low oil absorption, giving the latter better working qualities. Alumina hydrate imparts good finish or gloss to an ink.

Magnesium Carbonate (Magnesia)

This is a precipitated mixture of hydrated magnesium carbonate and magnesium hydroxide. It is used similarly to alumina hydrate. It gives a flatter finish to an ink, is slightly less transparent and has a little lower oil absorption.

Blanc Fixe

This is precipitated barium sulfate. It is much less transparent than alumina hydrate and magnesia and has a much lower oil absorption due to its high specific gravity. It has much poorer working qualities and is only used as a cheap extender. It dries with a very dull finish.

Gloss White

Gloss White is made by coprecipitating a mixture of alumina hydrate and blanc fixe. It usually contains 25% of alumina hydrate and 75% of blanc fixe, by weight. It has much better working qualities than a corresponding dry mixture of alumina hydrate and blanc fixe and is used as an economical and inert base in printing inks.

OPAQUE WHITE PIGMENTS

Lithopone

This is a mixture of zinc sulfide and barium sulfate. It has good opacity and hiding power and is used to a great extent for this purpose in printing inks.

Zinc Oxide

Zinc Oxide or zinc white has good opacity and is greatly used for opaque mixing whites and other opaque inks.

Titanium Oxide

This is the most opaque pigment available today. It is also the most inert and stable, making it safe to use with any of the printing ink vehicles and pigments. It can be baked without discoloration.

The above white pigments are used in conjunction with the colored pigments as indicated by the various formulae.

Wax Offset Compound

1. Beeswax	22
2. Petrolatum Amber	20
3. Mutton Tallow	5
4. Paraffin Oil	22
5. Kerosene	10
6. Naphtha (High Flash)	4

Melt (1), (2), (3) and (4) and stir until dissolved. Turn off heat and work in (5) and (6).

Soft Wax Non-Offset Compound

No. 1 Lithographic Varnish	35
Soft Cup Grease	35
Paraffin Wax	10
Beeswax	20

---

* Ink, Intaglio

1. Gilsonite	22
2. Petroleum Naphtha	51
3. Glue	5
4. Water	22

(1) is dissolved in (2) and (3) in (4) heating moderately. Emulsify the two with vigorous stirring.

---

* Intaglio Printing Ink

Grind together in ball mill

A. Nitrocellulose (½ sec.)	154
Isopropyl Alcohol	54
Diethyl Phosfate	50
Lampblack	50
Alcohol	400
Toluol	34
B. Nitrocellulose (½ sec.)	123
Alcohol	43
Triphenyl Phosfate	17
Acetone	440
Toluol	267
Tricresyl Phosfate	17
Castor Oil	16
Milori Blue	120

---

* Intaglio Inks, Water Resistant

One hundred parts by weight of rosin are dissolved in one hundred parts by weight of benzine. The solution is then emulsified in an aqueous solution of 160 parts by weight of sodium resinate (rosin soap) and 800 parts by weight of water. This emulsion is then mixed with 200 to

280 parts by weight of pigment and ground into an intaglio ink.

One hundred parts by weight of dammar may be dissolved in 100 to 150 parts by weight of benzol. The solution is then emulsified with a solution of sixty parts by weight of rosin soap in 600 parts by weight of water. The emulsion is thereafter mixed with 160 to 220 parts by weight of pigment and ground into an ink. Paste colors may be used in the place of dry colors.

It is possible to use a brownish colored pigment which is made in the form of an emulsion of petroleum pitch or asphalt. Coloring matters or pigment pastes plus diluents may then be added to the emulsion. Thus a solution of one hundred parts by weight of petroleum pitch (melting point 100 to 120° Co) in one hundred parts by weight of benzol is emulsified with 200 parts by weight of rosin soap of 20 per cent concentration. Pigments are added for example in the form of a paste consisting of one hundred parts by weight of lithol red R and 300 parts by weight of water plus a neutral diluent.

Invisible Ink

Cobalt Chloride	3 dr.
Water	4 oz.
Glycerin	1 dr.

Ink, Invisible

Linseed Oil	1 dr.
Ammonia Water	20 dr.
Water	100 dr.

This ink leaves no visible stain on the paper, but when it is dipped in water, and while it is wet, the secret can be read. As the paper dries the writing again disappears.

Laundry Marking Ink

A. Soda Ash	1
Gum Acacia	1
Water	10
B. Silver Nitrate	4
Gum Acacia	4
Lampblack	2
Water	40

Wet cloth with solution A and dry. Write with solution B using a quill pen.

Silver Nitrate	6
Gum Acacia	6
Soda Ash	8
Distilled Water	15
Ammonium Hydroxide	8

Silver Nitrate	15
Copper Sulfate	35
Gum Arabic	20
Sal Soda	20
Distilled Water	80
Ammonium Hydroxide	50

A. Copper Chloride	85
Sodium Chlorate	106
Ammonium Chloride	53
Water	600
B. Aniline Hydrochloride	60
Glycerin	30
Gum Acacia	20
Water	130

Mix 1 part of A with 4 of B and use immediately as mixture does not keep. The marking is "fixed" by steaming it.

Aniline Black	7 gm.
Alcohol	200 cc.
Hydrochloric Acid	12 cc.
{ Shellac	10 gm.
{ Alcohol	800 cc.

Dissolve the shellac in alcohol and then stir in other ingredients.

Indelible Laundry Ink

1. Phenol	32 fl. oz.
2. Nitrobenzol	30 fl. oz.
3. Turpentine	12 fl. oz.
4. Nigrosine (Spirit Soluble)	3¾ lb.
5. Alum (Powd.)	6 oz.

Heat (1), (2) and (3) in enamel pot to 105° C.; turn off heat and add (5) and stir until dissolved; add (4) and stir until completely dissolved. Allow to stand 24 hours, filter and bottle. The above gives a black ink. By substituting other spirit soluble colors different shades are obtained.

Marking Ink

A water glass marking ink is made by cooking together fifty parts by weight of water glass, 38 to 40 degrees Bé. concentration, and twenty-five parts by weight of each of water and ground rosin. The cooking continues until a smooth soap solution is formed. Before this solution cools down, twenty parts by weight of carbon black are added. When the proportions used above are changed, so that equal parts by weight of water glass water and rosin are used, and when this soap solution is mixed with twenty-five parts by weight of carbon black and seventy-five parts by weight of mineral black, a so-called marking india ink is obtained. This ink may then be compressed into tablets and dried. When moistened with a wet brush, the color is transferred to the same and hence the ink can be used for marking purposes with or without stencils.

Marking Ink, Waterproof

A waterproofing marking ink is made by heating almost to the boiling temperature a mixture of seventy parts by weight of water, five parts by weight of ammonia, 0.910 specific gravity, and twenty-five parts by weight of pulverized, red acaroid resin. The mass is constantly agitated while being heated. Then sufficient ammonia is added in small proportions, until the resin is completely dissolved, that is the undissolved part from the first cooking is brought into solution. The solution, still in the hot state, is then passed through a very fine sieve or through a hair cloth. The sieved mass is then mixed with one-half part by weight of acid green, three parts by weight of bluish or violet-tinted nigrosin, three parts by weight of sulphonated castor oil and 0.1 part by weight of tri-cresol. In order to make the ink somewhat thicker in consistency, a little shellac size or casein solution is added. If the acaroid resin solution becomes too thick, this is generally due to the use of too much shellac size or casein solution or ammonia.

Acid Proof Marking for Quartz Thermometers

A sharply defined, clearly visible marking unaffected by acids and alkalis, and permanent up to 1000°, is made by coating the surface with a polymerized material (tung or linseed oil, Japan varnish, etc.), lightly scratching, etching with dil. HF at 80–90°, and filling the marking with a pigment (grey-black) of CuO, sand, and glycerin (I), which is fixed by gently heating to drive off the (I) and afterwards to a temp. sufficient to fuse the pigment. A green pigment consists of Pb_2O_3 (5 pts.), SiO_2 (1 pt.), and Co_2O_3 (35 pts.) suspended in (I).

Marking Ink

Without the use of stencils may be made by mixing four parts by weight of haematin and twelve parts by weight of white dextrin in eighty parts by weight of warm water. The solution is then allowed to cool and is separately mixed with three solutions, each prepared in the warm state with two parts by weight of water, containing in one case 0.3 part by weight of crystalline sodium carbonate, in a second 0.3 part by weight of potassium dichromate and in a third 0.3 part by weight of potassium binoxalate. In each case the mixture is vigorously agitated. Then ten parts by weight of pine soot or carbon black, double calcined, are uniformly moistened with alcohol, and this mass is then slowly mixed with the above-mentioned solutions. There is also added 0.1 part by weight of tri-cresol and the entire mixture is then passed through a fine screen. When the aforementioned ingredients are to be used in the finely pulverized condition, then thirty parts by weight of vine black (Frankfurt black) are used in place of the carbon black or pine soot and no alcohol is used. All the ingredients are well mixed and then screened and a marking powder is thus obtained, which is very well suited for marking packages after it has been dissolved in eight to ten times the quantity of hot water.

Marking Ink

Inks for marking packages, boxes, bales, etc., for shipment are discussed. Two of the formulas given contain rosin: (1) Waterglass (38–42° Bé) 50, H_2O 25 and ground rosin 25 parts are cooked together and 20 parts of lampblack added. (2) Medium hard stearin pitch 30 and rosin pitch 25 parts are melted together and coal tar light oil 40 and lampblack 5 parts are added.

Marking Ink

Asphaltum	1 lb.
Coal Tar	4 lb.
Benzol	3 qt.

Marking Ink

Castor Oil	7
Rosin Oil	1
Methyl Violet	2
Alcohol	90

Blue Marking Ink

Shellac	2
Gum Acacia	2
Borax	2
Aniline Dye	sufficient
Ultramarine Blue	sufficient
Water	26

Ink, Meat Branding

14 lb. of spirit soluble nigrosine is dissolved in a warmed mixture of Glycerin 28 lb., Glycopon AA,* 10 lb., Acetic

Acid Glacial 12 lb. Cool and add 136 lb. alcohol.

Meat Stamping Inks

A. Red

Carmine	16
Ammonium Hydroxide	120
Glycerin	45

Stir until dissolved then stir in

Dextrin	20

B. Blue

Pure Food Blue Dye	30
Dextrin	20
Glycerin	82
Water	70

Inks for Metals

An iron marking black can be obtained by mixing thirty parts by weight of medium hard stearin pitch with twenty-five parts by weight of rosin pitch, forty parts by weight of coal tar light oil and five parts by weight of carbon black. The two pitches are first melted together, the molten mass removed from the flame and then very carefully mixed with the light oil or crude benzol. Great care must be taken to avoid the mass running over or the benzol or light oil catching fire. Then the carbon black is added after first being passed through a fine screen. This ink is very well suited for marking metal containers and sheet metal and in fact or all purposes where the ink does not penetrate into material and hence must possess a marked tendency to adhere firmly to the surface of the same.

Mimeograph Ink Base

1. Lampblack (Best Grade)	10.5
2. Violet Toner	1.1
3. Aluminum Hydrate Light	3.8
4. Long Varnish	1.1
5. Castor Oil	65.5
6. Lanolin	18.0

Mix (1), (2) and (3) dry and add (4) and (5) and continue mixing until uniform; add (6) and mix until thoroughly incorporated. Then grind on a four roll mill. This base ink is too heavy for direct use and is thinned down with castor oil to suit.

Mimeograph Ink

Lampblack (Best Grade)	6.4
Violet Toner	0.6
Aluminum Hydrate (Light)	2.2
Long Varnish	0.6
Castor Oil	78.5
Lanolin	11.7

Follow same procedure as for mimeograph ink base.

Ink, Mimeograph

Shellac	4 oz.
Borax	4 oz.
Water	30 oz.

Boil the above until dissolved; stir in

Nelgin	8 oz.

Add with stirring

Water	20 oz.

Then work in

Lampblack or other pigment	5 oz.

Outdoor Ink

Shellac	12.5
Alcohol	22.5
Cresol Tech.	15
Nigrosine Base	5

* Ink, Recording Instrument

Water Soluble Eosin	1 oz.
Formic Acid	2 cc.
Glycerol	5 cc.
Water	1 gal.
Am. Hydroxide	to make alkaline

Rubber Stamp Ink

Aniline Red	20
Glycerine	6
Molasses	3
Boiling Water	80

Black Stencil Ink

Paris Paste is thinned down wit water and rapid stirring to the consistency desired.

If a waterproof ink is desired th water is replaced by a rubless wax emulsion or borax shellac solution.

Ink for Use on Metals

Copper Sulphate	10 g.
Hydrochloric Acid, Conc.	4 g.
Ammonium Chloride	8 g.
Gum Arabic	4 g.
Lamp Black	2 g.
Water	10 g.

Typewriter Ribbon Ink

Petroleum Oil	108
Peerless Carbon Black	25– 30
Oleic Acid	20
Toner (Oil Soluble)	10

Grind until uniform.

Stamp Pad Ink

Glycerin	5 lb. 6 oz.
Water	4 lb. 2 oz.

Varm to 150° F. and add

Methyl Violet	6⅔ oz.

lowly while stirring. Allow to cool and tand for a few days and filter.

Magenta	4 oz.
Acetic Acid	4 oz.
Water	1 qt.
Alcohol	1 pt.
Glucose (43° Bé)	1 pt.
Glycerine	2 qt.

Add the dye slowly with stirring to the ıixture of other ingredients. Warm and tir until dissolved. Allow to stand a ew days and filter. For violet and green ıks acetic acid may be used as above; or other colors leave out acetic acid.

---

* Ink, Sheep Marking

Pot. Permanganate	4
Trisodium Phosfate	5
Dextrin	1
Water	to suit

---

* Sausage Marking Ink

Spar Varnish	10– 20
Paraffin Wax	1– 2
Petrolatum	20– 38
Chinawood Oil Varnish	70– 85
Pigment	127–138

Stamping Ink

Denatured Alcohol	1 part
Spirit Soluble Aniline Dyestuff	1–3 parts
Glycerine	4–5 parts

Mix thoroughly in water bath at 100–30° F. Allow to cool.

Use.—Apply to inking pads or as stenling ink.

---

* Ink, Stencil

Rosin Oil	120
Turkey Red Oil	90
Mineral Oil	10
Water	60
Carbon Black	17

Ink, Sympathetic

A solution of oxal-molybdic acid yields ı "ink" the characters made with hich are invisible in the lamp-light, or . weak daylight, but which, exposed to rong sunlight or electric arc light, sudenly appear in deep indigo blue. The id is prepared by adding to a boiling lution of molybdic acid one of oxalic id, also boiling, letting cool, and reovering the crystals which form. Dislve these in cold water to make the "ink." A sheet of paper immersed in the solution and dried in the dark becomes blue when exposed to the sun. If written on with a pen dipped in plain water, the letter will appear white on a blue ground. If the paper be held close to a hot fire, the blue becomes black. Similarly, the blue letters that appear on a white ground, if strongly heated, become permanently brown or black.

Ink, Invisible or Sympathetic

1. Make a five or ten per cent solution of cobalt chloride in soft or distilled water. When marks are made with this on paper it is not noticeable when dry at ordinary temperature; on heating the paper, *blue-green* lines will appear.

2. Writing or a drawing made with a ten per cent solution of lead acetate (or sugar of lead) in water will turn *black* if exposed to hydrogen sulfide, or if a weak solution of ammonium sulfide is brushed gently over it.

3. Writing made with a five or ten per cent solution of ammonium or potassium thiocyanate in water will turn a deep *red* if brushed gently or sprayed with a dilute solution of ferric chloride.

Transfer Ink

1. Ultramarine Blue	50
2. Gum Mastic	30
3. Beeswax	10
4. Petrolatum	10

Melt (3) and (4), work in (1) and mix with melted (2).

Waterproof Drawing Inks

Yellow

Fresh Bleached Shellac	28 gm.
Borax Crystallized	7 gm.
Water	1000 cc.

Dissolve the above by warming and stirring; then add with stirring

Erythrosine Yellow	1 oz.

By substituting the following dyes in a like amount the corresponding shades are obtained:

Orange—Brilliant Orange R
Yellow—Chloramine Yellow
Green—Brilliant Milling Green B
Blue—Wool Blue G Extra
Violet—Methyl Violet B
Brown Benzamine Brown 3GO

Dye Toners for Printing Inks

Distilled, Low Titre, Oleic Acid	50 parts
Oil Soluble Dye	50 parts

(Such as Victoria Blue Base, Methyl Violet, etc.)

These are heated together and ground over rolls until a heavy paste is formed.

Ink Eradicator for Tracing Cloth

Turpentine	17
Pumice Dust	53
Petrolatum	14
Paraffin	16

Ink for Zinc

Copper Acetate	1
Ammonium Chloride	1
Water	15
Lampblack	½

Copper Sulfate	1
Pot. Chlorate	1
Water	36

Blue Copying Pencil

Aniline Blue (Water Soluble) Powder	2 kg.
Water	4 kg.

Dissolve by heating; then cool and add

Gum Tragacanth Powder	20 gm.

and stir until dispersed; now add

Milori Blue (Powder)	4/7 kg.
Kaolin (Powder)	3 3/7 kg.

Make acid with sulfuric acid; allow to stand overnight and neutralize with soda ash. Extrude the leads and dry for a few days. Rub off crystals which have formed on leads, by means of a damp rag. Dry in an oven and clean off crystals again in same way. Repeat until more crystals form on drying.

Redissolve in a similar amount of water to which has been added the following filtered solution.

Sugar	80 gm.
Albumen	20 gm.
Water	120 gm.

then add with stirring

Indigo-Carmin	500 gm.

and heat on a water bath until of a doughy consistency.

The Milori Blue and Kaolin should first be mixed together with water to form a slurry and ground wet and dried and powdered. To this is added and thoroughly mixed in

Sulfuric Acid	½ kg.

The finished lead is waxed or greased to protect it from atmospheric moisture.

Colored Pencil Leads

Ammonium Hydroxide	2
Shellac	3
Venice Turpentine	1
Prussian Blue or other pigment	6
Clay or Chalk	4

The pigment are ground to a fin paste with water; the shellac is dissolve in the ammonia. The Venice turpentin is rendered fluid by short heating. Th clay is worked to a smooth slurry wit water and pressed through muslin an dried and powdered. Mix everything tc gether in a mill until the consistency i that of a thick dough. This is then fe into a pressing machine of the macaror type with openings of the size require The extruded leads are placd in a dry ing oven for drying.

Red Indelible Lead

Rosin Soap	60 gm.
Water	6 kg.

Dissolve with heat and add

Shellac	40 gm.

Stir in

Ponceau-Creosot	2 kg.

and

Albumen	40 gm.
Gum Tragacanth	40 gm.
Water	120 gm.

Mill in

Cinnabar Powd.	2 kg.
Kaolin Powd.	2 kg.

Extrude through press and dry.

* Stencil Sheet

The resins used in the following fo mulae are phenol formaldehyde chin wood oil types which are hardened b heat.

16 parts of resin varnish, formed b dissolving 45 parts of resin in tl solvent
9 parts peanut oil
3½ parts of aluminum stearate.

The second formula is as follows:

16 parts of resin varnish
10 parts of peanut oil
6 parts of diethylene glycol monoeth ether
4½ parts aluminum stearate.

The third formula is as follows:

16 parts of resin varnish
10 parts corn oil (refined)
16 parts of diethylene glycol monoeth ether
4½ parts aluminum stearate.

Experiments have shown that in the second and third formulae given above, that diethylene glycol monobutyl ether may be substituted in the same proportions for the diethylene glycol monoethyl ether.

The invention also contemplates the use of other non-volatile derivatives of glycol which may serve well in this use. The invention further contemplates in all of the above formulae, before the mixtures are finally prepared, that a coloring agent be added such as victoria blue base or other suitable coloring material in sufficient quantity to render the coating opaque so that the cut letters of the stencil will reveal the white fibres of the backing sheet by the color contrast.

To prepare stencil paper according to the first formula, it has been found well to first mix 9 parts of peanut oil and 3½ parts of aluminum stearate and add to this the desired quantity of coloring material which has been previously dissolved in an appropriate amount of solvent, which is preferably 30 parts of any one or any mixture of alcohols, benzols or esters, but which may by way of example be toluol and ethyl or butyl alcohol mixed in proportions of approximately 20 to 25 parts respectively.

This mixture is added at a temperature of about 45 to 50 degrees centigrade and while maintained at this temperature, 16 parts by weight, of the resin varnish containing substantially 45 per cent of the solid resin is then added and the product stirred.

Suitable base sheets such as yoshino paper are then coated in the usual way, with the coating solution prepared as above outlined, the sheets being immersed or floated upon the coating bath. When the sheets have been thoroughly covered with the solution they may be drawn across a straight edge or wire to remove the excess material, if this is necessary, and then hung up to dry. The drying operation serves to volatilize part of the solvent.

After the drying operation, the sheet is placed in an oven the temperature of which may be from 90 to 150 degrees centigrade but preferably substantially 140 degrees centigrade, and baked for a period of 30 to 60 minutes. After baking, the sheets are removed and cooled at which time the stickiness present during the heat treatment disappears. The product at this time is substantially insoluble by virtue of the reaction produced by the heat treatment involving the constituent parts of the potentially reactive resin.

In preparing stencil sheets, according to the second and third formulae, the process is exactly the same, except that the oil, either peanut or corn oil, as the case may be, the glycol derivative and the aluminum stearate are mixed together prior to the introduction of the solvent and coloring agent.

It has been found in practice that a suitable base paper such as yoshino paper, coated and treated in the above indicated manner produces a greatly improved stencil paper. The coating as prepared is quite insoluble in oils and inks and yet is of such character that it may be cut by a stylus or cutting type to form very accurate letters.

It is further found that stencil paper thus prepared when cut, is very durable and will withstand the wear of a large number of reproductions without causing blurs or imperfect letters. Furthermore, paper so produced is unaffected by weather conditions, remaining in good workable condition at all times and is not subject to being dried out or hardened by exposure to the atmosphere nor is it subject to being cracked in a manner to pass ink when it is folded or creased in ordinary usage.

---

* Stencil Sheet Coating

Water	130
Sulfo Turk C	.40
Tricresyl Phosfate	20
Ultramarine Blue	10
Oleyl Alcohol	10
Gelatin	8
Beechwood Flour	6
Myricyl Alcohol	5
Glycerol	3
Dinitro Toluene	0.5

Stencil Paper

1. A stencil sheet coating composition containing the following substances in substantially the proportions specified:

Aluminum Stearate	2 parts
(45% Solution) Phenol Formaldehyde Resin	16 parts
Chlorinated Naphthalene	14 parts
Corn Oil	13 parts

INSECT, RODENT AND WEED DESTROYERS

* Agricultural Insecticide

An emulsifying composition comprises casein, gamboge, ethyl alcohol, and soda. As an illustrative example of such preparation, 0.84 pounds of casein is slowly poured into about 2 gallons of cold water, and stirred until thoroughly wet and soaked, then 0.63 pounds of dehydrated sodium carbonate is added, stirring until all the casein is well in solution. Six gallons of denatured alcohol is then added, and 1.67 pounds of powdered gamboge. The gamboge is added slowly with constant stirring. Enough water is then added to make up a total of 20 gallons. The resultant solution should be clear and of a dark cherry color, and contain not over 0.1 to 0.2% of free alkali. Instead of denatured ethyl alcohol, other available alcohols may be used, for instance methyl alcohol or the higher primary or secondary alcohols. With this composition, the oil to be emulsied is incorporated, preferably by slow additions, with agitation. Most oils emulsify therein readily. Heat may be applied if quicker emulsification is desired. For petroleum oil for example, with a specific gravity of 0.891, a proportion of 1 part by volume of the foregoing composition to 5 parts of the oil affords a satisfactory product. Such emulsion will contain about 83.3 per cent of oil, making up to a consistency about that of lard at the same temperature. Such a product, even after standing in a warm place for months is free from separation. Emulsions customary heretofore, under the same conditions all show separation, with oil at the top and water on the bottom.

---

* Agricultural Insecticide

An insecticide and fungicide comprises an oil-in-water emulsion of the 150°–300° C. distillate fraction of crude shale oil, in which the oil globules are at least as large as approximately 4 microns in diameter. The process of preparing this comprises mixing with water and a 150°–300° C. distillate fraction of crude shale oil an emulsifying agent in an amount which is incapable of producing a uniform emulsion composite when the mass is agitated at normal atmospheric temperature but is effective for producing a uniform emulsion composite when the mass is heated to approximately 80° C. and vigorously agitated, heating the mass to approximately 80° C. and vigorously agitating the heated mass. As a typical example of the method employed in preparing an insecticide and fungicide according to the present invention, 600 cc. of shale oil kerosene which is rich in nitrogen bases and contains saturated and unsaturated hydrocarbons are mixed with approximately 400 cc. of water carrying approximately 2½ grams of sulfite waste liquor calculated on the dry basis. In commercial practice it is most convenient to use the sulfite waste liquor in the commercial form containing 50 per cent solids and in this case 600 cc. of shale oil kerosene would be mixed with 395 cc. of water and 5 cc. of the sulfite waste liquor. The mass prepared in either of the ways mentioned is then heated to approximately 80° C. and thereafter vigorously stirred to form an emulsion composite.

* Insecticide, Agricultural

	Ex. I	Ex. II
	Per Cent	Per Cent
Glue	1.00	1.20
Water	20.00	24.00
White oil, 60 viscosity (100° F)	......	65.46
White oil, 80 viscosity (100° F)	77.77	......
Alcohol	......	0.25
Sodium compound of preferentially oil-soluble sulfonates (55 per cent)	1.23	1.64
Lead arsenate	......	9.45
Toluol	Trace	Trace

The composition may be very readily emulsified or thinned with water; for example passage through a rotary pump is usually sufficient for the production of a very stable emulsion.

For use as a spray, the thick emulsion is diluted with between about 10 and 100 volumes of water, a good distribution

All formulae preceded by an asterisk (*) are covered by patents.

being obtained with slight stirring. It is preferred to dilute the thick emulsion with about thirty to fifty volumes of water, which gives an emulsion of approximately 2% oil content.

Insecticide Spray, Agricultural

Shale Oil Kerosene	600 c.c.
Water	390 c.c.
Sicapon	10 gm.

Heat between 70–100° C. and stir vigorously to emulsify. The above is a concentrate and is diluted to 60 liters for actual spraying.

Agricultural Spray

Nicotine	1.20
Soap	20.20
Water	75.20

Agricultural Spray

Anthracene Oil	75
Fish Oil Soap	3
Water	22

* Agricultural Spray

Mineral Oil	2–2½ gal.
Diglycol Oleate	2–3 oz.
Trihydroxyethylamine Linoleate	½–1 lb.
Water	100 gal.

Add in the order above and beat vigorously. The above emulsion is quick breaking and spreads easily.

* Larvicide, Agricultural

Nicotine Sulfate	5–10
Sugar	2–5
Water	2–8
Diatomaceous Earth	75–100

* Pyrethrum Extract

Ground Pyrethrum Flowers	30 lb.
Ethylene Dichloride	20 gal.

Percolate the above and finally squeeze out the wet mass. If a concentrate is desired the extract is heated to drive off the solvent. There remains an okoresin which may be dissolved in kerosene or other distillate.

Seed Disinfection

Formalin vapor is generated by boiling a formalin soln. contg. 1 part of 40% in 100 parts water and the seed is exposed 1–10 min. Tests in 4 widely sepd. areas for 4 yrs. have given efficient control of oat smut (*Ustilago avenae*) and wheat bunt (*Tilletia caries* and *T. foetens*) in every case and the cost is extremely low. The germinability of the seed grain is not impaired.

Insecticide Spray
(Agricultural Quick-Breaking)

Diglycol Oleate	2 lb.
Pyrethreum Extract (Mineral Oil or Kerosene)	50 lb.

Mixing the above together gives a concentrated spray base free from alkalies. The active principle of pyrethreum is thus unaffected. Burning due to alkali is also eliminated.

The above concentrate emulsifies readily on stirring in water with a pump. It is "quick-breaking" when sprayed on the foliage.

Insecticidal Dust

Sulfur	60.00
Nicotine	1.90
Lead Arsenate	10.00
Arsenic	2.00
Talc	28.00

* Insecticide, Nicotine

An anhydrous insecticide base consists of

Nicotine	1622
Oleic Acid	2002
Soda Ash	150

Water is added to this for any dilution desired for spraying.

* Floatable Powdered Insecticide

Quicklime	300
Paste Copper Arsenite	300

Mix thoroughly and when heat begins to develop add

Stearic Acid	12

Mix thoroughly and grind well before use. Other insecticides such as arsenates, pyrethrum, derris, "nicotine" dust, etc. may be used. They should first be made into pastes with water.

* Weed Killer

Sodium Chlorate	1.8
Calcium Chloride	1.2

Vegetable Weevil, Insecticide for

Sodium silicofluoride when used as a dust (about 30–40 lbs. per acre) gives good results.

Weevils, Killing Corn

Fumigation with CS_2 is recommended. Approx. 1 lb. of CS_2 is used to 100 cu.

ft. of space to be fumigated. If the contact period exceeds 36 hrs., permination is injured. Optimum results were obtained at temps. of 75–90° F. in closed bins.

* Herbicide

Water	67 lb.
Salt	19 lb.
Sulfuric Acid	8 lb.
Iron Sulfate	3 lb.
Hydrochloric Acid	3 lb.

The above is diluted with water and sprayed on the weeds.

Spray, Horticultural

0.84 pounds of casein is slowly poured into about 2 gallons of cold water, and stirred until thoroughly wet and soaked, then 0.63 pounds of dehydrated sodium carbonate is added, stirring until all the casein is well in solution. Six gallons of denatured alcohol is then added, and 1.67 pounds of powdered gamboge. The gamboge is added slowly with constant stirring. Enough water is then added to make up a total of 20 gallons. With this composition, the oil to be emulsified is incorporated, preferably by slow additions, with agitation. Most oils emulsify therein readily. Heat may be applied if quicker emulsification is desired. For petroleum oil for example, with a specific gravity of 0.891 a proportion of 1 part by volume of the foregoing composition to 5 parts of the oil affords a satisfactory product. Such emulsion will contain about 83.3% of oil, making up to a consistency about that of lard at the same temperature. Such a product, even after standing in a warm place for months is free from separation.

For horticultural spraying, a petroleum oil emulsion as indicated, would ordinarily be used at a spraying strength of 2% oil. This would be obtained from the preparation referred to in the above example by diluting 2.4 gallons of the emulsion to 100 gallons with water. When sprayed, a highly satisfactory coating on the vegetation is had, with a minimum loss from run-off or drip, and at the same time the oil is well protected against damaging tender foliage.

Bordeaux Mixture

The following is the method of making Bordeaux Mixture for horticultural spraying. The customary wash is known as "4-4-50," and the official formula and instructions are as follows:

Copper Sulphate (98 per cent.)	4 lb.
Best Quicklime (in lump form)	4 lb.
Water	50 gal.

The copper sulphate should be dissolved in a small wooden vessel at the rate of 1 gal. of water per lb. of sulphate (iron or tin vessels must not be used). The lime should be slaked to a fine paste with a little water in another vessel, and water added gradually to make a milk, and finally diluted in a large barrel to the requisite amount (46 gal.). The 4 gals. of copper sulphate may now be poured slowly into the diluted milk of lime and the mixture stirred thoroughly during the process. The two components of the mixture may be kept separately for a long time, but, after mixing, the spray fluid should be used as soon as possible—at all events, within 24 hours. When used on a large scale it may be convenient to make up a stock of each ingredient which may be diluted down and mixed as required. For this purpose, 50 lb. of copper sulphate may be dissolved in 50 gals. of water and 50 lb. of lime, slaked and diluted to 50 gals. of milk of lime. Each gallon will then represent 1 lb. of copper sulphate and 1 lb. of lime. When required for use, the contents of the barrels should be thoroughly stirred and the requisite number of gallons taken out and diluted according to the above formula. For a 50-gallon barrel, for instance, 4 gals. of lime-milk should be removed and diluted with 42 gals. of water, and when thoroughly stirred and strained the 4 gals. of copper solution may be added slowly. The addition of refined sugar (2 oz. to 50 gals.) is useful in delaying flocculation.

Treeband Composition

Sulfur Flowers	6
Linseed Oil	75

Heat 1½ hrs. with stirring until uniform. Cool and thin with cottonseed oil.

* Tree Spray

The basic ingredients of this improved spray are oil and hydrated lime. The oil is preferably a highly refined petroleum oil having a viscosity of about 50 to 150 sec. Saybolt at 100° F. and a specific gravity of about 0.89 to 0.81. The hydrated lime is preferably very finely powdered and it may be of commercial purity. Other solids, such as talc, gypsum and bentonite may be used instead of or combined with the lime, but it has been found that hydrated lime gives the best results. A preferred formula for this tree spray is: hydrated lime, 15 lbs.; white oil, ½ gal.; casein, .38 lbs. The oil is thoroughly mixed with the solid ingredients in any suitable apparatus, for example, the mixture may be passed through screens of about ten mesh so that the solids will absorb a uniform amount of the oil. The above proportions may be varied throughout a considerable range, but for practical purposes, the amount of oil used should not be sufficient to yield a sticky or pasty mass. Instead of using casein in the above formula, about 2 per cent of gum arabic or an equivalent amount of gum ghatti, dehydrated sulfite liquor or other materials which have an emulsifying action may be used. The powder will keep indefinitely, and since it is relatively dry no preservative is necessary for the casein. The dry powder may be shipped to the orchards in paper bags and may be mixed with water to form a spray, about 20 pounds of powder being incorporated in about 50 gals. of water. The amount of water will, of course, vary within wide limits, and it is usually desirable to proportion the mixture so that the oil content in the final spray will be from ½ to 2 per cent. When this emulsified mixture is sprayed onto the foliage, it does not cling thereto in large drops but it spreads evenly, adheres closely to the leaves and gives a smooth, uniform coverage which is superior to that obtained by prior sprays. This is a particularly important feature because it insures the effectiveness of the spray and it makes frequent and/or repeated spraying unnecessary. Heretofore sulfur has been avoided in oil sprays because of its injury to plant foliage. Ordinarily sulfur, when combined with oil, causes the leaves to burn and curl up. It has been discovered that sulfur in amounts to about 15 per cent (calculated on the dry powder basis) may be mixed with this spray and used on sensitive foliage with safety. It has been found that other toxic ingredients, such as lead arsenate, monochloronaphthalene and their equivalents, may also be employed. For instance, 0.1 per cent monochloronaphthalene (calculated on the diluted spray) has been added to my spray and has been found effective without apparent injury to sensitive plant tissue. There has been developed no theory to explain why normally injurious elements such as sulfur, are rendered non-injurious when applied with a lime-oil spray of the character described, but experiments have proven this to be a fact. This improved spray has been found to be particularly effective for combating the oriental fruit moths which infest the sensitive leaves of peach trees. It may also be used against codling moths, red spider, aphids, scale insects and other insect and fungus pests. The term "refined oil" as used in the claims includes relatively viscous oils which have been treated to remove the unsaturated hydrocarbons or other compounds which may be injurious to foliage.

Pine Oil Insecticides

Yarmor Steam-distilled Pine Oil is rapidly displacing such ingredients as methyl salicylate, citronella, lemon oil, safrol and oil of wintergreen in household insecticides for it possesses a pronounced germicidal value, aside from its pleasant perfume odor.

1. Formulae

A.	Pyrethrum Extract	1 qt.
	Gasolene-kerosene	5 qt.
	Citronella	1 oz.
	Yarmor Pine Oil	6 oz.
	Paradichlorbenzene	8 oz.
B.	Pyrethrum Extract	1 qt.
	Gasolene-kerosene	5 qt.
	Paradichlorbenzene	4 oz.
	Cedarwood Oil	3 oz.
	Yarmor Pine Oil	3 oz.
	Methyl Salicylate	2 oz.
C.	Pyrethrum Extract	1 qt.
	Gasolene-kerosene	5 qt.
	Yarmor Pine Oil	5 qt.

Fungicide

A composition consisting of 95 per cent dusting sulfur and 5 per cent by weight of either of the following dry and finely ground substances: aluminum hydroxide, zinc oxide, or hydroxide, aluminum sulfate or zinc sulfate, or the basic sulfates of these metals, or other non-hygroscopic salts formed from an anion, the hydroxide of which is amphoteric, etc., was found to be much superior to straight sulfur dusts, and at least equal to the most efficient lime-sulfur liquid sprays without having any of the drawbacks of the latter.

---

* Fungicide

Light Petroleum Oil	18
Gum Arabic	1
Gum Ghatti	1
Phenol	2
Sodium Polysulfide	1½–7

Fungus Killer

Copper Carbonate	36
Copper Sulfate	3
Sulfur	58

* Fumigation Composition

Liquid HCN (25–50%) with chloropicrin (2–3%, as warning agent) is absorbed by a granular material (6–20-mesh) consisting of calcined infusorial earth.

Fruit Spray Residue, Removing

Treatment with 0.3–1% Hydrochloric Acid at 95–105° F. gives good results in removing arsenical residues.

* Green Lead Arsenate

A method of making a green colored insecticide comprises reacting a water-soluble chromate with lead hydrogen arsenate so that some of the lead hydrogen arsenate is converted into a lead chromium arsenic compound and adding ferric ferrocyanide so that the lead chromium arsenic compound formed and the ferric ferrocyanide cooperate to produce a green-colored insecticide having increased fungicidal and adhesive properties and without an appreciable increase in water solubility. To 1680 lbs. of litharge in suspension in water add 150 lbs. of nitric acid (36° Baumé) follow this with 20 lbs. of sodium dichromate. To this add 1380 lbs. of arsenic acid (61 per cent AS_2O_5). After reaction between the litharge and the arsenic acid has taken place add 15 lbs. ferric ferrocyanide (Prussian blue). The yield on a dry basis will be around 2585 lbs. of dry green lead arsenate.

Argentine Ant Poison

This poison consists of a syrup, attractive to the insects, containing from one to two tenths of one per cent of the chemical element arsenic in the form of sodium arsenite. In view of the uncertain purity of commercial sodium arsenite, it is advisable to prepare the chemical in solution from arsenious oxide, a stable, standard compound universally obtainable and of known poison strength. The poisoned syrup prepared from this material is not immediately fatal to the worker ants, but instead is carried by the insects to the nests, where the queen and brood are killed.

Inasmuch as the syrup does not keep very well without a preservative, it is perhaps better to make up a small supply each time it is used. In order that such a plan may be convenient, a "stock solution" of sodium arsenite is made up. This does not ferment and if kept in a well-stoppered bottle will not deteriorate appreciably. The stock solution is mixed as desired with thin syrup.

One ounce arsenious oxide (common "white arsenic")

¾ ounce sal soda crystals (if the soda has crumbled down into a fine white powder, use only ⅜ ounce)

Boil the above ingredients together with about one pint of water in a granite-ware pan. Do not use aluminum or galvanized vessels. After the arsenic is practically all dissolved, add enough water to make the total volume of the solution *one quart.* Sometimes the arsenic is not quite pure, and leaves a little cloudiness which will settle over night, and which does no harm anyway. Mix thoroughly, bottle and label POISON. At the time the syrup is desired for use, mix the Stock Solution as above prepared with honey according to the following figures:

Stock Arsenic Solution	1 fl. oz.
Thin Honey	1 pt.

Method of Use.—Soak pieces of excelsior in the syrup, place in cans; cover with loose-fitting lids, and place outfit in path of ants.

Note. Ants seem to like straight honey best. If economy is desired, cane syrup may be substituted for a part of the honey ingredient.

Ant Repellent

1 lb. sugar in 1 qt. of water
125 grams arsenate of soda

Boil and strain.
Add spoonful of honey.

Ant Poison

Sugar	1 lb.
Water	1 pt.
Thallium Sulfate	27 gr.
Honey	3 oz.

Stir together and heat to a boil. Avoid vapor when boiling.

Ant Destroyer

Tartar Emetic	1 lb.
Sugar Powd.	1 lb.
Arsenic Sulfide Powd.	½ oz.

Ants, Carpenter, Destroying

Bore sloping hole at top of infested wood and pour in a mixture of equal parts of carbon disulfide and carbon tetrachloride. The heavy liquid and its vapor will sink down and permeate crevices.

Another method is to dissolve one pound paradichlorbenzene in two quarts of kerosene and spray this solution.

Ants, Preventing Entry of

Sprinkle Clovel or Oil of Sassafras at entrances. Ants do not like these odors and will not enter.

Ant Powder

Sodium Fluoride	78
Pyrethreum Powd.	8
Starch	14

Fire Ant, Insecticide for

Thallium Sulfate	2 oz.
Sugar	5 lb.
Honey	½ lb.
Water	4½ pt.

Ant Poison

Thallium sulphate has been found effective in exterminating in 3 or 4 weeks small red ants in houses, where arsenic compounds had previously failed. The following mixture was used:

Water	1 pt.
Sugar	1 lb.
Thallium Sulphate	27 gr.
Honey	3 oz.

The whole is brought to the boil and well stirred.

Fire ants in Texas were exterminated by the use of a syrup containing 2 oz. of thallium sulphate in 4½ pints of water, 5 lb. of sugar and ½ lb. of honey being added, and when dissolved the whole made up to a gallon with water. Four teaspoonfuls of this are poured on a moistened sponge which is placed in a box near the ant nest.

Insecticide

Naphthalene	2 lb.
Oleo-resin Pyrethrum	2 oz.
Methyl Salicylate	2½ pt.
Deodorized Kerosene	6¼ gal.

Dissolve the first two ingredients in the kerosene by mixing or shaking and add the methyl salicylate.

Insecticide (Bed Bugs)

Cresol	3 fl. oz.
Dichlorobenzene	13 fl. oz.

Use one pint of this mixture to five pints kerosene.

Insecticide, Bed Bug

Kerosene	90
Clovel	5
Cresol	1
Pine Oil	4

Bed Bug Exterminator

Insect Powder	150
Colocynth	50
Phenol	50
Oil of Turpentine	100
Alcohol	1000

Macerate the crude drugs in the alcohol for eight days, express, and filter, then add the phenol and oil.

Bed Bug Killer

Kerosene	96–98
Phenol	4–2

Use as spray in cracks and on springs.

Insecticide for Mexican Bean Beetle

Spray with

Barium Silicofluoride	5 lb.
Water	50 gal.

Insecticide, Cabbage Maggot

Calomel	4
Gypsum Powder	96

---

* Insect and Mildew-Proofing Canvas

A process has been patented for the treatment of fabrics with thallium salts in such a way as to render them water, moth, mould, and insect-proof. The process is said to be suitable for the treatment of textiles such as tent canvas, and in addition to imparting the properties mentioned above, is claimed to render the materials more durable.

Two solutions are required, A and B. An example of A is as follows:

	Parts
Soap	15
Casein	10
Carrageen	5
Bentonite	30
Petrolatum	500
Water	2400

Solution B consists of a solution of a thallium or certain other metallic salt, one part of the salt being dissolved in about 40 parts of water.

The soap, casein, carrageen, and water are beaten together, a small amount of borax having been added to render the casein soluble. The temperature should be from 140° to 200° F. Into the emulsion which is formed, the bentonite is slowly sifted with constant stirring, and when incorporated, the melted petrolatum is added, the liquid being continuously beaten during the operation.

The canvas to be treated is immersed in and thoroughly wetted with A, wrung out and passed into B, washed in water to remove excess of metallic salts, again wrung, and dried. The finished goods are said to be entirely without any greasiness, and to be of a good color.

Cattle Spray

Kerosene Extract of Pyrethrum Flowers	8 parts
Steam-distilled Pine Oil	10 to 15 parts
Petroleum Oil (40 to 65 secs. viscosity)	to make 100 parts by volume

The kerosene extract is made at the rate of five pounds of flowers to a gallon of oil. The kerosene used should be highly refined so as to be as nearly non-irritant as possible. One may purchase ready made extract from the previously mentioned companies. Pine Oil is the repellent in the formula. *Steam-distilled pine oil is more repellent to flies and less irritating to the skin than the cheaper destructively distilled pine oil.* If necessary the latter may be used at the rate of 20 to 25 parts per hundred.

Cattle Louse Insecticide

Dust with

Sodium Fluoride	1
Diatomaceous Earth	1

Cattle Parasiticide

Precipitated Chalk	40
Rock Salt	60
Pine Tar	2
Copper Sulfate	1

Make into plastic mass with water; cast into blocks and dry.

Pine Oil Cattle Sprays

The axiom "contented cows produce more milk" has been the basis for considerable research work on pine oil cattle sprays.

Various cattle sprays are being marketed, differing in ingredient content, but producing comparable results in combating warble and horse flies. There are also a few pine oil cattle sprays on the market that have outstanding merit. These sprays could be materially improved by the addition of more pine oil as evidenced by the subsequent data.

A series of four sprays were subjected to identical conditions for a period of time at an agricultural college and a city sanitation department.

The sprays were composed of the following ingredients, all figures computed on a volume basis:

	Formula No. 1	Formula No. 2A	Formula No. 2B	Formula No. 3
(a) Heavy-bodied Paraffin Oil....	15%	20%	.....	30%
(b) Kerosene Ext. of Pyrethrum....	5%	8%	8%	8%
(c) Yarmor Pine Oil..	25%	30%	30%	50%
(d) Long-time Burning Oil........	55%	42%	.....	12%
(e) Petroleum Distillate..........	.....	.....	62%	.....
	100%	100%	100%	100%

The product is prepared by simple mixing of the ingredients. Care must be taken that the ingredients are not allowed to absorb water as this may produce a cloudy product. The cloudiness is easily removed, however, by filtration through kieselguhr or like material.

(a) The heavy-bodied paraffin oil is obtainable from any oil refinery and should conform to the following specifications:

Bé. or A.P.I. Gravity	28.0
Specific Gravity 60° F.	0.88725
Flash Point	350° F.
Fire Point	405° F.
Viscosity at 100° F.	90 to 100 Saybolt units
Pour Point	30° F.
Color	No. 2 Tag-Robinson

(b) The kerosene extract of pyrethrum calls for a concentration of the extract from five pounds of flowers per gallon of kerosene. Lethane may also be used as a replacement product for kerosene extract of pyrethrum.

(c) Yarmor Pine Oil conforms to specifications—eighty-two per cent secondary and tertiary alcohols.

(d) The long-time burning oil is obtainable from any oil refinery and should conform to the following specifications:

Bé. Gravity at 60° F. 40.0–45.0

This fraction of oil is a shade heavier than kerosene.

(e) The petroleum distillate is obtainable from any oil refinery and should conform to the folowing specifications:

Flash, Cleveland Open Cup 260° F.
Fire, Cleveland Open Cup 300° F. Minimum

Viscosity, Saybolt Thermo at 60° F.	1000 to 1500
Color, 18″ Lovibond	5.0 Maximum
Cloud Test	32° F. Maximum
Unsaturation	4% Maximum
A.P.I. Gravity	36.5 and 38.5°

These sprays were originally tested according to the Peet-Grady Method and the results are tabulated for Formulae No. 2A and No. 2B.

Formulae No. 2A and No. 2B

	Down in 10 min.	*Dead after 24 hrs.*
Test No. 1	100	70
2	98	72
3	98	75
4	100	66
5	98	55
6	99	62
7	100	49
8	97	47
9	100	71
	99	63

These sprays were later tested on a practical scale at an agricultural college and a city sanitation department. The comments are indicative of what to expect when they are applied in the field.

Formulae No. 2A and No. 2B received the unanimous vote as being the most effective and most presentable products of the four. They possessed the following characteristics:

1. Burning or blistering of hides—negative
2. Odor—mild odor of the pine forest
3. Tainting of milk—negative if sprayed 30 min. before milking time and usual care exercised.
4. Clarity—free from suspended matter
5. Color—dark amber
6. Repellency—three to six hours
7. Volatility—relatively slow drying
8. Kill—63%
9. Knock-down—99%
10. Matting of hair—negative
11. Healing properties—the pine oil content promotes healing of open wounds and cuts.

Results of field tests may be duplicated provided no deviations are made in raw materials specified.

Cockroaches, Exterminant for

(1)

	Parts by Weight
Powdered Borax	4
Flour	2
Chocolate Powder	1

(2)

	Parts by Weight
Powdered Borax	10
Insect Powder	1
Starch	1

(3)

	Parts by Weight
Kieselguhr	22
Sodium Fluoride	40
Sodium Chloride	10

The ingredients in the finest powder are thoroughly mixed and the powder sprinkled about runs of the insects.

(4) Freshly burnt plaster of Paris and fine oatmeal (dry) in equal parts are thoroughly mixed and the powder is dusted around places infested by roaches.

Insect Powder (Cockroach)

Powdered Borax	8 lb. 10 oz.
White Hellebore	8 oz.
Dalmation Powder	8 oz.
Ground Cloves	4 oz.
Cayenne Pepper	2 oz.

Roach Poison

Sod. Fluoride	50
Flour	50

Roach Powder

Sodium Fluoride	65
Pyrethrum	30
Starch	5

Earthworm Poison

Corrosive Sublimate	1 oz.
Water	75 gal.

Sprinkle ground with this solution which is unharmful to plant life; vegetation should be sprinkled with water after this treatment.

Fly Spray

This is made by macerating 500 gms. of pyrethrum with 4 liters of kerosene (followed by expression) after 24 hours. Perfume by adding 90 cc. of methyl salicylate to each 4 liters of solution.

Pyrethrum	240 gm.
Kerosene	2000.0 cc.
Gasoline	2000.0 cc.
Napthalene	30.0 gm.

Macerate the pyrethrum in the petroleum liquids for 48 hours, then strain, express and then add the naphthalene.

Fly Spray

Deodorized Kerosene	89
Methyl Salicylate	1
Pyrethrum Powd.	10

Percolate a few times and filter.

Fly Catching Mixture

Rosin	56
Ester Gum	1
Heavy Mineral Oil	40

Melt together and stir until dissolved. Remove from heat and stir in

Glycerine	2½
Honey	1½

Fly Paper

Rosin	32
Rezinel No. 2	20
Castor Oil	8

Heat above and stir until uniform. Apply hot to suitable paper.

Increasing rosin content gives a heavier faster drying coating. Decreasing rosin gives a thinner stickier coating which remains sticky for longer periods.

Fly Paper

Water	21
Glucose	16
Sod. Silicate	11
Glycerin	½

First soak coated paper in a weak alum solution; dry and then coat with above.

Fly Paper

Rosin	32 gm.
Flexoresin E1	20 gm.
Castor Oil	8 gm.

Melt together, and dip paper into warm mixture.

Fly Paper Composition

A. Rosin	118
Rozolin	70
Paraffin Wax	10
B. Rosin	125
Rosin Oil	30
Rozolin	40
C. Rosin	100
Pine Oil	30
Rosin Oil	30
Thin Mineral Oil	30
Glyceryl Bori Borate	4
Glycerin	2

Beet Fly, Spray for

Eggs and pupae are not greatly harmed by contact insecticides. The larvae may be killed by 5–6% $BaCl_2$ soln. or 0.15% nicotine spray (40 gallons per acre, min.), but it is more advisable to destroy the flies with a spray contg. 0.3–0.4% NaF and 2% sugar.

Warble-Fly, Control of

Good results are gotten by spraying with

Soft Soap	¼ lb.
Water	1 gal.
Derris Powder	½ lb.

Bracken, Eradication of

Spray with 1% solution of sod. chlorate.

Moth Spray

Camphor 10, naphthalene 40, capsicum 100, oil of cloves 10, turpentine 100 and alc. 900 parts are macerated for 48 hrs. and strained.

Mothproofing Solution

For textiles—non-staining

Sod. Aluminum Silicofluoride	0.52
Water	98.48

* Mothproofing Composition

Chlorxylenols (mixed)	3–5
Trinitroisobutylxylene	3–5
Magnesium Carbonate	94–90

Mothproofing

Sodium Fluoride	0.5
Sodium Taurocholate	0.2
Carbon Dioxide to saturation point of water	
Water	100

Textile Mothproofing

Paranitro Chlorbenzol	10–20
Paradichlorbenzol	80–90

Codling Moth Bands

Bands are treated with a solution obtained by heating

Beta Naphthol	1 lb.
Red Engine Oil (300 sec.)	1½ pt.
Aluminum Stearate	½ oz.

Codling Moth Spray

Nicotine tannate kills by contact the mature eggs and young larvae of the codling moth. It remains toxic to the larvae for at least 21 days, and is more toxic as a stomach poison than $PbHAsO_4$.

The spray-tank mixt. is prepd. from U. S. P. tannic acid and free nicotine (50% soln.), the quantities being 4 parts tannic acid dissolved in 1600 parts water to which 1 part of nicotine soln. is added. The mixt. is compatible with S, but not with soap, lime-S soln., Ca caseinate or other alk. or acid substances.

Nematodes, Spray for Combating

Carbon Bisulfide	68
Rosoap	8
Water	26

Agitate violently and dilute 1: 50 with water before use. Formaldehyde may be added to control fungus pests.

Peach-Borer (lesser), Control of

Paradichlor Benzol	1 lb.
Crude Cottonseed Oil	2 qt.

Other oils are not as satisfactory as cottonseed oil.

Rodent Poison

Strychnine	0.55
Saccharine	0.15
Flour	98.30

Strychnine	0.35
Anise Oil	0.15
Sugar	20.50
Flour	79.00

Non-Poisonous Rat Destroyer

Gypsum	100
Rye Flour	300

Dry thoroughly in oven and add 0.1 oil of anise. Keep in air-tight containers.

Mouse Exterminator

Barium Carbonate	100
Oatmeal	300
Saccharin	1
Water	enough

Make a stiff dough, force through a coarse sieve, and dry in an oven.

"Silverfish," Poison for

White Arsenic	30 gm.
Flour	500 c.c.
Water	to make paste

Snail Killer

Ferrous Sulfate	20
Ferric Sulfate	20
Copper Sulfate	45

Field Mouse Poison

Whole Wheat	125 lb.
Thallium Sulphate	1½ lb.
Hot Water	6 qt.
Starch, Dry	½ lb.
Glycerin	½ pt.

The thallium sulphate is dissolved in the hot water, and to this is added the starch, previously mixed with a little cold water. The clear starch paste thus made is boiled for 2 to 3 minutes, the glycerin is added and the mixture boiled for a short time and then incorporated with the wheat.

A simple rat poison consists of a tapioca flour paste, containing 2½% of thallium sulphate, and spread on slices of bread. Another bait which has been used successfully is made as follows: ¼ oz. of thallium sulphate is dissolved in a large tea cup of boiling water and half a cupful of corn syrup, and 12 oz. of peanut butter are added. Thin slices of bread from two loaves are well covered with this mixture and cut into small squares. Tablespoonful doses of these squares are placed in the tracks of the vermin.

Bed Bug Spray

Lysol	1 oz.
Carbon Tetrachloride	75 parts
Refined Kerosene	25 parts

Mix. Sure death to bugs.

Moth Killer

(For Upholstered Furniture)

Ethylene Dichloride	74 parts
Carbon Tetrachloride	25 parts
Paradichlorbenzene	1 part
and Diglycol Oleate	1 part

Insect Exterminator

Kerosene, Refined Grade	1 gal.
Pyrethrum Powder, Best Grade	½ lb.
Paradichlorbenzene	1 lb.
Perfume	sufficient

* Electrical Insulating Compound

The following formulae may be used as a covering or lute as well as for molding into forms:

(a)	Molasses	20 lb.
	Litharge	50 gm.

Stir until homogeneous. Allow to stand until desired degree of hardness or plasticity is reached.

(b)	Nitrobenzol	100 gm.
	Manganese Resinate	50 gm.
	Molasses	40 lb.

The amounts in above formula may be varied to produce different consistencies.

---

* Electric Insulation

Elec. insulating coating compns. are prepd. on a base of urea–CH_2O resin, mixed with materials with which solid solns. are formed. Example: 33 parts of a 46% soln. of urea–CH_2O resin in n–BuOH are mixed with a soln. of nitrocellulose 15, blown castor oil 30, and rosin 6 parts in AcOBu 25, EtOH 75 and PhMe 75 parts, by wt. Coatings of this compn. are very adherent, elastic and durable after drying for 8 hrs. at 75°. They are resistant to mineral acids and to the action of transformer oils at high temps.

---

*Electrical Insulation

The following composition is suitable for transformers, capacitators, cables, etc.

Crude Scale Wax	80
Petrolatum	10
Mineral Oil	10

---

* Electrical Insulation Lining

Glue	1
Water	24
Sulfoturk C	2
Mica	5
Sod. Silicate	5

* Liquid Electrical Resistance

A H_2O–tube resistance for high voltages is provided with means for cooling the liquid so that its resistance is maintained substantially const. Instead of H_2O an aq. soln. of $CuSO_4$ or Manganni's liquid, contg. mannitol 121, H_3BO_3 41 and KCl 0.06 g. per l., may be used.

---

* Non-Drying Plastic Conductor

Glycol Bori Borate	20–30
Water	5
Carbon Black	10

Insulating Coating, Electrical Filament

Layers of a satd. soln. of $Al(NO_3)_3$ of d_{27} 1.4 mixed with 3–10% by wt. of SiO_2 are applied to a filamentary W wire, and the wire is heated after each successive layer is applied to convert the $Al(NO_3)_3$ to Al_2O_3. The wire is subsequently heated to a temp. above the m. p. of SiO_2 but below the crystn. point of W to form a hard homogeneous insulating coating.

---

* Insulating Tape, Electrical

Tape is treated with following at 165°.

Carnauba-Montan Wax	40–50
Rosin	32–40
Castor Oil	10–28

---

* Insulation Composition, Coil Impregnation

Rosin	70
Asphalt	30

Apply at 160–175° C.

---

* Insulation, Heat

Flake Mica	800
Flour Paste	100
Phenol	1

Put between strips of cotton or jute and wrap around steam pipes.

All formulae preceded by an asterisk (*) are covered by patents.

* Heat Insulation

Portland Cement (quick setting)	15–40%
Mineral Wool	40–65
Diatomaceous Earth	20–30

* Heat Insulator

A material weighing approx. 20 lb. per cu. ft. comprises the set product of a mixt. formed from calcined gypsum 2 lb., $Al_2(SO_4)_3$ 3 oz., $CaCO_3$ 1.5 oz., soap 4 g., talc 8 g. and water 26 fl. oz

* Heat Insulating Material

Glass Wool or Mineral Cotton	80
Asbestos	18
Plaster or strong Glue	2

* Refrigerator Insulating Compound

Latex	13 gal.
Bentonite	60 lb.
Trihydroxyethylamine Abietate	2 lb.
Water	1 gal.

* Cable Oil, High Tension

The following is used for saturating high tension paper wound cables.

Mineral Oil	85–90 lb.
Rosin	10–15 lb.
Rubber	0.2–0.5 lb.

Sanctuary Oil

Rape Seed Oil	3 lb.
Cotton Seed Oil (Winter Pressed)	1 lb.
Linseed Oil	1 lb.

* Stove Wick

Pumice Powdered	4
Charcoal Powdered	1
Coke Powdered	1
Sand Powdered	1
Grit Powdered	1
Rosin Powdered	⅛
Silicate of Soda	2

Water sufficient to make paste. Press into shape and vitrify by heat to drive off all volatile matter.

LACQUERS, PAINTS, VARNISHES, STAINS

Nitrocellulose Lacquers

These lacquers may be divided into two parts:—volatile and non-volatile constituents. Under the former may be classed the liquids used to carry the solids into solution. The non-volatile matter consists of nitrocellulose, gums or resins, and a plasticizer.

A film of nitrocellulose alone is not satisfactory for most uses, as it lacks adhesion, is stiff and brittle, lacks flexibility and elasticity; and as a result of this, it will split or peel off the surface. Nitrocellulose has a high viscosity, and a lacquer solution will not contain as much solids for the coating as a paint or varnish of like viscosity.

Resins are used to give a lacquer more solids without increased viscosity, greater adhesion, more gloss and sometimes greater hardness. The resins used are both natural and synthetic. The former class contains such well known materials as rosin, shellac, dammar, kauri, copals, sandarac, mastic, and elemi. A legion of names may be mentioned in the latter class. But we will confine ourselves to the most representative and popular members of each kind. In this class are found ester gum, bakelite, beckacite, amberols, lewisols, and the rezyls and teglacs.

Lacquer films become hard and brittle with age. To overcome the cracking and peeling of a brittle film due to the expansion, contraction, or bending of the coated surface, a plasticizer is incorporated into the lacquer. These materials may be oils, such as castor oil, blown castor oil, blown rape seed oil, OKO oil, and lacquer linseed oil. A very important class is the high boiling esters which are solvents for the cotton and many times for the resins. In this class will be found the ethyl, butyl and amyl esters of the phthalates, tricresyl phosphate, tri-

All formulae preceded by an asterisk (*) are covered by patents.

phenyl phosphate; just to mention a few of the most common ones in use. These plasticizers are non-volatile and will remain in the film for a very long time. They tend to form solid solutions with the nitrocellulose. A very important class and coming to the fore are the resin-plasticizers. In this class will be found ethyl or methyl abietate, beckolac 1308, paraplex 5B as those most popular to-day.

By the use of the term solvents, we mean those liquids that are used to dissolve the nitrocellulose. Solvents are classified as low boilers and high boilers. Each class performs a certain function. Low boilers are used to carry the cotton into solution, provide volatility for the lacquer, and also give the initial set for the film. Usually the low boiler is a faster solvent for the cotton than the high boiler. The most popular member of this class is ethyl acetate. The high boilers provide smooth flow, prevent blush, orange peel and give homogeneity to the film. In this class are found butyl acetate, amyl acetate, butyl proprionate, ethyl lactate, butyl lactate, and the cellosolves.

Latent solvents are compounds or liquids that are not solvents for cotton by themselves. But they become so, by the mere addition of a solvent. In this class are the methyl, ethyl, butyl, propyl, and amyl alcohols.

In the making of a solvent mixture or thinner for a lacquer, other liquids are used, such as benzol, toluol, xylol, solvent naphtha, and also special petroleum naphthas. These do not dissolve cotton, and also lower the solvent power of a solvent when mixed with them. This class of liquids is called diluents, and though they are excellent solvents for a great many of the resins, we will call them diluents as they are not solvents for the nitrocellulose. They give bulk to the mixture, aid in keeping the resins in solution, help balance the formula, and also lower the price.

In the compounding of lacquers, certain standard or stock solutions are used; nitrocellulose or cotton solutions, and the resin solutions. They are blended in various proportions, a plasticizer and the solvents added to bring it to the desired viscosity or concentration.

The nitrocellulose solutions are usually made to contain a definite amount of ounces to the gallon, or to hold a certain amount of cotton in the gallon of solution. Or else it may be cut according to the percentage formula, as a 20%, 25%, or 35% solution.

Cotton Solution No. 1

Dry ½sec Cotton	25 %
Den. Alcohol	10.7%
Butyl Acetate	16.1%
Toluol	32.1%
Ethyl Acetate	16.1%

This solution contains 2 pounds of dry cotton in the gallon of solution. The solution weighs 8.3 pounds per gallon.

Cotton Solution No. 2

Dry ½sec Cotton	35.8 lb.
Ethyl Acetate	24.8 lb.
Toluol	24.2 lb.
Ethyl Alcohol	15.2 lb.

This solution is a 36% cut, and contains approximately 59.5 ounces of dry cotton in the gallon.

Cotton Solution No. 3

Dry 70sec Cotton	1.13 lb.
Alcohol	.51 lb.
Benzol	3.10 lb.
Ethyl Acetate	3.00 lb.

This yields one gallon of solution of a high viscosity cotton.

Cotton Solution No. 4:—To 24 ounces of film scrap add one gallon of solution of 25% Ethyl Acetate; 25% Alcohol, 16% Toluol, and 34% Bayway Solvent No. 55.

Resin solutions are cut from 4 to 14 pounds of resin to the gallon of solvent, or else as a 50/50 cut of resin and the solvent. The solvents used are generally benzol, toluol, xylol, alcohol, and ethyl acetate. In general, different resins will require different solvents. Some manufacturers cut their resins in a thinner to insure greater compatibility with the cotton solutions. Ester gum, Lewisol, beckacite, amberol are dissolved in one gallon of toluol or thinner. The proportions are 8 pounds of the resin to one gallon of the solvent. Elemi gum is dissolved in an equal weight of solvent. For Kauri gum, dissolve 40 pounds of the resin in 60 pounds of a solution of 85% denatured alcohol and 15% ethyl acetate. Dammar Solution is made by dissolving 80 pounds of dammar in a mixture of 20 pounds of ethyl acetate and 40 pounds of petroleum naphtha of boiling range between 80 and 130° C. When completely dissolved add 100 pounds of ethyl alcohol, agitate for a while and allow to settle overnight for a thorough dewaxing. The shellac solution may be the ordinary 4 or 5 pound cut of shellac in alcohol.

A good solvent should possess high solvent power, offer excellent blush resistance, give good flow, make for excellent compatibility and a thoroughly homogeneous film, and should be fast in its action. The formulae listed below may be used for solvents and reducers to thin the various stock solutions, when incorporating them with the other ingredients for a lacquer for sale or use.

Solvent	No. 1	No. 2	No. 3	No. 4	No. 5
Toluol	65%	60%	50%	50%	70%
Ethyl Acetate	10%	15%	15%	15%	15%
Den. Alcohol	....	....	15%	10%	5%
Butyl Alcohol	15%	15%	....	5%	....
Butyl Acetate	....	....	....	20%	....
Amyl Acetate	....	....	13%	....	....
Amyl Alcohol	....	....	7%	....	....
Cellosolve	5%	5%	....	....	5%
Butyl Cellosolve	5%	5%	....	....	5%

The following formulae contain the main elements of a good thinner for general use, namely

1—Good solvent power.
2—Good blush resistance.
3—Proper speed of evaporation.
4—Low cost.

Solvent (Thinner)	No. 1	No. 2	No. 3	No. 4	No. 5
Petroleum Naphtha	44%	....	....	20%	30%
Toluol	....	50%	70%	40%	32%
Ethyl Acetate	22%	18%	15%	10%	....
Ethyl Alcohol	12%	12%	10%	10%	10%
Butyl Acetate	....	....	....	10%	23%
Butyl Alcohol	....	....	....	10%	5%
Amyl Acetate	22%	20%	....	....	....
Amyl Alcohol	....	....	....	....	....
Butyl Cellosolve	....	....	5%	....	....

Wood Lacquers

In a general run of wood lacquers, one will be called upon to supply a sanding sealer, high gloss clear, flat lacquers, rubbing or polishing lacquers, and various specialties as required by the trade such as alcohol proof lacquer, and rubbed effect lacquer.

SANDING SEALER:	No. 1	Non-Volatile Dry Basis	No. 2	Non-Volatile
Cotton Solution No. 1	1 qt. or 2 lb.	½ lb.	4 lb.	1 lb.
Cotton Solution No. 4	1 qt. or 2 lb.	5 oz.		
Resin Solution	1 pt. or 1 lb.	½ lb.	2 lb.	1 lb.
Dibutyl Phthalate	⅛ lb.	⅛ lb.	½ lb.	½ lb.
Blown Castor Oil	⅛ lb.	⅛ lb.		
Zinc Stearate (R. B. H.)	1 lb. paste	32½% solids	1 lb.	32½% solids
Solvent No. 3	1 qt.		1 qt.	

The resin in No. 1 is amberol No. 801 and in No. 2 is Lewisol No. 2. Each solution is made by cutting 8 pounds of the respective resin in 1 gal. of a cheap thinner.

Clear Lacquers

A high gloss clear can be made by taking

Cotton Solution No. 4 1 gal.
Cotton Solution No. 1 ½ gal.
Ester Gum Solution 1 gal.
8 pounds resin to 1 gal. thinner
Amberol Solution ½ gal.
(8 pounds resin to 1 gal. thinner as above.)
Blown Castor Oil ½ lb.
Solvent No. 5 1 gal.

Below we will give a table of various wood lacquers. In this table will be found the non-volatiles. By the use of the standard solutions of cotton, resin and solvents as given above these formulae may be compounded. The addition of solvent and amount will be left to the individual, to meet his specific problem of price and quality.

Clear Lacquers

½ Sec. Nitrocellulose	2	2	2	2	2	2
Dammar Solution	5	2	...	..	..	..
Ester Gum Solution	..	3	1	3	2	4
Kauri Solution	..	..	1½	..	..	..
Amberol Solution	..	..	...	2	..	2
Lewisol Solution	..	..	...	..	3	..
Blown Castor Oil	½	¼	½	¼	¼	¼
Dibutyl Phthalate	½	..	¼	½	..	¼
Tricresyl Phosphate	..	½	...	..	½	..

Flat Lacquer

Cotton Solution No. 1 2 lb.
Cotton Solution No. 3 ½ lb.
Amberol Solution 1 lb.
Zinc Stearate (RBH) 1 lb.
Tricresyl Phosphate ¼ lb.
Solvent No. 4 to one gallon.

Rubbing or Polishing Lacquer

Cotton Solution No. 1 4 lb.
Cotton Solution No. 3 1 lb.
Lewisol Solution 1 lb.
Dibutyl Phthalate ½ lb.
Solvent No. 4 to one gallon.

Alcohol Proof Lacquers

Cotton Solution No. 1 1 gal.

Amberol Solution	1 qt.
Paraplex 5B	2 lb.

Solvent No. 3 to spraying consistency.

Cotton Solution No. 1	4 lb.
Lewisol Solution	2 lb.
Dibutyl Phthalate	12 oz.
Solvent No. 3	1 qt.

By combining the flat and gloss lacquers in varying proportions, any desired effect of semi-gloss, satin finish or rubbed effect may be obtained.

Wood Enamels (Pyroxylin)

In a discussion of the pigmented enamels two factors must be considered. The ability to grind the pigment in the plant, or must the ground pigment be bought from an outside source. For the former we will list below some represented grinds in a plasticizer and gum solution. These will be explained in detail and the difference from the mill ground product shown.

	Pigment, Lbs.	Blown Castor Oil, Lbs.	D.B.P., Lbs.	Ester Gum, Sol., Lbs.	Lewisol Sol., Lbs.
Black	10	16	8	12	12
White	60	8	4	12	12
Red	40	26½	13½	18	..
Blue	45	22	11	9	6
Orange	80	14	6	..	..
Yellow	67	15	5	9	..
Green	58	13½	6½	3	6
Indian Red	68	14	6	9	..

To make these all equal to 100 pound basis add enough toluol to make 100 pounds. This will also thin the mixture to the proper grinding consistency for a roller mill. For a ball or pebble mill slightly more thinning will be required.

The R.B.H. pigments are dispersed in a medium consisting of ½ second nitrocellulose in a solvent mixture. These lacquer pigments will be found to be of a uniform dispersion, excellent covering power, smooth, and may be obtained in any quantity from a gallon can to a fifty gallon drum. In the use of the R.B.H. pigments additional plasticizer must be added to compensate for the added cotton and pigment. It will also be found necessary to carefully watch the resin content for gloss lacquers as these pigments have a tendency to flatten a lacquer.

In the formulation of a wood enamel, a good clear lacquer is usually taken as the base and the pigment grind added to this to meet the required specification for covering power. Sometimes more resin is added to bring up the gloss. If flattening is desired a zinc stearate mixture is added. The base clear used will depend on the price of the enamel. If a cheap enamel is being formulated, a base clear high in ester gum will be indicated. Also the viscosity may be increased by the use of high viscosity cotton or the film solution. For the better grade enamels, the lower viscosity cotton is used to give more solids, and the better resins increased, such as amberol, lewisol, beckacite, and the rezyls. These resins will also give the tougher and more flexible film.

Metal Lacquers

These lacquers are used as a protective and ornamental coating on all class of metal objects, such as, brass goods, plated ware, and even iron and steel, and some of the newer alloys. When the purpose is to protect the highly polished surface against tarnishing, the lacquer is made of a rather high viscosity cotton, as this type will give a tougher film than ½ second cotton. The film is thin and almost imperceptible. The resin used is usually low in acid number and of a very pale color. The low acid number being required so as not to attack the metal coated. The resin will add to the adhesion of the lacquer.

High Viscosity Cotton	4	4	4	4
Elemi Solution	–	–	2	–
Dammar Solution	–	1½	–	1
Lindol	1	–	–	–
Dibutyl Phthalate	–	1	1	1
Blown Castor Oil	–	–	–	1
Ester Gum Solution	–	–	–	1

Clear finishing lacquers for metal and automobile work may be included in this class.

Dry Pyroxylin	10 parts
Rezyl 19	20 parts
Dibutyl Phthalate	5 parts

Dry Pyroxylin	10 parts
Rezyl 113	30 parts
Dibutyl Phthalate	3 parts

Dry Pyroxylin	6
Ester Gum	1¼
Blown Castor Oil	1½
Dibutyl Phthalate	1½

For the enamels for metal, we again refer to the grinds given under wood enamels and follow the same system of incorporating the pigment. That is, take a clear base, and add sufficient pigment to reach the requirements for good covering power. In this class of material it is advisable to increase the plasticizer, for better flexibility and better adhesion.

Automobile Lacquers

This class of lacquer deserves a special division and a complete line of formulae will be given to cover the entire requirements.

Primer Surfacer.—This type of material should possess excellent adhesion, extreme flexibility and toughness, dry quickly, high filling power, and be easily sanded by the dry or wet paper in either water or naphtha.

To 2 pounds of dry ½ sec. cotton add 12 lb. of grind of

40	lb. Keystone Filler
20	lb. Lithopone
10	lb. Talc or Barytes
40	lb. Beckolac No. 1308
6½	lb. Blown Castor Oil
3½	lb. Dibutyl Phthalate

in 1 gal Butyl Acetate

Polishing Black.—High solids, good covering power, good color, excellent flow, easy rubbing and must come to a high polish with the least amount of Rubbing. To

1	lb. dry ½ sec. cotton
½	lb. dry 30 sec. cotton

add 2 lb. of the following pigment grind

10	lb. Super Spectra Black
15	lb. Blown Castor Oil
15	lb. Tricresyl Phosphate
2½	lb. Butyl Stearate
15	lb. Lewisol Solution
42½	lb. Toluol

make up to two gallons with an extremely good solvent.

High Gloss Black.—This lacquer should possess high gloss of a lasting quality, good coverage, good color, excellent flow and smoothness and be able to stand the wear of the sun's rays.

Dry ½ sec. cotton	5 lb.
Dry 15 sec. cotton	3 lb.
Ester Gum	3 lb.
Lewisol	9 lb.
Lindol	2 lb.
Blown Castor Oil	2 lb.
Black Grind (above)	10 lb.
Solvent q.s.	10 gal.

Leather Lacquers

Leather lacquers or leather dopes are used in the manufacture of artificial leather and split leather. The solvents are quick drying. These lacquers are usually made from a medium to a high viscosity cotton. They contain castor oil and other oils as plasticizers and no resins. The resins are not used as they tend to detract from the flexibility. The usual starting point in this work is to begin with the plasticizer equalling the dry cotton. The plasticizers that may be recommended for this work are numerous. The old favorites are blown castor, raw castor oil, blown rapeseed oil and treated linseed oil. The newer ones are ADM 100, butyl acetyl ricinoleate, beckolac 1308 and hydroresin.

Bronzing Lacquer

A special grade of nitrocellulose is usually used for this type of material. It is called bronzing cotton and has a viscosity of from 30 to 40 seconds. Resins are not used as the free acid may cause the powder to turn. A formula that has been tested and used is:

Dry Pyroxylin	4	parts
Dibutyl Phthalate	1¼	parts
Bronze Powder	5	lb.
Solvent	5	gal.

Specialty Lacquers

A lacquer in vogue today for decorating purposes is the crystal lacquer. This material depends on the action of naphthalene to crystallize and of a cotton solution and at the same time not affect the strength of the film.

Cotton Solution No. 1	15	lb.
Cotton Solution No. 3	5½	lb.
Naphthalene Flakes	4	lb.
Cyclohexanone	6½	lb.
Amberol Solution	2	lb.
Tricresylphosphate	½	lb.
Amyl Acetate	5	lb.

Fill to 10 gal. with solvent.

A "matt" lacquer for the furniture trade may be made by taking:

5 lb. Cut White Shellac	2½	lb.
A. S. Solution Cotton	½	lb.
Raw Linseed Oil	2	oz.
Blown Castor Oil	2	oz.
Acetone	1	pt.
Toluol	1	pt.

Fill to gal. with denatured alcohol.

Nail Polish Lacquer (Clear)

Cotton Solution No. 1	32	oz.
Cotton Solution No. 3	16	oz.
Dammar Solution	16	oz.

Tricresyl Phosphate	16 oz.
Butyl Cellosolve	16 oz.

C.P. Acetone to one gallon (1 qt.).

The above may be colored to suit.

All the formulae given above though having proved their practical use by standing the test of sale and resale to consumers are only offered as a starting basis for one's problem. In each trade there are individual requirements, and it is up to the skill and ingenuity of the compounder to adapt or change his formulae to meet these requirements.

Olive Green Dipping Enamel

Carbon Black	10
Chrome Green (25%)	192
Boiled Linseed Oil	63
Varnish	15
Benzine	13

* Anti-fouling Lacquer

Low-viscosity nitrocellulose 10.5, resin 7.6, mercuric resinate 1.5, Paris green 1.5, pigment 6.0, castor oil 6.5, butyl acetate 21.6, butyl alcohol 7.2, ethyl acetate 8.6, denatured alcohol 15.0, methyl alcohol 3.9, benzene 8.4 per cent. The permeability of the films to water is decreased by increasing the percentage of gum, whereas it may be increased by increasing the proportion of softener or cellulose derivative.

*Lacquer Black Coating

Pyroxylin	10
Castor Oil	10
Ethyl Acetate	30
Benzol	35
Alcohol	14.5
Nigrosene	0.5

* Lacquer Coating, Non-inflammable

Cellulose Acetate	12
Monoethylin Palmitate	12
Triphenyl Phosfate	6
Acetone	8.5
Ethyl Acetate	7.5
Alcohol	6.0
Toluol	16.0
Cellosolve	17.6
Acetone Oil (90°–150° C.)	16.0
Diacetone Alcohol	2.5

Non-Blushing Lacquers

¼ sec. Pyroxylin	10
Hydro Resin	2½
Blown Castor Oil	4½
Ethyl Acetate	10
Butyl Acetate	15
Toluol	27
Den. Alcohol	17
Naphtha	14

½ sec. Pyroxylin	10
Ester Gum	10
Hydro Resin	3
Blown Castor Oil	3
Butyl Alcohol	8
Toluol	43
Butyl Acetate	20
Ethyl Acetate	3

Pearl Wood Lacquer

18 oz. ½ second Nitrocellulose
8 oz. High Viscosity Nitrocellulose
6 oz. Dammar Gum-Pale
6 oz. Shellac
2 pt. Butyl Acetate
1 pt. Butyl Alcohol
¼ pt. Amyl Acetate
4 pt. Toluol
3 oz. Dibutyl Phthalate
4 oz. Pearl Essence

White Lacquer Enamels

(1) Nitro-cotton Solution:
- 10 parts Nitro-cotton No. 6—dry
- 30 parts Butyl Acetate
- 10 parts Toluol
- 10 parts Ethyl Acetate

The ingredients are mixed and the cotton dissolved.

(2) Pigment Paste:
- 10 parts Alftalate 222 A 100 per cent.
- 10 parts Toluol
- 20 parts Titanium Dioxide (100 per cent Titanium White)

The paste is ground finely on a mill.

(3) 60 parts nitro-cotton solution are mixed thoroughly with 40 parts of pigment paste, and the enamel then diluted with the above-mentioned solvent mixture to brushing, spraying or dipping consistency.

Nitrocellulose Lacquers

(a)

4.5 parts 222 A. Alftalate
12 parts Nitro-cotton No. 6
36 parts Butyl Acetate
23 parts Ethyl Acetate
24.5 parts Toluol

(b)

14 parts 222 A Alftalate
14 parts Nitro-cotton No. 6
16 parts Butyl Acetate
25 parts Ethyl Acetate
6 parts Industrial Methylated spirit
6 parts Butanol
19 parts Benzol

(c)

18 parts 222 A Alftalate
6 parts Nitro-cotton No. 6
15 parts Butyl Acetate
7 parts Industrial Methylated spirit
7 parts Butanol
31 parts Toluol
8 parts Benzol
8 parts Ethyl Acetate

(d)

17 parts 222 A Alftalate
16 parts Nitro-cotton No. 6
19 parts Butyl Acetate
30 parts Ethyl Acetate
8 parts Industrial Methylated spirit
8 parts Butanol
22 parts Benzol

The above lacquers differ from each other chiefly in their contents of alftalate in proportion to nitro-cotton. The higher the alftalate content the greater the filling property and elasticity. The above solvent mixtures should only be regarded as examples. They may, of course, be changed in the usual way for nitrocellulose lacquers. It must, however, always be remembered that alftalate 222 A is insoluble in methylated spirit.

* Lacquer, Shellac Ester

The following formulae have unusual elasticity and gloss and possess good adhesive properties and excellent durability and resistance to the actinic rays.

(a)

Pyroxylin (wet)	11
Butyl Ester of Bleached Shellac	20
Butyl Acetate	24
Toluol	40

(b)

Pyroxylin	9
Dammar (dewaxed)	5.9
Butyl Ester of Bleached Shellac	10
Alcohol	16
Ethyl Acetate	2.5
Petroleum Distillate (80–130° C.)	4.9
Butyl Acetate	24.4
Dibutyl Phthalate	1

* Lacquer Thinner

Ethyl Acetate	20–40%
Ethyl Alcohol	32–70%
Ethylene Dichloride	10–28%

Lacquer Thinners

A

Butyl Acetate	20
Ethyl Acetate	10
Denatured Alcohol	10
Toluol	60

B

Butyl Acetate	25
Ethyl Acetate	15
Butyl Alcohol	10
Toluol	50

* Undercoat, Lacquer

Shellac	2–4 lb.
Dibutyl Phthalate	2–10 oz.
Denatured Alcohol	1 gal.

Imitation Chinese Lacquer

Alcohol	1 gal.
Shellac	4 lb.
Sealing Wax	4–16 oz.

Different colored sealing waxes produce different colored lacquers.

* Non-Gelling Lacquers

I

	Parts
Half Second Cotton	12
Dammar	12
Dibutyl Phthalate	8
Zinc Oxide	30
Tartaric Acid	.3–1.5

In 100 parts of a solvent mixture consisting of:

	Per cent by volume
Ethyl Lactate	20
Butyl Acetate	10
Toluol	70

II

	Parts
Half Second Cotton	12
Ester Gum	9
Tricresyl Phosphate	6
Zinc Oxide	20
Sodium Tartrate	.3–1.5

In 100 parts of a solvent mixture consisting of:

	Per cent by volume
Isopropyl Lactate	40
Xylol	60

III

	Parts
Half Second Cotton	12
Glyptal Resin ("Rezyl 12")	20
Dibutyl Phthalate	3
Titanium Dioxide	20
Zinc Oxide	10
Tartaric Acid	.1–.5

In 100 parts of a solvent mixture consisting of:

	Per cent by volume
Isobutyl Lactate	42
Naphtha (boiling range 140–190° C.)	58

IV

	Parts
Half Second Cotton	12
Dammar	12
Dibutyl Phthalate	8
Zinc Oxide	30
Tartaric Acid	.3–1.5

In 100 parts of a solvent mixture consisting of:

	Per cent by volume
Normal Butyl Lactate	37
Naphtha (boiling range 150–200° C.)	63

V

	Parts
Half Second Cotton	12
Ester Gum	9
Tricresyl Phosphate	6
Zinc Oxide	30
Sodium Tartrate	.3–1.5

In 100 parts of a solvent mixture consisting of:

	Per cent by volume
Ethyl Oxyisobutyrate	30
Butyl Acetate	10
Toluol	60

VI

	Parts
Half Second Cotton	12
Dammar	12
Dibutyl Phthalate	8
Titanium Dioxide	20
Zinc Oxide	10
Tartaric Acid	.1–.5

In 100 parts of a solvent mixture consisting of:

	Per cent by volume
Butyl Oxyisobutyrate	20
Ethyl Acetate	10
Butyl Acetate	10
Toluol	60

---

* Non-Gelling Metallic Lacquers

A typical non-livering composition consists of (in parts by weight): Cellulose nitrate 8.5, tricresyl phosphate 20, gold bronze 17.5, ethyl acetate 31.5, benzol 60, and the citric or tartaric acid 0.14 to 6.8. Other pigments to which this invention refers include Vandyke brown, red oxide of iron, iron blues, and chrome yellow. The addition of an acid of the nature specified above, preferably dissolved in a solvent for the base material, to compositions which have already livered is effective in delivering them, i.e., restoring them to their original condition and preventing further livering. For this purpose the acid preventive agent is used in the same proportions as indicated.

Artificial Flower Pearl Lacquer

40	oz.	High Viscosity Nitrocellulose
1½	pt.	Cellusolve Acetate
½	pt.	Dibutyl Phthalate
1	qt.	Butyl Acetate
1.2	lb.	Glyptal
2½	gal.	Toluol
1½	gal.	Ethyl Acetate
32	oz.	Pearl Essence

Pearl Dipping Solution

3	lb.	High Viscosity Nitrocellulose
4½	gal.	Amyl Acetate
8	oz.	Pearl Essence

Pearl Enamels

1 pt. Lacquer Enamel (Black, Blue, Red, etc.)
7 pt. Outdoor Durable Clear Lacquer
8 oz. Pearl Essence

---

* Non-Chalking Lacquer Coating

Undercoating: Half-second nitro-cotton 10 oz., ester gum 5 oz., blown castor oil 8 oz., dissolved (to 1 gallon) in a mixture of ethylene glycol monoethyl ether 25 per cent, toluene 37, xylene 23, and ethyl alcohol 15 per cent; the pigment may be 5 oz. of carbon black and 0.7 oz. of Prussian blue. Intermediate coat: Half-second cellulose nitrate 20 oz., tricresyl phosphate up to 14 oz., dissolved in a mixture of equal volumes of butyl acetate and toluene to make 1 gallon of solution. Top coating: Low-viscosity cellulose acetate 20 oz., resin 0–20 oz., plasticizer 8–18 oz., dissolved in 1 gallon of a mixture of ethyl acetate 25, acetone 30, ethyl lactate 25, and ethyl alcohol 20 per cent. Tests have shown that whereas ordinary cellulose nitrate lacquer coatings will chalk and bloom within two or three months of exposure and cellulose acetate directly over nitrate will blister and peel after several months of severe weather conditions, the combined (triple) coating described above will remain in good condition for two years or more when exposed to equally severe weather conditions.

---

* Lacquer Pigment Base

The process may be carried out as follows: 15 lb. of nitrocellulose (viscosity ½ sec. American) in the alcohol-damp condition and 15 lb. of alcohol (or appropriate amount of other liquid, such as benzol, toluol, or xylol) are kneaded together until the excess of alcohol is taken up by capillarity; 86 lb. of pigment (e.g. a blend of 25 per cent of titanium dioxide on a barytes base) is added, and the kneading and mixing operation is continued until all the pigment is thoroughly wetted (about half an hour). Ethyl acetate (5 lb.) is then added and the kneading resumed until the pigment particles are sufficiently dispersed as indicated by visual tests; this occupies an hour or more. The product is plastic or putty-like and may be sold as such, and may be diluted for use with 5 lb. of ethyl acetate, 2 lb. of alcohol, and 10 lb. of toluol. It is possible to mix all the ingredients together at once to form the putty-like mass, but the procedure described above gives better results since the viscosity is more easily controlled.

Pyroxylin and Rubber Lacquer

Pyroxylin	10
Rubber	5
Ethyl Crotonate	100

---

* Lacquer, Quick Drying

Pyroxylin	10.5
Denatured Alcohol	4.5
Butyl Acetate	26.5
Ethyl Acetate	6.0
Butanol	5.0
Toluol	26.0
Dibutyl Phthalate	5.0
Glycol Abietate	16.5

Paper Lacquer

Dry nitrocellulose, 100 lb.; rezyl 11, 250 to 300 lb.; tricresyl phosphate, 50 to 100 lb.; and paraffin wax, 4 to 8 lb. Extra wrappings in cardboard containers are sometimes rendered unnecessary by coating one or both surfaces of the container with the foregoing type of coating. Rezyl lacquer coatings are suggested also for washable and other wallpapers.

---

* Paper Lacquer

The following lacquer gives a brilliant surface to paper or cardboard. It likewise renders it water-proof.

Pyroxylin	16
Ethyl Acetate	20
Butyl Acetate	7½
Butyl Lactate	7½
Octyl Phthalate	15
Alcohol	10
Dammar (de-waxed)	4
Albertol	3½
Ester Gum	2½
Toluol	14

---

* "Pearl" Lacquer

Silky Lead Iodide	25
Pyroxylin	10–15
Lacquer Thinner	100–150

---

* Bronze Lacquer, Non-Thickening

Pyroxylin	7
Dibutyl Phthalate	3
Butyl Acetate	10
Ethyl Acetate	30
Butyl Alcohol	10
Ethyl Alcohol	35
Bronze Powder	5

To the above when homogeneous, is added water 5 parts, slowly with stirring.

---

* Crackle Lacquer Base

This "crackle base" consists of a metallic soap, such as an aluminium soap, mixed by grinding or otherwise with a solvent such as ethyl acetate and preferably also, during the grinding, with a small quantity of pyroxylin to give body to the mixture. A suitable composition consists of aluminium stearate 25, ethyl acetate 74.5, pyroxylin 0.5 per cent. Other aliphatic alcohols or esters may be used as solvent, but hydrocarbons are not suitable, as they tend to cause the base to gel during storage. The crackle base should be added to the ordinary cellulose nitrate lacquer in such proportion that the finished product contains 10–15 per cent by weight of the metallic soap. In thinning the mixture of lacquer and crackle base it is desirable to use ethyl acetate or other readily volatile solvent in order to accelerate the speed of drying.

* Crystallizing Lacquer

About 12 lb. of nitro-cotton and 25 lb. of salicylic acid are dissolved in a mixture of acetone 45, ethyl acetate 45, and butyl alcohol 10 per cent to produce a liquid of specific gravity about 0.95 or 0.96. The composition is applied to paper, leather, or other base and the solvent allowed to evaporate at about 60° to 85° F. When crystallization is complete the coated product is passed through a warm solution of sodium borate or sodium phosphate, whereby more or less of the salicylic acid is dissolved out according to the period of immersion. In place of the above alkaline treatment it is possible to remove the salicylic acid by passing live steam through the paper. When such coatings are applied to wood it is preferable to wash the product with borax solution or benzol sufficiently to strike through to the wood and dissolve from it a certain amount of the natural gum or resin, thereby accentuating the grain; a protective coating of varnish, etc., should then be applied.

Gloss Furniture Lacquer

Gals.	Pts.	Lbs.	Material	Wt. %
28	7.63	213.64	Cotton Solution......	48.76
10	8.45	84.55	Lewisol No. 3 Solution.	19.30
2	8.65	16.30	Dibutyl Phthalate....	3.72
6	7.51	45.06	Butyl Cellosolve......	10.28
3	7.29	21.87	Butyl Acetate.........	4.99
3	6.76	20.28	Butyl Alcohol........	4.63
6	6.07	36.42	Lactol Spirits A......	8.32
		438.12		100.00

Tube B.—Gardner Holdt @ 80° F.
Sp. Gr. .921 @ 80° F.

This lacquer, to quote a finisher, "flows like a varnish." It, therefore will rub down with a minimum of labor, which leaves more lacquer *on the work*. It is very tough and three months of exposure facing south at 45° to the horizontal did not damage it.

Cotton Solution

Gals.	Pts.	Lbs.	Material	Wt. %
		193.00	Wet Cotton..........	28.09
22	7.36	161.92	Ethyl Acetate........	23.57
46	7.22	332.12	Toluol..............	48.34
		687.04		100.00

Yield 90 Gallons of Solution
Weight, 7.63 Lbs. per Gal.

This solution contains 1½ lbs. of dry cotton in each gallon of solution (or 19.66% by wt.). The 193 lbs. of wet cotton is a standard weight drum and is composed of 135 lbs. of dry cotton and 58 lbs. of alcohol.

Lewisol No. 3 Solution

Gals.	Pts.	Lbs.	Material	Wt. %
		8.00	Lewisol No. 3........	52.56
1		7.22	Toluol..............	47.44
		15.22		100.00

Yield, 1.8 Gals. Weight, 8.45 Lbs. per Gal.
Each gallon of solution contains 4.4 lbs. of gum

* Wrinkle Finish Lacquer

A wrinkle finish is produced by applying to a lacquer film a mixture of liquids, e.g., AcOBu, AcOEt and PhMe, having a solvent action on the film. The lacquer is prepared from dry nitrocellulose 7, chinawood oil 9, Ca resinate 10, AcOBu 40 and PhMe 34 parts.

Tinting Lacquers, Shellacs, Etc.
(Light Yellow to a Ruby Red Color)

Resublimed iodine added in the proportion of 2 grams of iodine to 1 gallon of lacquer or shellac will produce a clear golden yellow color that is fast.

This yellow color can be deepened by the addition of more iodine to a point when it begins to take on a clear ruby red color at about 50 grams per gallon. This color is also fast.

Air-Plane Wing Dope
(Non-inflammable)

A formula used in England is as follows:

Acetate of Cellulose	350 gr.
Triphenyl Phosphate	50 gr.
Acetone	2,500 ccm.
Benzol	1,200 ccm.
Alcohol	1,200 ccm.
Benzylic Alcohol	100 ccm.
	5 l.

The effect, characterized by a higher flaming point and by retardation, may perhaps be augmented by the use of chlorhydrocarbons in heavy proportion:

Acetate of Cellulose	150 gr.
Glyceryl Phthalate	100 gr.
Dichloride of Ethylene	600 ccm.
Methylated Spirits	200 ccm.
Methyl Glycol	100 ccm.
Acetate of Methylglycol	100 ccm.
	1 l.

Addition of Pigments

The addition of pigments, oxide of zinc for instance, still further decreases inflammability. Metallic salts applied to the cloth as the first step would act as retarding agents, but they are not used as the dope would adhere less firmly to the cloth. In this connection, it must be noted that the presence of a non-saponifiable substance, such as petrol, in the cloth completely prevents the adherence of dope.

The aeroplane wings are brushed with the acetate of cellulose solution. Pads or other machines are not much used for the cloth, as the solution is so volatile. After drying a second and even a third coat is given.

The dry dope should stick tightly on the tissue, like the skin of a drum, and should resist changes of temperature, wet weather and sunlight. It is recommendable to protect it by means of a varnish, generally with a base of nitrocellulose, to which pigments are added to decrease very considerably its inflammability. This protecting varnish can be prepared as follows:

Viscous Solution of Nitrocellulose	118 kgs.
Castor Oil	23 kgs.
Acetone	90 l.
Amyl Acetate	67 kgs.
Methylated Spirits	67 kgs.

Airship Fabric Dope

The rubberized fabric composing the gas bags of airships is also treated with Pyroxylin dope as follows:

Amyl Acetate	21%
Mutyl Acetate	36%
62° Gasoline	28%
Denatured Alcohol	2%
Castor Oil	8%
Pyroxylin	5%

Air Plane Dope

To harden and increase the tensile strength of fabric used in airplane construction:

Pyroxylin	8 oz.
Solvent	1 gal.

The solvent consists of the following:

Ethyl Acetate	44%
Amyl or Butyl Acetate	22%
Denatured Alcohol	2%
Benzol	32%

* Anti-fouling Composition

Petrolatum	5
Heavy Lubricating Oil	5
Rosin	2½
Paraffin	2½
Salt	1

Paints

Paints are surface coatings consisting essentially of pigments ground in vehicles of drying oils and varnishes. The quantity and type of pigments determine the color, hiding value and to a large extent the body or consistency of the material. They may also influence the drying time as well as the life of the paint.

The vehicle portion, both as to quantity and type, influences essentially the life, gloss, flexibility and drying time of the material. It consists of drying oils, gums, varnishes, dryers and volatile matter.

Dryers are metallic soaps of fatty acids, such as Co, Pb, and Mn, compounds of linoleic and abietic acids, known as linoleates and resinates. These are the important metals used for dryers. More recently, other organic acids have been used in place of the fatty acids, particularly naphthenic acid. The naphthenates are quite commonly used at present.

Volatiles, such as turpentine, solvent naphtha, varnolene, benzine, etc., are used merely to give fluidity in order to permit application by spraying, brushing and dipping.

Typical paint formulas follow:

For exterior use where surfaces are exposed to atmospheric conditions.

1. White House Paint

White Lead	210 lb.
Zinc Oxide	60 lb.
Asbestine	30 lb.
Refined Linseed Oil	12 gal.

Grind and add

Varnolene	1 gal.
Linseed Oil	7 gal.
Liquid Dryer (containing 5% Mn and 5% Pb metal)	1 gal.
Yield	27 gal.

2. Black

Lamp Black	30	lb.
Litharge	8	lb.
Whiting	52	lb.
Asbestine	60	lb.
Raw Linseed Oil	25	gal.

Grind and add

Mixed Dryer (containing about 5% each of Pb and Mn and 1% Co)	3	gal.
Linseed	11	gal.
Yield	53¼	gal.

3. Green

Chrome Green	75	lb.
Barytex	75	lb.
Silica	75	lb.
Asbestine	75	lb.
Linseed Oil	22	gal.
Grind and add		
Dryers Mixed	1¾	gal.
Varnolene	1½	gal.
Linseed	11	gal.
Yield	47¼	gal.

In grinding the pastes above add the oils first into the mixer and while mixing follow with the pigments. After the grind, the remaining vehicles are added.

Other Colors:

For light tints such as ivory, cream, buff, gray, light brown, light green, and light blue, use the white house paint formula and add small quantities of colors in oil to the finished product to obtain the required shades. The colors in oil most generally used for ivory, cream, buff, gray, and light brown are raw and burnt umbers, lamp black, chrome yellows, ochers, and red oxides.

For light blue, use either prussian or ultramarine blue and lamp black, chrome yellow and red oxide, depending upon shade required. These are the most usual combinations but others may be used. It depends entirely upon the shades required.

Bright red or vermilion, use Formula as the above black or green, substituting Toluidine red for the colored pigments, leaving the rest of the formulas the same. Because of the price, toluidine is little used. Para Toner is generally substituted.

Red Lead

Red Lead	1,000	lb.
Linseed Oil	10	gal.
Grind and add		
Linseed Oil	5	gal.
Kettlebodied Linseed Oil	10	gal.
Varnolene	1¼	gal.
Lead, Manganese Dryer	1¼	gal.
Yield	41½	gal.

Metal Protective Paint

Zinc Dust Paint

Zinc Oxide	250	lb.
Zinc Dust	750	lb.
Linseed Oil	10	gal.
Grind and add		
Linseed Oil	5	gal.
Kettlebodied Linseed Oil	10	gal.
Varnolene	1¼	gal.
Lead Manganese Dryer	1¼	gal.

Outside House Paints are also made in paste form and sold as such. The user reduces them gallon to gallon with linseed oil and adds about 1 pint of Pb-Mn Dryer.

PASTE PAINTS

Zinc Oxide

Zinc Oxide	415	lb.
Refined Linseed	11	gal.
Yield	500	lb.

Red Lead

Red Lead	465	lb.
Raw Linseed Oil	4½	gal.
Yield	500	lb.

White Lead

Corroded White Lead	430	lb.
Refined Linseed Oil	6½	gal.
Yield	500	lb.

Both the white ready mixed and paste paints are made also by combining White Lead, Zinc Oxide, Titanox and TiO_2 with inerts in various proportions. Lithopone is sometimes included and although claims are made for these pigments whether used alone or in combination, the Pb-Zn combination seem to give best durability for exterior purposes. For hiding, TiO_2 titanox and lithopone are best in the order named.

INTERIOR PAINTS

White Flat Wall Paint

Lithopone (high oil absorption)	400	lb.
Asbestine	100	lb.
Refined Linseed Oil	7½	gal.
3 Hour Kettle Bodied Linseed Oil	2½	gal.
60% Limed Rosin Soln. in Varnolene	2½	gal.
Varnolene	5	gal.
Grind and add		
Varnolene	15	gal.
Pb-Mn Dryer	⅝	gal.
Yield	45	gal.

Eggshell

Low Oil Lithopone	400	lb.
Asbestine	50	lb.
Whiting	50	lb.
50 Gal. Ester Wood Oil Varnish	17½	gal.
3 Hr. Kettle Body Linseed Oil	7½	gal.
60% Limed Rosin Soln.	5	gal.
Grind and add		
Varnolene	5	gal.
Mixed Dryer	1⅜	gal.
Yield	51⅞	gal.

Gloss

Low Oil Lithopone	375	lb.
Zinc Oxide	125	lb.
Refined Linseed Oil	12½	gal.
3 Hr. Kettle Bodied Oil	10	gal.
Grind and add		
3 Hr. Kettle Bodied Oil	5	gal.
60% Pale Ester Gum Soln.	16¼	gal.
Mixed Dryer	1⅞	gal.
Varnolene	11¼	gal.
Yield	70	gal.

Tint as above under house Paints, before painting.

Wall Sealer

Silica	20	lb.
Asbestine	10	lb.
50 Gal. Ester Wood Oil Varnish	3	gal.
Grind and add		
50 Gal. Ester Wood Oil Varnish	7	gal.
Blown Linseed Oil	2	gal.
Varnolene	2	gal.
Mixed Dryer	½	gal.
Yield	16	gal.

used on walls for reducing porosity.

Wall Wash for Neutralizing Free Lime on Fresh Plaster Walls

Zinc Sulphate	1	lb.
Water	1	gal.

Floor Paint

Lithopone	150	lb.
Zinc Oxide	50	lb.
22 Gal. Varnish *	8	gal.
Grind and add		
No. 22 Gal. Varnish *	16	gal.
Varnolene	3	gal.
Mixed Dryer	½	gal.
	32¾	gal.

* Ester gum-wood oil varnish may be used. Preferably however use a partial phenol-formaldehyde condensation gum variety such as paranol or amberol.

Tint to required color with colors ground in Varnish. For colors note discussion under outside paints.

Quick Drying Enamels

Same as floor paints except use only the phenol-formaldehyde type. Also have it a little longer in oil, about 27.

Enamel for Walls and Wood Work

Low Oil Lithopone	350	lb.
Zinc Oxide	25	lb.
3 Hr. Kettlebodied Linseed Oil	12	gal.
Light Ester Wood Oil	5	gal.
Grind and add		
Light Ester Wood Oil 50% Varnish	15	gal.
Dammar Soln. in Varnolene	1	gal.
Varnolene	10	gal.

Varnishes

Varnish is a gum cooked in a drying oil and thinned with volatile solvents. Dryers are added in the form of metallic compounds during the heating process or they are added as metallic linoleates and resinates after the varnish is made. (Other organic compounds of these metals are also used such as the naphthlanates.)

The presence of Pb, Mn, and Co in solution accelerates the drying of varnishes very materially. They act as oxygen carriers, absorbing oxygen from the air and surrendering it to the oils, which combine with it to form a hard rubbery material.

Gums impart hardness to a varnish film, and oils impart flexibility. The "longer" a varnish the more flexible it is. This length is measured by the number of gallons of oil used per 100 lb. of gum, 50 gal., 25 gal., 10 gal., etc., denoting the addition of the corresponding gallons of, say, combined linseed and china wood oils to 100 lb. of gum.

The most common gum used is ester gum, the glyceryl compound of abietic acid or rosin. Limed rosin is also used extensively but gives more discoloration and is not as neutral as the ester. Neutrality is important, particularly when used in paint formulation when such basic pigments are used as White Lead and ZnO. An acid varnish may result in coagulation or "livering" of the paint caused by metallic soap formation.

Gloss Oil

W. W. Rosin 100 lb.

Melt and heat to 450 F. and add slowly

Hydrated Lime (stir when adding) 7 lb.

Raise temp. to 550 F. continue stirring for about 15 minutes. Draw from fire, let temp. drop to 400 and add slowly while stirring

Varnolene 10 gal.

Centrifuge while hot.

Yield 20 gal.

---

50 Gal. Rosin Varnish

W. W. Rosin 100 lb.

Melt and heat to 450° F. and add slowly while stirring

Hydrated Lime 7 lb.

Raise to 550 and hold 10 minutes, add slowly

China Wood Oil 43 gal.

Heat to 520 and add

Litharge 5 lb.

Raise to 570 and let cool to 550. Hold 20 min. and add

3 Hr. Linseed Oil 7 gal.

Heat to 535 and add

Mn Resinates 3 lb.

Draw from fire and add

Varnolene 60 gal.

Centrifuge while hot 120 gal.

---

25 Gal. Rosin Varnish

W. W. Rosin	100	lb.
Hydrated Lime	7	lb.
China Wood Oil	21	gal.
Litharge	3	lb.
3 Hour Linseed Oil	4	gal.
Mn Resinate	1½	gal.
Varnolene	35	gal.
	70	gal.

---

50 Gal. Ester Varnish

Ester Gum 100 lb.

China Wood Oil 42 gal.

Melt and heat to 520

PbO and heat to 570 5 lb.

Drop to 550 hold for ½ hour.

Add

3 Hr. Bodied Linseed Oil 8 gal.

Raise to 540 and add

Mn Resinate, draw from fire 3 lb.

Varnolene 60 gal.

120 gal.

Ester Cut

Ester Gum 500 lb.

Varnolene 25 gal.

Heat to about 400 carefully in 200 gal. kettle. Draw from fire and add

Varnolene 25 gal.

Yield 100 gal.

Similarly varnishes of any length can be made.

---

4 Hour Varnish (Partial Phenol-formaldehyde Type of Resin

China Wood Oil 21 gal.

25% Phenol-formaldehyde condensation gum like paranol or amberol 100 lb.

Heat to 500° F. and add

PbO. Heat to 550 and hold for about 20 min. 3 lb.

Add

3 Hour Bodied Linseed Oil 4 gal.

Mn Resinate. Heat to 530 and draw from fire 2 lb.

Add

Xylol 10 gal.

Varnolene 25 gal.

70 gal.

---

40 Gal. Phenol-formaldehyde Type of Gum

Resin (Durez 500 Gum Plastic) 100 lb.

China Wood Oil 32 gal.

Heat gum and oil to 460 F. in 20 min. Add

3 Hour Linseed Oil 8 gal.

Hold for body for about 20–30 minutes and add

Cobalt Linoleate (5¾% metal) 1 lb.

Lose heat to 425, add

Xylol 24 gal.

Varnolene 30 gal.

---

25 Gal. Ester Varnish

Ester	100	lb.
China Wood	21	gal.
Litharge	3	lb.

3 Hour Linseed Oil	4	gal.
Mn Resinate	1½	gal.
Varnolene	35	gal.
	70	gal.

Dammar Cut

Dammar	500	lb.
Varnolene	25	gal.

Dissolve as under Ester Cut above.

Varnolene	25	gal.
	100	gal.

There are many other gums that may be used, particularly the innumerable synthetics, but the above illustrate the general type of formula. Many of the modern synthetic gums are really complete varnishes and need be merely dissolved and driers added in order to make a finished product.

The main type of synthetics may be divided into two parts: 1 Phthalic anhydride glycerine condensation products and 2, Phenol-formaldehyde condensation products. Fatty acids are always incorporated with these materials and thus the gums really contain oils and the finished product in many cases are in reality varnishes and may be so used.

The first type, the phthalics, are best used when light color is required, but they do not dry hard through unless applied in a very thin film. They tend to remain soft underneath. They are particularly good in white baking enamels where discoloration is not permissible.

The phenolics are excellent for fast drying and give excellent dry, hard films. They however discolor badly, particularly on baking.

For exterior purposes (spar varnishes) the long oil 50 gal. type is used. For interior the shorter 25 gallon type. A 25 to 30 gal. ester varnish is generally sold as a general purpose varnish for floors, furniture, etc.

Up to a certain point drying of all varnishes can be hastened by adding driers, cobalt being a top or surface drier while manganese and lead are through driers. Excessive driers, however, hasten the deterioration of the film and may cause wrinkling, particularly in baking. A proper balance should always be sought. The quantity of metal should be determined empirically. Based on solid content, lead is used up to about .1%, Mn .05%, Co up to .05%. Of course these ratios can vary greatly with individual requirements.

Speed of drying depends largely also on the length of the varnish, the shorter drying faster.

Baking Enamels, primer and undercoats can be formulated after the manner of floor paint and 4 hour enamels. Each particular problem requires its own special formula and must be made up largely empirically. Certain fundamental facts of course should be known such as increase in pigment content increases the flatness of the finish; increase in non-volatile oil and gums increase the gloss; the longer the varnish the more flexible the film and also the softer; Phenolics give harder films than phthalics and in general less gloss; certain pigments such as toners do not stand excessive baking, that is high temperature and long baking. Also dryers must be used in much smaller amounts with the latter than in air drying paints.

Interior Enamel I

Pigment	40%
Vehicle	60%
	100%

Pigment

Zinc Oxide, French Process	100%

Vehicle

Heat Bodied Linseed Oil	60%
Mineral Spirits	12%
Turpentine	25%
Lead-Cobalt Liquid Drier	3%
	100%

Interior Enamel II

Pigment	47%
Vehicle	53%
	100%

Pigment

Lithopone	80%
Zinc Oxide, French Process	20%
	100%

Vehicle

Heat Bodied Linseed Oil	50%
Dammar *	10%
Turpentine	8%
Mineral Spirits	30%
Cobalt Liquid Drier	2%
	100%

* Dammar dissolved in part of Mineral Spirits.

Interior Enamel III

Pigment	34%
Vehicle	66%
	100%

Pigment

Lithopone	100%

Vehicle

Limed Rosin	20%
China Wood Oil	35%
Linseed Oil	10%
Above cooked together and reduced with	
Mineral Spirits	33%
Cobalt Liquid Drier	2%
	100%

Interior Flat Paint I

Pigment	65%
Vehicle	35%
	100%

Pigment

Lithopone	85%
Extenders *	15%
	100%

* Extenders for interior flat paints include asbestine, talc, silica, whiting, china clay, barytes.

Vehicle

Limed Rosin	8%
Linseed Oil	7%
China Wood Oil	25%
The above cooked together and reduced with	
Mineral Spirits	58%
Lead-Cobalt-Manganese Liquid Drier	2%
	100%

Interior Flat Paint II

Pigment	65%
Vehicle	35%
	100%

Pigment

Lithopone	80%
Zinc Oxide	5%
Extenders	15%
	100%

Vehicle

Refined Linseed Oil	30%
Blown Linseed Oil	6%
Limed Rosin †	4%
Mineral Spirits	57%
Lead-Cobalt-Manganese Liquid Drier	3%
	100%

† Limed Rosin dissolved in part of Mineral Spirits.

Interior Gloss Paint I

Pigment	60%
Vehicle	40%
	100%

Pigment

Lithopone	65%
Zinc Oxide	20%
Extenders *	15%
	100%

* Extenders for interior gloss paints include whiting, barytes, china clay, asbestine.

Vehicle

Heat Bodied Linseed Oil	65%
Mineral Spirits	32%
Lead-Cobalt Liquid Drier	3%
	100%

Interior Gloss Paint II

Pigment	55%
Vehicle	45%
	100%

Pigment

Lithopone	80%
Extenders	20%
	100%

Vehicle

Limed Rosin †	20%
China Wood Oil †	25%
Refined Linseed Oil	25%
Mineral Spirits	27%
Cobalt Liquid Drier	3%
	100%

† Limed Rosin and China Wood Oil cooked together and reduced with Mineral Spirits.

Interior Gloss Paint III

Pigment	52%
Vehicle	48%
	100%

Pigment

Lithopone	90%
Asbestine	10%
	100%

Vehicle

Refined Linseed Oil	45%
Blown Linseed Oil	10%
Limed Rosin ‡	7%
Mineral Spirits	35%
Lead-Cobalt Liquid Drier	3%
	100%

‡ Limed Rosin dissolved in part of Mineral Spirits.

Exterior House Paint I

Pigment	67%
Vehicle	33%
	100%

Pigment

White Lead	70%
Zinc Oxide (Amer. Process)	20%
Extenders *	10%
	100%

* Extenders for exterior paints include barytes, asbestine, silica.

Vehicle

Raw Linseed Oil	80%
Kettle Bodied Linseed Oil	5%
Turpentine	11%
Lead-Manganese Liquid Drier	4%
	100%

Exterior House Paint II

Pigment	64%
Vehicle	36%
	100%

Pigment

Lithopone	40%
Zinc Oxide, 35% Leaded	45%
Extenders	15%
	100%

Vehicle

Raw Linseed Oil	83%
Kettle Bodied Linseed Oil	7%
Mineral Spirits	5%
Lead-Manganese Liquid Drier	5%
	100%

Exterior House Paint III

Pigment	65%
Vehicle	35%
	100%

Pigment

White Lead	40%
Titanox B	20%
Zinc Oxide, Amer. Process	25%
Extenders	15%
	100%

Vehicle

Raw Linseed Oil	80%
Kettle Bodied Linseed Oil	5%
Mineral Spirits	11%
Lead-Manganese Liquid Drier	4%
	100%

Exterior House Paint IV

Pigment	63%
Vehicle	37%
	100%

Pigment

Zinc Sulphide	25%
White Lead	15%
Zinc Oxide, 35% Leaded	40%
Silica	10%
Asbestine	10%
	100%

Vehicle

Raw Linseed Oil	80%
Kettle Bodied Linseed Oil	5%
Turpentine	5%
Mineral Spirits	6%
Lead-Manganese-Cobalt Liquid Drier	4%
	100%

Black Stoving Enamels or Baking Japans

These are applied by dipping, brushing or spraying and are stoved at 150° F. to 400° F. from 1 to 4 hours according to the nature of the japan. Egg shell gloss or flats are made by adding vegetable black in sufficient quantity to give the desired result and thinned down with volatile thinner.

General Method of Procedure

The japans are made by cooking linseed oil with litharge, red lead and black oxide of manganese (or burnt umber) for about five hours at 450° F. to 475° F. The dryers are gradually taken up and the oil oxidized to an almost solid mass. This is known as lead oil. Stearine pitch, together with a bone pitch, to increase blackness, are added to the hot mass and thoroughly cooked for two to three hours until they are all completely amalgamated. It is then thinned down with kerosene and tar spirits, strained and tanked until impurities have settled out. Sometimes a half to one ounce of Prussian blue to the gallon is added during heating. This increases opacity and in parts increased hardness and drying to the oil. These japans are used for the cycle and bedstead trade, also as insulating varnish for impregnating armature and field coils of motors and dynamos, also transformer and magnet coils.

Black Stoving Enamel

Gilsonite Selects	100 lb.
Manjak	10 lb.
Linseed Oil	10 gal.
Burnt Umber	5 lb.
Kerosene	16 gal.
Tar Spirits	16 gal.

Stove at 300° F. for four hours.

Black Stoving Enamel

Stearine Pitch	100 lb.
Rosin	20 lb.
Raw Linseed Oil	50 gal.
Flake Litharge	24 lb.
Manganese Dioxide	2 lb.
Kerosene	20 gal.
Tar Spirits	40 gal.

Stove at 300° F. for four hours.

Black Varnish (Cycles)

Prepared Pitch	37.5 parts
Boiled Linseed Oil	31.5 parts
Petroleum	12.5 parts
White Spirit	18.5 parts

Stove at 180° C.

Black Stoving Enamel

Stearine Pitch	34 parts
Asphaltum	11 parts
Boiled Linseed Oil	22 parts
Turpentine	13 parts
White Spirit	20 parts

Stove at 120° C.

Air Drying Black Enamels and Varnishes

Formula A

Asphaltum	100 lb.
Boiled Linseed Oil	4 gal.
Red Lead	2 lb.
Manganese Dioxide	1 lb.
White Spirit	20 gal.

The White Spirit is added to the mixture of the other materials.

Formula B

Asphaltum	100 lb.
Boiled Linseed Oil	2 gal.
White Spirit	14 gal.

Brunswick Black A

Asphaltum	100 lb.
Dark Rosin	80 lb.
Litharge	2 lb.
Manganese Dioxide	1 lb.
White Spirit	18 gal.

Brunswick Black B

Asphaltum	30 lb.
Dark Rosin	100 lb.
Slaked Lime	4 lb.
Boiled Linseed Oil	3 gal.
Litharge	2 lb.
Manganese Dioxide	1 lb.
White Spirit	30 gal.

Brunswick Blacks are only for indoor use such as for coating iron work and are too brittle for outdoor use.

Berlin Black

Berlin Blacks are air drying enamels which give a mat or eggshell finish.

Brunswick Black	12 gal.
Vegetable Black	20 lb.
Turpentine	6 gal.

Wood Paints

No. 1 Paint. Weight per gallon 14.8 lb.

Pigment	62%	
Lithopone		50%
35% Leaded Zinc Oxide		40%
Silica		5%
Asbestine		5%
Vehicle	38%	
Raw Linseed Oil		80%
Kettle Bodied Oil		8%
Naphtha		7%
Turp, Drier		5%

The above paint was reduced for primer by the addition of one quart of raw linseed oil and one quart of turpentine to one gallon of paint.

No. 2 Paint. Weight per gallon 11½ lb.

Pigment	44%	
Titanox B		70%
Titanium Dioxide		15%
Zinc Oxide		15%

Vehicle	56%	
* Phenol Rosin Varnish		75%
Boiled Linseed Oil		12%
Turpentine		6%
Xylol		3.4%
Solution		2.6%
Drier		1.0%

* The Phenol Rosin Varnish was made up (by weight) as follows:

Phenol Rosin	13.0%
Wood Oil	45.0%
Heavy Naphtha	42.0%

This paint was reduced for priming purposes by the addition of one-half gallon raw linseed oil and one-half pint of turpentine to one gallon of paint.

No. 3 Paint. Weight per gallon 11½ lb.

Pigment	43%	
Titanox B		70%
Titanium Dioxide		15%
Zinc Oxide		15%

Vehicle	57%	
* Phenol Ester Varnish		77%
Boiled Linseed Oil		12%
Turpentine		5.4%
Solution		2.5%
Drier		3.1%

* The Phenol Ester Varnish consisted of:

100% Phenol Formaldehyde

		By Weight
Type—Resinoid	25%	19.1%
Ester Gum	71%	
Rosin	4%	
Wood Oil	67%	35.0%
Bodied Linseed Oil (Body Q Oil)	33%	
Heavy Naphtha		37.1%
Xylol		2.8%
Turpentine		6.0%

Reduction of the No. 3 paint for priming purposes was accomplished by adding one-half gallon raw linseed oil and one-half pint of xylol to one gallon of paint.

No. 4 Paint. Weight per gallon 11½ lb.

Pigment	43%	
Titanox B		70%
Titanium Dioxide		15%
Zinc Oxide		15%

Vehicle	57%	
* Phthalic Anhydride Varnish		83.5%
Boiled Linseed Oil		11.4%
Drier		2.4%
Solution		2.7%

* Phthalic anhydride varnish percentages by weight:

Glycerol Phthalate Linseed Acid Resin		42.5%
Heavy Naphtha	90%	57.5%
Pine Oil	10%	

Reduction of this paint for priming purposes was effected by the addition of one-half gallon of raw linseed oil to one gallon of paint.

Flat Lacquer Paste
(All by Weight)

½″ RS Cotton—dry basis	4	oz.
Aluminum Stearate	16	oz.
Dibutylphthalate	1	oz.
Ethyl Alcohol, including alcohol in cotton	10	oz.
Ethyl Acetate	13½	oz.
Butyl Acetate	3	oz.
Butyl Alcohol	4	oz.
Toluol	13½	oz.

Grind 18 hours in a one-gallon porcelain mill with stone pebbles. The above gives proper size batch for such a mill. The mill should be one-half full of one-inch flint pebbles.

Clear Gloss Lacquer
(By Weight)

½″ RS Cotton—dry basis	7½%
Pale Dewaxed Dammar—solid basis	4½%
Dibutylphthalate	3 %
Blown Castor	1½%
Methyl Alcohol	4 %
Ethyl Alcohol, including that in cotton	7½%
Butanol	6 %
Ethyl Acetate	8 %
Butyl Acetate	18 %
Toluol	40 %

EXTERIOR WOOD PAINTS

Formula No. 1—Priming Coat
(New Outside Wood)

Materials	Soft Paste	Heavy Paste
White-lead	100 lb.	100 lb.
Pure Linseed Oil	4 gal.	4 gal.
Pure Turpentine	1¾ gal.	2 gal.
Pure Drier	†1 pt.	†1 pt.
Gallons of Paint		9 gal.
Coverage (700 sq. ft. per gal.)		6,300 sq. ft.

The addition of a very small amount of lampblack-in-oil to this formula results in a more even and perfect appearing job after the subsequent coats have been applied.

It is especially important that the priming coat be mixed and applied properly. It is the foundation for all succeeding coats of paint and unless it secures a firm and lasting anchorage the coats that follow will merely be lying on the surface and will cause endless trouble. More than ordinary care in the mixing and brushing on of the priming coat will provide good insurance against future trouble.

The painter may use his own judgment in using a smaller quantity of oil for woods which are less absorbent such as southern yellow pine, white spruce, Alaska cedar and cypress.

† When boiled oil is used, reduce drier to ½ pint.

Formula No. 2—Second Coat
(New Outside Wood)

Materials	Soft Paste	Heavy Paste
White-lead	100 lb.	100 lb.
Pure Linseed Oil	¾ gal.	1½ gal.
Pure Turpentine	1½ gal.	1½ gal.
Pure Drier	1 pt.	†1 pt.
Gallons of Paint	5⅝ gal.	6 gal.
Coverage (800 sq. ft. per gal.)	4,500 sq. ft.	4,800 sq. ft.

Where light-colored paint is being mixed, it is good practice to tint the body coat approximately the shade of the final coat as it will afford better hiding power.

Formula No. 3—Third Coat
(New Outside Wood)

Materials	Soft Paste	Heavy Paste
White-lead	100 lb.	100 lb.
Pure Linseed Oil	* 2¼ gal.	* 3 gal.
Pure Turpentine	1 qt.	1 qt.
Pure Drier	1 pt.	†1 pt.
Gallons of Paint	5⅞ gal.	6¼ gal.
Coverage (800 sq. ft. per gal.)	4,700 sq. ft.	5,000 sq. ft.

Repainting Outside Wood.—Two coats usually are enough on wood which has been painted before, the old paint serving as a priming coat.

Before repainting, scrape off all loose and peeling paint and touch up the bare spots and defective places with paint mixed according to Formula No. 4 and then apply two coats as follows:

* In sections where dirt discoloration or mildew is prevalent, particularly on exposures not subjected to direct sunlight, better results will be obtained by reducing the linseed oil content by one-half gallon and increasing the turpentine by one pint.

Although turpentine has been specified in Formulas 2, 3, 4 and 5 many painters are using a flatting oil instead with excellent results. They find it improves the paint's brushing and flowing qualities.

† When boiled oil is used, reduce drier to ½ pint.

Formula No. 4—First Coat
(Repainting Outside Wood)

Materials	Soft Paste	Heavy Paste
White-lead	100 lb.	100 lb.
Pure Linseed Oil	2 gal.	2 gal.
Pure Turpentine	1¾ gal.	2 gal.
Pure Drier	†1 pt.	†1 pt.
Gallons of Paint		7 gal.
Coverage (800 sq. ft. per gal.)		5,600 sq. ft.

This coat will hide the old surface better if it is tinted to about the color of the final coat. If a white job is wanted the addition of a very small amount of lampblack-in-oil to this formula will result in a more even and perfect appearing job after the final coat has been applied.

Formula No. 5—Second Coat
(Repainting Outside Wood)

Materials	Soft Paste	Heavy Paste
White-lead	100 lb.	100 lb.
Pure Linseed Oil	* 3 gal.	* 3 gal.
Pure Turpentine	— gal.	1 qt.
Pure Drier	†1 pt.	†1 pt.
Gallons of Paint		6¼ gal.
Coverage (800 sq. ft. per gal.)		5,000 sq. ft.

Paint Ingredients in Tabular Form.—For convenience and ready reference, the previous formulas are tabulated later following which will be found the same formulas reduced to the basis of one gallon of paint.

† When boiled oil is used, reduce drier to ½ pint.

* In sections where dirt discoloration or mildew is prevalent, particularly on exposures not subjected to direct sunlight, better results will be obtained by reducing the linseed oil content by one-half gallon and adding one pint of turpentine to this formula.

Painting Porch and Other Floors.—The same precautions must be taken in preparing to paint a floor as in the preparation of any other surface. If the old paint is rough and scaly or thick and gummy, the floor should be cleaned down to the wood by planing, burning and scraping or by the use of a liquid paint remover. If a remover containing lye or other strong alkali is used, the surface must be brushed afterward with a coat of strong vinegar to neutralize all remaining traces of alkali and then thoroughly washed with water. Make sure that every part of the floor is firm and solid. After sandpapering and cleaning, the floor is ready for painting.

Priming Soft Wood Floors.—If the floor is of white pine, poplar, hemlock, or other soft wood, use the following formula for the first coat:

Formula No. 6—Priming Coat
(Soft Wood Floors)

Materials	Soft Paste		Heavy Paste	
White-lead	100	lb.	100	lb.
Pure Linseed Oil	3	gal.	3	gal.
Pure Turpentine	2¾	gal.	3	gal.
Pure Drier	†1	pt.	†1	pt.
Gallons of Paint			9	gal.
Coverage (700 sq. ft. per gal.)			6,300	sq. ft.

In applying use a brush well filled with paint and brush out well. One cause of sticky floor paint is flowing the paint on so thick that it does not dry thoroughly underneath, and then hurrying too much with the other coats.

After the priming coat is dry, all joints, cracks, nail-holes and other defects should be filled with a good white-lead putty. The putty should be firmly pressed into the joints or holes and smoothed over with a putty knife. When the putty is entirely dry, sandpaper.

† When boiled oil is used, reduce drier to ½ pint.

Priming Hard Wood Floors.—New hard wood floors—oak, maple, ash, yellow pine or walnut—are not often painted but, if they are to be painted with white-lead, use the following first-coat formula:

Formula No. 7—Priming Coat
(Hard Wood Floors)

Materials	Soft Paste		Heavy Paste	
White-lead	100	lb.	100	lb.
Pure Linseed Oil	2	gal.	2	gal.
Pure Turpentine	2¾	gal.	3	gal.
Pure Drier	†1	pt.	†1	pt.
Gallons of Paint			8	gal.
Coverage (700 sq. ft. per gal.)			5,600	sq. ft.

† When boiled oil is used, reduce drier to ½ pint.

The priming coat is the most important. A first-class foundation saves material and labor in repainting.

Body and Finishing Coats.—For the body or second coat and the finishing or third coat on new floors, whether the wood is soft or hard, use the two formulas that follow. These same formulas should be followed in repainting wood floors with two coats.

Formula No. 8—Second Coat
(Wood Floors)

Materials	Soft Paste		Heavy Paste	
White-lead	100	lb.	100	lb.
Pure Linseed Oil	½	gal.	½	gal.
Pure Turpentine	2¼	gal.	2½	gal.
Pure Drier	½	pt.	½	pt.
Gallons of Paint			6	gal.
Coverage (800 sq. ft. per gal.)			4,800	sq. ft.

Formula No. 9—Third Coat
(Wood Floors)

Materials	Soft Paste		Heavy Paste	
White-lead	100	lb.	100	lb.
Pure Linseed Oil	—		½	gal.
Pure Turpentine	1	gal.	1	gal.
Pure Drier	½	pt.	½	pt.
Floor Varnish	1	gal.	1	gal.
Gallons of Paint	5¼	gal.	5½	gal.
Coverage (800 sq. ft. per gal.)	4,200	sq. ft.	4,400	sq. ft.

For porch floors a varnish should be used that will withstand outside exposure. Where dark colored paint is used, thin tinting colors with turpentine to paint consistency before adding to the paint.

Two things to keep in mind throughout the work are: first, vigorous brushing to spread out each coat to the utmost; second, allowing each coat at least four days to dry.

Underside of Porch Floors.—Porch floors require protection against moisture from the damp space beneath the porch. This space is frequently left without sufficient ventilation. If the soil is damp the porch floor cannot help absorbing a great deal of moisture, which is almost certain to cause blistering and peeling of paint. To prevent trouble of this sort give the underside of the floor, also the tongue and groove edges of the boards, a coat of paint mixed as follows:

Formula No. 10
(Underside Porch Floors)

Materials	Soft Paste		Heavy Paste	
White-lead	100	lb.	100	lb.
Pure Linseed Oil	3¼	gal.	4	gal.
Pure Turpentine	2	gal.	2	gal.
Pure Drier	1	pt.	†1	pt.
Gallons of Paint	8⅝	gal.	9	gal.
Coverage (700 sq. ft. per gal.)	6,038	sq. ft.	6,300	sq. ft.

† When boiled oil is used, reduce drier to ½ pint.

Colored Exterior Paint.—All formulas given so far in this book make white paint. Where colored paint is wanted it can be made simply by adding tinting colors of the proper shade in the right amounts. The tinting colors are known as "colors-in-oil" and can be bought in tubes or in cans wherever you buy your white-lead.

While there is hardly a limit to the number of tints and shades that may be produced by adding colors to white-lead paint, some colors have a tendency to fade rather quickly on exposure to sunlight and should be avoided unless, as is sometimes the case, this faded, weathered appearance is desired for special architectural reasons. Formulas for making a number of desirable colors are printed later. Any of these colors can be varied indefinitely by increasing or decreasing the amount of tinting materials specified.

Most of the color formulas given call for the use of two or more tinting materials but it should be remembered that simpler colors may be made with but one coloring material. Lamp-black, added in varying amounts to white-lead paint, produces a range of pleasing grays; chrome yellow will produce creams, yellows and buffs; chrome green will make shades of green; and venetian red provides a variety of pinks.

Since there is no standard of tone or tinting strength for colors-in-oil of various manufacture, all formulas for producing colored paint must necessarily be approximate. Chrome yellows and ochres, for example, are particularly subject to variation in both strength and tone.

As explained under "Tinting" the tinting colors should be added to the batch of paint before the final thinning. Never pour in all at once the entire quantity of color specified. Add the color gradually and note its effect as it is being stirred into the paint. Stop when the right shade is reached even if you have used less than the formula calls for. On the other hand, you will have to provide more color if the specified amount fails to bring the batch to the shade wanted. Should you accidentally mix too much color in the paint it will be necessary to add more white-lead, properly thinned.

When a formula calls for large amounts of tinting color, it is necessary to provide an extra quantity of thinners to avoid changing the consistency of the paint. This extra color should be thinned before mixing in. Dump the color into a pail and bring it to paint consistency by stirring in linseed oil and turpentine (equal quantities of each).

Permanence of Colors.—The colors which follow are grouped according to their relative permanence. Of course, all colors are subject to some fading but those classified as "permanent" are less likely to show noticeable change on exposure than those requiring tinting materials of a more fugitive type. The latter colors are grouped as "fairly permanent" and "not permanent."

Formulas for Exterior Colors.—The following formulas, with the exception of Numbers 1017, white-lead. If you are tinting a batch of paint which contains more or less than 100 pounds of white-lead, simply increase or decrease the quantity of coloring material proportionately. For the approximate amount of color to be used in tinting one gallon of paint, read the section entitled "Tinting One Gallon of Paint."

Permanent

Fawn—No. 1001
9 oz. Raw Umber

Buff—No. 1002
9 oz. Raw Umber
1½ lb. Raw Sienna

Rose Buff—No. 1003
9 oz. Raw Umber
1½ lb. Raw Sienna
13 oz. Burnt Sienna

Cafe-au-lait—No. 1004
9 oz. Raw Umber
1½ lb. Raw Sienna
13 oz. Burnt Sienna
2 oz. Lampblack

Tan—No. 1005
8 lb. Raw Sienna

Drab—No. 1006
8 lb. Raw Sienna
4 lb. Raw Umber

Golden Brown—No. 1007
8 lb. Raw Sienna
7 oz. Venetian Red

Ivory—No. 1008
13 oz. French Ochre

Ash Gray—No. 1013
2 oz. Lampblack

Lead Gray—No. 1016
8 oz. Lampblack

Fairly Permanent

Colonial Yellow—No. 1009
13 oz. French Ochre
1½ lb. Medium Chrome Yellow

Jade—No. 1011
1¼ lb. Medium Chrome green

Putty—No. 1014
2 oz. Lampblack
3 oz. Medium Chrome Yellow

Silver Green—No. 1015
2 oz. Lampblack
3 oz. Medium Chrome Yellow
12 oz. Medium Chrome Green

Not Permanent

Ceiling Blue—No. 1010
2 oz. Chinese Blue

Opal—No. 1012
1¼ lb. Medium Chrome Green
8 oz. Chinese Blue

Dark Colors.—These colors are used chiefly for sash and blinds and require no white-lead. Each formula is complete in itself, the thinners being shown with each color. Formulas Nos. 1 and 2 should be used for the priming and second coats respectively, on new unpainted wood and Formula No. 4 for the first coat on repaint jobs when the following colors are used as the finishing coat. The addition of lampblack to the above formulas (on the basis of 8 ounces of lampblack to each 100 pounds of white-lead) will provide a satisfactory ground color.

Red No. 1017—(Permanent)
No White-lead
20 lb. Venetian Red
10 lb. Indian Red
1¼ gal. Pure Linseed Oil
1 pt. Pure Turpentine
1 pt. Pure Drier

This will make about 2¾ gallons of paint which will cover approximately 2,200 square feet, one coat.

Green—No. 1018—(Permanent)
No White-lead
10 lb. Chromium Oxide
1 qt. Pure Linseed Oil
½ pt. Pure Turpentine
½ pt. Pure Drier

This will make about a gallon of paint which will cover approximately 800 square feet, one coat.

Brown—No. 1019—(Permanent)
No White-lead
10 lb. French Ochre
3 lb. Venetian Red
½ lb. Lampblack
3 qt. Pure Linseed Oil
1 pt. Pure Turpentine
½ pt. Pure Drier

This will make about 1½ gallons of paint which will cover approximately 1,200 square feet one coat.

Painting Wood Shingles on Side of House.—Paint for wood shingles used as siding should be prepared as follows:

For priming coat use Formula No. 1.

For the second coat use:

Formula No. 11—Second Coat
(Wood Shingles as Siding)

Materials	Soft Paste	Heavy Paste
White-lead	100 lb.	100 lb.
Pure Linseed Oil	1¼ gal.	2 gal.
Pure Turpentine	1 gal.	1 gal.
Pure Drier	1 pt.	1 pt.
Gallons of Paint	5⅝ gal.	6 gal.
Coverage (600 sq. ft. per gal.)	3,375 sq. ft.	3,600 sq. ft.

For the third coat use Formula No. 3.

Staining Wood Shingles and Rough Siding.—A small amount of tinting material, sufficient to stain the shingles or siding to the desired color, should be added to a mixture of the following oils:

⅓ Flatting Oil
⅔ Pure Boiled Linseed Oil

In order to obtain the desired color it is necessary only to add the proper tinting colors-in-oil to the above oil mixture. The color formulas which follow give the amounts of colors-in-oil required for each gallon of the oil mixture to produce some of the more common colors. These are but a few of the many colors obtainable.

Gray
12½ lb. White-lead
½ oz. Lampblack

Deep Red Brown
3 lb. Dark Indian Red

Bright Red
4 lb. Venetian Red

Green
1½ lb. Chromium Oxide
or
3 lb. Medium Chrome Green

Blue
4½ lb. White-lead
1½ lb. Prussian Blue
8 oz. Lampblack

Note.—While creosote oil sometimes is used for staining shingles and rough siding it is not needed to produce a good, penetrating stain and is very likely to cause trouble if the surface is painted in the future. Creosote stains beneath a coat of paint are apt to "bleed" through and cause discoloration and spoil an otherwise good job.

Helpful Hints in Mixing and Applying Paint.—1. Be sure to mix plenty of paint, both for body and trim. It is better to have some left than to run short, especially if you are using a colored paint. There will be no waste, for the left-overs are useful for painting cellar stairs, roof valleys or gutters and various odd jobs where the color of the paint makes no material difference. The body and trim color left-overs may be used for such work and a little lampblack added to the batch to produce a neutral shade.

2. Be sure to put the tinting colors in the paint before the final thinning. The colors should first be thinned to paint consistency and added to the mix after the white-lead has been broken up in the case of heavy paste white-lead, or before the final thinning if soft paste white-lead is used. To put in the colors in their paste form or in dry form is to invite streaking when the paint is brushed out.

3. Strain your paint before using it. Stretch a double thickness of cheesecloth or a fine wire screen over a tub or pail and pour your freshly mixed paint through it. This will remove small lumps of color, skins and other foreign matter that may have fallen into the mixing tub. Straining the paint also adds to its spreading qualities.

4. Benzine and kerosene should never be used as a substitute for turpentine. Mineral oil and other non-drying oils have no place in paint. Avoid them.

5. Use only the best liquid drier, made by some well-known manufacturer.

6. Knots and sappy streaks in new wood should be shellacked, after the priming coat is applied, with pure shellac varnish, brushed out very thin. When the lumber is extremely knotty, less oil and more turpentine may be used than the formula calls for, as too much oil on the knots causes later coats to draw and check.

7. Do no outside house-painting in extremely cold, frosty or damp weather. Painting may be done in winter if care is taken to choose periods when the temperature is favorable (not lower than 50° F.) and surfaces are dry.

8. Moisture is paint's worst enemy. Wood in new buildings almost always contains a good deal of moisture. Let the wood dry out thoroughly before painting. Never put more than the priming coat on the outside of a house until the plaster inside is thoroughly "bone dry." Oil and water will not mix and paint applied over a damp surface may eventually peel.

9. Be equally careful when repainting. Wait for dry weather and examine the surface carefully for moisture before painting.

10. The surface to be painted should be smoothed down before the new paint is applied. If the old paint was white-lead and linseed oil only a light sanding and dusting off will be needed. If hard, brittle paint was used it may be necessary to scrape the surface or perhaps remove the old paint with a gasoline or acetylene torch and scraper. Do not paint over loose or scaling paint. Be sure to brush off all the dust and dirt that has collected on the drip-caps over windows and doors, as well as on the window headers and sills. If not removed, the dust and dirt will mix with the fresh paint and cause streaking.

11. Use plenty of "elbow grease." Brush the paint well into the pores of the wood and do not allow it merely to flow from the brush. It is doubly important to brush the priming coat in closely.

12. For putty use only pure white-lead (either soft paste or heavy paste) thickened to putty consistency with dry whiting. With this putty fill all nail-holes, cracks, knot-holes, dents and other defects in the surface. These places should be filled tightly after the priming coat is dry. Putty containing petroleum and marble dust often mars an otherwise good painting job by making yellow nail-holes and cracks.

13. Preparations of cheap shellac, rosin, etc., are likely to cause knots to show yellow.

14. It is well to mix the paint 48 hours before being used but do not put in the drier or all the turpentine until just before application. Paint should not be allowed to stand for long periods unless it is kept in fully sealed, air-tight containers; otherwise it will become fatty.

15. Two coats of paint, properly mixed and well brushed out, are always better than one thick, heavy coat.

16. In the case of linseed oil substitutes it is sometimes claimed that they are "just as good."

Interior Wall Paints

Preparing the Surface.—It is always advisable to allow plaster at least six months to dry and season thoroughly before attempting to paint it. Fresh plaster contains free alkali which has a tendency to keep paint from drying properly and to cause colors to bleach out.

A good many people do not care to let their walls go unpainted for six months. In such cases, painters oftentimes artificially "age" the new plaster by treating the surface with a solution made by dissolving two pounds of zinc sulphate in one gallon of water. After this solution is applied, sufficient time is allowed for the plaster to dry before priming.*

* In the case of the priming coat, figure 800 square feet per gallon. Also for the second and third coats, if turpentine is to be used.

Before applying any paint, be sure that the plaster or old paint is clean and smooth. Go over the wall very lightly with fine sandpaper or a wide putty knife to remove grit and any loose plaster or paint, taking care not to scratch the surface.

Fill all cracks and holes with patching plaster. The proper filling of cracks is essential to a good-appearing and permanent paint job on plaster. The plaster, to be filled properly, should be first cut out in the shape of an inverted V or triangle.

The edges of the opened crack should be soaked with water to aid the patching plaster in forming a bond with the old wall.

Interior Wood Painting

All loose dust and dirt should be removed before painting. If the surface is excessively dirty or covered with grease, it should be washed. This is especially true of kitchen, bathroom and laundry walls and ceilings.

Walls that have been calcimined should be washed off with sponge and warm water before applying the priming coat.

It is frequently possible to paint successfully over wallpaper provided there is but one layer on the wall and that layer in fairly good condition. All sections of loose paper should be torn away and if there are any cracks underneath, they should be repaired with patching plaster and the seams rubbed with No. 0 sandpaper. Painting is then done as if on bare plaster.

Some wallpapers contain bleeding colors. When any light paint is applied over them the oil in the paint dissolves the color and discoloration results. This can be stopped usually, by the application of two thin coats of shellac over the priming coat. If this difficulty is anticipated it would be well to test a little light paint on the dark colors and if bleeding results it would probably be easier to remove the paper than to apply the two coats of shellac.

If the paper is textured in a pleasing manner it need not be removed but it should be remembered that textures cannot be hidden completely with paint and if the texture is displeasing, the paper should be removed.

If there is more than one layer of paper on the wall, or if the paper is extremely loose or if there is considerable plastering to be done, it would be better to remove all the paper using a broad knife or similar tool after saturating the paper with warm water. The plaster should then be washed to remove all traces of paste.

Formula No. 12—Priming Coat
(Interior Plaster)

Materials	Soft Paste		Heavy Paste	
White-lead	100	lb.	100	lb.
Pure Boiled Linseed Oil	3	gal.	3	gal.
Floor Varnish	2	gal.	2	gal.
Pure Turpentine	1¼	gal.	1½	gal.
Gallons of Paint			9½ gal.	
Coverage (600 sq. ft. per gal.)			5,700 sq. ft.	

Formula No. 13—Second Coat
(Interior Plaster)

Materials	Soft Paste		Heavy Paste	
White-lead	100	lb.	100	lb.
Pure Turpentine	1¼	gal.	1½	gal.
Floor Varnish	¾	gal.	¾	gal.
Pure Drier	½	pt.	½	pt.
Gallons of Paint			5¼ gal.	
Coverage (700 sq. ft. per gal.)			3,675 sq. ft.	

Formula No. 14—Third Coat, Flat Finish
(Interior Plaster)

Materials	Soft Paste		Heavy Paste	
White-lead	100	lb.	100	lb.
Pure Turpentine	1¾	gal.	2	gal.
Floor Varnish	1	pt.	1	pt.
Pure Drier	½	pt.	½	pt.
Gallons of Paint			5 gal.	
Coverage (800 sq. ft. per gal.)			4,000 sq. ft.	

Formula No. 15—Third Coat, Eggshell Finish
(Interior Plaster)

Materials	Soft Paste		Heavy Paste	
White-lead	100	lb.	100	lb.
Pure Turpentine	¾	gal.	1	gal.
Floor Varnish	1¼	gal.	1¼	gal.
Pure Drier	½	pt.	½	pt.
Gallons of Paint			5¼ gal.	
Coverage (700 sq. ft. per gal.)			3,675 sq. ft.	

Enamel Finish.—When a prepared enamel is to be used as the finishing coat, the priming and second coats should be mixed according to formulas No. 12 and No. 13. Then follow with enough coats of formula No. 13 to make a ground which will not only completely hide the surface but will be flat and uniform. The finish of prepared enamel may then be applied over this ground.

Colored Interior Paint.—The preceding formulas covering the painting of interior plaster surfaces produce white paint. If colored paint is desired, the white paint can be readily tinted by the addition of proper tinting colors before all the thinners are added, as explained under "Tinting." See also the section in "Colored Exterior Paint" which gives some valuable pointers on the selection and use of colors-in-oil.

Formulas for Interior Colors.—The following formulas are based on the use of 100 pounds of white-lead. For smaller or larger amounts of white-lead simply decrease or increase the quantity of coloring material accordingly. See also "Tinting One Gallon of Paint."

Formula No. 16—Third Coat, Oil Gloss Finish
(Interior Plaster)

Note.—The following formula should be used only as a base for dark colors, as light-colored paint containing considerable raw linseed oil will yellow badly when used on interiors. Where a light-colored gloss finish is required, follow Formula No. 17.

(a) Materials	Amounts
Heavy Paste White-lead	100 lb.
Pure Linseed Oil	3 gal.
Flatting Oil	¼ gal.
Pure Drier	1 pt.
Gallons of Paint	6¼ gal.
Coverage (800 sq. ft. per gal.)	5,000 sq. ft.

or

(b) Materials	Amounts
Heavy Paste White-lead	100 lb.
Pure Linseed Oil	3 gal.
Pure Turpentine	¼ gal.
Pure Drier	1 pt.
Gallons of Paint	6¼ gal.
Coverage (800 sq. ft. per gal.)	5,000 sq. ft.

Warm Gray—No. 1020
9 oz. Raw Umber

Lemon Ivory—No. 1021
2 oz. Medium Chrome Yellow

Shell Pink—No. 1022
2 oz. Medium Chrome Yellow
4 oz. Venetian Red

Rose Gray—No. 1023
2 oz. Medium Chrome Yellow
4 oz. Venetian Red
1 oz. Lampblack

Buff—No. 1024
3½ lb. French Ochre

Peach—No. 1025
3½ lb. French Ochre
2 oz. Venetian Red

Silver Gray—No. 1026
1 oz. Lampblack

Light Blue—No. 1027
1 oz. Lampblack
7 oz. Chinese Blue

Canary—No. 1028
8 oz. Medium Chrome Yellow

Pistachio—No. 1029
8 oz. Medium Chrome Yellow
1½ oz. Medium Chrome Green

Stippling.—This is one of the most useful methods a painter can employ to give unusual beauty to an interior wall job. A stippled effect is produced simply by striking the wet surface, before the paint has set, with a special type of brush known as a wall stippling brush. The ends of the bristles "pick up" the paint resulting in a uniform pebbly surface that eliminates all possibilities of brushmarks or surface blemishes of any kind.

Since a paint coat to be stippled can be applied with less attention to even brushing, this method adds practically nothing to the labor time required for the job. At the same time it adds greatly to the finished effect.

One hundred pounds of heavy paste white-lead thinned with 2 gallons of flatting oil (or turpentine) makes a paint suitable for stippling. If a heavier stipple is desired the quantity of flatting oil may be reduced accordingly.

Special Wall Finishes.—Many people prefer walls decorated in one color and without doubt in many cases good taste dictates this treatment. Others prefer blended, mottled or figured wall effects and these are frequently suitable. Some owners think they must give up the sanitary and other advantages of paint when anything but a plain unfigured finish is desired. This is a great mistake. Quite a number of very beautiful and highly decorative blended, mottled and figured wall effects are obtainable with paint made of white-lead and flatting oil. Moreover, with these effects are still retained ease of cleaning, sanitary qualities and rich texture.

Plain walls are desirable where simplicity is indicated, where care must be taken not to detract from pictures or in large formal rooms where a certain severity is required. But there are many cases where the use of special finishes is not only in excellent taste but preferable. To meet this demand, there are described below and on the following pages some of the blended, mottled and figured wall effects obtainable with paint.

Crumpled Roll Finish.—To produce this finish, select two harmonious colors differing enough in tone to offer a pleasing contrast.

The ground or second coat, using the second coat formula, should be tinted to match one of the colors selected and should be applied in the regular way and allowed to dry. Then the finishing coat is brushed on, a workable section at a time, and "rolled" as described below while still wet. Prepare the finishing coat according to the third coat flat finish formula and tint it to match the second color chosen.

The "rolling" or mottling is done with a double sheet of newspaper or other absorbent paper crumpled tightly into an elongated wad seven to eight inches in length. Newly printed newspapers should not be used because the printing ink may come off the paper and spoil the appearance of the wall.

Starting at the top left-hand corner of the freshly painted surface and rolling diagonally downward, turn the roll of crumpled paper over and over with the fingers, pressing it firmly against the wall to keep it from slipping.

Continue the rolling to the bottom of the wall and repeat for the next strip, permitting the end of the roll of paper to just overlap the edge of the previous strip.

New rolls should be substituted when the paper becomes so saturated with paint as to leave an indistinct impression.

After a wall has been rolled it should be examined. All blank or missed spaces should be patted with the crumpled paper, and all blurs touched up and rerolled while they are still wet.

Care should be taken to apply no larger section of the finishing coat than can be conveniently rolled before it sets up.

The principal problem involved in a treatment of this type lies in the selection of the two colors to be used. Such colors as ivory for a ground and tan for a finishing coat combine nicely, as do salmon pink and pale smoke gray, and buff and light gray.

If considerable difference exists between the colors selected for use, an effect may be expected that is sharper and more clearly defined than in the case of two colors which are more or less similar. Just as a dark finish may be employed over a light ground, in the reverse way a light finish may be employed over a dark ground.

It must, however, be kept in mind that as only about one-third of the ground coat shows through, the finishing coat is the one which determines the dominant color of the decorative effect.

In new work the second coat should be tinted to the desired ground color, while the third coat should be colored in a sufficiently different manner to show a proper degree of contrast when removed by rolling in the manner previously described. On repaint work, however, the side wall color already in place, if in good condition and free of grease and dirt, may be employed as the ground, and in such an instance the single finishing coat to be applied over it should be tinted with proper reference to the ground so that the desired degree of difference will be apparent.

Experiment with this finish will show that the size of the figure is determined by the closeness with which the paper selected for use is crumpled. Paper crumpled loosely will produce a more or less widely spaced effect, while closely crumpled paper will produce an exceptionally uniform treatment.

Where a three-tone finish is desired, another coat of flat paint, tinted to a third color, should be applied over the two-tone effect and then rolled as previously described.

The crumpled roll finish should not be attempted on rough-finished surfaces since the high points of the plaster will prevent the paper from reaching the paint in the depressed portions, thus leaving an indistinct pattern.

Stencil Finish.—Whether a decorative note of color is required over an entire side wall or simply in small spots here and there in the panels, the stencil offers a ready means of supplying it. It is also invaluable as a quick method of securing a frieze or panel border where moldings are missing.

Although a stencil can be applied with ease, there are two points which should not be overlooked in connection with its application. First, care should be taken to avoid the use of a too thin paint as a stencil color. The paint should be of paste consistency, thinned slightly with flatting oil, and should be applied with a brush carrying very little paint. Second, care should be taken actually to compare the stencil color directly against the ground over which it is to be applied, since those colors in the immediate vicinity of the stencil will influence and seem to change its color characteristic.

Tiffany Finish.—This finish, which was originated by the famous Tiffany Studios of New York City, is sometimes called a blended or glazed finish. To prepare a surface for the tiffany finish it should first be brought up to the ground color selected by adding the required amount of tinting materials to Formula No. 14. This coat should be allowed to dry thoroughly. Over this should be brushed a coat of straight flatting oil, taking care to cover no larger area than can be conveniently worked—about twenty-five square feet.

While the flatting oil is still wet, the glazing colors should be applied here and there. Some of the colors-in-oil used for tinting paint are better adapted to glazing work than others. Raw and burnt sienna, raw and burnt umber, rose lake, cobalt and chinese blues and lampblack are most frequently used as glazing colors. The last two mentioned should be used very sparingly since they exhibit a tendency to ''strike in'' and unless care is taken a spotty effect may result.

The colors should be blended one into another with a wad of cheesecloth, using either a circular or a figure 8 motion. High lights should then be wiped out here and there to permit the ground color to show through and the work finished by tamping with a ball of cheesecloth.

The method as outlined above applies of course to smooth finish plaster, but equally interesting effects on this same order may be obtained on rough finish plaster, provided the glazing colors when applied are blended into one another by tamping with a stippling brush.

Shaded Tiffany Finish.—The shaded tiffany differs from the regular tiffany in that the coloring, instead of being the same all over, gradually gets darker down the wall, being very light at the ceiling line. This interesting decorative effect is often employed as a treatment for alcoves, side wall panels or for vaulted ceilings to give the appearance of increased height.

An appropriate flat ground color, prepared according to the third coat flat finish formula is selected, applied and allowed to dry. Next a coat of straight flatting oil is brushed on to cover as much of the surface as can be easily worked at one time.

While the flatting oil is still wet, the glazing colors should be applied near the top of the wall in small spots, considerably removed from one another. Farther down the wall, the spots should be made larger and, as the baseboard is approached, should be more closely spaced.

As explained under ''Tiffany Finish,'' the colors should be blended into one another with a ball of cheesecloth with a faint suggestion of wiped high lights, through which the ground color is barely visible.

The work should then be finished by tamping with a ball of clean cheesecloth starting at the top of the wall.

The plain shaded effect, which is produced by using but one glazing color, is rendered in the same way except that the color gradation should be as even as possible with no attempt made to suggest high lights by wiping through to the ground color beneath. The ground should be permitted to show only at the top of the wall.

Paint Blend.—This finish employs the same blending principle as the tiffany, except that tinted flat paint is used instead of flatting oil and colors.

While the ground, prepared just as for the tiffany by using Formula 14, tinted to the desired color, is still wet, the blending is done with paint mixed to the same formula (No. 14). The necessary quantity of paint for the blending is divided into two or more batches and these parts tinted to different but harmonizing colors. These colors, in well-assorted groups, are spotted over the wet ground and then, before the paint has set up, smoothly blended into each other by tamping with a stippling brush. The effect produced is very similar to the tiffany.

The principal advantage of this finish is the fact that the painting and the blending can be accomplished at the one time instead of, as in the tiffany, having to wait until the ground coat is dry before doing the blending.

Polychrome Finish.—The polychrome or multi-colored finish is interesting for use where spots of color are required to accentuate certain moldings composed of individual units such as the egg and dart, bead, floral motifs, etc., that may be present in the interior. It is, as a general rule, most satisfactory for use as an added touch of decoration where a plain one-tone treatment has been employed on side wall and ceiling.

This finish is best obtained by applying to the various units composing the molding several different colors which have been extended into tints by the addition of white-lead. These tints should be quite light and nearly equal in value. Tinting parts of the molding in certain of these light colors offers a particularly effective treatment for large rooms, since it lends a colorful touch to an interior that might otherwise appear cold and uninteresting.

Should the effect appear too bright it can be toned down, when the paint is dry, by the application of a thin glaze coat as described below, under "Antique Finish."

Two-Tone Glaze or Antique Finish.—This method of finishing the plain one-tone wall, or some more elaborate decorative treatment, is indispensable where the colors used need to be softened and a rich depth of tone added to the work.

The effect is obtained by first preparing a thin semi-transparent glaze composed of flatting oil to which tinting material has been added to produce the depth of tone required. Apply this glaze over the dry finishing coat and then, while the glaze is still wet, wipe lightly over it with a ball of clean cheesecloth. This operation will remove a certain amount of the glaze, permitting enough to remain on the surface to give an antique effect.

Wiped Stencil Finish.—A coat of straight flatting oil is applied over a dry, flat, one-tone ground coat prepared according to Formula No. 14, and tinted to the desired color. On this wet surface the glazing colors are spotted unevenly. The colors are then blended one into another until a tiffany finish is produced.

While the tiffany is still wet the stencil selected for use should be placed firmly against the surface and the glaze appearing through the openings of the stencil should be removed by wiping with a ball of cheesecloth. This allows the ground color to show through.

The ease with which an error can be corrected by simply glazing over the spot and rewiping through the stencil can be seen.

There are many interesting possibilities with this finish. When the stencil is placed against the wall, the glaze may be wiped out clean to show a clear-cut pattern or it may be wiped lightly to show a faint and somewhat indistinct outline. In the latter case, care should be taken to wipe clean the edge of the area appearing through the stencil openings. This operation permits a small amount of the glazing color to remain in the center of each figure, to harmonize with the remainder of the glazing color used on the side wall.

Another interesting treatment is secured by wiping clean the areas appearing through the stencil openings and then applying, in the regular stencil manner, some of the clear glazing colors used in originally spotting the wall for the glazed effect. This will naturally produce a stencil in complete harmony with the remainder of the side wall since the same colors are used.

The wiped stencil is, of course, appropriate for use only on plaster having a smooth finish. Obvious difficulties would be encountered in endeavoring to wipe clear the surface of a rough-finished ground.

Striping.—Where a simple method of treatment is required to lend a distinctive air to an interior which has been painted in a plain one-tone effect, striping may be used with good results. Striping is simply a narrow banding line of some harmonizing color of greater strength than that applied on the side wall.

For general use this line should perhaps be three-quarters of an inch in width outlining all window frames, door frames, and running parallel with any other interior trim.

The striping line should be applied direct to the side wall a few inches out from the wood trim, the distance depending largely on the width of the stripe which is, in turn, determined by the size of the room. The usual distance is about three to four inches for a three-quarter inch stripe.

Striping is also employed where imitation stone effects are required as a method of marking their outline.

Panel Effects with Paint.—Large interior surfaces are sometimes found that would appear far more interesting if paneled than if left in large unbroken areas.

Striping or stenciling with paint to produce panels offers a simple solution of the problem. In laying off the side wall in panels, considerable discretion should be exercised in order that the panels may be interesting in shape. As a general rule, panels should be taller than they are wide in order to lend an atmosphere of height to the interior. When panels have been outlined and the decorative panel treatment carried out, a solid striping line of color or a stencil border should be applied to frame properly each panel. The width of the border is dependent on the panel size.

Sponge Mottle Finish.—In the sponge mottle finish the colors chosen for the ground and mottling coats should differ sufficiently to show the desired degree of contrast in the finished effect.

A flat ground, properly tinted, should first be applied and allowed to dry. Prepare this ground according to Formula No. 14; use this formula also for the mottling coat.

Now cut a coarse fibre sponge in half in order to make a flat surface, soaking one of the halves in water to soften the fibres and then wringing it out carefully.

To do the mottling, lightly press the flat side of the sponge into some of the mottling coat paint, previously spread on a board, and then tamp the wall with it here and there. Go over the entire surface in this way, making no attempt to follow a set pattern. Much of the charm of the sponge mottle finish is lost if the sponge markings are placed in straight lines and at fixed intervals.

More than one mottling color may, of course, be employed. Use a separate sponge for each color.

A beautiful and changeable effect may be secured by using an eggshell gloss (third coat, eggshell finish), over a flat ground coat. By tinting both the ground and the mottling coats to the same color an effect of tracery may be obtained due to changes in the angle of reflected light.

Combination Effects.—All the special wall finishes described on the foregoing pages are subject to interesting variations and many may be used with excellent results in combinations one with another. A little experimenting will disclose innumerable possibilities. For example, the two-tone crumpled roll finish serves as an excellent background over which to apply a sponge mottle or stencil, giving an elaborate and highly decorative treatment.

White-Lead and Oil Plastic Paint.—The trend is away from excessively rough surfaces as wall finishes, but modified or low-relief textures are gaining in popularity. This latter type of textural effect can be produced readily with a white-lead and oil plastic paint. Such a paint is made with materials that the painter always has in his shop, is relatively low in cost and gives a durable finish that can be kept clean by washing.

The resulting paint, although heavy, will brush out with comparative ease, after which it may be manipulated or textured with a brush, whiskbroom, sponge or any other means.

A plastic paint prepared as described may be tinted while it is being mixed, or may have colors-in-oil worked into it while it is still wet on the wall. Such a paint sets up overnight and can easily be glazed to lend additional color to the surface if such a procedure is desired.

White-lead and oil plastic paint may be applied to any surface that is in condition to receive paint—plaster, wall board, fabric wall coverings, brick, concrete, wood and glass. In the case of fabric wall coverings, all loose or slack fabric should be pasted or nailed in place with nails driven through tin disks. One coat of plastic paint, which is sufficient for all ordinary texturing, will completely hide small defects and nail heads.

When the plastic paint is to be applied to new plaster walls, it is recommended that the walls first receive a priming coat of wall primer. If the walls have been previously painted with an oil paint, and are in satisfactory condition for repainting, the plastic finish may be applied direct.

Use an ordinary four-inch wall brush and coat only a workable section at a time. If too large an area is covered before the texturing is begun, the paint may be difficult to manipulate.

Plastic Textures.—Paint prepared according to Formula No. 18 may, when applied, be textured to produce interesting and highly decorative effects.

For a wall effect of modified texture, apply a coat of paint mixed as follows:

Formula No. 18—Plastic Paint

Materials	Soft Paste		Heavy Paste	
White-lead	100	lb.	100	lb.
Dry Whiting	44	lb.	22	lb.
Flatting Oil	1¾	gal.	1½	gal.
Pure Drier	¼	pt.	¼	pt.
Gallons of Paint	7¼	gal.	5¼	gal.
Coverage (160 sq. ft. per gal.)	1,160	sq. ft.	840	sq. ft.

If soft paste white-lead is used, thin the whiting with the flatting oil and mix thoroughly with the white-lead, adding the drier and such tinting colors as may be required.

If heavy paste white-lead is used, add half the flatting oil to the white-lead and use the remainder to thin the whiting. Then mix the two batches together thoroughly, adding the drier. Tinting colors may also be put in if desired.

A gallon of white-lead and oil plastic

paint will cover from 100 to 220 square feet, the difference in spreading rate depending upon the thickness of film required to produce the desired texture. The maximum coverage of 220 square feet to a gallon represents a spreading rate beyond which the plastic paint would be too thin for producing even the most modified relief effect. The minimum coverage of 100 square feet to the gallon represents a spreading rate which, if further reduced, will not give overnight drying, due to the heaviness of the texture. An average coverage of 160 square feet per gallon should be estimated in figuring costs on plastic lead paint.

Basket Weave.—Drag the wide edge of a whisk broom down over the paint about six inches, until a square is formed. Then place the broom immediately below, and at the left edge of the square, and draw it horizontally across the wall until the right edge of the motif above is reached. Repeat the first process below the horizontal markings. When this pattern is laid over an entire wall the effect resembles a basket weave and makes an interesting modern design for small rooms or for the tea room, shop or studio.

Fan Swirl.—Starting at the top of the wall, place a whisk broom against the wet plastic paint and give the wrist slightly more than a half turn to the right to produce a circular effect. Repeat the process, making another similar figure at the right of the first one. The whisk broom is held in horizontal position. The bristles at the right act as the axis upon which the broom is turned. After several of these fan-shaped swirls have been executed, a second series should be worked below the first and just close enough to enable the sweep of the whisk broom to carry the pattern up over the lower part of the first line.

The Fan Swirl texture is particularly striking if a glaze is added to accentuate the high points.

Grass Cloth.—The beauty of the Grass Cloth effect depends as much on the colors used as on the texture. A coat of tinted plastic paint is first brushed on in the usual way. While this coat is still wet, spots of plastic paint of various colors are applied here and there. A whisk-broom is then drawn vertically across the surface so as to blend the colors.

Another way to produce the Grass Cloth finish is as follows: Put on a coat of tinted plastic paint. Then texture this with a whisk-broom in the manner described and, when dry, glaze it with colors thinned with flating oil.

Weave Moderne.—This effect is produced simply by drawing a whisk-broom through the plastic paint at various angles. The broom sweeps should be fairly long and overlap so as to form an interesting series of interlacing diagonal lines. Particularly effective results may be had with this effect by glazing with gold, silver, bronze or some other metal color.

Water Wave.—Beginning at the top of the wall, draw a whisk-broom or paint brush slowly downward, at the same time moving it from left to right to produce a series of wavy lines.

Vein Relief.—To produce this effect, simply strike the wet plastic paint sharply all over with the flat side of a four-inch wall brush.

Swirl Overlay.—There are two ways of forming this interesting figure. One is to place the flat side of a coarse fibre sponge against the plastic paint, pulling the sponge sharply away after a quarter twist of the wrist. The second method is to use, in place of the sponge, a flat block of wood about six inches square and an inch thick. With either tool the markings should be made so that the swirls overlap.

Gothic Scroll.—A serving spoon is the tool used in producing this pattern. The bowl of the spoon is pressed against the wet plastic and moved spiral-fashion. The outer sweep of the spiral should be six or eight inches in diameter, the spiral becoming smaller as it approaches the central point from which the spoon is lifted. A second spiral, overlapping the first, is then added and the process continued to form an all-over treatment.

Waving Reed.—First drag a graining comb horizontally across the plastic paint. Then, using the rounded end of the handle of a paint brush or putty knife, make upward curving lines a foot to a foot and a half long. All the lines should have the same general curvature and taper off at the point to resemble reeds bending slightly before the wind. The "reeds" should interlace to provide a uniform all-over pattern. The use of a glaze will bring out the texture strikingly.

Thatched Reed.—This effect is obtained by drawing the rounded end of the handle of a brush or putty knife through the plastic paint to establish vertical and diagonal markings, closely interlaced. These, in the final finish, should suggest the matted effect of closely woven thatch. The texture is emphasized if a glaze is applied.

Willow Twig.—This design is made by placing a rolling pin against the plastic paint and simply rolling the pin upward.

Fretted Texture.—Just tamping the wet plastic paint uniformly with a coarse fibre sponge produces the fretted texture.

Bamboo Effect.—First, tamp the wet plastic paint uniformly with a coarse fibre sponge. Then, with a length of rounded stick, such as a pencil or piece of half-round molding, press in the bamboo-like marks. These markings should be sloped uniformly to the right or left but no attempt made to produce an even design.

Pine Needle Texture.—The background of this effect is produced by tamping the wet plastic paint uniformly with a coarse fibre sponge. The "needles" are then formed by tamping the paint with a wood block around which heavy cord has been wound in fan shape. The block should be about four inches square wrapped with six or seven turns of cord so that the turns are together at one end of the block, thus forming the fan shape.

Palette Blend.—The Palette blend is produced by brushing on a coat of plastic paint in the regular way and then applying spots of plastic paint of another color while the all-over coat is still wet. This done, the two colors are blended together by placing a straightedge against the surface at various places and giving the tool a quarter twist. The staightedge may be celluloid, wood or metal. Care should be exercised to hold it very lightly against the surface so that too much plastic paint is not piled up. The two colors used should give a good contrast. About three times as much paint will be needed for the undercoat as for the spots.

Travertine.—First apply a cream-colored plastic paint uniformly over the surface. Then press a sponge lightly here and there, evenly distributing the sponge markings and spacing them from four to eight inches apart. The markings should measure about three inches in width and be longer horizontally than vertically. Such markings can readily be made by grasping the sponge tightly. A straightedge is finally drawn lightly across the textured plastic paint from left to right so as to smooth down all points raised in stippling.

After the textured paint has set, it is marked off into blocks. This is accomplished by cutting parallel lines spaced about a quarter of an inch apart and then lifting out the plastic paint between the lines.

It is customary to use a thin glazing coat in the case of the Travertine effect. The liquid glaze may be made with flatting oil, burnt umber and burnt sienna.

Caenstone.—This texture is secured simply by stippling cream-colored plastic paint in a uniform manner with a stippling brush and then glazing. The blocking off is done in the same way as in the case of the Travertine effect.

Tapestry Effect.—This effect is obtained by dragging a graining comb through the plastic paint to give a series of vertical lines and then striking the paint lightly here and there with a sponge or a wad of paper. Glazing with gold, silver or bronze gives a rich, beautiful finish.

Painting Fabric Coverings.—To overcome defects in plaster walls or to anticipate others which it is feared may develop, plaster walls are sometimes covered with muslin or a specially prepared fabric of some kind which is then painted. No difficulties are encountered in painting such fabric coverings. The painting is done in the regular way just as if plaster were being painted, and the finished job is practically indistinguishable from ordinary painted plaster. If the fabric has been previously treated with a size, no priming coat is necessary.

Painting Wall Board.—Composition wall board, which is used on many interiors to take the place of plaster, may be painted with satisfactory results. Such surfaces may be treated like plaster walls and the painting should be done in accordance with the recommendations given for painting plaster.

Washing Painted Walls. — Walls painted with white-lead can be cleaned, without harm, provided the following procedure is employed.

A workable portion of the wall should be sponged with a good white soap solution, the work progressing from the baseboard toward the ceiling. This section should then be rinsed with clear water and the adjoining section cleaned in the same manner. The white soap solution should effectively remove ordinary dust and dirt which accumulates on most walls.

In certain public buildings, the walls receive severe mechanical injury and become badly soiled, and it is sometimes necessary to use a solution stronger than that containing only white soap. Some of the washing powders, which do not contain an excessive amount of alkaline material, prove very effective in such cases. Cleaning powders that contain a certain amount of abrasive material will naturally wear down the paint film regardless of how hard it may be and their use should be avoided whenever possible. A little experimenting will enable one to determine just how strong a soap solution is necessary to produce the desired results without injuring the paint film by either chemical or mechanical action.

Painting New Inside Wood.—The following formulas are for white paint. If the paint is to be colored, tint it as explained.

Formula No. 19—Priming Coat (New Inside Wood)

(a) Materials	Soft Paste	Heavy Paste
White-lead	100 lb.	100 lb.
Flatting Oil	2¾ gal.	3 gal.
Pure Linseed Oil	3 gal.	3 gal.
Pure Drier	1 pt.	1 pt.
Gallons of Paint		9 gal.
Coverage (800 sq. ft. per gal.)		7,200 sq. ft.

or

(b) Materials	Soft Paste	Heavy Paste
White-lead	100 lb.	100 lb.
Pure Raw Linseed Oil	3 gal.	3 gal.
Pure Turpentine	2¾ gal.	3 gal.
Pure Drier	1 pt.	1 pt.
Gallons of Paint		9 gal.
Coverage (700 sq. ft. per gal.)		6,300 sq. ft.

As on outside wood, the painter may exercise his discretion in reducing the quantity of linseed oil for woods which are less absorbent such as southern yellow pine, white spruce, Alaska cedar, hemlock and cypress. The amount of flatting oil and drier should be increased correspondingly.

Formula No. 20—Second Coat (New Inside Wood)

(a) Materials	Soft Paste	Heavy Paste
White-lead	100 lb.	100 lb.
Flatting Oil	1¾ gal.	2 gal.
Gallons of Paint		5 gal.
Coverage (900 sq. ft. per gal.)		4,500 sq. ft.

or

(b) Materials	Soft Paste	Heavy Paste
White-lead	100 lb.	100 lb.
Pure Turpentine	1¾ gal.	2 gal.
Pure Drier	½ pt.	½ pt.
Gallons of Paint		5 gal.
Coverage (800 sq. ft. per gal.)		4,000 sq. ft.

Formula No. 21—Third Coat, Flat Finish (New Inside Wood)

(a) Materials	Soft Paste	Heavy Paste
White-lead	100 lb.	100 lb.
Flatting Oil	1¾ gal.	2 gal.
Gallons of Paint		5 gal.
Coverage (900 sq. ft. per gal.)		4,500 sq. ft.

or

(b) Materials	Soft Paste	Heavy Paste
White-lead	100 lb.	100 lb.
Pure Turpentine	1¾ gal.	2 gal.
Floor Varnish	1 pt.	1 pt.
Pure Drier	½ pt.	½ pt.
Gallons of Paint		5 gal.
Coverage (800 sq. ft. per gal.)		4,000 sq. ft.

Formula No. 22—Third Coat, Eggshell Finish

(a) Materials	Soft Paste	Heavy Paste
White-lead	100 lb.	100 lb.
Flatting Oil	1¾ gal.	2 gal.
Wall Primer	8 gal.	8 gal.
Gallons of Paint		13 gal.
Coverage (900 sq. ft. per gal.)		4,700 sq. ft.

or

(b) Materials	Amounts
Heavy Paste White-lead	100 lb.
Pure Turpentine	1½ gal.
Floor Varnish	¾ gal.
Pure Drier	½ pt.
Gallons of Paint	5 gal.
Coverage (700 sq. ft. per gal.)	3,500 sq. ft.

Formula No. 23—Third Coat, Oil Gloss Finish (New Inside Wood)

Note.—The following formula should be used as a base for dark colors only, as light-colored paint containing considerable raw linseed oil will yellow badly when used on interiors. Where a light-colored gloss finish is required, follow Formula No. 17.

(a) Materials	Amounts
Heavy Paste White-lead	100 lb.
Flatting Oil	¼ gal.
Pure Linseed Oil	3 gal.
Pure Drier	1 pt.
Gallons of Paint	6¼ gal.
Coverage (800 sq. ft. per gal.)	5,000 sq. ft.

or

(b) Materials	Amounts
Heavy Paste White-lead	100 lb.
Pure Linseed Oil	3 gal.
Pure Turpentine	¼ gal.
Pure Drier	1 pt.
Gallons of Paint	6¼ gal.
Coverage (800 sq. ft. per gal.)	5,000 sq. ft.

Enamel Finish.—When a prepared enamel is to be used as the finishing coat, the priming and second coats should be mixed according to Formulas No. 19 and No. 20. Then follow with a sufficient number of coats of Formula No. 20.

Formula No. 24—First Coat Over Shellac
(Special Interior Wood Finish)

(a) Materials	Soft Paste	Heavy Paste
White-lead	100 lb.	100 lb.
Flatting Oil	1¾ gal.	2 gal.
Gallons of Paint		5 gal.
Coverage (900 sq. ft. per gal.)		4,500 sq. ft.

or

(b) Materials	Soft Paste	Heavy Paste
White-lead	100 lb.	100 lb.
Pure Turpentine	1¼ gal.	1½ gal.
Floor Varnish	¼ gal.	¼ gal.
Pure Drier	¼ pt.	¼ pt.
Gallons of Paint		4¾ gal.
Coverage (800 sq. ft. per gal.)		3,800 sq. ft.

Old woodwork should be rubbed smooth with sandpaper until all gloss has disappeared. Then apply one coat of paint mixed according to Formula No. 24.

When the first coat on either new or old work is dry and hard, putty all defects such as knot-holes, dents, cracks, etc., with putty made by stiffening heavy paste white-lead to putty consistency with dry whiting.

From this point new and old work should be treated alike. When the first coat is dry, rub it down with No. 0 sandpaper. Repeat coats of Formula No. 24 as many times as are necessary to bring the surface to clear white with no dark places showing through, always sanding between each coat.

Next apply one coat of high-grade white enamel. After this is dry, rub it down with pumice and water. Then apply a second coat of the same enamel and finish with rotten stone and sweet oil. Polish finally with a chamois.

This completes the full-gloss finish.

For a silk finish, rub down the last coat with fine pumice and water.

To obtain an ivory effect, tint the last coat with just enough raw sienna to turn it off the white, before applying the enamel. The enamel coats must be tinted in like manner.

Interior Wood Stains

Staining Interior Wood.—In staining new interior wood a coat of liquid composed of equal parts of raw linseed oil and turpentine, particularly if the wood is soft, should first be applied to make an even foundation for the stain. If this precaution is not taken, the stain will strike in here and there, appearing dark in some spots and light in others. When this coat is dry, the stain should be applied over it. After the stain has been on the surface for 5 or 10 minutes wipe off the surplus with a dry rag or waste.

Stain Formulas
(Natural Wood)

(a) 2 qt. Flatting Oil
2 qt. Pure Raw Linseed Oil
1 qt. Pure Drier

—or—

(b) 2 qt. Pure Raw Linseed Oil
2 qt. Pure Turpentine
1 qt. Pure Drier

To this may be added colors-in-oil, in the approximate proportions outlined below, to obtain the required color.

Cherry

2 lb. Burnt Sienna
1 lb. Raw Sienna

If the burnt sienna has more of a brown than a fiery red tone, omit the raw sienna but use three pounds of burnt sienna instead of two.

Mahogany

2 lb. Van Dyke Brown
1 lb. Rose Lake

Vary the proportions of the above colors to get the depth desired for this stain.

Light Oak

2 lb. Raw Sienna
½ lb. Raw Umber

If the raw sienna is inferior in staining power, omit the raw umber and use three pounds raw sienna.

Dark Oak

2 lb. Raw Sienna
¾ lb. Burnt Umber
Small amount Burnt Sienna

Walnut

6 lb. French Ochre
1 oz. Venetian Red
1 oz. Lampblack

For graining colors the tinting materials given under "Staining," for the particular wood to be imitated, should be thinned to brushing consistency with

3 parts Pure Turpentine
2 parts Pure Raw Linseed Oil
1 part Pure Drier

This paint should be applied over the dry ground and, while still wet, should be dragged, combed, or otherwise figured, in imitation of natural wood graining.

Painting Interior Floors.—There are two kinds of floors that require painting—new floors laid with soft wood such as hemlock or white pine; old floors that have become worn, scratched, stained or otherwise marred. New floors of hard wood, such as oak, ash, maple or yellow pine may be painted, if desired, but waxing or varnishing or staining makes a handsomer finish.

Success with newly painted floors depends chiefly upon the choice of right materials and knowing how to use them. In fact, the only important particular in which the film of floor paint needs to differ from that on a window frame, door or the side of a house is the finish. The priming coat must anchor firmly into the wood, it must dry thoroughly and the outer coat must become hard before the floor is used.

Other Finishes for Hard Wood Floors.—For hard wood floors that are not to be painted, four kinds of treatment may be named—oiling, shellacking, varnishing and waxing. The processes overlap more or less and vary according to the kind of wood. The treatment selected should also depend upon the way the floor is to be used. A few fundamentals may be stated.

Open-grained hard woods, such as oak, birch, ash or walnut, should be treated first with a good silex paste filler. Close-grained hard woods, like maple or cherry, require no filter. Yellow pine, owing to the pitch it is likely to contain, should first have a thin coat of shellac to prevent the pitch from blistering later coats.

Good silex paste fillers may be purchased ready to apply. Or an excellent one may be made by mixing the finest silex, or silica, with equal parts of pure linseed oil, pure turpentine and best japan drier, so as to form a medium paste. Reduce this paste to a fairly thin mixture with turpentine only, allowing the filler to stand for a time. In some cases it is possible to add the colors-in-oil, with which the wood is to be stained, directly to the filler. This is good practice. Brush across the grain of the wood with a stiff, stubby brush that will work the paste well into the pores. One coat makes a fair job, but two coats make a better one, filling up the checks which the first coat did not fill.

After the filler has dried for about an hour, rub briskly across the grain of the wood with coarse burlap or excelsior to remove surplus filler left on the surface.

The purpose in using fillers is to fill the pores of open-grained wood, and to prevent darkening by the excessive absorption of varnish or other material used for the finish.

Oil Finish.—Oiling, no doubt, is the most durable finish for a floor, though it requires frequent going over. One effect of oil is to darken considerably the natural color of the wood. For a floor oil use three parts of pure boiled linseed oil to one part of turpentine. When boiled oil cannot be obtained take four parts raw oil, one part turpentine and one part drier. Stir frequently while using; apply with a strong, stiff brush; rub well into the wood. Clean off all surplus oil not taken up by the wood. An oiled floor should be wiped frequently with an oiled cloth. Oily rags are liable to take fire spontaneously and should be burned.

Shellac Finish.—This treatment gives a fairly lasting finish if the floor is not to have very rough usage. Three or four coats of shellac, thinned down with good quality denatured alcohol, are recommended for either soft or hard wood floors.

Refinishing Old Floors.—The proper time to take care of a floor is when the first bare spot appears. Then all that is necessary is to scrub thoroughly, apply a coat of floor varnish or paint to such places as show wear and, when dry, go over the entire floor.

To bring a badly worn floor back to its original state of perfection requires considerable work and ingenuity. There are two good methods by which this can be done. One is to remove the old finish and then scrape the wood with a carpenter's steel floor scraper. This scraping and subsequent sandpapering brings the wood back to its original condition and all that is then necessary is to fill, stain and varnish or paint as a new floor. This is a somewhat expensive proceeding, however, and many people prefer to do the work in the following way:

1. Apply a good liquid paint and varnish remover. Cheap soda solutions discolor the wood. Cover ten or twelve boards at a time, the entire width of

the room. When finish has softened, remove most of the film with a broad knife, finishing up with coarse steel wool dipped in remover.

If the floor is not badly discolored, a thorough washing up with denatured alcohol will be sufficient for the final cleaning. If bleaching is required, however, a hot saturated oxalic acid solution (as much acid as the quantity of boiling water will dissolve) should be applied over the entire floor. If there are some spots that do not bleach out after ten minutes, apply more of the hot solution to these places until the entire floor is uniform in color. Then, remove excess acid with warm water and sponge and allow to dry.

Sometimes, when there are only a few dark, worn places in the floor, it is only necessary to apply the bleaching solution to these spots, cleaning up the rest of the floor with alcohol.

2. Sandpaper with No. 1½ grade, rubbing with the grain of the wood. Wipe up the loose dust carefully and then refinish in the manner desired. It will not be necessary, of course, to use filler.

Painting Stucco, Concrete, Brick, Etc.

Preparing Stucco or Concrete. — Stucco, concrete work and the mortar in brick or stone work should be allowed to stand and dry at least a year before paint is applied. If painted within a year, it may be aged artificially by washing with a solution made by dissolving two pounds of zinc sulphate in one gallon of water or with ordinary carbonic acid water.

Boiled linseed oil should be used as specified wherever possible, especially on stucco and concrete. If boiled oil is not available, raw oil and drier may be used.

Formulas for New Work.—For painting stucco, concrete, brick or stone, apply three coats of paint mixed according to the following formulas:

Formula No. 25—Priming Coat (Stucco, Concrete, Brick, Stone)

Materials	Soft Paste	Heavy Paste
White-lead	100 lb.	100 lb.
Pure Boiled Linseed Oil *	3 gal.	3 gal.
Spar Varnish	2 gal.	2 gal.
Pure Turpentine	1¼ gal.	1½ gal.
Gallons of Paint		9⅜ gal.
Coverage (200 sq. ft. per gal.)		1,875 sq. ft.

Formula No. 26—Second Coat (Stucco, Concrete, Brick, Stone)

Materials	Soft Paste	Heavy Paste
White-lead	100 lb.	100 lb.
Pure Linseed Oil	2 gal.	2 gal.
Pure Turpentine	1¼ gal.	1½ gal.
Pure Drier	†1 pt.	†1 pt.
Gallons of Paint		6½ gal.
Coverage (400 sq. ft. per gal.)		2,600 sq. ft.

Formula No. 27—Third Coat, Gloss Finish (Stucco, Concrete, Brick, Stone)

Materials	Soft Paste	Heavy Paste
White-lead	100 lb.	100 lb.
Pure Linseed Oil	3 gal.	3 gal.
Pure Turpentine	—	1 qt.
Pure Drier	†1 pt.	†1 pt.
Gallons of paint		6¼ gal.
Coverage (600 sq. ft. per gal.)		3,750 sq. ft.

* If pure boiled linseed oil is not available, use pure raw linseed oil and add 1½ pints pure drier.

† When boiled oil is used, reduce drier to ½ pint.

Formula No. 28—Third Coat, Flat Finish (Stucco, Concrete, Brick, Stone)

Materials	Soft Paste	Heavy Paste
White-lead	100 lb.	100 lb.
Flatting Oil (or turpentine)	1¾ gal.	2 gal.
Gallons of Paint		5 gal.
Coverage (600 sq. ft. per gal.)		3,000 sq. ft.

Semi-Flat Finish. — An excellent semi-flat finish on brick, stone, concrete and stucco can be secured by applying over the second coat one or two coats of paint made according to Formula No. 22, substituting spar varnish for the floor varnish listed in the formula.

For brick-red finish on outside brick, thin the color with flatting oil.

Painting Concrete Floors.—The foregoing priming coat—Formula No. 25—may be used in priming concrete floors, substituting floor varnish for the spar varnish listed. The second and third coats must be made to produce a harder finish than is necessary in the case of concrete walls, as floors are subjected to much more severe usage. The following formulas will produce the hard finish needed:

Formula No. 29—Second Coat (Concrete Floors)

Materials	Soft Paste	Heavy Paste
White-lead	100 lb.	100 lb.
Pure Linseed Oil	½ gal.	½ gal.

Pure Turpentine	2¼ gal.	2½ gal.
Pure Drier	1 pt.	1 pt.
Gallons of Paint		6 gal.
Coverage (400 sq. ft. per gal.)		2,400 sq. ft.

Formula No. 30—Third Coat (Concrete Floors)

Materials	Amounts
Heavy Paste White-lead	100 lb.
Pure Turpentine	1½ gal.
Floor Varnish	4 gal.
Gallons of Paint	8¼ gal.
Coverage (600 sq. ft. per gal.)	5,000 sq. ft.

When the third coat is dry the floor should be finished by applying a coat of wax or a high-grade floor varnish. The third coat should be tinted with a little lampblack to match the natural color of concrete.

After the priming coat is dry all cracks and other defects in the floor should be filled with a good putty. The putty should be firmly pressed into the cracks and smoothed over with a putty knife.

Two things to keep in mind throughout the work are: first, vigorous brushing to spread out each coat to the utmost; second, allowing each coat at least four days to dry. One cause of stickiness on floors is flowing the paint on so thick that it does not dry thoroughly underneath, and then hurrying too much with the other coats.

Metal Painting

Preparing the Surface.—To obtain the best results with red-lead, care should be exercised in applying as well as mixing the paint. A vital point is to clean off all loose rust, dirt and other foreign material before commencing to paint. Wire brushes and scrapers will be found to be effective in removing rust and scale. The sand blast will give good results and is strongly recommended, but thorough scraping and brushing will usually be satisfactory. Rust, the great enemy of iron and steel, is an accelerator of further rusting when it is loose enough to retain moisture. If rust is allowed to remain it will work disaster even after the paint has been applied. Besides, rust and dirt are likely to cause peeling.

Number of Coats.—Three coats of paint are necessary on all outside work. Two coats will do for metal indoors. In no case will one coat of paint completely cover bare metal. To the naked eye, the metal may appear to be covered but under the microscope it is another story. Many small pinholes and air bubbles will be found. Even a second coat will not absolutely cover all these pinholes. A third coat is really necessary. Of course, the more the paint is brushed out, the more the pinholes and air bubbles are worked out. Plenty of good brushing effort is essential to a first-class job.

Mixing the Paint.—Paint is made with paste red-lead exactly as white-lead paint is made with heavy paste white-lead, by simply adding linseed oil a little at a time and stirring constantly with a wooden paddle. Dry red-lead is mixed with oil in the same manner, the only difference being that it is less easy to incorporate with the oil.

If the paint is to be tinted, "break up" or soften the red-lead first with just enough linseed oil to make a workable paste; then add the coloring material and finally the remainder of the oil. When drier is used, put it in after the coloring material and before adding the final oil.

Applying the Paint.—Steel and iron should never be painted during wet weather nor when covered with dew or frost. Early morning painting during the late summer months is not recommended as a usual thing. It is always better to wait until the sun has had time to dry everything out. It is bad practice to attempt painting in freezing weather.

Red-lead paint can best be applied with a round or oval brush. Be sure to use plenty of paint, covering the surface well and not attempting to make a gallon of paint go too far. Pay particular attention to bolts, rivet heads, edges and corners, as they are more subject to destructive influences than perfectly flat surfaces.

The priming coat is the most important. Extra care and precaution should be taken during its application.

Allow plenty of time between coats for the previous coat to dry thoroughly. A week is not too long, especially for the priming coat.

Formula No. 31—Priming Coat (Exterior and Interior Metal)

Materials	Paste Red-lead	Dry Red-lead
Red-lead	100 lb.	100 lb.
Pure Linseed Oil (See Note Below)	2⅜ gal.	3⅝ gal.
Pure Turpentine	1 pt.	1 pt.

Pure Drier	1	pt.	1	pt.
Gallons of Paint	4⅞	gal.	5¼	gal.
Coverage (800 sq. ft. per gal.)	3,900 sq. ft.		4,200 sq. ft.	

Formula No. 32—Second Coat (Light Brown)

(Exterior and Interior Metal)

Materials	Paste Red-lead		Dry Red-lead	
Red-lead	100	lb.	100	lb.
Pure Linseed Oil (See Note Below)	2½	gal.	3¾	gal.
Pure Lampblack-in-oil	12	oz.	13	oz.
Pure Turpentine	1	pt.	1	pt.
Pure Drier	1	pt.	1	pt.
Gallons of Paint	5	gal.	5½	gal.
Coverage (800 sq. ft. per gal.)	4,000 sq. ft.		4,400 sq. ft.	

Note.—If genuine boiled linseed oil is available, we advise the use of one-third boiled oil to two-thirds raw oil. In this case, omit the drier.

The lampblack is added to the red-lead for the second coat to change the color of the paint to a light brown, which enables the painter to see readily if any places have not been covered properly. Moreover, a slightly shaded second coat facilitates the inspection of the final coat in the same way.

Formula No. 33—Third Coat (Dark Brown)

(Exterior and Interior Metal)

Materials	Paste Red-lead		Dry Red-lead	
Red-lead	100	lb.	100	lb.
Pure Linseed Oil	3⅝	gal.	5	gal.
Pure lampblack-in-Oil	6	lb.	6½	lb.
Pure Turpentine	1	pt.	1	pt.
Pure Drier	1	pt.	1	pt.
Gallons of Paint	6¾	gal.	7⅜	gal.
Coverage (800 sq. ft. per gal.)	5,400 sq. ft.		5,900 sq. ft.	

Dark Finishes.—Where a dark color is desired other than the browns secured by shading red-lead with lampblack, decorative finishes such as greens and black, are obtainable by simply adding tinting materials to red-lead.

Formulas for tinting paste red-lead light and dark green and black follow:

Formula No. 34—Third Coat (Light Green)

(Exterior and Interior Metal)

Materials	Amounts
Paste Red-lead	100 lb.
Pure Linseed Oil	5½ gal.
Medium Chrome Yellow-in-oil	30 lb.
Chinese Blue-in-oil	12 lb.
Pure Turpentine	1 pt.
Pure Drier	1 pt.
Gallons of Paint	9¾ gal.
Coverage (800 sq. ft. per gal.)	7,800 sq. ft.

Formula No. 35—Third Coat (Dark Green)

(Exterior and Interior Metal)

Materials	Amounts
Paste Red-lead	100 lb.
Pure Linseed Oil	4 gal.
Medium Chrome Yellow-in-oil	12½ lb.
Chinese Blue-in-oil	7½ lb.
Pure Turpentine	1 qt.
Pure Drier	1 qt.
Gallons of Paint	7½ gal.
Coverage (800 sq. ft. per gal.)	6,000 sq. ft.

Formula No. 36—Third Coat (Black)

(Exterior and Interior Metal)

Materials	Amounts
Paste Red-lead	100 lb.
Pure Linseed Oil	14 gal.
Lampblack-in-oil	52 lb.
Chinese Blue-in-oil	16 lb.
Pure Turpentine	½ gal.
Pure Drier	½ gal.
Gallons of Paint	24⅜ gal.
Coverage (800 sq. ft. per gal.)	19,500 sq. ft.

Intermediate shades of green and brown may be secured by varying the amount of coloring matter used. Where the formulas given are altered to any great extent, however, be sure that the amount of linseed oil used is increased or decreased accordingly.

Light Finishes.—In cases where decorative finishes are desired other than the dark ones obtainable by tinting red-lead, use second and third coats of pure white-lead paint tinted to the required color, for either exterior or interior work. Where considerable additional tinting material is required, add linseed oil and turpentine equal to one-half the weight of the tinting material. White-lead and linseed oil are especially adapted for use over red-lead and linseed oil because linseed oil dries much the same with the two pigments, and therefore makes a homogeneous film.

The following white-lead second and final coats will be found to give good results generally, over a priming coat of red-lead:

Formula No. 37—Second Coat
(Exterior Metal)

Materials	Soft Paste		Heavy Paste	
White-lead	100	lb.	100	lb.
Pure Linseed Oil	¾	gal.	1½	gal.
Pure Turpentine	1½	gal.	1½	gal.
Pure Drier	1	pt.	1	pt.
Gallons of Paint	5⅝	gal.	6	gal
Coverage (800 sq. ft. per gal.)	4,500 sq. ft.		4,800 sq. ft.	

Formula No. 38—Third Coat
(Exterior Metal)

Materials	Soft Paste		Heavy Paste	
White-lead	100	lb.	100	lb.
Pure Linseed Oil	2¼	gal.	3	gal.
Pure Turpentine	1	qt.	1	qt.
Pure Drier	* 1	pt.	* 1	pt.
Gallons of Paint	5⅞	gal.	6¼	gal.
Coverage (800 sq. ft. per gal.)	4,700 sq. ft.		5,000 sq. ft.	

* Under poor drying conditions, such as cold or humid weather, the amount of drier should be increased, not to exceed twice the amount called for by the formula.

A very attractive light gray, which will in one coat (if applied fairly heavy) hide the red-lead undercoating, can be obtained with the following formula:

Formula No. 39—Third Coat
(Light Gray)
(Exterior Metal)

Materials	Soft Paste		Heavy Paste	
White-lead	100	lb.	100	lb.
French-ochre-in-oil	8	oz.	8	oz.
Lampblack-in-oil	4	oz.	4	oz.
Pure Raw Linseed Oil	3½	gal.	3½	gal.
Pure Turpentine	—		1	qt.
Pure Drier	1	pt.	1	pt.
Gallons of Paint			6¾	gal.
Coverage (800 sq. ft. per gal.)			5,400 sq. ft.	

Where white or an exceptionally light tint is desired on interior work over a red-lead priming coat two coats of white-lead paint should be used to obscure totally the red-lead undercoat. In such cases, apply Formula 41 for the second coat, adding about one ounce of lampblack if the final coat is to be white or an exceptionally light tint. The practice of adding lampblack should be followed also on exterior work. For the final coat, use Formula 42 or Formula 43, according to finish desired.

Painting Metal Ceilings.—Painting metal ceiling with red-lead or white-lead paint will practically eliminate the most common trouble experienced with interior sheet-metal work of this type, the formation of rust spots.

Where the ceiling is to be finished in white or a very light tint, it is recommended that all the coats, including the priming coat, be of white-lead.

For priming, use the following:

Formula No. 40—Priming Coat
(Interior Metal)

Materials	Amounts
Heavy Paste White-lead	100 lb.
Pure Linseed Oil	2 gal.
Pure Turpentine	1 gal.
Pure Drier	1 pt.
Gallons of Paint	6 gal.
Coverage (800 sq. ft. per gal.)	4,800 sq. ft.

The second coat should be mixed as follows:

Formula No. 41—Second Coat
(Interior Metal)

(a) Materials	Amounts
Heavy Paste White-lead	100 lb.
Flatting Oil	2 gal.
Gallons of Paint	5 gal.
Coverage (900 sq. ft. per gal.)	4,500 sq. ft.

or

(b) Materials	Amounts
Heavy Paste White-lead	100 lb.
Pure Turpentine	2 gal.
Pure Drier	1 pt.
Gallons of Paint	5 gal.
Coverage (800 sq. ft. per gal.)	4,000 sq. ft.

If a flat finish is desired, the third or final coat should be made as follows:

Formula No. 42—Third Coat,
Flat Finish
(Interior Metal)

(a) Materials	Soft Paste		Heavy Paste	
White-lead	100	lb.	100	lb.
Flatting Oil	1¾	gal.	2	gal.
Gallons of Paint			5 gal.	
Coverage (900 sq. ft. per gal.)			4,500 sq. ft.	

or

(b) Materials	Soft Paste		Heavy Paste	
White-lead	100	lb.	100	lb.
Pure Turpentine	1¾	gal.	2	gal.
Floor Varnish	1	pt.	1	pt.
Pure Drier	½	pt.	½	pt.
Gallons of Paint			5 gal.	
Coverage (800 sq. ft. per gal.)			4,000 sq. ft.	

If an eggshell finish is preferred, use the following for the third coat:

Formula No. 43—Third, Eggshell Gloss Finish

(Interior Metal)

(a) Materials	Soft Paste	Heavy Paste
White-lead	100 lb.	100 lb.
Flatting Oil	¾ gal.	1 gal.
Floor Varnish	1¼ gal.	1¼ gal.
Gallons of Paint		5¼ gal.
Coverage (800 sq. ft. per gal.)		4,200 sq. ft.

or

(b) Materials	Soft Paste	Heavy Paste
White-lead	100 lb.	100 lb.
Pure Turpentine	¾ gal.	1 gal.
Floor Varnish	1¼ gal.	1¼ gal.
Pure Drier	½ pt.	½ pt.
Gallons of Paint		5¼ gal.
Coverage (700 sq. ft. per gal.)		3,675 sq. ft.

(b) Materials	Amounts
Heavy Paste White-lead	100 lb.
Pure Turpentine	1½ gal.
Floor Varnish	¾ gal.
Pure Drier	½ pt.
Gallons of Paint	5 gal.
Coverage (700 sq. ft. per gal.)	3,500 sq. ft.

Painting Galvanized Iron.—No paint can be recommended to stand up satisfactorily on galvanized iron at all times because the coating left by the galvanizing process has a tendency to repel paint. Sometimes the paint takes hold properly right away; other times considerable difficulty is encountered in making the paint adhere.

It has been the experience of practical painters that paint made of pure red-lead and linseed oil gives good results most consistently. The best results are obtained after the galvanized iron has been exposed to the weather at least six months.

Apply three coats of paint mixed according to the following formulas:

Formula No. 44—Priming Coat

(Galvanized Iron)

Materials	Amounts
Paste Red-lead	100 lb.
Pure Raw Linseed Oil	2⅜ gal.
Pure Turpentine	1 pt.
Pure Drier	1 pt.
Gallons of Paint	4⅞ gal.
Coverage (800 sq. ft. per gal.)	3,900 sq. ft.

Formula No. 45—Second Coat

(Galvanized Iron)

Materials	Amounts
Paste Red-lead	100 lb.
Pure Raw Linseed Oil	2½ gal.
Lampblack-in-oil	12 oz.
Pure Turpentine	1 pt.
Pure Drier	1 pt.
Gallons of Paint	5 gal.
Coverage (800 sq. ft. per gal.)	4,000 sq. ft.

Third Coat

(Galvanized Iron)

Mix the third coat similar to the second coat except where a decorative finish is desired other than the slightly shaded red-lead color. In the latter case, substitute one of the tinted red-lead finishing coats.

Painting Radiators.—Pipes and radiators never before painted should first be cleaned thoroughly with wire brushes to remove all traces of rust, dirt and grease. Then apply a priming coat of red-lead paint based on Formula No. 44.

In the case of pipes and radiators that have been painted before and that show some defect such as blistering or peeling, the old finish should be removed and the foregoing priming coat applied. If the old finish shows no defects, the priming coat may be omitted.

In the painting of pipes and radiators the decorative requirements of the room should be considered. The finish may be in aluminum or bronze, or in some light tinted paint which will harmonize with the color scheme of the room.

In the painting of pipes and radiators the decorative requirements of the room should be considered. The finish may be in aluminum or bronze, or in some light tinted paint which will harmonize with the color scheme of the room. The metallic powders, if these are used, should be thinned to suitable painting consistency with a mixture of one part good varnish and two parts flatting oil. This makes an excellent bronzing liquid.

If a light-tinted flat paint is decided upon, apply a second coat, tinted to approximately the color desired in the finishing coat, based on Formula No. 41. Then follow with the finishing coat tinted to the desired color and mixed according to the above formula or, if a semi-gloss finish is desired, according to Formula No. 43. When a full gloss is desired, a good prepared enamel may be employed for the finishing coat.

Ample time should be permitted to elapse between coats so that each may dry and harden thoroughly before the next is applied. If it is possible to permit the steam to pass gradually through the pipes between coats, the drying may be hastened in this way. However, the steam should not be turned on full. If the pipes are submitted to sudden heating, the coating will undoubtedly be affected.

It should also be kept in mind that almost all light tints show a tendency to darken slightly due to heat. This should be taken into consideration when the color is selected.

Boat Painting

The practice in painting boats is regulated largely by one thing—the type of craft. If a boat is a yacht or a launch, the owner aims to keep it always clean and bright. Its appearance is a matter of pride with him. Hence the handsomest job obtainable is none too fine, and coat upon coat of paint is often applied in order to get an unusually fine finish.

A rowboat, on the other hand, is not a show boat. While the possessor of one or a fleet of them wants a job that looks well, only an ordinarily good finish is called for.

When it comes to canoes an altogether different problem is presented. A high-class finish is wanted, but it is not obtained in the same way, because a canoe is usually built of canvas.

For present purposes, therefore, boats have been classified into three groups: Power and Sail Boats; Row Boats; Canvas Canoes. In this order, directions for painting them are taken up.

Power and Sail Boats.—The outside of the hull, deck-house and some parts of the interior are proper subjects for the paint brush. Some of these parts should receive attention at least every year.

Preparing the Surface.—If the wood is new, dust it off carefully and cover all knots and sappy streaks with orange shellac. The shellac can be made by thinning dry orange gum shellac with good quality denatured alcohol, proportioned on the basis of three pounds of shellac to one gallon of alcohol, or the liquid shellac may be purchased as "3 pound cut pure orange shellac." Brush the shellac on thin. If it is put on too thick the paint will alligator, leaving the knots bare.

Painting the Hull.—Prime the new wood with a thin coat of paint mixed as follows:

Formula No. 46—Priming Coat (Boat Exterior)

Materials	Soft Paste		Heavy Paste	
White-lead	100	lb.	100	lb.
Pure Linseed Oil	4	gal.	4	gal.
Pure Turpentine	1¾	gal.	2	gal.
Pure Drier	†1	pt.	†1	pt.

Gallons of Paint	9 gal.
Coverage (700 sq. ft. per gal.)	6,300 sq. ft.

† When boiled oil is used, reduce drier to ½ pint.

After the priming coat has dried thoroughly, fill all cracks, nail-holes, dents and other defects in the surface carefully with putty. The hardest and most serviceable putty is that based on white-lead. It should consist of white-lead, either soft or heavy paste, stiffened to putty consistency with dry whiting.

Use sandpaper to smooth down the rough places. Then apply a second coat of paint, mixed as follows:

Formula No. 47—Second Coat (Boat Exterior)

Materials	Soft Paste		Heavy Paste	
White-lead	100	lb.	100	lb.
Pure Raw Linseed Oil	1¼	gal.	1¼	gal.
Flatting Oil (or Turpentine	1	gal.	1¼	gal.
Pure Drier	1	pt.	1	pt.

Gallons of Paint	5½ gal.
Coverage (800 sq. ft. per gal.)	4,400 sq. ft.

Repeat the second coat as many times as desired. Many boatmen put on five or six coats brushed out very thin. Without question this is the best practice, as a number of thin coats produces much better results than the same thickness of film produced by putting on two or three thick coats.

Finish with a coat of paint mixed as follows:

Formula No. 48—Finishing Coat (Boat Exterior)

Materials	Amounts
Heavy Paste White-lead	100 lb.
Flatting Oil (or Turpentine)	2 gal.
Spar Varnish	½ gal.

Gallons of Paint	5½ gal.
Coverage (800 sq. ft. per gal.)	4,400 sq. ft.

The preceding formula gives a "flat" or glossless finish, which wears much better under exposure to the water than a glossy paint rich in oil.

Painting Deck, Spars and Outside of Cabin.—Use the same formulas for the priming and second coats on the deck, spars and outside of the cabin as for painting the hull. Then apply the following finishing coat. Be sure to allow plenty of time between coats for the preceding coat to become dry, at least forty-eight hours.

Formula No. 49—Gloss Finishing Coat (Boat Exterior)

Materials	Soft Paste	Heavy Paste
White-lead	100 lb.	100 lb.
Pure Raw Linseed Oil	2¼ gal.	3 gal.
Pure Turpentine	1 qt.	1 qt.
Pure Drier	1 pt.	1 pt.
Gallons of Paint	5⅞ gal.	6¼ gal.
Coverage (800 sq. ft. per gal.)	4,700 sq. ft.	5,000 sq. ft.

Painting the Interior.—New woodwork inside of cabins, saloons, etc., should first receive a thin coat of good orange shellac. Sandpaper the shellac when dry. Putty all nail-holes and joints. Then apply a priming coat mixed as follows:

Formula No. 50—Priming Coat (Boat Interior)

(a) Materials	Soft Paste	Heavy Paste
White-lead	100 lb.	100 lb.
Flatting Oil	1¾ gal.	2 gal.
Gallons of Paint		5 gal.
Coverage (900 sq. ft. per gal.)		4,500 sq. ft.

or

(b) Materials	Soft Paste	Heavy Paste
White-lead	100 lb.	100 lb.
Pure Turpentine	1¾ gal.	2 gal.
Floor Varnish	¼ gal.	¼ gal.
Pure Drier	¼ pt.	¼ pt.
Gallons of Paint		5¼ gal.
Coverage (800 sq. ft. per gal.)		4,200 sq. ft.

Follow with a second coat, mixed as follows:

Formula No. 51—Second Coat (Boat Interior)

Materials	Soft Paste	Heavy Paste
White-lead	100 lb.	100 lb.
Flatting Oil (or Turpentine)	2¼ gal.	2½ gal.
Gallons of Paint		5½ gal.
Coverage (900 sq. ft. per gal.)		4,950 sq. ft.

If an eggshell gloss is desired, apply a finishing coat mixed as follows:

Formula No. 52—Finishing Coat, Eggshell Gloss (Boat Interior)

Materials	Soft Paste	Heavy Paste
White-lead	100 lb.	100 lb.
Flatting Oil (or Turpentine)	¾ gal.	1 gal.
Floor Varnish	1¼ gal.	1¼ gal.

(If turpentine is used, add ½ pt. pure drier.)

Gallons of Paint	5¼ gal.
Coverage (800 sq. ft. per gal.)	4,200 sq. ft.

Note.—If an extra fine finish is desired, draw the oil from the white-lead in the case of all three coats.

If a gloss finish is desired, a prepared enamel may be used for the finishing coat, or a gloss finish may be made by thinning 3 pounds of white-lead with sufficient turpentine to make a thick paste and then thoroughly mixing it with 1 gallon of high grade floor varnish.

Tints.—The finishing coats specified for the hull, the deck, the spars and the outside and inside of the cabin make white paint. Where a colored paint is desired, tint the final coat in usual way.

Painting Metal Parts.—Iron and steel hulls, masts or other metal parts of a vessel should be painted with two coats of red-lead, thinned according to the following formula:

Formula No. 53 Metal Work on Boats)

Materials	Amounts
Paste Red-lead	100 lb.
Pure Raw Linseed Oil *	2⅜ gal.
Pure Turpentine	1 pt.
Pure Drier	1 pt.
Gallons of Paint	4⅞ gals.
Coverage (800 sq. ft. per gal.)	3,900 sq. ft.

* If genuine boiled oil is available, use one-third boiled and two-thirds raw oil, omitting the drier.

On ornamental parts, finish with white-lead tinted to suit. Below the waterline, finish with anti-fouling, if desired.

Repainting.—In repainting, use the same formulas given for painting new work, except that the priming or first coat may be omitted. Old coats should be well smoothed down and the surface dry before new coats are applied.

Row Boats.—Do not attempt to paint immediately after taking the boat from the water. Let it dry out thoroughly. No matter how good a paint is it will not stick to a wet surface.

Neither will paint adhere properly to a boat's bottom that is covered with dirt, water plants, marine animals, etc.

Clean off all such accumulation by scraping or scrubbing.

Stop up all leaks before applying any paint. Cracks and seams can be filled up with caulking cotton soaked in paste white-lead, nail-holes with bits of pine, and very small leaks with white-lead putty.

Paint applied over an uneven surface is bound to present a bad appearance. Where the old paint is rough, sandpaper it down smooth and touch up all bare spots before applying the first coat.

After heeding the foregoing directions, apply two coats of paint, inside and outside, mixed according to the following formula:

Formula No. 54

(Row Boats—Exterior and Interior)

Materials	Amounts
Heavy Paste White-lead	25 lb.
Pure Turpentine	½ gal.
Spar Varnish	½ pt.
Pure Drier	1 gill
Gallons of Paint	1¼ gal.
Coverage (800 sq. ft. per gal.)	1,000 sq. ft.

If a colored paint is wanted, tint the last coat. The addition of a very little lampblack or dropblack will produce a gray. A little chinese blue will make a light blue. (For other colors follow tinting directions using only one-quarter of the quantity of ingredients called for, as Formula No. 54 is based on 25 pounds of white-lead instead of 100 pounds.)

The finish produced by two coats of paint mixed according to Formula No. 56 will be "flat" or lustreless. If an eggshell gloss is desired, use Formula No. 54, modified by the use of an additional pint of spar varnish, for the finishing coat.

Canvas Canoes.—When the paint is so badly cracked and broken that the canvas shows through in places, it is best to remove the old coat entirely by means of a paint remover and start anew. After the old paint is off, sandpaper the surface and apply a coat of paint composed of:

Formula No. 55

(Canoes)

Materials	Soft Paste	Heavy Paste
White-lead		
Pure Turpentine	¾ pt.	1 pt.
Spar Varnish	½ pt.	⅓ pt.
Pure Drier	1 gill	1 gill
Gallons of Paint		¼ gal.
Coverage (700 sq. ft. per gal.)		175 sq. ft.

Tint as desired.

The above formula should make enough paint for the first coat on one canoe. Put the paint on thick and work it well into the canvas by careful brushing. When dry, sandpaper the surface and then apply two coats of japan color thinned with spar varnish and just enough turpentine to make the paint brush out smooth. One pint of japan color and one pint of varnish should be sufficient to do the work.

If the old paint on a canoe is in good condition, the white-lead paint need not be applied. Simply sandpaper the old coat down smooth and apply the two coats of japan color and varnish.

To refinish the inside of a canoe, sandpaper the old varnish thoroughly and put on one coat of good spar varnish. One pint of varnish should be sufficient.

Patching.—To mend a hole in a canoe, insert a piece of canvas beneath the torn part, pasting the patch on with a little white-lead and rubbing varnish, and clinching it to the ribs of the canoe with brass or copper tacks. Very small holes can be fixed by plugging them with white-lead stiffened slightly with whiting.

White Enamel Paint, Outdoor

1.

	Parts
Albertol 177 C Extra Pale	100
Linseed Stand Oil Extra Pale	400
Thickened Wood Oil Extra Pale	100
Cobalt (calculated as metal)	0.4
White Spirit	200–300

The albertol is dissolved in the white spirit either in the cold, or at a temperature of 50° C. (112° F.), and the stand oils, driers and the remainder of the white spirit added to this solution. The finished varnish is then ground with zinc white. To obtain a still better white color, it is advantageous, instead of using zinc white alone, to use 75 per cent zinc white and 25 per cent titanium white.

Another very usual procedure is to grind the white pigment with a corresponding quantity of linseed stand oil to form a thick paste. The remainder of the oils, the resin solution, the driers and the diluents are added to this white paste.

2. Decorators' Varnish

	Parts
Albertol 177 C.	100
Linseed Stand Oil	90
Thickened Wood Oil	30

Cobalt (calculated as metal)	0.12
Diluents	125–175

The stand oils are mixed together, and the albertol dissolved therein at a temperature of 150° C. (302° F.). As the temperature falls, the cobalt drier and finally the diluents are added. According to the paleness desired, albertol 177 C extra pale, pale or dark is used.

3. Long Oil Outdoor Varnish

100 parts Albertol 177 C are dissolved at a temperature of 150–160° C. (302–320° F.) in
100 parts Linseed Stand Oil. When solution has taken place, further
165 parts Linseed Stand Oil and
85 parts Thickened Wood Oil are added. The temperature is then again for a short while raised to 100° C. (212° F.). Finally,
0.35 part Cobalt (calculated as metal) is to be added, and then
200–275 parts Diluents.

If the American method is preferred, see example No. 7.

4. Flatting Varnish

100 parts Albertol 201 C are cooked with
70 parts Linseed Stand Oil at 240–260° C. (464–500° F.) until a small test of the batch, thinned out with double the normal proportion of diluents, and cooled down under the tap, remains quite free from cloudiness.
30 parts Thickened Wood Oils are then added and the temperature again raised to 240° C. (464° F.); after again carrying out the dilution test described above.
0.1 part Cobalt (calculated as metal) and
100–150 parts Diluents are added at falling temperature.

5. White Tin-printing Enamel

(May also be used as a white indoor enamel).

100 parts Albertol 201 C Extra Pale
90 parts Linseed Stand Oil Palest
20 parts Thickened Wood Oil Palest
0.075 part Cobalt (calculated as metal)
125–175 Thinner

Proceed as in 4 above.

White Enamel Paint Indoor

Zinc White	80
Titanium White	20
Varnish	120

Grind together thoroughly and thin to brushing consistency.

White Enamel Paint, Tin Printing

Lithopone or Titanium White	100
Varnish	100–140

Thin to viscosity desired.

It is recommended that a stoving temperature of 100° C. (212° F.) be not exceeded.

6. Decorator's Varnish

100 parts Albertol 201 C.
90 parts Linseed Stand Oil
30 parts Thickened Wood Oil
0.12 part Cobalt (calculated as metal)
120–175 parts Diluents

Procedure exactly as in the case of example No. 4.

7. Quick-drying Outdoor Varnish by the American Method. (Also suitable for Boat and Finished Varnish.)

100 parts Albertol 201 C are heated with
250 parts Raw Wood Oil under constant stirring, as rapidly as possible, to a temperature of 275° C. (527° F.), and then removed from the fire. Owing to internal heating, the temperature continues to rise. Therefore
16 parts Lead Resinate are added immediately.

Preparation of the Lead Resinate:

8 parts litharge are dissolved in 100 parts of rosin at 240° C. (464° F.).

To cool the batch,
50 parts Linseed Stand Oil are added when the lead resinate has been taken up. Then
0.09 part Cobalt, and finally
150–300 parts White Spirit are added.

No dilution test is necessary.

Water Paints

A.	Potato Starch	10
	Cold Water	30
	10 Bé Caustic Soda	10

Mix the starch with cold water and add the caustic slowly in a thin stream till a transparent thick liquid is obtained.

B.	90-Mesh Lactic Casein	6
	Water	20
	20 Bé Caustic Soda	10

Soak the casein in the warm water, not over 130° F., and add the caustic whilst stirring.

C.	Medium Congo Copal	20
	Linseed Oil	50
	White Spirit	30
	Manganese (as Resinate)	.1

Linseed Oil Varnish as above	80
Water	150

Grind in the required amount of pigment with the oil varnish and then stir in the water. Run the three solutions together through a Hurrel Homogenizer and the resulting emulsion will be stable for a year. If for export to a hot country, it is advisable to add a litle preservative, e.g., metachlor-paracresol.

Irish moss is sometimes used in order to obtain a high viscosity in paste distempers and so keep the pigment from settling. It is usually dissolved beforehand to form a very thick jelly and then added. One well-known brand of distemper on the market is composed of an anhydrous basis of

Chalk	84.0 %
Blanc Fixe	1.5 %
Zinc Oxide	.25%
Brunswick Green	7.4 %
Dextrine	5.0 %
Irish Moss	1.1 %
Nitrobenzene	.05%

the whole being so adjusted as to contain approximately 90% of water.

Silicate Water Paint

Sod. Silicate	40
Pot. Silicate	25
Asbestine	15
Pigment (High Density)	20

Dilute with sufficient water before use.

A paint similar to this, but containing much less pigment, may be used for coating electric light bulbs, which should first be cleaned with care or trouble will be experienced with adhesion. The following modification works more smoothly and gives a better coating, but is not so durable or waterproof.

$Na_2O.3.3SiO_2$ (S.G.1.4)	20%
Rice Starch	5%
Pigment	20%
Water	55%

Fireproof Paint

Aluminum Powder	1 lb.
Sodium Silicate 22° Bé	1 gal.

Water Paint

Double Boiled Oil, with Driers	50	} 100
Water	45	
Sodium Silicate	5	
Pigment		50

The oil, which may be diluted if required with 120° F. flash white spirit, should be added to the aqueous phase in a slow stream with rapid and vigorous stirring.

The oil may be replaced with, for example, latex, and paints can be made on the following lines:

1.	Sodium Silicate	10
	Ammonia	10
	Water	10
	Zinc Oxide	5
	Sulphur	3
	Zinc Dimethyldithio Carbamate	.5
2.	60% Latex	100
	Whiting	200
	Spindle Oil	60
	Glue	5

The two solutions are made separately as indicated, and mixed. The ratio of silica is not mentioned, but presumably $3.3SiO_2$ is indicated. The more alkaline varieties of sodium silicate cause precipitation of latex by reason of hydrolysis. If, however, ammonia be added to the solution this increases the OH ion concentration and prevents splitting of the silicate, so that the latex is thickened and rendered stable. Aluminium sulphate also thickens latex by precipitation of the protein portion, giving a butter-like product.

1.	Pale Boiled Oil	45 lb.
	Rosin	45 lb.
	White Spirit	25 lb.

Melt the rosin in the oil and dilute while hot with the white spirit. Then grind in the pigment.

2. Casein	120 lb.
Water	600 lb.
Borax	24 lb.
Ammonia	3 lb.
10% Potassium Bichromate	30 lb.
Mirbane	3 lb.

Dissolve the casein by steeping it in the water at 130° F., then add the borax and the ammonia. Allow to cool and add the bichromate solution. By vigorous shaking emulsify the mirbane with twice the amount of the casein solution just prepared, and add the milky product to the balance. Then mix the oil into the casein solution, using a whisk or colloid mill. It should be noted that while 90-mesh casein is usually selected on account of its speedy solubility, it is much better to use 30-mesh casein as this contains fewer grits (from the grinding stones) and its viscosity is more uniform.

Water Soluble Shellac Solution

(1) To 5 parts of sulfonated rape oil add 1 part of sodium hydroxide. Warm in a water bath until the excess water has been evaporated.

(2) Dissolve 3 parts of No. 1 in 36 parts of water.

(3) Add 5 parts of a 20% ammonia solution to the 39 parts of No. 2.

(4) To 44 parts of No. 3 add 25 parts of flaked orange shellac and agitate in a mechanical churn until solution is complete. Under normal conditions this will require about 6 hours.

The resultant heat should dissipate about 22 parts of the water so that the completed mixture will contain approximately 4½ lb. of shellac per gallon of mixture.

Matt Finish Distemper

A typical formula for a matt distemper of this type with good covering power and resistance to water is casein 10 per cent, lime 10 per cent, clay 10 per cent, lime-proof pigment 10 per cent, and chalk 60 per cent. The purpose of the clay is to keep the other pigments in suspension and to aid in the brushing of the paint.

Oil-bound Distemper

(1) casein 30 kg., water 150 litres, borax 3.5 kg., phenol 1.0 kg.; (2) formalin 2.0 litres, water 5.0 litres; (3) rosin 15 kg., boiled oil 15 kg., white spirit 10 litres. The casein is soaked in the warm water, the borax added first and then the phenol. This solution is allowed to stand for 24 hours, and the ingredients in the second list are then added and after mixing hot those in the third. It is well known that pigments grind better in oil than in water and it is a great advantage to grind the pigment into the oil medium in the third list before emulsifying it with the casein solution. The proportion of pigment usually incorporated is about six to eight times the total weight of the medium.

Water Paint

Trihydroxyethylamine Linoleate	0.6
Glue	10
Water	32
Varnish	16
Naphtha	4
Sodium Ortho Phenyl Phenate	0.1

Paint, Oil Emulsion

Trihydroxyethylamine Linoleate	0.3
Glue	5
Water	16
Linseed Oil Varnish	8
Phenol	0.2

Procedure for the above oil emulsion paints is to dissolve the water soluble materials and heat together with stirring until free from lumps. The oil, varnish or other water insoluble material is run in slowly while stirring vigorously with a high-speed mixer. Best results are obtained by not too long mixing and occasional rest periods.

Railroad Water Tank Paint

Protecting the interiors of steel water tanks from rust and corrosion is often a troublesome problem because the paint or other protective material is nearly always under water, and frequent repairs or repainting mean putting a tank temporarily out of service. Therefore the method successfully used by the Union Pacific System should be of interest to all with similar problems.

This 10,000-mile rail system has 260 steel water storage tanks at 230 stations. They vary in capacity from 6,000 to 1,000,000 gal. and run up to 100 ft. in diameter at main terminals where maximum daily consumption is 1,400,000 gal. The total storage capacity is 31,300,000 gal. and represents an investment of several million dolars. Probably no railway system encounters a greater variety of climatic and water conditions than the Union Pacific. Its painting jobs are therefore put to severe test and the problem of protecting the large investment is of great importance.

Steel tanks are given a shop coat of ready-mixed red lead paint inside and out. After erection the exterior is given a brown and a black coat, both being mixtures of red lead and lampblack, with lampblack increased in the black coat.

Interiors receive three coats in addition to the shop coat. The first field coat is brown and is made by adding 10 oz. of lampblack paste, 6 fluid oz. of japan drier, and 2 lb. of finely powdered litharge to 1 gal. of ready-mixed red lead paint. A second field coat, light brown, has the same composition as the first with the exception of the lampblack paste, 5 oz. of which are used instead of 10. The third field coat, red, is the same, with all lampblack omitted.

The litharge passes a No. 325 sieve with total residue on the sieve not exceeding 1 per cent by weight. The ready-mixed red lead paint pigment contains 88 per cent of red lead by weight, which must run not less than 94 per cent true red lead. The lampblack paste is 25 per cent pure lampblack by weight, balance pure linseed oil. Addition of the litharge gives an extraordinarily hard paint film that does not become unduly soft by continued soaking. The ready-mixed red lead paint contains 6 per cent by weight of pigment.

Tanks are inspected annually and painted at intervals of from four to ten years, depending upon local conditions. To avoid interruption of water service, a set of three 8,000-gal. steel tanks with demountable steel trestle support is conveyed on flat cars to the vicinity of the paint job as a temporary storage plant. The permanent tank is drained and the steel cleaned, sometimes by sandblasting, but more generally by scraping and wire brushes. Brush painting is usually used. The paints described cover about 00 sq. ft. per gal. with the brush method.

Through experience it has been learned to watch closely the following vital items:

1. The steel work must be dry and temperature conditions favorable when paint is applied.
2. The paint must be thoroughly mixed at the start and frequently stirred.
3. A rather high proportion of pigment is desirable especially on interior surfaces.
4. Each coat must be brushed out to a thin film.
5. Litharge is to be used in each field coat for interior surfaces.
6. Proper intervals of time must be allowed for the drying of each coat.

Outside Wood Paint

	Priming Coat		*Second Coat*	
Soft Paste White-lead	100	lb.	100	lb.
Pure Linseed Oil	2	gal.	3	gal.
Pure Turpentine	1¾	gal.	—	
Pure Drier *	1	pt.	1	pt.
Gallons of Paint	7	gal.	6½	gal.
Coverage, one coat	5,600	sq. ft.	5,000	sq. ft.

* When boiled oil is used, reduce drier to ½ pt.

Structural Paint

	First or Intermediate Coats		*Top or Finishing Coats*	
Blue Lead in Oil, Paste	100	lb.	100	lb.
Raw Linseed Oil	2⅜	gal.	3	gal.
Turpentine or Paint and Varnish Manufacturer's 48° to 50° naphtha	1¾	gal.	2	qt.
Drier (rosin free)	1	qt.	1	qt.
Approximate Paint Produced	7¼	gal.	6⅔	gal.
Weight per Gallon, Approximately	17.8	lb.	18½ to 19½	lb.

Paint for Interior Plaster

Priming Coat

Soft Paste White Lead	100 lb.
Pure Boiled Linseed Oil	3 gal.
Floor Varnish	2 gal.
Pure Turpentine	1¼ gal.
Gallons of Paint	9½ gal.
Coverage (600 sq. ft. per gal.)	5,700 sq. ft.

Second Coat

Soft Paste White Lead	100 lb.
Pure Turpentine	1¼ gal.
Floor Varnish	¾ gal.
Pure Drier	½ pt.
Gallons of Paint	5¼ gal.
Coverage (700 sq. ft. per gal.)	3,675 sq. ft.

Third Coat—Flat Finish

Soft Paste White Lead	100 lb.
Pure Turpentine	1¾ gal.
Floor Varnish	1 pt.
Pure Drier	½ pt.
Gallons of Paint	5 gal.
Coverage (800 sq. ft. per gal.)	4,000 sq. ft.

Third Coat—Eggshell Finish

Soft Paste White Lead	100 lb.
Pure Turpentine	¾ gal.
Floor Varnish	1¼ gal.
Pure Drier	½ pt.
Gallons of Paint	5¼ gal.
Coverage (700 sq. ft. per gal.)	3,675 sq. ft.

Black Walnut Stain

Gilsonite	2 lb.
Turpentine	2 lb.

Ebony Stain

Nigrosine (water soluble)	16 lb.
Oxalic Acid	7 lb.
Water	640 lb.

Clear Shingle Stain

Creosote Oil	1 gal.
Kerosene	1 gal.
	2 gal.

Colored Shingle Stain (Red)

Red Oxide	45 lb.
Asbestine	15 lb.
Linseed Oil	3 gal.
Grind and add	
Creosote Oil	12 gal.
Kerosene	12 gal.
	29½ gal.

Similarly other colored shingle stains can be made by changing the colored pigments.

* Mahogany Stain

The method of producing a fadeless mahogany stain, which consists in mixing with the steam extracted water insoluble extract of quebracho wood sufficient hot concentrated alkali solution to produce a pH value of about 11 to 12, and digesting with sufficient added hot water to produce a pH value between 7.0 and 8.5 in the final product.

Traffic or Road Marking Paint

I. Cold Cut Method for Traffic Paint:

Cumar V	100 pounds
Kettle Bodied Linseed Oil	4 gallons
Xylol	3 gallons
V. M. and P. Naphtha	18 gallons

Cobalt Linoleate Solution—or Naphthenate Cobalt Drier No. 42—Equivalent to 1¼ pounds—.009 lb. Cobalt Metal.

Procedure: Cut the Cumar by agitating in a power mixer or tumbling barrel with 3 gallons of Xylol and 18 gallon of V. M. and P. Naphtha. This ma require 2–4 hours. When solution ha been completed add the linseed oil an the Cobalt Drier.

The following grinds are suggeste Pigments may be added according t specific requirements. These grinds ca be made conveniently in a pebble mil

	No. 1 Grind		No. 2 Grind	
Above Vehicle	40%	by weight	40%	by weig
Titanium Pigment	42%		33.3%	
Asbestine	18%		14.7%	
Diatomaceous Earth			12.0%	

Following the grind the batch i thinned 50% by weight with a mixtu of 85% V. M. and P. Naphtha and 15 Xylol. Grind No. 2 dries at a fast rate.

II. Varnish Type:

Varnish A

Cumar V 1	100 lb.
China Wood Oil	33 gal.
Glycerine	18 lb.
Litharge	2½ lb.
Cobalt Acetate	½ lb.
Mineral Spirits	60 gal.

Cooking Method: Carry China Woo Oil and Glycerine to 400° F. Add 3 pounds Cumar meanwhile running hea rapidly to 560° F. Withdraw from fi and hold for first string from stirre Chill with remaining Cumar. Body (i necessary) by holding around 500–480 F. until a sample cooled on tin gives good string. Cook in Cobalt Acetat Cool to 450° F. or below and thin.

Cumar Cut B:

Cumar	100 lb.
Xylol	3⅓ gal.
V. M. and P. Naphtha	13⅓ gal.

This is a cut of 6 pounds of Cuma to the gallon of thinner.

The solution is made by agitatin Cumar and the thinners in mechanic mixer or tumbling barrel for 2–3 hour The following grinds are suitable:

	Grind 3		Grind 4	
Lithopone	840	parts by weight		part by weig
Titanium Lithopone	...		840	
Asbestine	360		360	
Cumar Cut B	300		300	
Cobalt Linoleate Paste Drier (5% Cobalt)	10		10	
Varnish A	500		500	

These grinds are made in a pebb mill and are further thinned with 1 parts by weight of a mixture of 80 V. M. and P. and 20% Xylol.

Varnish Type II road paint dries at slower rate than the cold cut type, b as a better covering power. Grind type 4 is suggested for application over asphalt.

Vehicle for Ready Mixed Aluminum Paint:

Some manufacturers find it desirable to offer aluminum paints with the aluminum powder already mixed with the vehicle. This practice is not generally advocated but it may be said that a fair measure of success has been realized with some vehicles in which aluminum powder has been mixed and which has undergone limited storage.

Cumar V 1	100 lb.
China Wood Oil	5 gal.
*Kettle Bodied Linseed Oil	5 gal.
Xylol	15 gal.
V. M. and P. Naphtha	15 gal.
Cobalt Resinate (3½% Metal)	1 lb.

* Linseed Oil Bodied 3 hours at 575° F.

Cooking Method: China Wood Oil and Linseed Oil are carried to 400° F. at which point 50 pounds Cumar are added. The temperature is carried to 450° F. and is held until a good body is attained. This is determined by testing samples cooled on tin until a stiff button is obtained.

In experiments this vehicle has been mixed with two pounds aluminum flake per gallon and has, in our observation, given good flaking results upon standing several months.

Vehicle for Aluminum Paint for Exposure to High Temperatures:

The formula given below is suggested for an aluminum liquid which is to be exposed to high temperatures. In many cases, since temperatures and other conditions vary, the varnish maker will have to vary his formulations to meet special conditions:

Cumar W	100 lb.
China Wood Oil	2½ gal.
Light Cold Pressed Menhaden Oil	2½ gal.
Xylol	5 gal.
V. M. and P. Naphtha	20 gal.
Cobalt Resinate (3½%)	0.8 lb.

Cooking Method: Carry China Wood Oil to 400° F. Add 50 pounds Cumar and bring heat up rapidly to 565° F. Hold for a short time and then check with the fish oil and Cumar. Hold at 500° F. (re-heating if necessary) for about 10 minutes. Then add drier and thin.

Vehicles for Aluminum and Bronzing Liquids

The following formulae are types of vehicles which experiments have indicated as being suitable for use for aluminum coatings.

In most cases it is desirable to add 1¾ to 2 lb. aged aluminum flake to each gallon of liquid.

Vehicle for Outside Aluminum Paints:

A. Spar Type

34	gal.	China Wood Oil
10	gal.	Kettle Bodied Linseed Oil *
25	lb.	Rosin
7	lb.	Litharge
75	lb.	Cumar V
6½	oz.	Cobalt Acetate
65	gal.	Mineral Spirits

* Linseed Oil is bodied at 575° F. for 3 hours.

Cooking Method: Heat China Wood and Rosin with a fast fire to 400° F. While still on the fire add 25 pounds Cumar. When the temperature of 475° is reached begin adding litharge while stirring rapidly. The Litharge may be dusted in or mixed with China Wood Oil to a fluid consistency prior to addition. The heat is checked only slightly during the Litharge addition and during this operation it is necessary to whip down the foam. The fire is then raised to bring the temperature rapidly to 575° F. This point should be reached within 25 to 30 minutes of the start. The kettle is withdrawn from the fire at this point and held until the temperature gains 585–595° F. This requires only a minute or two. The heat is checked with the linseed oil followed by the Cumar. Stir rapidly and the temperature drops below 500° F. Hold between 500–480° F. until a sample cooled on tin gives a moderate body.

Add Cobalt Acetate at 480° F. Cook until acetate fumes cease, cool and thin.

The addition of 15–25% Xylol increases flaking effect.

The addition of 15–25% of Coke Oven distillate (Xylol or Hi Flash Naphtha) increases the flaking effect of the liquid.

B. Cumar—Phenolic Resin Type—

The following varnish involves the use of Cumar with the oil reactive phenolic resins. The usual low cooking temperatures may be used:

China Wood Oil	25 gal.
Oil Reactive Phenolic Resin	20 lb.

Fused Lead Resinate 5 lb.
Cumar
V 3 M.P. 260–270° F.
or
W 1 M.P. 300–320° F. 75 lb.
Mineral Spirits 45 gal.
Liquid Drier *

Cooking Method: Run China Wood Oil, Phenolic Resin and Lead Resinate to 400° F. Add one-half of the Cumar and carry to 480° F.–500° F. Hold for body at this temperature. Chill with the remaining Cumar. Cool, thin and add liquid drier.

* It is recommended that enough liquid driers be added to give a concentration of .02% to .03% Cobalt Metal on the weight of the oil. Cobalt Linoleate or Naphthenate Drier solution may be used.

C. Cold Cut Type

Where the user wishes to prepare a cold cut aluminum vehicle for outside use, he may use the following formulation as a guide in his work. It must be realized that such a formula as given below will not be as durable as the spar types, but will give suitable service in many cases.

100 lb. Cumar V 2
20 gal. V. M. and P. Naphtha
5 gal. Xylol
10 gal. Kettle Bodied Linseed Oil
Liquid Cobalt Drier (Equivalent to 0.03% Cobalt on weight of oil)

Procedure: Cut Cumar by agitating with the V. M. and P. Naphtha and Xylol for several hours. When completely dissolved add the Linseed Oil and Cobalt Drier.

Vehicle for Interior Aluminum Paints:

50 lb. Cumar V 2
2 gal. Xylol
10½ gal. V. M. and P. Naphtha
2 gal. Kettle Bodied Linseed Oil
1 pt. Japan Drier

Wall Sealers

The following formula may be used by the paint and varnish manufacturer in developing a good wall sealer.

Varnish II

China Wood Oil	30 gal.
Kettle Bodied Linseed or Perilla Oil *	3 gal.
Cumar W 1	88 lb.
N Rosin	12 lb.
Powdered Basic Lead Carbonate (White Lead)	3¼ lb.
Mineral Spirits	50 gal.

* Kettle Bodied at 575° F. for 3 hours.

Procedure: Heat China Wood Oil and Rosin in kettle quickly to 400° F. Add about 25 pounds of Cumar and run rapidly to 565–570° F. Withdraw from the fire. Hold until the temperature reaches 580° F. Chill the batch with the Linseed (or Perilla) Oil and 45 pounds of Cumar. Stir as the temperature drops to about 525–520° F. Add the white lead and stir until taken up. Add the remaining Cumar and cool to about 490° F. Hold between 490° F. and 475° F. for approximately 30 minutes or until a sample cooled on tin gives a one inch string or more. Cool below 450° F. and thin.

The following grind is suggested.

Paste No. 1

Titanox C	1000 lb.
Varnish II	388 lb.
Total	1388 lb.

Grind on stone mill.

Reduction of Paste No. 1

Paste No. 1	1388 lb.
Varnish II	253 lb.
Mineral Spirits	229 lb.

Liquid Cobalt Drier: Add equivalent of 0.03% cobalt metal on the weight of the oil.

If it is desirable to make a less expensive pigment combination it is possible to replace 25% of the Titanox in the above grind with inerts. A combination of 10% Asbestine and 15% Whiting can be used for this purpose.

* Glazing Composition

Whiting	15
Asbestine	15
Asbestos Fiber	5
Aluminum Powder	9
Linseed Oil Boiled	30
Naphtha	26

Candy Glaze

Shellac (arsenic free)	4 lb.
Alcohol	6.5 lb.
Isopropyl Acetate	2.4 lb.

Candy Glaze

Copal Bold Chips	6 lb.
Isopropyl Alcohol (98–99%)	12 lb.
Isopropyl Acetate	2 lb.

* Acid Resistant Paint

Asbestos Fibre	28
Aluminum Silicate	44
Barium Sulfate	28
Stearin Pitch	10
Petroleum Asphalt	15
Mineral Asphalt	10
Naphtha	200

Antifouling Paint

a. Rosin	2 lb.
Lithopone	1 lb.
Naphtha	160 lb.

b. Chrome Green	1 lb.
Lithopone	2 lb.
Rosin	3 lb.
Naphtha	160 lb.

c. Chrome Green	21 lb.
Rosin	12 lb.
Naphtha	160 lb.

First apply a coat of (a) and when dry apply a coat of (b). When this has dried apply (c).

---

* Paint, Automobile Top

Carbon Black	16
Calcium Phosfate	77
Calcium Carbonate	3
Sodium Silicate	2
Water	2
Rosin	103
China Wood Oil	223
Naphtha	359

Auto Top Dressing

Mix a solution of benzol and asphalum to the consistency of milk and to each pint of the resulting mixture add about two or three tablespoons of linseed oil. The linseed oil is added to make the dressing more flexible.

Blackboard Paint

Carbon Black	15 lb.
Shellac	14 lb.
Prussian Blue	1 lb.
Lithopone	1 lb.
Powdered Carborundum	7 lb.
Drier Liquid	16 lb.
Alcohol	130 lb.
Linseed Oil Boiled	7½ lb.

Bridge Paint

Undercoats

25 lb. Dry Red Lead
½ gal. Raw Linseed Oil
½ gal. Boiled Linseed Oil
1 gill Petroleum Spirits

Finish Coats

100 lb. Commercial Hard Paste White Lead Carbonate
2 gal. Raw Linseed Oil
2 gal. Boiled Linseed Oil
2–2⅓ oz. Chinese Blue in Oil
19 oz. Burnt Umber in Oil

These quantities make about 7 gal. of paint.

---

* Paint, Cement

Hydrated Lime	43
Hydraulic Cement	19.5
Talc	12.0
Metronite	11.5
Salt	6.5
Mica	5.0
Gum Arabic	1.6
Gum Karaya	0.5
Irish Moss	0.1
Calcium Stearate	0.3

This is used as a cold water exterior paint.

Cement Water Paint

50	lb. White Portland Cement
5	lb. Gypsum
4½	lb. Calcium Chloride
½	lb. Hydrated Lime
60	lb.

Mix intimately in pebble mill. Stir about 7 to 8 lb. of the above into 1 gal. of water and paint over wet surface. When paint sets up, wet down with ordinary tap water.

Cold Water Paint, Outside

Whiting	55 lb.
Clay	15 lb.
Dextrine	2 lb.
Casein	12 lb.
Lime	15 lb.
Trisodium Phosfate	1 lb.
Corrosive Sublimate	1 oz.

Ten pounds of the above are used with 1 gal. of water.

Enamel Paint Remover

Benzene (90° Bé)	50
Alcohol	25
Acetone	10
Nitric Acid	10
Sulfonated Oil	5
Beeswax	1

---

* Enamel Paint (Outdoor)

White Lead	50–75
Zinc Oxide	25–50
Barium Fluoride	5–10

China Wood Oil	10–15
Linseed Oil	5–10
Turpentine	10–20
Manila Copal	5–10
Alcohol	50–70
Ethyl Acetate	30–50

Flexible Paint for Marking or Stencil Work

Adheres well to rubber goods. Can be hot pressed into fabrics.

Gutta Percha	60
Colored Pigment	40

The colored pigment is milled into the Gutta Percha on a roll mill. Pigments such as vermilion, cadmium sulphides, ultramarine, etc., may be used. Organic color lakes are also satisfactory. On account of the smaller quantity of lake needed, the difference should be made up with blanc-fixe.

The mixed compound is dissolved in solvent naphtha with slight warming. A 20% solution gives good coverage and may be sprayed easily.

Freight Car Paint

Iron oxide paste, containing 25% linseed oil, 100 lb.; rezyl 110, 42 lb. and xylol 18 lb.; liquid drier, 8 lb.; naphtha or mineral spirits, 59 lb.; total, 227 lb. or 23¼ gal. The liquid drier should contain 1 lb. of lead linoleate and ½ lb. of manganese linoleate dissolved in turpentine or coal-tar naphtha. A still more rapid-drying and enduring paint can be made by grinding the pigment in a solution of rezyl 110 instead of using an oil paste.

Galvanized Iron, Treatment before Painting

Some people, before painting it, wash the galvanized metal with vinegar. This is said to be good. Others scrub it well with burlap wet with benzine. Scrubbing the surface with soap and sand can be recommended. The best method seems to be, however, to leave the galvanized metal exposed to the weather for a few months.

Still others report good results from washing the well-cleaned surface with a one per cent solution of copper chloride, acetate or sulphate. The solution is left on for a time and then brushed off before painting is attempted. A few months of exposure is probably better, however, even than this treatment.

Light sand-blasting is also said to have been used for cleaning galvanized iron and putting it in condition to take paint. No doubt this would accomplish the purpose.

Even in the case of perfectly clean zinc, it is not easy to get paint to stick always. No paint yet invented adheres to it as well as in the case of iron or wood. What chemists call "the surface tension" is different. Not that any good paint invariably all comes off. Generally most of it stays on but that is not very satisfactory.

If galvanized iron is weathered and then well cleaned, there is seldom any trouble encountered when the paint is red-lead. Probably most of the difficulties in painting galvanized surfaces are traceable to improper preparation done by not too expensive labor. This is why weathering, which does not skip anything, is best.

Paint Grinding

A small percentage of Oleic Acid materially helps the grinding of Carbon Black.

Heat Resisting Paint

Powdered Graphite	1	lb.
Lampblack	1	lb.
Black Oxide of Manganese	0.33	lb.
Japan Gold Size	0.33	pt.
Turpentine	0.50	pt.
Boiled Linseed Oil	0.33	pt.

Mix together until a uniform consistency is obtained.

High Light Reflecting Paint

The following formulae are suggested for obtaining proper illumination in interiors and providing desirable paints that can be washed repeatedly:

100 lb. Pure White Lead (heavy paste)
2 gal. Flatting Oil

or

100 lb. Pure White Lead (heavy paste)
2 gal. Pure Turpentine
1 pt. Floor Varnish
½ pt. Pure Drier

They may be tinted as follows (Quantities are per 100 lb. white lead):

Ivory White	— 3 oz. French Ochre
Cream	— ¾ lb. French Ochre
Light Buff	— 3 lb. French Ochre

Priming Coat

100 lb. Pure White Lead Soft Paste
2¼ gal. Pure Boiled Linseed Oil
2 gal. Spar Varnish
1½ gal. Pure Turpentine

Makes about 9 gal.

Second Coat

100 lb. Pure White Lead Soft Paste
2¼ gal. Pure Raw Linseed Oil
1 gal. Spar Varnish
1 gal. Pure Turpentine
1½ pt. Pure Drier

Makes about 7¾ gal.

Third Coat

100 lb. Pure White Lead Soft Paste
2 gal. Pure Turpentine
½ pt. Pure Drier

Makes about 5¼ gal.

---

* Paint, Hydrocarbon Resistant

Minium	10
Litharge	2
Glycerol (30° Bé.)	3
Sod. Silicate (36° Bé.)	9

This paint resists water, oils, cold and heat.

---

* Paint, Iron Protective

Zinc Chromate	12.5
Basic Lead Chromate	12.5
Sublimed Blue Lead	25
Magnesium Silicate	50
Linseed Oil	60
China Wood Oil	20
Turpentine	10
Drier	5
Petroleum Naphtha	5

---

* Latex Paints

Latex (50% solids)	50%
Kieselguhr	16½%
Lithopone	40%
Lime	2%
Zinc Oxide	8%
Sulphur	3%
Barytes	5%
Soap	½%

The whole of the fillers are ground wet with 40 parts of water to form a thick cream, and then added to the latex. The film may be vulcanized after application.

Paints made on the following formula do not coagulate, "ball-up" or pull off.

Casein Solution:

Casein	80 oz.
Borax	12 oz.
Water	480 oz.

Add to this a mixture of .880 ammonia 90 cc. and saturated phenol solution 10 cc.

Pigment Paste:

Casein Solution	7 pt.
Water	9 pt.
Lithopone	50 lb.

Paint:

Preserved Latex	16 pt.
Casein Solution	14 pt.
Pigment Paste	32 pt.

Heat Sensitive Paints

The Double Iodide of Silver and Mercury

Silver Iodide	5 parts
Mercuric Iodide	1 part

This compound mixed with shellac and painted on thin strips of steel change from a very bright yellow to a deep red as the temperature increases.

Luminous Paints

	Parts by Weight
White	
Luminous Calcium Sulphide	20
Zinc Oxide	10
Barium Sulphate	10
Varnish	30
Yellow	
Luminous Calcium Sulphide	20
Barium Sulphate	5
Barium Chromate	4
Varnish	25
Yellow	
Luminous Calcium Sulphide	20
Barium Sulphate	5
Orpiment	4
Varnish	25
Red	
Luminous Calcium Sulphide	20
Barium Sulphate	5
Realgar	4
Varnish	25
Green	
Luminous Calcium Sulphide	20
Barium Sulphate	5
Ultramarine Blue (French)	3
Cobalt Blue	3
Varnish	28–30

Violet

Luminous Calcium Sulphide	20
Barium Sulphate	5
Violet Lake	2
Varnish	25

Luminous Paint

Barium Sulfate	34 lb.
Indian Lake	22 lb.
Madder Lake	23 lb.
Luminous Calcium Sulfide	76 lb.
Varnish	73 lb.

Luminous Paint

The following are two formulas for luminous paint giving a yellow glow:

	I	II
Strontium Carbonate	100	100
Sulphur	100	30
Potassium Chloride	0.5	—
Sodium Carbonate	—	2
Sodium Chloride	0.5	0.5
Manganese Chloride	0.4	0.2

The mixture is heated in a crucible for three-quarters of an hour at about 1,300° C. The more permanent variety of luminous paint used for watch hands consists of zinc sulphide activated with radium bromide.

* Marine Paint

Coal Tar	1 gal.
Sodium Cyanide	5 oz.
Cement	1 lb.

Structural Metal Paints

The Three Principal Paint Formulas

Full Weight Formula

Dry red-lead....	33	lb....	79.23%
Raw linseed oil...	1 gal. =	7.75 lb....	18.61%
Turpentine......	½ pt. =	0.45 lb....	1.08%
Drier...........	½ pt. =	0.45 lb....	1.08%
Total.........	1.58 gal. =	41.65 lb....	100.00%

Weight of one gallon, 26.4 lb.

Equivalent Paste Red-lead Formula

Paste red-lead.	2.232 gal. =	100 lb..	86.54%
Additional oil..	1.851 gal. =	14.34 lb..	12.36%
Turpentine....	0.116 gal. =	0.835 lb..	0.55%
Drier.........	0.116 gal. =	0.835 lb..	0.55%
	4.315 gal. =	116.01 lb..	100.00%

Medium Weight Formula

Dry red-lead....	28	lb....	76.40%
Raw linseed oil..	1 gal. =	7.75 lb....	21.14%
Turpentine.....	½ pt. =	0.45 lb....	1.23%
Drier..........	½ pt. =	0.45 lb....	1.23%
Total.......	1.51 gal. =	36.65 lb...	100.00 %

Weight of one gallon, 24.2 lb.

Equivalent Paste Red-lead Formula

Paste red-lead.	2.232 gal. =	100 lb...	83.09%
Additional oil..	2.352 gal. =	18.23 lb...	15.14%
Turpentine....	0.147 gal. =	1.06 lb...	0.88%
Drier.........	0.147 gal. =	1.06 lb...	0.88%
	4.878 gal. =	120.35 lb...	100.00%

Light Weight Formula

Dry red-lead....	25	lb...	74.30 %
Raw linseed oil..	1 gal. =	7.75 lb...	23.03 %
Turpentine.....	½ pt. =	0.45 lb...	1.335%
Drier..........	½ pt. =	0.45 lb...	1.335%
Total........	1.47 gal. =	33.65 lb...	100.00 %

Weight of one gallon, 22.2 lb.

Equivalent Paste Red-lead Formula

Paste red-lead.	2.232 gal. =	100 lb...	80.8 %
Additional oil..	2.75 gal. =	21.31 lb...	17.2 %
Turpentine....	0.17 gal. =	1.23 lb...	0.88%
Drier.........	0.17 gal. =	1.23 lb...	0.88%
	5.32 gal. =	123.77 lb...	100.00%

STRUCTURAL METAL PAINTS

Tinted Paint Formulas

Light Brown

(28 lb. Pigment to 1 gal. Oil)

Dry Red-lead Formula

Dry red-lead....	28	lb....	75.98%
Paste lampblack.	⅕ lb. =	0.2 lb....	0.54%
Raw linseed oil..	1 gal. =	7.75 lb....	21.04%
Turpentine.....	½ pt. =	0.45 lb....	1.22%
Drier..........	½ pt. =	0.45 lb....	1.22%
	1.53 gal. =	36.85 lb....	100.00%

Weight of one gallon, 24.15 lb.

Paste Red-lead Formula

Paste red-lead.	2.232 gal. =	100 lb..	82.58 %
Paste lampblack	0.082 gal. =	0.75 lb..	0.62 %
Raw linseed oil.	2.352 gal. =	18.23 lb..	15.05 %
Turpentine....	0.147 gal. =	1.06 lb..	0.875%
Drier.........	0.147 gal. =	1.06 lb..	0.875%
	4.96 gal. =	121.1 lb..	100.000%

Weight of one gallon, 24.42 lb.

Note: Any red-lead paint may be tinted light brown by adding two ounces of paste lampblack to each gallon of paint, or three-quarters of a pound of paste lampblack to each 100 pounds of paste red-lead.

Black

100	lb. paste red-lead..............	2.23 gal.
52	lb. paste lampblack.............	5.5 gal.
16	lb. paste Prussian blue.........	1.6 gal.
108.5	lb. raw linseed oil.............	14.0 gal.
3.6	lb. turpentine..................	0.5 gal.
3.6	lb. drier.......................	0.5 gal.
283.7	lb.	24.33 gal.

Weight of one gallon, 11.7 lb.; contains 3.8 lb. dry red-lead.

Dark Brown

100	lb. paste red-lead.............	2.23 gal.
6	lb. paste lampblack............	.67 gal.
28.1	lb. (3 ⅝ gal.) linseed oil........	3.63 gal.
.90	lb. (1 pint) turpentine.........	.12 gal.
.90	lb. (1 pint) drier..............	.12 gal.
135.9	lb.	6.77 gal.

Weight of one gallon, 20 lbs.; contains 13.7 lb. dry red-lead.

Light Gray

Heavy Paste Formula

100	lb. paste white-lead (heavy paste)	2.85 gal.
0.25	lb. paste lampblack	.028 gal.
0.5	lb. paste French ochre	0.033 gal.
31.0	lb. (4 gal.) raw linseed oil	4.00 gal.
0.90	lb. (1 pint) turpentine	.125 gal.
0.90	lb. (1 pint) drier	.125 gal.
133.55	lb.	7.16 gal.

Weight of one gallon, 18.6 lb.

Soft Paste Formula

100	lb. paste white-lead (soft paste)	3.23 gal.
0.25	lb. paste lampblack	.028 gal.
0.5	lb. paste French ochre	.033 gal.
25.2	lb. (3 ¼ gal.) raw linseed oil	3.25 gal.
0.90	lb. (1 pint) turpentine	.125 gal.
0.90	lb. (1 pint) drier	.125 gal.
127.75	lb.	6.79 gal.

Weight of one gallon, 18.8 lb.

Light Green

100	lb. paste red-lead	2.23 gal.
30	lb. paste chrome yellow med.	1.25 gal.
12	lb. paste Prussian blue	1.2 gal.
42.625	lb. raw linseed oil	5.5 gal.
0.90	lb. turpentine	0.125 gal.
0.90	lb. drier	0.125 gal.
186.4	lb.	10.43 gals.

Weight of one gallon, 17.9 lb.; contains 8.9 lb. dry red-lead.

Dark Green

100	lb. paste red-lead	2.23 gal.
12.5	lb. paste chrome yellow med.	0.522 gal.
7.5	lb. paste Prussian blue	0.75 gal.
31.0	lb. raw linseed oil	4.0 gal.
1.80	lb. turpentine	0.25 gal.
1.80	lb. drier	0.25 gal.
154.6	lb.	8.00 gal.

Weight of one gallon, 19.3 lb.; contains 11.6 lb. dry red-lead.

Paint, Oil Emulsion

(1)	Linseed Oil	9
	Water	16
	Alum	1
	Glue	4
	Chlorphenol	0.1
	Sulfo Turk C	0.5

---

(2)	Potato Starch	10
	Water	30
	Casein	6
	Varnish	80
	Water	170
	Am. Oleate	3

---

(3)	Casein	3
	Water	30
	Borax	0.35
	Phenol	0.2
	Formaldehyde	0.2
	Rosin	1.5
	Pale Boiled Linseed Oil	1.5
	Naphtha (V.M. & P.)	2.0
	Sulfo Turk C	0.3

Olive Drab Paint

White Lead (ground in raw linseed oil)	6 lb.
Raw Umber (ground in raw linseed oil)	3 lb.
Chrome Yellow (ground in raw linseed oil)	½ lb.
Linseed Oil Raw	1 pt.
Turpentine	½ pt.
Japan Drier	¼ pt.

Outside White Paint Base

Lithopone (high oil absorption)	250 lb.
Zinc Oxide	250 lb.
Asbestine	105 lb.
Refined Linseed Oil	18¼ gal.
Bodied Linseed Oil	6¾ gal.
Varnolene (Naphtha)	2½ gal.

---

* Paint, Outside

White Lead	37.5
Keene's Cement	12
Tartaric Acid	0.5
Linseed Oil	42.5
Turpentine	5
Japan Drier	1.5
Paraffin	0.5
Carbon Tetra Chloride	0.5

Outside White Paint

Material	Pounds
Carbonate White Lead	41.0
Zinc Oxide	20.5
Asbestine	7.3
Linseed Oil	25.8
Turpentine and Driers	5.4
	100.0

Pounds	
21.8	Titanox B.
21.8	Basic Carbonate White Lead
12.4	Zinc Oxide
6.0	Asbestine
31.9	Linseed Oil
6.1	Turpentine, Varnolene and Driers

Pounds	
24.0	Lithopone (Albalith)
24.0	Zinc Oxide (American Process) XX
6.0	Asbestine
6.0	Silica
30.9	Alkali refined or mechanically refined Linseed Oil
2.5	Kettle Bodied Linseed Oil
6.6	Turpentine, Varnolene and Driers

Pounds
- 25.5 AX1 Lithopone
- 28.7 35 per cent Leaded Zinc Oxide
- 4.8 Asbestine
- 4.8 Silica
- 29.9 Refined Linseed Oil
- 2.6 Kettle Bodied Linseed Oil
- 1.8 Drier
- 1.9 Thinners

Cheap Outside White Paint

Lithopone	300 lb.
Paris White	200 lb.
Asbestine	130 lb.
Refined Linseed Oil	7 gal.
Refined Fish Oil	7 gal.
Limed Gloss Oil	11¾ gal.
Varnolene (Naphtha)	5¼ gal.
Kerosene	5¾ gal.
Liquid Japan Drier	2¼ gal.
Spar Varnish	3 gal.
Water	4¾ gal.

Where colored paint is desired raw oils and dark gloss oil may be used with suitable pigments replacing all or part of the above pigments.

---

* Fresh Plaster, Painting On

The following composition when applied to fresh plaster acts as a moisture absorbent and permits of the application of paint at once.

Rosin or Shellac	20–60
Titanox	5–20
Zinc Oxide	5–10
Denatured Alcohol	25–50
Xylene	50–75

Paint, Cold Water

Casein	10
Lime	10
Chalk	60
Clay	10
Pigment	10

To the above dry mixture water is added just before use.

---

* New Plaster Wall Size

Copal	25
Alcohol	30
Xylol	60
Lithopone	10
Titanox	5
Zinc Oxide	10

Varnish Formula No. LV-112
40-gal. Long

- 92½ lb. Lewisol No. 2
- 13 gal. China Wood Oil
- 16 gal. Linseed Oil (bodied 4½ hrs. at 590° F.)
- 2¾ lb. Lead Acetate
- 8 gal. Linseed Oil (bodied 4½ hrs. at 590° F.)
- 58 gal. Varnolene or Oleum

Run Lewisol No. 2, China Wood Oil, and 16 gal. Linseed Oil to 450° F. in 15 minutes and add Lead Acetate. Run to 565° F. in 10 minutes and hold for sign of string. Check with 8 gal. Linseed Oil, hold at 500° F. for 11½ minutes, and reduce at about 450° F.

After cold or after grind add Cobalt in the proportion of .03% based on the weight of the oil.

Rubbing Varnish

- 100 lb. Lewisol No. 2
- 20 lb. Hardened Rosin (800 lb. Rosin, 64 lb. Lead Acetate, 40 lb. Lime)
- 10 gal. China Wood Oil
- 5 lb. Powdered Litharge
- 2½ lb. Zinc Sulfate
- 8 gal. Dipentene
- 30 gal. Benzine
- 4 lb. No. 49 Drier

Directions:

- 4 gal. CW Oil and H Rosin run to 510° F.
- 4 gal. More China Wood Oil added and run to 540° F.
- 2 gal. China Wood Oil, Litharge, Zinc Sulfate and the
- 100 lb. Lewisol No. 2 added and run to 500° F. Hold for 20 minutes to hard pill. Cool and reduce.

Varnish

- 77 lb. Lewisol No. 2
- 29 gal. China Wood Oil
- 2¾ lb. Lead Acetate
- 8 gal. 4-hr. Bodied Linseed
- 56 gal. Varnolene or Oleum

Run Lewisol No. 2 and China Wood Oil to 450° F. in 15 minutes. Add Lead Acetate and run to 565° F. in 10 minutes. Hold for signs of string (about 1 minute). Check with Bodied Linseed Oil and hold for 3 minutes at 500° F. Cool to 450° F. and reduce.

Varnish

- 84 lb. Lewisol No. 18
- 16 lb. WW Gum Rosin
- 2¾ lb. Lead Acetate
- 3 lb. Harshaw's No. 42 Cobalt
- 29 gal. China Wood Oil

4 gal. Heavy Bodied Linseed Oil (bodied 4½ hrs. at 590° F.)
8 gal. Dipentene
76 gal. Varnolene

Run the Rosin and China Wood Oil to 450° F. in 15 minutes. Add Lead Acetate and run to 565° F. in 8 minutes. Check with Heavy Bodied Linseed Oil, stir and add Lewisol No. 18. Stir until all in solution. Run to 500° F., hold for body if necessary, cool to 450° F. and reduce. Not as durable as No. 2, but easier to handle.

Varnish Formula No. LV-89

25-gal. Long

13 gal. China Wood Oil
2 gal. Bodied Linseed Oil (4½ hours at 590° F.)
50 lb. Lewisol No. 2
10 lb. Prepared Rosin
1 gal. Dipentene No. 122
30 gal. Varnolene or Oleum

Run China Wood Oil and Lewisol No. 2 to 425° F. slow (20 minutes). Stir continually, run to 520–530° F. in 14 minutes. Hold for string, in this case 9 minutes, check with Bodied Linseed Oil, Prepared Rosin, cool to 450° F. and reduce.

Varnish Formula No. LV-93

25-gal. Long

This varnish is recommended where permanency of white, waterproofness, good flow and color, and very fast dry are desired, but where it is not necessary to pass the severe fume closet test.

7½ gal. China Wood Oil
25 lb. Lewisol No. 2
5 lb. Prepared Rosin *
¾ lb. Litharge
15 gal. Varnolene or Oleum

Run 6½ gal. Wood Oil and 17 lb. Lewisol No. 2 to 575° F. in 15 minutes (held for 1 minute). String and check immediately with 1 gal. China Wood Oil, Litharge, Prepared Rosin, and balance of Lewisol No. 2. Drop heat to 475° F., hold for 10 minutes at 475–450° F. for signs of string and reduce.

Note the varnish must be checked immediately at first sign of string at 575° F.

* The Prepared Rosin for the above is made by heating 800 lb. Rosin with 32 lb. Lead Acetate and 25 lb. Lime.

After the grind or before the varnish is put up add .35% Cobalt based on the non-volatile content of the varnish. Yield, 25¼ gallons.

The above gallons are "U. S. gallons."

Varnish Formula No. LV-107

40-gal. Long

50% Solids

Approximate Body F—Gardner-Holdt Scale

92½ lb. Lewisol No. 2
29 gal. China Wood Oil
2¾ lb. Lead Acetate
8 gal. Linseed Oil (bodied 4½ hrs. at 590° F.)
58 gal. Varnolene or Oleum

Run Lewisol No. 2 and China Wood Oil to 450° F. in 15 minutes and add Lead Acetate. Run to 565° F. in 10 minutes and hold for signs of string (not over 45 seconds). Check with Bodied Linseed Oil, hold 3 minutes, and reduce at about 450° F.

After cold or after grind add Cobalt in the proportion of .035% based on the weight of the oil.

This varnish dries in from 2 to 4 hours. Yield 104¾ gallons.

The above gallons are "U. S. gallons."

Varnish Formula No. LV-111

40-gal. Long

92½ lb. Lewisol No. 2
21 gal. China Wood Oil
8 gal. Linseed Oil (bodied 4½ hrs. at 590° F.)
2¾ lb. Lead Acetate
8 gal. Linseed Oil (bodied 4½ hrs. at 590° F.)
58 gal. Varnolene or Oleum

Run Lewisol No. 2, China Wood Oil and 8 gal. Linseed Oil to 450° F. in 15 minutes and add Lead Acetate. Run to 565° F. in 10 minutes and hold for signs of string (not over 45 seconds). Check with 8 gal. Linseed Oil, hold at 500° F. for 7 minutes, and reduce at about 450° F.

After cold or after grind add Cobalt in the proportion of .03% based on the weight of the oil.

Ester Gum Mixing Varnish

(L.V.-151)

22½ gal. China Wood Oil
22½ lb. Imperial Ester Gum No. 8

Heat to 525° F. and hold for string and add 45 lb. Imperial Ester Gum No. 8, 2½ lb. Red Lead, 3¾ lb. Ground Litharge, and gain to 550° F. and add 6 gal. LV-150 Oil.

Stir well, and, if necessary, hose to 500 and let cool to 425°.

Add Oleum.

LV-150 Oil

30 gal. China Wood Oil
30 lb. W. G. Rosin

Gain to 525° F. and hold for string and add 30 lb. W. G. Rosin, 30 gal. Superior Linseed Oil and stir well and gain to 545° and add slowly 15 lb. Ground Litharge.

Stir for 15 minutes and let cool and tank.

The above gallons are "U. S. gallons."

Sanding Sealer

Gals.	Pts.	Lbs.	Material	Wt. %
8		61.04	Cotton Solution...	41.70
3		25.35	Lewisol Solution...	17.32
	6	5.83	Zinc Stearate Base.	3.98
	2	2.16	Dibutyl Phthalate.	1.47
1	4	10.83	Butyl Acetate.....	7.40
1	4	10.14	Butyl Alcohol.....	6.93
1		6.76	Denatured Alcohol.	4.62
4		24.28	Lactol Spirits A...	16.58
20		146.39		100.00

SAND IN TEN MINUTES

Cotton Solution

Gals.	Pts.	Lbs.	Material	Wt. %
		193.00	Wet Cotton.......	28.00
22	7.36	161.92	Ethyl Acetate.....	23.57
46	7.22	332.12	Toluol...........	48.34
		687.04		100.00

YIELD 90 GALLONS OF SOLUTION
WEIGHT 7.63 LB. PER GAL.

This solution contains 1½ lb. of dry cotton in each gallon of solution (or 19.66% by wt.).

The 193 lb. of wet cotton is a standard weight drum and is composed of 135 lb. of dry cotton and 58 lb. of alcohol.

Lewisol Solution

Gals.	Pts.	Lbs.	Material	Wt. %
		8.00	Lewisol 1.2 or 18..	52.56
1		7.22	Toluol...........	47.44
		15.22		100.00

WEIGHT 8.45 LB. PER GAL.

Zinc Stearate Base

Gals.	Pts.	Lbs.	Material	Wt. %
		75.00	Zinc Stearate.....	25.72
25		180.50	Toluol...........	61.90
5		36.10	Toluol*..........	12.38
		291.60		100.00

YIELD 37½ GALLONS—WEIGHT 7.78 LB. PER GAL.

* Grind in pebble mill four hours and rinse out with Toluol.

White Enamel

129 lb. Titanox B
43 lb. Zinc Oxide
120 lb. LV-89 Lewisol No. 2 Varnish
200 lb. Grind
164 lb. LV-89 Lewisol No. 2 Varnish
15 fl. oz. Drier

Drier

100 gal. 25 lb. Harshaw Lead No. 45
75 lb. Varnolene
31 gal. 25 lb. Harshaw Cobalt No. 42
75 lb. Varnolene

White Enamel

129 lb. Titanox B
43 lb. Zinc Oxide
120 lb. LV-111 Lewisol No. 2 Varnish
200 lb. Grind
164 lb. LV-111 Lewisol No. 2 Varnish
15 fl. oz. Nuodex Cobalt

White Enamel

129 lb. Titanox B
43 lb. Zinc Oxide
120 lb. LV-112 Lewisol No. 2 Varnish
200 lb. Grind
164 lb. LV-112 Lewisol No. 2 Varnish
15 fl. oz. Nuodex Cobalt

White Enamel

129 lb. Titanox B
43 lb. Zinc Oxide
120 lb. LV-115 Lewisol No. 2 Varnish
200 lb. Grind
164 lb. LV-115 Lewisol No. 2 Varnish
15 fl. oz. Nuodex Cobalt

Varnish Formula No. LV-66

Approx. 22-gal. Long

This varnish is recommended where extreme waterproofness, weather resistance and ability to resist yellowing out of direct light are not required, but where it is desirable to pass very severe gas tests.

144 lb. Lewisol No. 2
16 lb. W G Rosin
2¾ lb. Lead Acetate
3 lb. No. 42 Drier (Harshaw)
15 gal. Kellogg KVO Linseed Oil
15 gal. China Wood Oil

4	gal. Heavy Bodied Linseed Oil 4½ hours at 590° F.
8	gal. Gum Turps
76	gal. Varnolene or Oleum

Run the Wood Oil and the Kelloggs KVO Linseed to 450° F. in 15 minutes. Add Lead Acetate and heat to 525° F. in 7 minutes. Hold at 525° F. for 10 minutes. **Immediately** add the rosin and Lewisol No. 2 and the Heavy Bodied Linseed Oil. Stir well and heat to 500° F. and hold for 50 minutes. Cool to 400° F. and reduce, adding the Cobalt after the grind in proportion of .035% Cobalt as metal based on the weight of the oil. This varnish dries in from 2 to 4 hours depending, of course, on conditions. Yield, 140 gallons.

The above gallons are "U. S. gallons."

Four Hour Varnish

The following formula using Nevindene is suggested where rapid drying is desired in a medium oil varnish. The Limed Rosin is used to assist kettle manipulation, to prevent drier precipitation and to keep the Nevindene completely dissolved. To obtain maximum speed of drying no Linseed Oil is used.

Medium Oil Varnish

Nevindene	81	lb.
Limed Rosin (5%)	13	lb.
No. 1 Fused Lead Resinate	6	lb.
China Wood Oil	25	gal.
No. 1 Cobalt Drier	1	gal.
No. 1 Manganese Drier	⅜	gal.
Mineral Spirits	44	gal.

0.60% Lead Metal based on weight of China Wood Oil.

0.03% Cobalt Metal based on weight of China Wood Oil.

0.011% Manganese Metal based on weight of China Wood Oil.

Procedure

Heat the Wood Oil to 400° F. and add 13 lb. of Limed Rosin and 40 lb. of Nevindene. Run the batch so as to get to the top heat of 565° F. in approximately 30 minutes from the start of the cook. Hold at 565° F. until a few drops "spun" on glass "pick up" 12 to 15 inches before "breaking." Chill with the Lead Resinate and balance of 41 lb. of Nevindene to cool around 495° F. Hold here for a syrupy body but do not "string" the varnish. As soon as the desired body is obtained add enough Mineral Spirits to completely "check" the batch. Add the liquid driers at 350° F.

Remarks

This varnish is a so-called "four hour" varnish. It is highly water and alkali resistant. Samples have been maintained at a temperature of 30° F. for 7 days without showing precipitation.

Cobalt Drier

W. W. Rosin	100	lb.
Refined Linseed Oil	100	lb.
Cobalt Acetate	16	lb.
Mineral Spirits	35	gal.

Heat Rosin and Linseed Oil to 350° F. and add Cobalt Acetate slowly. Keep the temperature rising. When nearly all the Acetate has been added, the mixture may crystallize but in raising the temperature to 500° F. it will again become liquid. Add the balance of Acetate if not already added and hold at 500° F. until all acetic acid fumes have been eliminated. Cool to 390° F. and add Mineral Spirits.

This drier contains one ounce of Cobalt Metal per gallon.

Manganese Drier

W. W. Rosin	100	lb.
Refined Linseed Oil	100	lb.
Manganese Acetate	15½	lb.
Mineral Spirits	35	gal.

The procedure in making this drier is the same as that described for the Cobalt drier.

This likewise contains one ounce of Manganese Metal per gallon.

Short Oil Varnish

Neville Hard Resin	100	lb.
China Wood Oil	10	gal.
Mineral Spirits	25	gal.
No. 1 Cobalt Drier	1¼	gal.

0.10% Cobalt Metal based on weight of China Wood Oil.

Procedure

Heat the Wood Oil to 400° F. and add 30 lb. of Hard Resin. Run the batch so as to get to the top heat of 565° F. in about 35 minutes from the start of the cook. Hold at 565° F. until a few drops "spun" on glass "pick up" about 24 inches before "breaking." Chill with the balance of 70 lb. of Hard Resin. The batch should not show a "string" at any stage. If desired, just enough of the Resin may be added to "chill" to 490° F. and the kettle held here for a final "stout" body. The Resin must all be in solution when the kettle has cooled to 425° F. The Mineral Spirits should be added as soon as all of the Resin has dissolved. Add the Liquid Drier at 350° F.

Remarks

This varnish serves to illustrate the use of straight Cobalt Drier with Neville Resin. For many purposes it will be desirable to replace some of the Cobalt Drier with Fused Lead Resinate.

Medium Oil Varnish

Neville Hard Resin	84	lb.
No. 1 Fused Lead Resinate	6	lb.
Ester Gum	10	lb.
China Wood Oil	20	gal.
Bodied Linseed Oil	3	gal.
Mineral Spirits	42	gal.
No. 1 Cobalt Drier	1	gal.
No. 1 Manganese Drier	3/8	gal.

0.60% Lead Metal based on weight of China Wood Oil.
1.20% Lead Metal based on weight of Linseed Oil.
0.03% Cobalt Metal based on weight of China Wood Oil.
0.06% Cobalt Metal based on weight of Linseed Oil.
0.011% Manganese Metal based on weight of China Wood Oil.
0.022% Manganese Metal based on weight of Linseed Oil.

Procedure

Heat the Wood Oil to 400° F. and add 10 lb. Ester Gum and 40 lb. of Hard Resin. Run the batch so as to get to the top heat of 565° F. in approximately 30 minutes from the start of the cook. Hold at 565° F. until a few drops "spun" on glass "pick up" 12 to 15 inches before "breaking." Chill with the Lead Resinate, the Hard Resin (the 44 pounds that have been "held out") and enough of the Linseed Oil, if necessary, to cool to approximately 495° F. Hold here for a syrupy body but do not "string" the varnish. Add balance of Linseed Oil, if any, and follow at once with the Mineral Spirits if necessary to further "check" the batch. Add the liquid driers at 350° F.

Remarks

When freshly made, this varnish may show some "silking," but ageing for one or two days usually eliminates it. Under good conditions, this varnish will permit of the application of two coats a day. Here again, faster drying may be obtained by increasing the drier content, particularly the Cobalt.

Four Hour Varnish

A variation is given below. Three gallons of China Wood Oil have been replaced by three gallons of Bodied Linseed Oil. This gives a film with slightly more flexibility.

Nevindene	81	lb.
Limed Rosin (5%)	12½	lb.
No. 1 Fused Lead Resinate	6½	lb.
China Wood Oil	22	gal.
Bodied Linseed Oil	3	gal.
No. 1 Cobalt Drier	1 1/16	gal.
No. 1 Manganese Drier	3/8	gal.
Mineral Spirits	44	gal.

0.60% Lead Metal Based on weight of China Wood Oil.
1.20% Lead Metal based on weight of Linseed Oil.
0.03% Cobalt Metal based on weight of China Wood Oil.
0.06% Cobalt Metal based on weight of Linseed Oil.
0.01% Manganese Metal based on weight of China Wood Oil.
0.02% Manganese Metal based on weight of Linseed Oil.

Procedure

Heat the Wood Oil to 400° F. and add the Limed Rosin and 40 lb. of Nevindene. Run the batch so as to get to the top heat of 565° F. in approximately 30 minutes from the start of the cook. Hold at 565° F. until a few drops "spun" on glass "pick up" 12 to 15 inches before "breaking." Chill with the Lead Resinate and enough Nevindene to cool to around 495° F. Hold here for a syrupy body but do not "string" the varnish. As soon as the desired body is obtained, add any remaining Nevindene and enough Mineral Spirits to completely "check" the batch. Add the liquid driers at 350° F.

Remarks

This varnish is a so-called "four hour" varnish. It is highly water and alkali resistant. Samples have been maintained at a temperature of 30° F. for 7 days without showing precipitation.

Method: The China Wood Oil is heated to about 470° F. and 75 pounds Cumar added with stirring while on the fire. The temperature is run up to about 530° F. and the kettle is withdrawn and held until a drop of the oil on cold glass sets to a hard button. The balance (25 pounds) of Cumar is added with stirring. The temperature falls below 500° F. The kettle is put back on the fire and heated to about 510° F. It is held for 15 to 30 minutes until sufficient body is attained as indicated by a drop of the melt cooled on glass. In this varnish it should give a hard button. The batch is cooled and the cobalt linoleate is added. Thinning is started at 450° F. or below. It should be noted that at no point in this operation is the China Wood Oil cooked so that it strings from the stirrer.

This formula is successful except where elasticity is of utmost importance in which case a longer oil varnish may be used.

Cumar in Concrete Paints

The following varnish A may be used for general purpose alkali resisting varnishes or as a vehicle for concrete paints. However, varnish B is more satisfactory where greater elasticity and ease of grinding are required.

Varnish A

China Wood Oil	20 gal.
Cumar V	125 lb.
Mineral Spirits	35 gal.

Method: Put China Wood Oil in kettle, run very quickly (12–16 minutes) to 400° F. and add 100 pounds of Cumar. Carry the heat rapidly to 56° F. (this point should be reached within 20–25 minutes of the start) and withdraw the kettle from the fire as the temperature gains 570–575° F. Do not allow the batch to string, but check with the remaining 25 pounds of Cumar. This must be stirred in rapidly. It will be necessary to cool from this point by running a stream of water on the kettle until the temperature is just below 500° F. Body the batch between 500–450° F. as it cools. The varnish can be reheated to 480–490° F. if necessary. The body is estimated by cooling a sample of the melt on tin. The batch is thinned at 450–430° F.

A Cobalt Japan (Equivalent of 1 lb. of 5% Cobalt Linoleate) is added later.

Varnish B

China Wood Oil	30	gal.
Cumar V	100	lb.
Litharge (Sublimed)	7½	lb.
Cobalt Acetate	½	lb.
Mineral Spirits	60	gal.

Method: Put 25 gallons China Wood Oil in the kettle, carry to 400° F. then add 25 lb. Cumar. Run quickly (within about 15–17 minutes of start) to 485° F. Check fire, and gradually stir in litharge. When the litharge is in, boost the fire to reach 590° F. This takes 5 to 7 minutes. Take off fire at 590° F. and gain 600° F. which temperature is reached quickly. Chill at once with 5 gallons of China Wood Oil and follow at once with the Cumar. Stir rapidly and the temperature drops below 500° F. Hold at 455–475° F. for the proper body (about 40 minutes to an hour is required). Sometimes it is necessary in this operation to place on the fire to maintain the temperature. Add the Cobalt Acetate, around 460° F. cool to 450–440° F. and thin. When intended as a grinding vehicle it is better to add the cobalt as a liquid drier after grinding.

Note: The excess of litharge, added to restrain the rate of oil polymerization at the elevated temperature, forms a cloud of insoluble lead drier which requires some time to settle. White lead or Lead Acetate in equivalent amount can be used instead of the litharge.

Concrete Silos, Varnish for Interior of

This simple coating is suggested as a wash coat for concrete silo interiors since it will resist the alkaline action of the concrete and the organic acids and other reactive liquids which, generated in the ensilage, have a destructive action on the concrete.

Cumar V-3	100 lb.
Xylol	5 gal.
V. M. and P. Naphtha	15 gal.

Dissolve the Cumar by agitation with the solvent mixture in a vessel provided with a mechanical mixer or in a tumbling barrel. The solution possesses a comparatively low viscosity.

Stir in about 300 pounds of Portland cement and apply with a heavy brush. It will be understood that if a glaze coat is required less cement will be used. If a flatter finish is desired a greater amount of cement can be added.

The mixture is applied with a heavy brush.

Alkali Resisting Varnish

Where a varnish of maximum alkali resistance is desired the following formula is suggested.

China Wood Oil	10–12 gal.
Cumar W	100 lb.
Cobalt Linoleate of 5% Metal Content (or equivalent)	8 oz.
Mineral Spirits	28 gal.

VARNISH

Medium Oil—China Wood Oil (High Cooking Temperature)

China Wood Oil	20 gal.
Cumar V	125 lb.
Mineral Spirits	35 gal.

Method: Put China Wood Oil in kettle, run very quickly (12–16 minutes) to 400° F. and add 100 pounds of Cumar. Carry the heat rapidly to 565° F. (this point should be reached within 20–25 minutes of the start) and withdraw the kettle from the fire as the temperature gains 570–575° F. Do not allow the batch to string but check with the remaining 25 pounds of Cumar. This must be stirred in rapidly. It will be necessary to cool from this point by running a stream of water on the kettle until the temperature is just below 500° F. Body the batch between 500–450° F. as it cools. The varnish can be reheated to 480–490° F. if necessary. The body is estimated by cooling a sample of the melt on tin. The batch is thinned at 450–430° F.

A Cobalt Japan (Equivalent of 1 lb. of 5% Cobalt Linoleate) is added later.

Long Oil—China Wood—Linseed—Rosin Type

China Wood Oil	20 gal.
Linseed Oil	10 gal.
Cumar V	100 lb.
Rosin	20 lb.
Cobalt Linoleate (about 5% metal)	1.5 lb.
Mineral Spirits	40 gal.

Method: The China Wood Oil and Rosin are heated to about 535° F. and drawn off the fire. When the oil strings the Linseed Oil is added to chill. The Cumar is then added and the kettle put back on the fire, heated to 500–510° F., held for 15 to 30 minutes or until sufficient body has been obtained. Add drier, cool and thin at 450° F. or below.

Long Oil—China Wood Oil—Cumar Spar Type with Litharge

China Wood Oil	30 gal.
Cumar V	100 lb.
Litharge (Sublimed)	7½ lb.
Cobalt Acetate	½ lb.
Mineral Spirits	60 gal.

Method: Put 25 gallons China Wood Oil in the kettle, carry to 400° F. then add 25 lb. Cumar. Run quickly (within about 15–17 minutes of start) to 485° F. Check fire, and gradually stir in litharge. When the litharge is in, boost the fire to reach 590° F. This takes 5 to 7 minutes. Take off fire at 590° F. and gain 600° F. which temperature is reached quickly. Chill at once with 5 gallons of China Wood Oil and follow at once with the Cumar. Stir rapidly and the temperature drops below 500° F. Hold at 455–475° F. for the proper body (about 40 minutes to an hour is required). Sometimes it is necessary in this operation to place on the fire to maintain the temperature. Add the Cobalt Acetate, around 460° F., cool to 450–440° F. and thin.

Note: The excess of litharge, added to restrain the rate of oil polymerization at the elevated temperature, forms a cloud of insoluble lead soap which requires some time to settle. White lead or Lead Acetate in equivalent amount can be used instead of the litharge.

Long Oil—China Wood—Spar with Rosin and Litharge (Regular 34 gallon type)

China Wood Oil	34 gal.
Cumar V	70 lb.
Rosin	30 lb.
Litharge	7½– 8 lb.
Cobalt Acetate or equivalent Cobalt Linoleate or Manganese Resinate	5.3 oz.
Mineral Spirits	60–65 gal.

Method: Run China Wood Oil and Rosin to 465–470° F. and add Litharge while stirring down foam. Carry quickly to 575–585° F. and pull from the fire while it gains 600° F. Check at once with the Cumar which quickly lowers the temperature to about 535° F. Chill here with hose to about 515–510° F. and gain body as the batch slowly cools to 480° F. Add Cobalt Drier at 480° F. Cool to 450–440° F. and thin. It is not desirable in any case to allow the China Wood Oil to string.

Note: For better flowing results use 31 gallons China Wood Oil and 3 gallons of kettle-bodied Linseed Oil. For longer oil batches use 34 gallons of China Wood Oil as given above and chill at 600° F. with 3 to 6 gallons of Linseed Oil.

Long Oil—China Wood Oil—Spar with a Resinate (25 gallon Quick Drying)

China Wood Oil	25 gal.
Cumar V	75 lb.
Fused Lead Resinate (5% Lead Content)	25 lb.
Cobalt Linoleate (6.5% Metal)	12 oz.
Mineral Spirits	50–55 gal.

Method: Run the China Wood Oil to 300° F. and add the Fused Lead Resinate, then carry temperature quickly to 560° F. and withdraw from the fire. Allow it to gain 575° F. Hold a moment and chill immediately with 75 pounds of

Cumar. Stir rapidly and the temperature drops to 510–515° F. Allow the varnish to gain body as it cools from this point. It is important to gain a good body so that when the batch is thinned with 50–55 gallons Mineral Spirits it will have an F. or G. (Gardner-Holt) body. It is not good practice to string the Cumar Varnish, therefore the progress of the bodying of the oil is noted by withdrawing samples from the stirrer and testing on pieces of tin.

Fused Zinc Resinate with a small percentage of lead can be used instead of the Fused Lead Resinate in the above formula. Limed Rosin can also be used if approximately 1½ pounds litharge is added at 460° F. on the up-heat.

Rather than cook the Cobalt drier into the batch, some varnish makers prefer to add the Cobalt in the form of a liquid drier.

Short Oil—China Wood Oil Alone

China Wood Oil	12 gal.
Cumar V	100 lb.
Cobalt Linoleate of 5% Metal Content (or equivalent)	8 oz.
Mineral Spirits	28 gal.

Method: The China Wood Oil is heated to about 470° F. and 75 pounds Cumar added with stirring while on the fire. The temperature is run up to about 530° F. and the kettle is withdrawn and held until a drop of the oil on cold glass sets to a hard button. The balance (25 pounds) of Cumar is added with stirring. The temperature falls below 500° F. The kettle is put back on the fire and heated to about 510° F. It is held for 15 to 30 minutes until sufficient body is attained as indicated by a drop of the melt cooled on glass. In this varnish it should give a hard button. The batch is cooled and the cobalt linoleate is added. Thinning is started at 450° F. or below. It should be noted that at no point in this operation is the China Wood Oil cooked so that it strings from the stirrer.

Short Oil—China Wood Oil with a Holding Agent

China Wood Oil	10 gal.
Cumar W or Cumar V	90 lb.
Fused Lead Resinate (Metal content about 5%)	10 lb.
Cobalt Acetate	2 oz.
Mineral Spirits	18–20 gal.

Method: The China Wood Oil and Lead Resinate are put into the kettle and the heat is carried rapidly to 575–580° F. This point should be reached within 20–25 minutes of the start. The kettle is withdrawn from the fire at this point and the temperature is allowed to gain about 590° F. Do not allow the batch to string but check with 65 pounds of Cumar and stir rapidly. The temperature drops to 500–480° F. Put kettle on fire and heat to 500 or 510° F. Cook at 500–470° F. until a sample cooled on glass gives a hard button. Gradually add the remaining Cumar without allowing the temperature to be reduced too much. Add the Cobalt Acetate at 470° F. and hold until it is taken up. Cool and begin thinning at 430–410° F.

Instead of using Fused Lead Resinate, untreated Rosin can be added to the China Wood Oil at the start and at 450° F. to 470° F. about 1½ pounds of powdered Litharge dusted in, while the oil is stirred rapidly. From this point the upheat is continued and the remaining procedure is followed.

Medium Oil—China Wood—Linseed Oil (Low Cooking Temperature)

China Wood Oil	15½	gal.
Refined or 3½ hour Kettle Bodied Linseed Oil	2½	gal.
Cumar V	100	lb.
Cobalt Linoleate (5% Metal Content)	12	oz.
Mineral Spirits	34	gal.

Method: The China Wood Oil is heated to about 470° F. and about 74 pounds Cumar added while still on the fire, with sufficient stirring to prevent the Cumar from sticking to the bottom of the kettle (as local overheating would darken the varnish). This should require not over 12 minutes. The temperature is then run up to about 535° F., the kettle withdrawn from the fire and held for 15 minutes, or until a drop on a cold glass plate sets up to a fairly hard button. The cooking under any circumstances should not be continued so far that the oil begins to string from the stirrer. The balance of the Cumar (25 pounds) and the Linseed Oil are added with stirring to check the heat. It is important to get the proper body without stringing and this method has been found to be both easy and safe. The final bodying is conducted at 500–480° F. until a sample tested on glass indicates that the correct body has been obtained. The Cobalt is then added and after cooling below 450° F. the batch is thinned. This varnish has excellent lustre and is hard and tough.

This varnish can be improved in drying

time by the addition of two pounds of Lead Resinate or Lead Linoleate with the Cobalt Linoleate.

75-Gallon Rosin Varnish Formula

I Wood Rosin 100 lb.
run to 450° F. and add
Hydrated Lime 7 lb.
run to 560° F. and add in a slow stream with stirring
Raw China Wood Oil 37½ gal.
and
Raw Linseed Oil 9½ gal.
run to 590° F. and add
Raw Linseed Oil 28 gal.
Sublimed Litharge 8 lb.
run to 510° F. and cook at this temperature until proper body is obtained (about 4 hours). Reduce with
Turpentine 40 gal.
Varsol 45 gal.
in which is dissolved
Cobalt Linoleate Paste Drier 4 lb.

25-Gallon Ester Gum Varnish Formula

Ester Gum	40	lb.
China Wood Oil	9	gal.
Bodied Linseed Oil	1	gal.
Litharge	1	lb.
Manganese Acetate	4	oz.
Cobalt Acetate	1	oz.
Turpentine	5	gal.
Mineral Spirits	10	gal.

Heat 9 gal. Wood Oil and 35 lb. Ester Gum to 400° F. Add 1 lb. Litharge. Raise quickly to 580° F., gain 590 off fire; hold for light string from stirring rod. Add immediately 5 lb. Ester Gum and 1 gal. of Bodied Linseed Oil (3 hrs.). At 440° F. add driers. Then thin.

50-Gallon Ester Gum Varnish Formula

Ester Gum	36	lb.
China Wood Oil	13	gal.
Perilla Oil	1½	gal.
Bodied Linseed Oil	3½	gal.
Litharge	1 lb. 11	oz.
Manganese Acetate	6	oz.
Cobalt Acetate	1½	oz.
Turpentine	13½	gal.
Mineral Spirits	12	gal.

Heat Wood Oil, Perilla Oil and Ester Gum to 400° F. Add Litharge. Quickly raise to 580–590° F. off fire. Hold for light string. Add Bodied Linseed Oil. At 440° F. add driers and reduce.

75-Gallon Ester Gum Varnish Formula

Ester Gum	40	lb.
China Wood Oil	15	gal.
Bodied Linseed Oil	15	gal.
Litharge	3	lb.
Manganese Acetate	½	lb.
Cobalt	⅛	lb.
Turpentine	22	gal.
Mineral Spirits	10	gal.

Heat Wood Oil and Ester Gum and 5 gal. Linseed Oil to 400° F. Add Litharge. Raise quickly to 580°. Gain 590. Hold for light string. Add balance of Linseed Oil. Reheat to 500° F. until 2″ to 3″ string established from stirring rod. Cool to 440° and thin.

25-Gallon Amberol F-7 Varnish Formula

Amberol F-7	90	lb.
Lead Resinate	10	lb.
China Wood Oil	22	gal.
Medium Bodied Linseed Oil	3	gal.
Mineral Thinner	45	gal.
Liquid Cobalt Drier	¼	gal.
Liquid Manganese Drier	⅛	gal.

Cook the Amberol F-7 and China Wood Oil to 560°, check with 3 gallons of Linseed Oil and hold for body at 500°. Pull from fire. Add Lead Resinate and then to G body. Add Liquid Driers.

50-Gallon Amberol F-7 Varnish Formula

Amberol F-7	95	lb.
China Wood Oil	35	gal.
Medium Bodied Linseed Oil	15	gal.
Fused Lead Resinate	8	lb.
Cobalt Acetate	6	oz.
Manganese Acetate	4	oz.
Mineral Thinner	73	gal.

Cook the Amberol F-7, China Wood Oil and 5 gal. of Linseed Oil to 560°. Check with 10 gal. of Linseed Oil and hold at 500° F. for body. Pull from fire. Add Lead Resinate and when all is in, add Cobalt and Manganese Driers. Thin at 450° F.

75-Gallon Amberol F-7 Varnish Formula

Amberol F-7	100	lb.
China Wood Oil	47	gal.
Med. Bodied Linseed Oil	28	gal.
Fused Lead Resinate	7½	lb.
Mineral Thinner	102½	gal.

Liquid Cobalt Drier sufficient to give metallic cobalt equal to .03% of the oil content.

Cook the Amberol F-7, China Wood Oil and 14 gal. of Linseed Oil to 550° F. Check with 14 gal. of Linseed Oil, hold for body at 500° F. Pull from fire. Add Lead Resinate and thin to body F. Add Liquid Driers.

25-Gallon Amberol 226 Varnish Formula

Amberol 226	100 lb.
China Wood Oil	25 gal.
Mineral Thinner	38 gal.

Liquid Cobalt Drier sufficient to give Cobalt Metal equal to .03% of the oil content.

Cook the Amberol and China Wood Oil to 460° F. and hold for body. Thin with Mineral Thinner to Body F. Add Liquid Cobalt Drier.

50-Gallon Amberol 226 Varnish Formula

Amberol 226	100 lb.
China Wood Oil	50 gal.
Mineral Thinner	61½ gal.

Liquid Cobalt Drier containing sufficient Cobalt Metal to equal .03% of the oil content.

Cook the Amberol and China Wood Oil to 460° F. and hold for body. Thin to Body F and add Liquid Cobalt Drier.

75-Gallon Amberol 226 Varnish Formula

Amberol 226	100 lb.
China Wood Oil	75 gal.
Mineral Thinner	92 gal.

Liquid Cobalt Drier containing sufficient Cobalt Metal to equal .03% of the oil content.

Cook the Amberol and China Wood Oil to 460° F. and hold for body. Thin to Body F and add Liquid Cobalt Drier.

25-Gallon XR-254 Bakelite Varnish Formula

China Wood Oil	22½ gal.
Varnish Grade Linseed Oil	2½ gal.
(Spencer Kellogg and Sons Superior Oil)	
Bakelite XR-254	100 lb.
Mineral Spirits (Varsol)	27½ gal.
Dipentene	10 gal.

Procedure: Oils and resin in kettle to 450° F. in 25 minutes. Hold at 450° F. for 22 minutes and thin immediately. Driers: To each gallon of the above varnish add 1½ fl. ounces XK-1092 Liquid Cobalt Drier and one ounce XK-944 Lead Manganese Drier. (See 75-254 Bakelite for drier formulae.)

50-Gallon XR-254 Bakelite Varnish Formula

China Wood Oil	25 gal.
Bakelite XR-254	50 lb.
Mineral Spirits	22½ gal.

Procedure: Oil and resin in kettle to 450° F. in 30 minutes. Hold at 450° F. for 20 minutes and thin at once. Driers: To each gallon of the above varnish add 1½ fl. ounces XK-1092 Cobalt Drier and ½ fl. ounce XK-944 Lead Manganese Drier. (See 75-254 Bakelite for Drier Formulae.)

75-Gallon XR-254 Bakelite Varnish Formula

China Wood Oil	25 gal.
Bakelite XR-254	33⅓ lb.
Mineral Spirits	24 gal.
Dipentine	2¾ gal.

Procedure: Oil and resin in kettle to 450° F. in ½ hour. Hold at 450° F. for 35 minutes and thin at once. Driers: To each gallon of the above varnish add 2 fl. ounces of XK-1092 Cobalt Drier and 2 fl. ounces of XK-944 Lead Manganese Drier.

XK-944 Lead Manganese Drier

This drier is prepared by dissolving 2.7 lb. of lead manganese Soligen in 1½ gallons of mineral spirits by warming. One fluid ounce of this drier contains approximately one gram of lead and .21 gram manganese as metal.

Varnish Formula

Rezyl No. 113 solution containing 40% by weight of Rezyl No. 113, 30% Xylene and 30% "Hi-Flash Naphtha." Driers are present as linoleates equivalent to 0.4% Lead, 0.05% Manganese and 0.03% Cobalt based on the weight of the Rezyl.

Varnish Formula

Solution containing 35% solids, *i.e.*, 17½% each of Rezyl No. 113 and No. 1102, plus 32½% Xylene and 32½% "Hi-Flash Naphtha," all percentages by weight. Driers present are 0.3% Lead, 0.04% Manganese and 0.02% Cobalt based on the total weight of the combined Rezyls.

Baking Varnish for Wrinkle-Finish on Metal

Manila Gum	2½ lb.
Tung Oil	2½ pt.
Raw Linseed Oil	½ pt.
Zinc Sulphate	3 oz.
Lead-Manganese Drier	3 oz.
Turpentine	½ pt.
Varnolene	4 pt.

Melt gum to 625° F., cool to 575°. Heat again to 640° F., cool to 600°. Heat again to 650° F., cool to 600°. Heat again to 610° F.

Heat oils separately to 375° with the zinc sulphate, add to gum, then add drier; heat to 560° F., cool and add thinner at 375° F.

---

* Light Fast Colored Varnish

Example 1.—In 100 parts of commercial spirit varnish (containing as essential part a resin, for instance shellac) there is dissolved 0.5 part of perchloric acid (concentrated). There is thus obtained a varnish which can be colored fast to light. By using 0.25 part of Malachite green crist, there is produced for example a beautiful green coloring fast to light.

Example 2.—In 100 parts of warm commercial spirit varnish there are dissolved 0.25 part of Victoria blue B, highly concentrated, whereupon 0.5 part of concentrated nitric acid is added. The varnish is of blue color fast to light.

Bookbinder's Varnish

Venice Turpentine	5 kg.
Bleached Shellac	11 kg.
Alcohol	35 kg.

Anti-Rust Varnish

Cumarone China Wood Varnish	25 parts
White Spirit	15 parts
Lead Chromate	¼ part

Varnish, Anti-Skinning Agent for

The addition of 0.1% guaiacol diminishes "skinning."

Amberol Varnish

K-12-A Amberol	90	lb.
Limed Rosin	10	lb.
Wood Oil	11	gal.
Kettle Bodied Linseed Oil	4	gal.
Lead Acetate	2	lb.
Mineral Spirits	22½	gal.

Heat the Amberol, 5½ gallons of wood oil and one gallon linseed oil to 560° F. Hold for five minutes. Add remainder of wood oil and gain 540° F., check with rosin, add lead acetate, linseed oil and reduce.

Bakelite Varnish

XR-254	100	lb.
Wood Oil	23	gal.
Improved Raw Linseed	2½	gal.
Cellosolve	6	gal.
Toluol	2	gal.
High Flash Naphtha	4	gal.
Mineral Spirits	33	gal.
Cobalt Drier (Resinate)	7/32	gal.

The XR-254 and wood oil were run to 480° F. in 30 minutes, held 45 minutes and the linseed oil added. The batch was then pulled from the fire held for body and reduced.

Bottle Varnish

Rosin	65
Ceresin	5
Japan Wax	5

Melt and stir until uniform. While stirring and heating add slowly

Barytes (Powder)	25

Allow to cool to 90° C. and add slowly with stirring

Alcohol	2

taking care that it does not boil off. Other pigments may be used in place of barytes. This varnish is applied hot. It may also be used for bottle cork capping.

---

* Bakelite Type Varnish

Resins of the phenol-aldehyde or of the glyptal type, capable of being hardened, are mixed with an equal wt. of rosin, or other non-hardening resin, and the mixt. is heated at 200° for 30–60 min. The resulting resin is very sol. in turpentine and oils to give a varnish which dries in 8–10 min. The rosin serves to render the synthetic resin permanently sol. Varnish may be made directly, for example, as follows: ceresol 100, 40% HCHO 100, rosin 100, hexamethylenetetramine 1–1.5 and chinawood oil 200 parts are heated together under reflux for 2 hr. The H_2O is then distd. off. Heating is continued at 250° for 1 hr. and 10–20 parts of Pb or Mn tungstate are added. Turpentine may be added.

Bakelite-Nevindene-Ester Gum All Round 50 gallon Utility Spar Varnish

(To compete with the lower priced Albertols)

Nevindene	10 lb.
Ester Gum	80 lb.
Bakelite XR-821	10 lb.
China Wood Oil	50 gal.
Mineral Spirits (Sunoco)	60 gal.
Solvent Naphtha (2-50-W)	10 gal.
Metallic Cobalt in the form of Cobalt, Linoleate or other Soluble Form	13 gm.

Heat the Nevindene, Bakelite Resin, Ester Gum and China Wood Oil to 470° F. in 30 minutes. Hold for at least 30 minutes for a string of about 3 inches cold from glass. Check with all of the

Solvent Naphtha and part of the Mineral Spirits to 350° F., or less. Add the driers and remainder of thinners.

Length	50 gal.
Body	E
Color	5+
Non-volatile	50%
Drying Time	4 hr.

Note: In order to render this Varnish free from gas check it must be held for not less than 30 minutes at 470° F., as the proportion of Bagelite Resin is comparatively small.

This Varnish will compare favorably with varnishes made with any of the Albertols costing 4c to 5c per pound more than the combined Resins herein.

It will have greater elasticity and durability in as much as the Phenolformaldehyde is reacted with the oil in place of having been previously reacted with the Ester Gum.

It also has the further advantage of being cooked at a low heat.

Bakelite-Nevindene Varnish for Maximum Adhesion

The following Varnish represents a Bakelite Varnish containing all the resin of acid, alkali and water resisting characteristics, and probably represents the maximum in adhesion for this type of Varnish. The addition of Nevindene adds to the film hardness and improves the adhesion to a greater extent than in the similar type straight Bakelite Varnish. It materially reduces the cost.

Nevindene	20	lb.
Bakelite XR-254, XR-820, or XR-821	65	lb.
China Wood Oil	48	gal.
Mineral Spirits (Sunoco)	44½	gal.
Solvent Naphtha (2-50-W)	5	gal.
Metallic Cobalt in the form of Cobalt Linoleate, or other Soluble Form	45	gm.
Bakelite XR-302	30	lb.

Heat the Bakelite XR-254, Nevindene and China Wood Oil to 450° F., in 30 minutes. Hold for exactly 9 minutes by the clock. Check with all the Solvent Naphtha and part of the Mineral Spirits to 350° F. Add driers and remainder of thinners. At as low a temperature as possible, preferably cold, add the XR-302.

Length	50 gal.
Body	E
Color	5
Non-Volatile	60%
Drying Time	4 hr.

In addition to its lower cost than the Resin for which it is substituted, Nevindene also permits the use of more thinners, and in this respect further reduces the cost.

The China Wood Oil is heated only for a sufficient length of time to render it free from gas check and is as free from jell formation as it is possible to make.

This Varnish is not as sensitive to driers as usual and will not skin in the container. It has improved gloss and flow.

Bakelite-Nevindene Floor Varnish

This is an all round Floor Varnish and Floor Enamel vehicle. On account of its great water, acid and alkali resistance, it is particularly suitable as a vehicle for concrete floors, and as a Wall Sealer. When made properly, it has good gloss and dries quickly to a very hard film surface.

When used as a vehicle for pigments, the acid number should be increased by the substitution of 3 to 5 pounds of Rosin in place of part of the Nevindene. It must be remembered that neither Bakelite nor Nevindene have any appreciable acid number.

Nevindene	80	lb.
Bakelite XR-821	20	lb.
China Wood Oil	25	gal.
Mineral Spirits (Sunoco)	21½	gal.
Solvent Naphtha (2-50-W)	6	gal.
Metallic Cobalt in the form of Cobalt Linoleate or other Soluble Form	6.5	gm.

Note: If preferred, a mixture of Cobalt and Lead Linoleate of equivalent strength may be used in place of straight Cobalt.

Heat the Nevindene, Bakelite Resin and China Wood Oil to 470° F. in 30 minutes. Hold for 20 to 25 minutes for a firm 3 to 4 inch string from glass. Check with all the Solvent Naphtha and part of the Mineral Spirits to 350° F. or less. Add the drier and remainder of thinner.

Length	25 gal.
Body	E
Color	6−
Solids	60%
Drying Time	2–4 hr.

The Varnish has a tendency to yellow over a period of time, but when used with the usual floor colors, this is of no consequence.

Typical Blended Oil Esterified Rosin Mixing Varnish

W. W. Rosin	125	lb.
Glycerine, sp. gr. 1.26	12½	lb.
Zinc Dust	6	oz.

Cobalt Resinate	5	lb.
Tung Oil	25	gal.
Heated Treated Linseed Oil (Stand Oil)	9	gal.
White Spirit	50	gal.

Melt the rosin and the glycerine with 5 gallons of tung oil. Heat the mixture to 350° F. Add the zinc dust and raise the temperature slowly to 600° F. Allow to cool to 280°–300° F. In another pot heat and mix the stand oil and tung oil rapidly to 550° F. Remove from the fire, when the temperature will rise to 570°–580° F. Check with base and cool to 500° F. immediately. Allow to cool to 470° F. and hold to 2-inch string. Add drier before adding white spirit for thinning.

Typical Example of an Enamel Varnish, Using Modified Phenol Formaldehyde Resin

Hard Resin	100	lb.
Varnish Linseed Oil	12	gal.
Stand Oil	15	gal.
Thickened Wood Oil	12	gal.
White Spirit	30	gal.
Turpentine	5	gal.
Cobalt metal in suitable liquid drier such as cobalt resinate	.25	lb.

Heat the varnish oil to 400° F. and add resin gradually. When all the resin is in, raise the temperature to 500° F., and test for stability of mix by thinning the sample with double the quantity of thinners mentioned in the formula. The thinning should be done after the sample has been cooled by immersion in water. Now add thickened oils and maintain temperature at 450° F. until varnish remains clear after thinning test. Allow to cool to 400° F. Add driers and thin out according to formula.

* Crystallizing Varnish

Glyptal (Chinawood Fatty Acids Type)	19	lb.
Blown Chinawood Oil	38	lb.
Liquid Cobalt Drier	5.5	lb.
Solvent Naphtha	9.5	lb.
High Boiling Gasoline	28	lb.

Varnish, Electrical Conducting

Varnish	54
Lithopone	37.8
Lampblack	8.2

Varnish, Emulsion

1. Proflex	5
2. Water	50
3. Varnish (4 hour)	40

Allow (1) and (2) to soak ½ hr. and warm and stir until all particles disappear. Put in a vessel fitted with a high-speed mixer and run (3) into it slowly, while stirring vigorously. Stir until uniform.

Varnish, Flat

Linseed or Chinawood Oil	15–30%
Calcium or Aluminum Stearate	15–30%
Kerosene 40° Bé.	33–40%
Naphtha	Balance

Hard Cold Made Varnish

Bleached Shellac	20	lb.
Sandarac	38	lb.
Pale Manila Gum	32	lb.
Rosin WW	10	lb.
Denatured Alcohol	16	gal.
Carbon Tetrachloride	4	gal.

Mix in tumbling barrel until dissolved.

* Varnish, Insulating

Rosin 1,000, metallic aluminium a little quantity, glycerine 50–150, anhydride of sodium sulphide or anhydride of sodium selenide 5–30, and tung-oil 1,500, which has been previously stirred up with 0.1–0.5% of anhydride of sodium sulphide or anhydride of sodium selenide are mixed together and treated at 240–300° C. When they have sufficiently reacted upon themselves, color, pigment or other suitable plastic matter is added or not added to the mixture according to the requirement of the circumstance, and the mixture is properly diluted with a solvent, *e.g.*, turpentine oil.

Insulating Varnish

Cumarone Resin	30	parts
Ester Gum	16	parts
Wood Oil	114	parts
White Spirit	132	parts
Kerosene	57	parts
Linseed Oil	48	parts
Cobalt Acetate	0.05	part

Orange Shellac Varnish

T. N. Orange Shellac	200	lb.
Alcohol	40	gal.
Powd. Oxalic Acid	20	oz.

Tumble in barrel for 6–8 hrs. until dissolved; strain through cheese-cloth.

Long Oil Outdoor Varnish

100 parts albertol 201 C are cooked in 100 parts Linseed stand oil at a temperature between 240 and 260° C. (464–

500° F.), until a small test of the batch, thinned out with double the normal proportion of diluents and cooled down under the tap, remains quite free from cloudiness. Then further

225 parts linseed stand oil (in two portions) are added, and the whole is mixed with

125 parts thickened wood oil. After each addition of oil is made, the batch is again brought to a temperature of 240° C. (464° F.), and in this way any slight turbidity which may be produced when adding the oil is eliminated. At the conclusion of the cooking process the dilution test described above is again carried out, in order to make quite sure that the albertol is completely dissolved.

0.45 part cobalt (calculated as metal) is added at falling temperature, and finally

200–300 parts diluents are added.

For the higher temperatures which are necessary in the case of Albertol 201 C we recommend to work with enamel- or aluminium-kettles in good condition, for the contact of the batch with iron in the heat causes strong darkening of the varnish.

Quick Drying Rubbing Varnish

Beckacite Extra Hard	300	lb.
Chinawood Oil	22½	gal.
Thinner	75	gal.
Liquid Drier	2½	gal.

Directions: Heat gum and Chinawood Oil to 565° F. This operation takes approximately 45 minutes. Remove kettle from fire and material automatically rises in temperature to 575° F. Cool material to about 375° F. and add thinner. Then add about 3 gallons of liquid drier.

Liquid Drier

Rosin	60 lb.
Cobalt Acetate	40 lb.
Mineral Spirits	100 gal.

Quick Drying Floor or Interior Varnish

Beckacite Extra Hard	200	lb.
Chinawood Oil	30	gal.
Heavy-bodied Oil	7½	gal.
Mineral Spirits, depending on body desired	75–85	gal.

Directions: Heat gum and the Chinawood Oil to 565° F. This operation takes approximately 45 minutes. Remove kettle from fire and material will automatically rise in temperature to 575° F. Hold heat at 575° until liquid attains desired body. At this point chill with 7½ gal. of Heavy-bodied Oil. Allow material to cool to about 375° F. and thin with Mineral Spirits. When cold add about 4 gallons of lead manganese liquid drier. This formula makes approximately 145 gallons.

Quick Drying Spar Varnish

Beckacite Extra Hard	160	lb.
Chinawood Oil	50	gal.
Heavy-bodied Oil	10	gal.
Mineral Spirits, depending on body desired	75–85	gal.

Directions: Heat gum and Chinawood Oil to 565° F. This operation takes approximately 45 minutes. Remove kettle from fire and material will automatically rise to 575° F., at which time add the Heavy-bodied Oil. To chill back and prevent polymerization, cool material to about 375° F. and add thinner. Then add about 3 gallons of liquid drier. This formula makes approximately 165 gallons.

Heavy-Bodied Oil

One part raw wood oil, three parts bleached Linseed. Heat to 565° F. for 2½ hours.

Typical Resinate Varnish

W. W. Rosin	150 lb.
Lime	9 lb.
Manganese Linoleate	1 lb.
Tung Oil	40 gal.
White Spirit	75 gal.
Turpentine	5 gal.

Melt the rosin and add the lime, and heat the mixture to 525° F., holding for 15 minutes. After adding the tung oil, heat to 350° F. and stir in the litharge. Heat to 490° F. and hold for pill, about 1 to 1½ hours, then add the manganese linoleate; cool and reduce.

Rubber Shoe Varnish

Limed Rosin	10 lb.
Stearin Pitch	30 lb.
Asphalt	30 lb.
Coal Tar	10 lb.
Benzol	100 lb.
Light Naphtha	20 lb.

Allow to settle and decant before using.

Short Oil Varnish (Wood Oil)

Cumarone Resin	100 lb.
China Wood Oil	12 gal.

Thinner	28 gal.
Cobalt Linoleate	14 oz.

Thinner consists of equal parts of white spirit and solvent naphtha.

Note: In using cumarone resins in oil varnishes do not use oxide drier powders, *e.g.*, litharge, as owing to the neutrality of cumarone, the drier is liable to be precipitated.

The oil is heated to 100° C. The cumarone resin is added gradually in small portions and the mix kept at 100° C. for 2 hours. It is then brought up to 250° C. for fifteen minutes, removed from the fire and allowed to cool to 180° C. The drier is then added, and after further cooling add the thinner.

Short Oil Varnish (Linseed Oil)

Cumarone	100	lb.
Linseed Oil	10	gal.
White Spirit	25	gal.
Cobalt Liquid Drier	1¼	gal.

Medium Oil Varnish

Indene Resin	81 lb.
Limed Rosin	13 lb.
Fused Lead Resinate	6 gal.
Cobalt Liquid Drier	1 gal.
Manganese Drier	3 pt.
White Spirit	44 gal.

Long Oil Varnish (Linseed)

Cumarone	86 lb.
Linseed Oil	45 gal.
White Spirit	72 gal.

Driers as above.

Spar Varnish

Cumarone Resin	100	lb.
Rosin	20	lb.
Linseed Oil	10	gal.
Cobalt Linoleate	2½	lb.
Thinners	40	gal.

Straw Hat Varnish

Elemi	50 lb.
Rosin	45 lb.
Sandarac	30 lb.
Shellac	5 lb.
Castor Oil	12 lb.
Alcohol	860 lb.

Transfer Varnish

Gum Mastic	6 lb.
Rosin	12 lb.
Sandarac	25 lb.
Limed Rosin	1 lb.
Venice Turpentine	25 lb.
Alcohol	75 lb.

Violin Varnish

Gum Sandarac	78 lb.
Gum Elemi	31 lb.
Gum Mastic	98 lb.
Castor Oil	48 lb.
Alcohol	980 lb.
Venice Turpentine	20 lb.

Water Shellac Varnish

Borax	20 lb.
Shellac	60 lb.
Water	167 lb.

Warm with stirring until dissolved.

* Varnish, Water Resistant

Tung Oil	100
Cresol	120
Formaldehyde	120
Rosin	50
Pyridine	1
Chlorinated Naphthalene	70–120

Melt together with stirring and gradually raise temperature to 140° C. Cool and thin with following solvent

Toluol	90
Xylol	250

Whitewash

The following will give equal results. Dissolve six pounds of trisodium phosphate in two gallons of water. Soak ten pounds of casein in four gallons of water for two hours, or until soft, add to the first solution and dissolve. Stir to smoothness twenty-five pounds of whiting and fifty pounds of hydrated lime in seven gallons of water. When the mixtures are cold, slowly add the first solution to the lime, stirring continuously. Dissolve five pints of formaldehyde in three gallons of water and just before use add it slowly to the whitewash, stirring hard. Do not make more than can be used in one day.

Whitewash (Without Glue)

Dissolve 15 lb. salt in 7½ gallons water and add slowly with stirring 50 lb. hydrated lime.

* Plastic Paint

Calcined Gypsum	100
Paper Pulp	1–8

The above is mixed with water as a texture coating and may be "stippled" by a brush or sponge.

* Plastic Paint

Ground Calcined Sulfate	40–60
Ground Mica	15–35
Asbestos Powd.	10–15

Casein 100 mesh	8–10
Slaked Lime	5–7

* Paint, "Raised Surface"

Crude Crepe Rubber	10
Trichlorethylene	80
Tetrachlorethane	20
Ethyl Acetate	25
Methanol	15

Powdered mica, aluminum or pigments may be dusted on surface while wet to give a "raised" or relief effect.

* Caking of Crystals, Prevention of

Fine asbestos fibre up to 5% is mixed in to prevent caking of crystalline materials.

* Roof Paint

Coal Tar	20
Gasoline	5
Alcohol	1

* Paint, Rust Proofing

For use on metals submerged in water.

Gilsonite Paint	98.6
Sodium Alumino Silicate (Finely Ground)	0.9
Mercuric Chloride (Finely Ground)	0.5

Structural Steel Paint

Dry Red-Lead, 20 lb	=0.273 gal.
Raw Linseed Oil, 5 pt.	=0.625 gal.
Turpentine, 2 gills; Liquid Drier 2 gills	=0.125 gal.
	1.023 gal.

100 pounds of heavy paste white-lead, 4 ounces of paste lampblack and 8 ounces of French ochre, with 4 gallons of raw linseed oil and a pint each of turpentine and drier. The lampblack with the white-lead produces a light gray which the ochre, being a pale yellow color, turns into a slightly warmer tint.

Though the paint is just off the white, its slight deepening by adding the lampblack and the ochre causes it to be sinsibly more opaque. One coat of this gray will "cover" or conceal the brilliant scarlet of red-lead, which one coat of pure white will not do. Some put a further coat of white, or a light color, over the gray.

The finishing coat used on the Philadelphia-Camden highway bridge was a substantial gray paint weighing 20.5 pounds per gallon. The paint was mixed on the following basis: 100 pounds paste white-lead, 1.5 pounds paste lampblack, 0.1 pound paste Chinese (Prussian) blue, 1 gallon raw linseed oil, 2 gallons boiled linseed oil, 1 quart turpentine, 1 quart drier.

* Shellac Paint, Metallic

(Non-gelling)

Bleached Shellac Solution	25 lb.
Copper Bronze Powder	3 lb.
Malic Acid	0.2–1.5 lb.
Tricresyl Phosfate	0.5 lb.

Ship Paint

The experts in charge of dry-dock work on the Atlantic coast have found satisfaction in repainting work done with the following formula:

Paste Red-Lead	100 lb.
Raw Linseed Oil	1½ gal.
Japan Drier	1 qt.
Turpentine or Mineral Spirits	1½ qt.
	4¼ gal.

Paints, Phosphorescent

A paint having a green-blue phosphorescence contains $Sr(OH)_2$ 20.7, S 8.0, MgO 1.0, Na_2CO_3 3.0, Li_2SO_4 1.0, colloidal Bi 6.0 cc. (0.3 g. in 100 cc. H_2O). One with a reddish glow contains BaO 40.0, S 9.0, Li_3PO_4 0.7, $Cu(NO_3)_2$ 3.5 cc. of a 0.4% alc. soln.

* Paint, Plastic

Dead-burnt gypsum or Keene's cement is ground wholly or completely to 325-mesh and mixed with 1–5% of starchy material, 1–6% of gum arabic (20-mesh) or other H_2O-sol. gum, and a hydration accelerator, *e.g.*, alum. The paint can be applied with a brush or trowel and may also contain fillers.

Paint Base for Textiles

Lithopone	75 gm.
Linseed Oil, Boiled	15 cc.
Oil of Turpentine	10 cc.

Working Formula:

Lithopone	25 oz.
Linseed Oil, Boiled	5 oz.
Oil of Turpentine	3⅓ oz.

Put in cornucopia to make design on cloth. Before it is thoroughly dry, shake on gold dust or steel beads or similar material. Remove excess with a blower.

* Water Paint

Am. Linoleate	7
Glue	13
Water	600

Allow to soak overnight and heat and stir to dissolve; cool. Run in slowly with stirring

Varnish	150
Rosin	80
Turpentine	70
Pigment	to suit

* Water Paint

A compn. to be applied to old water-paint coatings before applying a new coating is prepd. as follows: Wax 3, pitch 15, and benzine 10 parts are heated together on the water bath, and wood meal 5, NaOH soln. (sp. gr. 1.32) 4, and chalk 3 suspended in water 60 parts are added, the whole being stirred and then poured through a fine sieve. Washing or scraping of the old coating is rendered unnecessary.

* Water Paint for Stucco

A paint suitable for use on cement stucco is formed of white portland cement 50, hydrated lime 50, NaCl 7, Ca stearate 3 and sucrose 2 parts, ground dry in a ball mill with any desired coloring matter. Al stearate may be substituted for Ca stearate and some other modifications may be made in the compn.

Cheap White Paint

Whiting	105	lb.
Barytes	105	lb.
Lithopone	200	lb.
Zinc Oxide	20	lb.
Raw Linseed Oil	12½	gal.
Blown Linseed Oil	3	gal.
Liquid Drier	2	gal.
Naphtha	11½	gal.
Turkey Red Oil	1	gal.
Water	7	gal.

Grind pigments in oil and then mix in other liquids.

Liquid Paint Drier

1. Rosin W. W.	200	lb.
2. Calcium Hydroxide	16	lb.
3. Lead Acetate (Powd.)	16	lb
4. Chinawood Oil	8	gal.
5. Manganese Borate	2	lb.
6. Benzine	98	gal.
7. Kerosene	9	gal.

Melt (1) and (2) and strew (3) over surface. Heat slowly raising temperature to 450° F. and heat until odor of acetic acid is gone. Mix (4) and (5) and stir into above and mix thoroughly while heating. Raise temperature to 540° F. stirring and beating down foam. Cool to 460° F. and add Kerosene while stirring. When cooled to 240° F. add benzine with stirring.

This gives a practically colorless quick drier.

Wood Paint Primer

Pigment	65.6%	
Basic Carb. Lead		60%
Zinc Oxide		20%
Titanox B		19%
Aluminum Bronze Pwd.		1%
Vehicle	34.4%	
Raw Linseed Oil		40%
Boiled Linseed Oil		30%
Turpentine		16½%
Solvent Naphtha		10%
Drier (Pb. Mn.)		3½%
Weight per Gallon	16.7 lb.	

25-Gallon Rosin Varnish Formula

I Wood Rosin	50 lb.
Raw China Wood Oil	25 gal.
Hydrated Lime	2 lb.

heat to 550° F. (to 570° F. off fire). Check with

I Wood Rosin 50 lb.

add

Sublimed Litharge 6 lb.

allow to cook at 500° F. for 1½ hours, cool and reduce with

Turpentine	20 gal.
Varsol	20 gal.
Cobalt Linoleate Paste Drier	4 lb.

50-Gallon Rosin Varnish Formula

I Wood Rosin 100 lb.

run to 450° F. and add

Hydrated Lime 6 lb.

run to 560° F. and add slowly with constant stirring

Raw China Wood Oil	37½	gal.
Raw Linseed Oil	10	gal.

heat to 550° F. (to make 575° F. off fire). Check with

Linseed Oil 2½ gal.

Sprinkle on top of batch

Sublimed Litharge 4 lb.

allow to cook down to 450° F. and reduce with

Turpentine	30	gal.
Varsol	30	gal.

in which has been dissolved

Cobalt Linoleate Paste Drier 6 lb.

Paint and Varnish Remover

Benzol (90%)	3 gal.
Denatured Alcohol	2 gal.
Paraffin Wax	1 lb.

* Paint and Varnish Remover

Caustic Soda	10.45
Sod. Silicate (40–42° Bé)	9.14
Water	69.55
Copperas	0.71
Flour	10.15

Paint and Varnish Remover

Benzol	50
Methanol	25
Acetone	15
Gasoline	10
Paraffin Wax	2½

Paint and Varnish Remover

Gasoline	50
Benzol	15
Acetone	35
Paraffin	3

Paint and Varnish Removers

Trisodium phosphate and sodium metasilicate will quickly and easily remove varnish. They will also work on paint if not too old or too thick. Use 1 lb. to 1 gallon of boiling water. Mop or brush on, and let stand 20 to 30 minutes. Then rub off and rinse well with water.

Wood Bleaches

As a wood bleach sodium perborate is probably superior to any of the others now used (including the old stand-by oxalic acid). It has the great advantage over the acid bleaches that it can be mixed directly with sodas and alkalies, since it is stable in alkaline solution. A soluble silicate should be present as a stabilizer. A good mixture is 90% sodium metasilicate and 10% sodium perborate. Some of the metasilicate may be replaced by trisodium phosphate. This is a combination paint and varnish remover and wood bleach. Use 1 lb. to 1 gallon of boiling water. Mop or brush on, and let stand 20 to 30 minutes. Then rub off and rinse well with water.

Wood, Plastic

Nitrocellulose	15–20
Ester Gum	5– 9
Castor Oil	1– 5
Wood Flour	15–30
Lacquer Thinner	79–66

Wood Filler Powder

Silica Powder	200 lb.
China Clay	32 lb.
Linseed Oil	44 lb.
Turpentine	40 lb.
Liquid Drier	24 lb.

Acid Proof Wood Stain

Solution A

Copper Sulfate	12½
Pot. Chlorate	12½
Water	100

Solution B

Anilin Oil (Light)	15
Hydrochloric Acid (Conc.)	18
Water	100

The wood surface must be freed thoroughly from paint, varnish, grease and dirt. Heat solution A to a boil and give wood two coats while hot, allowing first coat to dry before applying second. Apply two coats of solution B in the same way. When surface is thoroughly dry wash well with soap and water. Dry and rub well with linseed oil.

Wood Stains, Non-Grain Raising

Water or Spirit Soluble Dye	4– 6 oz.
Ethylene Glycol	15–25 oz.

Heat on water bath until dissolved; cool and add

Methanol	1 gal.

* Putty

Marble Dust	10
Whiting	70
Linseed Oil	2
Mineral Oil	15
Asbestos Powd.	2.5
Machine Oil	0.5

Preparing Zinc for Painting

A practical formula is: 135 grams sodium dichromate, 400 cc. nitric acid, 600 cc. sulfuric acid, and 20 liters water. Contrary to most etching solutions, this gives an even crystalline ground which will not show under a paint. A brown scum usually appears on the surface when the metal comes from this solution. However, immersion for about a minute in a dilute nitric and sulfuric acid solution readily removes this scum. The plate is then washed free from acid and dried. This drying is important. The water must either be wiped off by means of sawdust or any other absorbing medium, or be displaced by dipping the plate into a lacquer thinner that is sufficiently miscible with water so as to allow the plate to dry free from contact with water.

This process has the same disadvantage as sandblasting in that it is often quite impractical to apply the finishing material immediately after treatment.

Oil Soluble Stain

Red Mahogany

Sudan Red	2 oz.
Pylakrome Black No. 319	3 oz.
Azo Orange 30	1 oz.

Dissolve in two gallons benzol.

Brown Mahogany

Azo Oil Yellow 408	2 oz.
Pylakrome Oil Green 430	½ oz.
Sudan Red	1 oz.
Azo Orange	2½ oz.

Dissolve in two gallons benzol.

Walnut

Azo Oil Yellow 408	7 gm.
Sudan Red	½ gm.
Pylakrome Green 430	1 gm.
Azo Orange	4 gm.

Dissolve in one pint of benzol.

Oak

Azo Yellow	15.5 gm.
Pylakrome Black 319	.5 gm.

Dissolve in two pints of benzol.

The above also soluble in waxes, acetone, turpentine and lacquers.

Synthetic Resin Finishes

Oxidizing rezyl solutions make excellent vehicles for aluminium-bronze finishes for either interior or exterior work, the powder being mixed just prior to application. For general decorative work, rezyl 114 is recommended as giving a quick and hard-drying gloss. Rezyl 1102 is exceptionally resistant to heat, hence well adapted for use on steam pipes, radiators and the like, as well as for prolonged baking at high temperatures. For oil refinery and filling station equipment, aluminium finishes made from rezyl 1102 are recommended, because resistant to petrol. Typical formulas follow:

Rezyl 114, 100 lb. and coal-tar naphtha 100 lb. (J and 33 lb. of xylol); mineral spirits, 70 lb.; lead linoleate, 2 lb.; cobalt linoleate, ¾ lb.; aluminium-bronze, 70 lb.; total, 342¾ lb. or 38¾ gal. This is an air-drying finish for brush application.

A baking finish for spray application is made as follows: Rezyl 1102, 100 lb.; xylol, 150 lb. (same as Solution A, 250 lb.); toluol, 150 lb.; cobalt linoleate, ½ lb.; aluminium-bronze, 70 lb.; total, 470½ lb. or 55¼ gal.

A harder and quicker-drying, but somewhat brittle, vehicle for indoor use can be obtained by blending rezyl 114 with cumarone resin. Rezyl 113 in equal parts of coal-tar naphtha and mineral spirits is recommended as an aluminium-bronze vehicle for outdoor use. Its adhesion, toughness, rapid drying, durability make it superior to the long-oil spar varnish ordinarily used for this purpose. It works more easily than rezyl 1102, dries a trifle more slowly but forms a more flexible film, and hence is well adapted for all types of exposed metal work. Rezyl 110 dries somewhat more slowly than rezyl 113, but brushes more easily and permits of the use of mineral spirits with aluminium-bronze for priming wood, for which its elasticity, adhesion and durability recommend it. When used in metal paints, the vehicle should contain 10 per cent of coal-tar naphtha and 10 per cent raw linseed oil to insure proper floating and leafing of the aluminium-bronze. Although rezyl 1103 is still slower drying than rezyl 110, it makes aluminium paints with excellent working qualities for brush application.

Good adhesion and elasticity make the oxidizing rezyls excellent for quick-drying undercoats. The following are typical formulas in addition to the primer formulas already given:

Baking primer: Iron oxide, 150 lb.; rezyl 110, 100 lb. and xylol, 43 lb.; V. M. and P. naphtha, 155 lb.; lead linoleate, 2 lb.; manganese linoleate, ¾ lb.; total, 450¾ lb. or 38¼ gal. For best results, this primer is applied in a thin film and baked at least one hour at 200° F.

A surfacer which has given good results in both air-drying and baking is formulated as follows: Iron oxide, 50 lb.; lithopone, 50 lb.; black mineral filler, 300 lb.; silica, 100 lb.; rezyl 114, 100 lb. and xylol, 100 lb.; mineral spirits, 50 lb.; turpentine, 30 lb.; lead linoleate, 2 lb.; manganese linoleate, ¾ lb.; total, 782¾ or 57¾ gal. Several coats of this surfacer can be applied in rapid succession, and the whole film baked hard in one operation. It has good water-resistance, elasticity and toughness, yet sands easily and lacquer can be applied over it without lifting.

Fused Manganese Resinate

Rosin	200 lb.
Manganese Dioxide	25 lb.

Heat Rosin to 310–330° F. and add dioxide slowly with careful stirring.

Raise temperature to 430° F. and then to 485° F. at which point all but five pounds of the dioxide should have been added. The addition of the last five pounds should not be made until a chilled sample is of a clear amber color. Stir until thick; remove from heat and shovel into cooling forms.

Limed Rosin

Rosin	200 lb.
Slaked Lime	10 lb.

Heat Rosin to 480° F.; remove from heat; sprinkle lime on surface and stir in gradually. Heat again to 550–580° F. Allow to cool to 480° F. and pour into forms.

* Phthalic Anhydride Varnish Resin

By cooking a mixture of two parts phthalic anhydride, two parts glycerol and four parts linseed oil fatty acids for 6 hours at 325–400° F., and then continuing the reaction for the same period and at the same temperature, but with the addition of another two parts of phthalic anhydride, resins with the above-mentioned qualities are produced. When incorporated with driers, varnishes and enamels in which these resins are the vehicles, dry to hard tough flexible films in 4–6 hours. Without driers they bake at 200° F. for 2 hours to hard coatings with excellent outdoor durability.

Waterproof Shellac

Scrap Celluloid	20 oz.	A.
Methylated Spirits	2¼ pt.	A.
Acetone	2¼ pt.	A.
Camphor	1½ oz.	A.
Benzole	2 pt.	A.
Orange Lac	8 lb.	B.
Methylated Spirits	1 gal.	B.
Butyl Alcohol	1 gal.	B.
Benzole	1 part	C.
Methylated Spirits	1 part	C.
Acetone	1 part	C.
Butyl Alcohol	1 part	C.

Mix the above separately, and take three parts of A. to seven parts of B.

If the mixture is to be sprayed, use C. as a thinner. It would then be necessary to prevent frothing by the addition of Glycol or Butanol in the proportion of 1 gallon to 30 gallons of the mix.

Water Solution of Shellac

100 gm.	Water
8 gm.	Ammonium Hydroxide
2 gm.	Glycerine
20 gm.	Bleached Wax-free Shellac

The water, ammonium hydroxide and glycerine are first mixed together. The shellac is then added. The mixture is allowed to stand for one hour or longer. It is then heated on a water bath to 150° F., whereupon a clear solution is produced. This material is useful as an inexpensive varnish. This material may be improved by substituting Aquaresin (G M) in place of glycerine.

* Transfers

A suitable paper sheet is first impregnated, as by means of immersion or spraying, with a material to act as an ink-absorption minimizer and ink softener. This material is a liquid mixture including one or more volatile solvents, one or more oils, fats or waxes, and phenol. Various formulae have in practice been found satisfactory. A recommended formula is the following:

Toluol	6 gal.
Kerosene Oil	2 gal.
Neat's Foot Oil	2 gal.
Phenol	7 lb.

Another formula giving good results is the following:

Ethylene Dichloride	5¼ gal.
Carbon Tetrachloride	1¾ gal.
Petroleum Jelly	14 lb.
Phenol	7 lb.

After the paper is impregnated, the volatile solvents should be completely or substantially completely evaporated; it being recommended that the impregnated paper be allowed to season for from one to several days. The phenol left in the paper is for the purpose of acting later, at the time of heat and pressure transfer, as a dissolving or softening agent for the pigmented ink laid down on the paper at the time the paper is printed to form the new transfer sheet. The oily or greasy material remaining in the paper after the evaporation of the volatile solvents restrains such ink against other than minute absorption by or penetration into the paper.

The paper selected is preferably fairly smooth to accept good clean printing; but such paper is not necessarily heavily sized or calendered or otherwise specially finished.

The printing may be executed with ordinary printing equipment and by any of the usual printing methods; for instance, lithography, typography or rotogravure may be successfully employed. However, in order to secure best results, the printing inks used should be somewhat different from those of customary composition. Ordinary printing inks include oil varnish, which will dry within a relatively short time, forming a considerable bond with the fibers of the paper; and consequently these inks are not of maximum efficiency in carrying out the present invention. Instead there is recommended an ink having a richly pigmented content; preferably so rich as to give a stiff paste were not some slow evaporating solvent incorporated.

The ink is thus richly pigmented, and yet is brought down to the proper consistency, that is, the usual consistency of an oil varnish printing ink by the addition of such a solvent as benzyl alcohol or ethylene glycol monomethyl ether. Such ink may be conveniently made up of the following:

Color Pigment	3 lb.
Linseed Oil Varnish	4 lb.
Copal Resin	½ lb.
Dibutyl Phthalate	1 lb.
Benzyl Alcohol	½ lb.

Another very satisfactory ink for use in connection is made up of the following:

Color Pigment	3 lb.
Blown Castor Oil	5 lb.
Cumarone Resin	½ lb.
Ethylene Glycol Monomethyl Ether	1½ lb.

In making the ink, thorough grinding is important if not essential.

Best results are obtained when the printing is so executed that neither too much nor too little ink is supplied. The feeding of the ink should be so regulated that the solids of the design will have a good ink coverage, but there should not be supplied surplus beyond this to such an extent that smearing of the half-tones of the design will occur.

After the paper is printed, the solvent content of the ink slowly evaporates, allowing the remainder of the ink to set but without drying completely from the top surface of the printing down to the paper. The ink remains thus only partially dry apparently for an indefinite period. The richly pigmented ink residue left on the paper as a result of the printing operation is only loosely connected with the fibers of the paper, and if pressure or friction is applied, the deposited ink may have a large portion thereof easily removed, but not so easily as to be capable of being accidentally smudged by lightly slipping friction such as might occur in ordinary transport and handling. In other words, said richly pigmented ink residue remains somewhat soft, yet has a certain toughness and pliability, or self-sustaining quality; which result is obtained by the addition of a proper amount of resinous material, such, for instance, as specified in the ink formulae given.

The base material is desirably, if not essentially, treated in such manner as to carry an ink-transfer accelerator at the time of heat and pressure transfer. While the phenol residue in the paper is activated by the heat of the transfer step to soften the ink, the accelerator acts to intensify such softening. Thus the accelerator acts in conjunction with the phenol residue in the paper, thereby to hasten transfer of the ink to the base material. The accelerator, further, acts as a binder to hold the transferred ink on the base material—yet without any undesirable binding action on or adhesive cling to the paper of the transfer sheet.

The practical value of this ink-transfer accelerator will be appreciated, when it is explained that a fair transfer may be occasionally effected even when the ink used for the printing of the transfer sheet is ordinary printing ink rather than a special ink as hereinabove described.

It is recommended, however, that such special ink be employed in every case; since always in transfer work the very finest possible results, and as uniform results as possible, are desirable.

If the transfer is to be made to a plain base material, or one not previously lacquered, said ink-transfer accelerator may comprise, a solution of phenol in a volatile solvent or solvents. The base material is sprayed or otherwise coated with such solution, and while such coating is still moist, the printed face of the transfer sheet is laid against the coated side of the base material, and the heat and pressure transfer effected. Thus, at the time of heat and pressure transfer, the phenol and its still unevaporated solvents on the base material, are applied to and squeezed under pressure and in the presence of heat against the printing of the transfer sheet and against the paper carrier,—this carrier having, as aforesaid, not only a phenol content, but also a residue of oily or greasy matter. A recommended formula for said solution is the following:

Toluol	6½ gal.
Benzyl Alcohol	2½ gal.
Phenol	7 lb.

Another formula for said solution giving good results is the following:

Ethylene Dichloride	6 gal.
Carbon Tetrachloride	2 gal.
Benzol	2 gal.
Phenol	7 lb.

If the transfer is to be made to a base material previously coated with a pigmented or clear lacquer (for instance, nitrocellulose lacquer), the ink transfer accelerator to be carried by the base material may be provided by modifying said lacquer. Excellent results in this connection are obtained when a surplus of oil, as castor oil, is added to the lacquer. Such surplus oil content of the lacquer, at the time of the heat and pressure transfer is liberated and driven out of the heated lacquer coating, and is taken up and absorbed by the paper carrier and the ink thereon. This surplus oil is similarly liberated and similarly acts, when, as is preferred, the lacquer coating is dried before the transfer; the surplus oil in this case being liberated as soon as the lacquer coating becomes thermoplasticized. The liberated oil acts, in conjunction with the phenol residue in the paper, as an ink-transfer accelerator pursuant to the invention. That is to say, this liberated oil acts to intensify the softening action of the phenol in the paper carrier on the ink of the latter, in the presence of the heat of the transfer; so that, here also, a perfect transfer is effected.

Further, in the case of a lacquered base material, the surplus oil in such lacquer serves another useful purpose as will now be explained. Many ordinary commercial lacquers, when used as a base coating for a base material, have a tendency, as the result of a heat and pressure transfer, to stick to the paper carrier of the transfer sheet and thereby make removal of the latter difficult if not impossible. When an ordinary commercial lacquer has a surplus oil content pursuant to the invention, this sticking trouble is completely overcome.

A recommended formula for the new lacquer is the following:

Butyl Acetate	6 gal.
Toluol	3½ gal.
Camphor	8 oz.
Soluble Cotton	50 oz.
Castor Oil	1 gal.

Another satisfactory lacquer formula is the following:

Ethyl Acetate	4½ gal.
Ethyl Lactate	1½ gal.
Ethylene Dichloride	3 gal.
Cellulose Acetate	75 oz.
Castor Oil	½ gal.
Tricresyl Phosphate	½ gal.

Excellent results are obtained when the heat and pressure are applied for about from five to ten seconds; the applied temperature is about 200° F. and the pressure is about 100 lb. per square inch.

On removal of the stack from the press, the paper sheet may be immediately stripped off by manual pull without the use of water or solvents, easily, and without blurring or smudging the transferred printing. Then the base material may be finished in any desired way, as by applying a coating of lacquer or the like, thereby to set the transferred ink.

1. Ordinary Composition for Transfers.

Parts by Weight	
100	Rosin
30	Beeswax
30	Gold Bronze or Pigment

2. Indelible Marking Composition—Blacks.

Parts by Weight	
100	Stearic Acid
150	Induline Base

3. Indelible Marking Composition—Colors.

Parts by Weight	
100	Cumar Light
25	No. 4 Litho Varnish
8	Mineral Oil
2½	Cobalt Drier
30	Permanent Pigment

4. Permanent Marking Composition.

Parts by Weight	
100	Cumar Light
50	Processed Rapeseed Oil
35	Bronze or Pigment

5. Water Soluble Transfer Composition.
   1. Printing Compound.
      - a. Glycerine — 100 by wt.
      - b. Gum Arabic — 40 by wt.
      - c. Color (Dye or Pigment) — 25 by wt.
   2. Dusting Material.
      - a. Gum Tragacanth Powder

6. Embroidery Composition for Transfers.

Parts by Weight	
16	**Cumar**

4 Rosin
4 Canauba Wax
2 Stearic Acid
8 Ultramarine Blue
31.2 Titanox Ground
8.8 Litho Varnish Ground

7. Leather Composition for Transfers.

Parts by Weight
100 Shellac—Orange or White
50 Venice Turpentine
40 Pigment

8. Indelible Transfer Ink.

Parts by Weight
100 Cumar
10 Varnoline
10 No. 4 Litho Varnish
20 Turkey Red Oil
20 Dyestuff (Induline Base)
30 Permanent Pigment

9. Flexible Marking Composition.

Parts by Weight
100 Light Cumar
55 Processed Rapeseed Oil
30 Rubber Latex
45 Vermilion

10. Fugitive Transfer Composition.

Parts by Weight
100 Rosin
10 Beeswax
1 Cobalt Drier
25 Gold Bronze

11. Water Fugitive Transfer Composition.

Parts by Weight
1 Mutton Tallow
1 Cocoa Butter
4 Paraffine
6 Rosin
Sufficient quantity—Pigment

Laboratory Table Finish

A black acid proof stain is made as follows:

Apply 2 coats of hot aq. soln. contg. 4% copperas, 4% blue vitriol and 8% $KMnO_4$. Rub off the excess of the second coat and apply 2 coats of aq. 12% aniline and 18% concd. HCl. When dry apply a coat of linseed oil.

Turpentine Jelly

Aluminum Stearate	40 lb.
Turpentine	20 gal.

In a steam jacketed kettle put the turpentine and add the stearate a little at a time stirring to incorporate it uniformly. Allow to stand overnight and then heat to 150° F. while stirring; keep heat until a clear jelly forms. Stop heating when desired consistency is attained.

---

* Acid Proof Coating

For use on tanks, pipes, roofing, etc.

Portland Cement	40 lb.
Mica	5 lb.
Sulfur	50 lb.
Aluminum Powder	5 lb.

Mix and heat together until uniform.

---

* Bituminous Coating

A compn. suitable for coating or surfacing purposes or incorporating in road-making materials comprises low-temp. tar with a fatty pitch dispersed therein and an addn. of CaO or other alk. compd. adapted to accelerate hardening. In an example 2 parts stearin pitch is heated with 5 parts shale oil at not over 60 lb. per sq. in. to 150° for 4 hrs. to give a soln. which is dild. with 40 parts shale oil and stirred into 250 parts tar warmed to 50°. The cooled product may be mixed with gravel and slaked CaO.

Butter Tubs, Coating For

To eliminate woody odor in butter, the inside of tubs is sprayed with

Casein	50
NaOH	4
Water	170

followed by 4% formaldehyde.

Cellulose Coatings

After treatment with a dilute mineral acid at a moderate temperature, cellulose (in the form of cotton fibre, rags, or waste) can be disintegrated and reduced to a fine powder. In the latter condition it is capable of even dispersion in a dilute adhesive medium, such as nitrocellulose solution, drying oil or starch. A paint for metal or wooden surfaces can be obtained, for example, by incorporating twenty parts of the powdered disintegrated cellulose with a clear solution of nitrocellulose plasticized with tricresyl phosphate. Similarly, the new material can be mixed with viscose solution to form a paste-like product, which can be applied as a paper coating.

---

*Concrete Coating

Thirty-eight parts of rosin are melted with 1.9 parts of Zn chromate and added at 220° to 32 parts of a mixt. of the oil of Dryandria cordata and boiled linseed oil. Thirty parts of thinner and drier are added. To 60 parts of this varnish are added 40 parts of titanox or ZnO pigment.

* Corrosion Resistant Coating

Coatings for preventing corrosion on metals contain, e.g., stearic pitch 105.3, orthophosphoric acid 4.54 kgs., solvent naphtha 155 l. and petroleum 100.8 l.

*Pipe Coating

Pitch	100 lb.
Mica Powder	10–30 lb.

Heat and stir until uniform; apply hot.

Pipe Line, Coating for Petroleum

Among a great variety of compns. protecting against corrosion, best results were obtained with a mixt. of 50% clay and 50% of an asphalt m. 80°. This mixt. was applied to 8-inch lines connecting Baku with Batum. Pipes of smaller diam. should be coated with a mixt. having a higher content of clay.

Protecting Coating for Wax Finishes

Copal Varnish	6 lb.
Boiled Linseed Oil	6 lb.
Turpentine	10 lb.

Mix above together, and apply a thin coat to the wax finish. This will protect it from damp without dulling the finish.

* Rubber Pyroxylin Coatings

Nitrocellulose Solution (commercial duco)	50 cc.
Latex	20 cc.
Nitrocellulose Thinner	50 cc.
Water	100 cc.
Castor Oil	2 cc.

A satisfactory composition for this purpose may be made up with rubber cement according to the following formula:

Nitrocellulose Solution (commercial duco)	50 cc.
10% Rubber Cement	20 cc.
Thinner	50 cc.

The thinners or solvents used for nitrocellulose products which may be used in the above compositions are, amyl acetate, ethyl acetate or butyl acetate. Benzol and alcohol mixture which is a common solvent for nitrocellulose and rubber may be used.

In using these nitrocellulose compositions the leather is first treated or impregnated with the waterproofing composition containing rubber and after the waterproofing treatment is completed a coating of the nitrocellulose composition or dressing is applied to the surface of the leather. When the solvent in the nitrocellulose composition evaporates a surface finish remains on the leather which is not impaired by flexing the leather and which gives to the leather a smooth finished appearance and the "feel" which is a desirable characteristic of leather when used in articles such as shoes and other kinds of footwear.

* Wall Coating

Mica	49
Clay	30
Casein	18
Alum	2
Cream of Tartar	1

Color to suit.
Mix with hot water and apply.

* Wrinkled Finish Coating

Glycerol	75
Phthalic Anhydride	148
Linoleic Acid	85
Tung Oil	85

This mixture is heated at a temperature of about 230–250° C. for a period of about one-half to one hour, until a sample on cooling yields a non-sticky or only very slightly sticky mass. The heating operation is preferably carried out in a non-oxidizing atmosphere such as may be obtained by passing a stream of carbo dioxide or nitrogen or the like through or over the reaction mass.

When the reaction is complete, the resin composition is cooled to about 150° C. and is thinned with coal-tar naphtha (boiling point 160–200° C.) until a solution is obtained containing about 40% resin. A liquid drier such as linoleate, or resinate, is added in amounts sufficient to give a metallic cobalt equivalent of about 0.02 to 0.1%, based on the weight of resin. The solution is then ready for use and may be applied to a surface in any suitable manner, such as by brushing or flowing the solution thereon. The coating is preferably heated to a temperature of about 100° C. for one hour, whereupon there is obtained a light colored adherent film, having a wrinkled finish and being of superior hardness and durability, and being substantially insoluble in the usual solvents.

Filler for Cast Iron

This material is used to fill in the rough surfaces on cast-iron motor blocks, engines, machine-parts, etc., to obtain smooth surface, before enamel or lacquer is applied.

Japan Varnish	1½ gal.
Spar Varnish	½ gal.
Keystone Filler	4 lb.
Aluminum Silicate Flake	20 lb.

Filler for Automobile-Body Work

Rubbing Varnish	2 gal.
Blown Linseed Oil	¼ gal.
Japan Varnish	¾ gal.
Keystone Filler	4 lb.
Sublimed White Lead	4 lb.
Aluminum Silicate Flake	20 lb.

* Crack Filler

Silex	2
Lacquer	4
Cornstarch	3
Wheat Flour	3
Glue Powder	2

* Milk Bottle Caps

Heavy paper is impregnated in molten mixture as follows:

Carnauba Wax	80
Rosin	18
Sulfur	2

Dispersions of Casein and Shellac

Casein and shellac are animal products which are acidic in character and hence combine with alkaline reagents. The products formed by reaction with Triethanolamine are similar to soaps in that they form colloidal dispersions with water. Partial neutralization of the casein and shellac will, like the partial neutralization of stearic acid and rosin, produce sufficient soap to emulsify the remainder of the material. The greater the amount of Triethanolamine used, the more nearly colloidal and clear will be the solution.

As a rule, between 5 and 15 per cent of Triethanolamine by weight of the casein or shellac produces an excellent dispersion in water. One formula in use takes two ounces of Triethanolamine to one pound of casein and one gallon of water. The Triethanolamine not only produces a uniform solution, but protects the casein from decomposition and makes it somewhat more flexible. A similar product can be made by melting shellac with Triethanolamine and dissolving in boiling water. It is sometimes advisable, however, to carry along some alcohol with the water to give a clear solution. For example, shellac treated with 10 per cent of Triethanolamine is completely soluble in 50 per cent alcohol.

Treating Concrete Oil Tanks

It is recommended that all concrete oil-storage tanks should be treated with silicate when first built. Concrete needs water to obtain its final set. If oil is put on it, the oil drives out the water which is needed for the curing. Oil therefore prevents the full curing of the concrete. To protect the concrete from the oil, the tanks should have a treatment with silicate similar to that for waterproofing concrete. The process recommended is as follows:

1st coat, 1 part of silicate and 3 parts of water
2nd coat, 1 part of silicate and 2 parts of water
3rd coat, 1 part of silicate and 1 part of water
4th coat, 1 part of silicate and 1 part of water

Precipitated Cobalt Linoleate (Drier)

A.	1. Linseed Oil	50 gal.
	2. Caustic Soda (76%)	80 lb.
	3. Water	32 gal.
	4. Water Boiling	166 gal.
B.	5. Cobalt Acetate	250 lb.
	6. Water Boiling	100 gal.

Dissolve (2) in (3). In another vessel mix (1) and ¾ of the mixture of (2) and (3) mix thoroughly and allow to stand two days. Heat while stirring until liquid and add 10 gal. hot water. Bring to a boil whipping down foam. Cool by addition of cold water if foam cannot be controlled. Test with phenolphthalein; if alkaline continue boiling; if neutral add part of remaining caustic soda solution and boil until a sample on glass sets clear. The finished soap should be but faintly alkaline.

Dissolve (5) in (6) and heat to a boil; run the above soap solution heated to a boil into it slowly while stirring until precipitation is completed. Allow to cool over night and draw off water. Wash the precipitate thoroughly with hot water.

Lead Drier

Lead Tungate as a vehicle and drier for quick drying paints is prepared as follows:

Litharge 30
or Basic Lead Carbonate 35
is added slowly with stirring to
China Wood Oil—Fatty Acids 100
heated to 300° F. Stir until uniform.

* Nitrocellulose Emulsion

Ten parts by weight of nitrocellulose in the form of low viscosity nitrocotton was added with stirring to 50 parts of hexalin acetate, the mixture being heated to about 80° C. to facilitate conversion into a homogeneous liquid solution. To the solution was then added 10 parts of a 10% aqueous gum tragacanth solution which was prepared by soaking the hard, horn-like raw gum in water, for about twenty-four hours, and then heating to effect its solution. The 10% solution of gum tragacanth was a semi-solid paste and was readily disseminated throughout the cellulose nitrate solution to produce a homogeneous composition. Water was then gradually added and mixed into the solution, the solution taking up or absorbing the water substantially without precipitation of nitrocellulose until about 150 parts had been added, whereupon a change of phase occurred and the nitrocellulose solution became dispersed as fine, discrete particles in the aqueous medium. The dispersion was of a pastelike consistency, and when spread as a thin layer on glass and then dried, resulted in a continuous, translucent film.

* Urea Resin Stoving Finishes

The initial water soluble condensation product of urea and formaldehyde (when reacted in the proportion of one gram of the former and five cc. of the 40 per cent solution of the latter) has been discovered to yield a highly resistant end product on treatment with salicylic acid. The latter, in solution in a suitable organic solvent mixture (e.g., ethyl alcohol, butyl alcohol and ethyl lactate), is incorporated with the aqueous solution of the initial urea-formaldehyde compound and enters into reaction during the stoving operation. The compositions present features of interest as protective coatings for articles of non-ferrous metals, including brass and aluminum. Application may be by dipping or spraying, and the film is finally hardened by stoving for twenty minutes at 135° C., when the coating passes into the insoluble state. Successive coats can therefore be applied without danger of re-softening provided the stoving operation is carried sufficiently far. Even very thin hardened films of the composition are claimed to exhibit prolonged resistance to sulfur compounds, air, moisture, salt spray, alcohol, acetone and perspiration.

Paint Remover

5 gal. Benzol
3 gal. Ethyl Acetate
2 gal. Butyl Acetate
3½ lb. 122° M. P. Paraffin
½ oz. Nitrocellulose

Dissolve nitrocellulose in acetates. Dissolve paraffin in benzol. Mix two.

Waterproof Show Card Ink

Hydromalin 13.8 lb.
Carnauba Wax 25 lb.

Heat together for ½ hr. at 120–140° C. Turn off heat and dissolve with stirring.

Any oil soluble dye 0.3 lb.

When temperature has fallen to 100° C., add while stirring vigorously,

Distilled Water, Boiling 178 lb.

Stir until uniform.

* Varnish for Wax Coated Surfaces

The varnishing of surfaces coated with paraffin wax, especially, has been well nigh impossible because of the length of time required for drying.

The following cold varnish dries very rapidly on waxed surfaces:

Ester Gum 25
Acetone 75

Allow to stand overnight and stir before using. While this varnish separates in two layers and is cloudy, it will give a clear film and should not be filtered. No other solvents or proportions of ingredients will produce as good results.

Wood-Oil Stand-Oil (Thickened Wood-Oil)

A batch of raw wood-oil, in preferably not over 10 gallons at a time, is heated as rapidly as possible (within 20 to 30 minutes) to 260° C. Pull off from the fire at this temperature. The temperature will rise automatically (polymerization) and as soon as a temperature of 280° C. is reached, the reaction is stopped by the addition of 3 to 5 gallons of cold thin linseed stand-oil, so that the temperature falls to 240° C. As soon as the consistency of a thin stand-oil is reached, further thickening is stopped by pouring the oil into a cold large flat vessel.

The oil thus cooled, now serves for the quick cooling of the next batch of thickened wood-oil. It is best to retain from the cold thickened wood-oil as much as is necessary for the next batch. In the course of time, the linseed oil content of the thickened wood-oil will decrease, and eventually be eliminated altogether, leaving *pure thickened wood oil.*

Unfortunately, raw wood-oil is sometimes found on the market, that even at a temperature of 260° C. will not by itself cause internal further rise of temperature nor polymerize further. It is, therefore, recommended to make a preliminary test of each new delivery of wood-oil. This test is to show exactly when internal heating takes place. If at 260° this internal heating does not set in, it will be necessary to raise the temperature a further 5° to 10° C., or at any rate, as high as is required until internal heating commences. In this case, the cooling down with cold oil is not carried out at 280° C., but at 285° C. or 290° C. It is, therefore, necessary to establish beforehand the proper temperature for the preparation of the wood-oil stand-oil, because the heating of the wood-oil has to be carried out quickly, in one action. The heating must not be interrupted in any way, as even by a temporary slight cooling or keeping at the same temperature for even a moment, the wood-oil would be gelatinized by the renewed heating.

Medium Long-Oil Varnish for Inside and Outside Use

Ingredients.—100 lb. Albertol 209L, 100 lb. Varnish linseed oil, 125 lb. Linseed stand-oil, 75 lb. Thickened wood-oil, 0.3 lb. Cobalt (calculated as metal), 150–200 lb. Thinners.

Procedure.—The varnish linseed oil is heated to 150–200° C. (302–392° F.) and the Albertol gradually fed into the hot oil, at such a rate that no accumulation of undissolved Albertol takes place. When the Albertol is all in, the batch is heated to 240–260° C. (464–500° F.) and the temperature maintained until a small test taken from the batch and thinned down with double the proportion of thinners after cooling down to normal temperature under the tap, shows no signs of cloudiness. When this point is reached, and not before, the two thickened oils are added, and the heating of the batch is continued at 200–220° C. (392–428° F.) until a further test remains quite bright, when tested as just described. The batch is then allowed to cool down and during the cooling, first the driers, then finally the thinners are added. In using a fluid siccative containing 2% metallic cobalt, the quantity required for the above recipe is 15 lb.

Enamel Varnish

Ingredients.—100 lb. Albertol 111L, 160 lb. linseed stand-oil, 40 lb. thickened wood-oil, 0.15 lb. cobalt (calculated as metal), 125–175 lb. thinners.

Procedure.—The previously prepared thickened oils are mixed and heated to 150° C. (302° F.). The Albertol is then gradually fed into the hot oils at such a rate that no accumulation of undissolved Albertol takes place. When all the Albertol is in, the batch is allowed to cool, and during the cooling, first the driers and finally the thinners are added. When using a liquid cobalt drier containing 2% metallic cobalt, 7.5 lb. of the liquid drier are required for this formula.

Long-Oil Boat Varnish (Yacht Varnish, Marine or Submersible Varnish, Non-Spotting Outside Varnish)

Ingredients.—100 lb. Albertol 111L, 300 lb. thickened wood-oil, 0.3 lb. cobalt (calculated as metal), 100–300 lb. thinners.

Procedure.—The previously prepared thickened oil is heated to 150° C. (302° F.) and the Albertol is gradually fed into the hot oil, at such a rate that no accumulation of undissolved Albertol takes place. When all the Albertol is in, the batch is allowed to cool, and during the cooling, first the driers, then finally the thinners are added. If a fluid cobalt drier, containing 2% metallic cobalt be employed, 15 lb. of it will be required for this batch.

LEATHER, HIDES, SKINS, FURS

Chrome Tan Calf Finish

Gelatin		1½ oz.
Casein		1½ oz.
Borax		¾ oz.
Shellac		2 oz.
Dextrine		3 oz.
Water	to make	1 gal.
Pigment	to suit	

Kip Butt Finish

Shellac	6 oz.
Gelatin	6 oz.
Soap	4 oz.
Water	1 gal.

Percentage of pigments as required.

Leather Rolls, Coating for

Red Lead	2.5 oz.
Clovel	2.5 oz.
Lampblack	2 oz.
Glycerol	2.5 oz.
Gelatin	1.5 lb.
Acetic Acid	1 gal.

Patent Leather Softening Emulsion

Castor Oil	4 parts
Casein	4 parts
Methylated Spirits	1 part
Benzol	1 part
Water	50 parts
Preservative	A trace

Imitation Leather Dressing

A transparent dressing for imitation leather may be made as follows:

½ second dope solution (nitrocellulose approximately 30%)	19 lb.
Wood alcohol	33 gal.
Castor Oil	2 qt.
Amyl Acetate	13 gal.

Should a colored dressing be desired a proper dye may be added to the above solution to obtain desired shade.

* Artificial Leather Base

A. Water	100
Acetone	400
Pyroxylin	500
Tricresyl Phosfate	120
Castor Oil	250
B. Crepe Rubber	50
Benzol	500

Solution A and B are allowed to swell separately and then milled together until homogeneous.

Leather Finishes

Unpigmented finishes, known as seasonings are applied in dilute solutions to the grain side, leaving a very thin flexible film, sufficiently hard to take a polish when the leather is glazed. That is when the leather is rubbed on glass or agate.

Egg Albumen Finish (for light colored leather)

Egg Albumen	1.5 parts
Milk	4.5 parts
Water	94.0 parts

The above are thoroughly mixed together. This film becomes insoluble to water when exposed to light and air over a period of time. A much more rapid method of rendering it insoluble is by ironing the skin or by treating it with a dilute solution of a metallic salt which does not react with the tannin of the skin.

Note: In making the above mixture, care must be taken not to exceed 130° F. otherwise the albumen will coagulate.

Blood Albumen Finish (for glazed black leather)

Blood Albumen	10 to 18%
Nigrosine	1%
Glycerine	½%
Milk	10%
Water to make 100%	

The skin is also ironed to render the film insoluble.

Temperature of mixing should not exceed 130° F.

All formulae preceded by an asterisk (*) are covered by patents.

Casein Finish

Only lactic casein should be used, and not rennet casein.

Casein	2 %
Borax	0.35%
Water	90 %
Milk	10 %

The casein is added to the warm milk and water at about 130° F. and the borax is stirred in afterwards. Formaldehyde is added as a fixative. The formaldehyde (less than 10%) must be added cold, very slowly in a thin stream with constant agitation to the cold casein solution, otherwise it will cause the casein solution to gel.

Nitrobenzene is added as a preservative.

Coloring Leather Black

Make a thin paste of Paris Paste and water and rub into the leather. When dry coat with a bright drying wax emulsion or shellac solution. This gives a permanent non-fading black.

Cellulose Finish for Patent Leather Splits

After the usual rolling and smoothing processes, the splits are brushed free from dust. They are then given two priming coats and a final gloss finish.

Priming Coat:

Celluloid	100 gm.
Amyl Acetate	100 gm.
Ethyl Acetate	50 gm.
Acetone Alcohol	300 gm.
Fusel Oil	300 gm.
Solvent Naphtha	100 gm.
Alcohol	100 gm.
Castor Oil	125 gm.
Mineral Dye (Umber)	50 gm.

The celluloid is dissolved in the mixture of amyl acetate, ethyl acetate and acetone alcohol. The dye is dissolved in the castor oil and a little of the solvents. It is then milled and added to the dissolved celluloid together with the rest of the solvents. The mixture is blended in a mill and applied to the splits by brush and dried at 35° C. When dry, the leather is pressed and a second coat of primer is given. When dry, the flesh side of the splits is wetted down and the grain side pressed with a grain-patterned plate. It is then sprayed with the final gloss finish.

Gloss Finish:

Celluloid	100 gm.
Amyl Acetate	100 gm.
Ethyl Acetate	150 gm.
Acetone Alcohol	300 gm.
Fusel Oil	200 gm.
Solvent Naphtha	200 gm.
Alcohol	200 gm.
Castor Oil	100 gm.

Solution of above is effected similar t the priming coat.

Leather Finish

Dissolve

1 oz. Nigrosine sol. in spirit in a mixture of
3 gills spirit shellac solution and
¾ gill acetine by heating on the water bath, allow to cool and filter.

Spirit Shellac Solution

is prepared by dissolving
8 oz. shellac in
1 gallon methylated spirit by heating on the water bath, filter, and allow to cool.

The leather is brushed over once o twice with this solution and after dryin polished with a cloth with or without th application of cream.

Leather Finish

A typical example of wax pigmen finish—a russet finish—is as follows:

Boil 40 lbs. grey carnauba wax wit 4 lbs. caustic soda and 5 gallons of wate for at least 8 hours, making to origina volume with water, until saponificatio is complete; often a further boiling i necessary. Then add the following pig ments:

Venetian Red	3 lb.
Raw Umber	11 lb.
Brown Acid Dye	2 lb.

and more water as required.

Artificial Leather Dope

High grade for hand finishing.

8 oz. Pyroxylin (30–40 second viscosity)
1 qt. Butyl Acetate
1 pt. Amyl Acetate
1 pt. Butanol
2 qt. Toluol or solvent Naphtha
1 oz. Acetanilid
2 oz. Camphor

Cheaper grade of Artificial Leather Dope

26 oz. Pyroxylin
2 pt. Ethyl Acetate
1 pt. Methyl Acetate
1 pt. Denatured Alcohol
4 pt. Benzol
2 oz. Camphor

Castor or Rapeseed Oil to be used as plasticizer for both of above.

Pyroxylin artificial leather is made from a cotton fabric, upon which has been built up a plurality of coats of mixtures of oils pyroxylin and plasticizers together with pigments to give the desired color. When the desired thickness has been attained the material is run through an embossing machine where, under proper conditions, the desired grain effect is impressed into the fabric. If a hard finish is desired a nitrocellulose coating with a minimum of oil is applied as a final measure. But since from 3 to 30 coats are applied it is probably economical to use low grade dope for the intermediate coats and a high grade one for the first two coats and the last two or three coats. The dope itself is applied by a blunt knife operated by a machine. For this reason they are rather viscous. The manipulation of the solvent formulae to give the desired qualities together with cheapness is a very specialized art and each manufacturer cherishes what he conceives to be the best and cheapest formula. To avoid blushing when using cheap low boiling solvents use forced drying under heated drying tunnels at a temperature of 150° to 200° F.

Split Leather

Split leather is technically treated the same as cotton cloth, but has the added advantage of it being possible to correctly call it "leather" and a compensating cost from splitting with that of only requiring three coats whereas 6 to 30 coats are used on cotton. Because of the irregular shape of the hide the dope is applied by hand with a 2½" by 6" swab and since it is brushed it is necessary to use high boiling point solvents and, in the case of black or patent leather, each coat is pumiced smooth to remove all flow and brush marks.

---

* Artificial Leather

A suitable cloth is coated with a composition consisting of 1 pt. of nitrocellulose, 4–5 pts. of linseed oil (blown with air at about 250° so that its viscosity at 25–30° is 60–75 sec. as measured by the time taken by a steel ball of 0.25 in. diam. to fall through 12 in.), and a pigment dissolved in a mixed solvent (*e.g.*, EtOAc 30, C_6H_6 30, methylated spirits 40 pts.); linseed oil may be replaced by other drying or semi-drying oils, and a drier may be incorporated.

---

* Imitation Leather Finish

A 9:1 mixture of tung and linseed oils is heated to 249°, then allowed to cool to 243°, PbO is added, and the temp. maintained at 238–243° for 30 min. Mn and Co linoleates, and synthetic or natural resin, *e.g.*, Amberol (B.S. 1, light), South Sea gum, are then added, the temp. is restored to 218°, and the batch thinned out with a mixture of heavy and light petroleum naphthas.

Leather Stain Remover

A solution for removing stains from the flesh side of leather is composed of the following:

Water	250 cc.
Oxalic Acid	3 gr.

Waterproof Boot Dressing

Spermaceti	3 oz.
Raw India Rubber	6 dr.
Tallow	8 oz.
Hogs Lard	2 oz.
Amber Varnish	5 oz.

Leather (Matt) Finish

Dissolve 1 lb. of white Borax Chip soap in 4 qts. of water and add to it 2 qts. of sulphonated castor oil, and boil until you get a perfect soft soap or emulsion. Add to the above a solution made from 4 oz. of flaxseed thoroughly leached in 2 qts. of water, and then add 6 oz. of gelatine dissolved in 2 qts. of water, and 4 oz. of logwood crystals should be added in the dry powdered form, 1 lb. of lamp black and 4 oz. of direct black. When all is together in the kettle you should boil for about one hour, then add sufficient cold water to make a total of 3 gallons and then heat to about 125° F., and stir well until mixture is perfectly smooth.

The above should be boiled in the steam jacket or over the fire; it cannot be done with the steam pipe on account of the water from the exhaust.

The above gives an excellent oily finish, and if your chrome matt leather should feel too rich or oily on the face, you might reduce the amount of sulphonated castor oil used. The gelatine is used to make the finish adhesive and by the use of a larger amount a brighter finish will be produced, particularly when the leather has been ironed. The lamp black gives the matt calf the dull appearance desired.

The weights as given above are as follows:

2 qt. sulphonated castor oil.
4 oz. flaxseed.
6 oz. gelatine.
4 oz. logwood crystals.
1 lb. lamp black.
4 oz. direct black.
1 lb. white Borax chip soap.

Finishing of Black Vegetable Tanned Calfskins

After tannage, the goods are well washed, struck out, equalized and retanned in sumac at 30 degrees C., for 100 skins about 50 or 60 pounds of sumac are dissolved in hot water. The goods go into a sumac bath which has been used for a previous pack and stay there for 24 hours and then go to a new bath, then follow horsing up, setting out, boiling of the grain with a clear cod oil, hanging in the air, striking out of the flesh and fat-liquoring of the flesh with a mixture of degras and cod oil, too much grease should be avoided in order to produce the brilliant grain. In place of the above the goods may be fat-liquored in the drum giving 10 to 12 per cent. of fat-liquor. The following is a good recipe:

For 100 pounds of goods—
5 pt. of cod liver oil,
5 pt. of moellon,
5 lb. of Marseilles soap,
5 oz. of Borax and 100 pints of water.

After fat-liquoring, they are hung up, struck out on flesh and grain and dried out, stored for a few days and then blacked. If pure iron black be employed, a solution of logwood to which it is well to add a little potassium bichromate and sodium carbonate is applied to the grain and made to penetrate. When the logwood has penetrated, the solution of iron is similarly applied.

Excellent results can be obtained with aniline blacks, which are simple to apply by passing the solution over the goods with a brush. It is preferable to use a basic dye rather than an acid black when it is to be applied with a brush.

After dyeing, they are given a light coat of oil on the grain, partially dried, boarded in several directions, laid in pile over night, cleaned on the grain with a little barberry juice, dried and rubbed with a soft flannel. To obtain a brilliant finish a light coat of finish should be applied after the barberry juice, e.g., a solution of 10 per cent. of blood albumen in water should be applied, the goods dried, glazed on the machine, boarded, again given a coat of finish, dried, glazed and finally boarded again.

Dyeing Chrome Side Leather Black

Recipe No. 1 Logwood and Bichromate of Potash. For dyeing 100 lbs. of leather, washed and shaved ready for coloring, use:

Logwood Crystals	1½ lb.
Extract of Fustic	4 oz.
Borax	3 oz.

Boil the logwood and Borax, until dissolved, in 6 gallons of water. Then dissolve the fustic paste in 2 gallons of hot water and stir it into the logwood liquor. Then add enough cold water to make 12 gallons of dye. In a pail dissolve:

Bichromate of Potash 1¼ oz.

Put the leather into the drum with 3 or 4 gallons of water and run the drum five minutes to wet the leather. Then pour the prepared logwood liquor at 12° Fahr. into the drum and run the leather in it twenty minutes. Next pour the bichromate of potash solution into the drum and continue the drumming for ten minutes, when the process should be complete. Wash the leather in three changes of water; then fat-liquor it.

Recipe No. 2. Logwood and Titanium Salts. Prepare a logwood solution by boiling logwood and Borax, then add fustic paste and have 12 gallons of the dye as described in Recipe No. 1. In a little hot water in a pail dissolve for 100 lbs. of leather:

Titanium Potassium Oxalate 6 oz.

Run the leather in the logwood liquor twenty minutes. Then add the titanium solution to the liquor and run the drum fifteen minutes. A good black results. The leather should then be washed, fat-liquored and finished.

To get the black deeper into the leather or through it, drum it in palmetto extract, then in 3 ozs. of titanium salts in solution; add the logwood, and after twenty minutes, pour in 3 more ozs. of titanium salts in solution, and after running the drum ten minutes longer, wash and fat-liquor the leather.

Recipe No. 3. Logwood and Nigrosine. This process colours the flesh blue and the grain black. For each dozen sides, dissolve 8 ozs. of nigrosine in hot water, and drum the leather in the solution twenty minutes or until the color is well taken up. Then drain off all the water and fat-liquor the leather with a suitable fat-liquor, after which, black the grain on a table by brushing in logwood and copperas or logwood and bichromate of potash, first applying the logwood and then the copperas or other striker. When the grain has become black, wash it, set it out, apply a coat of oil, and hang the leather up to dry.

When leather is drummed in a logwood liquor containing Borax until the color is taken and then spread on a table or run through a machine and blacked upon the grain, it dries out with blue flesh and black grain.

After leather is dyed with logwood and striker, it should be very thoroughly washed before it is dried and finished, to get rid of all the dye liquor.

It is considered by some tanners conducive to a better color to run the leather in a solution of palmetto extract or of gambier before giving it the dye. A good method is to apply palmetto liquor, say 2 lbs. to 100 lbs. of leather, then to drum the leather in an alkaline logwood fustic liquor, and then to develop the color with a solution of titanium salts as described in Recipe No. 2.

Good results are also secured by fat-liquoring the leather first, then running it in gambier or palmetto and afterwards dyeing with logwood and striker. A better black, as to color and durability, is obtained by using titanium salts in place of iron liquor.

Methods of Dyeing Goat Skins Black

Dyeing with Logwood and Titanium Salts. For 100 lbs. of shaved skins use:

Logwood Crystals	1½	lb.
Extract of Fustic Paste	4	oz.
Borax	3	oz.

Boil the logwood in 6 gallons of water until dissolved; then add the fustic paste and stir thoroughly; run in enough cold water to make 12 gallons of liquor. Add the Borax and then color the skins by drumming them in the dye until the logwood is taken up. The temperature of the liquor should be 120° Fahr. While the skins are running in the dye, dissolve in a pail of hot water for each 100 lbs. of skins:

Titanium Potassium Oxalate 6 oz.

When the twenty minutes are up pour this solution into a drum and drum the skins ten or fifteen minutes longer. Then wash them in warm water and fat-liquor them.

Dyeing with Logwood Acetic Acid and Nitrate of Iron. To color 100 lbs. of skins use:

Logwood Crystals	1½	lb.
Black Nigrosine	1	oz.
Borax	4	oz.
Acetic Acid	1½	oz.
Nitrate of Iron	3	oz.

Boil the logwood in a few gallons of water; add the Borax and enough water to make 12 gallons of liquor. In a pail of hot water dissolve the nigrosine. Run the skins in the logwood liquor for ten minutes; add the nigrosine and run ten minutes longer. Then dissolve the acetic acid and nitrate of iron in 2 gallons of water. Pour the solution into a drum and run the latter fifteen minutes. Then drain the liquor out of the drum, wash the skins in two or three changes of water and then fat-liquor them. The temperature of the dye liquor should be 120° Fahr.

Dyeing with Logwood and Copperas. For each 100 lbs. of skin to be dyed, prepare a logwood liquor by boiling in a few gallons of water:

Logwood Crystals	1½	lb.
Fustic Paste	4	oz.
Borax	4	oz.

Drum the skins in this liquor, of which there should be 12 gallons at a temperature of 120° Fahr. for twenty minutes. In the meantime dissolve in 3 gallons of boiling water:

Copperas	2	oz.
Bluestone	½	oz.

Add cold water to the solution to reduce the temperature to 100° Fahr. When the twenty minutes are up, pour the solution into a drum and allow the latter to rotate fifteen minutes. Then remove the skins from the drum, wash them in two or three changes of warm water and finally fat-liquor them.

Dyeing Kangaroo Skins Black

Recipe 1. For each 100 lb. of skins, dissolve by boiling in 10 gallons of water:

Logwood Crystals	1½	oz.
Fustic Paste	4	oz.
Borax	3	oz.

Add 5 gallons of cold water to the liquor and use it at 125° Fahr. Drum the skins in it for twenty minutes. While the drum is running, dissolve in a pailful of hot water:

Bichromate of Potash 1 oz.

Pour this solution into the drum and run the drum ten minutes. Then drain the liquor out of the drum and wash the skins in three changes of water. They are then ready to be fat-liquored.

Recipe 2. A good color can be obtained with logwood and titanium salts in the following manner: For every 100 lbs. of skins, boil until dissolved in 10 gallons of water:

Logwood Crystals	1½ lb.
Fustic Paste	4 oz.
Borax	3 oz.

In another tub dissolve in 10 gallons of hot water for every 100 lbs. of skins:

Titanium Potassium Oxalate 5 oz.

Put the skins and half of the titanium solution into the drum and run the drum ten or fifteen minutes; then pour the logwood liquor in and run the drum fifteen minutes; finally, to develop the colour, pour in the rest of the tianium solution and run the drum ten minutes longer. Wash the skins and finish them, but have 1 lb. of titanium salt in the barrel of seasoning and no copperas. The logwood liquor should be increased to 15 gallons by the addition of 5 gallons of cold water and used at a temperature of 125° Fahr.

Blacking Chrome Sole Leather

When the leather is blacked first and then stuffed, it is taken, a side at a time, slicked out smooth on a table and given a coat of logwood liquor, then a coat of striker, next another coat of logwood and more striker, after which it is washed, run through a wringer and put into condition for stuffing.

The logwood liquor is made of 6 lb. of logwood crystals and 2 lb. of Borax in 50 gallons of water. The striker is made of 7 lb. of copperas and 5 lb. of blue vitriol in 50 gallons of water, although any other good striker may be used.

Coloring Chromed India-Kips

An excelent colour is secured by using the following process:

For each 100 lbs. of leather ready to be coloured, boil in 10 gallons of water, 1½ lbs. of logwood crystals and 4 oz. of Borax, then stir into the liquor 4 oz. of fustic paste. Use this liquor at 125° Fahr. Drum the leather in it for one half-hour; then pour into the drum a solution of 5 oz. of titanium postassium oxalate in a pail of hot water and run the drum fifteen minutes longer. If the leather has not been fat-liquored, it should next be washed and then fat-liquored, oiled and dried. The grain should be well struck out, and oiled with a mixture of one part olive and three parts paraffin oils. Drying should be done somewhat slowly; and when dry the leather should be dampened, staked and tacked.

Chrome Liquor

The chrome liquor can also be made by dissolving ten pounds of sodium bichromate in two gallons of water, and adding to this liquor ten pounds of sulphuric acid. Then add to the solution six pounds of syrup glucose at intervals allowing the agitation to subside before adding another portion. This liquor should be diluted to 45 Bé., and fifteen pounds of it will tan one hundred pounds of skins. The dry skins, after they have been washed back, can be also chrome tanned with six pounds of tanolin dissolved in two gallons of boiling water. Drum the skins in the salt water solution ten minutes, then add the chrome liquor in portions of one-third at a time at intervals of one-half hour, drumming for two hours. Then dissolve and pour into the drum eight ounces of sodium bicarbonate and drum one-half then add six ounces more of the sodium bicarbonate and drum another hour. After the leather has been drained at least twelve hours it is washed and neutralized with Borax.

Fat-Liquor for Chrome Side Leather

No. 1. Put 10 lbs. of palmetto, fig or other good soap into a clean barrel with 10 gallons of water. Boil with steam until dissolved. Then take four gallons of neatsfoot oil and cut it by stirring into it a few ounces of Borax dissolved in hot water. Add the oil to the soap and boil again; then add 6 lbs. of moellon degras and boil until the liquor is thoroughly emulsified. Run in enough water to make 40 gallons of fat-liquor. Four gallons of this emulsion may be used for each dozen sides.

No. 2. For 100 lbs. of heavy grain chrome leather:

Fig Soap	1 lb.
German Degras	3 lb.
Neatsfoot Oil	3 lb.
Sod Oil	3 lb.
Borax	4 oz.

Boil the first three ingredients in 6 gallons of water for one half-hour. Then add the sod oil and Borax and stir thoroughly. Add water to make 12 gallons of liquor, which may be applied to the leather at any temperature between 125 and 140° Fahr. If the leather is greasy, wash it in a warm solution of Borax. If the fat-liquor is not fully taken up by the leather, pour in the drum 4 ozs. of salts of tartar dissolved in 3 gallons of hot water and run the drum fifteen minutes longer. The grain should receive a good coat of cod or neatsfoot oil before the leather is dried.

Fat-Liquor for Chrome Glove Leather

The following is given as especially suitable for glove leather:

Olive Chip Soap	12 lb.
Glauber's Salt	3 lb.
Borax	2 lb.
Sod Oil	5 gal.
Cod Oil	3 gal.
Neatsfoot Oil	1 gal.

Boil the first three ingredients for one-half hour; then add the oils and boil again about one-half hour; then fill up the barrel to make 50 gallons of fat-liquor. Use 7 lbs. of this fat-liquor for 100 lbs. of leather. Dilute it with hot water and use at 125° Fahr., drumming the leather in it for forty minutes.

Fat-Liquor

An excellent fat-liquor for chrome glove skins is made of one pound of soap, eight ounces of neatsfoot oil, one and a half pounds of egg yolk and two ounces of Borax for one hundred pounds of leather. The soap and Borax are boiled and dissolved in a few gallons of water; the oil is then added and the mixture thoroughly stirred. A few gallons of cold water are added to reduce the temperature to 90 degrees, when the egg yolk is added and the liquor thoroughly stirred is used at a temperature of 120 degrees. There should be twelve gallons of it. The preparation of sheepskins is about as follows: They are dewooled with a paint of sodium sulphide and lime or one made of lime and red arsenic; limed for a few days in clear white lime.

Fat-Liquor for Sheep Leather

Put 10 lbs. of potash soap into a clean barrel with 10 gallons of water, and boil and stir it until it is dissolved. Into 4 gallons of best neatsfoot oil stir 4 ozs. of Borax dissolved in a quart of boiling water, taking care to stir thoroughly to cut the oil. Put the oil into the soap solution and stir thoroughly. Then run in enough cold water to make 50 gallons of fat-liquor. The user can, if he desires, add 10 lbs. of egg yolk to the oil and soap solution, but not until it has been cooled down to 75° Fahr. with cold water.

Leather Heavily Fat-Liquored

Chrome leather that has been so heavily fat-liquored that the grain is greasy should be given a sig before the logwood liquor is applied with a brush. The object of this treatment is to cut the grease out of the grain so that the logwood can penerate the grain. For this purpose a warm solution of Borax is very beneficial. The strength of the solution must depend upon the condition of the leather.

Dressing Oil as an Alkali Fat-Liquor with Borax

Mix in a wooden tank arranged with open steam coil. Use 3 to 5% of Borax in ⅓ of water to ⅔ of oil. Heat the mixture with steam. When the Borax is thoroughly dissolved and the mixture stirred up, you will have a splendid fat-liquor which may be used as soon as it is cold.

Caution

It is very necessary in mixing this oil to use a wooden tank or a lead lined tank, and an open steam coil, as an iron tank or dry heat would have the effect of darkening the oil.

Fat-Liquoring

For 100 lbs. of skins take:

2 lb. of Marseilles Soap,
1 lb. of Neatsfoot Oil,
35 gr. of Borax,
4 to 5 gallons of water at 60 degrees C. and drum for 40 minutes.

The skins are now passed through water and if the shade is not sufficiently black they can be darkened further. For 50 skins take 5 gallons of logwood infusion, and pass the skins through it three times. Then the skins are immersed in a bath of sulphate of copper. Rinse the skins in water, set out by machine, apply a light coat of neatsfoot oil and hang up to dry.

Finishing is as usual with the following, which will produce a fine lustre:

4 litres of logwood infusion,
250 cc. of ox blood,
500 cc. of milk,
300 gr. of barberry juice,
13 whites of egg,
60 gr. ammonia,
60 gr. alcohol,
12 gr. sulphate of iron,
2 litres of nigrosine solution,
500 cc. of gall nut infusion.

Recipe for Fat-Liquor

Cook seven pounds olive chip soap and seven pounds fig soap in 25 gallons of water. Add one pound of powdered

Borax and cook until cut, then let cool to 120° and add six gallons egg yolk. Fill the barrel with cold water to make fifty gallons. This will fat-liquor sixty-five to seventy horse hides, kip or cow hides. The leather should be run for one hour in stuffing mill, which should be kept at 120°.

Olive Oil, being a vegetable oil, produces lasting effects on leather. It does not evaporate, spew or become gummy. The lasting effects of this oil have long been known, but because of its high cost it has been used only on the fine grades of leather.

Degreasing Before Dyeing

After the fat-liquored skins have laid in pile for about twelve hours, they are degreased by brushing over on the grain with 3 per cent. Borax and 2 per cent. good white soap, made into a solution with 95 parts soft water. After washing the grain with this, the skins are rinsed in warm soft water, dried for dyeing, or in some cases sponged over with linseed mucilage which retards the fixing of the colour, and keeps the grain a uniform shade but it must be allowed to dry before the dye is applied. Some dyers prefer to dye before fat-liquoring, because less dye is required, and if acid dyes are used sulphuric acid may be used in the dye-bath. It must be remembered, however, that fat-liquoring subsequent to dyeing removes a lot of the dyestuff. Where the dyeing follows the fat-liquoring under no circumstances must sulphuric acid be allowed in the dye-bath, as this will precipitate free fatty acid on the leather and cause uneven dyeing.

Formula for Producing Plump Leather

Soaking.

Dissolve five pounds of Borax in hot water and add it to 1,000 gallons of water and soak hides from 24 to 48 hours changing the water, if necessary, where the hides are very dirty. Have your stock as clean as possible before it is put into the limes.

Liming.

The best method depends somewhat on the kind of stock being made. Starting stock in new and strong limes and finishing up with weak ones makes the leather very plump as well as soft, but it is preferable to start in weak limes and finish up in strong ones. Extreme plumping at the start tends to weaken the fibres of the leather.

To Give a White Flesh Side to Calf Leather

After tanning with sumac, the skins are dried and shaved. They are then fulled very soft, dyed on the grain side only, racked, stretched over a frame and dried. When the grain side is finished, the flesh side is pumiced, coated with the white dressing and glass papered. This white dressing is made as follows: For a dozen skins, 2 pounds of Spanish white and 12 ounces of white tallow soap are stirred together with the white of 12 eggs and 2¾ gallons of water.

The skins after a thorough cleansing are repeatedly coated with a mixture of 100 parts of glycerine, 0.2 of salicylic acid, 0.2 of picric acid, and 2.5 of Borax. They are then nearly dried and impregnated in a dark room with a solution of bichromate of potash, after which, drying is completed and both sides given a coat of shellac varnish.

Variety of Useful Shades

By increasing or decreasing the quantity of blue or black in the medium and dark browns a large variety of useful shades can be obtained. For the light shades of brown the yellows are used as the shading agents.

To finish colored leather, take:

Egg Albumen	6 oz.
Glycerine	2 oz.
Borax	2 oz.
Shellac	4 oz.

Dissolve the albumen in lukewarm water, then dissolve the shellac with the Borax and add to the albumen together with the glycerine, and use cold. It is always advisable to add to the seasoning mixture a little dyestuff of the same kind as that used for dyeing the skins. The above ingredients will be found enough to make 10 gallons of seasoning. The method of procedure, after applying the season, is the same as for blacks, except that if a finishing oil for colors is found desirable, special attention must be given to the selection of the right quality. The oil used should not be greasy; its consistency should be thin, and it should rapidly disappear into the leather. The object of its use is merely to lubricate and soften the grain, and only a light application is necessary.

Blue or Purple Coloring

In some cases the sides are dyed blue or purple on the flesh. This is performed by dissolving 6 pounds of logwood paste and 2 pounds of Borax and a small quantity of blue aniline in warm water, heated to the boiling point. If a purple-black is required, an additional quantity of Borax and a small quantity of blue aniline should be added, the quantities to be regulated by the shade required. The sides are run in this liquor for 20 minutes.

The finish is made up in the following way: For 10 gallons of season, 4 ounces ruby shellac, 2 ounces ammonia, 2 ounces haematin, 6 ounces nigrosine, 3 ounces chrome leather black and 2 ounces glycerine are used. First dissolve the shellac in water, to which has been added the ammonia; then dissolve haematin, nigrosine and chrome leather black; stir the whole together until fairly cold. Give the sides a coat of this and air off, then glaze; then another coat, and glaze again, after which they can be grained and are ready to be sent out.

Tanning Fur Skins

Cut off the useless parts of the skin, and then soften it by soaking, so that all flesh and fat may be scraped from the inside with a blunt knife. Soak the skin next in warm water for an hour, and during that time mix equal quantities of Borax, saltpetre and Glauber salts with enough water to make a thin paste. About half an ounce of each ingredient will give enough for an opossum skin, and proportionately more will be required for larger ones. When the skin has soaked in the warm water, lift it and spread it out flat, so that the paste may be applied with a brush to the inside of the skin; more paste will be required where the skin is thick than where it is thinner. Double the skin together, flesh side inwards, and put it in a cool place for twenty-four hours, at the end of which time it should be washed clean, and treated in the same way as before with a mixture of one ounce of sodium carbonate (washing soda), one-half ounce Borax and two ounces hard, white soap; these must be melted slowly together without being allowed to boil. The skin should then be folded together again, and put in a warm place for twenty-four hours. After this, dissolve four ounces alum, eight ounces salt, and two ounces sodium carbonate (baking soda) in sufficient hot water to saturate the skin; the water used should be soft, preferably rain water. When this is cool enough not to scald the hands, the skin should be immersed and left for twelve hours; then wring it out and hang it up to dry. The soaking and drying must be repeated two or three times, till the skin is soft and pliable, after which it may be rubbed with fine sandpaper and pumice stone to obtain a smooth finish.

For Tanning White Goat Skins

4 oz. Sulphate of Alumina
4 oz. Sulphate of Potash
2 oz. Borax

If the skins are very greasy, use 3 oz. of salt petre for a driver. All alum skins should be dried out in the air and dampened by sprinkling a little water on the flesh. Roll them up and allow them to stand for a day or two then arm crutch them.

For Tanning Snake Skins

A combination tannage is best. The ingredients are salt, alum, gambier and common flour. One third of each of the chemicals and one pound of flour. Cover with about a couple of gallons of water. Add about five ounces of Borax to make the skins soft.

The Graining Process

Graining is an art well understood in morocco finishing, and therefore it is not necessary to describe it in much detail. The skin is sometimes bruised on the flesh before graining, but the general idea is to get as pronounced a marking as possible, and this is done by crossing and recrossing in the ordinary way. Morocco graining, however, is a process in which there is every opportunity for the workman to use his brains and experience, and for this reason none but the best workmen are usually employed for this purpose.

After graining in the damp state, the skins are aired off, and a coat of the following season carefully applied. Dissolve 4 to 6 ounces best orange shellac in hot water, and add 2 ounces of Borax, making up the whole into six gallons of finish.

Blacking Kangaroo Calf and Sides

After stock has been stuffed and dried out it should lay some days to mellow down. Then yellow back it. Take a 50-gallon barrel and put in:

10 gal. of Water
1 gal. Neatsfoot Oil
1 lb. Sal Soda
1 lb. of Borax
5 lb. of Turmeric

Boil well, then fill up barrel with cold water in drum and 12 pails of mixture and turn 15 to 20 minutes, then take out and color as follows:

1 pail of Sig.

1 pail of Logwood Liquor
1 pail of Black
A brush for each

Imitation English Oak

To make an imitation English oak or to bleach dark leather, submerge same in a solution composed of

4 ounces Borax
4 ounces Oxalic Acid

thoroughly dissolved in

1 gallon of water.

Deliming Hides and Skins

Crocodile, Lizard and Python Skins. The dehydrated skins are restored by soaking in cold water softened with 8 lbs. of borax per 1,000 gallons, worked over the beam on the flesh side, and limed to loosen the scales and separate the fibers. The skins are given 10 to 15 days in fairly mellow lime liquors (no sodium sulphide), and hauled daily. The strength of the lime liquor is maintained by small additions of lime paste on alternate days. The scales are removed with an unhairing knife, swollen flesh detached, and the pelts washed in a paddle with running water. Deliming is accomplished in the same vessel, using 2 lbs. of boric acid for 100 lbs. of skins, and paddling for about 2½ hours. Finally the skins are washed for 20 minutes in clear water.

Tanning Reptile Skins

Dehydrated skins are soaked in water (8 lbs. Borax per 1000 gallons); worked over beam on flesh side and treated with sat. lime solution to loosen scales. Weak lime liquors are used now to treat skins for 10 days. Remove scales by knife and wash pelts in running H_2O. Then delime with 2 lbs. boric acid per 100 lbs. skins, paddling for two to three hours; then a clear H_2O wash.

Bates stock immersed in water containing Fastan to bring gravity to 1° Bé. On next day strengthen liquor to 2° Bé. On 3rd day, strengthen to 3° Bé. Remove skins on fourth day and place in Hypo bath for 24 hours. Then wash and fat liquor.

Home Tanning of Leather and Fur Skins

Preparation of the hide or skin for tanning may be started as soon as it has been taken off the animal, drained, and cooled from the body heat. Overnight will be long enough. If tanning is not to be started at once or if there are more hides than can be handled at one time, the hides may be thoroughly salted and kept for from three to five months. The hides must never be allowed to freeze or heat during storage or tanning. Some tanners state that salting before tanning is helpful. It can do no harm to salt a hide for a few days before it is prepared for tanning.

The directions here given have been prepared for a single heavy cow, steer, or bull hide weighing from 40 to 70 pounds or for an equivalent weight in smaller skins, such as calf or kip skins. The heavy hides are best suited for sole, harness, or belting leather. Lighter hides weighing from 20 to 40 pounds should be used for lace leather.

Preliminary Operations

Before it is tanned a hide or skin must be put through the following preliminary operations. As soon as the hide or skin has been put through these processes, start the tanning, following the directions given for the particular kind of leather desired.

Slaking Lime

Put from 6 to 8 pounds of burnt or caustic lime in a clean half barrel, wooden tub, or bucket, with a capacity of at least 5 gallons. Use only good-quality lime, free from dirt and stones; never use air-slaked lime. To the lime add about 1 quart of water. As the lime begins to slake add more water, a little at a time, to keep the lime moist. Do not pour in enough water to quench the slaking. When the lime appears to be slaked, stir in 2 gallons of clean water. Do all this just as in making whitewash. Slake the lime on the day before the soaking of the hide is begun, and keep the limewater covered with boards or sacks until ready to use it.

If available, fresh hydrated lime, not air-slaked, may be used instead of the burnt or caustic lime. In this case use from 8 to 10 pounds in 4 or 5 gallons of water.

Soaking and Cleaning

If the hide has been salted, shake it vigorously to remove most of the salt. Spread it out, hair side down, and trim off the tail, head, ears, all ragged edges, and shanks.

Place the hide, hair side up, lengthwise, over a smooth log or board, and, with a sharp knife, split it from neck to tail, straight down the backbone line, into two half hides, or "sides." It will be more convenient in the later handling, especially when the hide is large, to then split each side lengthwise through the "break," just above the flanks, into two strips, making the strip with the backbone edge about twice as wide as the belly strip. Thus a whole hide will give two sides or four strips. If desired, small skins need not be split. In these directions "side" means side, strip, or skin, as the case may be.

Fill a 50-gallon barrel with clean, cool water. Place the sides, flesh side out, over short sticks or pieces of rope and hang them in the barrel of water. Let them soak for two or three hours. Stir them about frequently to soften, loosen, and wash out the blood, dirt, manure, and salt. The sticks or pieces of rope may be held in place by tying a loop of cord on each end and catching the loops over nails in the outside of the barrel near the top.

After soaking for about three hours take out the sides, one at a time, and place them, hair side up, over a "beam."

A ready-made beam can be bought. A fairly satisfactory one may be made from a very smooth slab, log, or thick planed board, from 1 to 2 feet wide and 6 to 8 feet long. The slab or log is inclined, with one end resting on the ground and the other extending over a box or trestle so as to be about waist high.

With the side lying hair side up over the beam, scrub off all dirt and manure, using if necessary a stiff brush. Wash off with several bucketfuls of clean water.

Turn the side over, flesh side up, and scrape or cut off any remaining flesh. Work over the entire flesh side with the back edge of a drawing or butcher knife, held firmly against the hide, while pushing away from the body. Wash off with one or two bucketfuls of clean water. This working over should always be done.

Refill the soak barrel with clean, cool water and hang the sides in it as before. Pull them up and stir them about frequently until they are soft and flexible. Usually a green or fresh hide needs to be soaked for not more than from 12 to 24 hours and a green salted hide for not more than from 24 to 48 hours.

When the sides are properly softened—that is, about like a fresh hide or skin—throw them over the beam and thoroughly scrape off all remaining flesh and fat. It is of the greatest importance to remove all this material. When it can not be scraped off, cut it off, but be careful not to cut into the hide itself. Even should there appear to be no flesh to take off and nothing seems to be removed, it is necessary to thoroughly work over the flesh side in this way with the back of a knife. Finally wash off with a bucketful of clean water.

The side must be soft, pliable, and clean all over before being put into the lime, which is the next step.

Liming

Wash out the soak barrel. Pour in all of the slaked lime; nearly fill the barrel with clean, cool water; and stir thoroughly. Place the sides, hair side out, again over the short sticks or pieces of rope, and hang them in the barrel so that they are completely covered by the limewater. See that the sides have as few folds or wrinkles as possible and also be sure that no air is trapped under them. Keep the barrel covered with boards or bags. Pull up the sides and stir the limewater three or four times each day until the hair will come off easily. This takes from 6 to 10 days in summer and possibly as many as 16 days in winter.

When thoroughly limed, the hair can be rubbed off readily with the hand. Early in the liming process it will be possible to pull out the hair, but the hide must be left in the limewater until the hair comes off by rubbing over with the hand. For harness and belting leathers leave the hide in the limewater for from 3 to 5 days after this condition has been reached.

Unhairing

When limed, throw the side, hair side up, over the beam, and, with the back edge of a drawing or butcher knife, held nearly flat against the side, push off the hair from all parts. If the side is sufficiently limed, a curdy or cheesy layer of skin rubs off with the hair. If this layer does not rub off, the side must be returned to the limewater. After removing the hair, put the side back in the limewater again for another day, until any fine hairs that may remain can be easily scraped off. Now thoroughly work over the grain or hair side with a dull-edged tool to "scud" or work out as much lime, grease, and dirt as possible.

Fleshing

Turn the side over and "scud" it again, being sure to remove all fleshy matter. Shave down to the hide itself, but be careful not to cut into it. Remove the flesh by scraping and by using a very sharp knife, with a motion like that of shaving the face.

Now proceed as directed under "Bark-tanned sole and harness leather," "Chrome-tanned leather" or "Alum-tanned lace leather," depending upon the kind of leather desired.

Wastes from Liming

The lime, limewater, sludge, and fleshings from the liming process may be used as fertilizer, being particularly good for acid soils. The hair, as it is scraped from the hide, may be collected separately, and, after being rinsed several times, may be used in plastering. If desired, it can be thoroughly washed with many changes of water until absolutely clean and, after being dried out in a warm place, can be used for padding, upholstering, insulation of pipes, etc.

Bark-tanned Sole and Harness Leather

Deliming

After the sides have been put through the unhairing and fleshing operations, rinse them with clean water. Wash the sides in cool, clean water for from six to eight hours, changing the water frequently.

Buy 5 ounces of U. S. P. lactic acid (or 16 ounces of tannery 22 per cent lactic acid). Nearly fill a clean 40 to 50 gallon barrel with clean, cool water, and stir in the lactic acid, mixing thoroughly with a paddle. Hang the sides in the barrel and leave them there for 24 hours, pulling them up and stirring frequently.

Take out the sides, work over or "scud" them thoroughly, as directed under "Unhairing," and hang them in a barrel of cold water. Change the water several times, and finally leave them in the water overnight.

If lactic acid can not be obtained, use a gallon of vinegar instead.

Tanning

The sides are now ready for the actual tanning. From 15 to 20 days before this stage will be reached weigh out from 30 to 40 pounds of good-quality, finely-ground oak or hemlock bark and pour onto it about 20 gallons of boiling water.

Finely-ground bark, with no particles larger than a grain of corn, will give the best results. Simply chopping the bark into coarse pieces will not do. Do not let the tan liquor come in contact with iron vessels. Use the purest water available. Rain water is best.

Let this bark infusion stand in a covered vessel until ready to use it. Stir it occasionally. When ready to start tanning, strain off the bark liquor through a clean, coarse sack into the tanning barrel. Fill the barrel about three-quarters full with water, rinsing the bark with this water so as to get out as much tannin as possible. Add 2 quarts of vinegar. Stir well. Place the sides, from the deliming, over sticks, and hang them in this bark liquor with as few folds and wrinkles as possible. Move the sides about and change their position often in order to get an even color.

Just as soon as the sides have been hung in the bark liquor, again soak from 30 to 40 pounds of ground bark in about 20 gallons of hot water. Let this second bark liquor stand until the sides have become evenly colored, or for from 10 to 15 days. Take out of the tanning barrel 5 gallons of liquor and pour in about one-quarter of the second bark liquor. Also add about 2 quarts more of vinegar and stir it in well. Five days later add another fourth of the tan liquor only (no vinegar). Do this every 5 days until the second bark liquor is used up.

The progress of the tanning varies somewhat with conditions and can best be followed by inspecting a small sliver cut from the edge of the hide. About 35 days after the actual tanning has been started a fresh cut should show two dark or brown narrow streaks about as wide as a heavy pencil line coming in from each surface of the hide.

At this stage weigh out about 40 pounds of fine bark and just moisten it with hot water. Do not add more water than the bark will soak up. Pull the sides out of the bark liquor and dump in the moistened bark, keeping in the barrel as much of the old tan liquor as possible. Mix thoroughly and while mixing hang the sides back in the barrel. Actually bury them in the bark. All parts of the sides must be kept well down in the bark mixture. Leave the sides in this bark for about six weeks, moving them about once in a while.

At the end of six weeks pull the sides out. A cutting should show that the tanning has spread nearer to the center. Pour out about half the liquor. Stir the bark in the barrel, hang the sides back, and fill the barrel with fresh, finely ground bark. Leave the sides in for about two months, shaking the barrel from time to time and adding bark and water as needed to keep the sides completely covered.

At the end of this time the hide should be evenly colored all the way through, without any white or raw streak in the center of a cut edge. If it is not struck through, it must be left longer in the wet bark, and more bark may be needed.

For harness, strap, and belting leather the sides may be taken out of the bark liquor at this stage, but for sole leather they must be left for two months longer.

When fully tanned through the sides are ready for oiling and finishing.

Oiling and Finishing

Harness and belting leather.—Take the sides from the tan liquor; rinse them off with water; and scour the grain or hair side thoroughly with plenty of warm water and a stiff brush. Then go over the sides with a "slicker," pressing the slicker firmly against the leather while pushing it away from the body. "Slick" out on the grain or hair side in all directions. For harness, belting, and the like this scouring and slicking out must be thoroughly done.

A slicker can be made from a piece of copper or brass about one-fourth inch thick, 6 inches long, and 4 inches wide. One long edge of the slicker is mounted in a wooden handle and the other long edge is finished smooth and well rounded. A piece of hardwood, about 6 inches square, 1½ inches thick at the head, and shaved down wedge-shape to a thin edge, will also serve as a slicker.

While the sides are still damp, but not very wet, go over the grain or hair side with a liberal coating of neat's foot or cod oil. Hang up the sides and let them dry out slowly. When dry, take them down and dampen well by dipping in water or by rolling them up in wet sacking or burlap.

When uniformly damp and limber, evenly brush or mop over the grain or hair side a thick coating of warm dubbin. The dubbin is made by melting together about equal parts of cod oil and tallow or neat's-foot oil and tallow. This dubbin when cool must be soft and pasty, but not liquid.

Hang up the sides again and leave until thoroughly dried. When dry, scrape off the excess tallow by working over with the slicker. If more grease in the leather is desired, dampen again and apply another coating of the dubbin, giving a light application also to the flesh side. When again dry, remove the tallow and thoroughly work over all parts of the leather with the slicker. Rubbing over with sawdust will help to take up any surface oiliness.

If it is desired to blacken the leather, this must be done before greasing. A black dye solution can be made by dissolving one-half ounce of water-soluble nigrosine in 1¼ pints of water, with the addition, if handy, of several drops of ammonia. Evenly mop or brush this solution over the dampened but ungreased leather and then grease as directed in the preceding paragraph.

Sole leather.—Take the sides from the tan liquor and rinse them thoroughly with clean water. Hang them up until they are only damp and then apply a good coating of neat's foot or cod oil to the grain or hair side. Again hang them up until they are thoroughly dry.

When repairing shoes with this leather it is advisable, after cutting out the piece for soling, to dampen and hammer it down well, and then, after putting it on the shoe, to make it waterproof and more serviceable by setting the shoe in a shallow pan of melted grease or oil and letting it stand for about 15 minutes. The grease or oil must be no hotter than the hand can bear. Rubber heels should not be put in oil or grease. The soles of shoes with rubber heels may be waterproofed in the same way, using a pie pan for the oil or grease and placing the heels outside the pan. Any good oil or grease will do. The following formulas have been found satisfactory:

Formula 1:	Ounces
Neutral Wool Grease	8
Dark Petrolatum	4
Paraffin Wax	4
Formula 2:	
Petrolatum	16
Beeswax	2
Formula 3:	
Petrolatum	8
Paraffin Wax	4
Wool Grease	4
Crude Turpentine Gum (gum thus)	2
Formula 4:	
Tallow	12
Cod Oil	4

Chrome-tanned Leather

For many purposes chrome-tanned leather is considered to be as good as the more generally known bark or vegetable-tanned leather. The chrome process, which takes only a few weeks as against as many months for the bark-tanning process, derives its name from the use of chemicals containing chromium or "chrome." It is a chemical process requiring great care. It is felt, however, that by following exactly the directions here given, never disregarding details which may seem unimportant, a serviceable leather can be produced in a comparatively short time. The saving in time seems sufficient to justify a trial of this process.

Deliming

After the sides have been put through the unhairing and fleshing operations rinse them off with clean water.

If sole, belting, or harness leather is to be tanned, soak and wash the sides in cool water for about six hours before putting them into the lactic acid. Change the water four or five times.

If strap, upper, or thin leather is to be tanned, put the limed white sides into a wooden or fiber tub of clean, lukewarm (about 90° F.) water and let them stay there for from four to eight hours before putting them into the lactic acid. Stir the sides about occasionally. Be sure that the water is not too hot. It never should be so hot that it is uncomfortably warm to the hand.

For each large hide or skin buy 5 ounces of U. S. P. lactic acid (or 16 ounces of tannery 22 per cent lactic acid). Nearly fill a clean 40 to 50 gallon barrel with clean, cool water, and stir in the lactic acid, mixing thoroughly with a paddle. Hang the sides in the barrel, and leave them there for 24 hours, plunging them up and down occasionally.

For light skins, weighing less than 15 pounds, use only 2 ounces of U. S. P. lactic acid in about 20 gallons of water.

If lactic acid can not be obtained, use 1 pint of vinegar for every ounce of lactic acid. An effort should be made to get the lactic acid, however, for vinegar will not be as satisfactory, especially for the medium and smaller skins.

After deliming, work over both sides of the side as directed under "Unhairing."

For sole, belting, and harness leathers, hang the sides in a barrel of cool water overnight. Then proceed as directed under "Tanning."

For thin, softer leathers from small skins, do not soak the sides in water overnight. Simply rinse them off with water and proceed as directed under "Tanning."

Tanning

The tanning solution should be made up at least two days before it is to be used—that is, not later than when the sides are taken from the limewater for the last time.

Remember that this is a chemical process and all materials must be of good quality and accurately weighed, and that the specified quantities of water must be carefully measured.

The following chemicals are required: Chrome alum (chromium potassium sulphate crystals); soda crystals (crystallized sodium carbonate); and common salt (sodium chlorid).

For each hide or skin weighing more than 30 pounds use the following quantities for the stock chrome solution:

Dissolve 3½ pounds of soda crystals (crystallized sodium carbonate) and 6 pounds of common salt (sodium chlorid) in 3 gallons of warm, clean water in a wooden or fiber bucket. The soda crystals must be clear or glasslike. *Do not use the white crusted lumps.*

At the same time dissolve, in a large tub or half barrel, 12 pounds of chrome alum (chromium potassium sulphate crystals) in 9 gallons of cool, clean water. This will take some time to dissolve and will need frequent stirring. Here again it is important to use only the very dark, hard, glossy, purple or plum-colored crystals of chrome alum, not the lighter, crumbly, dull lavender ones.

When the chemicals are dissolved, which can be told by feeling around in the tubs with a paddle, pour the soda-salt solution slowly in a thin stream into the chrome-alum solution, stirring constantly. Take at least 10 minutes to pour in the soda solution. This should give one solution of about 12 gallons which is the *stock chrome solution.* Keep this solution well covered in a wooden or fiber bucket, tub, or half barrel.

To start tanning, pour one-third (4 gallons) of the stock chrome solution into a clean 50-gallon barrel and add about 30 gallons of clean, cool water; that is, fill the barrel about two-thirds full. Thoroughly mix the solution in the barrel and hang in it the sides from the deliming. Work the sides about and stir the solution frequently, especially the first two or three days. This helps to give the sides an even color. It should be done every hour or so throughout the first day. Keep the sides as smooth as possible.

After three days, temporarily remove the sides from the barrel. Add one-half of the remaining stock chrome solution, thoroughly mixing it with that in the barrel, and again hang in the sides. Move the sides about and stir the solution three or four times each day.

Three days later, once more temporarily remove the sides. Pour into the barrel the rest of the stock chrome solution, thoroughly mixing it with that in the barrel, and again hang in the sides. Move the sides about and stir frequently as before.

After the sides have been in this solution for three or four days, cut off a small piece of the thickest part of the side, usually in the neck, and examine the freshly cut edge of the piece. If the cut edge seems to be evenly colored greenish or bluish all the way through, the tanning is about finished. Boil the small piece in water for a few minutes. If it curls up and becomes hard or rubbery, the tanning is not completed and the sides must be left in the tanning solution for a few days longer, or until a small piece when boiled in water is changed little if at all.

The foregoing quantities and directions have been given for a medium or large hide. For smaller hides and skins the quantities of chemicals and water can be reduced. For each hide or skin weighing less than 30 pounds, or for two or three small skins together weighing not more than 30 pounds, the quantities of chemicals may be cut in half, giving the following solutions:

For the soda-salt solution, dissolve 1¾ pounds of soda crystals (crystallized sodium carbonate) and 3 pounds of common salt (sodium chlorid) in 1½ gallons of clean water.

For the chrome-alum solution, dissolve 6 pounds of chrome alum (chromium potassium sulphate crystals) in 4½ gallons of cool, clean water.

When the chemicals are dissolved pour the soda-salt solution slowly into the chrome-alum solution as already described. This will give one solution of about 6 gallons which is the *stock chrome solution*. For the lighter skins tan with this solution, exactly as directed for medium and large hides, adding one-third, that is, 2 gallons, of this stock chrome solution each time, and begin to tan in about 15 gallons instead of 30 gallons of water. Follow the directions already given as to stirring, number of days, and testing to determine when tanning is completed. Very small, thin skins probably will not take as long to tan as will the large hides. The boiling-water test is very reliable for showing when the hide is tanned.

Washing and Neutralizing

When the sides are tanned, take them out of the tanning solution and put them in a barrel of clean water. The barrel in which the tanning was done can be used after it has been thoroughly washed.

When emptying the tanning barrel be sure to carefully dispose of the tanning solution. Although not poisonous to the touch, it probably would be fatal to farm animals should they drink it, and it is harmful to the soil.

Wash the sides in about four changes of water. For medium and large hides, dissolve 2 pounds of borax in about 40 gallons of clean water and soak the sides in this solution overnight. For hides and skins weighing less than 25 pounds, use 1 pound of borax in about 20 gallons of water. Move the sides about in the borax solution as often as feasible. After soaking overnight in the borax solution, remove the sides and wash them for an entire day, changing the water five or six times. Take the sides out, let the water drain off, and proceed as directed under "Dyeing black," or, if it is not desired to blacken the leather, proceed as directed under "Oiling and finishing."

Dyeing Black

Water-soluble nigrosine.—One of the simplest and best means of dyeing leather black is the use of nigrosine. Make up the dye solution in the proportion of one-half ounce of water-soluble nigrosine dissolved in 1¼ pints of water. Be sure to get water-soluble nigrosine. Evenly mop or brush this solution over the damp leather after draining as already directed and then proceed as directed under "Oiling and finishing."

Iron liquor and sumac.—If water-soluble nigrosine can not be obtained, a fairly good black may be secured with iron liquor and sumac. To make the iron liquor, mix clean iron filings or turnings with one-half gallon of good vinegar and let the mixture stand for several days. See that there are always some undissolved filings or turnings in the vinegar. For a medium or large hide put from 10 to 15 pounds of dried crumbled sumac leaves in a barrel containing from 35 to 40 gallons of warm water. Stir well and when cool hang in it the wet, chrome-tanned sides. Leave the sides in this solution for about two days, pulling them up and mixing the solution frequently. Take out the sides, rinse off all bits of sumac, and evenly mop or brush over with the iron liquor. Rinse off the excess of iron liquor and put the sides back in the sumac overnight. If not black enough the next morning, mop over again with iron liquor, rinse, and return to the sumac solution for a day. Take the sides out of the sumac, rinse well, and scrub thoroughly with warm water. Finally wash the sides for a few hours in several changes of water.

While both of these formulas for dyeing have been given, it is recommended that water-soluble nigrosine be used whenever possible, as the iron liquor and sumac formula is somewhat troublesome and may produce a cracky grain. After blackening, proceed as directed under "Oiling and finishing."

Oiling and Finishing

Thin leather.—Let the wet tanned leather from the dyeing, or, if not dyed, from the neutralizing, dry out slowly. While it is still very damp go over the grain or hair side with a liberal coating of neat's foot or cod oil. While still damp tack the sides out on a wall or tie them in frames being sure to pull them out tight and smooth, and leave them until dry. When dry take down and dampen well by dipping in warm water or by rolling them up in wet sacking or burlap. When uniformly damp and limber go over the sides with a "slicker," pressing the slicker firmly against the leather, while pushing it away from the body. "Slick" out on the grain or hair side in all directions.

After slicking it may be necessary to "stake" the leather. This is done by pulling the damp leather vigorously back and forth over the edge of a small smooth board about 3 feet long, 6 inches wide, and 1 inch thick, fastened upright and braced to the floor or ground. The top end of the board must be shaved down to a wedge shape, with the edge not more than one-eighth inch thick and the corners well rounded. Pull the sides, flesh side down, backward and forward over this edge, exactly as a cloth is worked back and forth in polishing shoes.

Let the sides dry out thoroughly again. If not sufficiently soft and pliable, dampen them with water, apply more oil, and slick and stake as before. The more time given to slicking and staking, the smoother and more pliable the leather will be.

Thick leather.—Thick leather from the larger hides is oiled and finished in a slightly different manner. For harness and strap leather, let the tanned sides, dyed if desired, dry down. While they are still quite damp slick over the grain or hair side thoroughly and apply a liberal coating of neat's foot or cod oil. Tack on a wall or tie in a frame, stretching the leather out tight and smooth, and leave until dry. Take the sides down, dampen them with warm water until limber and pliable, and apply to the grain side a thick coating of warm dubbin. This dubbin is made by melting together about equal parts of cod oil and tallow or neat's foot oil and tallow. When cool it must be soft and pasty, but not liquid. If too nearly liquid, add more tallow. Hang up the sides again and leave them until thoroughly dried. When dry, scrape off the excess tallow by working over with the slicker. If more grease in the leather is desired, dampen again and apply another coating of the dubbin. When again dry, slick off the tallow and thoroughly work over all parts of the leather with the slicker. Rubbing over with sawdust helps to take up surface oiliness.

Chrome-tanned leather is stretchy, so that in cutting the leather for use in harness, straps, reins, and similar articles it is best to first take out most of the stretch.

Chrome leather for shoe soles must be heavily greased, or, in other words, waterproofed, unless it is to be worn in extremely dry regions. Waterproofing may be done after repairing the shoes by setting them in a shallow pan of oil or grease so that just the soles are covered by the grease. The soles should be dry before they are set in the melted grease. Melted paraffin wax will do, although it makes the soles stiff. The simple formulas given are satisfactory for waterproofing chrome sole leather.

Alum-tanned Lace Leather

Deliming

After the sides have been put through the unhairing and fleshing operations, rinse them off with cool, clean water for from six to eight hours, changing the water frequently.

Buy 5 ounces of U.S.P. lactic acid (or 16 ounces of tannery 22 per cent lactic acid). Nearly fill a clean 40 to 50 gallon barrel with clean, cool water and stir in the lactic acid, mixing thoroughly with a paddle. Hang the sides in the barrel and leave them there for 24 hours, pulling them up and stirring them about frequently. Take out the sides, work over or "scud" thoroughly, as directed under "Unhairing," and hang them in a barrel of cool water. Change the water several times, and finally leave them in the water overnight.

If lactic acid can not be obtained, use a gallon of vinegar instead.

Tanning

While the sides are being delimed, thoroughly wash out the barrel in which the hide was limed. Put in it 15 gallons of clean water and 12 pounds of ammonia alum or potash alum and stir frequently until it is completely dissolved.

Dissolve 3 pounds of washing soda (crystallized sodium carbonate) and 6 pounds of salt in 5 gallons of cold, clean water in a wooden bucket. The soda crystals must be clear and glasslike. *Do not use white crusted lumps.*

Pour the soda solution into the alum solution in the barrel very, very slowly, stirring the solution in the barrel constantly. Take at least 10 minutes to pour in the soda solution in a small stream. If the soda is poured in rapidly the solution will become milky and it will not tan. The solution should be cool, and enough water to nearly fill the barrel should be added.

Hang each well-washed side from the deliming in the alum-soda solution. Pull up the sides and stir the solution six or eight times each day. Do not put the bare hands in the liquor if they are cut or cracked or have sores on them.

After six or seven days remove the sides from the alum-soda solution and rinse well for about quarter of an hour in clean, cold water.

Oiling and Finishing

Let the sides drain and dry out slowly. While still very damp go over the grain or hair side with a liberal coating of neat's-foot or cod oil. After the oil has gone in and the sides have dried a little more, but are still slightly damp, begin to work them over a "stake." The time to start staking is important. The sides must not be too damp; neither must they be too dry. When light spots or light streaks appear on folding it is time to begin staking. Alum-tanned leather must be thoroughly and frequently staked.

Staking is done by pulling the damp leather vigorously back and forth over the edge of a small, smooth board, as described. The sides must be staked thoroughly all over in order to make them pliable and soft, and the staking must be continued at intervals until the leather is dry.

When dry, evenly dampen the sides by dipping them in water or by leaving them overnight covered with wet burlap or sacks. Apply to the grain or hair side a thick coating of warm dubbin. This dubbin is made by melting together about equal parts of neat's-foot oil and tallow or cod oil and tallow. When cool, the dubbin must be soft and pasty but not liquid. If too nearly liquid, add more tallow. Leave the greased sides, preferably in a warm place, until dry. Scrape off the excess tallow and again stake the sides. If the leather is too hard and stiff, dampen it evenly with water before staking.

After staking, go over the sides with a "slicker," pressing the slicker firmly against the leather, while pushing it away from the body. Slick out on the grain or hair side in all directions.

Alum-tanned leather almost invariably dries out the first time hard and stiff. It must be dampened again and restaked while drying. In some cases this must be done repeatedly and another application of dubbin may be necessary. By repeated dampening, staking, and slicking the leather can be made as soft and pliable as desired.

Tanning Fur Skins

Much of the value of a fur skin depends upon the manner in which it is handled in the raw state. After the animal has been caught, every effort should be made to follow the best practices in skinning and curing, in order to obtain a skin of the greatest possible value. Certain trade customs also must be followed to secure the top price. Fur skins as a protection are a necessity for those living in cold climates, but comparatively few are used for this purpose. Most of the fur skins are made into articles which are more or less of a luxury, and as such are valued largely by their appearance and finish which an inexperienced worker can seldom make sufficiently pleasing. Furthermore, raw fur skins are valuable, and, if well cared for, usually find a ready market. Nevertheless, the spread between the prices paid for raw furs and those demanded for finished fur articles is enormous. No doubt, this spread in many instances inspires the attempts at home manufacture.

An inexperienced person should not try to tan valuable fur skins or large hides, such as cattle, horse, or bear, for making into coats, robes, or rugs. The risk of damage or of an unsatisfactory product, as measured by the usual standards of finish and appearance, is too great. The difficulties in properly handling large hides make the chances of success remote, except by those having suitable equipment and experience. Moreover, tanning the skin is only one step in the production of the finished article. After being tanned, all skins must be tailored, many must be dyed, and small ones must be matched, blended, and sewed together. All these operations require experience and practice to secure the attractive appearance desired by wearers of furs. Some of the operations, such as those of bleaching and dyeing, are so highly specialized that their undertaking should not even be considered by an amateur. From the standpoint of serviceability and usefulness, inexperienced persons might meet with a fair degree of success in tanning and tailoring fur skins, but few can ever hope to make a fur piece or garment which will compare favorably in appearance with the shop or factory product.

Alum-tanned leather almost invariably dries out the first time hard and stiff. It must be dampened again and restaked while drying. In some cases this must be done repeatedly and another application of dubbin may be necessary. By repeated dampening, staking, and slicking the leather can be made as soft and pliable as desired.

Tanning Fur Skins

No formulas for tanning are foolproof and success can be attained only by close observation, plenty of work, and the exercise of care and patience. All skins are not treated just alike. In fact, each skin has its own peculiarities, which only experience can show how to treat. Some skins are tough and fairly thick and will stand mistreatment; others are very thin and tender and are easily ruined. Some are fat and greasy and require thorough working out of the grease; others do not. An inexperienced person should experiment with the least valuable skins. If a number of skins of the same kind are to be tanned, one or two of the poorest should be tried first.

Soaking and Fleshing

The first step is to get the skin thoroughly softened, cleaned, and free from flesh and grease.

Split the tail the entire length on the underside. If the skin is "cased," split it neatly down the middle of the belly. Soak it in several changes of clear, cool water. When the skin begins to soften, lay it on a beam or smooth pole and begin working over the flesh side to break up the adhering tissue and fat. All dried skins have a shiny, tight layer of tissue. This tissue must be broken up and entirely removed, which is best done by repeated alternate working and soaking.

A good tool for scratching the tissue is a metal edge of any kind, such as a drawing knife or an ordinary knife with dull saw teeth or notches filed in it. Working over with these dull teeth scratches or breaks up the tissue so that it can be scraped off after further soaking.

At the same time the grease and oil are worked out of the skin. This operation is of the utmost importance. It is utterly useless to start tanning until all the tissue and grease have been removed and the skin is uniformly soft and pliable, without any hard spots.

The time of soaking depends upon the condition of the skin. Some skins require only about two hours, while others need a much longer time. Very hard skins often must be thoroughly dampened, rolled up, fur side out, and put away in a cool place overnight to soften. While a skin must be soaked until soft, it should not stay wet longer than necessary, as the hair may start to slip.

In fleshing and scraping, care also must be taken not to injure the true skin or expose the hair roots, especially on thin skins.

When the soaking is well advanced and the skin is getting in good shape, work it in lukewarm water containing an ounce of soda or borax to the gallon. Soap also may be added. This treatment promotes softening, cleans the skin, and cuts the grease.

Work again over the beam and finally rinse thoroughly in lukewarm water. Squeeze out most of the water, but do not wring the skin. Without further drying, work the skin in gasoline, using several changes if very much dirt and grease are present. Squeeze and hang up the skin for a few minutes.

The skin should now be ready for tanning. When painting or pasting of the tan liquor on the flesh side only is included in the directions for tanning, it is best to dry out the hair or fur side first by working in sawdust. In this way any heating of the fur side while the skin is tacked out is avoided, as are also matting and stiffening of the fur. If while drying out the fur, the flesh side becomes too dry, it must be evenly dampened with a wet cloth before applying the tan-liquor.

Combination Tannage

A combination tannage is a combination of mineral and vegetable tanning. It has an advantage over the salt-acid or salt-alum processes in giving a soft and flexible skin, as well as a more lasting tannage.

One of the most popular and successful formulas for a combination tannage is: A pasty mixture of alum, salt, gambier, and flour, with or without glycerin or olive oil, is made as follows: Dissolve 1 pound of aluminium sulphate and 1 pound of salt together in a small quantity of water. Dissolve 3 ounces of gambier or Terra Japonica in a little boiling water. Mix the two solutions and make up to 2 gallons with water. As this solution is used, mix it with enough flour to make a moderately thin paste. If the skin has a hard texture and lacks natural grease, thoroughly mix a little olive oil or glycerin with the paste.

Soak, soften, and clean the skin as previously described and tack it out flat and smooth, flesh side up. Apply from two to three coatings of the paste, depending upon the thickness of the skin. Only thick skins require three coatings. Each coating should be about one-eighth inch thick and should be applied at intervals of a day. Between applications the skin should be kept covered with sacking or paper. Scrape off most of the old coating before putting on a new one. After the last coating has been applied, spread out the skin uncovered or hang it up to dry slowly.

When practically dry, wash off the flour paste, rinse for several minutes in water containing an ounce of borax to the gallon, then in water alone. Squeeze out most of the water. Put the skin over a beam and slick it out well on the flesh side with the back of a knife or edge of a wooden slicker, thus working out most of the water. Again tack the skin out smoothly, flesh side up, and apply a thin coating of any animal fat, fresh butter being particularly good, or a nondrying oil, such as neat's foot, castor, or olive oil. Glycerin or a soap may be used instead of the grease or oil. If the skin originally was very greasy, it may not be necessary to apply any oil.

When nearly dry, but still slightly damp, begin to work the skin in all directions, stretching it from corner to corner and working the flesh side over a stake or a wooden edge, such as the back of a chair or piece of board clamped in a vise.

The time to begin working is important and is best judged from experience. The skin must not be too wet; neither must it be too dry. The appearance of a few light spots or a light streak on folding is a good indication of the time to start working the skin.

Work the skin in all directions back and forth, as if shining shoes with a cloth. The skin may also be worked this way through smooth metal rings. Much of the success in getting a soft skin lies in this repeated working, which must be done *while the skin is drying out, not after it is dry.* If the skin is not soft enough when dry, it must be evenly dampened and worked again while drying. This may be repeated several times if necessary.

After softening and drying out it is well to give the skin a hasty bath in gasoline. If the skin is greasy, this must be done. This also helps to deoderize some skins, such as those of the skunk.

Finally, to clean and brighten the tanned skin, tumble or work it repeatedly in dry, warm sawdust, preferably hardwood sawdust, or bran or cornmeal. Clean these out of the fur by gentle shaking, beating, combing, and brushing.

The flesh side may be smoothed if necessary by working over a sandpaper block. This also helps to further soften the skin. If desired, the thicker sections of the skin may be made thinner and more flexible by shaving off some of the skin or hide.

Salt-Alum Tannage

The salt-alum process, an old method for fur-skin tanning, is widely used. It is considered slightly better than the salt-acid tannage, being a little more permanent and, when properly carried out, giving skins which have a little more stretch and flexibility. It often happens, however, that alum-tanned skins come out stiff and hard and must be repeatedly worked and sometimes retanned.

A salt-alum tanning solution may be made up in the following proportions: 1 pound of ammonia alum or potash alum, dissolved in 1 gallon of water; 4 ounces of washing soda (crystallized sodium carbonate) and 8 ounces of salt, dissolved together in one-half gallon of water. When dissolved, pour the soda-salt solution very slowly into the alum solution while stirring vigorously.

The skin, cleaned and softened as previously described, may be tanned by immersion in this solution for from two to five days, depending upon its thickness. Because of the action of alum on some furs it may be best, as a general rule, to apply the tanning liquor as a paste to the flesh side only.

Mix the tan liquor as used with sufficient flour to make a thin paste. Add the flour in small quantities, with a little water, and mix thoroughly to avoid lumps. Tack the skin out smoothly, flesh side up. Apply a coating of the paste about one-eighth inch thick and cover the skin. The next day scrape off most of the paste and give another coating. Apply altogether, at intervals of a day, from two to three coatings, depending upon the thickness of the skin. Only thick skins should need as many as three treatments. Leave the last coating on for three or four days. Finally scrape off and rinse clean in water, putting in about an ounce of borax to the gallon of water. Rinse at last in water only.

Work over the beam to remove most of the water. Stretch the skin out flat and sponge over the flesh side with a thin soap paste. After this has gone in, apply a thin coating of oil. Leave the skin stretched out to dry, and while it is still damp, work and stake as described, wetting and working repeatedly if necessary. Finally, clean in gasoline and sawdust and finish as described above.

Salt-acid Tannage

One of the oldest processes of tanning requires various mixtures of common salt and sulphuric acid. Tanning, or, more correctly speaking, tawing, by this means is open to the objection that sulphuric acid must be used very cautiously, and must be completely neutralized to prevent later damage to the skin. Skins tanned with salt and acid also show a tendency to become damp and clammy in wet weather and, if repeatedly subjected to wetting, lose their tanned effect.

A salt-acid tanning solution may be made up in the following proportions: For each gallon of water use 1 pound of common salt and one-half ounce of concentrated sulphuric acid. Dissolve the salt and carefully pour in the acid with stirring. This tan liquor must be made and used in jars or wooden vessels, *never in metal containers of any kind.* (When pouring in the acid, do not inhale any more of the fumes given off than is necessary, and also be careful not to get any of the strong acid on the skin or clothing.) As soon as the acid-salt solution has cooled, it is ready for use.

Put the cleaned, softened skin in the solution so that it is entirely covered and leave it for from one to three days, depending upon its thickness. During this time stir the skin about frequently. If desired, the solution may be painted on instead. In this case, tack out the skin smoothly, flesh side up, paint over with the solution, and cover the skin with well-dampened sacking or cloth. At the end of six hours, paint over it again. With thicker skins, give one or two more applications of the solution about six hours apart, keeping the skin covered between applications. After the last application, hang up the skin or spread it, flesh side up, without cover, and let it dry.

After tanning, either by immersion or by painting, rinse the skin in clear water and squeeze out most of the water, but do not wring it. Then work the skin for about 10 minutes in a solution made up in the proportion of an ounce of borax in a gallon of water, and finally rinse well in clear water and squeeze.

Work over the skin with a slicker to remove most of the water, tack it out flat, flesh side up, and apply a thin coating of grease or oil. Leave the skin stretched to dry, and while still damp work and stake as described.

Finally clean in gasoline and sawdust, and finish by shaking, beating, sandpapering, brushing, and combing.

One Bath Fastan-Chrome Stock Liquor

10 parts of dry one bath chrome dissolved in 100 parts H_2O; cool to 90° F. and add 100 parts Fastan.

To ⅓ of this liquor add 100 parts H_2O containing HyPo (15% on weight of pelts) add pelts and drum for one hour. Then add another ⅓ of stock solution and drum for two hours; and then add last ⅓ and drum for one hour or longer.

To bleach chrome tanned leather, adjust the Ph of leather to 3.5 or 4.0 by treating in H_2O at 95° F. Then add dissolved oxalic acid so as to have 1% in the solution; drum, 20 minutes, wash in 95° F. H_2O for 30 minutes, then wash in cool H_2O.

Float the stock in a drum in 10 gallons cold H_2O per 100 lbs. leather. Dissolve 15% HyPo separately and add 10% Fastan.

Add this mix to drum in three equal portions at intervals of ten minutes and run for two hours.

Remove stock, wash for 20 minutes at 110° F. and float in 110° F. H_2O using 5 gallons per 100 lbs. stock, fat liquor with sulphonated Cocoanut Oil and 5% TiO_2 for ¾ hour.

Fulling of Skins and Hides

1% Paraldehyde (on wet weight of skins). Drum the pelts for one-half hour or without agitation, for several hours. If greater degree of swelling is desired use greater per cent.

Sheep skin skivers are tanned by treating for 3 hours at 85% F. in a bath of 20% Fastan and 15% "HyPO" (based on wet weight skinvers). After tanning, Fat Liquoring consists of 3–5% sulfonated oil (Castor or Cocoanut).

Stuffing Leather

10 to 20% of Bentonite is emulsified in a sulphonated Castor solution (10% on weight of leather or a 2 to 3% solution).

Fat liquors for leather. 2 parts Sulphonated Neats Foot Oil and 1 part straight Neats Foot Oil.

---

* Fur Skin Tanning

The washed skins are chrome tanned in the usual way in a bath containing

5 lb. of chrome alum dissolved in 10 gal. of water at 70° F. and paddled for 2 hours. A solution of 3 lb. of washing soda is then run in and the process continued for a further 2 hours. The skins are then left for 12 hours, rinsed, and then washed in a bath containing ½ per cent. borax on the weight of the skins.

The patented process is that to the above chrome solution, 60 grm. of formaldehyde are added to every 10 litres of chrome solution. After proper tanning, the skins are rinsed and while moist are subjected to treatment with chloride of lime, being worked for 15 minutes in the cold in a solution of 120 grm. hydrochloric acid (32° Tw.) per 10 litres. Then, without rinsing, they are transferred to a bath containing the clear solution left from suspending 2 to 4 grm. chloride of lime in 10 litres of water over half an hour. They are then replaced, again without rinsing, in the acid bath for 15 minutes and finally rinsed in a bath containing 1 to 2 per cent. sodium thiosulphate, rinsed and finally hydroextracted. They are then ready for dyeing.

---

* Leather Substitute

Cotton flannel napped on both sides is impregnated with a solution of rubber containing rubber 70, resin 3, ZnO 20, pigment 7%, dried, smoothed under tension, and vulcanised, if desired. One side of the material is then starched, and coated with a rubber mix containing about 30% of cotton flocks, which is vulcanised, together with the unvulcanised impregnating rubber, and neutralised with anhyd. NH_3. Additional coatings are then applied which are starched, vulcanized, and neutralised, and finally the article is dusted with talc.

---

* Leather, Substitute

Cotton or jute is prepd. in the form of a nappe of regular thickness and dipped into a liquid bath contg. resin 10, aq. NH_3 200, $PhNH_2$ 100, water 800 g. and latex 10 l., pressed, dipped into an aq. bath contg. ACOH and afterward dried to give a leather-like substance.

---

* Fur, Carroting

The following solution prevents yellowing and the fur is given better felting properties.

First make up a mercuric nitrate solution by mixing 80 parts of a 40% nitric acid solution with 20 parts of metallic mercury. This gives a solution containing about 32% of mercuric nitrate, 20% of free nitric acid and 48% of water. This solution is about the same as the mercuric nitrate carroting solution ordinarily used.

Then take one half gallon of the above solution, dilute it with three gallons of water and add thereto about one and one half gallons of water containing about one half ounce of ammonium fluoride. To the resulting solution then add two and one half gallons of a commercial peroxide solution (usually about 3% strength) and finally two and one half gallons of water.

Fur Dye Mordants

1. 1 gr. $K_2Cr_2O_7$ / 0.5 gr. cream of tartar / 0.1 gr. $CuSO_4 .5 H_2O$ } to 1 liter of water
2. 4 gr. $CuSO_4 .5 H_2O$ / 2 cc. $CH_3COO H$ (30%) } to 1 liter of water
3. 4 gr. $FeSO_4 7 H_2O$ / 2 cc. $CH_3COO H$ (30%) } to 1 liter of water

---

* Dehairing Hides

H_2SO_4 (6%)	1000
Silver Nitrate	0.05

Immerse skins in above at 60° C. The separated hair may be washed and used for making felts.

---

* Hide Depilatory

Water	5270 lb.
Sod. Sulfide (30%)	50 lb.
Glucose	25 lb.
Maltose	18 lb.
Lactic Acid	0.4–0.65%

Felting Liquid

Mercury	1.2 lb.
Nitric Acid	2.5 lb.

Let stand in cool place until the mercury is dissolved.

Warning—Do not inhale the fumes. Then add a mixture of 5¼ lbs. nitric acid (40%) in 60 lbs. water. Lastly add 33 lbs. Perhydrol (100 vol. peroxide) and use at once.

Warning—Do not inhale fumes.

Special Felting Liquid

Mercury	2.4 lb.
Nitric Acid	5.0 lb.

Let stand in cool place until the mercury is dissolved. Then add a mixture of 10½ lbs. nitric acid in 50 lbs. of water. Lastly add 33 lbs. perhydrol and use at once.

---

* Felting Animal Hairs

To enhance capacity for felting animal hairs are treated with

Am. Silicofluoride	4 lb.
Water	1000 lb.

Tanning Liquid

Material of vegetable origin such as wood waste, bark, seed husks, straw, peat, etc., is treated with 25–40% H_2SO_4 or HCl with heating in the presence of Na_2SO_4 or NaCl. The soln. is dild. with water to an acidity of 1–1.5%, then nearly neutralized with soda, and used to tan hides.

* Sole Leather, Tanning

The cleaned limed skins are treated for 24 hrs. in following bath:

Pot. Chromate	250 gm.
Boric Acid	200 gm.
Formic Acid	180 gm.
Glycerol	50 gm.
Water	100 qt.

* Tanning Agent, Synthetic

Three mols. of resorcinol or pyrogallol are condensed with 1–1.5 mol. of BzH or a substitution product thereof at atm. or raised temp. in an aq. medium in the presence of a small quantity of acid. The reaction is stopped, by neutralizing the acid, as soon as (or shortly before) the reaction mixt. gives a neg. $FeCl_3$ test for resorcinol or pyrogallol. Water-sol. products useful as tanning agents are obtained.

* Fur Carroting Solutions

1.	Hypochlorous Acid	13–50
	Sulfuric Acid	15–60
2.	Hydrogen Peroxide	20–100
	Sulfuric Acid	17–85

CEMENT, CONCRETE, STONE AND MATERIALS OF CONSTRUCTION

Acid-Proofing Creamery Floors

Paraffin (150° F.)	4
Turpentine	1
Toluol	16

Warm and stir until uniform. Pour into cans and allow to "set." Spread on floor and allow to penetrate for 24 hrs. At the end of this time the residual layer should be driven into the concrete by heat. A free flame should not be used due to fire hazards; hot irons will be found safe and effective in forcing the paraffin into the pores and capillaries of the finish for some distance below the surface.

After either treatment, the floor should be given a good waxing with any standard floor wax suited for this purpose. As the wax film is worn away through use, it is replaced by a fresh coating with the use of a polishing machine. Neither of these methods of acid-proofing creamery floors will change the color of the finish appreciably.

* Alabaster and Gypsum, Hardening

Articles made of the above and similar materials are given the hardness and appearance of marble by impregnating with after first drying at 150–200° F.

Water	1 gal.
Sod. Silicate	1 lb.
Magnesium Sulfate	1 oz.
Fused Calcium Chloride	1 oz.

They are then polished and rubbed with linseed oil.

* Asphalt Emulsion

Melt together 350 pounds of Asphalt and 6½ pounds of Pine Oil, keeping at a temperature between 145 and 175° F. In another container dissolve 20 pounds of Rosoap and 1 pound Caustic Soda in 150 pounds of water and heat to the same temperature. Run the Asphalt into the water solution slowly while beating vigorously. The type of beating necessary is that gotten from a high speed turbo mixer or colloid mill.

All formulae preceded by an asterisk (*) are covered by patents.

* Asphalt Emulsion

Asphalt 300, rosin 75, boiled chinawood oil 150, boiled linseed oil 150 and Na_2CO_3 7.5 parts, which is dissolved in hot H_2O. When cool 3.4% of ammonia is added. Any further desired amt. of H_2O may be added.

Bituminous Composition

(for roads, floors, tennis-courts, etc.)

Sand	75–86
Bitumen	11–15
Fire Clay	3–10

Bitumen Emulsion

An emulsion for road making contains Spramex bitumen 48, water 49.5, oleic acid 2 and calcined Na_2CO_3 0.5%. The bitumen is warmed at 95–98°, and the oleic acid added. The water is heated separately with the Na_2CO_3, and the two liquids are introduced into the emulsifier. Another emulsion contains Spramex bitumen 50, mineral oil 2–2.5, resin soap 1.5–2, KOH 1, and water 45%. The bitumen is melted and the mineral oil added during agitation. The water is heated to boiling, and in it are dissolved the soap and KOH. The liquids are mixed at 95°. With more bitumen there must be added 1–2% of glue, starch, gelatin or Na silicate, during or after emulsification.

* Slag Brick

The brick is composed of granulated blast-furnace slag 200–300, cement 50–100, pulverized $CaCl_2$ 3–8, $Pb_3(BO_3)_2$ 1–6, and pigments 5–10 pts. It is moistened and pressed.

Brickwork, Painting

Use any good quality outside paint. The first coat should seal the pores of the brick; for this the paint is thinned with turpentine and boiled linseed oil, and many painters also add varnish. The second coat is not thinned so much, and for the third the paint is used as it comes in the can.

* Brick, Weatherproofing

A coating for brick, stucco, cement or iron consists of

Cod Liver Oil Crude	1 gal.
Beeswax Crude	12 oz.
Glacial Acetic Acid	4–12 oz.

Coloring Cement Gray

Paris Paste	8
Cement or Plaster	100
Water	sufficient

The Paris Paste is dispersed in the water by rapid stirring.

If a darker color is desired the percentage of Paris Paste is increased.

Concrete or Mortar

How to Figure Quantities

Quantities of Cement, Fine Aggregate and Coarse Aggregate Required for One Cubic Yard of Compact Mortar or Concrete

Mixtures			Quantities of Materials				
Cement	F. A. (Sand)	C. A. (Gravel or Stone)	Cement in Sacks	Fine Aggregate		Coarse Aggregate	
				Cu. Ft.	Cu. Yd.	Cu. Ft.	Cu. Yd.
1	1.5	...	15.5	23.2	0.86	....	....
1	2.0	...	12.8	25.6	0.95	....	....
1	2.5	...	11.0	27.5	1.02	....	....
1	3.0	...	9.6	28.8	1.07	....	....
1	1.5	3	7.6	11.4	0.42	22.8	0.85
1	2.0	2	8.3	16.6	0.61	16.6	0.61
1	2.0	3	7.0	14.0	0.52	21.0	0.78
1	2.0	4	6.0	12.0	0.44	24.0	0.89
1	2.5	3.5	5.9	14.7	0.54	20.6	0.76
1	2.5	4	5.6	14.0	0.52	22.4	0.83
1	2.5	5	5.0	12.5	0.46	25.0	0.92
1	3.0	5	4.6	13.8	0.51	23.0	0.85

1 sack cement = 1 cu. ft.; 4 sacks = 1 bbl. Based on tables in "Concrete, Plain and Reinforced," by Taylor and Thompson.

For Chemical Advisors, Special Raw Materials, Equipment, Containers, etc., consult Supply Section at end of book.

Materials Required for 100 Sq. Ft. of Surface for Varying Thicknesses of Concrete or Mortar

C. = Cement in Sacks.
F. A. = Fine Aggregate (Sand) in Cu. Ft.
C. A. = Coarse Aggregate (Pebbles or Broken Stone) in Cu. Ft.

Quantities may vary 10 per cent either way depending upon character of aggregate used. No allowance made in table for waste.

Proportion	1–1½			1–2			1–2½			1–3		
Thickness in inches	C.	F. A.	C. A.	C.	F. A.	C. A.	C.	F. A.	C. A.	C.	F. A.	C. A.
⅜	1.8	2.7		1.5	3.0		1.3	3.2		1.1	3.4	
½	2.4	3.6		2.0	4.0		1.7	4.3		1.5	4.4	
¾	3.6	5.4		3.0	6.0		2.5	6.3		2.2	6.8	
1	4.8	7.2		4.0	7.9		3.4	8.4		3.0	8.9	
1¼	6.0	9.0		4.9	9.9		4.2	10.5		3.7	11.1	
1½	7.2	10.8		5.9	11.9		5.1	12.7		4.4	13.3	
1¾	8.4	12.6		6.9	13.9		5.9	14.7		5.2	15.7	
2	9.6	14.4		7.9	15.8		6.8	16.9		5.9	17.7	
	1–2–2			1–2–3			1–2½–3½			1–3–5		
3	7.7	15.4	15.4	6.5	13.0	19.3	5.5	13.6	19.1	4.3	12.8	21.3
4	10.2	20.4	20.4	8.6	17.2	25.8	7.3	18.1	25.4	5.7	17.0	28.4
5	12.8	25.6	25.6	10.8	21.6	32.2	9.1	22.6	31.8	7.1	21.3	35.5
6	15.4	30.7	30.7	12.9	25.8	38.6	10.9	27.2	38.2	8.5	25.6	42.6
8	20.6	41.0	41.0	17.2	34.4	51.6	14.6	36.4	51.0	11.4	34.1	57.0
10	25.6	51.2	51.2	21.5	43.2	64.4	18.2	45.3	63.5	14.2	42.5	71.0
12	30.7	61.4	61.4	25.8	51.6	77.2	21.8	54.5	76.3	17.0	51.1	85.1

* Concrete, Building Blocks

Cork Scrap	1–4 parts by volume
Cement	1 part
Sand	1–4 parts

Mix dry and gage with water to form a sticky plastic mass, which is then cast in forms.

* Concrete, Curing

Evaporation of water from freshly laid concrete is prevented by coating with a gel consisting of

Bentonite	100
Magnesium Oxide	2
Mineral Oil	10
Water	to suit

* Mortar, Road

Dry Sand	66–72
Cement	30.5–21.5
Iron Oxide	0.8– 1.5
Lime	0.5– 1
Calcium Fluoride	0.8– 1.5

* Lime Mortar, Hardening

A mixt. formed of $MgCl_2 \cdot 6H_2O$ 3.5, $MgSO_3 \cdot 6H_2O$ 2, hydrated lime 50 and plaster of Paris 50 parts is suitable for a hard interior wall plaster.

Masonry and Wall, Waterproofing

Tallow	10
Linseed Oil Bodied	5
Paraffin	1
Naphtha	32
Drier Liquid	0.13

Dustproofing Concrete Floors

"Concrete Special" silicate of soda is recommended for this purpose. It is a syrupy solution. Technically, it is 42.25° to 42.75° Baumé, with a ratio of sodium oxide to silica of 1: 3.25. It is diluted as noted below, and applied to the surface of the concrete after it has set. After the concrete is in place, it is desirable to wait at least two weeks before applying the silicate, and four weeks is still better. Also the silicate treatment may be satisfactorily applied to clean concrete at any later time; it is especially good on old concrete.

The diluted "Concrete Special" silicate soaks into the concrete, and a chemical reaction takes place which hardens the surface and makes it more dense.

Method of Application

In ordinary cases it will be found satisfactory to dilute each gallon of the silicate with four gallons of water. The resulting five gallons may be expected to cover 1000 square feet of floor surface, one coat. However, the porosity of floors varies greatly and the above statement is given as an approximate value for estimating purposes.

The floor surface should be prepared for the treatment by cleaning free from grease, spots, plaster, etc., and then thoroughly scrubbed with clear water. To get the best penetration the floor should be thoroughly dry, especially before the first application, and if practical it is well to let it dry for several days before the first scrubbing. . . . The solution may be applied with a mop or hair broom and should be continuously brushed over the surface for several minutes to obtain an even penetration. An interval of twenty-four hours should be allowed for the treatment to harden, after which the surface is scrubbed with clear water and allowed to dry for the second application. Three applications made in this manner will usually suffice, but if the floor does not appear to be saturated by the third application a fourth should be applied.

Acid Resistant Concrete

The same treatment with silicate of soda that is recommended for dust proofing is remarkably serviceable in rendering concrete resistant to acid. It works by filling the pores of the concrete with a material that is acid-proof. Concrete itself is rapidly attacked by acids, but when thus protected by an acid-proof filler, it has considerable acid-resistance. For example, a block of concrete was prepared with the silicate treatment applied to one end and not to the other. Concentrated hydrochloric acid was poured over the block. The acid ate rapidly into the untreated end leaving it friable and sandy. The treated end was only slightly affected.

Along this line, therefore, the silicate treatment has frequently done good service where old floors had to be used. The treatment is useful also for protection against dilute acids, and against organic acids. In some cases repeated silicating, perhaps once a year, may be desirable.

Cement Patches

In patching or resurfacing concrete "Concrete Special" silicate of soda can be used to insure a good bond between the old and new cement.

To refill a hole it should be chipped out clean and somewhat under-cut. The fresh surface should then be painted with "Concrete Special" silicate full strength. Neat cement should then be dusted over the surface and worked in with a broom or stiff brush. The new concrete can then be applied in the usual manner.

For resurfacing, the concrete should be roughened with a pick, all loose particles removed and the floor wet thoroughly with water over night. Immediately before the new surface is applied the old one should be washed with a freshly prepared mixture of 10 pounds of neat cement with one quart of "Concrete Special" in fourteen quarts of water. This mixture should be brushed in well and followed at once with the surface layer.

Concrete Efflorescence, Removal of

Where efflorescence occurs, it may be dissolved by a dilute solution of muriatic acid (1 part of concentrated acid to 10 parts of water). In using this treatment the surface of the concrete is wetted before applying the acid and is thoroughly washed after the acid treatment.

The length of time required for the acid solution to dissolve efflorescence will depend upon the amount of the latter. In most cases, the acid can be washed off within three or four minutes. It is best not to leave the acid solution on longer than four minutes, for it may etch the colored concrete. If some deposit still remains after the first application, a second can be made. The acid solution should be brushed on smoothly, using the least amount possible for each application.

Efflorescence also can be removed with a solution of equal parts of paraffin oil and benzine rubbed vigorously into the surface when the concrete is dry. This treatment also improves the wearing qualities of the surface by filling the pores and bringing out the color more uniformly. It is frequently applied to concrete surfaces for these reasons only.

Concrete, High Early Strength

Increasing the time of mixing will increase early strength. For concrete cured at normal temperatures, increasing the mixing time from 1 minute to 2 minutes will add about 100 pounds per square inch to the strength at three days. About 200 pounds per square inch are added by increasing the mixing time from 1 to 5 minutes.

Concrete that is to attain high early strength should be kept damp at a temperature of 70 degrees Fahrenheit or above, beginning soon after it is placed. Concrete cured below 70 degrees hardens more slowly and it is not likely to have high strength at an early age.

The admixtures commonly used to increase the rate at which concrete hardens are calcium chloride and calcium oxychloride. These materials may be used within certain limits to hasten hardening and to increase early strengths of concrete.* The quantities of admixtures should not exceed from 2 to 4 per cent of calcium chloride or 7 to 10 per cent of calcium oxychloride by weight of the cement.

The calcium chloride is dissolved in the mixing water before adding it to the other materials in the mixer. Most contractors make up a solution of known concentration, adding the desired amount to each batch. Thus, if it is desired to use 2 pounds of calcium chloride per sack of cement a solution containing 1 pound per quart can be made, 2 quarts of the solution being added to the mixture for each sack of cement in the batch. It is important to remember that this solution is to be regarded as part of the mixing water.

* There is evidence to show that calcium chloride and similar compounds do not react in the same manner with all brands of portland cement. Trial batches of the brand of cement and the brand of accelerator proposed to be used should be made up and rate of hardening at the specified temperature noted before proceeding with their use in important work.

Concrete Floor Hardeners

The fluosilicates of zinc and magnesium, when dissolved in water, have been used with fair success for hardening defective concrete finish. In making up the solutions, ½ pound of the fluosilicate should be dissolved in one gallon of water for the first application and 2 pounds to each gallon for subsequent applications. The concrete floor must be clean and free from plaster, oil, paint or other foreign substances, otherwise the solutions will not penetrate sufficiently to react. For the same reason the surface must be absolutely dry. After the floor has dried, the second application may be made. About 3 or 4 hours are generally required for absorption, reaction and drying. In this treatment, with the average floor, one gallon of the liquid will cover approximately 130 square feet. Care should be taken to mop the floor shortly after drying to remove incrusted salts, otherwise white stains may be formed.

Sodium Silicate Treatment

When sodium silicate is used, it is applied in a 20% solution in two or more coats twenty-four hours apart. Ordinarily the sodium silicate requires considerable time to dry before the floor can be used. Commercial sodium silicate varies in strength from 30 to 40% solution. It is quite viscous and requires thinning with water before it will penetrate the floor. It has been found satisfactory to dilute each gallon of the silicate with three gallons of water. Each gallon of the resulting solution will cover approximately 200 square feet of floor surface. The floor should be thoroughly cleaned of all foreign matter, and should be dry before the first application of the silicate solution.

Aluminum Sulphate Treatment

This treatment consists in one or more applications of solutions of aluminum sulphate to the clean, dry surface. The solution is made up in a wooden barrel or stoneware vessel and the water should be acidulated with not more than one teaspoonful of commercial sulphuric acid for each gallon of water. The sulphate does not readily dissolve and requires occasional stirring for a few days until the solution is complete. About 2½ pounds of the powdered sulphate will be required for each gallon of water and one gallon of the solution should cover about 100 square feet of floor surface. For the first treatment the solution may be diluted with twice its volume of water. Twenty-four hours after this application the stronger solution may be used, and twenty-fours should elapse between subsequent applications.

Zinc Sulphate Treatment

This treatment consists of the application of about 16% solution of zinc sulphate made acid with a teaspoonful of commercial sulphuric acid to every gallon. The mixture is applied in two coats, the second coat being applied four hours after the first. The surface should be scrubbed with hot water and mopped dry just before the application of the second coat. This treatment gives the floor a darker appearance.

Concrete

Recommended Proportions of Water to Cement and Suggested Trial Mixes

Kinds of Work	Add U. S. Gals. of Water to Each Sack Batch if Sand is			Suggested Mixture for Trial Batch			Materials per Cu. Yd. of Concrete *		
					Aggregates			Aggregates	
	Very Wet	Wet	Damp	Cement Sacks	Fine Cu. Ft.	Coarse Cu. Ft.	Cement Sacks	Fine Cu. Ft.	Coarse Cu. Ft.
5-Gallon Paste for Concrete Subjected to Severe Wear, Weather or Weak Acid and Alkali Solutions									
Colored or plain topping for heavy wearing surfaces as in industrial plants and all other two-course work such as pavements, walks, tennis courts, residence floors, etc.	4¼	Average Sand 4½	4¾	1	1	1½	10	12	15
					Maximum size aggregate ⅜″				
One-course industrial, creamery and dairy plant floors and all other concrete in contact with weak acid or alkali solutions.	3¾	4	4½	1	1¾	2	8	14	16
					Maximum size aggregate ¾″				
6-Gallon Paste for Concrete to be Watertight or Subjected to Moderate Wear and Weather									
Watertight floors such as industrial plant, basement, dairy barn, etc. Watertight foundations. Concrete subjected to moderate wear or frost action such as driveways, walks, tennis courts, etc. All watertight concrete for swimming and wading pools, septic tanks, storage tanks, etc. All base course work such as floors, walks, drives, etc. All reinforced concrete structural beams, columns, slabs, residence floors, etc.	4½	Average Sand 5	5½	1	2¼	3	6¼	14	19
					Maximum size aggregate 1½″				
7-Gallon Paste for Concrete Not Subjected to Wear, Weather or Water									
Foundation walls, footings, mass concrete, etc., not subjected to weather, water pressure or other exposure.	4¾	Aver. Sand 5½	6¼	1	2¾	4	5	14	20
					Maximum size aggregate 1½″				

* Quantities are estimated on wet aggregates using suggested trial mixes and medium consistencies—quantities will vary according to the grading of aggregate and the workability desired.

It may be necessary to use a richer paste than is shown in the table because the concrete may be subjected to more severe conditions than are usual for a structure of that type. For example, a swimming pool ordinarily is made with a 6-gallon paste. However, the pool may be built in a place where soil water is strongly alkaline in which case a 5-gallon paste is required.

Recommended Mixtures for Several Classes of Construction

Intended primarily for use on small jobs

Kind of Work	Gallons of Water to Add to Each One Sack Batch			Trial Mixture for First Batch			Maximum Aggregate Size
	Dry Sand and Pebbles	Moist Sand and Pebbles	Wet Sand and Pebbles	Cement	Sand	Pebbles	
				Sacks	Cu. Ft.	Cu. Ft.	Ins.
Foundation walls which need not be watertight, mass concrete for footings, retaining walls, garden walls, etc.	7½	6	5	1	3	5	2
Watertight basement walls and pits, walls above grounds, dams, lawn rollers, hand tamper, shoe scrape, hot beds, cold frames, storage and cyclone cellar walls, etc.	6½	5	4¼	1	2½	3½	1½
Water storage tanks, well curbs and platforms, cisterns, septic tanks, watertight floors, sidewalks, stepping stone and flagstone walks, driveways, porch floors, basement floors, garden and lawn pools, steps, corner posts, gate posts, piers, columns, chimney caps, concrete for tree surgery, etc.	5½	4¼	3¾	1	2	3	1
Fence posts, clothes line posts, grape arbor posts, mail box posts, etc., flower boxes and pots, benches, bird baths, sundials, pedestals and other garden furniture, work of very thin sections.	4½	3¾	3½	1	2	2	¾

Concrete and Cement Waterproofer

A quantity of naphtha is heated to a temperature of approximately 80° C. and aluminum stearate in the ratio of 2 to 10 parts by weight of stearate to 100 parts of naphtha is added to the hot naphtha. The two materials are then agitated until a complete solution of the stearate in the naphtha is effected. A quantity of anhydrous acetic acid, equivalent to 0.3% to 1.5% by weight of the solution, is then added and the resulting mixture is thoroughly agitated. The product thus obtained is a clear solution having a specific viscosity Engler at 0° F. of 15 to 45 seconds per 100 cc. which can be stored without fear of gelling occurring at ordinary atmospheric temperatures and which may be applied to the substance to be waterproofed by means of a brush, spray or other device, and good penetration be obtained.

Acid Wash for Concrete Surfaces

Aluminum Chloride (Commercial)	1 lb.
Water	10

To be flushed over concrete surface and washed off with clean water.

Cement Accelerator

Commercial Calcium Chloride	4 lb.
Water	96 lb.

The above to be used as gauging water for concrete.

---

* Cement Coloring

Carbon Black	30
Iron Oxide	4
Water	100
Pine Oil	1

---

* Cement, Retarding Settling of

The set of portland cement is retarded by adding 0.25–1.5% Manganese Sulfate.

---

* Cement Coating

A coating compn. has approx. the following compn.: white portland cement (waterproof) 62.5, high-Ca hydrated lime 31.1, Irish moss (powd.) 0.1, NaCl 6.0, $Ca(C_2H_3O_2)_2$ 0.3%.

Cement Floor Hardener

Magnesium Fluosilicate	1 lb.
Water	15 lb.

The above to be flushed over a cement urface. Wash with clean water to remove soluble salts.

Hydraulic Cement

Portland Cement	90 lb.
Aluminum	2 lb.
Ferro Silicon	8 lb.

Cement Preservative

Chinawood Fatty Acids	10 lb.
Paraffin Wax	10 lb.
Kerosene	40 gal.

Cement, Resistant to Calcium Chloride Solutions

Aluminum Oxide	40
Lime	40
Iron Oxide or Silicon Dioxide	15
Calcium Chloride	1

* Cement, Slow Setting

A compn. is formed of MgO 2, alum 2, NaOH 3, NaCl 1, lime 100, water 67, a pigment such as whiting 2 and a filler such as sand 125 parts.

* Cement Size

The size contains 10 pts. of amorphous SiO_2, 6 pts. of china clay, 5 pts. of talc, 0.8 pt. of pigment, and the reaction product from Portland cement 48, $CaCO_3$ 10, alum 4.7, lactic casein 5.5, sulphuric casein 5.5, borax 1.8, and $Ca(OH)_2$ 3.5 pts.

Removal of Paint from Stone Surfaces

Paper pulp (old newspapers, cement sacks or stock pulp) is prepared by shredding in water by means of a steam jet. Excess water is drained off, 10–15% washing soda is added to the pulp, followed by sufficient fireclay (or lime), to render the mass plastic. Apply as a poultice to the surface to be treated; allow to remain 24 hours.

The poultice can usually be stripped off easily at the end of the above period. In obstinate cases, repeat treatment.

Last traces of pigment are removed by scrubbing with a bristle brush with clear water.

Removal of Pitch, Asphalt, Etc., from Stone Surfaces

Soak one or two thicknesses of blotting paper with carbon bisulphide. Lay over stain and apply a heated flat iron or similar heat retaining body. Remove iron when cool. The bituminous material will be found to be largely or wholly absorbed by the blotting paper. Repeat treatment in case of only partial removal.

Note: As carbon bisulphide is inflammable, the above treatment should not be attempted in the vicinity of sources of ignition.

Cement Coated Wire

To increase the holding power of fastening devices made from wire, the latter is supported as a coil on a rotating mandrel dipped into one of the following mixtures.

1.	Chinawood Oil	30
	Ester Gum	20
	Naphtha	50
2.	Rosin	15
	Calcium Hydroxide	0.9
	Lead Oxide (PbO)	0.3
	Manganese Dioxide	0.2
	Chinawood Oil	33.6
	Naphtha	50.0

* Dry Rot, Prevention of

Wood or cloth is impregnated with following to prevent dry-rot and for waterproofing:

Copper Abietate	15 lb.
Amyl Alcohol	250 lb.

Dissolve with heat and add

Turkey Brown Oil	250 lb.

To above add slowly while beating vigorously

Water	5000 lb.
Sod. Acetate	235 lb.

* Fireproof Construction

The following is used in fire-proof building construction:

Volcanic Cinders	45 lb.
Cement	20 lb.
Gypsum	5 lb.
Hydrated Lime	5 lb.
Finely Ground Pumice	25 lb.

Water sufficient for setting.

Flooring, Rubber Composition

A compn. may be used formed of rubber 4, cork powder 12, S 1, ZnO 2, whiting 1, French chalk 2 and paraffin 2 parts, with coloring substances as desired.

Industrial Flooring Composition

Alpha Gypsum	10–77
Asphalt	4–36
Sand or Gravel	0–86

Terrazzo Floor Finish

1. Base Slab

The surface of the base slab shall be struck off reasonably true at a level not less than $2\frac{1}{4}$ inches below the required finish grade.

2. Aggregates

No fine aggregate or sand shall be used in the terrazzo finish. The coarse aggregate shall be (insert here the kind and color of marble chips desired). The coarse aggregate shall be graded in three sizes: $\frac{1}{8}$ inch, $\frac{1}{4}$ inch and $\frac{1}{2}$ inch.

3. Mixtures

The mortar base for the terrazzo finish shall be mixed in the proportions of one part of portland cement to 3 parts of clean, coarse sand, mixed with not more than 6 gallons of water per sack of portland cement.

The terrazzo mixture shall be one part of portland cement and 3 parts of stone chips.

Not more than 4 gallons of mixing water, including the moisture in the aggregate, shall be used for each sack of portland cement in the mixture.

4. Consistency

The terrazzo concrete shall be of the driest consistency possible to work with a sawing motion of the strike-off board or straight-edge. Changes in consistency shall be obtained by adjusting the proportions of aggregate and cement. In no case shall the specified amount of mixing water be exceeded.

5. Placing

Before placing the mortar base and the terrazzo finish, the surface of the structural concrete slab shall be covered with a uniform layer of fine sand $\frac{1}{4}$ inch thick, and covered with an approved tar paper.

The mortar base shall be at least $1\frac{1}{4}$ inches thick and shall be screeded to an even surface $\frac{3}{4}$ of an inch below the finished floor level.

Metal dividing strips about $1\frac{1}{2}$ inches wide, at least 20 gauge, shall be inserted in the mortar or supported on the slab to conform to the designs specified by the architect. The top of the strips shall be at least $\frac{1}{32}$ of an inch above the finished level of the floor.

When in the opinion of the engineer the mortar base has hardened sufficiently to withstand rolling, the terrazzo mixture shall be placed to the level of the tops of the dividing strips.

6. Finishing

After striking off to the finished level the concrete topping shall be rolled length and crosswise so as to secure thorough compaction of the stone chips and cement paste. Additional stone chips of the larger size shall be spread over the topping during rolling until 85 per cent of the finished surface shall be composed of stone. Immediately after rolling, the surface shall be floated and troweled once. No attempt shall be made to remove trowel marks.

After the terrazzo concrete has hardened enough to prevent dislodgments of aggregate particles, it shall be ground down with an approved type of grinding machine shod with free, rapid cutting carborundum stones to expose the coarse aggregate. The floor shall be kept wet during the grinding process. All material ground off shall be removed by squeegeeing and flushing with water.

Air holes, pits and other blemishes shall then be filled with a thin grout composed of neat cement paste. This grout shall be spread over the surface and worked into the pits. After all patch fillers have hardened for seven days the floor surface shall receive a second or final grinding to remove the film of cement paste and to give the floor a polish. It shall then be thoroughly washed and all surplus material removed.

7. Curing and Protection

All freshly placed concrete shall be protected from the elements and from all defacements due to building operations. The contractor shall provide and use when necessary tarpaulins to cover completely or enclose all freshly finished concrete.

If at any time during the progress of the work the temperature is, or in the opinion of the engineer will, within twenty-four (24) hours, drop to 40 degrees Fahrenheit, the water and aggregate shall be heated and precautions taken to protect the work from freezing for at least three (3) days.

As soon as the concrete has hardened to prevent damage thereby, it shall be covered with at least one (1) inch of wet sand, or other covering satisfactory to the engineer, and shall be kept continually wet by sprinkling with water for at least seven (7) days.

8. Cleaning

After removing all loose material, the finish shall be scrubbed with warm water and soft soap, and mopped dry.

---

* Marble, Imitation

An imitation marble slab is formed from a mixt. contg. Keene's superfine white cement 64, marble or alabaster powder 160, pure lime 1 and alum 1 part and may be colored by immersion in water on which coloring matter is floated and rendered translucent by immersing in oil. It may be further dipped in "liquid bronze" and lacquer after drying and wax-polished.

---

* Paving Material

About 100 parts of a fuel oil contg. asphalt is heated to about 105° and mixed with 5–10 parts of a metallic salt of a higher fatty acid such as Al oleate (which is preliminarily mixed with a portion only of the oil), the mixt. is allowed to cool to about 50° and there is then added about 120–40 parts of S and the materials are further thoroughly mixed.

---

* Road Surface, Bituminous

Road surfaces which have been sprayed with bitumen, tar or their emulsions or the like are further treated with a material formed by mixing gravel 1000 (or a similar quantity of broken stone or blast-furnace slag) with bitumen, tar or tar oil 4–5 parts.

Road Composition, Bituminous

Compns. which may be spread hot or pressed into bricks or the like are formed of peat moss, coconut fiber or similar material 10, stone and sand 62 and 28% of a binder comprising fuel oil, hard pitch such as that derived from petroleum residuum and native bitumen (suitably in the relative proportions of 2: 9: 4).

---

* Paving Composition, Cold

Tar 100, asphalt 25 and milk of lime (about one-third of which is lime and two-thirds water) 40 parts are stirred together, or a similar mixt. may be prepd. contg. 10–12% hydrated lime added as dry lime (water and "road metal" being later added).

---

* Plaster, Sound Adsorbent

Furnace Slag (12–20 mesh)	300
Plaster of Paris	125
Microcosmic Salt	2
Sod. Benzene Sulfonate	1
Water	85

Board, Plaster or Wall

Portland Cement	67
Ground Stone	109
Shredded Sugarcane Fiber	24

---

* Building Plaster

Dry Slaked Lime	15–30%
Limestone Dry (Powd.)	75–85%

The limestone should be of 50–100 mesh. The addition of

Plaster of Paris	5–15%

may be made

---

* Patching Plaster

Plaster of Paris	32
Dextrin	4
Volcanic Ash	4

Patching Plaster

Plaster of Paris	32
Dextrin	4
Pumice Powder	4

---

* Plaster, "Nailable"

A plaster through which nails may be driven without splitting consists of

Plaster of Paris	100 lb.
Fine Cotton Fibres	20–50 lb.

---

* Plaster, Magnesium Finishing

A stock mixt. is formed of powd. magnesite 1 and powd. brick or stone 2–7 parts, with or without pumice powder, and 3 parts of this stock mixt. is used with 1 part of $MgCl_2$ soln. of suitable concn. Water and pigments such as Fe or Zn oxide also may be added.

---

* Plastic Roofing Compound

Asphalt	34
Kerosene	26
Asbestos	40

---

* Artificial Slate

Artificial slate, especially for use in the manuf. of writing tables, is made of a mixt. of about 4 parts cement, 9–10 parts powd. slate or slag, 5–6 parts K_2SiO_3, and one or more metal salts such as $Ca(No_3)_2$, $Ba(NO_3)_2$, $K_4Fe(CN)_6$, etc. Coloring matter may also be added. The mixt. is kneaded into a paste by a fatty soln. or emulsion (as of K soap) in water, pressed and laid down in moist air.

* Slate, Writing

Cardboard is coated with

Alum	1
Titanium Dioxide	3
Pot. Silicate	1
Pumice Stone	1
Linseed Oil	1
Gasoline	2

Light, Stone-like Product

Silocel or Kieselguhr	45 parts
Portland Cement	45 parts
Color Pigment	10 parts

Thoroughly mix all to get a uniform powder. Then add 70 parts of water, and work in well. The resulting mix is a damp, pulverent, crumbly agglomeration. This mixture put into molds and subjected to a pressure of approximately 200 lb. per sq. in. will yield a shaped body which will set to a stone like mass in a few hours. The density of the mass is approximately 1.3 times that of water. The dry powder will keep indefinitely, but once the water is added the shaping must take place within an hour. After setting, curing in a damp atmosphere for a few days will materially increase the strength.

If more water is used than above, a paste will result. The paste need not be molded under pressure, but the resulting set product is much denser.

* Stone, Artificial

Alum Rock	59
Alabaster Powd.	22
Alabaster Plaster	17
Water	2

The alum rock is fused; the other ingredients are added and the mass is molded.

* Synthetic Stone

Shale (325 mesh)	16.7
Calcium Hydroxide	10.0
Water	5.3
Limestone Aggregate (Powd.)	68.0

Mix thoroughly and compress at 2500 lb. pressure in polished molds. Remove; allow to stand for an hour. Put in autoclave with saturated steam at low pres sure for 1–2 hours. Turn off steam an allow to cool for one hour.

* Stucco Composition

Portland Cement	28
Silica Dust	28
Limestone	10
Asbestos Flour	6
Titanium Dioxide	4
Boiled Linseed Oil Mineral Spirits Turpentine	to make to a working consistency

* Tennis Court Composition

A layer of broken brick and clinke of a granule size between 0.1 and 0.6 mm. is covered with a layer of compn. contg. gravel 15, coarse CaO 5, rock salt 5, cement 5 and brick dust 70%.

Tile and Floor Composition

Asphalt Emulsion	1.75
Cement	1
Crushed Rock	5

* Concrete Tiles and Blocks

Rubber latex (suitably 0.5–2.0 pints per cu. yard of concrete) is added to a mixt. such as one comprising sand 3 and portland cement 1 part and the compn. may be reënforced with bamboo or other material.

Structural Tile

Calcium Carbonate (Marble Dust, Fine)	15 parts
Powdered Glass	4 parts
Magnesium Oxide (Heavy)	8 parts
Magnesium Chloride Solution (Sp. Gr. 1.19 @ 25° C.)	13 parts

Mix powders and make a thick paste with the solution of magnesium chloride. Pour into paraffined molds on a hard shiny surface. Let stand till dry.

* Wall and Ceiling Composition

A mixt. of trachyte tuff 57, sand (washed free from clay) 34, portland cement 5.7 and soda 2 parts is used with sufficient water to cause the material to set.

* Wall Efflorescence, Prevention of

The parts are scaled, washed with water contg. 4% formaldehyde and coated with cement or lime mixed with water also contg. 4% formaldehyde. A top coating of plaster is mixed with 5% alum and water contg. 4% formaldehyde.

---

* Waterproof Construction Material

Suitable for walls, roofs, roads, etc.

Coal Tar	5–85
Rubber	1–25
Granite Dust	5–85

Warm together and mill until uniform. This may be molded into bricks or sheets.

---

* Wood Preservative

Wood is protected against fungi and insects by 0.1 *N* Na_3 AsO_3 or 0.2 *N* NaF, is mixed with a relatively concd. soln. of an alk. earth or heavy metal salt, *e.g.*, 8 *N* $CaCl_2$, with or without a protective colloid. The mixt. does not form a ppt. until after a certain time, and meanwhile the wood is given a single impregnation with the mixt.

---

* Wood Preservative

Copper Chips	4%
Zinc Chips	1.5%
Cresol	7.5%
Ammonium Hydroxide	7%
Water	80%

Dissolve by continued agitation.

---

* Rot-proof Fibre Board

Such a board is manufactured from a pulp consisting of

Fibre	1000
Water	50,000
Rosin Size	20
Zinc Arsenite	3
or Creosote	50

---

* Wood Impregnating Liquid

Neutral Mineral Oil	15
Mineral Spirits	36
Liquid Manganese Drier	10
Gloss Oil	10
Turpentine	20
Orthodichlorbenzol	5
Chrome Alum	1
Oil Soluble Dye	0.5

Wood Preserving Composition

A preservative compn. is prepd. by mixing water 3–4 qt., lime 1–3 lb., African gum 1.5–2 lb., a mineral or vegetable oil such as paraffin oil 0.75–1.0 qt., creolin 0.5–2 oz., paraffin oil 3–4 oz., HCl 1.5–2 oz. and "a poison" 4–8 oz.

---

* Wood Preservative

A method of making wallboard or like article comprises precipitating zinc-meta-arsenite in the presence of a plant fiber, and shaping and drying the fiber. The fiber, after felting on the Fourdrinier and before drying is treated successively, by spraying or otherwise, with two solutions, as hereinafter described, which react quickly to deposit zinc-meta-arsenite upon and throughout the mass of fiber. Excess water may then be pressed out and the wall-board dried as usual. No corrosive substances are liberated during the drying, and the dried wall-board is found to be quite uniformly impregnated with microscopic crystals of zinc-meta-arsenite. The two solutions above mentioned are prepared as follows: Solution A: An aqueous solution of a soluble zinc salt, preferably zinc sulphate, is prepared, preferably without heating. A suitable concentration is 15 per cent, calculated as zinc sulphate. Solution B: Arsenious oxide is dissolved in water by heating to the boiling point for some time, preferably in the presence of small proportions of soda ash, sodium hydroxid or sodium arsenite, which act to facilitate and accelerate the solution of arsenious oxide in water. The resulting solution is cooled to room temperature, and after cooling sodium bicarbonate is dissolved in it, in proportions as indicated below. A suitable concentration is 4 per cent As_2O_3. The reaction which occurs when these two solutions are mixed in proper order may probably be represented by the following equation:

$$ZnSO_4 + As_2O_3 + 2NaHCO_3 = Zn(AsO_2)_2 + Na_2SO_4 + H_2O + 2CO_2.$$

The zinc sulphate and arsenious oxide are preferably used in the proportions indicated by the above equations, that is to say in equimolecular proportions. The proportion of sodium hydrogen carbonate used is preferably approximately 95 per cent of the quantity indicated by the equation.

---

* Timber Preservative

Fuel Oil	85
Asphalt	10
Naphthalene	5

Heat together to dissolve and apply hot.

---

* Timber Preservative

15 grams of copper resinate are dissolved in 250 grams of amyl alcohol, and 250 grams of turkey-red oil are

added; the mixture is made up to 5 liters bulk by emulsification in water containing 235 grams of crystallized sodium acetate.

* Artificial Wood

Sawdust	100
Manganese Dioxide	1–3
Linseed Oil Boiled	2–8

* Wood, Cigar Box

The wood is extracted with acetone to remove resins.

* Lumber, Synthetic
(For Wall Board)

Sawdust	80–90%
Hardened Synthetic Resin	10–20%

The resin may be melted or dissolved and mixed thoroughly with the sawdust under pressure and heat.

Wood, Metal Coating

Wood, stone, textiles, paper, etc., are coated with the following which is first melted, cooled, ground and taken up with water.

Metal (Powder)	40– 70
Paraffin Wax	60– 90
Graphite (Powd.)	60– 90
Precipitated Chalk	100–150
Sod. Silicate	180–220
Casein	40– 70

* Writing Surface, Washable

A base of cardboard, wood or the like is coated with a pulp prepd. by dry mixing marble cement (Keene's cement) 20 and zinc white 50 parts, and then incorporating linseed oil varnish 15 and turpentine oil 15 parts into the mixt. The coated base is air-dried for 3–4 days, and the surface is then polished.

* Floor Covering Material

Dry rubber contg. 15–35% of S is thoroughly mixed with small pieces of cork to give a product contg. 50–85% of cork. The mixt. is molded, hot-vulcanized under high pressure and cooled before releasing the pressure. Pigments and fillers may be added to the mixt.

* Roofing Composition

Pitch	25–40
Asphalt	25–50
Cork	25–50
Asbestos	5–10

* Roofing Felt, Impregnation for

A compn. is used which is relatively more fluid than asphalt alone at elevated temps. and which comprises 60–98% of asphalt having a m. p. not less than about 32° together with 40–2% of paracoumarone resin.

Roof Coating

Asphalt	10 lb.
Varnolene	3 gal.
Short Fibered Asbestos	5 lb.

* Plastic Roofing Composition

Petroleum Asphalt	34
Kerosene	26
Asbestos	40

Wood Strengthener

A solution to help retain nails in wood is made as follows:

Rosin	1 lb.
Benzol	1 gal.

Nails are dipped in this solution, withdrawn, allowed to dry and they are then ready for use.

* Fireproof Wall Board

Fibrous wall board having finely ground mica interspersed among the fibers is fireproofed by impregnation with a soln. comprising NH_4 phosphate, H_3BO_3, $MgSO_4$ and water.

METALS AND TREATMENT

* Aluminum, Surface Hardening

A uniform hard coating of aluminum oxide is deposited if the aluminum is made the anode in an electrolyte containing 5–12% H_2SO_4.

---

* Recovering Aluminum from Foils

Paper-backed Al foil is packed to a *d* of 0.75–1.25 lb./cu. ft. and heated in a closed retort at 450–550° until all volatile matter is expelled; air is then admitted and the temp. kept const. until the C is burnt out. The residual Al is removed and melted with the usual fluxes.

---

* Aluminum Bronze Powder

Al is ground at 50° with stearic acid and a solvent, *e.g.*, naphtha, turpentine, C_6H_6, until a smooth paste is obtained from which the solvent is removed until the remainder contains 58% Al, 1% stearic acid, 1% Al stearate, and 40% solvent (preferably varnolene—a petroleum fraction, b.p. 152–207°).

Core Binder
(for aluminum castings)

Sharp Sand	45 lb.
Molding Sand	45 lb.
Rosin Powd.	2 lb.
Flour	1 lb.

or

Sharp Sand	71 lb.
Molding Sand	25 lb.
Rosin Powd.	4 lb.

Spray with molasses water and bake at 325° F. Remove from oven and coat with soapstone. Return to oven to dry.

Core Oil

1. Tung Oil	10 gal.
2. Linseed Oil	20 gal.
3. Mineral Oil	20 gal.
4. Varnish "foots"	5 gal.
5. Benzine	5 gal.
6. Rosin	200 lb.
7. Lime Slaked	6 lb.
8. Litharge	7 lb.
9. Manganese Dioxide	3 lb.

Melt 1 and 6, stir in 7, 8 and 9.

Heat to 500° F. for 20 minutes. Add 2 a little at a time and keep at 400° F. for 20 minutes. Raise temperature to 480° F. and keep there for two hours. Cool to 300° F. and add with stirring 3, 4 and 5.

---

* Corrosion Inhibitor, Auto Radiator

To prevent corrosion of cooling radiators containing glycerine or glycols, 0.2% of dimethylmorpholine or ethanolmorpholine is used.

---

* Cast Iron, Strong Malleable

An annealing pot is charged with white iron castings and with 4–15% of Fe oxide, and maintained at a temp. of 900–980° for 20–50 hrs. to graphitize the free cementite; the temp. is then lowered to between 730° and 650° and maintained at such range for 10–50 hrs. to graphitize pearlitic cementite and effect decarbonization of the white cast iron.

---

* Coating Iron Sheets

Fe or steel sheet (etc.) is plated with Sn, then coated with an asphaltic-base enamel, and baked at 230°. The Sn prevents flaking of the resulting japan finish.

---

* Sticking Iron Sheets, Prevention of

Rolled and pickled ferrous sheets and plates are treated with 0.001–0.2% sod. chromate solution prior to annealing. This prevents sticking of piled sheets in box annealing.

---

* Electro-deposited Metal, Stripping

To facilitate removal of deposited metal, the mandrel is coated with a 0.05–0.5% soln. of beeswax in turpentine to which 1% of CS_2 also is added.

Metal Annealing Bath

Sod. Chloride	30 lb.
Pot. Sulfate	44 lb.
Pot. Carbonate	21 lb.
Borax	5 lb.

All formulae preceded by an asterisk (*) are covered by patents.

* Case Hardening of Tools

The tool is heated to 600–700°, sprinkled with $K_4Fe(CN)_6$, again heated to 800–900°, and cooled in a mixture of 250 g. of coal dust, 75 g. of $K_4Fe(CN)_6$, 500 c.c. of oil, 125 c.c. of H_2O, and 250 g. of powdered slate.

Bright Tin Finish for Screws

Use the following tin solution to produce a tin deposit on your work:

Aluminum Sulphate	2 oz.
Cream Tartar	2 oz.
Tin Crystals	½ oz.
Water	1 gal.

Use a zinc container for the solution; place the screws in the pan and boil for 45 minutes. A new solution is necessary for each batch of work. If the deposit is not bright enough, tumble the screws in an oblique tumbling barrel, using clean hardwood sawdust.

Carbonizing Steel

The steel blanks are tumbled, burred and tumble finished previous to carbonizing and are then placed in the revolving drum of the carbonizing machine and ¾ pints of carbonia oil with ½ bushel of Burnt Bone added. The drum is closed securely, gas turned on and heated to 700–750 degrees F. for 3 hours. The heat is turned off and the drum allowed to run for 2 hours to cool off. The contents are removed and sifted and tumbled in ½ bushel of No. 2 Granulated cork and 2 pints of japan oil for 5 minutes; then dried and cleaned by tumbling in ½ bushel of sawdust for 5 minutes to put on a high polish.

Bake at 120 degrees F. for 8 to 10 hours to harden oil.

Gum for Parting Punch from Die

1 lb. Beeswax
½ lb. Rosin
¼ lb. Venice Turpentine

---

* Casting Magnesium, Molds for

Examples 1 and 2

Molding Sand	approx.	93 to 97 parts
Sulfur	approx.	6 to 2 parts
Boric Acid	approx.	1 1 part
Di-ethylene Glycol, 40 per cent aqueous solution		Sufficient

Example 3

Molding Sand	approx. 94 parts
Sulfur	approx. 2 parts
Anthracene	approx. 3 parts
Boric Acid	approx. 1 part
Di-ethylene Glycol, 40 per cent aqueous solution	Sufficient

Example 4

Molding Sand	approx. 93 to 91 parts
Borate of Glycol	approx. 3 to 5 parts
Naphthalene	approx. 4 4 parts
Water	Sufficient

Example 5

Molding Sand	approx. 98 parts
Boric Acid	approx. 2 parts
Ethylene Glycol, 40 per cent aqueous solution	Sufficient

Example 6

Molding Sand	approx. 95 parts
Naphthalene	approx. 4 parts
Boric Acid	approx. 1 part
Ethylene Glycol, 40 per cent aqueous solution	Sufficient

Example 7

Molding Sand	approx. 97 parts
Ammonium Bisulfate	approx. 2 parts
Boric Acid	approx. 1 part
Ethylene Glycol, 40 per cent aqueous solution	Sufficient

Example 8

Molding Sand	approx. 93 parts
Ammonium Bisulfate	approx. 2 parts
Naphthalene	approx. 4 parts
Boric Acid	approx. 1 part
Ethylene Glycol, 40 per cent aqueous solution	Sufficient

Example 9

Molding Sand	approx. 98 parts
Boric Acid	approx. 2 parts
Ethylene Glycol, 25 per cent aqueous solution	Sufficient

Example 10

Molding Sand	approx. 93 parts
Naphthalene	approx. 4 parts
Ammonium Bisulfate	approx. 2 parts
Boric Acid	approx. 1 part
Ethylene Glycol, 60 per cent aqueous solution	Sufficient

Example 11

Molding Sand	approx. 97 parts
Ammonium Bisulfate	approx. 2 parts
Boric Acid	approx. 1 part
Ethylene Glycol, 25 per cent aqueous solution	Sufficient

Example 12

Molding Sand	approx. 90 to 88 parts	
Organic Borate	approx. 3 to 5 parts	
Sulfur	approx. 2	2 parts
Naphthalene	approx. 3	3 parts
Ammonium Bisulfate	approx. 2	2 parts
Water		Sufficient

Example 13

Molding Sand	approx. 98 to 97 parts	
Borax (sodium tetraborate)	approx. 2 to 3 parts	
Aqueous Solution containing 15 per cent sulfuric acid and 40 per cent ethylene glycol		Sufficient

* Mold, Magnesium Casting

(a)	Molding Sand	920
	Water	60
	Sulfur	1
	Boric Acid	1
	Ammonium Chloride	2
	Ammonium Sulfate	2
	Ammonium Bromide	2
(b)	Molding Sand	92
	Water	6
	Sulfur	2

* Casting Mold Coating

Graphite Powd.	4
Core Oil	2
Gasoline	1.7

* Casting Oxidizable Metals, Mold for

Molding Sand	93
Glycol Bori-Borate	3
Naphthalene	4
Water	sufficient

* Molding Sand, Self Hardening

Ordinary molding sand is mixed with 3–10% of a mixture formed from the following:

40° Bé. Sod. Silicate	80–90
Bentonite	2–7
Water	3–18

* Mould-Core Wash

Am. Nitrate	2
Silica Flour	6
Molasses	1.5
Asbestos	1
Water	4

Nickel Welding Wire

Silicon	0.2–4.0%
Titanium	0.05–2%
Magnesium	0.12–0.2%
Nickel	Balance

* Recovering Platinum

Flotation concentrates containing Pt are roasted to expel S, As, etc., cooled, mixed with 5% of NaCl, and treated with Cl_2 at 200–550° for 5 hr. The product is leached with 0.5–1% HCl, the Cu removed by agitation with $CaCO_3$, and the Pt metals are pptd. by Zn dust.

* Powder, Parting or Facing

Fine Coke, Coal or Graphite	100
Rosin Residue	200

Heat to 250° C. to drive off volatile matter.

Rust Proofing

Copperas is dissolved in water, 3.5 kg. per l., in an iron vat. The materials to be treated are suspended in this bath. The bath is heated to 95–98° and held there for about ½ to 1 hr.

* Parkerizing, (Rustproofing)

A mixt. of Mn and Fe^{++} salts gave the best results; the latter may be prepd. in quantity by dissolving 1 part of Fe filings in 10 parts of 65% H_3PO_4 at 100°. After filtration the soln. is cooled slowly and the large crystals formed are centrifuged and dried rapidly at 60°. The Mn salt (mixed with Fe) is prepd. similarly from ferromanganese, and if the bath is boiled before use partial hydrolysis occurs with the formation of $Fe(H_3PO_4)_2$ and free acid. The optimum ratio of Fe to Mn in the protective coating is 1: 1, which is formed by selective deposition from a soln. of ratio 2:1; consequently the bath must be constantly replenished with Mn. The standard concn. adopted was 35 lb. of $Fe(H_2PO_4)_2$ per 125 gals. of H_2O and the temp. was maintained at 99°. Cu phosphate in small quantity acted as an accelerator.

Rustproofing Iron

The article is cleaned by sand-blasting or pickling in acid and plated with a thin layer of Zn from a bath contg. NaCN 4, $Zn(CN)_2$ 5, NaOH 4 oz., and a small amt. of Hg per gal., zinc anodes contg. 0.5% of Hg and a c. d. of 25 amp./sq. ft. being used at 5 v. After being washed well, the plated articles are dipped in a soln. contg. Ni chloride 4, NH_4Cl 6, NaCNS 2, and $ZnCl_2$ 0.5 oz. per gal. The black deposit thus obtained may be coated with lacquer or given an oil finish in the usual way.

* Rust, Removing

Ferrous-metal articles are immersed in a bath of NaOH 20 and $KMnO_4$ 1% at 190° F. to break down the resistance of oxide and scale to acid. The articles are then immersed in a hot pickling path of HCl, HNO_3 or H_2SO_4 for 5 to 10 min. Articles so treated are substantially free from etching.

* Steel Hardening Composition

Pot. Ferrocyanide	50–70
Soda Ash	7–12
Salt	15–25
Wood Charcoal	10–20

* Steel, Hardening and Tempering

Linseed oil is heated to about its b. p., 2 oz. resin per gal. of oil is added, and the metal to be treated is immersed in the oil and resin soln. until the metal acquires the temp. of the soln.; the metal is then removed from the soln. and covered with powd. resin, plunged into cold coal-oil and permitted to remain in the latter until the temp. of the metal and immersion bath are equalized.

Magnetic Chromium Steel, Heat Treatment of

The best magnetic properties of a steel contg. 1.3% C and 2.1% Cr are obtained by quenching from 850° in oil. The steel should not be held too long between 750° and 850°, as a change takes place in the double carbide. Incorrect heat treatment can be remedied by holding at 950–1000° for 1 hr., cooling in air, and then hardening.

Steel Parts, Preventing Corrosion of

Steel parts exposed to corrosive fumes are coated with

Lanolin	10
Naphtha	20

* Pickling Solution for Stainless Steel

10% (10% 60° Sulfuric Acid)
10% Copper Sulfate

Heated to 160–200° F.

* Steel, Toughening

Mild steel is heated above the Ac2 point, *e.g.*, to 760–780°, air-cooled until it just exhibits magnetic properties, and then quenched in cold H_2O.

* Annealing Chrome Steel

Low-C Cr–Ni or high-Cr steels are heated rapidly to 800–1000° and immediately quenched, the process being repeated several times until the desired properties are obtained. The process may be used also for softening an alloy of 51.3% Ni, 27.5% Mn, and 20.5% Cu, using a temp. of 900° and 10 alternations of heating and quenching.

* Ingot Molds

Ingot molds are made from Fe alloy contg. C about 2, Mn 0.6–1, P less than 0.1, S less than 0.1 and Si about 3%. Details are given of the manuf. of the alloy from crude Fe and steel.

MISCELLANEOUS

* Anti-Fogging Agent
(For Windows and Windshields)

Borax	125
Water	64,000
Elm Bark	1,000
Acacia	16,000
Alcohol	1,000

Anti-Fogging Compound

1. Glycerin	8 oz.
2. Pot. Oxalate	16 oz.
3. Spirits Turpentine	1 oz.
4. Camphor	⅛ oz.

Warm (1) and (2) until dissolved; add (4); remove from heat and add (3).

Anti-Fogging Liquid for Windshields

Glycerol	10–20
Albumen	1
Water	89–79
Phenol	0.1

* "Anti-Fogging" Liquid

The following is applied to windows or automobile windshields to prevent dimming by moisture.

Soap	1¼ lb.
Glycerol	½ oz.
Water	1 pt.
Acetone	1 oz.

Anti-Fog Windshield Liquid

Glycerol	10 oz.
Alcohol	⅛ oz.
Rose Water	6 oz.
Salt	0.06 oz.
Sulfur Powd.	0.06 oz.

Anti-Mist Liquid
(For Use on Glass)

Potash Coconut Oil Soap	120
Glycerin	60
Turpentine	8
Naphtha	3
Clovel	1

* Anti-Stick Coating Composition

Sod. Alginate	2
Soap	1
Glycerin	1
Glue	0.25
Water	to suit

The above solution is useful on sticky surfaces such as asphalt and soft resin coatings.

Algae Removal

In a swimming pool one pound of copper sulfate, or blue stone, to two million pounds of water destroys algae. This material is likely to be fatal to fish. The solid is placed in a sack and dragged back and forth across the pool to secure proper mixing. In computing amount needed, one gallon of water weighs eight and one-third pounds, or one cubic foot of water weighs 62.5 pounds.

* Air-Conditioning Water Treatment

A composition to be added to water used in air conditioning apparatus for preventing corrosion, congestion and rust in said apparatus consisting of the following ingredients and their reaction products: water approximately 6.78 per cent, caustic soda approximately 1.45 per cent, sodium bichromate approximately 1.16 per cent, soda ash, approximately 2.90 per cent, di-sodium phosphate approximately 1.63 per cent, sodium silicate approximately 86.08 per cent, and tannin .006 per cent.

Gasoline Carbon Looseners

There are in the market a number of gasoline addition agents for the removal of carbon. These are used in the following manner:

Add 4 oz. to five gallons of gasoline in tank or supply through manifold by attached cup.

The formulas for a few of these are:

1. Medium Oil	50%
Varnoline	50%
2. Medium Heavy Oil	50%
Light Paraffin	50%
Wintergreen Odor	0.2%
3. Kerosene or Varnoline	80%
Vaseline	20%

All formulae preceded by an asterisk (*) are covered by patents.

Brake Lining, Composition for

Crepe Rubber	14
Litharge	10
Barytes	34
Zinc Oxide	5
Carbon Black	3
Graphite	4
Sulfur	4
Asbestos Yarn	12
Brass Wire	14

---

* Anti-Freeze

Coffee 2 lb.

Is extracted with

Water 4 lb.

and added to following solution:

Calcium Chloride	260 lb.
Glycerin	1 gal.
Water to make	100 gal.

Anti-Freeze

Pints of anti-freeze per gal. of water for protection at:

	+10° F	0° F	−10° F	−20° F
Denatured alcohol 180° proof	3.4	4.9	6.5	8.3
Denatured alcohol 188° proof	3.3	4.7	6.0	7.7
Glycerine (USP) 95%	3.8	5.3	7.1	9.0
Radiator glycerine 60%	10.0	18.7	39.0	106.5
Ethylene glycol 95%	2.7	4.0	5.1	6.5

Specific gravity for protection at:

	+10°F	0°F	−10°F	−20°F	−30°F
Denatured alcohol	0.968	0.959	0.950	0.942	0.921
Glycerine	1.090	1.112	1.131	1.147	1.158
Ethylene glycol	1.038	1.048	1.056	1.064	1.069

---

* Anti-Freeze, Prevention of Foam in

Lard Oil	80 gm.
Triethanolamine	40
Sperm Oil	14
Gum Arabic	2
Butyl Carbitol	36
Water	94
Calcium Acetate	1.2
Dye	0.3

The above is added to 2 gal. Ethylene Glycol to produce an anti-freeze which is non-corrosive and doesn't foam.

---

* Anti-Freeze Alcohol, Corrosionless

0.1% Dibutylamine or diethanolamine is used with aqueous alcohol solutions to inhibit corrosion of iron.

---

* Anti-Freeze Liquid

Sod. Nitrate	98–99 lb.
Glue	1– 2 lb.

---

* Non-Corrosive Anti-Freeze Liquid

Methanol	74
Water	25
Borax	1

---

* Anti-Freeze Liquid, Non-Corrosive

Methyl Alcohol	70
Water	30
Sod. Antimony Tartrate	0.1

---

* Anti-Freeze Solution

Calcium Chloride	100 lb.
Glycerol	1 gal.
Slaked Lime	4 oz.
Water to make	17 gal.

---

* Anti-Freeze Solution

An aq. soln. of 260 lb. of $CaCl_2$ is mixed with glycerol 1 gal. and alc. 2 qt. and with a coffee ext. derived by boiling 2 lb. of coffee in water; the mixt. is dild. to 100 gal.

Anti-Freeze Solution

Denatured Alcohol	50
Methanol	10
Glycerin	30
Water	10

---

* Non-Sulfating Battery Acid

Distilled Water	755 cc.
Sulfuric Acid	240 cc.
Aluminum Sulfate	18 gm.
Am. Sulfate	4 gm.
Basic Aluminum Acetate	5 gm.

---

* Battery Box Composition

Ground Scrap Rubber	15
Ground Used Tire Fabric (de-rubberized)	7.5
Montan Wax	7.5
Carbon Black	1.5
Silica Powder	30
Blown Asphalt	22.5
Gilsonite	16

---

* Brake Fluid, Hydraulic

Triethylene Glycol	90
Sulfo Turk C	10

---

* Carbon, Removing Cylinder

Oil of cedar wood 10, acetone 50, benzene 15, alc. 24 parts are mixed with naphthalene in the proportion of 1 lb. $C_{10}H_8$ to each 6 gal. of the liquid.

---

* Electrical Potting Composition

The following is used on fixed electrical condensers and dry batteries.

Chlorinated Naphthalene 25–40 lb.
Gilsonite 20–30 lb.
Montan Wax 30–50 lb.

It has a flow point of 80–100° C.

* Engine Joint Seal

Aluminum Oxide 5
Blown Castor Oil 60

Heat with stirring at 150° C. until uniform and then stir in

Mica Powder 15
Asbestos Short Fibres 15

Gasket Compound

Asbestine Powd. 56
Copal Varnish 9 Cut 44

Grind in ball mill for 3 hours.

Engine Carbon Remover

Diethyl Formamide 1– 5
Benzol 49–40
Alcohol 50–55

Puncture Preventive, Tire

Bentonite 100
Magnesium Oxide 2
Asbestos Fiber 50
Water suitable quantity

Battery Terminals, Coating for

Diglycol Stearate 10
Water 300

Heat until melted and stir until dispersed. Run in slowly with stirring

Graphite Powd. 30–100

Tire Paint

Precipitated Chalk 40 lb.
Spanish White 20 lb.
Gilder's Whiting 15 lb.
Gum Tragacanth 10 lb.
Phenol Crude 10 oz.
Clovel 10 oz.

Allow gum to soak overnight in 7 gal. water; add phenol and pigments while stirring; if too thick add more water and then stir in the Clovel.

* Prevention of Frothing and Foaming in Aqueous Solutions

Add 1% by weight or less of a 2 to 10% solution of a metallic soap in pine oil. The soap may be aluminum, barium, calcium or zinc oleate or stearate.

* Depolarizer for Dry and Leclanché Batteries

Manganese Dioxide 80
Graphite 20
Silica Gel 1

Oriental Barometer

Cards, artificial flowers, etc., stained with cobalt chloride, change their color with the varying hygrometric state of the air; turning pink or red with much moisture, and blue when it is dry.

Belt Dressing Stick

Rosin 65 lb.
Tallow 6 lb.
Stearic Acid 1 lb.
Scale Wax 20 lb.
Castor Oil 2.0 lb.
Rosin Oil 0.5 lb.
Lanolin 4.2 lb.

Boiler Scale, Removal of

8–10% HCl is most suitable for Cu or brass app.; 5–10% HCO_2H, for Al or tinned metals; 15% AcOH, for Zn or galvanized iron.

* Boiler Compounds

Soda Ash 67 parts
50% Caustic Solution 10 parts
Powdered Chestnut Extract 20 parts
Water 3 parts

the chemicals mentioned being mixed and pressed into briquettes.

Soda Ash 55 parts
Sodium Aluminate 20 parts
Dextrine 8 parts
50% Caustic Solution 5 parts
Water 12 parts

these chemicals being mixed separately from those listed above and pressed into briquettes.

* Boiler Compound

Gallnuts (Powd.) 5 lb.
Pine Bark (Powd.) 2 lb.
Larch Bark (Powd.) 2 lb.
Rosin 1 lb.

100 gm. of the above are used per cubic meter of water.

Boiler Compound

Soda Ash 87
Trisodium Phosfate 10
Starch 1
Tannic Acid 2

* Box Toe Composition

Wood Pulp	60
Cotton Linters	30
Asbestos Fibre	10

Any thermoplastic sufficient to impregnate.

* Brake Lining

Asbestos Fiber	45
Magnesium Oxide	3.6
Rubber	18
Sulfur	9
Graphite	6
Litharge	6
Iron Oxide	3
Kaolin	9.1

* Brake Lining, Friction Material for

Black Clay	45
Zirconium Oxide	25
Feldspar	15
Agalmatolite	5
Magnesite	5
Kaolin	5

* Brake-Lining, Treatment for

To insure smooth brake action the lining is treated with

Ground or Deflocculated Graphite	1 oz.
Light Lubricating Oil	16
Naphtha	17
Carbon Bisulfide	19

* Hydraulic Brake Fluid

Hydra-acetyl Acetone	40
Castor Oil	60

* Brine Solution, Non-Corrosive

Calcium Chloride	40
Water	60
Zinc Chloride	0.4

* Caking of Powders, Prevention of

The addition of 1–5% of rice meal or flour prevents caking of powders and crystals.

* Arc Carbons

Calcium Fluoride	40
Strontium Fluoride	10
Sodium Silicate	5
Carbon Flour	45
Tar	sufficient to bind

* Carbon Electrode

A core is formed by baking

Calcium Fluoride	40
Strontium Fluoride	10
Sod. Silicate	5
Carbon Flour	45
Tar	sufficient to bind

* Carbon Remover

Isopropyl Ether	10
Propylene Dichloride	10
Ethylene Dichloride	10
Chloronaphthalene	8

* Carbon Deposits, Removing

The cylinders are heated a little above 65° C. and treated with following:

Aniline	25
Alcohol	25
Benzene	25
Naphthalene	25

* Catalyst, Oxidation

Platinum	0.3
Ferric Sulfate	1.0
Magnesium Sulfate	98.7

Flocculated Clay

A special flocculated clay of low acidity, which is especially suitable for compounding with rubber, is prepd., for example, as follows: 8 lb. of Florida clay are peptized in 6 gal. of soft H_2O by the addn. of 0.5 oz. of sodium pyrophosphate and 0.75 g. of rosin. A rosin soap forms. The impurities are allowed to settle and the clay is flocculated by adding 0.25 oz. of $Al_2(SO_4)_3$.

* Catgut and Tennis String, Preservative for

Gelatin 3 lb. is cooked with water 1 gal. up to a temp. of about 95° with addn. and stirring in of about 1 oz. of red oil and the further successive addn. of about 5 oz. glycerol and about 5 oz. of tallow oil while maintaining a temp. of about 95° during the prepn. of the compn.

* De-inking Newspaper

Print is removed from newspapers, etc., by immersing them in 7 times their weight of water containing 0.5% of Am. Linoleate or Sodium Sulforicinoleate and 0.5% of carbon tetrachloride or carbon disulfide.

* De-inking Newspaper

The paper is broken up and beaten for 5–7 min. at 100–107° (1 atm.) in a solution containing NaOH (10 pts.), $Na_2B_4O_7$ (25 pts.) Na_3PO_4 (4 pts.),

K_2CO_3 (1 pt.) in 60 pts. of H_2O. Ink pigments, etc., are removed by skimming and the pulp is then washed.

Demulsifier

Concentrated turkey red oil is a very efficient demulsifier and is used quite extensively in the oil fields for breaking petroleum emulsions. This material is made by slowly adding 10% of sulphuric acid (66° Baumé) to pale blown castor oil. The above is allowed to stand for two hours. It is then added to four times its volume of a half of one per cent water solution of sodium chloride and mixed thoroughly. After about twenty-four hours the water will be precipitated, whereupon the same is decanted and the remaining sulphonated castor oil is neutralized with ammonium hydroxide.

* Light Elastic Compound

Wood Charcoal Dust	35
Cotton Linters	10
Crepe Rubber	55

The above is to be worked on a rubber mill and may be vulcanized if desired.

* Electric Lamp "Getter"

The following formula is used for coating lamp filaments for "cleaning-up" deleterious residual gases.

Cryolite (Silk Bolted)	200 gm.
Red Phosphorus	30 gm.
Alcohol	230 cc.

Mix the above in a ball mill for half an hour. Place in a tray; evaporate alcohol and dry at 110° C. Sift through 200 mesh screen and bottle.

The method of application of this admixture to the filament is as follows:

Three hundred grams of metallic granulated zinc (between 20 and 40 mesh) or other suitable non-absorbent material are placed in a casserole and approximately 3 grams of diethyl phthalate is added thereto. The admixture is stirred until the zinc particles are uniformly coated with the diethyl phthalate. The binder is retained upon the surface of the zinc particles by surface tension effects. One hundred grams of the dry getter admixture is then added and the casserole and contents rotated so that the binder wetted zinc particles pick up the dry powder and cause it to adhere to the surfaces thereof by the adsorption of the diethyl phthalate.

The finished getter material will substantially comprise a quantity of round pellets made up of a core of zinc surrounded by a quantity of getter mixture.

These pellets are placed in a tube or other chamber together with the coiled filaments to be gettered and the tube and contents vigorously agitated. This forces the getter mixture into the core of the coil. The diethyl phthalate present acts as a binder and causes the getter mixture to pack so that future handling of the coils will not easily displace the getter.

During the preheating and baking out of the lamp the diethyl phthalate is vaporized completely away from the mixture. The remaining getter material comprising phosphorus in relatively coarse condition, superficially coated with inert non-inflammable cyrolite (or sodium ferric fluoride) withstands a higher baking temperature than the finely divided phosphorus getter composition heretofore employed.

* Electrode, Arc Lamp

Zirconium Fluoride	10
Carbon	30
Calcium Fluoride	60

* Electrolytic Condenser

An electrolyte for above consists of

Sod. Stearate	20
Diethylamine Stearate	10
Glycerol	95
Water	5

Embalming Fluid

Glycerin	250
Formaldehyde	1565
Pot. Nitrate	150
Borax	40
Boric Acid	120
Dark Red BA Dye	0.4
Water	2800

Embalming Fluids

Solution of Formaldehyde	11	lb.
Glycerin	4	lb.
Sodium Borate	2½	lb.
Boric Acid	1	lb.
Potassium Nitrate	2½	lb.
Solution of Eosin, 1%	1	oz.
Water enough to make	10	gal.

The sodium borate, boric acid and potassium nitrate are dissolved in 6 gallons of water; the glycerin is added, then the solution of formaldehyde, and lastly the solution of eosin, and the necessary amount of water.

Another formula in vogue is as follows:

Thymol	15 gr.
Alcohol	½ oz.

Glycerin	10 oz.
Water	5 oz.

* Embalming Fluid

Sodium Hydroxide	4
Water	100
Glycerol	20
Sodium Nitrate	5
Sodium Oleate	4
Sodium Orthophenylphenol	10
Sodium Oxalate	3
Color to suit.	

To the above add

Formaldehyde	32

before using.

* Tissue Filler, Embalmers

Celluloid Scrap	1
Methanol	7
Castor Oil	½
Oil Cloves	⅛
Triacetin	10
Ethyl Methyl Ketone	7
Eosin	to suit color

Embroidery Treatment

Cotton cloth is saturated with

Alum	1
Aluminum Chlorate (20%)	3
Water	17

Dry in air and embroidery is then worked on cloth. Then dry in oven at 80° F. Chlorine is liberated and attacks cotton so that latter may be brushed off from embroidery.

Oxidation of Ether, Prevention of

To every lb. of anaesthetic ether add 2 gm. Hydroquinone.

Fire Extinguisher

A fire extinguisher is absolutely necessary in the laboratory if the workers are to be protected. Manufactured extinguishers are rather expensive, but the following substitute is very efficient.

The metal part of a burned out electric light bulb is removed. The tube used to seal the bulb is dipped in carbon tetrachloride, and the tube broken. The vacuum draws the liquid into the bulb. The break is sealed with wax. Fire extinguishing "bombs" of this type may be put in convenient places about the laboratory.

* Chimney Fire Extinguisher

Pot. Chlorate	15.02
Sulfur	19.12
Zinc Powd.	49.14
Am. Chloride	5.46
Mag. Carbonate	2.74
Pot. Dichromate	1.70
Wood Pulp	4.82
Paraffin Wax	2.00

Flower Gardens (Chemical)

6 tablespoonfuls of salt
6 tablespoonfuls of bluing
6 tablespoonfuls of water
1 tablespoonful of ammonia water,

and pouring, after thorough mixing, over a clinker, a piece of coke or of brick in a broad bowl or dish. After the clinker (or coke or brick) has been wet with the liquid, drop on it a few drops of mercurochrome solution or of red ink or green ink. But do *not* use iodine, because this reacts with ammonia water to form the dangerously explosive nitrogen iodide, a black powder which is safe as long as it is wet but explodes with a loud report from very slight shock when it is dry. After the materials have been brought together, a coral-like colored growth soon begins to appear on the clinker. This increases rapidly.

The growth also tends to form on the edges of the dish and will climb up and over them unless they have been rubbed with vaseline. The growth will not extend beyond the vaseline.

The "depression flower garden" is a capillary phenomenon involving the tendency of ammonium salts to "creep." The saturated solution deposits crystals around its edges and upon the clinker where the evaporation is greatest. The crystals are porous and act like a wick, sucking up more of the solution by capillary action. The solution thus sucked up evaporates to produce more crystals, more wick, and more growth. The addition of a little more ammonia water to the dish will produce more growth after the first growth has stopped. Or the whole may be allowed to dry and may then be kept without further change.

The "mineral flower garden" which florists sometimes sell or display in their windows, depends upon an entirely different principle, that of osmosis or of osmotic pressure. A solution of sodium silicate or "water glass" is poured into a jar or globe, and crystals of readily soluble salts of certain metals which form colored and insoluble silicates are thrown in and allowed to sink to the bottom. Growths resembling marine plants spring up from these crystals and in the course of a few minutes climb rapidly upward through the liquid, often branching and curving, producing an effect which might lead one to believe that he sees exotic algae growing in an aquarium. The experiment works best if the solution of water glass is diluted to a specific gravity of about 1.10.

Ferric chloride produces a brown growth; nickel nitrate, grass green; cupric chloride, emerald green; uranium nitrate, yellow; cobaltous chloride or nitrate, dark blue; and manganous nitrate and zinc sulfate, white.

Freezing Mixture

A mixture of 230 g. of NH_4CNS, 30 g. of NH_4Cl, and 300 cc. of H_2O produces a fall of temp. from 15° to −19°. Increase of NH_4Cl content reduces the cooling effect, which is thus well under control.

Gelatine Capsules

Gelatine	8 parts
Water	8 parts
Sugar	2 parts
Glycerine	2 parts
Gum Arabic	1 part

Proceed as in printers rollers composition.

Gelatine Sheets

Gelatine	3 lb.
Water	5 lb.
Glycerine	5 oz.

Allow gelatine to soak in water until soft and dissolve on water bath. Add Glycerine and color solution if colored sheets are desired. Pour measured amount on polished plate glass that has been placed in absolutely level position so that the solution will not run off. Distribute solution evenly over surface by hand or with a fine comb. Allow to rest until gelatine has thoroughly set and then place in rack to dry. When dry, remove from glass by cutting to edges with sharp blade and lifting gradually off glass.

The thickness of the sheet depends on the amount of solution poured on glass. About 12 oz. of above sol. poured on glass 20×24 will give a sheet 3/1000 thick.

Glue Composition for Plaster Casting Molds

Powdered Hide Glue	1 part
Glycerine	1½ parts
Water	1 part
Sugar	½ part
Finely Powdered Silica	1 part

Proceed as in printers rollers composition.

Gems, Synthetic

Titanium Tetrafluoride	0.2
Beryllium Oxide	0.5
Iron Oxide	10
Aluminum Oxide	500
Magnesium Powder	100

Fuse together in a crucible and allow to cool slowly.

To Drill Holes in Glass

By taking a good steel drill and wetting with a saturated solution of camphor in oil of turpentine, holes may be rapidly and easily drilled through the thickest plate glass.

Frosting of Glass, Prevention of

(1) H_2O_6, NaOH soln. 6.5, palm oil 12 and rosin 1 part are cooked together. Thirty parts of H_2O are added and the cooking continued. Twenty parts of H_2O and 25–30 parts of glycerol are then added. The prepn. can be put up in cans. (2) A prepn. which can be put up in tubes is prepd. from soft soap 65, glycerol 30 and turpentine 5 parts. (3) Twenty parts of paraffin, 10 parts of wood oil and 70 parts of turpentine are used in making a non-hygroscopic compn.

Glycerine Jelly for Microscope Mounting

Water	3 oz.
Glycerine	3½ oz.
Gelatine	½ oz.
Carbolic Acid	1 dr.

Dissolve the gelatine in the water, and when dissolved add glycerine and carbolic acid. Warm for 15 minutes stirring continuously the whole time. Do not heat above 75° C. Allow to cool and on solidification drain off surplus water. Keep jelly in a cool place in an air tight jar.

Biological Fixing Fluid

These new fluids have been developed as the result of intensive research and are more or less free from the above difficulties. Materials fixed in them remain soft and will not harden when placed in a 70 per cent alcohol solution. In addition, all common stains may be used.

Two of the solutions are given as follows:

Cupric-paranitrophenol Fixing Solution

60 per cent Alcohol	100 cc.
Nitric Acid, sp. gr. 1.41–1.42	3 cc.
Ether	5 cc.
Cupric Nitrate, Crystals	2 gm.
Paranitrophenol, Crystals	5 gm.

This fluid is perfectly stable and is not limited as to duration of fixation, but has a slow penetration rate.

Cupric-phenol Fixing Solution

Stock Solution A

Distilled Water	100 cc.
Nitric Acid (as above)	12 cc.
Cupric Nitrate (as above)	8 gm.

Stock Solution B

80 per cent Alcohol	100 cc.
Phenol, Crystals	4 gm.
Ether	6 cc.

These solutions are perfectly stable and may be kept in glass stoppered bottles, but the mixture does not keep and for this reason the duration must not exceed forty-eight hours. For use, take: Solution A—one part; Solution B—three parts. In using either fixing solution wash the material in several changes of 70 per cent alcohol.

Artificial Perspiration

(Used in testing materials against defects from perspiration.)

Sodium Chloride	8 gm.
Acetic Acid	1 gm.
Butyric Acid	¼ gm.
Water	1000 cc.

---

* Permanently Neutral Formaldehyde

Commercial formaldehyde contains traces of formic acid and develops further amounts on standing.

In order to obtain a permanently neutral solution, it is only necessary to add to the commercial acid reaction formaldehyde a quantity of basic magnesium carbonate. Agitation or stirring may be used to effect intimate mixture. An excess of the salt does no harm. After neutralization the solution may be freed from the carbonate by filtration, decantation or other suitable means. Fifteen grams of hydrated basic magnesium carbonate is ample to neutralize six hundred grams of commercial formaldehyde solution. The excess may of course be used again. The use of a larger quantity does not change the hydrogen ion concentration which is found to be $1 \times 10^{-7.1}$. This is, for all practical purposes, neutral.

---

* Heat Producing Composition

Powdered Iron	17 oz.
Manganese Hydroxide	1 oz.
Graphitic Carbon	30 gr.
Ferric Chloride	30 gr.
Ferrous Sulfate	30 gr.
Manganese Chloride	30 gr.
Manganese Sulfate	30 gr.

On addition of water to the above, heat is generated.

---

* Heat Producing Composition

The following evolves much heat on addition of water.

Sod. Acetate	87
Sod. Hyposulfite	8½
Glycerol	3½
Calcium Chloride	1

---

* Heat Transfer Medium

Many substances have been used in the search for a suitable material for heat storage at high temperatures. Most organic materials decompose at comparatively low temperatures, and even diphenyl, which is one of the most suitable, is useless above about 900° F. The ideal heat-carrier should not decompose below say 1800° F., should be reasonable in price, non-corrosive, and as fluid as water within the widest possible range of temperature.

Recently a substance named ''N S fluid'' has been developed, which appears to approach very closely to this ideal. It consists of a mixture of inorganic salts of the general formula $R' Cl + R''' Cl_3$, *e.g.*, 1 mol. NaCl with 1 mol. $AlCl_3$, which solidifies to a homogeneous mass and at 302° F. liquefies into a well-defined solution which regulates its molecular ratio by expelling excess of $AlCl_3$ if present.

---

* Hydraulic Fluid

Water	10 gal.
Soda Ash	7 lb.
Soft Soap	2 lb.

Boil the above together and when dissolved run in with vigorous stirring

Lard Oil	2½ gal.

---

* Ice-Skating Rink, Artificial

A suitable floor is covered with

Sod. Hyposulfite	70
Borax	29
Alum	1

The surface may be covered with a mixture of powdered soap and stearic acid.

---

* Incense

An incense consists of redwood bark flour, 50.5 per cent by wt.; gum benzoin, powdered, 9.90 per cent by wt.; gum acacia, powdered, 16.50 per cent by wt.;

aromatic materials such as orris root, yara yara, rose leaves, vetiver, coumarin, etc., about 22.34 per cent by wt.; bergamot, oak moss, or other alcoholic extracts, 0.55 per cent by wt.; and saltpetre, 0.66 per cent by wt.

Aluminum Oleate

Distilled, Low Titre, Oleic Acid	282 lb.
Caustic Soda	40 lb.

The above forms 304 lb. Sodium Oleate.

Sodium Oleate	1824 lb.
Aluminum Sulphate	666 lb.

The yield of Aluminum Oleate should be 1740 lb.

Lead Oleate

Distilled, Low Titre, Oleic Acid	282 lb.
Caustic Soda	40 lb.

The above forms 304 lb. Sodium Oleate.

Sodium Oleate	608 lb.
Lead Acetate	379 lb.

The yield of Lead Oleate should be 769 lb.

---

* Packing, Oil Resistant

Crepe Rubber	40–50
Carbon Black	20–40
Cotton Linters	10–20
Glue	2– 5
Glycerol	1– 2
Diphonylguanadine	0.5– 1
Sulfur	2– 6

---

* Packing, Metallic Stuffing Box

Finely Divided Lead	90
Wool Grease	8
Graphite Powder	2

---

* Pectin, Soluble

Sod. Bicarbonate	5– 10
Tartaric Acid	12– 7
Pectin	10
Sugar	1000–1600

The above is a base for making jams and jellies.

---

* Printing Blankets, Preservative for

Carbon Black	5 lb.
Beeswax	0.6 lb.
Long Chinawood Oil Varnish	8 gal.
Japan Drier	3 qt.
Turpentine	2 gal.
Naphtha	7 qt.

Printers Rollers Composition

Powdered Hide Glue	1 part
Glycerine	1½ parts
Water	1 part
Sugar	½ part

Add glue and sugar to mixture of water and glycerine and stir well. Allow to stand until glue is thoroughly soaked and then place on water bath and melt. When mass is completely molten and all air bubbles have risen to surface, it is ready to be poured into molds.

Roller, Printers

Glue Highest Grade	20 lb.
Water	20 lb.

Soak ½ hr.

To this add

White Corn Syrup	40 lb.

Cook in double boiler for 2 hrs.

Add

Glycerine	16 lb.
Rezinel No. 2	1 lb.

Agitate with a high speed mixer until uniform and cast on a rubber core.

---

* Puncture-Sealing Compound

Castor Oil	½ lb.
Talc	1½ oz.
Wood Flour	1½ oz.
Water	½ lb.
Gum Arabic	¾ oz.
Benzol	¼ oz.
Clovel	1⁄16 oz.

---

* Radiator, "Stop-Leak" for

Flaxseed Meal	16⅔
Aluminum Powder	1½
Sod. Silicate	2½
Casein	2½

---

* Radiator Solution, Anti-Rusting

Saturated Soda Ash Solution	8
Saturated Copper Sulfate Solution	1

Eight ounces of above used to 5 gal. water.

---

* Refrigerant

Methyl Formate	90–95
Alcohol Anhydrous	5–10

---

* Refrigerant Leak Warning

Allyl Alcohol	0.5–1%

Chlorpicrin	0.5–2%

Either of the above is incorporated in the refrigerant. If any leak occurs it is quickly noticed.

* Shock Absorber Fluid

Glycerol	1
Caustic Potash	1
Water	75

Dissolve above in vessel fitted with a high speed mixer and while stirring rapidly run the following in slowly:

Red Oil	3–5
Methanol	1
Mineral Oil	19

Soluble Starch

Lintner Method: Potato starch is triturated with 7.5 per cent hydrochloric acid to a thin paste which is allowed to stand 7 days at 40° C. The modified starch is washed with cold water until the washings no longer redden litmus paper and is then expressed and dried.

Solomon Method: 100 gm. of starch is mixed with 1000 cc. of water in which 5 gm. of sulphuric acid has been previously dissolved. The mixture is then boiled for 2½ hours, after which the excess acid is removed by addition of barium carbonate and after filtration the filtrate is evaporated and the soluble starch is precipitated by addition of alcohol. The precipitate is collected and dried.

Leulier Method: 25 gm. of wheat starch is mixed with 100 gm. of alcohol (95%) containing 5 gm. of concentrated sulphuric acid. This mixture is heated in a flask provided with an invert condenser and is then boiled for 30 minutes. The modified starch is collected on a plain filter and is washed off with either cold water or alcohol until the washings no longer redden litmus paper. The washed starch is then dried.

---

* Soot Destroyer

Salt	285
Zinc Powder	14
Anthracite Coal Powd.	6
Hard Charcoal Powd.	3.5

Aluminum Stearate

Double Pressed Stearic Acid	284 lb.
Caustic Soda	40 lb.

The above forms Sodium Stearate, 306 lb.

Sodium Stearate	1836 lb.
Aluminum Sulphate	666 lb.

The yield of dry Aluminum Stearate should be 1752 lb.

Lead Stearate

Double Pressed Stearic Acid	284 lb.
Caustic Soda	40 lb.

The above forms Sodium Stearate, 306 lb.

Sodium Stearate	612 lb.
Lead Acetate	379 lb.

The yield of Lead Stearate, dry, should be 773 lb.

Zinc Stearate

Double Pressed Stearic Acid	284 lb.
Caustic Soda	40 lb.

The above forms Sodium Stearate, 306 lb.

Sodium Stearate	612 lb.
Zinc Sulphate	287 lb.

The yield of dry Zinc Stearate should be 631 lb.

Stiffeners for Toes of Shoes

Cumarone	12 lb.
Petroleum	1 gal.
Pine Oil	2 fl. oz.

---

* Thawing Composition

1 A thawing composition substantially consisting of an intimate mixture of grains of crystallized anhydrous magnesium chloride having the shape of thin tablets of a length not exceeding about 2 millimeters, and about one per cent of a finely powdered alkali chromate.

---

* Snow and Ice Melter

Salt	25
Am. Chloride	50
Mag. Sulfate	25

The above may be diluted with silica sand and water.

Copper Tubing, Bending

Fill tubing completely with molten lead and bend around wood form. When bent heat and drain out lead.

Ultra Violet Filter

A filter useful for absorbing ultra violet light in connection with fluorescence photography consists of a 2% solution of Sod. Nitrite in a glass cell 1 cm. in thickness.

Anti-Rot Compound for Wood

Sodium Fluoride	2 lb.
Water	98 lb.

X-Ray Screen, Fluorescent

Sodium Tungstate	29 gm.
Calcium Chloride	11 gm.
Sodium Chloride	58 gm.

The whole is intimately mixed, and heated in a crucible. The result of the reaction produced by the heat is that calcium tungstate is formed, which crystallizes out from the molten sodium chloride as the mass cools. After cooling, the mass is removed from the crucible and washed with water to dissolve the excess sodium chloride. The powder is then dried and sprinkled on a gummy sheet of stiff paper, and when dry makes a good fluorescent screen for experimental purposes.

Simple Azo Oil Dye

(I)

Ortho Toluidine	1	oz.
Sulphuric Acid	2	oz.
Sodium Nitrite	1½	oz.
Water. Ice.		

(II)

Beta Naphthol	¾	oz.
Caustic Soda	1½	oz.
Water. Ice.		

Procedure

1. Mix Ortho Toluidine and acid.
2. Add 8 ounces of water.
3. Add ice until temperature drops to 4° C.
4. Dissolve Sodium Nitrite in 6 ounces of water.
5. Add solution to Ortho Toluidine and acid slowly. Brown fumes will evolve. Stir until fumes cease. This should take about 10 minutes. This is part I.
6. Dissolve the Beta Naphthol in the Caustic Soda in 1 quart of water. Cool with ice to 10° C. This is part II.
7. To part II add part I stirring all the time. An orange colored precipitate will form. Filter, wash and dry.

Note: Make sure receptacles for both parts I and II are large enough.
Do not inhale fumes produced.

---

* Decolorizing Carbon

Pulverized bituminous coal is mixed with an aq. solution of an alkali salt, *e.g.*, Na_2CO_3 (30–40% of the coal), and a wetting agent, *e.g.*, sol. soap (1%). The H_2O is boiled off while continuing the agitation and the mixture is retorted at a red heat in absence of air and with avoidance of fritting. The powder may now be used after cooling, but is improved by further treatment in a pan the lid of which is adjusted to keep the mass just glowing by combustion, until NH_3 ceases to be evolved.

Sweeping compounds are usually made to contain a large percentage of filler, such as clean sand and sawdust. One well known mixture is made by dissolving 4 ounces of semirefined paraffin wax in 1 gallon of low viscosity lubricating oil; the wax being added to the hot oil. A dry mixture is prepared consisting of sawdust 20 pounds; clean sand 10 pounds, and salt three-quarters pound, and then the lubricating oil is thoroughly incorporated with this. While mixing these materials about 2 ounces of an odorizing oil, such as clovel or cedar, may be added.

---

* Thawing Composition Suitable for Use on Railway Switches, Etc.

Grains of cryst. anhyd. $MgCl_2$ having the shape of thin tablets of a length not exceeding about 2 mm. are mixed with about 1% of a finely powd. alkali chromate.

---

* Composition to Prevent Moisture Accumulation on Glass

To a satd. soln. of NaCl are added 50 g. of KNO_3, 25 g. of gelatin and 100 g. of 10% HCl, and the mixt. is boiled.

Radiator Solder

Flaxseed Meal	100
Aluminum Powder	1–2

Mix together until all the flaxseed is covered with Aluminum. When this is added to the water in a leaky automobile radiator, it swells and plugs up all leaks as the water circulates.

OILS, FATS, GREASES, LUBRICANTS, CUTTING OILS

* Graphite Lubricant

Graphite	85
Gum Tragacanth	10
Triethanolamine	1.6
Sod. Naphthionate	2
Water	250–400

Graphite Suspension

Diglycol Stearate	4
Water	100

Heat to 60° C. and remove heat and stir until a milky dispersion is formed. Add with stirring

Graphite	10–20

* Lubricating Grease Base

Japan wax and castor oil (1–1.2 pts. each) are melted and compounded with Al stearate (3 pts.) and the cooled base is pulverised and compounded with a mineral oil.

Cup Creases

Pressure

114 parts Fat
16 parts Quicklime
870 parts Petroleum Red Oil preferably 500 Visc. at 100° F. or over

No. 1

123 parts Fat
17 parts Quicklime
855 parts Petroleum Pale Oil 100 Visc. at 100° F.

No. 2

140 parts Fat
19 parts Quicklime
840 parts Petroleum Pale Oil 100 Visc. at 100° F.

No. 3

157 parts Fat
22 parts Quicklime
820 parts Petroleum Pale Oil 100 Visc. at 100° F.

No. 5

205 parts Fat
34 parts Quicklime
760 parts petroleum Oil

The weighed fat is placed in a steam jacketed kettle equipped with a paddle type agitator and a small portion of the Petroleum Oil, about half the volume of the fat, is added. Next the lime is hydrated and mixed with sufficient water to form a thin paste. The lime is added to the material in the kettle and the whole is cooked for several hours with continuous agitation. When a small portino of the soap on cooling is firm and brittle the remainder of the Petroleum Oil is added slowly to avoid chilling. The agitation is continued until a uniform grease without lumps is formed.

Locomotive Rod Cup Grease

35 parts Tallow
6.5 parts Sodium Hydroxide
50 parts Steam Refined Cylinder Oil
10 parts Water

Driving Journal Grease

40 parts Tallow
7 parts Sodium Hydroxide
45 parts Steam Refined Cylinder Oil
10 parts Water

Cup Grease

Lard Oil Extra No. 1	150 gal.
Inedible Tallow	300 lb.
Hydrated Lime	198 lb.
Western Mineral Lubricating Oil	900 gal.
Water	75 gal.
Oil Mirbane	6 lb.

Specifications

Mineral oil. Viscosity 180 to 100° F. (Saybolt) Color between 9 and 10½ (Robinson). Sp. Gravity 19 to 21.5 Baumé.

Extra No. 1 Lard. Color red or brown. Should not contain more than 7% free fatty acid calc., as oleic.

Inedible tallow. Clear, fresh and free from dirt. M. Pt. not less than 110° F. Free fatty acid not more than 5%.

All formulae preceded by an asterisk (*) are covered by patents.

Hydrated Lime. Finely powdered. Contain about 27% to 28% Water and 66.6% available CaO. (A.S.T.M. Tentative Standards.)

Procedure

Charge in an open steam jacketed kettle 50 gallons of the oil, slop cup grease, or tailings from the previous batch.

Charge into the kettle, the lard oil and tallow

Mix in separate vessel 198 lb. hydrated lime, 30 gallons mineral oil, 75 gallons water.

Add this mixture to kettle and start agitating paddles.

Turn on steam in jacket and bring temperature of contents of kettle to 300° F. in from 2 to 4 hours.

Allow lime soap formed to stand in kettle over night with steam on but without agitating.

Test soap to find if saponification is complete, and estimate water content. Soft pasty consistency indicates incomplete saponification or too much water. Stringy more or less transparent soap denotes excessive water evaporation. Soap should be firm and break evenly but should not crumble too readily (excess lime). Corrections should be made by adding from 1 to 5 gallons NaOH (20 B.) or in case of excess lime an appropriate amount of tallow.

Run in mineral oil till total oil in kettle is 400 gallons. Heat to 212° F.

Run in additional oil to 750 gallons.

Scrape down sides of kettle and add slowly 4 to 10 gallons water. Temperature should be 180° to 185° F.

Stir 10 minutes.

Add oil mirbane.

Run in remainder mineral oil (150 gallons).

Stir 15 to 20 minutes.

Fill at 160° F. to 170° F.

Grease produced is of medium consistency known as No. 3. Its melting point is 190° F. to 195° F. and consistency as taken by A. S. T. M. penetrometer at 77° F. is 180.

Manufacture of Cup Grease by Pressure Cooking

Cup Grease

Extra No. 1 Lard	150 gal.
Water	14 gal.
Western Pale Oil (180 Viscosity)	900 gal.
Powdered Hydrated Lime	198 lb.
No. 1 Inedible Tallow	300 lb.
Oil Mirbane	6 lb.

Procedure

Charge in closed pressure cooker of 12 barrels capacity, 30 to 50 gallons of oil or tailings from previous batch.

Run into pressure kettle, 150 gallons of lard and 300 lb. of melted tallow.

Mix in separate container:

Mineral Oil	30 gal.
Hydrated Lime	198 lb.
Water	10 gal.

Add this to kettle.

Close and fasten manhole or opening. Test for leaks by admitting compressed air till pressure of 15 lb. is reached.

Release pressure, close relief valve.

Turn on steam in jacket. Pressure of 100 lb. required. Rotate paddles at 38 r.p.m.

Maintain pressure kettle at temp. not less than 300° F. and 50 to 80 lb. pressure for 20 minutes.

In open steam jacketed mixer of 30 barrels capacity bring about 50 gallons of slop grease to temp. of 225° to 265° F.

Open valve on 12 barrel kettle and allow soap to be discharged under its own pressure into large open mixer.

Start paddles and begin adding oil till volume of oil is 400 gallons.

Bring to 212° F.

Oil added to 750 gallons.

Add 4 gallons water. Temp. 180° to 185° F.

Stir 10 minutes.

Add mirbane and balance of mineral oil.

Stir 15 to 20 minutes. Fill at 160° to 170° F.

Graphite Cup Grease

1. Graphite Cup Grease.

	Per cent by weight
Cup Grease No. 2	93.00
Medium Ground Graphite	2.00
American Talc	5.00

2. Graphite Lubricant.

Cup Grease No. 2	86.29
Steam Refined Cylinder Stock	6.80
Powdered Plumbago (Graphite)	6.91

3. Marine Graphite Grease.

Cup Grease No. 2	92.00
Fine Ground Graphite	8.00

4. Special Graphite Grease.

Hard Tallow	10.00
Dark Petrolatum	80.00
Fine Graphite	10.00

For Chemical Advisors, Special Raw Materials, Equipment, Containers, etc., consult Supply Section at end of book.

Slushing Oil (for foreign shipment)

Neutral 28° Paraffin Oil	4½ gal.
Anhydrous Lanolin	60 oz.

* Wool Lubricant

100 parts of olive oil or arachis oil or a mineral lubricating oil or free oleic acid or mixtures thereof are emulsified with 250 parts of water and 5 parts of the triethanolamine salt of the acid phosphoric ester of cetyl alcohol, if desired with the addition of 3 parts of glue powder, or of other animal or vegetable protective colloids, or of soaps or of an agent of the type of Turkey-red oil or of other sulphonation products of vegetable or animal fats or oils. The emulsions are then brought to the desired dilution with from 2 to 35 times their weight of water and are employable for example as oiling agents in making shoddy or in spinning fibrous materials or in brightening dyed fabrics.

Thread Grease

1 lb. Lanolin (dry)
2 oz. Vaseline

Melt No. 1 and No. 2 and add 3 oz. camphor.

Cordage Grease

Degras	30
Kerosene (Heavy)	60
Caustic Soda (36° Bé.)	10

Warm together and stir until uniform.

* Lubricating Grease, High Temperature

Mineral oil of a viscosity of not less than 90 sec. Saybolt at 100° is mixed and heated with an Al salt of a fatty acid such as Al stearate constituting 15–45% of the total mass, at temps. of about 70–125° and the mass is rapidly chilled to prevent reversion.

* Grease, Lubricating

A lubricating grease is manufd. by mixing 0.5% of rubber latex with 9.5% Al stearate and 90% hydrocarbon oil, and heating the mixt. with agitation to dehydrate the latex and produce a transparent homogeneous texture and subsequently cooling.

* Lubricating Grease, Gasoline Proof

Oleic Acid	2.5
Ammonium Linoleate	17.5
Glycerol	37
Lubricating Oil	35

* Grease, Lubricating

Cottonseed Fatty Acids	16
Crude Montan Wax	21.4
Slaked Lime	3.5
Caustic Soda	0.36
Heavy Black Mineral Oil	58.2
Water	2.7

Stainless Steel Lubricant

Lubricant for Drawing and Forming Stainless Steel

Heavy Drawing Compound	1 gal.
Hot Water	1 gal.
Lithopone	2 lb.
Flowers of Sulphur	½ to 1 lb.
Cresylic Acid	1 oz.

* Mill Grease

26.2 parts by weight of Rosin Oil
12.2 parts by weight of Tallow
59.0 parts by weight of Oil (500 second Saybolt at 210° F.)
3.2 parts by weight of Sodium Hydroxide

All of the tallow is mixed with one-half of the oil and all of the sodium hydroxide (which may be dissolved in a small amount of water) in a conventional kettle provided with heating and agitating means. This mixture is heated, with agitation, to a temperature of about 375 to 450° F., preferably about 400° F., until the tallow has been completely saponified.

The rosin oil is next added, the heating and agitation being continued, and the temperature is again raised to about 400° F. (375 to 450°) to effect a reaction between the excess alkali and the rosin-acid.

Finally, with continued heating and agitation, the other half of the oil is added and the temperature is again brought to 400° F. (375° to 450° F.). The mixture should be neutral or slightly alkaline at this point and if the reaction is acid, a calculated amount of sodium hydroxide should be added. The grease is then ready to be poured into moulds.

The mill grease prepared by this process is markedly different from and superior to greases formed by the usual process which consists in saponifying a mixture of fatty acid and rosin. Instead of a grainy, soft, low melting composition a smooth, clear, hard, elastic, high melting mill grease that shows unprecedented wear resistance is obtained.

Leather Stuffing

Ozokerite	6
Paraffin Wax	8
Rosin Oil	40
Mineral Oil	48

* Emulsified Fluid Lubricant

The soap base or emulsifier is first prepared. 300 pounds of elaine oil (commercial oleic acid) are heated in a kettle to 100°–110° C. and 300 pounds of water-white grade rosin are added, preferably in two equal portions, and the heating is continued until solution is complete. The solution is then cooled to about 95° C., and a solution of caustic potash containing 102 pounds of water and 47 pounds of 88–92% KOH (sufficient to saponify 73% of the elaine oil) is slowly added.

The contents of the kettle are heated from 95° to 101° C. for about an hour until the foam caused by the liberated carbonic acid disappears, and complete saponification has been effected between the caustic potash and a portion of the fatty acids. Then liquid is then cooled.

The cooled thick soapy liquid is next thinned, and the soap content is increased and rendered more effective as an emulsifier by adding an alcohol, such as denatured alcohol, and ammonia. The ammonia, however, is not, as a rule, added in sufficient amount to saponify all the fatty acids. If concentrated ammonia or a dilute water solution thereof is added directly to the soap solution, a stringy, ropy product, which is not an efficient emulsifier, will result. The ropiness can be prevented by first mixing the ammonia with denatured alcohol and then adding the mixture to the soap solution. Mix 98 pounds of 95% denatured alcohol with 31 pounds of ammonia (26° Bé.) and slowly stir the mixture into the soap solution. The resulting solution is a syrupy liquid containing ammonia and potash soaps and is used to mix with the neutral oil in making the emulsion.

The concentrated liquid emulsion is made by adding a solution of 21 gallons of neutral oil, such as paraffin oil, and 3½ gallons of the soap base, previously mixed together, to a dilute solution of soda ash. Preferably, add the above mixture to a solution made by adding 3.2 pounds of an alkaline compound, such as soda ash, to 24½ gallons of water. The soda ash is present in such quantity that it reacts with the remaining fatty acids and is preferably present in excess so as to act as a softening agent for the additional water that is added to the emulsion before it is utilized.

A concentrated emulsion may be made by adding the soda ash solution to a solution of the soap base in the oil in a crutcher, but when the ingredients are combined in this manner, the product gradually thickens when about ¾ of the dilute carbonate solution has been added. Then, after all the dilute carbonate solution has been added, the composition liquefies and produces an unstable emulsified liquid from which water and oil will separate in a few days.

A permanent concentrated emulsion, which will not separate, even though it is heated to the boiling point or cooled to the freezing point, may be formed if the ingredients are properly combined. Place the soda ash solution in the crutcher and gradually add the solution of soap base in oil to the soda ash solution while stirring. The resulting compound contains about 50% water, and the emulsion will not break on standing, heating or cooling.

The concentrated emulsion is used to make a suitably thinned emulsion such as cutting oil or emulsion. One volume of the concentrated emulsion may be diluted with 9 volumes of water for making a cutting compound.

---

* Lubricant

An oil such as a mineral oil is mixed with 3–5% of a Na, Zn or Ca soap and with 5–15% of an amide or anilide such as stearic, palmitic, oleic or arachidic anilide to increase the viscosity of the material.

---

* Lubricating Composition

Mineral Oil (300 Saybolt)	100
Stearic Toluide	5

Melt together at 230° F. and then cool quickly to congeal.

---

* Upper Cylinder Lubricant

A lubricant to be added to gasoline consists of

Gasoline	120
Benzol	15
Toluol	20
Camphor	50
Ether	20
Carbon Bisulfide	10
Castor Oil	40
Mineral Oil	50
Petrolatum	470
Clovel	5

---

* Lubricant, Journal-Box

Potash Coc. Oil Soap (40%)	100
Water	30
Neat's-foot Oil	3
Graphite	3
Cocoanut Oil	3
Sod. Stearate	6.25

The above prevents "hot-boxes" on railroad cars.

* Lubricant, Non-greasy

Cellulose Acetate	100
Diethyl Phthallate	100
Ethyl Lactate	100

Bicycle Chain Lubricant

12 kg. of rosin oil, 25.0 kg. of mineral oil, 1.0 kg. of 10° Bé. KOH, 4.5 kg. of $Ca(OH)_2$, 35.0 kg. of flake graphite and 22.5 kg. of mineral oil. The rosin oil and first portion of mineral oil are mixed and emulsified in the alk. soln. The $Ca(OH)_2$ and graphite are ground with the second portion of mineral oil, and well mixed with the emulsion.

Lubricant for Dies and Plates
(for moulded clay products)

No. 1.—Thoroughly mix, with both ingredients lukewarm, one part of Saponified Red Oil and five parts of kerosene.

No. 2.—Melt ten pounds of Double Pressed Saponified Stearic Acid to just above the melting point and add ninety pounds of kerosene with brisk agitation to obtain a thorough mixture.

Gun Lubricant

White Petrolatum	150
Bone Oil (acid free)	50

Graphite Grease

Ceresin	70
Tallow	70

Heat together to 80° C. and work in

Graphite	30

* Lubricant, Inorganic

The following formula gives a lubricant which is water soluble and not hygroscopic or deliquescent under ordinary conditions. It is particularly useful in systems carrying benzol, turpentine, oils and other water insoluble liquids.

Glacial Phosphoric Acid	100
Boric Acid	2
Orthophosphoric Acid	1¼

Heat to 122° C. and cool.

* Anti-oxident for Oils and Waxes

0.1% Tin Naphthenate is dissolved with heat in mineral oils and waxes to inhibit oxidation.

Boring Oil

A.	1. Oleic Acid	15
	2. Thin Mineral Oil	75
	3. Caustic Soda (40° Bé.)	5
	4. Alcohol	5

Warm 1 and 3 with stirring until uniform and while mixing vigorously run into it 2 and 4.

B.	Turkey Brown Oil	30
	Thin Mineral Oil	50
	Caustic Soda (20° Bé.)	10
	Alcohol	10

C.	Rozolin	18
	Thin Mineral Oil	74
	Caustic Soda (40° Bé.)	5
	Isopropyl Alcohol	5

D.	Naphthenic Acid	25
	Red Oil	25
	Thin Mineral Oil	100
	Caustic Soda (24° Bé.)	25
	Alcohol	25

E.	Rosin Oil	10
	Red Oil	10
	Thin Mineral Oil	70
	Caustic Soda (36° Bé.)	5
	Methanol	5

The above are mixed with water for use.

Rayon Lubricant (Partly Soluble Type)

70 to 80 parts Water White Mineral Oil

10 to 20 parts Mineral Seal Oil (used for cheapening cost of production)

10 to 15 parts Neat's-foot Oil (30 F. C. T. grade)

Adjust proportions to 100 parts.

Add in order named, agitate slowly and warm until thoroughly mixed.

Allow sample to stand for a short time to see if satisfactory.

Rayon Lubricant (Insoluble Type)

50 to 100 parts Water White Mineral Oil

0 to 50 parts Mineral Seal Oil

Adjust to viscosity and use desired for, with cost included in the final proportions used.

Open Gear Lubricant

A home-made mixture of ½ lb. white lead, ½ gal. cylinder oil and ½ lb. flake graphite makes an especially efficient lubricant for open gears, according to *Link-Belt Shovel News.* This mixture adheres well to the gears and can be painted on with a brush as required at intervals of about five hours. Cup grease may be substituted for the oil and the graphite may be omitted. Omission of the graphite is not advisable in warm weather.

Solid Lubricant

1. Rosin	9
2. Machine Oil	82
3. Caustic Soda (40° Bé.)	9

Melt 1 and 2 together and heat to 100° C. and run 3 into it slowly with stirring and raise temperature to 110–120° C.

* Candles, Non-sticking

To prevent candles from sticking to mold incorporate 5% of glycol or glycerin in wax mixture.

Valve Lubricant

Unaffected by gas and high temperatures.

1. Barium Stearate	50
2. Mineral Oil	40
3. Talc	10

Heat 1 and 2 together with slow mixing at 120–150° C. until dissolved; work in 3.

* Castor Oil, Sulfonating

Oils, fats and fatty acids and their mixts. (such as castor oil) are treated with H_2SO_4 of at least 1.80 sp. gr. and in a proportion of 45–100% the wt. of the material to be sulfonated. The acid is rapidly added with continuous stirring and the reaction mixt. is simultaneously cooled at least to 10–15° and the product is thereafter immediately washed in a salt soln., the latter is drawn off and the product is finally at least partially neutralized.

* Castor Oil, Thickening

HNO_3 of 80–85° Tw. is gradually added to about 80 times as much castor oil at a temp. of about 43°, and the temp. is allowed to rise to about 115° and maintained until "crackling" ceases.

Cutting or Spraying Oil

1. Mineral Oil	280 lb.
2. Miscibol	32 lb.
3. Oleic Acid	24 lb.
4. Water	15 lb.
5. Denatured Alcohol	10 lb.

Mix 1, 2 and 3 mechanically until dissolved. Heating speeds solution. Stir 4 into this and then add 5 with stirring. This produces a clear, stable, "soluble" oil.

If 70 parts water are added slowly, with stirring, to 10 parts of the above, a beautiful white stable emulsion results. The amount of water may be larger or smaller as needs require. This emulsion is useful as a lubricant, cutting oil, polish or agricultural spray.

Mineral Oil Softener

(For Use on Sulfur After-Treated and Vat Dye Blacks or Dark Blues. Also as a Soluble Cutting Oil)

Sulfonated Fish Oil (sperm 75%)	20 to 30 parts
Pale Paraffin Oil	70 to 80 parts

Adjust proportions on a 100 parts basis as to consistency desired. Heat fish oil until clear. Agitate and then add the paraffin oil cold and agitate again until the mix is clear. If desired for summer use the initial heating is sufficient, but for winter use it is desirable to heat a second time.

Test for use. A ten per cent solution in a blank solution should not separate on boiling. It is desirable that these tests be made on the material to be processed so as to see if any mineral oil marks show up on the finished material.

Dry Powdered Lubricant

Zinc Stearate	50
Talc	50

This is of advantage on machinery in mills where white goods are handled as this lubricant will not discolor goods.

Cutting Oil Emulsions

The term "cutting oil" is applied to soluble lubricating oils which are used as machine lubricants. In lathe and speed-tool operations the first requirement is a cooling medium which will carry heat away from the cutting edge. In addition, a certain amount of true lubricant is advantageous, and both of these requirements are satisfied by a dilute oil emulsion. With the proper oil and emulsifying agent, the corrosive action of the water is likewise decreased and rusting of steel prevented. In practice a soluble oil is used to produce a 5 to 25 per cent oil emulsion, and this is flowed over the cutting edge and continuously recirculated.

One of the most important requirements of a soluble oil for cutting is its dependability. It should not separate when left in open containers and it should always emulsify in water with only the simplest stirring methods. The resulting emulsion should also remain stable and uniform, a five per cent emulsion not separating oil in 24 hours. Soluble oils fulfilling these qualifications can be made with Triethanolamine. This agent, for one thing, permits the use of oils of high lubricating value which are otherwise difficultly emulsifiable. In addition it yields emulsions of such high dispersion and uniformity that lower concentrations of oil in water than are customary can be used with equal lubricating effect.

Another interesting application for soluble oils is in the lubrication of textile machinery. The elimination of ordinary oil spots from fabrics is usually an expensive hand operation. On the other hand, when the spot is caused by a soluble oil, it may be completely and readily removed in the regular scouring operation. If a stiffer lubricant, more of the texture of a grease, is desired, this can be made of any consistency by stirring thoroughly up to 20 per cent water into one of the soluble oils. Another way of making a soluble grease consists in melting 10 per cent of stearic acid into a lubricating oil, and then emulsifying this with an equal weight of hot water containing 4 per cent Triethanolamine.

Resin Soluble Cutting Oil

Rosin	7.5	lb.
100 visc. Spindle Oil	2.5	gal.
Oleic Acid	6.0	lb.
100 visc. Spindle Oil	5.5	gal.
32° Bé. Caustic Soda	4.0	lb.
Alcohol	2.1	lb.
Yield	10	gal.

Heat the rosin with the first portion of the spindle oil at a temperature of about 212° F. until the former is melted, then add the other ingredients in the order listed. The alcohol should be added when the batch has been cooled to room temperatures.

---

* "Cutting" Oils

The following formulae are used for cooling high-speed cutting tools.

Type A—Waterless Cutting Oils

1.	Rozolin	25
	Wool Grease	10
	Lard Oil	15
	Mineral Oil	50
2.	Rozolin	20
	Degras	5
	Mineral Oil	75

Type B—Soluble Cutting Oils

These are stirred while heating until saponification is completed.

1.	Red Oil	6
	Rozolin	8
	Caustic Soda (35° Bé.)	4
	Cellosolve	4
	Mineral Oil	78
2.	Rosin	5
	Rozolin	10
	Caustic Soda (35° Bé.)	3.6
	Butyl Cellosolve	3.4
	Mineral Oil	78
3.	Rozolin	12
	Paraffin Oil (28° Bé.)	81
	Caustic Soda (35° Bé.)	3.6
	Alcohol	3.4
4.	Castor Oil	10
	Rozolin	6
	Degras	1.7
	Mineral Oil	45.4
	Water	27.4
	Caustic Potash (35° Bé.)	9.5

The above are mixed with 3–10 times as much water before using. They are stable in presence of hard water.

Type C—Sulfur Cutting Oil

Rozolin	25
Rosin	5
Mineral Oil	30
Sulfo Turk C	20
Sulfur	15
Pine Oil	5

Heat with stirring at 350° F. until dissolved.

The above is dissolved in 4–20 times its volume of hot mineral oil for use.

---

* Cylinder Oil

Cylinder oil is made by heating to 400° F. heavy steam-refined lubricating oil 9, with Al stearate 0.3, asphaltic material 2 and lard oil 8.5 parts, cooling the mixt. and adding 80 parts of heavy, steam-refined lubricating oil.

---

* Dewaxing Lubricating Oil

The oil is mixed with about 2.5 times its quantity of a diluent comprising benzene 65, acetone 25–32 and naphtha 3–10% and the mixt. is chilled to about − 20°, the solidified wax is mechanically sepd. from the chilled oil, and the diluent is distd. from the dewaxed oil.

Drawing Oil

A. Rozolin 28
Caustic Potash (38° Bé.) 10
Thin Mineral Oil 64

B. Degras 40
Rosin 29
Rozoil 21
Caustic Soda (40° Bé.) 10

C. Tallow 10
Thin Mineral Oil 10
Japan Wax 1
Caustic Soda (40° Bé.) 4.2

* Fish Oil, Purifying

About 5% dry $Ca(OH)_2$ and 5% CaO are mixed with fish oils, agitated and filtered. By this treatment the fish oils are deodorized, decolorized and stabilized.

Increasing Viscosity of Oils

The viscosity of animal, vegetable or mineral oils is increased by dissolving therein 7–10% Ethyl Cellulose.

* Lubricating Grease Base

Lead Oleate	5
Castor Oil	15
Sperm Oil	5
Aluminum Stearate	75

* Insulating Oil, Refining

The oil is treated with 95–98% H_2SO_4 (15–50 vol.-%) for ½–2 hr., separated, neutralized, washed, and dried. It is then treated with absorbent material (0.5%), *e.g.*, SiO_2 gel, fuller's earth. The purified oil contains 0.1–6% of aromatic resinous compounds which act as antioxidizing agents. If these are deficient the oil may be blended, or the product (0.1–5%) obtained by treating turpentime with H_2SO_4, separating, neutralizing, and washing may be added.

* Penetrating Oil

The following is used for freeing rusted connections, bolts, etc.

Heavy Lubricating Oil	27–32%
Amyl Acetate	0.25–1%
Cottonseed Oil	2.0–2.5%
Kerosene	Balance

Penetrating Oil

Pine Oil	30
Blown Rape Seed Oil	30
Carbon Tetrachloride	10
Kerosene	100
Light Paraffin Oil	70

Oil, Penetrating

For freeing rusted bolts, screws, etc.

Kerosene	20
Mineral Oil Light	70
Secondary Butyl Alcohol	10

Pine Oil, Solidified

Trihydroxyethylamine Linoleate	1
Pine Oil	10
Water	8

Porcelain Mold Oil

Stearic Acid	24
Ozokerite	1
Paraffin Wax	3
Heavy Mineral Oil	82

Rayon Lubricating Oil

(insoluble type that can be used on the dipping whizzing method of oiling yarns)

70–80 parts of Water White Mineral Oil

20–30 parts Mineral Seal Oil

Warm water and mineral oil while stirring then add mineral seal, proportions may be adjusted to get the desired take-up of oil on the rayon yarns.

Soluble Oils

The name "soluble oil" has been given to a clear oil solution which emulsifies of itself when added to water. Such an oil possesses some advantages over an ordinary emulsion. In the first place, it has a good appearance since it is a clear, bright solution. Furthermore, because it contains little or no water, the user knows exactly what quantity of oil is being dealt with; and, in addition, its decreased volume on this account reduces handling, storage and shipping costs. Finally, in important applications, it can be used in a first step as an oil, its self-emulsifying properties being valuable in succeeding operations.

The usual oils to be put into soluble form are the mineral oils. The emulsifying action is brought about by dissolving a soap or similar compound in the oil. As a rule when sodium or potassium soaps are used, a large excess of fatty acid together with alcohol or other mutual solvents are necessary. On standing, the latter solvents are often evaporated, the entire soap then being thrown out of solution. Ammonium soaps are better as regards solubility in the oil, but possess no stability on standing. Naphthinic acids and sulphonated oils are soluble, but large quantities are required for emulsification, and both give the oil an acid reaction.

Because it is a mobile liquid which dissolves clearly in hydrocarbons, oils and most organic liquids, it is used for making soluble oils which emulsify readily in water. A few examples are given below:

A.
10 lb. Diglycol Oleate
50 Turpentine
100 Water
} a water in oil emulsion

B.
10 lb. Diglycol Oleate
50 Turpentine
100 Water
½ Caustic Soda
} an oil in water emulsion

C.
15 lb. Diglycol Oleate
60 Mineral Oil
75 Water
} a water in oil emulsion

Stainless Knitting Oil

75 to 80 parts Good Quality Mineral Oil (technical grade water white)
15 to 25 parts Good Quality Neat's-foot Oil (should show at least 30° F. c. t.)

Adjust proportions on a 100 parts basis as to viscosity desired.

Oil for Leather

Rozolin	10
Degras	10
Mineral Oil	82

---

* Linseed Oil, Substitute

Rosin, about	25 lb.
Animal Fat, about	12 lb.
Lubricating Oil, about 32° Bé.	3 lb.
Cobalt Acetate	¼ to 3 lb.
Litharge	2 to 10 lb.
Water, about	2 lb.
Kerosene	65 lb.

The ingredients (with the exception of the kerosene) are thoroughly mixed and heated to about 480° F. and then cooled, the kerosene being added and mixed after turning off the fire.

---

* Lubricating Oil, Chatterless

0.1–3% Lead Oleate or Sulforicinoleate is dissolved in the oil with stirring and heating.

---

* Lubricating Oil, Low Cold Test

At least 20 g./litre of $C_6H_4Me{\cdot}NH_2$, $C_6H_3Me_2{\cdot}NH_2$, or other alkyl-substituted NH_2Ph derivative is added to the oil to reduce its pour test.

---

* Lubricating Oil, Low Cold Test

Mineral Oil	100
Neat's-Foot Oil	4.5
Alum. Stearate	0.12

Heat together while stirring until uniform.

---

* Lubricating Oil, Low Cold Test

Up to 2% of xylylstearamide is added to a viscous hydrocarbon oil.

---

* Lubricating Oil, Reclaiming Used

Oil such as that which has been used in an engine or transformer is heated and impurities which settle out are removed; the oil is mixed with a fluid reagent comprising castor oil blended with about 10% of mineral oil of high sp. gr. at a temp. of about 70°, and the resulting mixt. is further heated and subjected to the action of superheated steam while stirring to drive off diluent material of non-lubricating character; about 3% of material such as trimethanolamine is added to assist in sepn. of the oil from the reagent, the mixt. is cooled to about 22°, the oil is sepd. from the reagent and deleterious matter by centrifuging, decolorizing material such as fuller's earth is added, and the resulting mixt. is passed through a straining medium such as an asbestos filter to clarify the oil. App. is described.

Mineral oil—Sulfonated-tallow emulsion suitable as finishing compound on cheaper goods where a slight odor is not objectionable, may be corrected by some special artificial odor. May be used for mercerized yarns.

50 parts 50% Sulfonated Tallow
10–15 parts Mineral Oil Softener
33–38 parts Water
2 parts Trisodium-phosphate (this should be dissolved up in the water)

Agitate while heating until solution reached and sample tested will emulsify satisfactorily in cold and lukewarm water.

---

* Mineral Oil Soluble Castor Oil

Castor oil is heat treated at a temp not exceeding about 310° until no more than 4.2% of the total wt. of the oil has been removed under atm. conditions, and the treated castor oil is mixed with mineral oil in various proportions.

---

* Mineral Oil Soluble Castor Oil

Castor Oil is heated to 200–280° C. with ½% of any of the following until a sample dissolves clearly in mineral oil.

Bauxite
Titanium Dioxide
Sod. Bisulfite
Phosforic Acid

Olive Oil, Bleaching

Dark oils are treated with a 12% soln. of tannic acid. From 1 to 4% of the acid is necessary, according to the color of the oil, and very thorough mixing of the oil and the soln. of the acid is required. A 5% soln. of citric acid also gives good results.

Palm Oil, Decoloring

The oil is heated to 90° in the presence of 0.01% Cobalt Resinate and air is blown through it for two hours.

Silk and Rayon Boil Off or Degumming Oil

50 to 70 parts 50% Sulfonated Castor Oil (No. 1)
Use necessary amount of 20% Caustic Soda Solution or slightly larger amount of Caustic Potash to make into liquid soap.
30 to 50 parts Sodium Silicate (best quality commercially, iron free)

Adjust proportions according to consistency required, after thoroughly agitating, allow to stand over 12 to 24 hours and drain off water layer.

Test. Try concentrated solution added to cold to hot water for solubility adding goods to be processed. Test their feel. If too stiff the amount of sodium silicate may be reduced accordingly.

The above formula may be used as a base and suitable solvents incorporated into the mixture that will remove oil spots, etc., in the heel, toe, or leg of hosiery, or material under process.

Tests. Solvents must be checked for miscibility and other necessary requirements for the particular use, especially boiling temperatures. (Those above 212° F. b. p. are best to use.)

* Textile Oil

An excellent oil for softening textile fibres and threads to be woven or knitted is made as follows:

Sulfo Turk C	12–15%
Oleic Acid	12–15%
Betanaphthol	2%
Light Mineral Oil	to make 100%

The above is especially good for rayon because it doesn't weaken the latter and washes off readily.

* Textile Oil, Soluble

Linseed Oil	100
Sod. Bisulfite (38° Bé.)	100

Heat at 60–90° C. with air until a sample is water soluble.

* Transformer Oil, Non-sludging

0.5% Cetyl Alcohol is added to transformer oil to inhibit sludging.

* Transformer Oils, "Non-sludging"

The addition of 0.1–1% Tetraethyl Lead to transformer oils eliminates "sludging" at high temperatures.

Transformer Oil, Stabilizing

The addition of 0.2–0.5% hydroxybiphenyl increases resistance to light, air and electrolytic action.

* Improving Transformer Oils

Transformer Oil	100 c.c.
Sod. Ethylate	0.03 gm.
Anhydrous Alcohol	0.50 c.c.

Heat to 120° C.

The above treatment gives an oil of increased insulating power.

Insoluble Oil Lubricant for Wool

10–20 lb. Lard Oil, No. 1 Quality
80–90 lb. Pale Paraffine (debloomed type) Oil

Mix cold with stirring, then heat until blended and add some type of artificial odor compound.

* Rancidity and Oxidation in Fats and Oils, Prevention of

The addition of maleic or fumaric acids or salts or esters of the same in amounts as low as 0.02% is effective in some cases.

Paper Size

40 lb. of cream sizing is boiled with 36 gallons of water. Then dilute with 185 gallons of water to get a solution suitable for calendar sizing. Mix 8 lb. of Hydrowax Cream plus ¼ lb. silicate of soda with 2 gallons boiling water and stir until uniform. Add this to the above solution and proceed as usual.

* Peanut Butter, Inhibiting Rancidity in
Peanut butter is mixed with 25% of its weight of crushed sesame seed.

---

* Valve Stem Packing

Calcium Carbonate (Powd.)	100
Graphite	35
Talc	5
Cottonseed or Other Oil	20

Grind until uniform.

---

* Heat Conducting Lubricant

Lubricants of improved thermal conductivity are made by incorporating 5–10% ammonium oleate or stearate.

---

* High Speed Bearing Lubricant

Mineral Oil	100 lb.
Zinc Oxide	50 lb.

---

* Flotation and Cutting Oil Base

I

100 parts of pine oil having a specific gravity of about 0.933 to 0.935, 50 parts of sulphur and 1/5 part of sodium carbonate are heated at the reflux temperature (about 180 to 190° C.) until the sulphur is combined. During the heat treatment, large amounts of hydrogen sulphide are given off but no water is liberated other than that which may be normally present in the pine oil initially used. The alkali may or may not be added in water solution. If no water is used it is preferable to have the alkali in a finally powdered state. As little as 1/10 part of sodium hydroxide in a 50% water solution gives the desired results.

After the sulphur is combined so that it will no longer precipitate on cooling, the heating may or may not be continued at higher temperatures, say about 200° C., until liberation of hydrogen sulphide ceases. In the event that the continued heating procedure is carried out, the amount of sulphur in the final product is about 20%. If the heating is not continued until evolution of hydrogen sulphide ceases the amount of combined sulphur in the finished product is about 23%.

The reaction product is given a single wash with an aqueous solution of caustic soda containing about 1% of caustic soda on the basis of the oil present. The washing is preferably carried out at about 100° C. to remove any corrosive sulphur compounds such as mercaptans and any occluded or dissolved hydrogen sulphide. The alkaline water is allowed to settle and decanted off. The oil may then be given successive washes with water at about 100° C. until the wash water is neutral to litmus paper or other indicator and gives no brown coloration with copper sulphate. The product is finally dehydrated to remove occluded or dissolved water. The dehydration is preferably carried out by heating the oil under reduced pressure at a temperature below 100° C.

The product is a non-corrosive material suitable for use as a cutting oil base and is free from objectionable odors. The product is also useful as a flotation agent in the separation of minerals from ore mixtures.

The amount of sulphur that may be held in permanent combination by this process will vary with the composition of the terpene oil used. The amount held in combination depends upon the proportion of unsaturated terpene hydrocarbons and tertiary terpene alcohols present in the terpene oil. For example, a standard grade "steam distilled pine oil" of specific gravity 0.933 to 0.935 containing from 10 to 15% hydrocarbons and about 60% tertiary alcohols will permanently combine with about 50% by weight of sulphur.

In order to obtain the maximum amount of sulphur in the final product by this process, it is necessary to start with about 100 parts by weight of the pine oil and 75 parts by weight of sulphur. When so large an amount of sulphur is used, it is advisable to use a little more alkali, for example, 1/2 part by weight.

II

100 parts of pine oil of 0.933 specific gravity are placed in a suitable vessel with 0.1 to 0.5 part of sodium carbonate in water solution. 0.5 part of copper hydroxide in paste form is added to the mixture. The mass is next heated until the water in the copper hydroxide paste has been boiled off. After boiling, 30 parts of technical sulphur are then added to the mass and a current of air is passed into the oil. The temperature is rapidly raised to 165° C. and then gradually to 195° C. over a period of from one to three hours. Air is continuously passed through the mass during the entire reaction period. The product is then cooled and may be washed in the manner described in Example I. The catalyst may be separated from the oil after the reaction by either decanting after settling or by filtration.

PAPER AND PULP

Preparation of Paper Pulps

Although chemists have produced paper pulps in the laboratory only an expert on paper can evaluate the products of these small scale experiments. The variation due to thermodynamic and mass action factors which can not be reproduced in the laboratory makes any comparison with products made on a factory scale extremely difficult. However, it is thought of some value to briefly outline the principal methods of production. A very wide variation in concentration, etc., is customary in plant practice but the following figures give a fair indication of recognized proceeding.

Groundwood

A flour of wood produced by grinding barked logs against stone. The process is purely mechanical.

Sulphite

Prepared by cooking wood chips at 70 to 80 lb. pressure 15 to 18 hours with a solution of sulphurous acid which has been passed through a tower of lime or dolomite. The final solution varies greatly but a total sulphur dioxide content, 4.5%, 3.5% free and the rest combined is considered good practice.

Sulphate (or Kraft)

Prepared by cooking wood chips at 120 to 140 lb. pressure about 8 hours with a solution of sodium hydroxide and sodium sulphide. The solution may have a formula approximating sodium carbonate, 11, sodium hydroxide 90, sodium sulphide 25 gm. per liter.

Soda

Prepared by cooking wood chips at 110 to 120 lb. pressure about 8 hours with a 6–8% sodium hydroxide solution.

Jute

Prepared by cooking cut burlap sacks (old bags) at normal or increased pressures with mild alkali such as 1–5% sodium hydroxide or 5–10% calcium hydroxide from 4 to 18 hours, washing and beating the product to pulp.

Rope

(Hemp or Manilla)

Prepared by cooking rope (old rope) as outlined for jute.

Note: There is more variation in method for production for the last two pulps than in the others. For instance there is one secret process which produces an excellent product, bleached, washed and ready for the beater continuously. All other methods are intermittent. The complete cycle is less than forty minutes. No other cycle is less than seven hours.

Principal Types of Paper

All papers are formed on a screen catching the suspended fibers and passing through the water. The resulting mat is dried by squeezing through felts and heating on hot cylinders.

Book

Chiefly prepared from sulphite and soda pulp.

News

About eighty per cent ground wood.

Wrapping

Sulphite, Sulphate, Jute, Rope, or mixtures.

Writing

May be old rag, but usually sulphite or sulphite and soda.

All formulae preceded by an asterisk (*) are covered by patents.

Minor Types

Waxed

A paper that has been run through paraffine.

Parchment

A paper that has been treated with concentrated sulphuric acid.

Glassine

A heavily beaten, unloaded paper. Supercalandered.

Grease Proof

Prepared as above, but not supercalandered.

Cellophane

Not technically a paper. A film of regenerated cellulose, cellulose nitrate or acetate.

Basic Weight

Paper is sold by basic weight. Official basic weight is the weight of 500 sheets, 25 by 40 inches. Trade custom basic weights vary. To convert from official to trade figures the following factors are useful.

Trade Name	Trade Size (inches)	Factor
Book	25×38	0.950
News	24×36	0.864
Wrapping	24×36	0.864
Writing	17×22	0.374

Determination of Basic Weight

To determine basic weight without the use of a special balance the ordinary analytical balance may be employed. The following formula will give the official basic weight.

$$\frac{\text{Gm.} \times 1.102 \times 1000}{\text{Total area in sq. inches}} = \text{Basic weight in lb.}$$

Obviously more than one sheet may be used and the total area becomes the area of one sheet multiplied by the number of sheets employed.

Stains Used in the Paper Trade

Herzberg Stain

A. Saturated Zinc chloride (at 70 deg. F.).

B. 5.25 gm. Potassium iodide, 0.25 gm. iodide, 12.5 c.c. water.

Add 25 c.c. of A to B. Mix vigorously. Allow the mixture to settle and decant the clear supernatant solution. Add a crystal of iodine.

Sutermeister Stain

A. 1.3 gm. iodine, 1.8 gm. potassium iodide made up to 100 c.c. with water.

B. Saturated calcium chloride (at 70 deg. F.).

To use this stain moisten the fibers with A. Absorb the excess solution with filter paper and moisten with B.

Phloroglucinol

5 gm. phloroglucinol in 250 c.c. of 50% hydrochloric acid. (One part conc. acid to one part water.) Avoid undue exposure to light.

To Prepare Paper for Identification

Take about one square inch of a representative sample of the paper and cover it with a small quantity of 0.5% sodium hydroxide solution. Transfer to a fine sieve and wash free of alkali. Transfer to a small bottle or heavy test tube and add glass beads. Shake vigorously to macerate the paper and dilute to about a 0.1% suspension. Dot a microscope slide with small portions drawn with a wide mouthed pipette. Absorb excess moisture with filter paper or dry in an oven at 105 deg. C.

Effect of Standard Stains

Phloroglucinol

This stain is especially useful for making quick tests for the presence of ground wood. There is no need to prepare a slide as a few drops on any sheet of paper not heavily coated will give a satisfactory test. In the presence of ground wood a deep wine-red color is instantly produced.

Note: A pale coloration is sometimes caused by the presence of poorly cooked jute or sulphite, but the characteristic stain of ground wood is so clearly defined there should be no confusion.

Herzberg Stain

The sample must be properly prepared.

Red—Linen, cotton, bleached hemp.

Blue—Well cooked wood.

Yellow—Ground wood, jute, unbleached hemp.

Sutermeister Stain

On properly prepared samples.

Red—Cotton, linen, hemp.
Blue—Soda.
Purple—Bleached or thoroughly cooked sulphite.
Green—Jute, poorly cooked unbleached sulphite.
Yellow—Ground wood.

To Strengthen Filter Paper

To harden filter paper so that it will stand considerable strain from a filter pump, it may be dipped in concentrated nitric acid (Sp. G. 1.42–1.43) for a few minutes. It should then be well washed with cold water. This treatment will produce a paper about ten times stronger than untreated filter paper and will not change its permeability nor increase the nitrogen content. The ash is slightly reduced.

To Waterproof Paper

Waterproofing is best accomplished by parchmentizing paper but this treatment leaves a surface that is too irregular to make a good writing surface. One part of any of the following to six parts of water are supposed to give a good waterproofed paper. Glue, gelatine, shellac or aluminum acetate. Excellent results are obtained by using one part of borax, five parts of shellac and ten parts of water. The mixture is brought nearly to the boil, but not boiled and kept hot until all the shellac has passed into solution. The paper may be dipped into the solution, or it may be applied with a wide brush. The surface is a satisfactory vehicle for ink or water color.

To Parchmentize Paper

Prepare a fifty per cent solution of sulphuric acid. Pass a water-leaf (unloaded) paper through the solution being careful that no air bubbles prevent even contact with acid. Each part of the paper should remain in contact with acid for about 5 seconds. Promptly plunge the paper into a large quantity of cold water. Then wash with a running stream of water from the faucet or a wash bottle with a wide-mouthed tip. Next wash with a weak solution of ammonia to remove the last trace of acid and finally wash with water to remove any ammonia. An excellent parchment-like effect is acquired by thick papers. However, there is an art in this and only experience can guide the operator in the length of time the paper should be in contact with the acid. If a longer time is required stronger acid may be used.

To Fireproof Paper

Prepare a solution as follows:

Ammonium Sulphate	8 gm.
Boric Acid	3
Borax	1.7
Water	100 cc.

The solution should be heated to 122 deg. F. and kept at this temperature. The paper is dipped in the solution and hung to dry. Wrinkles can be prevented by drying in a press, or the paper may be subsequently ironed.

To Remove Creases from Paper

Creases may be removed from even fine engravings if a little care is exercised. Place the sheet smoothed as far as possible by hand on a clean sheet of paper on top of a well-covered ironing board or similar surface. Cover with another clean sheet. Finally dampen a third sheet, place on top of the others and press with a moderately warm iron.

Temporary Tracing Paper

It is sometimes necessary to make a tracing on a regular sheet of writing or bond paper. Temporary translucence may be created by sponging the paper with benzine. As soon as the benzine evaporates the paper reverts to its normal condition. The last trace of odor can be removed with a draft of warm air. While still translucent the paper will take either pen or ink drawing without difficulty. The use of benzine provides a quick accurate method for tracing graphs.

PAPER COATINGS

Casein Glue

Casein	100 lb.
Water	50 gal.
Borax	17 lb.
Ammonia 26°	1 qt.

The casein is preferably soaked a few hours in the water, the borax dissolved in a little hot water—added, and the whole cooked to 160° F. till no undissolved particles of casein remain. Then the ammonia is added and the glue cooled.

Wax Emulsion

Carnauba Wax	50 lb.
Water	50 gal.
Soap	12 lb.

The soap is dissolved in the water and brought to boiling. The wax is added and boiling continued until all is emulsified. The emulsion is preferably stirred continuously until cold. The soap may be any good grade of washing soap free from rosin.

Yellow

Clay	50 lb.
Blanc Fixe Pulp (70% dry)	50 lb.
Chrome Yellow Pulp (50% dry)	125 lb.
Talc	12 lb.
Casein Glue	11 gal.
Carnauba Wax Emulsion	4 gal.

Blue

Prussian Blue Pulp (30% dry)	100 lb.
Violet Lake Pulp (35% dry)	75 lb.
Maroon Lake Pulp (35% dry)	75 lb.
Casein Glue	8 gal.
Carnauba Wax Emulsion	3 gal.
Talc	4 lb.

Pearl

Clay	50 lb.	Pulped together
Blanc Fixe Pulp	50 lb.	
Italian Talc	4 lb.	
Ultramarine Blue	5 lb.	
Water	4 gal.	
Casein Glue		12 gal.
Carnauba Wax Emulsion		4 gal.

Red

Red Pulp (40% dry)	200 lb.
Talc Italian	4 lb.
Casein Glue	12 gal.
Carnauba Wax Emulsion	6 gal.

White

Clay	300 lb.
Water	20 gal.
Italian Talc	18 lb.
Casein Glue	25 gal.
Carnauba Wax Emulsion	12 gal.

Paper Coating—Special for High Finish—White

Water	65 gal.
Soda Ash	3 lb.
Ammonia	4 gills
Satin White	440 lb.
Clay	650 lb.

Mix thoroughly and add the following solution

Water	50 gal.	
Casein	100 lb.	
Soda Ash	10 lb.	Dissolved in 3 gal. of hot water
Tri Sodium Phosphate	7 lb.	
Borax	5 lb.	
Ammonia		6 gills

Paper Coating—Friction Finish—Yellow

Casein	200 lb.
Borax	12 lb.
Ammonia	5 qt.
Water to make	150 gal.

Water	43 gal.
Talc	23 lb.
Clay	200 lb.
Blanc Fixe Pulp	390 lb.
Medium Yellow Pulp	18 lb.
Carnauba Wax Emulsion	16 gal.
Casein as above	32 gal.

Paper Coating—White—Soft Sized

Water	165 gal.
Clay	1300 lb.

Stir 15 min. in a rapid dissolver and add

Dry Casein 140 lb.

Stir 15 min. and add

Dry Borax 18 lb.

Stir 5 min. and add

Ammonia 4 qt.

Heat to 140° F. and stir till casein is dissolved and cool to room temp.

Strain before using.

If hard sized coating is desired, increase the amount of casein until the desired degree of sizing is obtained.

Coating for Paper

Sodium Silicate	30 gm.
Sodium Sulforicinoleate	20 gm.

Heat together on water bath and add 30 cc. boiling water.

Dip paper into this and draw out immediately. This gives a parchment like effect to the paper.

Keep the mix boiling for five minutes and dip second piece of paper into it. This gives a translucent paper.

Paper Coating

The relative amounts of clay, casein, and water in the coating slips were 100, 17.5, and 250 parts, respectively. The casein solution contained 100 parts of casein, 10 of borax, 5 of soda ash, and 2 of ammonia. The formula and method of preparation of the clay-casein mixture were as follows:

Clay slip: 3000 grams clay, 3825 ml. water, and 10 ml. concentrated ammonia. The clay was soaked overnight in 3000 ml. of water. The additional 825 ml. of water were used the next morning to wash the mixture into the agitator. The ammonia was then added, and the mixture agitated about 1.5 hours before the casein solution was added.

Casein solution: 525 grams casein in 2000 ml. of water; 52.5 grams borax in 750 ml. water; 26.25 grams soda ash (58 per cent sodium oxide) in 750 ml. water; 11 ml. concentrated ammonia in 164 ml. water. The casein was soaked in the 2000 ml. of water for one hour at room temperature and was stirred meanwhile. It was brought into solution by the addition of the three solvents in the order —borax, soda-ash, ammonia, each dissolved in the stated amount of water. After the solvents had been added, the solution was stirred for 1.5 to 2 hours, warmed to 57° C. (135° F.), and screened through a No. 200 sieve. The screened solution after being cooled to room temperature was mixed with the clay slip. The mixture was agitated for 1.5 to 2 hours, screened, and applied to the base paper in the coating machine.

Paper Coating
(Dull Black Velvet Finish)

Casein Solution (25–30% Casein)	100
Ivo Bone Black	75

Grind mixture in a suitable mill and use same day if no preservative is added.

Paris Paste, a colloidally dispersed carbon black can be used to give an intense black color. This has already been finely ground and goes into aqueous media readily.

Paper Coating

The following formula gives a solution which does not readily gel:

Lactic Casein	5.75%
Caustic Soda	.25%
Sodium Bicarbonate	.5 %
Blanc Fixe	33 %
Satin White	33 %
Water	27.5 %

The sodium bicarbonate should be added after solution of the casein by the caustic.

Clay works particularly well with casein solutions giving a smooth adherent film only a little less glossy than that given by satin white.

Blanc fixe and precipitated chalk are used for matt or semi-matt finishes. Neither clay nor blanc fixe gives such a waterproof coating as satin white. Depending on the effect desired and the nature of the casein and the fineness of grinding of the pigment, one part of casein will bind satisfactorily about 20 parts of barytes, 15 parts of chalk, 15 parts of coarse clay, 12 parts of fine clay, and 10 parts of satin white. In general, the greater the amount of mineral matter present the greater the gloss and the more readily does the paper receive the ink.

Paper Coating Composition

A typical formula is as follows:

Lactic Casein	9%
Borax	2%
Blanc Fixe	60%
Talc	1%
Special Soap Solution	4%
Water	24%

The soap solution is made by boiling together

Carnauba Wax	20.0 %
Potassium Carbonate	1.25%
Water	72.75%

A greater degree of water resistance can be secured by substituting for the borax one-sixth of its weight of caustic soda, and when solution is complete adding ammonium sulphate to the extent of one and one-half times the weight of caustic soda employed. All the additions are naturally made as solutions.

The most widely practised method of making up the finished coating solution consists, broadly, in first dissolving the casein in the manner stated above, reducing the temperature to 80–100° F. and adding it to a perfectly smooth thin slip of the mineral fillers and water. Any other components, such as foam reducers, are then added, and the batch taken to the coating machine.

Wax Emulsions for Paper Coating

A. Beeswax Emulsion

Yellow Beeswax	360	lb.
Caustic Potash	2	lb.
Ammonium Hydroxide	8¼	lb.

Boil with stirring for 1 hour. Add 150 gal. water and shut off heat.

Wax Emulsion for Paper Coating

B. Ceresin Emulsion

Ceresin	200 lb.
Stearic Acid	200 lb.
Caustic Potash	9 lb.

Ammonium Hydroxide	13 lb.
Water	200 gal.

Boil with stirring for 1 hour; bring up to 800 gal. with water while stirring.

Paper Coating Solution

1. Casein	500 lb.
2. Water	235 gal.
3. Borax	25 lb.
4. Tri-sodium Phosfate	32 lb.
5. Ammonium Hydroxide	10 gal.
6. Water	40 gal.

Soak (1) in (2) for an hour. Dissolve (3) and (4) in 25 gal. of (6) and add to casein. Heat to 150° F.; turn off heat, add (5) and balance of (6).

Paper Coating Solutions
Friction Finish

A. Clay Dry	100	lb.
Talc	2	lb.
Water	5	gal.
Dyestuff (4%)	4	lb.
Water	6	gal.
Casein Solution	6½	gal.
Beeswax Emulsion	6	gal.

B. Turkey Red Lake (33% dry)	300 lb.
Talc	2 lb.
Ammonium Hydroxide	1 qt.
Casein Solution	7 gal.
Beeswax Emulsion	5 gal.

Plate Finish

Clay (Dry)	100 lb.
Water	5 gal.
Dyestuff (4%)	4 lb.
Water	6 gal.
Casein Solution	10 gal.

Paper Coating Solutions
Litho Finish

Clay (Dry)	25 lb.
Blanc Fixe (Dry)	75 lb.
Water	11 gal.
Color	to suit
Casein Solution	15 gal.

Waterproof Coating

1. Ground Coat

Turkey Red Lake	300	lb.
Ammonium Hydroxide	2½	lb.
Casein Solution	20	gal.

2. Top Coat

Ammonia Casein Solution (1 lb. per gal.)	1 gal.
Orange Shellac Ammonia Solution	1 gal.
Water	1–2 gal.

Paper Coating Mixture

Dry Clay	0.9 kilo
Blanc Fixe Pulp (74% Solids)	0.9 kilo
Water	1.5 liter

Allow to stand overnight.

No. 5 Glue	0.2 kilo
Water	0.8 liter

Soak overnight and dissolve at 135° F.; cool. Stir for ½ hour.

---

* Paper and Wall Board, Fireproof

Pulp is beaten with an aq. soln. hav ing a d. of about 18° Bé. at about 38 and formed from a mixt. of $MgSO_4$ 24 borax 8, $(NH_4)_2SO_4$ 8 and dextrin 1. parts. The soln. used is formed from a mixt. of $MgCl_2$ 24, boric acid 4, NH_4C 4 and alco-glycerodextrin soln. 1.6 parts

---

"Glassine" Paper

Paper is coated with or dipped in th following:

Copal	100
Alcohol	300
Castor Oil	8–12

Glaze Paper

100 parts Carnauba Wax are melte together at 120–130° C. with 25 part curd soap, while stirring well. 900 part boiling water are then added while sti ring well, very slowly at first and the more rapidly, the whole being boiled u and stirred until cold.

---

* Greaseproofing Paper

The following treatment will rende paper moisture-, grease- and acid-proo

The paper is impregnated at 60–65 with an aq. solution containing (wt.-% pure gelatin 13, Irish moss 6.5, hide glu 3.25, glycerin 8.25, $COMe_2$ 0.375, NaOH 0.125, K alum 2, Na alum 1, 37% CH_2O 0.5. The dried paper is then treate with a solution of 1 lb. of 37% CH_2O and 1 oz. of glycerin in 1 gal. of EtOH

Imitation Parchment Paper

A small amount of Tricresyl Phosphat is added to a thin alcohol solution o bleached shellac. Paper dipped in th solution and dried will resemble parch ment, except that it will be very resistar to moisture.

Mimeograph Paper

The substance used for the coating consists of a mixture of hydrocarbons of the fatty series plus ozokerite, oleine, and palmitine.

The carrier for the coating is a light cellulose paper weighing about 12 gm. per square meter. This is placed on a metal plate, heated to 100° C. The coating is melted and painted on the surface with a soft sponge. The operation is done on the reverse side to the one on which the tracing is to be made. The molten coating penetrates the pores of the cellulose by dialysis and it thus becomes incorporated in a uniform manner which, when it comes into contact with the hot plate gives perfect glazing to that side of the sheet.

Formula for coating.

Tricosane	1250	parts
Ozokerite	55	parts
Oleine	32.5	parts
Palmitine	12.5	parts

* Paper, Moisture Proof

Paper is made from a mixt. of treated pulp 100, H_2O 2000 and salts of mixed fatty acids (such as linolic or oleic acid) 15 parts, and the paper is passed through a 10% aq. soln. of basic Al acetate and dried.

* Safety Paper

Paper is impregnated with

Alcohol	5	oz.
Water	2	qt.
Iodine	¼	oz.
Cobalt Nitrate	¼	oz.
Sod. Hyposulfite	⅛	oz.

"Safety" Paper

Paper treated to prevent fraudulent alteration and useful for checks, drafts, etc., is made by incorporating in it or coating it with a 10% water solution of a leuco indophenol and drying it. It is then passed through a bath containing 5 lb. of Manganous Sulfate per 20 gallons of water.

Paper Softener

Paper dipped in a 10% water solution of glycerine and dried will thereafter be very soft and cloth-like.

* Waterproof Heat Insulation Paper

Asbestos Fiber	93
Wood Cellulose	3
Starch	2
Iron Oxide	2
Sod. Silicate	0.5
Alum	0.25

Shellac Solution for Paper Waterproofing

In a wooden tank, fitted with steam injector place

Water	25	gal.
Orange Shellac	150	lb.
Ammonium Hydroxide	6½	gal.

Allow to stand overnight and then turn on steam until dissolved. Bring volume to 100 gal. with cold water. Two coatings of this solution are given to the paper.

RESISTANCE OF WRAPPING MATERIALS TO THE PASSAGE OF WATER VAPOUR

Materials Examined	Loss, in Grammes per Square Metre, in 24 Hours
Waxed paper	Down to 10
Waxed paper, after severe creasing	90 to 100
Coated viscose film	16 to 20
Viscose film	150 to 190
Coated glassine paper	100 to 150
Glassine paper	280
Vegetable parchment	185 to 320
Kraft papers	200 to 250
M.G. sulphite papers	Up to 480

Transparent Wrapping Material (Similar to Cellophane)

Ethyl Cellulose or Benzyl Cellulose dissolved in Ethyl Acetate and spread on a glass plate to dry will produce a perfectly transparent sheet with a high gloss. A small quantity of Tricresyl Phosphate or Dibutyl Phthalate will increase the flexibility of the same. This material may be colored as desired by the addition to the solution of Benzyl soluble dyes. The dyes are dissolved in Benzyl and added to the solution.

* Water and Flame Proof Paper

A mixture of asbestos fiber 93, wood cellulose 3, starch 2, Fe_2O_3 (or other metallic oxide) 2 pts., Na silicate 0.5, and alum 0.25 pt. is pulped with H_2O, sheeted, and dried, the metallic oxide combining with the silicate and alum constituents to form a binder for the fibers.

* Deinking Paper

The paper is treated with about 100 parts of water to each part of paper in an ordinary paper beater or hollander, the said water containing sufficient of the sodium hydroxide, sodium metasilicate and/or trisodium phosphate to give it the desired hydrogen ion concentration of pH 9.0 to 12.6. The duration of this beating is preferably about one hour. At the end of that time there is added for each 5 grams of paper present 10 ml. of a 1% soap solution or a 1% solution of the sulphonated oil in water, which is equivalent to 1/10 gram of soap to 5 grams of paper, or in the ratio of 1 part of soap or sulphonated oil to every 50 parts of paper. In other words, when treating a ton of paper, there would be required 1/50 ton or 40 pounds of soap or the said oil. The said sulphonated oil is quite dispersible in water and produces a sort of milky emulsion.

After the addition of the soap or oil, the beating is continued for about 20 minutes, until it is seen that the fibers have quite well separated from the fillers in the paper and the ink pigment. If the paper that is being treated does not contain any fillers, it is advisable, although not absolutely essential, to add a finely divided inorganic material such as finely powdered gypsum or land plaster, or a colloidal clay such as bentonite. The amount of such fillers added for this purpose may be on the order of from 10 to 25 pounds per ton of unfilled paper, such as newsprint, undergoing treatment. After adding this extraneous mineral matter, the beating is preferably continued a further 20 minutes so as to insure the gathering of the ink pigment upon the said mineral matter, so that it cannot redeposit upon the cellulosic fibers.

In either event, that is, when treating either coated or uncoated paper, the material is now ready for filtration to separate the cellulosic fibers from the pigment and fillers. This is accomplished by running it over a coarse sieve having meshes fine enough to hold the fibers but sufficiently coarse to let the fillers, ink pigment and dispersed ink vehicle pass through with the solution. It is preferred to do this without any suction, and the most advantageous manner of doing it is to pass the material over an ordinary screen such as is well known in the paper-making art. It is not advisable to employ any great degree of suction, as this tends to mat down the fibers and thereby causes physical entrapment of the pigment and fillers.

After the solution containing the suspended pigment and fillers, etc., is passed through the screen, the remaining cellulosic fibers may then be washed so as to insure the removal of any residual fillers and pigment, this being done either on the screen itself or, preferably, by transferring the wet mass of cellulosic pulp to a vessel containing clear water and thereupon again throwing this new mixture onto the screen. In localities where the water is hard and when soap has been used, it is advisable to employ a softened water or else to add sufficient alkali to the water so as to prevent the formation of insoluble calcium soap, which, if formed, would deposit itself upon the fibers and cause them to lose their brilliant white color. If such an alkaline washing solution is employed, this does not necessarily imply a waste of material, as the filtrate obtained from this washing step may, after correction for its hydrogen ion concentration, be employed for the first step in the treatment of a subsequent batch of printed paper. However, if the sulphonated oil is employed, this precaution will not be necessary.

In any event, and particularly when alkali is used in the wash water, the paper fiber is then further washed with pure water so as to remove the residual alkalinity thereof and produce a pulp suitable for the manufacture of new paper or paper product.

The present process has been particularly useful in the removal of colored printing ink from highly coated magazine stock and similar papers, it being well known that the removal of modern rotogravure and process inks presents a problem difficult of solution and a problem which was never presented to prior inventors, as in the past ordinary printing inks based upon merely a linseed oil base were the most commonly employed, whereas the modern printing inks often contain synthetic resins and dyestuffs which are by no means as easy to remove as the ordinary old-style printing inks.

The temperatures employed in the treatment, it may be stated, may be anywhere between room temperature and about 180° F. (87° to 90° C.), the latter temperature being particularly suitable. As the hydrogen ion concentration approaches 12.6, it will be advisable to use a somewhat lower temperature, say 160° F. (or from 60° to 75° C.). Heating the solution in this manner accelerates the action, but the process will work even at room temperatures, provided the agitation is sufficiently prolonged.

Acne Cream

Petrolatum White	10 lb.
Calamine	3 lb.

Dissolve following together separately by heating:

Camphor	1 oz.
Thymol	1 oz.
Menthol	1 oz.
Oil Rosemary	1 oz.
Methyl Salicylate	1 oz.
Oil Lavender	1 oz.
Resorcin	1 dr.
Betanaphthol	1 dr.

Mix all together cold.

Acne Lotion

Rose Water	5 gal.
Alcohol	2 gal.
Glycerine	1 pt.
Menthol	1 oz.
Phenol	2 oz.
Methyl Salicylate	1 oz.
Benzaldehyde F. F. C.	½ oz.
Zinc Oxide	2½ lb.
Calamine	2½ lb.
Boric Acid	1¼ lb.

Acne Ointment

Betanaphthol	2
Sublimed Sulfur	4
Balsam Peru	15
Petrolatum	15

Analgesic Balm

Lanolin Anhydrous	45 oz.
Yellow Petrolatum	25 oz.
Menthol	2 oz.
Ethyl Amino Benzoate	2½ oz.
Distilled Water	18 oz.
Oil Wintergreen	7½ oz.

(1) Triturate Ethyl Amino Benzoate with a portion of the Yellow Petrolatum until smooth. Gradually add the remainder of the Petrolatum and the Lanolin.

(2) Dissolve Menthol in Oil of Wintergreen and add the solution to No. 1, triturating until thoroughly mixed. Finally add the water and mix until homogeneous. Fill into tubes or jars.

Skin Ointment

Amber Petrolatum	270 oz.
Amber Liquid Petrolatum	78 oz.
Paraffin Wax	16 oz.
Lanolin Anhydrous	10 oz.
Zinc Oxide	12 oz.
Ethyl Amino Benzoate	8 oz.
Phenol	2 oz.
Oil Thyme	½ oz.
Thymol	¼ oz.
Oil Eucalyptus	½ oz.
Ichthyol	1 oz.

Mix oil, wax and fats together. Mix Zinc Oxide and E. A. B. and sift through No. 100 mesh sieve. Then add to melted oil mixture. Stir until cooled to about 50° C. to prevent powder from settling. Mix Phenol and Thymol with essential oils and warm to effect solution. Add to ointment at 45°–50° and stir well. Grind Ichtyol with a few pounds of the ointment and mix with bulk of ointment while still warm. Finally pass through ointment mill.

Note: For special treatment of burns add 4 oz. Picric Acid.

For acute eczema and other inflammatory conditions of the skin add 4 oz. Resorcin.

Anesthetic, Local

Ethyl Aminobenzoate U. S. P.	3
Benzyl Alcohol	5
Ether	10
Olive Oil	82

Antiseptic Solution

Boric Acid	25	g.
Thymol	1	g.
Eucalyptol	5	c.c.
Methyl Salicylate	1.2	c.c.
Oil of Thyme	0.3	c.c.
Menthol	1	g.
Ethyl Alcohol	300	c.c.
Purified Talc	20	g.
Water	to 1000	c.c.

Stir together and filter.

* Aspirin, Liquid

A solution of aspirin which does not

All formulae preceded by an asterisk (*) are covered by patents.

hydrolyze or decompose is made by using the following as a solvent.

Alcohol	10
Glycerin	10

Antiseptic Cure for Poison Ivy

Wash infected parts well with strong soap and water to remove poisonous oils.

Also use ether and chloroform or gasoline.

Then apply 5% solution ferric chloride mixed with 50–50 alcohol and water.

Pat generously on infected part.

Aspirin Tablets

Aside from other properties acetylsalicylic acid tablets must have good appearance and must dissolve rapidly in the stomach. Such tablets are made with base of 240 parts pulverized arrow-root starch and 240 parts heavy magnesium oxide. Base is well mixed and screened. Then it is moistened with solution of coconut oil, 10 parts in about 400 parts ether, and moistened mass screened again. Powder is spread on paper and ether evaporates. Acetylsalicylic acid, 2000 parts, are added and mixture carefully mixed to perfect homogeneity. Then it is mixed with acetone as required, about 30 parts to 250 parts powder. After drying and heating for 2 hours at 50° C., 2530 parts of the granulated mass are mixed with 30 parts pulverized agar-agar, 60 parts arrow-root starch and 80 parts pulverized talc. When unit of weight used is gram, 4000 tablets can be prepared from final mixture, each tablet weighing 0.7 gram and containing 0.5 gram of acetylsalicylic acid. To prevent powder from tablets from penetrating into lower die on tablet-making machine, latter is covered with cotton threads impregnated with paraffin oil.

Asthma Remedy

The following is smoked in a pipe or as a cigarette.

Powdered Grindelia Robusta	240 gm.
Powdered Jaborandi Leaves	240 gm.
Powdered Eucalyptus Leaves	120 gm.
Powdered Cubeb	120 gm.
Powdered Stramonium Leaves	450 gm.
Powdered Potassium Nitrate	360 gm.
Powdered Cascarilla Bark	30 gm.

Burn Treatment

Gum Tragacanth	30
Gentian Violet (1% sol.)	1000

Allow to swell; warm and stir. Applied to burns this leaves a thin moist, cooling, protective layer and rapid healing results.

Calamine Lotion

Calamine	8.00
Zinc Oxide	16.00
Glycerin	15.00
Lime Water	60.00
Rose Water q.s.ad.	120.00

Camphor Ice

Castor Oil	25 oz.
White Beeswax	15 oz.
Spermacetum	49 oz.
Camphor Powder	10 oz.
Ethyl Amino Benzoate	1 oz.
Carbolic Acid	20 gr.

Melt Castor Oil, Beeswax and Spermacetum together and add Camphor and Ethyl Amino Benzoate. Stir until dissolved. Then add Carbolic Acid and pour into molds.

Cream, Catarrh

Menthol	2
Eucalyptol	3
Oil of Pumilio Pine	3
White Beeswax	1
Hard Paraffin	6
White Soft Paraffin	85

The beeswax and white soft paraffin are melted together and stirred until nearly cold, and the medicaments, previously mixed, are incorporated while the mass is still soft. Some form of closed mixer is desirable, as otherwise there is considerable loss of menthol. The directions on the tube and its carton recommend the frequent use of the cream until the troublesome symptoms are relieved.

Pastilles, Catarrh

Gelatin	20.0
Glycerin	40.0
Sucrose	5.0
Citric Acid	2.0
Sodium Benzoate	0.2
Oil of Lemon	0.1
Solution of Carmine	sufficient
Triple Orange-flower Water	6.0
Distilled Water	to 100.0

The gelatin is soaked in one and a half times its weight of water until softened, the glycerin is added, and the mixture heated on a water-bath until the gelatin has dissolved and the weight has been reduced to 85. The acid and benzoate, dissolved in the orange-flower water, are added, then the oil of lemon and the carmine solution, followed by water to the required weight. The mass is strained through muslin while still hot. A trial pastille is poured and its weight ascertained. This weight is divided into the total weight of the mass and sufficient medicament is added for the number of pastilles that the mass is capable of making. Care should be taken to avoid the formation of air bubbles when stirring in the medicament. Metal moulds require to be very slightly lubricated with almond oil before the mass is poured.

Menthol and eucalyptus pastilles contain about 1/6 gr. of menthol and 1/2 m. of eucalyptol in each pastille.

Contraceptive Jelly

Water	76.85 c.c.
Sodium Chloride	3.00 gm.
Lactic Acid	2.00 gm.
Glycopon 4 A	15.00 gm.
Parachlormetaxylenol	0.10 gm.
Oxyquinoline Sulphate	0.10 gm.
Tragacanth Gum	2.75 gm.

Dissolve the lactic acid and sodium chloride in the water. Add the parachlormetaxylenol and oxyquinoline sulphate to the Glycopon 4 A. Warm till thoroughly dissolved, then add the tragacanth and stir till thoroughly mixed. To this, add the salt, and lactic acid solution slowly with hand stirring till cold. Allow to stand overnight, and stir the following day.

If a heavier jelly is required, reduce the amount of glycopon 4 A.

Corn Cures: are solutions of Pyroxylin, generally in mixtures of esters and alcohols to avoid the unpleasant hydrocarbon action on the body. An 8 oz. Pyroxylin solution in a mixture of 25% Butyl Acetate, 20% Butanol, 15% ethyl Acetate and 40% denatured alcohol characterizes them. The corn cures contain a small amount of Salicylic Acid and occasionally a trace of Hemp.

Corn Remedy

Acetone	168 oz.
Castor Oil	3 oz.
Venice Turpentine	6 oz.
Celluloid	10 oz.
Salicylic Acid	40 oz.
Ethyl Amino Benzoate	10 oz.

Dissolve the Salicylic Acid and Ethyl Amino Benzoate in the Acetone. Then add the Castor Oil and Venice Turpentine and finally the celluloid. Allow this mixture to stand, stirring it now and then until the Celluloid is completely dissolved. Then add sufficient Oil Soluble Chlorophyll to color it dark green.

Corn Removers

Solution of monochloroacetic acid in ratio of 1: 2 is suitable, but stronger solutions should not be used as they irritate skin. Another preparation contains 10 parts salicylic acid and 90 parts glacial acetic acid. This is thickened with mucilage containing 0.5 part gum tragacanth, 3 parts pectin, 3 parts glycerin and 43.5 parts water. About 5 parts of this mixture is used for thickening the preparation. Another composition contains 1 part glacial acetic acid, 8 parts lactic acid, 3 parts dried salicylic acid crystals and 8 parts of aforementioned thickener. Formic acid and carbolic acid, thickened with same thickener, may also be used.

Cough Lozenges

Extract Licorice	34 oz. 125 gr.
Powdered Cubebs	11 oz. 188 gr.
Ethyl Amino Benzonate	2 oz. 125 gr.

DENTAL PREPARATIONS

Dental Preparations

Dentists' Solution for Surface Anaesthesis

Glycopon S	80 fl. oz.
Ethyl Amino Benzoate	20 oz.
Oil Peppermint	50 minims.

Applied to the gums this solution desensitizes quickly and allows painless scaling. It also desensitizes the dentine and is therefore valuable in treating cavities.

Antiseptic Toothache Drops

Beechwood Creosote	15 oz.
Oil Clove	30 oz.
Cinnamic Aldehyde or Oil Cassia	20 oz.
Chloroform	30 oz.
Ethyl Amino Benzoate	5 oz.

Mix Creosote with oils and Chloroform then add Ethyl Amino Benzoate and stir until dissolved.

Toothache Gum

Yellow Beeswax	60 oz.
Venice Turpentine	10 oz.
Gum Mastic Powder	10 oz.
Ethyl Amino Benzoate	5 oz.
Dragon Blood Powder	10 oz.
Oil Clove	5 oz.

Melt Beeswax and Venice Turpentine together and add Gum Mastic. Stir until dissolved. Then add Ethyl Amino Benzoate and, when dissolved, Dragon Blood.

Stir until cooled to about 50° C. then add Oil Clove and mold into sticks.

* Plaster, Dental

(a)	Silica (finely ground)	60–80
	Alpha Gypsum	20–40
	Gum Arabic	0.5

(b)	Alpha Gypsum	95–99%
	Rochelle Salts	1–5 %

Dental Plate Adhesive

I.

Vanillin	0.5
Boric Acid Powd.	5.0
Powdered Acacia	
Powdered Tragacanth	
of each enough to make	100.0

II.

Powdered Acacia	
Powdered Agar-Agar	
of each	0.05
Powdered Tragacanth to make	10.00

In making these preparations, it is essential that all of the ingredients be in the form of a very fine powder.

Earache Oil

Oil Thyme	2 oz.
Oil Cajeput	2 oz.
Ethyl Amino Benzoate	3 oz.
Oil Apricot Kernel	93 oz.

Dissolve Ethyl Amino Benzoate in Apricot Kernel Oil by gently heating. When completely dissolved, allow to cool and add Oil of Thyme and Oil of Cajeput. Finally add sufficient Oil Soluble Chlorophyll to make it a light green color.

Acriflavine, Emulsion of

Acriflavene	0.5
Distilled Water (Hot)	25

Stir until dissolved.

In a separate container sterilize by heat

Lanolin	30 gm.

Allow it to cool and pour into a sterile mortar; add the acriflavine solution to this slowly while working it in with a pestle, not adding a further portion until the first is absorbed. Finally work in sterilized liquid paraffin oil a little at a time to make volume up to 500 c.c. This gives a permanent, sterile emulsion.

* Agar-Petrolatum Emulsion

In a steam-jacketed kettle heat under constant stirring:

Agar Agar Flakes	23.275	lb.
Sodium Benzoate	.75	lb.
Water	20.	gal.
Glycerin	5.	gal.

until the agar is dissolved.

Simultaneously in a steam-jacketed can prepare an emulsion from a strained solution of:

Acacia Gum Granular	27.75	lb.
Sodium Benzoate	.375	lb.
Water	13.125	gal.
and		
Liquid Petrolatum	50.	gal.

While the liquid petrolatum is being added to the Acacia solution, steam is circulating through the jacket of the can in order to heat the emulsion to about 60° C. As soon as a uniform emulsion is obtained it is transferred to the steam-jacketed kettle and mixed under stirring with the hot agar solution. Then the stirrer is shut off and the hot mixture by means of a pump is drawn from the kettle and strained through a 40 mesh strainer into two 50-gallon tin-lined copper cans, in which it is left undisturbed for forty-eight hours. The so obtained agar mush is soft and smooth, although of somewhat curdy appearance. It is transferred to a mixing tank and mixed with a strained solution of—

Acacia Gum Granular	27.75	lb.
Sodium Benzoate	.375	lb.
Water	13.125	gal.
and		
Liquid Petrolatum	37.5	gal.
Glycerin	9.18	gal.
Water, quantity sufficient to make a total of	175.	gal.
Flavoring Materials	1.	pt.

The mixture is circulated through a 40 mesh strainer and homogenized with 2500 lb. pressure.

The foregoing example is merely illustrative of the invention, which resides more particularly in the method and product involving the mixing of a hot solution of jell-forming substances with an immiscible ingredient such as oil at relatively high temperature, and the cooling of the mixture while it is quiescent, that is to say, avoiding stirring during the cooling of the mixture to room temperature.

* Castor Oil Emulsion, Laxative

Castor Oil	35
Milk of Magnesia	5
Water	60

Castor Oil Emulsion, Pharmaceutical

Sod. Hydroxide Sticks	9.8 gr.
Water	25 oz.
Castor Oil	6 lb. 5 oz.
Sodium Benzoate	100.5 gr.
Triethanolamine	288 minims.

Dissolve Sod. Hydroxide in water, add sodium benzoate and add Triethanolamine.

Mix with oil and run thru colloid mill. For flavor use sacharine (water soluble) vanilla and lemon extract.

---

* Pharmaceutical Castor Oil Emulsion

A prep. devoid of nauseous taste is prepared by emulsifying 35–80% of the oil in H_2O by the aid of 0.7% of NaOH or KOH, or of 5% of milk-of-magnesia.

Cod Liver Oil Emulsion

Cod Liver Oil	52 oz.
Water	88 oz.
Glycerine	10 oz.
Tragacanth	100 gm.
Oil of Sassafras	5 c.c.
Benzaldehyde	1 c.c.
Oil Coriander	1 c.c.
Oil Cardamom	2 c.c.
Tincture Vanilla	5 c.c.

Cod Liver Oil, Emulsion

Cod Liver Oil	26 oz.
Water	44 oz.
Tragacanth	50 gm.
Glycerin	5 oz.
Calcium Hypophosphite	½ oz.
Sodium Hypophosphite	½ oz.
Sacharine	.001 oz.

Flavoring

Benzaldehyde	7 parts
Oil Cassia	6 parts
Guaiacol	6 parts
Oil Sassafras	6 parts
Oil Wintergreen	16 parts

Use above mixture to taste.

Cod Liver Oil Emulsion

A

Gum Arabic	2 oz.
Tragacanth	½ oz.
H_2O	1½ qt.
Glycerine	8 oz.
Calcium Hypophosphite	½ oz.
Sodium Hypophosphite	½ oz.
Saccharin	1 gr.

B

Cod Liver Oil	29 oz.
Flavoring	

White Cod Liver Emulsion

"A"

Gum Arabic	15 gr.
Water	38 oz.
Calcium Hypophosphite	½ oz.
Sodium Hypophosphite	½ oz.
Glycerine	4 oz.
Saccharin	.001 oz.
Cod Liver Oil	26 oz.
Flavoring	As desired

"B"

Gum Arabic	225 gr.
Water	6 oz.
Glycerine	4 oz.

Mineral Oil and Agar Emulsion

1.

Liquid Petrolatum	30.0
Agar	1.0
Acacia	3.85
Tragacanth	0.75
Spirit of Chloroform	2.5
Tincture of Lemon Peel	2.0
Elixir of Saccharin	0.3
Phenolphthalein	1.5
Water, enough to make	100.0

Raise 60 c.c. of water to boiling point and dissolve the agar therein; heat the oil; mix the powdered gums in a suitable sized mortar and, with trituration, add hot oil. Mix, with constant stirring, the agar solution with the oil mixture; dissolve phenolphthalein in the mixed alcoholic preparations and add to the emulsion; to make up to desired quantity with warm water.

Mineral Oil and Agar Emulsion

"A"

Gelatin	27 gr.
Water	1800 c.c.

"B"

Mineral Oil	2000 c.c.

"C"

Agar	68 gr.
Gum Arabic	110 gr.
Gum Tragacanth	110 gr.
Oil of Orange	3 fl. dr.
Saccharin	10 gr.
Tinc. Vanillin	4 fl. dr.
Sodium Benzoate	2 dr.
Glycerine	4 fl. dr.
Water	2000 c.c.

Make up "A" mixture and add "B" to same, using electric stirrer. Run mixture through colloid mill, using .010 gap setting.

Prepare "C" mixture and add to

above. Run through colloid mill using .015 gap setting.

Add 960 grains phenolphthalein to each gallon of emulsion.

Mineral Oil and Agar Emulsion

2.

Agar	45 gr.
Boiling Water	8 oz.
Mineral Oil	8 oz.
Phenolphthalein	120 gr.
Flavoring Agent q. s.	

Dissolve the agar in the boiling water; heat the oil and mix the agar solution while hot; add the phenolphthalein and beat vigorously with an egg beater until cold. Add flavoring agents last. In respect to the latter, various oils may be used in this class of emulsion, such as cassia, wintergreen, etc., in suitable proportions.

In making these emulsions, acacia is used at times, in addition to agar, as it gives the emulsion a creamy appearance. It is important that liquid petrolatum of high viscosity be employed and to make such oil into a proper emulsion it is necessary to overcome the high viscosity by heating the measured quantity of liquid petrolatum on a water bath to about 50°C. The warm oil is then mixed with the hot agar solution as above.

Agar-agar of commerce contains 10 to 30 per cent. water. It should therefore be dried till it ceases to lose weight before using. To make a solution of agar it is necessary to boil for 30 to 40 minutes, making up the water lost by evaporation. An agar emulsion heavy enough to be permanent hardens to a stiff jelly, which cannot be poured out of a bottle unless the beaters are kept going till the product is cold.

* Laxative Mineral Oil Emulsion

1. Psyllium Seed	0.3 %
2. Agar	0.1 %
3. Gelatine	0.07%
4. Mineral Oil	70
5. Water	29.5

Swell 1, 2, 3 and 5 then boil; cool and filter add a preservative and run 4 in slowly while beating intermittently.

Paraffin Oil, Emulsion of
(for internal use)

Powdered Tragacanth	5 gm.
Moldex	1 gm.
Alcohol (90 per cent.)	10 gm.
Glycerin	150 gm.
Distilled Water	304.5 gm.
Mucilage of Gum Acacia	30 gm.
Liquid Paraffin Medicinal	500 gm.

If it is desired to include phenolphthalein in the emulsion, 10 gm. of phenolphthalein is added in the place of an equal amount of water in the above formula.

Emulsion of Liquid Petrolatum with Agar

Heavy Liquid Petrolatum	500.0 c.c.
Agar	5.5 gm.
Sugar	120.0 gm.
Acacia (fine powder)	30.0 gm.
Tragacanth (fine powder)	4.0 gm.
Tincture of Vanilla	8.0 c.c.
Tincture of Lemon	2.0 c.c.
Oil of Cassia	0.5 c.c.
Water, to make	1000.0 c.c.

Mix the agar and the sugar with 300 c.c. of boiling water and when they are dissolved strain the resulting solution and set it aside to cool. Triturate the powdered gums with the liquid petrolatum, then add the agar solution and whip the mixture with an egg beater. Finally add the tinctures and the oil and lastly enough water to make 1000 c.c.

Eye Lotions

The following two recipes are typical eye lotion preparations:

Boric Acid	50 gm.
Sodium Borate	50 gm.
Camphor Water	250 c.c.
Distilled Water to make	1,000 c.c.

Zinc Sulphate	2 gm.
Boric Acid	22 gm.
Camphor Water	250 c.c.
Distilled Water to make	1,000 c.c.

Eye Lotion

Zinc Sulphate	0.24
Sat. Sol. Boric Acid	180.00
Spanish Saffron (to color)	0.12
Alcohol	4.00
Camphor Water q.s.ad.	250.00

Eye Salves

Mercury oxide, red or yellow, is used mixed with wool grease, petrolatum or butter salve base. Special prepared salve bases may also be used. In one composition 1 to 3% yellow oxide of mercury (freshly precipitated) is mixed with 10% of anhydrous wool grease, 8% of distilled water, 2% of olive oil and white petrolatum to make 100%. In another composition one to 2.5% of collargol is mixed with 15% of distilled water, 12 parts anhydrous wool grease and white petrolatum to make 100%. Another formula contains 0.001 gram mercuric chloride, 0.05 gram boric acid, 5 drops olive oil and 5 grams white petrolatum. Also 5 parts red oxide of mercury are mixed with 3 parts opium and 100 parts fresh sweet butter; also one part red oxide of mercury is mixed with 29 parts white beeswax and 70 parts fresh sweet butter.

Foot Powder

Zinc Stearate	60 gm.
Alum Acetate	10 gm.
Menthol	½ gm.

Foot Powder

The ordinary old-time foot powder is composed principally of some such base as talc and starch, together with a little boric or salicylic acid. A modification of this old formula is as follows:

Salicylic Acid	6 dr.
Boric Acid	3 oz.
Powdered Elm Bark	1 oz.
Powdered Orris	1 oz.
Talc	36 oz.

Oxygen-liberating liquids and powders seem to be in favor for cleansing wounds and feet. A typical formula for such a powder is:

Sodium Perborate	3 oz.
Zinc Peroxide	2 oz.
Talc	15 oz.

Solutions for Perspiring Feet

Formic Acid	1 dr.
Chloral Hydrate	1 dr.
Alcohol, to make	3 oz.

Apply by means of absorbent cotton.

Boric Acid	15 gr.
Sodium Borate	6 dr.
Salicylic Acid	6 dr.
Glycerine	1½ oz.
Alcohol, to make	3 oz.

For local application.

Frost Bite Pencil

Camphor	25 parts
Iodine	50 parts
Olive Oil	500 parts
Paraffin, solid	450 parts
Alcohol	sufficient

Dissolve the camphor in the oil, and the iodine in the least possible amount of alcohol. Melt the paraffin and add the mixed solutions. When homogeneous, pour out into suitable molds.

Wrap the pencils in paraffin paper or tin foil, and pack in wooden boxes. By using more or less olive oil the pencils may be made of any desired consistency.

Gargle, Sore Throat

The preparation is used, diluted with two or three parts of water, either from a spray applied to nose and throat, or as a nasal douche from a nasal irrigator or syringe. Habitual users commonly inhale the solution into the nostrils from the palm of the hand. The preparation is also a most useful gargle for sore throats:

Sodium Bicarbonate	1.00
Borax	2.00
Sodium Benzoate	0.80
Sodium Salicylate	0.52
Menthol	0.03
Thymol	0.05
Eucalyptol	0.13
Oil of Pumilio Pine	0.05
Oil of Wintergreen	0.03
Alcohol (90 per cent.)	2.50
Glycerin	10.00
Solution of Carmine	0.52
Talc or Kaolin	sufficient
Distilled Water	to 100.00

The salts are dissolved in 80 of the water and the glycerin added. The other ingredients are dissolved in the alcohol and the alcoholic solution is triturated with the talc (about 5 per cent.), and the mixture added to the salt solution. The solution of carmine is added and the whole is filtered, distilled water being passed through the filter to produce the required volume. Filtration through talc or kaolin is essential to the production of a clear and bright solution.

Haemorrhoidal Suppositories

Ethyl Amino Benzoate	10 oz.
Bismuth Subgallate	10 oz.
Thymol Iodide	5 oz.
Cacao Butter, grated	75 oz.

Mix the powders with the Cacao Butter as directed in the USP and make the suppositories by the cold compression method. These suppositories relieve pain quickly, are astringent, antiseptic and granulating.

Influenza Remedies

1. Sodium Salicylate	3 oz. 287 gr.
Phenazone	1 oz. 362 gr.
Spiritus Ammonia Aromatic	6 fl. oz. 320 min.
Chloroform	40 min.
Water to Produce	1 gal.
Caramel as desired.	

The sodium salicylate and phenazone

are dissolved in the bulk of the water, together with sufficient caramel to give the necessary colour, which should be either dark brown or almost black, and the solution filtered. The chloroform is dissolved in the spirits and added to the clear aqueous solution, being made to bulk with water. The dose of the normal mixture is one fluid ounce.

2. Sodium Salicylate 3 oz. 287 gr.
   Liquor Ammonii Acetatis Conc. 1–7 2½ fl. oz.
   Ammonium Carbonate 480 gr.
   Water to Produce 1 gal.
   Caramel as desired.

The sodium salicylate, potassium bicarbonate, and ammonium carbonate are dissolved in the bulk of the water, with sufficient caramel to give a dark brown colour. The solution of ammonium acetate is added and the mixture filtered, and made up to the quantity with water. The dose of the normal mixture is one fluid ounce.

Cold and Influenza Mixture

Potassium Nitrate	2
Potassium Bromide	2
Spirit of Nitrous Ether	8
Strong Solution of Ammonium Acetate	4
Compound Tincture of Cardamom	2
Camphor Water	to 100

Label.—One tablespoonful to be taken every four hours.

Another mixture of similar type, but of more pleasant taste, is:

Ammonium Carbonate	3
Potassium Bicarbonate	6
Tincture of Ipecacuanha	12
Strong Solution of Ammonium Acetate	12
Chloroform Water	to 100

Label.—One tablespoonful to be taken every four hours.

Inhalants for Colds

No. 1

Chloroform	10 parts
Formaldehyde	5 parts
Ether	6 parts
Menthol	3 parts
Eucalyptus	3 parts
Lavender	4 parts
Isopropyl Alcohol	69 parts

Procedure: Dissolve menthol in chloroform, add the oils, ether and solvent. Mix and filter if necessary.

No. 2

Menthol	2 parts
Camphor	3 parts
Lavender	6 parts
Oil of Pine Needle	5 parts
Eucalyptus	3 parts
Rosemary	6 parts
Formaldehyde	4 parts
Acetone	71 parts

Procedure: Mix menthol and camphor, and warm until liquefied. Add the oils and the solvent. Mix and add the formaldehyde.

No. 3

Thymol	.5 part
Peppermint Oil	5 parts
Pine Needle Oil	3 parts
Formaldehyde	5 parts
Isopropyl Alcohol	86.5 parts

Procedure: Dissolve thymol in the oils, add solvent and formaldehyde. Mix thoroughly and filter.

No. 4

Phenol	1 part
Menthol	1 part
Camphor	.5 part
Thymol	.5 part
Eucalyptus	2 parts
Oil Clove	2 parts
Oil of Sassafras	2 parts
Chloroform	8 parts
Isopropyl Alcohol	83 parts

Procedure: Mix phenol, thymol, menthol and camphor and warm until liquefied. Add the solvents and finally the oils. Mix thoroughly and filter.

No. 5

Menthol	1 part
Camphor	1 part
Wintergreen	3 parts
Eucalyptus	2 parts
Oil of Pine Needle	3 parts
Formaldehyde	4 parts
Iodine (2½% sol.)	1 part
Isopropyl Alcohol	85 parts

Procedure: Mix menthol and camphor and warm until liquefied. Add the wintergreen, pine, formaldehyde, solvent and iodine. Filter.

No. 6

Phenol	1 part
Menthol	1 part
Camphor	1 part
Oil of Pine Needle	1 part
Sandalwood	1 part
Lavender	3 parts
Eucalyptus	2 parts
Strong Ammonia Solution	3 parts
Alcohol	87 parts

Procedure: Mix phenol, menthol and camphor and warm until liquefied. Dissolve the oils in part of isopropyl and add to above. Mix ammonia with rest of isopropyl and add to rest of mixture. Mix and filter.

Note: the formaldehyde solution recommended in the foregoing preparations is 37%. The pungency of any of the preparations can be toned down by the addition of sweeteners or by increasing the quantity of the more fragrant oils in each formula.

Vaginal Jelly

1. Gum Tragacanth	6
2. Glycerin	10
3. Water	100
4. Boric Acid	5

Mix 1 and 2 and add 3 and 4 slowly with stirring; let stand overnight.

Artificial Vaseline

Ceresin or Paraffin	15–20
White Mineral Oil	85–80

Wart Remover

A.	Salicylic Acid	2
	Glacial Acetic Acid	20

B.	Trichloracetic Acid	90
	Water	10

---

* Iodine, Colloidal

Finely divided or colloidal I is produced by treating cryst. I with 1% soln. of NaOH until a straw-colored liquid results, cooling, adding 10 g. of gum arabic or other protective colloid for each lb. of I, then adding rapidly strong HCl to ppt. all the I in the form of a brick-red ppt. (*i.e.,* finely divided colloidal I). The coarse crystals are filtered out, the filtrate is allowed to stand and concentrate and the supernatant soln. is drawn off to remove the NaCl.

Liniment

Camphor Oil	74 oz.
Oil Laurel, Expressed	10 oz.
Oleoresin Capsicum USP (VIII)	5 oz.
Ethyl Amino Benzoate	2 oz.
Camphor Powder	2 oz.
Oil Rosemary	2 oz.
Chloroform	5 oz.
Oil Mustard, USP	½ oz.

White Liniment

The following formula is said to yield a creamy white preparation of excellent penetrating power:

Ammonium Carbonate	30 gm.
Water	240 c.c.
Castile Soap	24 gm.
Hot Water	480 c.c.
Camphor	30 gm.
Alcohol	30 c.c.
Oil of Turpentine	q.s.

Dissolve the ammonium carbonate in 240 c.c. of water. Dissolve the soap in the hot water, then mix the two solutions. Dissolve the camphor in the alcohol and add to the first mixture; shake well. Now add the oil of turpentine in sufficient quantity to make a creamy emulsion, gradually adding and constantly shaking.

Lubricating Jelly

Fatty or oily substances, petrolatum, for example, have been used for lubricating surgical instruments, such as urethral sounds and vaginal specula. Latterly mucilaginous preparations are advised for this purpose, as they may be readily removed by washing with water. Such a preparation may be made as follows:

1.	Tragacanth, whole	48 gr.
	Carbolic Acid, liquefied	50 m.
	Glycerin	4 oz.
	Distilled Water	4 oz.

Mix the three liquids, pour upon tragacanth contained into a mortar or graduate, let stand for 12 to 24 hours or until the gum is thoroughly softened, then triturate or beat to a smooth paste. If desired still smoother, strain forcibly through cheese-cloth.

Dispense in an ointment jar.

2. A preparation now on the market is stated to contain the gelatin of Irish moss with oil of eucalyptus and formaldehyde. Such a preparation may be made by adding to 16 fluidounces of mucilage of Irish moss, 10 drops of oil of eucalyptus and 5 drops of formaldehyde solution.

Lubricating Jelly

1.	Karaya Gum	7.5 gm.
	Glycomel	10.0 c.c.
	Isohol	5.0 c.c.
2.	Water	100 c.c.
	Moldex	0.15 gm.

Bring 2 to a boil and stir until dissolved. Cool and add quickly to 1 while stirring. A heavy gel forms immediately. Air bubbles can be removed by keeping warm for a time. This gel spreads evenly on the skin and does not roll up. Various antiseptics and medicaments may be used to make vaginal and other jellies.

Menthol Pencil or Crayon

Menthol	100
Benzoic Acid	10
Eucalyptol	3

Melt together and cast in forms.

Migraine Salve

Ten parts beeswax and 46 parts anhydrous lanolin are melted and 180 parts distilled water added. Mass is well mixed and then mixture of 15 parts menthol, 16 parts methyl salicylate and 2 parts rosemary oil are worked in and uniform salve obtained. In another preparation 5 parts menthol are dissolved in 6.5 parts acetic ester, 4.2 parts absolute alcohol, 1.85 parts triple strength ammonia liquor and solution is worked up into salve with 45 parts anhydrous lanolin, 36.5 parts white petrolatum and perfumed with 0.5 part lavender oil and 1 part essence of eau de cologne.

Mosquito Cones

1. Powd. Charcoal	16	oz.
Nitrate Potassium	2	oz.
Carbolic Acid	1½	oz.
Insect Powder	8	oz.
Tragacanth Mucilage, a sufficient quantity		

Make into a stiff paste with the mucilage, and form into cones weighing about one ounce each.

2. Powd. Charcoal	16	oz.
Nitrate Potassium	2	oz.
Benzoin	4	oz.
Hard Tolu Balsam	2	oz.
Insect Powder	4	oz.
Tragacanth Mucilage, a sufficient quantity.		

Mosquito Powder

1. Oil Eucalyptus	1	oz.
Powdered Talcum	2	oz.
Powdered Starch	14	oz.

This powder is to be rubbed into the exposed parts of the body to prevent the attack of the insect.

2. Oil Pennyroyal	4	oz.
Powdered Naphthalin	4	dr.
Starch	16	oz.

Mix well and sift. This is to be used like the preceding.

Basic Formulae for Mouth Washes

It should be noted that terpeneless oils are best employed. For use, a few drops of the product are added to a glass of water.

Saccharin

Saccharin	52	gm.
Heliotropine	11	gm.
Peppermint Oil	50	c.c.
Cinnamon Oil	10	c.c.
Tincture of Cochineal	250	c.c.
Alcohol, q.s. to make	10,000	c.c.

Procedure.—Dissolve saccharin, heliotropine, peppermint, cinnamon, and cochineal in alcohol in the order stated. Mix well. Chill and filter.

Aromatic

Eucalyptol	10	oz.
Menthol	10	oz.
Clove Oil	5	oz.
Wintergreen Oil	1	oz.
Heliotropine	0.1	oz.
Acetic Ether	10	oz.
Chlorophyl Ether Soluble	2	oz.
Alcohol, q.s. to make	1,000	oz.

Procedure.—Add all ingredients to the alcohol one at a time, while mixing in the order given. Mix for three hours. Chill in a cooling tank to 40° F. and filter.

Salol-Thymol

Salol	10	oz.
Thymol	4	oz.
Oil of Lavender	1¼	oz.
Menthol	1¼	oz.
Benzoic Acid	20	oz.
Glycerin	500	oz.
Tincture of Cardamom	500	oz.
Alcohol	2,000	oz.

Procedure.—Dissolve salol, thymol, benzoic acid, and menthol in alcohol. Mix well. Mix tincture cardamom with glycerin, add to alcohol, add lavender. Mix four hours, chill and filter.

Mouth Wash

Benzoic Acid	1	lb.
Boric Acid	2	lb.
Borax	1	lb.
Alcohol	1½	gal.
Eucalyptus	3	fl. oz.
Oil of Thyme	1	fl. oz.
Oil of Wintergreen	2	fl. oz.
Water	15	gal.
Caramel Coloring	1¼	fl. oz.

The boric acid and borax are added to part of the water and dissolved by boiling. The solution is cooled by the addition of the rest of the water and left to become quite cold. The benzoic acid is dissolved in half the alcohol, and the essential oils in the remaining half, and the two mixed and added to the water solution. The caramel colour is added while stirring, and thorough mixing is continued for four hours.

Mouth Wash

Benzoic Acid	12 parts
Tincture of Rhatany	60 parts
Alcohol	400 parts
Oil of Peppermint	3 parts

A teaspoonful in a small wine-glassful of water.

Alkaline Mouth Wash

This is made as follows:

Potassium Bicarbonate	21.0 gm.
Sodium Borate	20.0 gm.
Sassafras Oil	1.0 c.c.
Thymol	0.5 c.c.
Eucalyptol	1.0 c.c.
Methyl Salicylate	0.5 c.c.
Cudbear	2.0 gm.
Alcohol	50.0 c.c.
Glycerin	90.0 c.c.
Magnesium Carbonate	10.0 gm.
Water	to 1,000 c.c.

Mix the potassium bicarbonate and sodium borate with 100 c.c. of water. When the effervescence ceases, add this solution to 500 c.c. of water. This is then added to the alcohol in which the essential oils have been previously dissolved. The tincture of cudbear and the rest of the water are next added with the magnesium carbonate. The whole is mixed thoroughly for 2 hours and allowed to stand for 48 hours, chilled, and filtered. Purified talc may be used in place of the magnesium carbonate.

Chloro-Phenol Mouth Wash

Benzoic Acid	4 oz.
Cinnamon Oil	8 oz.
Phenol	6 oz.
Chloroform	6 oz.
Alcohol	150 oz.
Oil of Peppermint	2 oz.
Glycerin	to make 400 oz.

Dissolve the benzoic acid in the chloroform, add the glycerin and mix. Dissolve the cinnamon, peppermint, and phenol in alcohol and mix the two solutions together. Mix for two hours, chill, and filter.

Resorcin Mouth Wash

Resorcin	50.0 gm.
Zinc Chloride	0.3 gm.
Menthol	5.0 gm.
Thymol	2.0 gm.
Eucalyptol	0.3 gm.
Camphor	0.3 gm.
Oil of Peppermint	0.5 gm.
Alcohol	250.0 gm.
Solution Hydrogen Dioxide	200.0 gm.
Water	to make 1,000 gm.

Dissolve the resorcin and zinc chloride in water, and the thymol, eucalyptol, wintergreen, menthol, and camphor in the alcohol. Mix the two solutions together, add the peroxide; stir for one hour, chill, and filter.

Astringent Mouth Wash

Zinc Chloride	1 gm.
Alcohol	12 c.c.
Eucalyptol	20 dr.
Oil of Cinnamon	2 dr.
Oil of Peppermint	3 dr.
Distilled Water to make	100 c.c.

Mouth Wash, Analgesic

Anaesthesin is used in analgesic mouth washes. This substance is easily soluble in alcohol and difficultly soluble in water and hence adheres to mucuous membrane with which it contacts. Two grams of anaesthesin are dissolved in 90 parts alcohol and 20 parts water are added. Peppermint oil, anise oil and clove oil may be added to finish preparation. Another product used in these mouth washes is ethyl paraphenol-sulfo-para-aminobenzoate in 2% solution. Novocaine hydrochloride may be used with addition of taste correctives. An interesting mixture is 800 parts tincture of pyrethrum, 40 parts tincture of Spanish pepper, 40 parts oil of clove, 20 parts menthol, 20 parts camphor and 80 parts chloroform.

Following mouth washes may be used for treating pain caused by cariotic teeth. Four parts red saunders are mixed with 2 parts guaiacum wood, 5 parts myrrh, 5 parts cloves and one part cinnamon bark. This mixture is digested with 290 parts 90% alcohol, filtered, and 0.1 part oil of clove and 0.1 part cinnamon oil. In another preparation 16 parts tincture of myrrh are mixed with 8 parts tincture of catechu, 4 parts tincture of guaiac, 4 parts tincture of rhatany, 3 parts tincture of cloves, 2 parts spirits of cochlearia, few drops cinnamon oil and 63 parts 50% alcohol. Simple preparation consists of 2 parts oil of black mustard and 30 parts spirits of cochlearia. Tannic acid is also used in these preparations. Thus 8 parts tannic acid are mixed with 5 parts tincture of iodine, 1 part potassium iodide, 5 parts tincture of myrrh and 200 parts rose water. Five parts tannic acid are also mixed with 5 parts tincture of pyrethrum, 4 parts lavender water, 40 parts 90% alcohol and 20 parts distilled water. Finally 6 parts tannic acid are mixed with 3 parts tincture of iodine, 6 parts tincture of myrrh, 70 parts 90% alcohol and 240 parts rose water.

Antiseptic Inhalant

Eucalyptol	20.0 c.c.
Menthol	7.5 gr.
Oil of Rosemary	10.0 c.c.
Oil of Pine Needles	10.0 c.c.
Oil of Lavender	3.0 c.c.
Oil of Jack Rose Comp.	2.0 c.c.
Brilliant Green	trace
Ethyl Alcohol (S. D.) q.s.	100.0 c.c.

Dissolve the menthol in the oils. Make a strong solution of brilliant green in alcohol. Use enough to give finished product a green tint. Add the remaining alcohol to make 100 c.c.

Antiseptic for Telephone Mouthpiece

1.	Stearic Acid	6.00
	S. D. Alcohol	20.00
2.	Sodium Hydroxide	1.35
	S. D. Alcohol	10.00
	Water	5.00
	Glycerin	5.00
	S. D. Alcohol	10.00
	Fluorescein	0.01
	Menthol	1.00
	Camphor	1.00
	Oil Eucalyptus	5.00
	Oil Lavender	5.00

Mix 1 and 2 at 60° C. Then add the remainder and before it cools pour into molds.

Mentholated Throat and Mouth Wash

Alcohol	4¾ gal.
Ethyl Amino Benzoate	12 oz. 350 gr.
Thymol	1 oz. 120 gr.
Eucalyptol	1 oz.
Oil Wintergreen	¾ oz.
Menthol	100 gr.
Boric Acid	3 lb.
Distilled Water	5¼ gal.

Dissolve Ethyl Amino Benzoate, Thymol, Eucalyptol, Oil Wintergreen and Menthol in Alcohol. Dissolve Boric Acid in hot distilled water, cool and filter. Add this aqueous solution slowly while stirring to the alcoholic solution and filter.

Zinc Chloride Mouth Wash

Tincture of Myrrh	2 fl. oz.
Thymol	5 gr.
Powdered Borax	½ oz.
Red Saunders	enough to color
Oil of Clove	5 dr.
Oil of Cinnamon	5 dr.
Zinc Chloride	4 gr.
Diluted Alcohol	1 pt.

Macerate three days with occasional shaking. Then filter.

Sterilizing Solution for Oral Mucous Membrane

Tincture Iodine	2 parts
Acetone	3 parts
Glycerin	1 part

---

* Ephedrine Nasal Spray

Ephedrine Hydrochloride	0.17–1.0
Gum Tragacanth	0.5–1.0
Water	99

Dissolve above and then beat in

Sod. Chloride	0.8
Ethylene Chlorhydrin	0.75
Eucalyptol	0.125
Phenol	0.062
Menthol	0.125

Antiseptic Oil Spray for Nose and Throat

Oil Sweet Almond	2 gal.
Ethyl Amino Benzoate	12 oz. 350 gr.
Thymol	3 oz. 360 gr.
Menthol	300 gr.
White Mineral Oil	8 gal.
Eucalyptol	3 oz.
Oil Wintergreen	2 oz. 120 minims.

Heat Oil Sweet Almond to about 70° C. and add Ethyl Amino Benzoate, Thymol and Menthol. Stir until dissolved. Then add slowly while stirring White Mineral Oil and then Eucalyptol and Oil Wintergreen.

Aseptic and Analgesic Dusting Powder for Wounds

Urea Crystals	80 oz.
Ethyl Amino Benzoate	5 oz.
Thymol Iodide	5 oz.
Boric Acid Powder	5 oz.
Bismuth Subgallate	5 oz.

Mix and grind in a ball or pebble mill and sift through a No. 120 mesh sieve. Fill into cans with sprinkler top.

Hay Fever Ointment

Lanolin Anhydrous	50 oz.
Yellow Petrolatum	25 oz.
Ethyl Amino Benzoate	5 oz.
Menthol	½ oz.
Epinephrin Solution 1-1000	2 oz.
Distilled Water	23 oz.

(1) Triturate Ethyl Amino Benzoate

and Menthol with a portion of the Yellow Petrolatum until smooth. Gradually add the remainder of the Petrolatum and the Lanolin.

(2) Mix Epinephrin Solution with Distilled Water and add this aqueous solution slowly under trituration to No. 1 and mix until homogeneous.

Mustard Ointment

Lanolin Anhydrous	45 oz.
Yellow Petrolatum	27 oz.
Ethyl Amino Benzoate	2 oz.
Camphor Powder	5 oz.
Distilled Water	20 oz.
Mustard Oil, USP	1 oz.

Triturate Ethyl Amino Benzoate and Powdered Camphor with portion of the yellow Petrolatum until smooth. Gradually add the remainder of the Petrolatum and the Lanolin. Then add the Mustard Oil and triturate until homogeneous. Finally add the distilled water and mix until perfectly smooth.

Zinc Ointment USP with Ethyl Amino Benzoate

White Petrolatum	65 oz.
Paraffin	15 oz.
Zinc Oxide	20 oz.
Ethyl Amino Benzoate	2 oz.

Melt Petrolatum and Paraffin together then add Zinc Oxide and Ethyl Amino Benzoate previously sifted through a No. 100 mesh sieve. Stir until cold and pass through ointment mill.

Haemorrhoid Ointment (Pile Ointment)

Yellow Petrolatum	53 oz.
Lanolin Anhydrous	30 oz.
Yellow Beeswax	5 oz.
Ethyl Amino Benzoate	5 oz.
Bismuth Subgallate	5 oz.
Thymol Iodide	2 oz.

Melt Yellow Petrolatum, Lanolin and Beeswax together and allow to cool. Mix the three powders and triturate with a portion of the ointment base until smooth. Then add gradually the remainder of the base and mix until ointment is homogeneous. Note: This ointment must not come in contact with iron as discoloration will result so only porcelain or wooden utensils should be used.

Stainless Iodine Ointment

Iodine, in moderately coarse powder	5 parts
Paraffin	5 parts
Oleic Acid	20 parts
Petrolatum	70 parts

Decolorized Tincture of Iodine

Iodine Crystals	50 parts
Potassium Iodide	25 parts
Stronger Ammonia Water	100 parts
Water	400 parts
Alcohol, a sufficient quantity to make	1000 parts

Protective Coating for Pills

Tincture Benzoin	50
Alcohol	50
Vanillin	0.2

Poison Ivy Lotion

Aluminum Sulphate Crystals	19½ oz.
Basic Secondary Lead Acetate	26 oz.
Distilled Water	60 oz.
Ethyl Amino Benzoate	2 oz.
Glycopons	50 oz.

Pyorrhea Astringent

Potassium Iodide	15 parts
Iodine Crystals	20 parts
Glycerin	25 parts
Zinc Phenolsulphonate	15 parts
Distilled Water, a sufficient quantity to make	100 parts

Refrigerant Counter Irritant

Menthol	10 parts
Iodine Crystals	10 parts
Chloroform	90 parts
Tincture Aconite, enough to make	480 parts

Granular Effervescent Cider Salt

I.

Powdered White Sugar	1800 gm.
Powdered Tartaric Acid	900 gm.
Powdered Sodium Bicarbonate	900 gm.

II.

Powdered White Sugar	3200 gm.
Powdered Sodium Bicarbonate	900 gm.
Powdered Tartaric Acid	820 gm.
Powdered Potassium Bitartrate	120 gm.

Cider Flavor

Amyl Alcohol	4
Chloroform	4

Amyl Acetate	4
Amyl Butyrate	4
Amyl Valerate	8

The cases given above are effervescent, but may not granulate easily. It will therefore be better to replace the 900 gm. of tartaric acid of the first recipe by 540 gm. of crystalline citric acid and 360 gm. of tartaric acid.

Smelling Salts

Phenol	1
Menthol	1
Camphor	2
Weak Solution of Iodine (2.5 per cent. v/v)	1
Oil of Pumilio Pine	1
Oil of Eucalyptus	1
Strong Solution of Ammonia	3
Ammonium Carbonate	90

The ammonium carbonate should be packed into the bottle, the strong solution of ammonia added, then the other ingredients, previously mixed. Sodium sesquicarbonate is sometimes substituted for ammonium carbonate.

Diabetic Tea

Two examples of such teas are as follows. Ten parts acacia flowers, 20 parts mistletoe, 15 parts water fennel seeds and 30 parts lady's mantle. Also 5 parts lovage root, 5 parts valerian root, 5 parts blackberry root, 35 parts herba herba anserinae and enough foliae myrtillae to make 150 parts in all.

Tobacco Cure

The following mouth wash is said to "cure" one of smoking.

Silver Nitrate	0.2
Oil Pepperment	0.1
Water	100

The mouth should be rinsed with this several times daily.

White Liniment

Am. Palmolate	18 lb.
Water	15 gal.

Stir until dissolved and add

Ammonium Hydroxide	4 gal.
Water	30 gal.

Stir mechanically and add slowly

Turpentine	12 gal.
Oil of Camphor	12 lb.
Cottonseed Oil	8 lb.

Stir 10–15 minutes and add

Am. Carbonate	20 lb.

Stir until uniform.

Athletic Liniment

Oil of Camphor	25 gm.
Emulsone B	3.5 gm.

Rub together in mayonnaise type mixer and add

Glycerin	7.5 gm.
Water	46.5 c.c.

Allow to soak for 1 hour and while beating add

Glycerin	7.5 gm.
Water	46.5

Beat intermittently for 1 hour.

This produces a heavy fluid emulsion which is very stable.

Tooth Paste Formula

Glycerine	41.0 parts
Pure Water	37.0 parts
Calcium Chloride	1.5 parts
Powdered Gum Tragacanth	2.0 parts
Powdered Neutral White Soap	15.0 parts
Calcium Sulfate	82.0 parts
Soluble Saccharin	0.2 parts
Oil Peppermint	2.0 parts

Procedure:

Mix the glycerine and gum tragacanth. Dissolve the calcium chloride in the water and add to the glycerine-gum tragacanth mixture, stir and let stand until the gum is thoroughly hydrated (approximately one hour).

Now mix all the powdered ingredients and sieve thru 40 or 60 mesh and add these and the essential oils to the elixir and mix until the paste is smooth.

The consistency can be changed as desired by adding more or less of the Calcium Sulfate but this should never be changed greatly.

A smoother, creamier paste will be produced if ground thru a paint or ointment mill before tubing.

(Acid) Tooth Paste Formula

Glycerine	200.0 parts
Flavor 6-143-C	9.6 parts
* Acid Solution	64.0 parts
Benzoic Acid	0.8 parts
Calcium Chloride	2.4 parts
Cerelose	40.0 parts
Powdered Gum Tragacanth	6.4 parts
Powdered Gum Karaya	7.2 parts
Calcium Sulfate	304.0 parts
Tricalcium Phosphate	90.4 parts
	724.8 parts

* The acid solution is made as follows: 5 parts each of citric, boric, and tartaric acids dissolved in 100 parts cold water.

Procedure:

(a) Mix the glycerine, flavor, acid solution benzoic acid, calcium chloride, and cerelose. Mix for 15 minutes.

(b) Mix the powdered gums, Calcium Sulfate and the Tricalcium Phosphate.

(c) Add (b) to (a) and mix at least two hours.

Mill through a paint or ointment mill before filling tubes.

Flavor is composed of 8.0 parts Oil Peppermint, 1/.1 parts Oil Spearmint, 0.3 parts Menthol and 0.4 parts Oil Cassia.

Tooth Paste

Glycerine	41.0 parts
Distilled Water	37.0 parts
Calcium Chloride	1.5 parts
Flavor	2.0 parts
Powdered Neutral White Soap	15.0 parts
Calcium Sulfate	82.0 parts
Powdered Saccharine	0.2 parts
Powdered Gum Tragacanth	2.0 parts
	180.7 parts

Procedure:

Mix the glycerine and gum tragacanth. Dissolve the calcium chloride in the water and add to the glycerine-gum tragacanth mixture, stir and let stand until the gum is thoroughly hydrated (approximately one hour). Now mix all the powdered ingredients and sieve thru 40 or 60 mesh and add these and the essential oils to the elixir and mix until the paste is smooth.

The consistency can be changed as desired by adding more or less of the Calcium Sulfate but this should never be changed greatly.

After mixing the paste should be ground thru a paint or ointment mill.

Flavor is composed of:

Oil of Peppermint	8.0 parts
Oil of Spearmint	4.0 parts
Menthol	0.2 parts
Thymol	0.2 parts

Tooth Paste

Precipitated Chalk	50 gr.
Powdered Soap	6 gr.
Glycerine	34 gr.
Saccharin	0.25 gr.
Water	2.26 gr.
Glycosterin	2.00 gr.
Oil Peppermint	1.00 gr.
Added Glycerine	2.00 gr.

Dissolve glycosterin in water and glycerine on water bath. Stirred mechanically while cooling. Added 15 gm. of the chalk and the soap slowly while stirring. Allowed to cool. Transferred to mortar and added rest of chalk, and oil of peppermint, and saccharin, and ground thoroughly.

Finally added the 2 gm. of glycerin, stirred and tubed.

---

* Tooth Paste

Gum Tragacanth	1
Pectin	5
Glycol	30
Water	63
Titanium Dioxide	37
Pepsin	22
Diethylaminoethyloleylamide Lactate	4
Glycerin	4
Flavor	2

Tooth Paste (Soapless)

Glycopon AAA	30	parts
Powdered Karaya Gum	0.3	parts
Powdered Tragacanth	0.3	parts
Glycosterin	3.0	parts
Crysalba (Swann Calcium Sulphate)	40	parts
Tricalcium Phosphate (Swann)	5	parts
Water	27	parts
Saccharin	.05	parts
Benzoic Acid	1	parts
*Flavor	.5	parts

*Flavor has the following composition:

Oil Peppermint	10	parts
Oil Spearmint	2	parts
Oil Cassia	.2	parts

Melt the Glycosterin. Mix the powdered gums with the Glycopon AAA. Add Benzoic Acid and Saccharin. Finally, the water. Mix for 5 minutes. Heat above melting point of Glycosterin and add to the latter with constant stirring. After mixing for about 5 minutes add the mixed Crysalba and Phosphate with stirring, until a smooth paste is produced. When the temperature is about 30° C. add the flavor mixture with stirring, and pour into tubes.

There seems no special difficulty in the preparation of this paste, and a smoother product will be obtained if the abrasives are mixed into the paste at a temperature sufficiently high to be above the melting point of the Glycosterin. And after all has been added it is passed thru an ointment mill. This paste does not seem to harden in the tube nor become friable after exposure for 24 hours. The flavor can of course be modified to suit individual taste.

Dental Cream

Precipitated Chalk	35%
White Neutral Soap	20%
Powdered Sugar	10%
Purified Talc	10%
Glycerin	25%
Oil Peppermint q.s.	

Mix the powders thoroughly together and then work into a paste with the glycerin. Add oil peppermint.

Tooth Powder Flavors

No. 1

Oil of Cinnamon	2.5 c.c.
Oil of Clove	.5 c.c.
Methyl Salicylate	8 c.c.

No. 2.

Oil of Spearmint	1 c.c.
Menthol	.5 gr.
Methyl Salicylate	8 c.c.

No. 3

Oil of Cinnamon	2 c.c.
Oil of Peppermint	2 c.c.
Oil of Clove	3 c.c.

No. 4

Oil of Anise	2 c.c.
Oil of Cinnamon	2 c.c.
Oil of Peppermint	1 c.c.
Methyl Salicylate	5 c.c.

Tooth Powders

Titanium Dioxide	115 gr.
Calcium Carbonate Heavy	600 gr.
Pulverized Neutral White Soap	100 gr.
Sodium Carbonate Monohydrated	140 gr.
Flavor (Oil of Wintergreen)	18 c.c.

Procedure: Rub up the oil with part of the calcium carbonate until finely dispersed. Add the other ingredients and mix thoroughly. Sift.

Calcium Carbonate	500 gr.
Tricalcium Phosphate	150 gr.
Calcium Chloride	20 gr.
Bicarbonate of Soda	50 gr.
Pulv. Neut. Soap	55 gr.
Confectioner's XXX Sugar	100 gr.
Flavor to Suit	8 gr.

Procedure: Mix the flavoring with the sugar thoroughly. Add the soap and mix again. Add the bicarbonate and the calcium chloride. Mix. Add the tricalcium and the chalk and mix thoroughly and sift.

Tooth Powders

Magnesium Carbonate	425 gr.
Precipitated Chalk	560 gr.
Sodium Perborate	55 gr.
Sodium Bicarbonate	45 gr.
Soap	50 gr.
Sugar	90 gr.
Methyl Salicylate	8 c.c.
Menthol	1 gr.
Oil of Cinnamon	2 c.c.

Procedure: Dissolve the menthol in the methyl, add the cinnamon and then add to and mix with sugar. Add soap, perborate and mix thoroughly. Add the chalk and mix and finally add the bicarbonate and the magnesium carbonate. Mix thoroughly and sift.

Precipitated Chalk	500 gr.
Tin Oxide	95 gr.
Tricalcium Phosphate	100 gr.
Soap	30 gr.
Sugar	50 gr.
Flavor to Suit	8 gr.

Procedure: Incorporate flavors with sugar, add soap and mix thoroughly. Add tricalcium and chalk, mix and add tin oxide.

Chalk	400 gr.
Calcium Chloride	60 gr.
Tri Calcium Phosphate	100 gr.
Pulverized Carnauba Wax	30 gr.
Soap	50 gr.
Talc	200 gr.
Sugar	100 gr.
Flavors	10 gr.

Procedure: Mix flavors with sugar add soap and mix thoroughly. Mix the wax with the talc thoroughly, add the tricalcium and mix thoroughly. Add the chalk and calcium chloride. Mix thoroughly and sift.

Cream of Tartar	350 gr.
Milk Sugar	300 gr.
Colloidal Clay	375 gr.
Flavor	8 gr.
Color if Desired (usually pink)	

Procedure: Add flavors to milk sugar and mix. Mix cream of tartar with the clay and mix both mixtures together thoroughly. Sift.

Tooth Paste Flavors

The following three formulas represent flavors to be used in preparing an herb toothpaste, a wintergreen and a cherry toothpaste.

1.

Oil of Arnica	1 part
Hyssop Oil	1 part
Oil of Thyme	1 part
Juniper-berry Oil	1 part
Calamus Oil	2 parts
Sage Oil	3 parts
Spearmint Oil	5 parts
Peppermint Oil	5 parts

2.

Oil of Wintergreen	8 parts
Peppermint Oil	10 parts
Aniseed Oil	10 parts
Cinnamon Oil	10 parts
Oil of Cloves	15 parts

3.

Oil of Cinnamon	4 parts
Anethol	8 parts
Oil of Cloves	4 parts
Oil of Bergamot	4 parts

Two flavors of the antiseptic type which may be used either in toothpastes or mouth washes are as follows:

1.

Oil of Anise	½ part
Eucalyptol	1 part
Menthol	1 part
Methyl Salicylate	½ part

2.

Oil of Anise	4 parts
Oil of Cassia	2 parts
Oil of Cloves	5 parts
Oil of Eucalyptus	8 parts
Oil of Lavender	4 parts
Oil of Peppermint	9 parts
Oil of Wintergreen	12 parts
Menthol	12 parts

Athlete's Foot Ointment

5% Flowers of Sulphur
95% Anhydrous Lanolin

Grind or triturate sulphur into lanolin and apply locally.

Mosquito Bite Ointment

Boric Acid Ointment U.S.P.	95%
Phenol	5%

Triturate phenol into ointment cold.

Gelatine in Pharmaceutical Preparations

When formaldehyde is added to a solution of gelatine a change is observed to take place which is dependent on the amount of formaldehyde added. When the amount added is less than .15% a viscous solution results, but insolubility is not obtained in such a solution until the gelatine has been permitted to dry out. In greater percentages formaldehyde produces a jelly that can not be remelted o. brought into solution again. This product is rubbery and possesses less strength when cold. If it is dried and powdered, the product is known as formogelatine and due to the antiseptic action of formaldehyde it remains sterile and is a germicide and is also employed as a surgical dressing.

Capsules

For use as containers of doses of medicines are made from gelatine. A strong solution of this gelatine is mixed with glycerine and a little sugar, and the whole kept at a temperature of around 120° F. An iron rod, the end of which is shaped exactly as the capsules required, is highly polished so that the gelatine when cool may be easily detached. The rod is dipped into the solution and then revolved in a drying chamber. The sections are removed as soon as solution has jellied and allowed to dry. In using them the two sections are made so that one fits down over the other like a cover.

For Coating Pills

Gelatine is used, the object in this case is to eliminate the taste of pill in swallowing and prevent evaporation of enclosed moisture. 1 part gelatine, 2 parts water are mixed with a little glycerine or sugar and the pills are coated by dipping. This amount of gelatine assists in overcoming the nauseating affects of the medicine.

Gelatine Cells for Ultrafiltration

For ultrafiltration of colloids, membranes are prepared by impregnating disks of hard filter paper or fat extraction thimbles with a solution of gelatine. A 2–10% solution of gelatine is used and the containing disk should be kept on water bath at a certain constant temperature during the impregnation. A porosity of the filter will vary with the temperature as well as with concentration. After removing disks from liquid they are allowed to drain, rotating constantly so as to prevent an excess of jell forming on one side. After ½ hour the papers are placed in a 2–4% formaldehyde solution for 24 hours to render insoluble, the temperature being maintained at around 10–15° C. The disks or thimbles are then rinsed in cold water and kept in water saturated with chloroform. For varying gradations in size of pore, concentrations of gelatine are varied from 2–10% the temperature still being kept constant.

PHOTOGRAPHY

* Flashlight Powder

Magnesium Powder	700–900
Sulfur	10– 18
Pot. Permanganate	100–140
Pot. Nitrate	70– 85
Magnesium Oxide (Calcined)	100–160
Charcoal	10– 30

All materials should be finely powdered before mixing, which, should be done most carefully.

Smokeless Flashlight Powder

Zirconium	28
Zirconium Hydride	7
Magnesium	7
Barium Nitrate	30
Barium Oxide	25
Rice Starch	5

Gelatin Film Cleaner

Alcohol	98–99
Diethyl Amine	2– 1

* Paper, Ultraviolet Sensitive

Unsized paper is treated with a solution of o-$C_6H_4(NO_2)$. CHO and an indicator (litmus or bromothymol-blue).

* Developer, Photographic

p-propylaminophenolhydrochloride	1
Sod. Thiosulfate	9
Pot. Carbonate	8
Water	200

Photographic Developer, Rapid

Soln. A: metol 5 g., hydroquinone 5 g., Na_2SO_3 100 g., H_2O 1 l.; soln. B: K_2CO_3 100 g., H_2O 1 l. Time of immersion of film in soln. A, 1 min., followed by immersion for ½ to ¾ min. in soln. B.

Photographic Developing Fixer

Metol	5– 10
Hydroquinone	15– 20
Sod. Sulfite	50– 80
Sod. Carbonate (Anhyd.)	30– 40
Caustic Soda	20– 30
Pot. Bromide	5– 10
Sod. Hyposulfite	250–300
Am. Picrate	3– 5
Water	1000

Photographic Film, Reclaiming

Forty kg. of discarded pieces of old film is washed for 15–20 min. in a soln. prepd. by heating to 70° 100 l. of H_2O and 0.7 kg. of NaOH. The alkali-contg. gelatin and Ag are drawn off and the celluloid is further washed with hot water before being used for other purposes The alkali is returned to the washer for treating another 40 kg. of film. After the alkali has been used on 80 kg. of film it is boiled with steam in a wooden vessel and HCl (d. 1.19) is added to complete coagulation of the gelatin. After settling, the supernatant liquid is removed, the ppt. is filtered and then ashed in a muffle at 500–600°. The dried substance is mixed with Na_2CO_3 1: 3 and heated in a crucible until it is liquid. To eliminate the admixtures, Ag is melted with KNO_3 until the surface is mirror-bright. The pptd. Ag_2S from the fixing soln. is treated in the same way except that it is melted with Fe, as $Ag_2S + Fe = FeS + 2Ag$. To regenerate fixing solution Ag is pptd. by Na_2S.

Photographic Negatives, Removing Water Spots from

For removing water-spot drying marks on negatives bleach in the following soln.: $K_2Cr_2O_7$ 1 g., H_2O 100 cc., HCl 2 cc. and redevelop with an elon-hydroquinone developer.

Photographic Negative Intensifier

The following formula is recommended for the intensification of underdeveloped transparencies: soln. (1): citric acid 10 g., distd. H_2O 500 cc., pyrogallol 7.5 g.; soln. (2): $AgNO_3$ 10 g., distd. H_2O 200 cc. For use mix 25 cc. of (1) with 20 drops of (2). Pour quickly over the

All formulae preceded by an asterisk (*) are covered by patents.

plate and immerse it in a tray of the soln. Fresh soln. is needed for max. intensification.

Photographic Negative Intensification

The negative may be intensified by mordanting dyes to the image. Bathe the well-washed negative for 5 min. in the following mordant bath: water 1000 cc., $CuSO_4 . 5H_2O$ 40 g., K citrate 40 g., AcOH (glacial) 30 cc., NH_4CNS, 20 g. Wash well, then bathe in the dye bath to the desired d.: Methylene blue (1% soln.) 287 cc., rhodamine (1% soln.) 333 cc., phosphine (1% soln.) 380 cc., AcOH (glacial) 10 cc. The bath is stated to give a neutral black tone.

Negative Reducer, Photographic

The following soln. is recommended: ferric NH_4 oxalate 40 g.; boric acid, 40 g.; Na thiosulfate, 200 g.; water, 1 l. For use, this is diluted with at least an equal vol. of water. It does not keep so well in the diluted form. If the reduced negative has a slight yellow color, it may be removed in a 1% $C_2H_2O_4$ soln.

* "Masking Cream," Photographic

Glycerol	6	oz.
Water	8	oz.
Whiting	32	oz.
Salt	½	oz.
Ocher	½	oz.

Photographic Printing on Cloth

For Ag images on cloth the following formula is recommended. Immerse the cloth in a 2% soln. of Iceland moss contg. $AlCl_3 . 6H_2O$, 2%. After drying, immerse in a sensitizing soln.: $AgNO_3$ 30 g., distd. H_2O 150 cc., $C_6H_8O_7 . H_2O$ 10 g. The sensitized fabric is dried in the dark and printed under a negative by using a printing frame. The image can be toned with Au thiocyanate, fixed and washed or toned with combining toning and fixing baths. For dye images the primulin method is recommended, which consists of: bathing the cloth for 10 min. in a warm soln. contg. 10 g. of primulin dissolved in 300 cc. of hot H_2O. The cloth is then washed and immersed in a bath consisting of: $NaNO_2$ 6.6 g., H_2O 1 l., HCl 15 cc. It will then be brownish red. After drying in a dark room the image is printed by the use of a contrasting transparency, the light-exposed areas bleaching out in printing. The image is washed and developed in a soln. of β-naphthol 3 g., NaOH 4 g., H_2O 300 cc. to give a red tone. For purple tones α-naphthylamine 4 g., water 200 cc., HCl 10 drops is used. After washing, dry slightly and iron the slightly moist material.

Printing Photographic, Single and Multiple Gum

Art papers are sized with 60–70 g. soft gelatin, 3–5 g. chrome alum and 1150 ml. water, to give clear highlights. Five parts Senegal or Sudan gum arabic is dissolved in 10 parts water, and filtered through linen. To this is added ¼ of its vol. 3% starch soln., to give a mat surface, and a few drops 10% phenol. For single gum, 1 part aquarelle or tempora color is mixed with 2 parts gum soln. and 3 parts 15% $K_2Cr_2O_7$ or $(NH_4)_2Cr_2O_7$ plus a few drops NH_4OH. This is brushed out on the paper and dried. If the proportion of gum is right the color will wash off without exposure by simple bathing ½ hr. in cold water. Exposure is 55° for thin, 65–70° for medium, 75–90° for dense, negatives, on Herlango Eder-Hecht neutral wedge photometer with Celloidin paper. Correct exposures develop in 1–2 hrs. in cold water, or 6–8 min. with sawdust and water and dil. K_2CO_3 at 30–35°. Cr salts are removed by bathing in 10% alum plus a few drops of glacial AcOH, 1–2 hrs., then wash well. For 3-step multiple gum, the proportions are: (1) the same as given above; (2) 1 g. of 4–8 parts dichromate, 0.5 part color; (3) 0.4 g. of 5 parts dichromate, 1 part color.

* Sensitizing Solution, Photographic

A sensitizing compn. especially applicable for fabrics, wood surfaces, tiles, etc., comprises $AgNO_3$ 300 grs., ferric ammonium citrate (green scales) 375 grs., citric acid 300 grs. and H_2O 1 pint. The soln. is applied in drops and the surface dried. The printed image is washed and fixed in hypo, various tones being possible by this treatment. Images on fabrics can be washed and ironed.

Sepia Toning, Improved Bleach for

What is said to be a much superior method of producing sepia prints is: The print to be toned is partially bleached in a soln. contg. 1 part in 4 of the following: $CuCl_2$ 240 g.; HCl 1 oz.; H_2O 20 oz. It is then redeveloped in normal metol-hydroquinone developer. The brown black thus produced is entirely free from double tones. Only a slight rinse between bleach and development and no fixing afterward are required. This bleach is also better than others in sulfide toning.

Toning Bath, Vanadium

Soln. A: distd. H_2O, 500 cc.; $FeCl_3 . 6H_2O$, 25.0 g.; $K_3Fe(CN)_6$, 1.0 g.; $KClO_3$, 2.5 g. Keep in a dark brown bottle in a dark place. Soln. B: VCl_4 as paste, 5 g.; HCl (concd.) 25%, 50 cc.; distd. H_2O, 200 cc. For use, take 100 cc. of A, 100 cc. distd. H_2O and 5–10 cc. of B. If the image fades during washing, the print should be bathed in 5% $C_2H_2O_4$.

Silver Toning Bath Stains, Removing

The chalky deposit which forms on sulfide-toned prints during washing with hard H_2O may be removed by means of dil. AcOH (2.5%). Mounted prints may be sponged with this soln. since the acid evaps. without harming the print.

PLASTICS (CELLULOID, CELLULOSE ESTERS, COMPOSITIONS), ETC.

COLORS FOR PLASTICS

Colors for Plastics

Mahoganies

Burnt Sienna	2.92
Black Oxide of Iron	.44
Deep Indian Red	.64
Resin	49
Wood Flour	49

Burnt Sienna, Dark	.8
Burnt Sienna, Very Dark	3.12
Black Oxide of Iron	.08
Resin	49
Wood Flour	49

Burnt Sienna	1.64
Black Oxide of Iron	.14
Deep Indian Red	.22
Resin	49
Wood Flour	49

Seal Browns

Burnt Sienna, Dark	1.85
Black Oxide of Iron	.1
Ultramarine Blue	.05
Resin	49
Wood Flour	49

Deep Indian Red	.75
Burnt Turkey Umber	1.75
Resin	49
Wood Flour	49

Red-Browns

Deep Indian Red	1.75
Burnt Turkey Umber	.75
Resin	49
Wood Flour	49

Deep Indian Red	1.50
Black Oxide of Iron	.5
Resin	49
Wood Flour	49

Blacks

Nigrosine Dye	1.4
Black Oxide of Iron	.6
Resin	49
Wood Flour	49

Olive Drab

Black Oxide of Iron	1.9
Yellow Oxide of Iron	.1
Resin	49
Wood Flour	49

Molding Powder

Asbestos Flour	147 parts
Chalk	147 parts
Clay	147 parts
Bakelite	30 parts
Cumarone	30 parts

Rubber Pyroxylin Mixture

A common solvent for rubber and pyroxylin is composed of following

All formulae preceded by an asterisk (*) are covered by patents.

Ethyl Butyrate
Propyl Propionate
Isobutyl Butyrate
or
Ethyl Oenanthate

Thus rubber and pyroxylin may be dissolved in these to form lacquers of special properties.

Sculptors Putty

Linseed Oil (Boiled)	15%
Fullers Earth	15%
Calcium Carbonate	70%

Mix all ingredients thoroughly.

* Imitation Porcelain Plastic

Molten chlorinated naphthalenes (approx. 50% Cl) are treated with 20–50% of talcum, kaolin, chalk, white lead, or of other insol. material or pigment. The masses soften when heated.

"Celluloid"—Non-Inflammable

Cellulose Acetate	119–180
Acetone	33– 48
Benzol	32– 52
Alcohol	14– 20

Cellulose Acetate

400 g. of AcOH and 7.5 g. of H_2SO_4 are added to 100 g. of air-dry cotton-wool (I), and 1–1.5 hr. later 250–280 g. of Ac_2O are added gradually, with const. agitation, maintaining the temp. at 20–30°. After complete dissolution of (I) (5–6 hr.) the triacetate is hydrolyzed at 15–35° during 20–70 hr. with a mixture of 52–56% AcOH and $\not>$ 7.5 g. of 95% H_2SO_4. H_2SO_4 may be replaced by $ZnCl_2$, in which case (I) should first be treated with 66.5% HNO_3 at 20–22° during 30 min., washed, air-dried, acetylated at 80° during 2.5 hr. using a 1 : 8 : 2.8 mixture of $ZnCl_2$, AcOH, and Ac_2O, and then hydrolyzed during 42 hr. at 15°. The product thus obtained possesses high viscosity, stability, and insulating properties, rendering it suitable for the manufacture of $COMe_2$-sol. insulating lacquers.

* Composition, Acid Proof

(Suitable for Storage Battery Boxes)

Cotton Flock	25
Infusorial Earth	25
Asphalt or Pitch	50

Cork Composition Binder

Casein	45
Borax	7
Water	120
Glycerine	76

Composition Ornaments

A pattern is carved out of wood and is covered by following composition to form a "die":

Oil of Tar	3 oz.
Soapstone	4 lb.
Emery Flour	4 lb.
Orange Shellac	6 lb.
French Chalk	4 oz.

Melt the shellac and add the oil of tar. Add the soapstone, mixing thoroughly. Mix separately the (dry) emery flour and French chalk; then pour this into the melted shellac and oil of tar, stirring thoroughly and vigorously. Place the pattern or "die" in a box, flat side down, and pour this mixture over same. When cool the result will be a mould into which can be cast the materials of which the ornaments or mouldings are composed.

The following composition has been tested and found excellent for mouldings and ornaments of this kind:

White Glue	13	lb.
Rosin	13	lb.
Raw Linseed Oil	⅓	qt.
Glycerine	1	qt.
Whiting	19	lb.

This mixture is prepared by cooking the white glue until it is dissolved. Then cook separately the rosin and raw linseed oil until they are dissolved. Add the rosin, oil and glycerine to the cooked glue, stirring in the whiting until the mass makes up to the consistency of putty. Keep the mixture hot.

Place this putty mass in the die, pressing it firmly into the same and allowing it to cool slightly before removing. The finished product is ready to use within a few hours after removal. Suitable colors can be added to secure brown, red, black or any other color.

In applying ornaments made of this composition to a wood surface, they are first steamed to make them flexible; in this condition they can be glued to the wood surface easily and securely. They can be bent to any shape, and no nails are required for applying them.

* Molding Composition, Celluloid

Pyroxylin	100
Tricresyl Phosphate	75– 80
Gypsum	300–350

* Phonograph Record Composition

Cellulose Acetate	250
Triphenyl Phosfate	50
Diacetone Alcohol	35
Triacetin	10
Acetone	1250
Aluminum Oxide	35
Asbestos Powder	sufficient to suit
Magnesium Oxide	sufficient to suit

* Bottle Caps, Gelatin

Elastic capsules are formed from a compn. consisting of a basic mixt. of gelatin 400 and glycerol 100 parts to which is added a mixt. of about 0.015–0.02 part of petroleum and about 0.03–0.04 part of benzoin-resin, and the capsules are hardened with a 3–5% formaldehyde soln. to which glycerol and alc. are added.

Gelatin Films, Hardening

Gelatin or other protein layers are rendered very insoluble by treatment with the following:

Formaldehyde	100 cc.
Pot. Carbonate	100 gm.
Water	1000 cc.

* Film, Non-Inflammable

Cellulose acetate 100, triphenyl phosphate 10–20, and diethyl phthalate 10–15 parts are closely combined with the aid of a solvent composed of acetone 85 and alc. 15 parts to form a flowable dope, which is cast on a suitable surface. The excess of solvent is evapd. and the resultant film stripped from the surface.

* Non-Inflammable Film, Prevention of Shrinkage

Such films are treated with dilute ammonia for six hrs. at 30–50° C. to prevent shrinking or wrinkling during development.

* Gelatin Sound Records

A gelatin soln. (which may be hardened with a dichromate) is poured on the original wax master record to form a film which when solid is mounted on a suitable backing such as a metal or celluloid plate. A compn. contg. gelatin 3, glycerol 15 and water 82% may be used.

* Cellulose Composition

Cellulose fibers are beaten in the following solution; the wet mass is shaped in forms and dried.

Magnesium Chloride	12
Magnesium Sulfate	12
Sod. Bicarbonate	12
Borax	2
Salt	2
Water	128

* Ivory, Artificial

Two parts of casein are dissolved in 12 parts of water to which a small addition of sodium hydroxide was added. To this casein solution add then a solution of sodium penta-sulfide which contains 3.7 parts of sulfur in 15 parts of water.

The decomposition of the poly-sulfide with 10% hydrochloric acid goes beyond the formation of colloidal sulfur and results ultimately in a flaky coagel of colloidal sulfur in casing. The acid is removed from the coagel by washing it with water; the greater part of the water retained by the casein is removed by pressing it at a relatively low pressure.

The obtained mass is now homogenized, for instance in a salve mill, and becomes thereby so far plastic that it may be transformed under high pressure into plates or other desirable forms. Application of heat greatly facilitates this transformation by pressure. The shaped masses are then finally dried at gentle heat and are hardened by heating them for 10 hours at a temperature of 90 to 95°.

* Plastic, Molding

500 grams of sugar, preferably brown cane sugar, is thoroughly mixed in about 1000 cubic centimeters of say 40% formaldehyde solution. When the sugar is completely mixed in the formaldehyde solution, heat is applied to the solution until a temperature of approximately 40° C. is attained. The solution is maintained at this temperature for about 10 minutes. After the expiration of this period, about 10 grams of sodium sulfite (Na_2SO_3) is added to the heated solution in small amounts while the temperature of the solution is increased to about 60° C. The temperature is preferably maintained at 60° C. for about 10 min-

utes. When this time has expired the mass of material is ready for use in the production of plastic masses and of cemented products.

The plastic material produced in this example sets and hardens by itself in the open air within a period of about 12 to 24 hours. Under the influence of heat, say at a temperature of 180° to 200° C., the plastic material sets in about 4 to 6 minutes to a relatively hard, strong, tough and resin-like solid.

If the aforesaid plastic material is used as a binder or cementitious substance for fibrous particles, such as sawdust or wood shavings, the plastic material is mixed thoroughly with said particles. The thus-treated mass may be permitted to set and harden in the open air but I prefer to subject the mass to heat and pressure. By applying a pressure of about several hundred pounds per square inch and a temperature of about 180° to 200° C. to the fibrous material containing the plastic substance, a dense hard board is obtained.

Benzyl Cellulose Plastic

Asbestos (Powd.)	300
Chalk (Powd.)	300
Clay (Powd.)	300
Benzyl Cellulose	125

A moulding pressure of 30–60 lb. per sq. in. is used.

---

* Plastic Composition

Ethyl Benzyl Cellulose	100
Triphenyl Phosfate	3
Tricresyl Phosfate	3
Benzol	200

The above is used for dental and other molding or modeling purposes.

---

* Celluloid Substitute (Non-Inflammable)

Camphor	35 parts
Rosin	65 parts

Fuse above at 130°–180° C. into a sticky condition in a closed vessel. The product formed by boiling 20 parts of waste floss silk in a 2% caustic soda solution is mixed in and the whole suspended in sufficient alcohol. Mix heated for 48 hours with 80 parts magnesium carbonate to saponify the rosin and convert the fiber into a state of colloidal solution. Finally the mass is kneaded with a roller at 70° C.

---

* Molding Composition, Thermoplastic

For example, take 100 pounds leather scrap, and soak this in water until the leather becomes softened or plumped. In practice, I ordinarily soak the leather for about 12 hours. I ordinarily add about 2% of urea or sulphuric acid to the water, for soaking or plumping.

After the leather has become sufficiently softened in the soaking vat, I remove it and place it in a steam-jacketed kettle, contained just sufficient water to cover the leather. The leather is then heated at a temperature of approximately 190° F. for about one-half hour. The temperature and the duration of heating should preferably be regulated so as not to produce complete disintegration or breakdown of the leather, which after heating should be gummy, and a sample placed between the fingers should pull out in long, fine threads. In the claims, the term "gum," refers to the tacky mass resulting from treating leather as above described. The mass remains tacky at room temperature.

After heating the leather, any water or solution standing in the kettle is run off.

The gummy leather is then conveniently run into a steam-jacketed mixer, containing the substances to be mixed with the leather gum, and which themselves have preferably been mixed previously to the addition of the gummy leather.

The mixture to which the gummy leather is added is preferably made up as follows: Naphthalene, amounting to 5% to 15% of the leather scrap, 10% for example (10 pounds in the example given above), is placed in the steam-jacketed kettle. There is also preferably added glycerine and ethylene-glycol, the relative amounts being variable and also varying, somewhat with the amount of naphthalene or equivalent used. As a representative proportion use glycerine to the amount of 10% of the weight of the leather scrap (10 pounds in the example given), and 1–2% ethylene-glycol (1–2 pounds in the given example). These are preferably mixed together in the mixer at a temperature of 200° F., and a thorough mixing of the said ingredients performed before adding the leather gum.

The leather gum may now be added to the mixed substances in the mixer and the mixing continued until a uniform mixture of the leather gum with the other substances is effected. When properly mixed, a sample should show a varnish-like homogeneous structure when placed on a glass plate.

Hexamethylene tetramine is preferably

added to the solution; preferably add an amount of this equal to about 2% of the scrap leather (2 pounds in the given example). This is preferably added to the solution or mixture of the gummy leather and the other ingredients in a steam-jacketed kettle, and the mixing continued to thoroughly incorporate the hexamethylene tetramine.

The contents of the mixer are then dried to expel moisture, and preferably broken up or pulverized. For drying, preferably use a vacuum drier, and temperatures of 170–180° F.

Fillers and coloring material may be added to the composition, either before or after drying; preferably the coloring material and fillers are added to the composition while the composition is in a plastic state in the steam-jacketed mixer.

The usual fillers, used in thermo-plastics, such for example as wood flour, asbestos, paper pulp, ground cork, etc., may be used.

The composition may be molded in heated molds. In molding articles with this composition, the usual temperatures of 300–350° F. or thereabout, and pressures of around 2,000 pounds per square inch, may be used.

The composition will soften in the molding dies, take the form and polish of the dies, and also undergo a change; and under the heating set or cure. When subsequently subjected to heat, the molded composition does not again become plastic.

The finished molded product is tough, possesses a good appearance, takes a polish from the mold, has high tensile strength and compression strength, and good di-electric properties and separates well from the mold. It is also water repellant.

---

* Thermoplastic

SeS_2, formed from 2 mols. of S and 1 mol. of Se by heating above the m. p. of Se (217°), is cooled, ground to a powder, fused at 125° with a filler (*e.g.*, asbestos, slate, Fe oxide, talcum, etc.), cooled, pressed cold and then converted into the hard state by subsequent curing for ½ hr. at 80–90°. The sulfide also acts as an excellent binder for cloth and paper and may be used for forming gears and insulating strips.

---

* Thermoplastic Composition

A.	Cellulose Acetate	100
	Chloroaniline	20–40
	Tritolyl Phosfate	10–15
B.	Cellulose Acetate	100
	Acetyl-o-anisidine	20–40
	Tritolyl Phosfate	10–15

Wax, Dental Impression

An impression material is prepd. b mixing and heating together a minera and drying oil mixt. 2.5–4.5, a beeswa and paraffin mixt. 1.5–2.5, Al stearat 2.5–3.5, rubber, gutta-percha or balat not more than about 0.06, starch 0.5–1. and glycerol not more than about 0.12 part.

---

* Plastic Insulation

A compn. which is waterproof, resis ant to acids and alkali and has elec. in sulating properties is composed of a unsaponifiable wax, such as paraffin wax ceresin or ozocerite, and rubber, gutta percha or balata, mixed to form a homo geneous mass. The wax forms 25–75% of the compn.

---

* Glass Substitute, Flexible

A transparent material which may re place glass for many purposes is mad by heating water (100 parts) to 45–50° adding gelatin (140), alc. (240), glycero (25), AcOH (25) and formol (30), stir ring to complete soln. and drying at be low 50°.

---

* Waterproof Plastic Coating

The following composition may be ap plied hot to waterproof cement, concrete etc.

Cumarone	100
Carnauba Wax	10
Rezinel No. 2	5
Marble Dust	to suit

Glue Composition

Indestructible mass for the manufac ture of ornaments, toys, etc. A har mass consists of 50 parts glue, 35 wa or rosin, 15 glycerine, and required quan tity of a metallic oxide of mineral colo A soft mass consists of 50 parts glue, 2 glycerine, 25 parts wax or rosin. Glu is melted in glycerine with the assistanc of steam and the wax or rosin added Mass poured in liquid state into moulds Degree of hardness of mass is increase by the addition of 30 to 35% zinc white

---

Printers Roller

A soft printer roller compound: Gelatine 32 parts, glue 4, softened in cold water and melted. To this add 4 glucose, 72 glycerine and 1 oz. methylated spirit. Whole mixed and cast in rollers. This is unaffected by temperature, retains its elasticity and does not shrink. Add formaldehyde to make glue insoluble in H_2O.

Composition for Printing Rollers

Ingredients	Composition "A"	Composition "B"
Glue	10 lb.	32 lb.
Molasses	0 lb.	12 lb.
Sugar	10 lb.	0 lb.
Glycerine	12 lb.	56 lb.
Isinglass	1½ oz.	0
India Rubber in Naphtha	0 lb.	10 lb.

PLATING

PREPARATION OF METALS FOR ELECTROPLATING

For the production of impervious adherent metal electrodeposits, the preparation of the articles for plating is of the greatest importance.

1. Polishing and Buffing

No general procedure can be given for all objects due to the large number of factors to be taken into account, such as composition of the object, shape, size, plate and surface finish desired, etc. The directions given here will be of a general nature, with some specific procedures for the common base metals iron and steel, and copper and brass. Treatises on the subject should be consulted for further information on these and other substances.

Naturally the smoothness and polish of the finished plate is greatly influenced by the same properties of the object before plating, particularly if the plate is thin, as is usually the case. Therefore, proper attention must be given to the operations of polishing and buffing the object before plating, and in some cases afterwards. The particular choice of cutting and finishing tools, abrasives, etc., is determined by the metal, the degree of finish on the final surface, etc.

For objects covered by a considerable amount of rust or millscale, sand-blasting or sand-rolling will greatly reduce the labor required for the final polishing. In sand-rolling the objects are rolled in steel barrels with abrasives such as sand, alundum, carborundum or emery mixed with water or oil. Where the number of objects is small a steel wire brush is best for removing coarse scale.

A certain amount of polishing should be used in all cases before plating, whether a high luster is desired or not. This is because the surface will be rendered more uniform, which will improve the quality of finish and corrosion resistance of the final plate. However, the polishing and subsequent treatments must be carefully studied and controlled in order not to weaken the surface layers with subsequent peeling after plating.

Under ordinary circumstances finishing is a two-step operation: "cutting down" to produce a smooth surface and "coloring" to produce a high final luster. It is often possible by proper choice of cloths, abrasives, speed of wheel, etc., to accomplish this with but two wheels, one for each step. However, in some cases more wheels are necessary for hard metals containing deep scratch marks, especially in the cutting-down step. Materials used for the wheels include muslin, flannel, felt, canvas, brushes of various kinds, leather and wood depending upon the nature of the material being polished, the coarseness of the abrasive, the finish desired, the preference of the polisher, etc. The abrasive composition is of much greater importance, since it is the medium doing the actual work. Excessive wheel wear means that the wheel is doing the work rather than the composition, and is due to improper choice or insufficient amount of composition. For efficiency the wheels are run at the maximum allowable speed. In some cases the limit is set by the material of or composition on the wheel, and in others by the material being buffed. Thus in the cutting-down step, where the abrasive is held on by glue a speed higher than 7,500 surface feet per minute will soften the glue and allow it to be torn from its setting on the wheel face. For soft metals on the other hand a speed this high generates enough heat to soften the metal and cause it to flow.

The first or cutting down step (often called simply polishing) is done by wheels faced with abrasive and glue. The abrasives used are either emery or artificial alumina, the latter being usually more desirable for most purposes. The glue should be the best quality hide glue; high viscosity, strength and flexibility being of prime importance. Application of the abrasive composition to the wheel is by rolling the wheel in a warm glue abrasive mixture and allowing to dry. If run at high speeds, polishing wheels should be faced with tallow to prevent burning.

The second or coloring step (often called simply buffing) is done by wheels faced with abrasive and grease. The abrasives used are of all kinds and grades, lime, silica, tripoli, emery, rouge, etc., being used. The melting point of the grease used will depend on the speed, a hard, high melting point grease being selected for buffing at high speeds. The grease should be of the saponifiable variety, because of the easier and quicker removal by alkaline cleaners.

For steel containing mill marks on which a high final luster is desired, the following combinations are suitable.

For very deep mill marks, two canvas wheels faced with glue and abrasive should be used. Suggested abrasive sizes are 120 and 220 mesh. These should be followed by one or two buffing steps on cloth wheels, depending upon the final finish desired.

In cases where the object is not deeply scratched to begin with, the following three-wheel combination offers advantages. One canvas wheel faced with glue and 180 mesh abrasive; one tampico brush wheel faced with fine emery paste; and one cloth wheel faced with chrome or steel rouge. The brush wheel offers the advantages of reaching backgrounds that cannot be reached with the usual polishing wheel, and of not requiring the glue-dressing step needed for the latter.

In going from one wheel to the nex the object should be rotated 90°, so th the new scratch marks are perpendicul to the old ones. The object must be ke on any one wheel until all the scrat marks of the previous step have be eradicated. If this takes an excessive long time, another wheel with an inte mediate grade of abrasive should be use

After polishing, the next step and t one of greatest importance is the clea ing of the article to be plated. The fo eign materials likely to be present metallic surfaces are of two classes: firs grease, dirt and organic substances; a second, oxides, scale, tarnish, and rust.

B. Removal of Grease

Grease of all kinds whether saponifiab or not can be removed by solution i organic solvents. In cases where the o jects are heavily coated with grease, cheap organic solvent such as gasoline, better a non-inflammable one such as ca bon tetrachloride or mixture containin it, should be used. However, this will n give complete cleansing, as the solvent o evaporation will leave a thin film grease, making another operation such dipping into fresh solvent necessary. Th latter is obviated in a recently designe apparatus, where the articles are su pended in the vapor above a boiling a paratus. The condensing solvent wash them free of grease, and since it is bein continually distilled, no second step necessary. A non-inflammable solve must be used in this case—trichlor ethylene has met with considerable fav recently because it does not hydrolyze a readily as carbon tetrachloride in th presence of moisture.

The common method of removin grease is by emulsification with alkalin solutions, which should be used as hot a possible. The detergents used in thes solutions are soap of all kinds, causti soda and potash, soda ash, trisodiu phosphate, sodium metasilicate, sodiu cyanide, borax, sodium sesquicarbonat sodium aluminate, etc., and all kinds o mixtures thereof. Sometimes finely d vided insoluble substances such as silic alumina, etc., are added. These are n fillers but help to clean either by scou ing of the surface or by adsorption o the dirt. Each plater, seller of platin supplies, etc., has a particular composi tion and procedure that he swears b Since the kind and degree of contamina tion of metallic surfaces vary conside ably in different plating shops, naturall certain particular mixtures used in co junction with a specific procedure wi ean more quickly than others. However, obably any hot alkaline solution will ork if given sufficient time. In general ther soap with one builder (alkaline lt) or a mixture of two alkaline salts used. The soap should be of a very luble variety so as to be quick and free nsing; fish oil soaps have been found ry satisfactory. Soda ash has been ed in the past as an alkaline soap ilder because of its cheapeness. Even day practically all commercial cleaners ntain much soda ash. However, it is ing gradually replaced by the more ficient detergents trisodium phosphate d sodium metasilicate. These seem to t more quickly not only because of gher alkalinity, but also due to specific nulsifying action. Caustic soda is used many mixtures; it cleans not only by s emulsifying action, but also by ponifying the fats present on the metal. Since any alkaline solution will have me saponifying action, the greases used the manufacturing and polishing opertions should be of the saponifiable ariety.)

Electrolytic cleaning is frequent prac-ce in plating shops. In this method an ectric current is passed through the bject, which is made one electrode in a ot alkaline solution. Usually the object made the cathode, both because of the reater gas evolution (hydrogen) which ives a scouring action, and the higher ee alkali concentration giving an in-reased cleaning action. Furthermore, as athode metals will not dissolve and some eduction of the oxides on the surface ay take place. The voltage applied hould be sufficient to produce a current ensity of 10 amp. per sq. ft. (1 amp. er sq. dm.) or greater. Any of the olutions used ordinarily for cleaning ay be employed; the alkali or alkaline alt content should be high to give good onductivity. Cleaners containing sus-ended solids should be avoided, as solids re often occluded to an electrode during lectrolysis. Iron bars or the containing ank may be used as anodes.

Special procedures must be used when he objects contain aluminum, zinc, tin r lead. For ordinary cleaning caustic oda or potash must be avoided as these ubstances will dissolve. In cathodic lectrolytic cleaning these will dissolve to ome extent in any case whether caustic s added or not, due to the formation of ree alkali at the cathode. Sometimes mall amounts of the zinc, tin or lead nay be redeposited from such cleaners, giving a film which will cause subsequent peeling of the electrodeposit. In such cases the object should be made the anode for short time, either in the same or in a separate bath. An alternative procedure is to use anodic cleaning. The mechanism of anodic cleaning is quite different from that of cathodic. In the latter, as stated above, the action is due to the bubbles of gas and the increased alkali concentration. However, with anodic cleaning the action is largely due to the etching (solution) of the surface. Since the impurities are on the surface only, they will thus drop off. Anodic cleaning is often used for brass and copper. Zinc should not be cleaned anodically as it is attacked so rapidly the surface blackens due to the finely divided metal formed.

A simple cleaning bath base may be made of the following:

8 oz. per gal. (60 g. per l.) Soda Ash (anhydrous sodium carbonate) (Na_2CO_3)

or

22 oz. per gal. (165 g. per l.) Washing Soda ($Na_2CO_3 \cdot 10H_2O$)

16 oz. per gal. (120 g. per l.) Trisodium Phosphate ($Na_3PO_4 \cdot 12H_2O$)

or

4 oz. per gal. (30 g. per l.) Sodium Metasilicate ($Na_2SiO_3 \cdot 5H_2O$)

To this should be added 1–2 oz. per gallon of soap and 1–2 oz. per gallon of caustic soda. If used electrolytically, most or all of the soap should be eliminated—0.1 oz. per gallon is sufficient.

For large scale production a double system will be found desirable. The greater part of the grease by solvent dip or by a strong hot soap solution; and then the object put into the electrolytic cleanser. Usually 3–4 minutes of the electrolytic cleaning is sufficient. When clean there should be a continuous film of water left on the object. Rinse thoroughly before proceeding with the pickling.

C. Removal of Oxides and Tarnish

Oxides, scale and tarnish are usually removed by solution in a suitable reagent, the process being usually called pickling. For iron and steel, sulfuric or hydrochloric acid is used; and for copper and brass sulfuric and nitric acids.

If the copper or brass is polished and clean, a short immersion in a "bright dip," composed of 425 ml. conc. H_2SO_4

and 75 ml. conc. HNO_3 in 500 ml. water is sufficient. For brass with appreciable amounts of oxide scales, a preliminary "scaling dip" in a solution composed of 375 ml. conc. H_2SO_4 and 75 ml. conc. HNO_3 in 550 ml. water should be used. The brass is dulled by the latter process and should subsequently be immersed in a bright dip.

For large scale treatment of iron and steel, sulfuric acid should be used because of its cheapness. The proper concentration is about 10% by weight (1 part conc. H_2SO_4 by volume to 16 of water). For smaller jobs hydrochloric acid is to be preferred because of its more rapid action. The concentration should be 7% by weight (5 parts commercial hydrochloric acid by volume to 32 of water). The time taken will depend naturally upon the amount of scale present and will vary from several minutes to an hour. These acids act not only by actual solution of the oxide, but also by attack of the metal with evolution of gas, which helps detach the scale. For objects with imbedded sand (from castings or sand blasting) hydrofluoric acid should be added to 4% by weight (1 part commercial hydrofluoric acid by volume to 16 of water). This will dissolve the silica.

After pickling thoroughly rinse the object and immerse immediately in the plating bath with the current on. The latter precaution is particularly important for acid plating baths to avoid partial solution of the metal before the current starts to flow. The exposure to the air of the prepared object should be a minimum, because the surface is unusually clean and particularly susceptible to oxidation.

D. Combination Procedures and Special Processes

In many cases some of these cleaning procedures can be combined or shortened. Thus if the metal has been highly buffed, the pickling step can be omitted. The oxides have been removed during buffing, and further oxidation prevented by the grease of the buffing composition. This grease may be removed either by solvent treatment of alkaline cleaning. Often a single solvent dip alone is satisfactory if the object is to be chromium plated, because the strongly oxidizing chromic acid bath will oxidize the traces of grease remaining. However, in some cases unsuccessful adhesion of the deposit occurs with this simplified treatment. This may be due to the presence of absorbed matter which is not removed by the solvent. In such cases the alkaline cleansers may yield better results, or a light scrubbing of the surface with Vienna lime may hel

In preparing highly polished brass f plating, the pickling step may be di pensed with by the addition of sodiu cyanide to the alkaline cleansing bat This will dissolve the traces of oxides an tarnish present. Cyanides should not b used for copper, as a film is formed whic is very difficult to wash off.

The pickling step induces the followin detrimental factors when used on ir and steel:

(1) Formation of surface carbon pr venting adhesion of the plate.

(2) Formation of hydrogen on the su face, which is occluded and adsorbed pr venting adherence and causing brittlenes The factors have caused the failure c plates (especially nickel) often in th past. The remedy found in recent yea (Madsenell process—patented) is degas fication. After pickling the metal to b plated is made the anode on a 12-vo circuit in concentrated sulfuric acid a room temperature. Usually a lea cathode is used. The current starts a about 5 amp. per sq. dm. and subside over a period of from 30 sec. to 10 mi to practically zero, when evolution of ga ceases. By this process the occluded an adsorbed gases and embedded oils an greases are removed. Although a passiv film of metal is probably formed, thi does not seem to be detrimental to th adhesion of the plate. An alternativ method is to use solutions of dichromate or chromic acid; old chromium platin baths serve admirably.

PLATING ROOM PROCEDURES

Nickel Plating—Still Tanks

Nickel solution:

Nickel Ammonium Sulphate	8 oz. per gal.
Nickel Sulphate	4 oz. per gal.
Boric Acid	2 oz. per gal.

Ph. value of above solution is kept a 5.8; nickel content, should be 3½ oz nickel per gal. Tanks used at room tem perature. Additions for nickel are mad by adding double nickel salt according t analysis shown. Practice is about 5 lb every ten days. Nickel anodes should b 99 plus, and maximum copper conten .30%. Amperage and voltage is limite to type of work, usually about 25 ampere and 6 volts for one hour.

Machine Nickel Plating

Nickel Sulphate	4 oz. per gal.
Nickel Ammonium Sulphate	12 oz. per gal.
Magnesium Sulphate	2 oz. per gal.
Boric Acid	3 oz. per gal.

Black Nickel Finish

ormula

Nickel Ammonium Sulfate	8 oz.
Sodium Sulfocyanate	2 oz.
Zinc Sulfate	1 oz.
Water	1 gal.

rocedure for Plating

Work is strung on racks.

Hung on mild alkaline solution to re-ove grease.

Wash in water.

All above work is done in the dip room.

The following work is finished in the uff room plating department.

Bright dip work is washed in milk alka-ne solution again before going through e following operations.

Buffed parts to be plated are first ipped and brushed with gasoline and ried in sawdust, after which they are ipped and brushed with milk alkaline olution.

Wash in water.

Dip in cyanide solution.

Washed with water.

Plate in black nickel solution.

Wash in cold water.

Wash in hot water.

Bright dipped parts are dried in saw-ust. Buffed parts are dried in hot box.

Cadmium Plating

ormula:

Sodium Cyanide	9 oz.
Cadmium Oxide	3 oz.
Sodium Hydroxide	2 oz.
Water	1 gal.

Use at room temperature using 8 to 10 mperes per sq. ft.

rocedure for plating:

Very greasy work is washed in gasoline nd dried in sawdust.

Wash and brush in milk alkaline olution.

Wash in water.

Dip in Muriatic Acid.

Wash in water.

Wash and brush in milk alkaline olution.

Wash in water.

Dip in Cyanide.

Wash in water.

Plate in cadmium solution from 20 inutes to 1½ hours depending on type of work and quantity of cadmium desired.

Wash in cold water.

Wash in hot water.

Dry in sawdust or hot box whichever the type of work requires.

Some work is rubbed with steel wool to brighten the metal finish.

Silver Plating

Silver bath formula:

Silver Cyanide	3½ oz.
Sodium Cyanide	5 oz.
Water	1 gal.

Silver strike formula:

Silver Cyanide	½ oz.
Sodium Cyanide	8 oz.
Water	1 gal.

Procedure for Plating:

Wash and brush in milk alkaline solution.

Wash in water.

Dip in Cyanide solution.

Wash in water.

Flash in silver strike at 6 volts.

Plate in silver bath for 30 min. at 2 volts.

Wash in cold water.

Wash in hot water.

Dry in hot box.

Stripping Solution

Stripping solution:

Whale Oil Soap	4 oz. per gal.
Sodium Hydroxide	8 oz. per gal.
Sodium Silicate	4 oz. per gal.
Cyanide Solution	2 oz. per gal.

Use at 212 degrees F.

Nickel Strip Solution

Nitric Acid	1 part
Sulphuric Acid	2 parts

Use at 40 degrees F.

Electric Cleaner

Mild Alkaline Solution 8 oz. per gal.

This solution is used with an E. M. F. of 6 to 12 volts, on work requiring exceptionally clean surface. It can be augmented by addition of stronger detergents but care must be used to prevent staining of colored work. Use at 200 degrees F.

Bright Dip

Sulphuric Acid 66° Baumé	68 oz. per gal.
Nitric Acid 42° Baumé	20 oz. per gal.

Hydrochloric Acid 24° Baumé	14 oz. per gal.
Water	40 oz. per gal.

Use at 40° F.

Blue Dip

Copper Carbonate stirred into concentrated ammonia until saturated. Use at temperature of 60 degrees C. Procedure for blue dipping brass is simply clean in potash, bright dip, blue dip, and hot water dry.

Brass and Bronze Plating

Formula for brass solution:

Copper Cyanide	4 oz.
Zinc Cyanide	1 oz.
Sodium Cyanide	6 oz.
Sodium Carbonate	2 oz.
Water	1 gal.

Temperature 90° F. Cathode current density 2.5 to 3 amperes per sq. ft.; 2 to 3 volts. Use rolled anodes, 80% copper, 20% zinc.

This solution will produce a good yellow deposit. If a green deposit is desired, for instance, such as is used for a flash deposit, in the novelty trade, previous to gold plating, use 1 ounce less of each, copper cyanide and sodium cyanide, and a small quantity of ammonium hydroxide.

As temperature plays a very important part in controlling a uniform deposit, it is advisable to have the tank equipped with a steam coil for proper regulation.

In operating a brass solution, it is well to keep in mind that a high current density tends to produce a deposit that is high in zinc; also, that the addition of ammonia or caustic soda to a brass solution has the same effect.

Bronze solution:

''Bronze plate'' (really a high-copper brass deposit) is generally produced in an alkaline solution, one similar to a brass solution, but with a higher copper content.

Copper Cyanide	4 oz.
Zinc Cyanide	½ oz.
Sodium Cyanide	5 oz.
Sodium Carbonate	2 oz.
Rochelle Salts	2 oz.
Water	1 gal.

Temperature 95° F. Cathode current density, 2 to 2.5 amperes per sq. ft.; 2 to 3 volts. Rolled bronze anodes, 90% copper, 10% zinc.

Temperature always plays a very important part in the control of this solution, so the tank should be equipped with a steam coil to keep the temperature constant. When rochelle salts are added to a bronze solution, better anode corrosion is obtained, and therefore, a more uniform deposit.

In replenishing the metal content of a brass or bronze solution, it is not advisable to make a stock from copper cyanide zinc cyanide and sodium cyanide, as it would be impossible to control the constituents in their proper proportion to produce a uniform color in the deposit A separate stock solution of the zinc salt and copper salt is recommended. They should be prepared by dissolving equal parts of copper cyanide and sodium cyanide, and zinc cyanide and sodium cyanide in water and placed in separate containers until wanted for use.

It is a known fact that when a zinc salt is added to a brass or bronze solution (and especially the latter), it takes considerable time before a uniform color of the deposit is obtained. This is probably due to the difference in potentials at which the two metals are deposited. It is by the formation of the double cyanides that it is possible to deposit these two metals from the same solution in different proportions.

Remarks on Brass and Bronze Solutions

Rochelle salts, when added to a brass or bronze solution, have the property of dissolving the oxides that form on the anodes, thereby permitting a more uniform deposit. One to two ounces per gallon is to be recommended.

It should be remembered that the factors that tend to make the zinc predominate in the deposits are a high zinc content, high current density, low free cyanide content, decrease in temperature, and the addition of ammonia or caustic soda to the bath.

When arsenic is added to a brass solution to produce a bright deposit, care should be used to avoid an excess as a light colored deposit will be the result To prepare the arsenic stock solution take two pounds of caustic soda and dissolve same into two quarts of cold water Then add one pound of white arsenic and when all has been dissolved, dilute to one gallon. One ounce of this stock solution is enough to add to each 100 gallons of solution. It is impossible to bright dip a piece of work that has been plated in a brass solution that contains an excess of arsenic. Arsenic should never be added to a bronze solution; neither should ammonium salts be added.

The free cyanide of a bronze solution is usually less than that of a brass bath The color desired should be regulated by the proportion of the copper and zinc salts used and the temperature at which the bath is operated.

Brass Plating on Steel (for rubber adhesion)

Copper Cyanide	4 oz.
Zinc Cyanide	1 oz.
Sodium Cyanide	6 oz.
Carbonate of Soda	2 oz.
Water	1 gal.

Temperature 80° F. to 85° F.

Cathode current density, 2.5 to 3 amps. per square foot. Rolled anodes should be used consisting of 80% Copper and 20% Zinc.

The work must be perfectly clean and it is necessary to maintain a regulated temperature and current density.

Bronzes, Restoration of Ancient

This article is made the cathode in 2% NaOH soln., and a weak current is passed for some hrs., a sheet-iron anode being used. In this way the incrustration is reduced again to metallic Cu, and the outer layers of dirt and loose sponge Cu are then readily removed by gentle brushing, this leaving a clean surface which usually shows all the original surface details. Malignant patina is due to the presence of Cu oxychloride in the corrosion products; the above electrolytic process effectively eliminates the patina, especially when the malignant salts impregnate the mass of the bronze. Another method which gives satisfactory results is to brush the parts affected with dil. Ag_2SO_4 soln., which converts the chlorides into insol. AgCl after being dried with blotting-paper, the surface is brushed with $Ba(OH)_2$ soln., which is allowed to dry, leaving a white powder, which is readily brushed away.

Plating Cadmium

For general purposes a soln. contg. Cd oxide 3.5, NaCN 10, Na_2SO_4 4.2, Ni sulfate 0.08 and lignin sulfonate 1% is recommended; for very bright plates the above figures should be modified to 6, 16, 6.6, 0.13–0.21 and 1.6%, resp. Both baths are operated at 15–50 amp./sq. ft., and at 25 amp./sq. ft. have a cathode current efficiency of 96%. Lime is said to be the best reagent for removing accumulations of Na_2CO_3.

---

* Plating Bath, Cadmium

Cadmium Hydroxide	48
Sod. Cyanide	120
Sod. Sulfate	60
Nickel Sulfate	1.5
Turkey Red Oil	12

Cadmium Plating

Formula for cadmium solution:

Sodium Cyanide	9 oz.
Cadmium Oxide	3 oz.
Caustic Soda	2 oz.
Water	1 gal.

Temperature 80° F. Cathode current density, 8 to 10 amperes per sq. ft.; 2 to 2½ volts. Use iron and cadmium anodes; one iron to three cadmium.

Remove cadmium anodes when solution is not in use.

In making the solution take ½ of the sodium cyanide, dissolve in hot water and then add the cadmium oxide. Dissolve balance of the sodium cyanide and caustic soda and add to the solution. Dilute with water to full volume.

The free cyanide content is a very important factor. It should equal the metal content, and for barrel plating it should be considerably higher.

Barrel plating cadmium solution usually contains twice the amount of chemicals used in the still solution.

Copper Plating

There are two types of solutions that are used for the deposition of copper, namely, the acid (sulphate) and the alkaline (cyanide) baths. Their use is dependent upon the class of work to be plated and the finish desired.

The cyanide solution is always used for depositing copper upon the ferrous metals, so as to prevent the deposition of copper by immersion which would be the result of the use of the acid bath on this class of work. There are two formulae for the cyanide solution, either of which will give satisfactory deposits—carbonate or cyanide.

Cyanide copper solutions:

Copper Cyanide	3½ oz.
Sodium Cyanide	4½ oz.
Carbonate of Soda	2 oz.
Hyposulphite of Soda	1/32 oz.
Water	1 gal.

Copper Carbonate	5 oz.
Sodium Cyanide	10 oz.
Hyposulphite of Soda	1/32 oz.
Water	1 gal.

Either solution should be operated at 100° F. to 110° F. Cathode current density 4 to 6 amperes per sq. ft., 1½ to 2 volts. Use rolled copper anodes. The free cyanide content of the bath should not be allowed to rise too high or else gassing will be produced at the cathode causing a blistered deposit. Enough cyanide should be used to keep the anodes fairly clean from the formation of basic copper salts, but not enough to prevent the dark discoloration which is produced by the use of the hyposulphite of soda. This discoloration usually disappears when the current is off for a few hours.

If the cyanide solution is operated at room temperature, a higher free cyanide content is necessary than at 110° F. With a metal content of approximately 2.50 oz. of metallic copper per gallon and operated at room temperature, a free cyanide content of 1 to 1.25 oz. per gallon will produce good results. If operated at 110° F. use a free cyanide content of .50 to .75 oz. per gallon.

Pitted deposits of copper are caused when the carbonate content becomes too high. When this occurs the carbonates may be precipitated from the solution by the addition of barium chloride. The precipitated carbonates are allowed to settle, the solution syphoned off, the carbonates removed from the tank, the solution is then replaced in the tank which is filled with water to proper solution level when the solution is ready for use.

It is not advisable to remove all of the carbonates, for without any carbonates a hard deposit will be produced.

Acid copper solution:

Copper Sulphate	28 oz.
Sulfuric Acid	3 to 5 fl. oz.
Water	1 gal.

Temperature 75° F. Cathode current density for still solution 10 to 15 amperes per sq. ft.; ¾ to 1 volt. Agitation of the cathode or of the solution allows the use of higher current density. Use rolled copper anodes.

Remarks on Copper Solutions

Bright deposits of copper from the cyanide solution may be obtained by adding to the bath lead carbonate which has been dissolved in a caustic soda solution. Agitation of the cathode is also necessary. The deposit from newly prepared cyanide solutions is usually hard and at times blistered. The addition of one or two ounces per gallon of caustic soda helps to overcome this condition.

Oxidized finishes are hard to produce uniformly from a cyanide solution that contains hyposulphite of soda.

More uniform bronze finishes are produced from an acid copper deposit. An excess of sulphuric acid in the acid solution produces a deposit that is hard and streaky; so will an exessive current density. The higher the sulphuric acid, the greater the conductivity of the bath.

A high acid content is indicated by the formation of copper sulphate crystals, especially when the temperature of the bath is below normal.

Coppering by immersion:

Copper Sulphate	1 to 2 oz.
Sulphuric Acid	½ to 1 oz.
Water	1 gal.

Where only a very thin film of copper is desired, the above solution will give good results. The work is free from grease by the usual cleansing methods and then immersed in the solution just long enough to become coated with copper. Rinse thoroughly in clean cold water and dry in sawdust.

---

* Copper Plating Bath

The bath contains $NaCu(CN)_3$, free NaOH, and Na K tartrate or citrate instead of free NaCN or its equiv., *e.g.*, $Cu(CN)_2$ 7.5–15, NaCN 3.2–7.5, NaOH 15–30, Na K tartrate 22–120 g. per liter. It is operated at 80–100°, using current at 6 volts.

Copper Electrotyping

The prepared graphited wax cases are "oxidized" and kept in starting tank for 2–5 minutes. They are then copperplated in

Copper Sulfate	210 gm. per liter
Sulfuric Acid	75 gm. per liter

Current density 110–140 amperes; 6 volts; temperature 85° F.

Copper Plating Glass

The following method is used for depositing silver upon glass, after which the silver may be copper plated:

The articles are freed from oil or grease, and placed in a dilute hydrofluoric acid solution to roughen the surface slightly; then rinsed in clean cold water; then they are ready for the silvering operation for which two solutions are necessary.

Solution No. 1.—Dissolve 90 grams of sugar in 250 c.c. of distilled water; add 4 c.c. of c. p. nitric acid and 175 c.c. of ethyl alcohol. Make up to 1 liter by diluting with distilled water.

Solution No. 2.—Dissolve 1.8 grams of silver nitrate in 100 c.c. of distilled water, and add ammonia drop by drop until the precipitate which forms is nearly redissolved; then add 0.9 gram of potassium hydroxide dissolved in 25 c.c. of water; and again nearly redissolve the precipitate by the addition of a few drops of ammonia.

Take 1 part of No. 1 solution and 9 parts of No. 2 solution; mix together thoroughly; and immediately immerse the glass articles into this mixture. The surface will be covered with a deposit of silver.

The deposit is quite adherent, and is a base for heavy deposits of silver or copper to be put on by electroplating.

Copper Plating Aluminum

The metal is cleaned with 10% NaOH saturated with NaCl, washed, dipped in 2% HCl, coated anodically (20–25 amp. per sq. dm. at 50–60 volts with electrode separation 6 cm. for 10 sec.–2 min. in 10% aq. $H_2C_2O_4$ or $CH_2Cl{\cdot}CO_2H$), washed, treated with Na_2CO_3 and $NaHCO_3$ (23 and 45 g. per liter) at 90–95° during 10–20 sec., and then coated with Cu from a neutral $CuSO_4$ or KCN–Cu bath.

Metalizing Non-Metallic Articles

Plastics, bone etc., are washed with naphtha to remove grease; dried and soaked in 3–4% aqueous quinol; then immersed in a solution of silver nitrate. Silver is deposited which may be polished. Other metals may be then plated thereon.

Gold Plating

1. Cyanide solution:

Metallic Gold as Fulminate or Cyanide	5 dwt.
Sodium Cyanide	2 oz.
Phosphate Soda	1 oz.
Water	1 gal.

Temperature 130 to 160° F.; 1 volt; 24 kt. gold anodes.

2. Chloride solution:

Gold Chloride	6 oz.
Hydrochloric Acid	10 oz.
Water	1 gal.

Room temperature; 2 to 3 volts.

In preparing the solution dissolve the gold chloride in dilute hydrochloric acid before adding it to the solution. The amount of free hydrochloric acid that the solution contains does not seem to make a great deal of difference in the operation of the bath, but it does have a decided effect upon anode. The greater the amount of free acid the faster the anode dissolves.

This solution is used where heavy deposits of gold are desired. The work is plated in the cyanide bath for a few minutes before placing in the acid bath.

3. Immersion gold solution:

Fulminate of Gold	4 dwt.
Yellow Prussiate Potash	12 oz.
Carbonate Soda	24 oz.
Caustic Soda	¼ oz.
Water	1 gal.

Solution should be boiled in a cast iron tank for an hour and allowed to cool to 180° F. before using.

If color is too light, it may be darkened by adding a very small amount of copper carbonate which has been taken up with yellow prussiate of potash.

4. Salt Water gold:

Yellow Prussiate of Potash	64 oz.
Sodium Phosphate	32 oz.
Sodium Carbonate	16 oz.
Sodium Sulphite	8 oz.
Gold as Fulminate	12 dwt.
Water	4 gal.

Boil for an hour and add to solution as required.

Solution is boiled for one hour, then diluted with water to make four gallons of solution. The solution is placed in a porous pot which is put in a tank that contains a saturated solution of sodium chloride heated to 190° F.

The porous pot is surrounded with a cylinder of zinc which is provided with a rest rod, on which the work to be plated is suspended in the gold solution.

The advantage of this type of solution over the cyanide solution is that a more uniform color may be obtained, although the deposit is not as rapid as with the cyanide solution, unless used with outside current. This is accomplished by connecting the zinc cylinder with the positive lead from the generator and the work rod with the negative lead. The amount of voltage is regulated with the class of work being done. If the work is wired or racked, 1 to 2 volts is sufficient. If basket work is being done, 5 to 6 volts give good results.

The solution is replenished from a stock solution:

Yellow Prussiate of Potash	16 oz.
Sodium Phosphate	8 oz.
Sodium Carbonate	4 oz.
Sodium Sulphite	2 oz.
Gold as Fulminate	1 oz.
Water	1 gal.

Green gold:

Metallic Gold as Fulminate or Cyanide	4 dwt.
Silver Cyanide	¼ dwt.
Sodium Cyanide	2 oz.
Water	1 gal.

Temperature 105° F.; 2 volts; 18 karat green gold anodes.

Dark or antique green gold solutions are produced by adding to the green gold solution a small quantity of lead carbonate that has been dissolved in caustic soda, and increasing voltage to 5 or 6. Agitation of the work produces best results.

White Gold

White gold and other karat gold solutions are best prepared by running the gold into solution with the porous pot method. This consists of making a cyanide solution of four ounces to a gallon of water which is to be the plating solution. Connect up tank for plating in the usual way. Place anodes on anode rod and on cathode rod suspend a porous pot which contains a fairly strong solution of sodium cyanide, 4 to 6 oz. per gallon. Into the porous pot suspend a sheet of copper, or better still a copper rod formed into a coil, and operate solution until the desired amount of gold has been dissolved from the anode. This can be readily determined by weighing the anode from time to time.

Rose gold solution:

Yellow Prussiate of Potash	4 oz.
Potassium Carbonate	4 oz.
Sodium Cyanide	¼ oz.
Gold as Fulminate	10 dwt.
Water	1 gal.

Temperature 175° F.; 6 volts. If a red color is desired, add small quantity of copper carbonate.

Cheap rose gold finish:

The work which must be brass is placed in the following dip until a smut is produced:

Copper Sulphate	16 oz.
Muriatic Acid	½ gal.
Water	1 gal.

Dissolve the copper sulphate in the water and then add the acid. The work should have a deep red smut which should be lightened somewhat by placing in a saturated salt solution for a few seconds. Plate in the regular fine gold solution, then relieve the high lights with bicarbonate of soda, replate in gold solution for a few seconds, dry and lacquer.

To remove fire scale after soldering on solid and karat gold, the work is pickled in a dip composed of: sulphuric acid 12 ounces, sodium bichromate 4 ounces, water 1 gallon; used hot.

It is then made the anode in the following solution:

Yellow Prussiate of Potash	2 oz.
Sodium Cyanide	8 oz.
Rochelle Salts	2 oz.
Water	1 gal.

Temperature 150° F. to 175° F., 6 volts, and lead cathodes.

Gold-Plating, Simple

The article to be plated, after being cleaned thoroughly is dipped into the following which has been previously boiled for an hour or so. This solution operates best at 140–150° F.

Yellow Prussiate of Potash	24 oz.
Sod. Carbonate	12 oz.
Caustic Soda	¼ oz.
Iron Sesquichloride	⅛ oz.
Gold Fulminate	3 pwt.
Water	1 gal.

When color of deposit becomes too red it is fortified by the addition of gold fulminate and boiling for an hour or so before use.

Gold (Colored) Plating

A.—Formula for rose gold solution:

Yellow Prussiate Potash	4 oz.
Potassium Carbonate	4 oz.
Sodium Cyanide	½ oz.
Gold as Fulminate or Cyanide	10 dwt.
Water	1 gal.

Use solution at a temperature of 175° F., with 6 volts.

Formula for smut green gold:

Gold as Fulminate or Cyanide	10 dwt.
Silver Cyanide	½ dwt.
Sodium Cyanide	6 oz.
Water	1 gal.

Dissolve a small amount of carbonate of lead with caustic soda in water, and add to the solution until smut is produced. Operate the solution at 100° F., with 6 volts, using 18 karat green gold anodes.

---

* White Gold Plating Solution

Pot. Gold Cyanide	4 gm.
Water	1000 c.c.

Nickel Formate sufficient to saturate water.

Iron Plating

Formula for iron solution:

Ferrous Chloride	40 oz.
Calcium Chloride	20 oz.
Water	1 gal.

Temp. 200° F.; current density 40 to

50 amp. per sq. ft.; 2 to 2½ volts; pH 1.5 to 2. Pure iron anodes.

This bath is used to produce heavy deposits of iron.

For thin deposits of iron use the following:

Dissolve 16 ounces of ammonium chloride in each gallon of water. Connect up tank, same as for plating, using cold rolled iron for anodes. On the cathode rod suspend some old plating racks or other work, and work solution with highest current density obtainable. After four or five hours of working the solution, there will be enough iron dissolved from the anodes and the solution will produce a deposit of iron. Operate solution at 80° F.; 1.5 to 2 amperes per sq. ft.; 1 volt.

Lead Plating

Formula for lead solution:

Lead Carbonate	20 oz.
Hydrofluoric Acid (50%)	32 oz.
Boric Acid	14 oz.
Glue	.025 oz.

To prepare the solution, place the hydrofluoric acid in a lead-lined tank and add the boric acid with constant stirring. When the boric acid is completely dissolved, the solution is allowed to stand until cool, when the lead carbonate is added in the form of a paste with water. The solution is allowed to settle when the clear solution is siphoned off and placed in the plating tank. The solution is then diluted to the proper volume with water and the glue added by dissolving the same in warm water. Mechanical agitation of the solution is essential.

A cathode current density of 10 to 20 amperes per sq. ft., 3 to 4 volts, and lead anodes are employed.

For thin deposits of lead, use the following:

Carbonate of Lead	2 oz.
Caustic Soda	6 oz.
Water	1 gal.

Lead anodes. Temperature 175° F.; 3 to 4 volts.

---

* Metal Plating, Non-electric

The cleaned metal is immersed in the following.

Thiourea	10
Mercuric Chloride	15
Water	1000

A coating of mercury is deposited which can serve as a base in electroplating.

NICKEL PLATING

Nickel Solutions

Many are the formulae for this solution, but they all contain double nickel salts, single nickel salts or both, some chloride salt and boric acid.

The constituents of the bath vary somewhat for the different classes of the base metal to be plated and there is no one solution that can be used and give ideal results on the different classes of work that require a nickel finish.

A nickel solution that has been used with good results on brass, copper and cold rolled steel is made of:

No. 1.	Double Nickel Salts	8 oz.
	Single Nickel Salts	4 oz.
	Boric Acid	2 oz.
	Sodium Chloride	2 oz.
	Water	1 gal.

Solution to be operated at 80° F.; 2 to 2½ volts; 6 to 8 amperes per sq. ft., and a pH of 5.8.

Depolarized nickel anodes 99%+ are recommended for use in this type of solution. Replenish the solution by the addition of single nickel salts.

For solutions that are operated at a higher temperature and a correspondingly higher current density, use:

No. 2.	Double Nickel Salts	8 oz.
	Single Nickel Salts	8 oz.
	Sodium Chloride	3 oz.
	Boric Acid	3 oz.
	Water	1 gal.

Temperature 110° F.; 2½ to 3 volts; 20 amperes per sq. ft., and a pH of 6. Depolarized nickel anodes 99%. Replenish by the addition of single nickel salts.

This solution can also be used for barrel plating at a temperature of 80° F. with very good results.

The low pH nickel solution has come into use recently where heavy deposits of nickel are desired. The solution should be operated at 150° F.; 3 to 3½ volts; with 50 amperes per sq. ft.; pH 2.

No. 3.	Single Nickel Salts	32 oz.
	Sodium Chloride	6 oz.
	Boric Acid	4 oz.
	Water	1 gal.

Nickel solution for die cast work:

No. 4.	Double Nickel Salts	10 oz.
	Sodium Chloride	7 oz.
	Sodium Sulfate	4 oz.
	Boric Acid	2 oz.
	Sodium Citrate	1 oz.
	Water	1 gal.

Operate solution at 75° to 80° F.; 2½

to 3 volts; 8 to 10 amperes per sq. ft.; pH 6.2 to 6.4.

Remarks on Nickel Plating

Bright deposits of nickel are obtained from No. 1 formula by the use of cadmium chloride or one of the prepared brighteners that are on the market. The pitting of nickel deposits is eliminated by adding hydrogen peroxide to the bath. Use from 1 to 10 cubic centimeters to each gallon depending upon the severity of the pitting.

Nickel solutions that are operated at 100° to 110° F. will plate faster and the deposit will be softer, although the deposit will be harder to nickel color. Solutions that are operated at low temperatures, 45° to 50° F. produce hard brittle deposits that have a tendency to peel and flake. This condition usually occurs during the winter months and raising the temperature will stop the trouble.

Defective nickel deposits may be stripped in a solution made of sulfuric acid 4 parts, water 1 part. Temperature 80° F., lead cathodes, 6 volts. If 3 or 4 oz. of copper sulfate per gallon are dissolved in the water before adding to the acid, the strip will not attack the base so readily.

Black nickel solution:

Double Nickel Salts	8 oz.
Sodium Sulphocyanide	2 oz.
Zinc Sulfate	1 oz.
Water	1 gal.

Temp. 80° F.; 1 volt; 1 to 1.5 amp. per sq. ft.; pH 6.

Work should be plated in white nickel solution for a few minutes or until the surface is completely covered with nickel and then placed in the black nickel solution.

Streaky deposits are caused by an excess of current, or a pH that is too low.

The addition of a small quantity of copper cyanide that is just dissolved in sodium cyanide will produce a darker deposit; 3 to 4 ozs. of copper cyanide is sufficient for 100 gallons of solution.

* Nickel Plating Bath

A bath for Ni deposition on printing plates is formed of Ni sulfate 82 lb., citric acid 27.25 lb. and water 375 gal., with addn. of KOH to the soln. until it is only slightly acid, then further addn. of K citrate 54 lb.

Aluminum, Nickel Plating on

The process adopted for castings and assemblies is: Stove at 315° for 1 hr.; polish; remove grease by trichloroethylene dip; boiling KOH dip, 15 sec.; wash; strong HNO_3 dip, 4 min.; wash; Ni plate, in $NiSO_4$ soln., for 30 min. at 15 amp./sq. ft.; wash and dry; stove at 482° for 15 min., starting up from cold. The yellowish tarnish on the Ni due to stoving can be removed by polishing or making the article anode in a strong H_2SO_4 soln. (d. 1.6) for 30–45 sec.

* Platinum Plating

$Na_2Pt(OH)_6$ (I) is prepared in a finely-cryst., readily sol. form by boiling aq. Na_2PtCl_6 with NaOH and treating the solution with an equal vol. of EtOH or $COMe_2$. The plating bath is made up by dissolving (I) in H_2O to give a 1% solution of Pt and adding Na_2SO_4, $Na_2C_2O_4$, and 0.2–2% of NaOH. The bath is operated at $>$ 40° (60–85°) at a c.d. of about 20 amp./sq. ft. Since the presence of SiO_2 in the bath produces poorly adherent, patchy deposits the salt is prepared in a Ni vessel and a similar vessel is used as the plating vat. Cu anodes plated with Pt may be used satisfactorily instead of the more costly Au or Pt sheets.

* Silver Plating Non-Conductors

The following is used for plating silver on non-conductors such as glass, ceramics, gas carbon, resins and other heat resisting materials.

100 grams of silver nitrate are dissolved in about a half liter of water and the solution so obtained is precipitated by addition of an excess of sodium hydroxide solution; the precipitated silver oxide is then washed until practically free from excess of alkali and other reaction products and is collected upon a filter. This gives about 70 gr. of silver oxide, which in the still moist condition is then ground up with 60 c.c. of mucilage or dissolved gum and the intimate mixture is treated with 20 gr. of glacial acetic acid while actively stirring. It will be noted that this quantity of acid is about half the calculated amount to convert the silver oxide present to acetate; consequently, its addition leaves much of the silver oxide unchanged and suspended in the mass of mucilage or dissolved gum. The silver acetate formed is then present in both true solution and colloidal dispersion and in intimate mixture with the same mass.

The relative proportions of silver acetate or other silver salt of an organic acid to the silver oxide present in the mass may be varied within quite wide limits bearing in mind that to obtain a good mirror-like deposit of silver the acetate should not be less than twenty per cent of the oxide and also bearing in mind that the higher the percentage of acetate present the higher the temperature required to produce the deposit. The proportion of acetate to oxide should not exceed ninety per cent.

Silver acetate is the most advantageous salt of silver to be used in the mixture, largely by reason of its solubility in water and the combustible nature of this salt but silver salts of other organic acids may be used if they are at least partly soluble in water or thoroughly dispersed.

In operation in full concentration or diluted with water to say about the consistency of thick cream, it can be painted or otherwise spread upon the surface to be silver plated and then by subjecting it to a moderate heat, say from a scarcely visible red heat 350 to 450° C. up to a bright red heat, say 900° C., the mixture is decomposed both the silver oxide and silver salt being converted to pure metallic silver, with complete elimination of all other ingredients of the mixture including the protective colloids. This decomposition is greatly facilitated by the oxygen given off from silver oxide, which brings about complete oxidation of the organic acid radical of the silver salt, and complete combustion of protective colloids originally present in the mixture.

In brief, silver oxide and silver acetate, at slightly elevated temperatures, mutually decompose each other and by simultaneous reduction of the former and oxidation of the latter yield pure silver as the only non-volatile residuum. The acetate of silver is the most advantageous salt in this connection because it is fairly soluble and hence more thoroughly permeates surfaces to which it is applied, although silver salts of other organic acids can be used if these are thoroughly dispersed in the protective colloid used.

Spotting, Prevention of Plating

After plating and rinsing, dry in an oven at a temperature of 400 to 450 degrees F. for several hours, then perform the final finishing operations. Still another method that has been used with some success is to rinse the work in a solution of 2 ounces of cream of tartar to the gallon of water, letting it remain in this rinse for 10 to 15 minutes, and then drying it after passing through cold and hot water rinses several times.

Silvering Mirrors

There are two methods of doing this, viz.: the hot and cold way. In the former method the glass to be silvered is cleaned thoroughly with wet whiting, then washed with distilled water, and prepared for the silver with a sensitizing solution of tin, which is well washed off immediately before its removal to the silvering table which is kept at a temperature from 35° to 40° C. The solution used is prepared as follows: in half a liter of distilled water 100 grams of silver nitrate are dissolved; to this there is added of liquid ammonia (sp. gr. 0.880) 63 grams; the mixture is filtered, and made up to 8 liters with distilled water, and 7.5 grams of tartaric acid dissolved in 30 grams of water are mixed with the solution. About 2.5 liters are poured over the glass for each superficial meter to be silvered. In about half an hour the silvered surface is cautiously cleaned by wiping with very soft chamois leather and the glass is treated a second time with solution like the first, but containing a double quantity of tartaric acid. After which the chamois is again used to remove all superfluous matter.

In silvering by the cold process two solutions are prepared. Silver nitrate 800 gm. and 1200 gm. of ammonium nitrate are dissolved in 10 liters of water and 1.3 kilos of pure caustic soda in 10 liters of water, and of each of these solutions 1 liter is added to 8 liters of water, which is allowed to rest till the sediment forms and then decanted. The second solution, invert sugar, is prepared by dissolving 150 gm. of loaf sugar with 15 gm. vinegar and 0.5 liter of water, and boiling this solution for half an hour. After cooling it is made up to 4200 c.c. with distilled water. For each square centimeter of glass to be silvered 15 c.c. of the silver solution are measured out, and from 7 to 10 per cent. of the sugar solution is added, both being stirred quickly together and poured over the cleaned glass. After about ten minutes the deposit of silver is complete and the exhausted solution may be carefully wiped off, the silvered surface washed off with distilled water and again treated with the mixed solutions to the extent of half the quantity used in the first application. The finished surface is wiped and washed off in the most careful manner and when thoroughly dry is coated with shellac or copal varnish. The glass to be treated should be absolutely clean and free from grease and the whole process requires much care to make it a success.

A more modern method is by reducing

the silver compound by the use of formaldehyde. A recipe of this type follows:

Silver Nitrate	1.6 gm.
Distilled Water	30.0 c.c.

Dissolve and of this solution take 8 c.c. add to it ammonia water, drop by drop, until the precipitate first formed is completely redissolved; then add 100 c.c. of distilled water. To this ammoniacal solution add 5 c.c. of 40 per cent. formaldehyde solution, mix quickly and then pour the mixed solutions upon the surface of the glass which is to be silvered. The entire operation of silvering should take about 2 minutes.

It is easy enough to write this description but the actual manipulation requires exquisite care. All forms of dirt and grease must be absent, even the trace of grease found naturally on the fingers. To successfully prepare a mirror will demand hours of preliminary practice.

Silver Plate on Glass

Clean the article from oil and grease. Place in a dilute solution of hydrofluoric acid to roughen the surface slightly, rinse in clean cold water. It is now ready for silvering. Two solutions are necessary.

Solution No. 1:

Pure Cane Sugar	90 gm.
Distilled Water sufficient to dissolve the sugar	
Nitric Acid (C. P.)	4 c.c.
Ethyl Alcohol	175 c.c.
Distilled Water to 1 litre.	

Solution No. 2:

Silver Nitrate	1.8 gm.
Distilled Water	180 c.c.

Add ammonia drop by drop until the precipitate which is formed is nearly redissolved. Then add

Potassium Hydroxide	9 gm.
Water sufficient to dissolve the potassium hydroxide.	

Add more ammonia drop by drop until the precipitate is nearly re-dissolved.

Take:

Solution No. 1	1 part by volume
Solution No. 2	9 parts.

Mix thoroughly, and immediately immerse the article. The surface will be covered with a deposit of metallic silver which is quite adhesive and serves as a base for further deposition of silver or copper.

To Copper Plate the Silvered Glass (Above) for Mirrors

It is necessary to have two copper sulphate solutions.

Solution No. 1 (Strike solution)

Copper Sulphate	8 oz.
Sulphuric Acid	1/4 oz.
Water	1 gal.

Current density 1 to 1½ amp. per square foot.

After the silver is covered with copper, the work is transferred to a regular acid copper solution as follows.

Solution No. 2.

Copper Sulphate	28 oz.
Sulphuric Acid	4 oz.
Water	1 gal.

Use cathode current density of 10 to 12 amp. per square foot.

Silvering Glass

Ammonium Hydroxide	1
Silver Nitrate	2
Water	3
Alcohol	3

Work in subdued light; dissolve; filter and mix with

Corn Sugar	1/4
Alcohol (25%)	10

Dip glass in this mixture and warm gradually to 70° C. when a mirror of silver is deposited.

* Silver Plating Compound

The product consists of an aq. $AgNO_3$ soln. to which is added sufficient Na_2CO_3 to obtain a milky ppt. of $AgNO_3$, 10–40% NaCl, 20–80% of a 50% $CaCO_3$ suspension, 1–20% abrasive and H_2O in sufficient amt. to produce a fluid mixt. $Na_2S_2O_3$ is ultimately added to produce a brilliant coating.

* Metallizing Patterns

The surface, *e.g.,* plaster of Paris, is impregnated with wax, the excess of which is removed, and the bared parts are moistened with a solution of $AgNO_3$ in an org. solvent containing a little H_2O. The surface is then rubbed with a 1:1 mixture of graphite and Cu powder to produce a Ag surface which can subsequently be plated with Cu.

Silver Plating Powder

Silver Nitrate	20
Am. Chloride	10

Sod. Bisulfate	40
Water	40
Pot. Carbonate	to make a paste

Keep in dark bottles.

Silver Plating

Formula for silver solution

1.	Silver Cyanide	3½ oz.
	Sodium Cyanide	5 oz.
	Ammonium Chloride	½ oz.
	Water	1 gal.
2.	Silver Chloride	3½ oz.
	Sodium Cyanide	8 oz.
	Ammonium Chloride	½ oz.
	Water	1 gal.

Either of the two solutions will give good results if operated at a temperature of 75° F. with a cathode current density of 4 or 5 amperes per sq. ft.; ¾ to 1 volt.

Solution 1 is generally used, but No. 2 is whiter.

Silver strike:

Silver Cyanide	½ oz.
Sodium Cyanide	8 oz.
Water	1 gal.

Use steel or carbon anodes; 6 volts.

Blue dip:

Bichloride of Mercury	1 oz.
Sodium Cyanide	6 oz.
Ammonium Chloride	1 oz.
Water	1 gal.

Brightener for silver solution:

Silver Solution	1 qt.
Sodium Cyanide	8 oz.
Carbon Bisulphide	1 oz.
Ether	1 oz.

To prepare the brightener place the carbon bisulphide and ether in a quart bottle and shake thoroughly. Dissolve the cyanide in the silver solution and fill bottle. Shake bottle from time to time until the carbon bisulphide is thoroughly dissolved and then filter.

One ounce of this stock solution should be sufficient for an addition to each 15 gallons of the regular plating solution. Care must be taken to avoid an excess or else the deposit will be rough and patchy. If an excess has been added, remove by raising the temperature of the solution to 140° F.

Silver strip solutions:

1.	Sodium Cyanide	12 oz.
	Caustic Soda	2 oz.
	Water	1 gal.

Reverse current with cold rolled steel as cathodes. Voltage 6 to 8. Agitate the work for a cleaner job.

2.	Sulfuric Acid	5 gal.
	Nitric Acid	1 gal.

Place crock that contains the strip in a hot water container. If all water is kept from the strip, brass or copper work will be attacked but very slightly.

Removing Fire Scale

To remove the fire scale from sterling silver use:

Nitric Acid	2 parts
Water	1 part

Use hot and agitate work.

Remove fire scale by reverse current with:

Sodium Cyanide	8 oz.
Water	1 gal.

Use hot and agitate work. Lead anodes; 4–6 v.

Bright dip:

Sulfuric Acid	2 gal.
Nitric Acid	1 gal.
Water	1 qt.

One ounce of muriatic acid for five gallons of above.

It is necessary to add water only when a new bright dip is made. Dip must be operated cold.

Matt dip:

Sulfuric Acid	1 gal.
Nitric Acid	1 gal.
Zinc Oxide	2 lb.

Operate hot and keep all water and chlorides from dip.

If the matt is coarse, add sulfuric; if too fine, nitric.

Burn Off Dip

If the work has been annealed, the fire scale should be removed in a hot sulfuric acid solution, 1 part acid, 3 parts water, rinsed in water and then placed in what is known as the "burn off" dip, made by using 2 parts of sulfuric acid, 1 part of nitric acid, and 5 parts of water.

The work is left in the "burn off" dip for five to twenty seconds, then rinsed in water, and bright dipped. If not bright enough, repeat the "burn off" and bright dip.

---

* Tin Plating

Formula for solution:

Sodium Stannate	12 oz.
Caustic Soda	1 oz.

Sodium Acetate	2 oz.
Hydrogen Peroxide (25 volume) or	⅓ oz.
Sodium Perborate	⅛ oz.
Water	1 gal.

The solution is operated at a temperature of 140° to 160° F.; 4 to 6 volts; 20 to 30 amperes per sq. ft.

The use of Hydrogen Peroxide or Sodium Perborate as an oxidizing agent is the greatest factor in controlling the character of the deposit as it prevents sponginess.

Small iron articles may be coated with tin in the following solution:

Tin Chloride	½ oz.
Aluminum Sulfate	2 oz.
Cream Tartar	2 oz.
Water	1 gal.

This solution is used in a copper tank which is lined with sheet zinc. The work should be clean and bright, and placed in iron wire baskets. If a large quantity of work is placed in the baskets, the work should be separated with perforated zinc sheets.

The solution is allowed to boil for 30 to 45 minutes and the addition of a very small quantity of sulfuric acid (about 1 drop to each gallon of solution) hastens the deposition of the tin deposit.

Immersion Tin—Caustic Soda Method

This method is used to tin by immersion, small brass or copper articles

Formula for Immersion Tin:

Caustic Soda	12 oz.
Stannous Chloride	4 oz.
Sodium Chloride	1 oz.
Water	1 gal.

The solution is placed in an iron tank which is heated with a steam coil. The bottom of the tank is covered with moss tin over which is placed an iron wire screen.

The work to be tinned is bright dipped or tumbled clean, placed in brass wire baskets and separated with sheets of perforated tin, placed in the solution at boiling temperature for 15 to 30 minutes, or until completely covered with tin. It is rinsed thoroughly in clean cold water and dried with the aid of hot water and sawdust.

The brightness may be increased somewhat by tumbling for a few minutes in hardwood sawdust.

Moss tin is prepared by melting the tin and pouring same into cold water at a slight elevation.

Tin Plating

The Na stannate plating soln. successfully used commercially has the compn.: Na stannate 32 oz./gal. and $SnCl_2$ 1/32 oz./gal., with anode c. d. not greater than 15 amps./sq. ft., cathode c. d. 15–45 amps./sq. ft., temp. 43–54° and 4–8 v. tank voltage. The Sn content is maintained by addns. of Na stannate. Very small addns. of Sn^{++} (as $SnCl_2$) are said to increase the throwing power of the soln. but too much to cause a powdery deposit. The soln. has a good throwing power and gives a good corrosion-resisting deposit.

Zinc Plating

The two types of zinc solutions that are in common use are the acid and alkaline solutions. The acid solution is usually preferred when cost is considered, as it can be made more cheaply, but the throwing power of this solution is lower than that of the cyanide bath.

Formula for acid zinc solution:

Zinc Sulphate	32 oz.
Ammonium Chloride	2 oz.
Sodium Acetate	2 oz.
Water	1 gal.

Temperature 80° F. Cathode current density, 15 to 20 amperes per sq. ft.; 3 to 4 volts.

Formula for cyanide zinc solution:

Zinc Cyanide	4 oz.
Sodium Cyanide	4 oz.
Caustic Soda	3 oz.
Water	1 gal.

Temperature 100° F. Cathode current density 10 to 15 amperes per sq. ft.; 2 to 3 volts.

Use pure zinc anodes in both solutions. Corn sugar may be used in the proportion of one ounce per gallon in either solution to obtain a finer structure of deposit.

Remarks on Zinc Solutions

The throwing power of the acid zinc solution is quite poor. The addition of one ounce of stannous chloride to a 100 gallon solution will improve the throwing power. An excess should be avoided, as it has a tendency to discolor the deposit. The pH is the most important factor to control in the acid solution. A pH of 3.5 to 4.5 using thymol blue as an indicator is about right. This should be maintained by adding the required sulphuric acid.

In the cyanide bath, the free cyanide is the most important factor to control. If the free cyanide is equal to the metal content best results will be had. An excess of free cyanides causes a bright, rough deposit.

Care should be used in drying zinc deposit to prevent stains. A thorough rinsing in clean cold water followed by hot water and hardwood sawdust is good procedure.

Zinc, Plating Nickel on

A cleaning soln.: Na silicate, 10 g./l., $+Na_3PO_4$, 30 g./l., operated at approx. the b. p. with just enough current to cause the article (cathode) to gas freely, was found to be best, as cleaning could be done in 0.5–3 min. without discoloration of the Zn. For picking, immersion for 0.5–1 min. in a soln. of 8% HCl was found best, etching, but not discoloring, the Zn. The importance of efficient rinsing between cleaning, pickling and plating, the avoidance of delay between pickling and plating, and the use of solns. for the prepn. and plating of Zn and its alloys only, are stressed. After varying the soln. compn. and conditions of operation considerably, the authors conclude the following soln. is best for the direct Ni-plating of Zn: $NiSO_4.7H_2O$, 75; $Na_2SO_4.10H_2O$, 200; NH_4Cl, 12, and H_3BO_3, 10 g./l., operated at room temp. with a mixt. of cast and rolled Ni anodes, at a $p_H = 6.0 \pm 0.2$ and an av. cathode c. d. = 10 amp./sq. ft. The soln. is said to become alk. on working, this necessitating daily addns. of H_2SO_4. A short initial "strike," at 30 amp./sq. ft., was first used but was found unnecessary after the bath had been worked for some time. Consistently good deposits of ductile Ni, which polished easily, were obtained from the above soln. It is suggested that the Ni deposit must be at least 0.00035 in. thick if it is to be serviceable.

* Plating Zinc-Tin on Iron

The plating bath comprises a solution of 81 g. of $ZnSO_4,7H_2O$, 3.5 g. of $SnCl_2$, and 150 g. of NaOH per litre to which are added 10 c.c. of sulphonated castor oil; it is operated at 6 volts and 10–20 amp./sq. ft., using anodes of 90:10 Zn–Sn alloy amalgamated with Hg (2%).

Black Finish on Brass

Solution No. 1.

Yellow brass may be colored blue black by immersion in a solution of water saturated with copper acetate to which ammonium carbonate has been added.

or

Solution No. 2.

Immerse in a solution of ammonium hydroxide which has been saturated with copper carbonate

or

Solution No. 3.

Immerse in

White Arsenic	12 oz.
Yellow Antimony Sulphide	¼ oz.
Water	1 gal.

or

Immerse in

Hyposulphite Soda	8 oz.
Acetate of Lead	4 oz.
Water	1 gal.

These solutions, except the one made up with copper carbonate, should be used hot. Immerse the work until proper color appears. The work should be finished with a coat of lacquer to prevent tarnishing.

Plating Baths

Basic recipes for still solutions have been developed for the guidance of the plater. However, the proportions of the constituents should be changed according to special requirements for individual needs. The following procedure is recommended for making up new solutions or replenishing old baths:

Fill the tank with one-third the amount of water required. Dissolve the Sodium Cyanide in this water, which should be at a temperature of about 50° C. (120° F.).

Then add the Metal Cyanide and stir until it is in solution. Finally add the balance of the ingredients and mix in the remaining two-thirds of water.

Brass Solution

Water	1 gal.
Sodium Cyanide (96–98%)	5½ oz.
Copper Cyanide	4 oz.
Zinc Cyanide	1 oz.
Soda Ash	2 oz.
Temperature	75–90° F.
Ratio Anode to Cathode Surface	2–1
Cathode Current Density	3–20 amp./SF
Voltage—Still Solution	3–5
Voltage—Barrel Solution	5–10
Anodes	Copper 80%, Zinc 20%

Copper Cyanide Solution

Water	1 gal.
Sodium Cyanide (96–98%)	4 oz.
Copper Cyanide	3 oz.

Soda Ash	1 oz.
Sodium Bisulfite	1 oz.
Ratio Anode to Cathode Surface	2–1
Temperature	68–120° F.
Cathode Current Density	10–25 amp./S.F.
Voltage—Still Solution	3–6
Voltage—Barrel Solution	8–12

Note:

1. For barrel plating double the proportions just given.
2. Hypo Soda can be used for brightening purposes in the concentration of 1/64 ounce per gallon when the deposit is not to be oxidized.
3. The reason for the addition of sodium bisulfite is to obtain better anode efficiency and better color of deposit.

Zinc Cyanide Solution

Water	1 gal.
Sodium Cyanide (96–98%)	3 oz.
Zinc Cyanide	5 oz.
Caustic Soda	4 oz.
Temperature	80–110° F.
Ratio Anode to Cathode Surface	1½ to 1
Cathode Current Density	15–30 amp./S.F.
Voltage—Still Solution	4–6
Voltage—Barrel Solution	8–12

Silver Cyanide Solution

Water	1 gal.
Sodium Cyanide (96–98%)	4½ oz.
Silver Cyanide	3 oz.
Temperature	Normal
Ratio Anode to Cathode Surface	1–1
Cathode Current Density	2–5 amp./S.F.
Voltage—Still Solution	½–1¼

Note: When making up a new solution, ⅛ ounce of ammonium chloride may be used.

Mercury Dip

Water	1 gal.
Sodium Cyanide (96–98%)	6 oz.
Bi-Chloride Mercury	½ oz.

Silver Cyanide Strike Solution

Water	1 gal.
Sodium Cyanide (96–98%)	8 oz.
Silver Cyanide	½ oz.
Caustic Soda	¼ oz.
Temperature	Normal
Voltage	6
Anodes	Sheet steel

Gold Cyanide Solution (Yellow)

Water	1 gal.
Sodium Cyanide (96–98%)	½–1 oz.
Sodium Gold Cyanide	½ oz.
Caustic Potash	⅛ oz.
Temperature	140–160° F.
Cathode Current Density	1–5 amp./S.F.
Voltage	2½

Lead Plating Iron Strips

The strip is passed in succession through vats contg. 50 and 70% HCl solns. and $ZnCl_2$ soln. plus a 2.5% soln. of NH_4Cl. Four kg. of the soln. contain in addition 1 part Hg, 2 parts $HgCl_2$ and 3 parts aqua regia. The strips are finally passed through a bath of molten Pb with 5% Sb.

POLISH, ABRASIVES, METAL CLEANERS

* Abrasive Compound

First produce two mixtures one of which consists of a potassium soap that is produced by heating and melting approximately thirty parts of stearic acid and adding, while heating and stirring, a solution of approximately six parts of potassium hydroxide and approximately twenty parts of water and then, after saponification has taken place, adding water to make one hundred parts.

The other mixture is produced by melting approximately five parts of a mixture consisting of approximately fifty per cent of beeswax and fifty per cent of japan wax with approximately ten parts of paraffin oil.

With this wax and oil combination is intimately mixed fifty parts, approximately, of the above described potassium soap mixture.

Then stir into this mass a mixture of approximately fifteen parts glycerine and approximately thirty parts of water.

To this combination is then added approximately seventy-five parts of silicon carbide and approximately twenty-five parts of electrically fused alumina.

All of these operations are performed in a water jacketed kettle at a temperature of about sixty degrees centigrade.

After agitating until the abrasive is thoroughly distributed throughout the mass, raise the temperature thereof until the water in the jacket is at a boil. These conditions are then maintained while continuously stirring until the mixture thickens to a stiff paste.

To this paste compound sometimes add a coloring pigment such as carbon black, and an essential oil, as methyl salicylate.

The above described abrasive compound is characterized by a very much slower rate of evaporation of its moisture content than is the case with those compounds of this class as heretofore produced.

Abrasive Polish

Abrasive (Tripoli, Silex, etc.)	40 lb.
Proflex	10 lb.
Suspendite	4 lb.

This is added to water with stirring. By varying the water used either a paste or liquid polish is formed.

Razor Strops, Abrasive for

Bauxite	42
Lard	42
Powd. Emery	15
Varnish	1

Aluminum Polish

1. Sapinone	1
2. Water	52
3. Oleic Acid	8
4. Ammonium Hydroxide	5
5. Alcohol	4
6. Infusorial Earth	20
7. Red Iron Oxide	8

Mix (2) and (3) and stir until uniform. Mix (6) and (7) and rub into a paste with part of (1), (2) and (5). Slowly add the balance and while mixing vigorously add mixture of (2) and (3).

---

* Cleaner, Aluminum (Non-Corrosive)

Tartaric Acid	99
Sodium Fluoride	1
Water	to suit

---

* Aluminum Cleaning Powder

Powdered Pumice	25
Powd. Calcined Silica	25
Sod. Sesquicarbonate	25
Trisodium Phosfate	10
Powdered Soap	10
Am. Chloride	5

Auto Polish

Paraffine Oil	5	gal.
Linseed Oil Raw	2	gal.
China Wood Oil	½	gal.
Benzol 90%	1	qt.
Kerosene	1	qt.

Odor to suit.

Mix oils together. Mix Benzol and Kerosene, then add to oils and stir thoroughly.

All formulae preceded by an asterisk (*) are covered by patents.

Auto Polish

Fullers Earth	4	oz.
China Clay	3	oz.
Kerosene	1¼	pt.
Mineral Oil	1¼	pt.
Turkey Red Oil	1	qt.
Ammonia Water (10%)	4	oz.
Water	2½	pt.
Formaldehyde (40%)	4	oz.
Glycerin	½	pt.

Automobile Polish

Carnauba Wax	9	lb.
Beeswax	4	lb.
Ceresin Wax	4	lb.
Naphtha	75	lb.
Stearic Acid	7	lb.
Triethanolamine	2.5	lb.
Water	75	lb.
Abrasive	25 to 60	lb.

Preparation

Add the Triethanolamine and stearic acid to the water, heat to 100° C. and stir to obtain a smooth soap solution. Then melt the waxes in the naphtha and, when the solution is about 85° to 90° C., add it to the hot soap solution. Stir vigorously until a smooth emulsion is obtained and then slowly until cold. If any separation occurs shortly after the emulsion has cooled, stir vigorously until the emulsion is creamy.

The method of adding the abrasive is dependent upon the type of abrasive used. An oil-absorbing abrasive should be well mixed with the hot oil solution before it is added to the soap solution, but an abrasive that absorbs water is best stirred into the finished emulsion. The latter type, like Bentonite, to the extent of 25 pounds, produces a paste with the above emulsion, while 60 pounds of the former, as Tripoli, makes a liquid polish.

Properties

This polish is non-destructive to lacquers. It is a cleanser and polisher combined and leaves a bright, hard film. It is applied by rubbing over the surface well to remove dirt and streaks and then polishing with a dry cloth.

Variations

The proportions of waxes can be changed depending upon the case of polishing required and the hardness of the final film. The naphtha and water contents can be varied slightly to change the consistency of the emulsion. When the primary use of this product is for polishing rather than as a cleaning and polishing combination, it will be more satisfactory without an abrasive.

Wax Automobile Polish

A.	Carnauba Wax	30 lb.
	Glyco Wax B	20 lb.
	Naphtha or Varnolene	68 lb.
	Turpentine	17 lb.
B.	Water	70 lb.
	Borax	10 lb.

Melt "A" together but do not heat above the boiling point of water. Meanwhile dissolve "B" while heating to a boil.

Run "A" into "B" *slowly* while stirring *vigorously.*

Motor Car Polishes

A good formula for a cleanser and polisher is:

Yellow Wax	20.0
Commercial Silica, Very Finely Powdered	40.0
Turpentine Substitute	40.0
Soft Soap	1.0
Water	5.0

Melt the wax and incorporate the powder, slowly adding the turpentine substitute, finally stir in the soap, previously dissolved in the water. Some may prefer it to be without the soap, but experience shows it to be worth its slight softening effect in yielding a higher and better polish. The paste may be tinted with ferric oxide.

Another formula is as follows:

Kieselguhr (Levigated)	11 parts
Silica (Levigated)	9 parts
Yellow Ochre	1 part
Red Ochre	1/10 part
Kerosene	16 parts
Soft Paraffin	2 parts
Powdered Soap	1 part

The following formula is suitable for polishing fabric bodies:

Oleic Acid	80.0
Liquid Paraffin	250.0
Potassium Hydroxide	16.0
Tragacanth	6.0
Water	to 1,000.0

Mix the oleic acid with the paraffin and slowly add the potassium hydroxide, previously dissolved in 200.0 of water. Soak the tragacanth in 500 cc. of water until fully absorbed, then heat to boiling, and when cool stir into the above emulsion.

Once a good surface has been produced by the above it is not an advantage to use too frequently, as frictional powders are bound to show the effect sooner or later if unwisely used. A thin film of wax once deposited on paintwork of the highly polished variety is best kept in condition by a hard wax polish. Beeswax is too soft, and the best for the purpose is Carnauba wax. This, however, is intractable and likely to crumble; it needs rubbing up with the cloth in order to soften it before applying. A modification enabling the polish to be easily applied and which does not modify in any way its polishing and surfacing effect is made as follows:

Grey Carnauba Wax	25.0
Japan Wax	5.0
Rosin	5.0
Melt and stir in	
Turpentine Substitute	60.0
Strain and add	
Solution of potash (1%)	5.0

This last addition has been found to give just sufficient saponification to prevent the paste crumbling. The preparation gives a highly polished hard surface, and where dirt and grease are not present its direct application forms a perfect protection of enamelled paintwork which can easily be kept clean with a dry cloth.

Automobile Polish and Cleaner

1. Celite (or other air-floated abrasive)	282 lb.
2. Isopropyl Alcohol	305 lb.
3. Glycerin	50 lb.
4. Naphtha	110 lb.
5. Oil of Camphor	105 lb.
6. Spindle (Mineral) Oil	555 lb.
7. Oxalic Acid	10 lb.
8. Suspensone	22 lb.
9. Water	1770 lb.
10. Emulsone B	10 lb.

"1," "8," "9," and "10" are mixed and allowed to stand over-night. Then add "3" and stir. Next add "7" and "2" and stir vigorously. Now add "4," "5" and "6" slowly while stirring vigorously. Continue stirring intermittently for 2 hours. Allow to stand overnight and stir for ½ hour the next day. If a thinner product is wanted reduce Emulsone B to 5 lb.

Auto Paste Wax Polish

Carnauba Wax	20
Beeswax	30
Japan Wax	30
Paraffin Wax	60
Turpentine	326

---

* Automobile Polish

Tartaric Acid	1.25
Oxalic Acid	1.25
Abrasive Mild	3.75
Suspendite	0.25
Mineral Oil	28.5
Water	65

Automobile and Floor Polish
(Wax Paste Type)—(Rubbing Type)

Yellow Beeswax	6 lb.
Ceraflux Tech.	16 lb.
Carnauba Wax	27 lb.
Montan Wax	8 lb.
Naphtha or Varnolene	89 lb.
Turpentine	10 lb.
Pine Oil	3 lb.

Melt together and pour into cans. Do not disturb until solidified. This makes an excellent auto polish of great durability and luster. Variations can be made to suit individual requirements.

Belt Dressing
(No. 1 Commercial Grade)

Castor Oil	40 parts
Cod Oil	40 parts
Neats-foot	40 parts

Mix thoroughly with heating if necessary.

Use: Clean belting to be dressed and apply dressing with brush or cloth. This is suitable where excess moisture or steam is present.

Belt Dressing

Tallow	10
Cod Oil	10

Brass Polish

Petroleum Spirits	30 parts
Ammonia	4 parts
Olein	10 parts
Tripoli Powder	50 parts
Methylated Spirits	10 parts
Water	20 parts

Brass Polish with Gasoline Base

Tripoli	1 lb.
Whiting	1 lb.
Prepared Chalk	1 lb.
Stearin	1 lb.
Gasoline	1 gal.
Oleic Acid	8 oz.

Dissolve the stearin in the gasoline, add the oleic acid and then stir in the powders, using care to keep them from forming in lumps. More or less stearin may be used to give any desired body, and the gasoline may be replaced in whole or in part with kerosene.

Brass, Refinishing Corroded

Saturate vinegar with salt and clean brass with this until all corrosion is removed. Polish with any good metal polish; wash; dry; wash with benzene to remove oil and grease; finish with spar varnish or lacquer.

Copper Cleaner

Oxalic Acid	1 oz.
Rotten Stone	6 oz.
Gum Arabic	½ oz.
Cottonseed Oil	1 oz.

Water sufficient to make paste.

Apply to small portion and rub dry with flannel.

Cellulose Friction Polishes

These are often "oil in water" type, and consist of emulsions of oil, gum or other emulsifier and water as lubricants to the friction polishing earths. Their great advantage is that they do not mark afterwards, and are free from a "film of wax" or other matter which can attract dust, but they wear away the enamel if used too frequently, and are not so waterproof as wax polishes. They should rub away to nothing on application, so that a polish ensures with the same rag.

Floss Powder	8 parts
Paraffin	8 parts
Methylated Spirits	2 parts
Glycerine	2 parts
Gum Tragacanth	⅛ part
Water	40 parts

Carborundum Suspension

Diglycol Stearate	4
Water	100

Heat to 60° C. and stir after turning off heat. Add with stirring

Carborundum Powder	4

Crocus Composition

Double Pressed Saponified Stearic Acid	11 lb.
Petrolatum	11 lb.
Edible Tallow	2 lb.
Crocus	165 lb.
Flint	23 lb.

"Dry Bright" Polish

Carnauba Wax	13.2 lb.
Oleic Acid	1.5 lb.
Triethanolamine	2.1 lb.
Borax	1.0 lb.
Water	108 lb.
Shellac	2.2 lb.
Ammonia (28%)	0.32 lb.

Melt the wax and add the oleic acid. The temperature should not be above 90° C. Using a hot water or steam jacketed kettle maintains a good temperature and prevents wax caking along the sides of the container. Add the triethanolamine slowly, stirring constantly. The solution should be *clear* at this point. Dissolve the borax in about a pint of boiling water and add to the wax solution to obtain a clear jelly-like mass. Stir for about 5 minutes. Add 92 pounds water, previously heated to boiling temperature, slowly with constant stirring. An opaque solution should be obtained. Cool. Add 16 pounds of water to the shellac and then the ammonia and heat until the shellac is in solution. Cool. Add this to the above wax solution and stir well to obtain an even mixture.

Properties

The above polish should give a clear film when applied to linoleum, mastic floors, etc., and one that is not too slippery. Shellac has been incorporated in the polish to cut down the slipperiness of a straight carnauba wax emulsion. It is necessary to use a good grade of light colored carnauba wax and the directions for making the polish must be carried out as described.

Variations

If 1.8 pounds of water soluble nigrosine is added to the water in the above formula, an excellent black leather polish can be made. By using stearic acid in place of oleic acid a thicker polish is obtained.

Dust-Cloth Fluid

Light Mineral Oil	3 gal.
Corn Oil	1 gal.
Clovel	3 oz.
Oil Soluble Yellow Color	to suit

Emery Grease

Double Pressed Saponified Stearic Acid	11 lb.
Edible Tallow	1 lb.

Paraffine	3 lb.
Petrolatum	1 lb.

Emery Paste

Double Pressed Saponified Stearic Acid	17 lb.
Oleo Stearine	2 lb.
Petrolatum	38 lb.
Japan Wax	3 lb.
Paraffine	26 lb.
Emery	300 lb.
Flint	100 lb.

Flatting Paste Emulsions

These are of the "water in oil" type, and consist essentially of oils, soap, and the friction or flatting powdered earths in fine form. They should be easy to work and yet not "scratch" the paint or varnish.

Tallow	20 parts
Soap	30 parts
Paraffin	18 parts
Water	20 parts
Waxes	8 parts
Turpentine	18 parts
Tripoli or Partly Brick Dust	60 parts

Mineral Oil Emulsion

Proflex	6 lb.
Water	60 lb.
Mineral Oil	50 lb.
Red Oil (Oleic Acid)	4 lb.

In using Proflex it should be strewn in the surface of the water which is being stirred with a high speed agitator. The oil or other water in soluble material is then run in slowly while stirring. The pigments or abrasives are then added in the same way.

Polish, Emulsion

Proflex	3 lb.
Water	17 lb.

Allow to soak for 15 minutes; stir until all particles are gone. Put into a mayonnaise type of mixer and, while beating add to it slowly

Mineral Oil	80 lb.

The above gives a white heavy cream which may be diluted with water to give a milky liquid.

Floor Oil

Mineral Oil	92
Turpentine	5
Beeswax	1
Shellac Wax	2

Dissolve waxes in mineral oil heated to 100° C.; cool and stir in turpentine.

Floor Oil, Low Priced

Light Mineral Oil	5	gal.
Automobile Engine Oil	½	gal.
Paraffin Wax	2	lb.
Clovel	½	pt.

* Oil, Floor (Non-Drying)

Mineral Oil	68
Oleic Acid	18
Ammonium Hydroxide	4
Pine Oil	10

Floor Polish

Carnauba Wax	30
Rosin	6

Heat above to 140° C., cool to 100° C. and add following with vigorous stirring which has been heated to 95–100° C.

Soap Flakes	10
Turpentine	1
Water	270

Floor Wax

Beeswax Yellow	5	lb.
Paraffin Wax	4	lb.
Soap Chips	3	lb.
Stearic Acid	3	lb.
Turpentine	3	gal.
Salts of Tartar	1½	lb.
Water	3½	gal.

Dissolve salts of tartar and soap in boiling water. Melt waxes in another container and heat to 200° F. when the boiling water soap solution is added slowly with vigorous stirring until homogeneous. Turn off heat and run turpentine in slowly with good stirring. Pack in cans when cold.

Finishing Floor Wax

Carnauba	5 lb.
Ozokerite	5 lb.
Turpentine	1 gal.
Gasoline	5 gal.

Heat gently until wax completely dissolves. Cool quickly.

Floor Wax, "Rubless"

Hydromalin	138 lb.
Carnauba Wax No. 2	250 lb.

Heat to 120–140° C. half hour. Cool to 100–105° C.

Add to the above slowly with stirring.

Natrex	30 lb.
Water	1780 lb.

Heated to 100° C. Keep as close to 100° C. as possible for 15 minutes.

This formula can stand additional water if a lower cost product is desired. The more water added, however, the lower the gloss will be. If a more water repellent product is desired the Natrex may be left out. When this is done, however, a lower gloss results.

Liquid Floor Wax (Rubbing Type)

Heat to 10 lb. of Glyco Wax B and 2 lb. Beeswax with 30 lb. Naphtha or kerosene until dissolved. Cool and stir thoroughly when thickening begins. Color yellow or orange with an oil soluble dye. This may be made thicker or thinner by varying the amount of wax.

Wood Floor Finish

Brush liberally with a mixture of three parts boiled linseed oil and one part turpentine; after a few minutes for soaking in, wipe up the excess. Two applications may be necessary, a day or two apart. This will darken the floor somewhat. For walnut tone, tint the oil with burnt umber ground in oil. Waxing can follow.

FURNITURE POLISHES

Wax Paste

Carnauba Wax	30 lb.
Beeswax	15 lb.
Ceresin Wax	15 lb.
Turpentine	26 lb.
Naphtha	24 lb.
Stearic Acid	8 lb.
Triethanolamine	4 lb.
Water	65 lb.

Liquid Wax

Carnauba Wax	10 lb.
Beeswax	4 lb.
Ceresin Wax	4 lb.
Naphtha	80 lb.
Stearic Acid	8 lb.
Triethanolamine	4.5 lb.
Water	200 lb.

Preparation

Melt the waxes and stearic acid and add the triethanolamine. Temperature should be about 90° C. Add the naphtha slowly so that a clear solution is maintained. Using a water or steam jacketed kettle prevents overheating and also caking of the waxes on the sides of the container. Add the boiling water to the naphtha solution and stir vigorously until a good emulsion is obtained and then slowly until the emulsion is cold.

Properties

Wax polishes of this type are used where a permanent finish is desired, as on woodwork, furniture, automobiles, etc. They require hard rubbing, but produce a polish of high luster. Triethanolamine stearate, being non-destructive to lacquer, is particularly indicated because of its ability to act as a cleanser as well as an emulsifier for the various constituents.

Furniture Polish

(Packages in glass only. No tin cans.)

200 gal.

Turpentine	8 gal.
Naphtha	30 gal.
Lt. Spindle Oil	49 gal.
Acetic Acid 36%	6 gal.
Water	100 gal.
Antimony Chloride	4 gal.
Gum Arabic	10 lb.
Gum Tragacanth	10 lb.
Perfume	1 gal.

Make up with water to 200 gallons and run through colloid mill.

Furniture or Auto Polish

Light Mineral Oil	1 gal.
Powd. Carnauba Wax	2½ oz.

Heat until wax is dissolved.

Furniture Polish

Yellow Ceresine	3 lb.
Japan Wax	1 lb.
Beeswax	2 lb.
Linseed Oil Raw	4 gal.
Turpentine	1 gal.
Paraffin Oil 28° gr.	1 gal.
Water	7 gal.
Carbonate of Potash	3 oz.
Soap Chips (Animal Fat Soap)	1 lb.

Mix the above thoroughly.

Cream Polish, Furniture

Carnauba Wax Bleached	6
Japan Wax	3½
Paraffin Wax	1½
Turpentine	12
White Curd Soap	3
Rosin Pale	2
Water	30
Clovel	Trace

Furniture or Auto Polish

1.	Blendene	10 parts by vol.
	Spindle Oil	60 parts by vol.
2.	Water	40 parts by vol.

Stir (1) with a high speed mixer. Add (2), stir five minutes. Blendene will give clear soluble oils with mineral oils, depending on grade, from two to six times its volume. The cruder the mineral oil, the higher percentage of oil will mix clear with Blendene. They emulsify readily on stirring in water.

* Furniture, Metal and Auto Polish

Nelgin	8 lb.
Water	126 lb.

Allow the above to soak a few hours, stir and then add the following mixture to it slowly with good stirring.

Light Mineral (Spindle) Oil	26 lb.
Blown Castor Oil	18 lb.
Varnolene or Solvent Naphtha	16 lb.
Lemenone Crude	16 lb.

This polish works exceptionally well on lacquered, painted or varnished metal surfaces.

Furniture Polish (Paste)

A.	Carnauba Wax	60
	Turpentine	60
	Stearic Acid	2
B.	Trihydroxethylamine Stearate	12
	Water	62

Heat (A) and (B) in separate vessels to 200° F. and run (B) into (A) slowly with vigorous stirring. Stop when homogeneous.

Furniture Polish (Liquid)

Carnauba Wax	6
Paraffin Wax	9
Ceresin	2
Naphtha	43
Turpentine	4
Stearic Acid	1
Trihydroxyethylamine Stearate	4.5
Water	130

Procedure—as above.

Furniture Gloss Oils

These are essentially emulsions of oil and gum in water. A little glycerine aids the ease of application.

Water	10 parts
Nut Oil	1 part
Mineral Oil	1 part
Acetic Acid	1/8 part
Gum Arabic	11 parts

Gas-Meter Diaphragm, Dressing for

Castor Oil	70
Linseed Oil Boiled	30

Glass Polish

1. Am. Linoleate	20
2. Orthodichlor Benzol	100
3. Water	200
4. Infusorial Earth	60

Dissolve (1) in (3) overnight and run (2) in while beating with high speed mixer. Then beat (4) in until uniform.

* Polish, Glass

Lard	10
Paraffin Wax	4
Naphtha	1
Glycerol	1

Glass Polish (Dry)

Precipitated Chalk	50
Kieselguhr	20
White Bole	30

Make into a slurry with water for use.

Glass Polish (Liquid)

White Bole	5
Vienna Chalk	10

Work into above

Oleic Acid	1
Denatured Alcohol	75

Then add while stirring vigorously

Water	20
Ammonium Hydroxide	15

Glass Polish

Whiting	54
Silica "Smoke"	18
Starch	15
Cream of Tartar	11
Magnesium Oxide	10
Infusorial Earth	2

For use make into a cream with water or benzine.

* Polish, Gold

Soap	20–25
Coconut Oil	1
Precipitated Chalk	25
Kieselguhr	8
Glycerol	40–45
Lemenone	1

* Gold and Silver Polish

China Clay	47
Precipitated Chalk	47
Am. Sulfate	5
Magnesium Powder	1

* Grinding and Polishing Compound

Silicon Carbide	10
Soap	20
Turpentine	20
Bentonite	20
Water	40

* Grinding Compound

Mineral Oil	15
Sulfo Turk C	15
Petrolatum	30
Silicon Carbide (150–220 mesh)	30
Emery (80–100 mesh)	10

Grindstones

Al_2O_3 is finely ground, made into a paste with a dil. acid, such as HCl, molded or pressed to the desired shape, dried and agglutinated at a temp. below 1600°.

* Household Cleaning Powder

Borax	24
Sod. Sesquicarbonate	50
Trisodium Phosfate	24
Sod. Silicate	2

Leather Polish

Carnauba Wax	11 lb.
Turpentine	16 lb.
Stearic Acid	3 lb.
Oil Sol. Nigrosine	2 lb.
Triethanolamine	1 lb.
Water	66 lb.
Water Sol. Nigrosine	1 lb.

Preparation

Dissolve the water soluble Nigrosine in the water, add the Triethanolamine and stearic acid and heat to boiling. Stir until a smooth soap solution is obtained. In a separate container, melt the carnauba wax in the turpentine and add the oil soluble Nigrosine. When this solution has reached a temperature of 85–90° C., add it to the soap solution. Stir vigorously to obtain a good dispersion of the wax and then stir slowly until the emulsion is cold.

Properties

This leather polish is a liquid cream which is readily applied to black shoes. It is excellent for removing grease and dirt and yields a bright waterproof finish. The use of Triethanolamine as the emulsifying agent eliminates any injurious solvent action on the leather.

Variations

If the Nigrosine is omitted from the above formula, the liquid is cream-colored and suitable for polishing light colored leathers. For tan and other colors, the appropriate dyes may be added. The substitution of naphtha for all or part of the turpentine decreases the odor and is sometimes desirable.

By changing the amount of water the consistency of this emulsion can be varied from a paste to a thin liquid.

Leather Belt Polish

A polish for unfinished edges of leather belting is composed of the following:

Water	1 gal.
Gum Tragacanth	2 oz.

Bismarck Brown Solution—in amount to obtain desired color.

Leather Dressing

Tallow	70
Petroleum Jelly	3.5
Diglycol Stearate	13
Beeswax	9
Rosin	2
Water	2

* Leather Dressing

Pyroxylin 100 sec.	1.7
Dibutyl Phthalate	0.8
Carnauba Wax	1.7
Titanium Dioxide	3.3
Ethyl Acetate	15.5
Butyl Acetate	10.3
Alcohol	66.7

Leather Dressings

One of the oldest and best known leather dressings consists of a soln. of 4 parts of rosin in 96 parts of C_6H_6 plus a trace of nitrobenzene. Another contains rosin 6, linseed oil 2, turpentine 4 and benzine 4 parts. A more complex prepn. consists of rosin 3 and EtOH 15 parts as soln. I and rubber latex 2, C_6H_6 15, turpentine 15 and CCl_4 10 parts as soln. II. Ceresin 5, stearin 2, soln. I 5 and soln. II 10 parts are heated together over a water bath. Three parts of K_2CO_3 in 30 parts of b. H_2O are added to make a dressing in emulsion form.

Leather Dressing

Cumarone	2 lb.
High Flash Gasoline	1 gal.
Castor Oil	¼ lb.

* Leather Finish

Prepare with stirring a first solution of borax, 17½ pounds; orange shellac flakes, 60 pounds; water, 40 gallons; prepare with heat and stirring a second solution, suspension or extension, of white neutral soap flakes, 6 pounds; carnauba wax, 19 pounds; water, 30 gallons. Mix in the ratio of from five to eight parts of the first solution to three parts of the second solution. The product is a smooth viscid paste, hard but flexible when the water of emulsion or solution has evaporated away, and not water-soluble thereafter to any practical extent.

---

* Leather, Preservative

Vaseline	62
Paraffin Wax	16
Lanolin	10
Am. Sulfoichthyolate	7
Neatsfoot Oil	5
Oil Birch Tar	to suit

---

* Leather Soles, Preserving

Larch Turpentine	80
Tallow (Beef)	6
Oil Birch Tar	4
Varnish	30

Leather Preservatives

A. Neatsfoot Oil (20° Cold Test)	20
Castor Oil	20

B. Lanolin Anhydrous	40
Neatsfoot Oil (20° Cold Test)	60

C. Neatsfoot Oil (20° Cold Test)	50
Lanolin Anhydrous	35
Japan Wax	20
Soap Chips	8
Water	90

Military Leather Paste Polish

Carnauba Wax	18
Candelilla Wax	2
Japan Wax	10
Paraffin Wax	2
Turpentine	20

Linoleum Polish

Carnauba Wax	1 lb.
Paraffin Wax	1 oz.
Yellow Wax	7 oz.
Turpentine	1 gal.

Metal Polish

Tank A

Dissolve thirteen (13) pounds of Oxalic Acid in forty (40) gallons of water. Heat to not more than 80° C. Add twelve (12) pounds of 26° Bé Ammonia.

Tank B

Mix twenty-five (25) pounds of Red Oil or Rozolin with twenty-five (25) pounds of Denatured Alcohol. Add twelve (12) pounds of 26° Bé Ammonia, to be warmed slightly to affect saponification.

Add contents of Tank A to Tank B while mixing. This can be done successfully in the cold, also with varying degrees of heat, but the mixture should not be too hot.

While adding Tank A to Tank B, Schulz Silica should be added slowly and the whole mixture stirred gently. The amount of Silica to be added ranges from 100 to 200 pounds to above proportions. 200 pounds are necessary if you desire a thicker and creamier polish. The above proportions produce approximately sixty to sixty-five gallons of polish.

Pine Oil Metal Polish

Although polishing powders are in use, metal polishes usually consist of some abrasive material in suspension in either a liquid or a semi-paste form.

The abrasive material should be selected with care in order not to scratch or otherwise mar the finishes on which the polish is applied. On very delicate finishes only the mildest abrasives should be employed such as rouge (iron oxide) or precipitated chalk (calcium carbonate). For dull surfaces siliceous materials are generally in use.

The Yarmor Steam-distilled Pine Oil is blended with the soap prior to the addition of the abrasive. The Yarmor Pine Oil softens the oxidizable and non-oxidizable material without injuring the surface. In addition, it gives body to the polish and helps hold the abrasive matter in suspension.

A typical formula is as follows:

Tripoli	20.00%
Oleic Acid	7.00%
Sodium Hydroxide (100%)	.50%
Yarmor	25.00%
Water	47.50%

This pine oil formula does the work fast and well and the polish holds a long time, spreads freely, wipes easily and leaves a fine finish. It is non-inflammable and does not possess any ingredients that injure metal surfaces.

Metal Polish

Naphtha	62	lb.
Oleic Acid	1	lb.
Abrasive	7	lb.
Triethanolamine	0.33	lb.
Ammonia (26°)	1	lb.
Water	128	lb.

Preparation

In one container mix together the naphtha and oleic acid to a clear solution. Dissolve the Triethanolamine in water separately, stir in the abrasive, if it is of a clay type, and then add the naphtha solution. Stir the resulting mixture at a high speed until a uniform creamy emulsion results. Then add the ammonia and mix well, but do not agitate as vigorously as before.

Properties

This polish has excellent cleansing properties and removes much of the dullness from metals by the solvent action of Triethanolamine. The emulsion is fairly stable and will not separate as when made from straight ammonia. In use, the metal is first gone over with this polish, which dries leaving a fine white coat. Rubbing with a dry cloth now brings out a high luster.

Variations

The choice of abrasive is very important in making a satisfactory metal polish, and the variety chosen depends upon the metal on which it is to be used. For fine metals, like silver, a jeweler's rouge or a precipitated chalk is used. For brass or nickel, a slightly coarser abrasive is valuable, such as the colloidal clay in the above formula, or a fine silica. A dye is often added to commercial polishes in addition to the other ingredients.

If a non-colloidal abrasive is to be incorporated, it should be mixed with the oleic acid and naphtha instead of with the water, and considerably higher proportions of acid and Triethanolamine will have to be used.

Metal Polish (Paste)

Palm Oil	20	lb.
Yellow Petrolatum	8	lb.
Paraffin Wax	4	lb.
Crocus "B"	12½	lb.
Silex Double Ground	12½	lb.
English Rottenstone Powd.	6	lb.
Bright Red Iron Oxide Powd.	2	lb.
Oxalic Acid	10	oz.
Clovel	8	oz.

Melt the first three items and when clear, while heat is on, add other items slowly while stirring until free from lumps; raise temperature, continuing stirring and run into cans.

Metal Polish

1. Ortho Dichlorbenzol	5
Naphtha or Mineral Spirits	20
Pine Oil	4
2. Trihydroxyethylamine Linoleate	2
Tripoli or Silex	50–75
Suspendite	9
Water	260
3. Ammonium Hydroxide	12

Add "1" to "2" with stirring and then stir in "3"; allow to stand overnight and stir before packaging.

This gives a polish which does not separate if made properly. If a thicker polish or paste is desired the Tripoli is increased and the liquids decreased.

Metal Polish

A. Ammonia 16°	12½	gal.
Alcohol	100	oz.
Oleic Acid	100	oz.

B. Oxalic Acid	10	lb.
H_2O	15	gal.
Ammonia 26°	4¼	gal.

For polish use

A	2½	gal.
B	1¼	gal.
H_2O	35¼	gal.
Air Floated Silex	97	lb.

Mix and run through colloid mill.

---

* Metal Cleaner

Zinc Powder	33.3
Sod. Acid Tartrate	100
Copper Oxide	10
Mineral Oil	to make paste

---

* Cleaner, Metal

Magnesite Powder	700	gm.
Mineral Oil	150	gm.
Oleic Acid	30	gm.
Denatured Alcohol	60	gm.
Sal Ammoniac	60	gm.
Thymol	0.2	gm.

Polish for Metal or Glass

Tallow	96
Whiting	32
Iron Nitrate	4

Warm and grind together.

* Metal Cleaning Pad

A fabric pad is filled with powdered

Calcium Carbonate	90 lb.
Soda Ash	8 lb.
Salt	2 lb.

Mixed Polish

Mixture 1

Carnauba Wax	8 parts
Montan Wax	8 parts
Paraffin Wax	4 parts

These are saponified in a hot solution of:

Potash	3 parts
Water	40 parts

Replace any evaporation with additional warm water. There is then added to this 20 parts of Turpentine.

Mixture 2

No. 1 Polish Black	4 parts
Water	20 parts

These should be milled together in a color mill until thoroughly dispersed.

While Mixture No. 1 is hot, add Mixture No. 2 slowly and with constant stirring. As it cools, the mass will slowly set to a paste. Before it is too stiff for flowing pour into suitable containers and set aside until cold.

These formulae may form the basis for any change which a particular manufacturer might wish to make. Other gums or resins may be substituted and the amounts of water or turpentine varied according to the final consistency desired.

In the formulae calling for carbon black to be ground into water, colloidal carbon would be of great advantage. This material is put on the market as Paris Paste and is a paste of carbon black in water containing 33⅓% carbon black. This paste may be diluted with water so as to give a concentration desired in the formula.

Nickel Silver Castings, Cleaning

If the nickel silver castings have any sand on them, it will be necessary to use a hydrofluoric acid pickle to remove the sand. This pickle is made by using 1 pint of 48 per cent hydrofluoric acid to each gallon of water. The pickle should be used cold, in a lead-lined tank, and care should be taken in handling the acid as it causes severe sores when it comes in contact with the body.

After the work is left in this pickle long enough to remove the sand, it should be rinsed in clean cold water, and then placed in a hot muriatic acid pickle, 1 part acid, 1 part water, to remove any oxidation. It is then immersed in a regular bright dip. This is made by mixing 2 parts sulfuric acid, 1 part nitric acid, and after this is made, add 1 quart of water and ¼ oz. hydrochloric acid to each gallon of the mixture. When it cools to room temperature it is ready to use. After bright dipping, pass the work through a cyanide dip made of sodium cyanide 6 ounces, water 1 gallon. Rinse in clean cold water, then in hot water, and dry in hardwood sawdust.

Buffing Nickel Polish

Double Pressed Saponified Stearic Acid	86 lb.
Paraffine	16 lb.
Edible Tallow	10 lb.
Japan Wax	3 lb.
Silex	376 lb.

Oil Polish

Mineral Oil	60 lb.
Naphtha	26 lb.
Turpentine	3 lb.
Stearic Acid	9 lb.
Triethanolamine	4 lb.
Methanol	4 lb.
Water	120 lb.

Preparation

Mix together the mineral oil, naphtha and turpentine and add the stearic acid. Heat the mixture to about 60° C. at which time the acid will dissolve to give a clear solution.

In a separate container mix the Triethanolamine, methanol and water and heat likewise to 60° C. Then add to this the first mixture and stir vigorously until the emulsion is smooth. Continue with gentle stirring until cool.

Properties

An oil polish of this type can be used both for furniture and automobiles. It can be rubbed dry to leave a glossy finish on the varnish or lacquer surface. Such a polish is more easily applied than a wax polish but it does not leave the same hard and permanent film.

Variations

The cleaning action of this polish can be increased with a slight alteration in formula; namely by the substitution of part of the mineral oil with kerosene or naphtha. Pine oil may also be substituted for the turpentine, or other solvent changes made. When this polish is to be used for lacquers, a fine abrasive is frequently added in small quantity.

Uses

Furniture and automobile polishes.

Soluble Oils, Cutting Oils, Polishes

A. (1) Rosoap 10 lb.
(2) Pine Oil 10 lb.
(3) Mineral (Paraffin) Oil 40 lb.

Mix (1), (2) and then add (3).

B. (1) Rosoap 31 lb.
(2) Rozolin 10 lb.
(3) Denatured Alcohol 4 lb.
(4) Mineral (Paraffin) Oil 159 lb.

The above oils give rich creamy emulsions with water.

Glaze for Paper, Wood or Metal

Casein	100	lb.
Borax	7–15	lb.
Trisodium Phosfate	7–15	lb.
Hexamethylene Tetramine	0.5– 8	lb.
Castor Oil	1– 5	oz.
Clovel	1	oz.

---

* Razor Hone

Carborundum Powder	4
Rubber	30
Factice	17
Red Iron Oxide	49

Mill together until uniform.

---

* Razor Paste

Bauxite	42 gm.
Raw Animal Fat	42 gm.
Powdered Emery	15 gm.
Liquid Varnish	1 gm.

Polishing Rouge

Double Pressed Saponified Stearic Acid	50 parts
Edible Tallow	25 parts
Camphor	3 parts
Paraffine Wax	2 parts
Fine Iron Oxide	20 parts

Shoe Cream

1. Trihydroxyethylamine Stearate 25 lb.
Beeswax 10 lb.
Candellila Wax 30 lb.
Carnauba Wax 40 lb.
Turpentine 20 lb.
2. Water 500 lb.

Heat (1) to 200° F. and in a separate pot heat (2) to 200° F. Run (1) into (2) slowly while stirring vigorously until cold. This gives a beautiful light cream. If a colored cream is desired dissolve some oil soluble dye in the wax mixture while it is melting.

White Shoe Cleaners, Paste
(For use in tubes)

Soap Flakes	10
Proflex	5
Water	35
White Pigment	150

White Shoe Cleaners, Liquid

A.	Soda Ash	1
	Rochelle Salts	2
	Titanox C	40
	Water	57

B.	Soda Ash	0.5
	Soap Flakes	3
	Lithopone	40
	Water	53
	Gum Arabic (50% Sol.)	4

Liquid Shoe Blacking

Nigrosine Base	8
Rozolin	17

Warm and stir until dissolved. Cool and add

Alcohol	24
Acetone	22
Benzol	42

Black Shoe Cream

Montan Wax Crude	15
Carnauba Wax Refined	15
Rosin	3
Caustic Potash	6
Soap Flakes	1
Water	156
Nigrosine (Water Soluble)	4

Shoe Cream, Neutral

Hydrowax Cream	50

Heat to 200° F. and to it add following solution warmed to 150° F. and stir until smooth.

Turpentine	29
Water	24
Proflex	3
Soap Flakes	1

White Shoe Dressing

Pipe Clay	450 gm.
Spanish Whiting	225 gm.
Flake White	180 gm.
Precipitated Chalk	115 gm.
Powdered Tragacanth	8 gm.
Phenol	4 gm.

Water to make a paste.

Shoe Polish, Paste

Carnauba Wax	20
Paraffin Wax	12

Heat to 200° F. and add to this slowly with good stirring while heating on a steam table

Turpentine	65
Carbon Black No. 1	2.5
Oil Soluble Black Dye	0.5

Stir until uniform.

Non-Caking Shoe Dressings

White shoe polishes, especially, have tendency toward cake formation of the pigments. This can be overcome by grinding the pigment with Aquaresin G.M. The latter forms a thin film around each particle of pigment. While this does not prevent settling, it does prevent formation of a hard cake and slight shaking distributes the pigment thoroughly.

Shoe Polish and Preservative

Carnauba Wax	2 parts by wt.
Beeswax	2 parts by wt.
Neatsfoot Oil	1 part by wt.

Heat by hot water bath (not over fire) till melted, and then add turpentine until a soft paste is obtained when the mixture is cold. This should be applied to the clean, dry leather with a rag or a piece of waste, and rubbed hard until no more polish is absorbed. Polish with a clean cloth. A higher polish will be obtained by reduction of the proportion of oil, but the leather will not be so well preserved.

* Shoe Uppers, Preserving

Larch Turpentine	10–32
Beef Tallow	45–55
Oil Birch Tar	8–14
Bone Oil	18–27

Shoe Dye

Shellac	12.7 kg.
Borax	3.2 kg.
Water	82.0 kg.
Carnauba Wax	6.3 kg.
Marseilles Soap	1.5 kg.
Potassium Carbonate	0.3 kg.
Nigrosin	12.0 kg.
Water	32.0 kg.

The shellac solution in borax and water is made first, the carnauba wax is emulsified in the soap, carbonate solution as above and the nigrosin and water added to it, it is then added to the shellac soln. with rapid agitation. Some ammonia may be added to prevent lumps.

Cold Polishing Dyes for Dressing Shoes

Carnauba Wax	7.5 kg.
Marseilles Soap	1.0 kg.
Potassium Carbonate	1.5 kg.
Water	79.0 kg.

Melt the carnauba wax, and add the heated mixture of the other ingredients. Stir rapidly, and add 11 kg. nigrosine previously dissolved in a small amount of the soap soln.

Dyeing "Shoe" Plush Brown

Four pieces of "shoe" plush weighing approximately 320 pounds are immersed in the dye bark at 120° F. and run for ten minutes or until thoroughly wet out. Two pounds of borax, seven pounds of trisodium phosphate and twelve pounds of olive soap are now added to the bath, which contains 800 to 850 gallons of water. The scouring is then continued at 120° F. for an additional 30 minutes. A 20 minute wash in a bath containing two pounds of trisodium phosphate follows. This wash is followed by three 15 minute rinses with water at 120° F. and one cold rinse. If soft water is not available, a small amount of soda ash is added to the first rinse to avoid the formation of any hard soap which would be extremely difficult to rinse out of the dense pile. The rinsing, even though it may seem too much, is vitally important to ensure the absence of all soap in the ensuing processes.

The cloth is dyed brown by running in a bath containing 30 pounds of potassium permanganate and 1 pound of zinc dust at 120° F. for one and a half to two hours. An addition of 5 to 10 pounds of potassium permanganate is usually necessary to obtain the desired depth of shade. Following the dyeing the cloth is rinsed at 160° F. with water made very slightly alkaline by the addition of one and a half pounds of trisodium phosphate. Two warm rinses complete the process.

Pure Turpentine Shoe Polish

Melt together the following:

Carnauba Wax	20
Paraffin Wax	12

In a separate vessel put the following:

Turpentine	65
No. 1 Polish Black	2.5
Oil Soluble Black Dye	0.5

Heat this to slightly above the melting point of the waxes. As soon as this point is reached, add the turpentine to the melted waxes, which should be just above their melting point. Stir vigorously and cool. The stirring should be continued during the cooling. As soon as it is cooled to a thin paste, pour into cans where it will further cool to a stiff paste.

Saponified Water-Wax, Shoe Polish

Mixture 1

Carnauba Wax	8 parts
Montan Wax	8 parts
Paraffin Wax	4 parts

These are saponified in a hot solution of:

Potash	3 parts
Water	50 parts

Replace any evaporation with additional warm water.

Mixture 2

No. 1 Polish Black	4 parts
Water	25 parts

These should be milled together in a color mill until thoroughly dispersed.

While Mixture No. 1 is hot, add Mixture No. 2 slowly and with constant stirring. As it cools, the mass will slowly set to a paste. Before it is too stiff for flowing pour into suitable containers and set aside until cold.

Shoe Cream, Black

A. Crude Montan Wax	18 kg.
Japan Wax	2 kg.
Carnauba Wax	4 kg.
Rosin	2 kg.

B. Water	260 kg.
98% Potash	6 kg.
Water-Soluble Nigrosin	12 kg.

Heat A and B separately to 95–100° C. and add B to A while stirring vigorously with an electric mixer.

Shoe Polish

1. Carnauba Wax 55 parts
   Crude Montan Wax 55 parts

are melted at 105–110° C.

Nigrosine Base 10 parts

dissolved in

Stearic Acid 20 parts

added, then

Ceresine 150 parts

and finally

Turpentine Oil 900 parts

The mass is filled at 45° C. (105° F.).

---

2. Carnauba Wax	65 parts
Crude Montan Wax	40 parts
Dyestuff Soluble in Oil	30 parts
Paraffin	110 parts
Ozokerite	10 parts
Turpentine Oil	760 parts

---

3. Carnauba Wax	65 parts
Crude Montan Wax	40 parts
Dyestuff Soluble in Oil	30 parts
Paraffin	40 parts
Ceresine	75 parts
Turpentine Oil	760 parts

It is recommended to use only stearic acid or crude Montan wax for dissolving the bases, as oleine or mixtures of crude Montan wax with oleine do not give such fine surfaces.

For Floor Polishes

1. Carnauba Wax	15 parts
Paraffin	26 parts
Ceresine	32 parts
Benzine	170–180 parts

Color to suit with any oil soluble color.

---

2. Carnauba Wax	60 parts
Paraffin	104 parts
Ceresine	128 parts
Turpentine	600 parts
Naphtha	100 parts

Shoe Polish

Beeswax	1 lb.
Ceresin Wax	1 lb.
Carnauba Wax	6 oz.
Turpentine	3 pt.
Yellow Soap	6 oz.
Oil Soluble Black Anilin	enough to color
Water	sufficient

Shave the soap and dissolve in the smallest possible quantity of water by means of heat, melt the waxes together, add the turpentine and stir well, then add the anilin dye and stir in the soap solution, continuing to stir until cold.

Shoe Polish

The basis of most paste polishes at the present time is beeswax. Sometimes some carnauba wax is used to give hardness to the polish and experiences indicates that a higher polish can be obtained where this ingredient is present. The turpentine in the polish serves to keep it soft and allows it properly to penetrate the leather, while the soap gives the necessary easy rubbing qualities. Knowing this, it is easy to modify any given formula so as to meet requirements. If, for instance, the gloss obtained is not high enough, it indicates that more wax should be used; if the polish dries out too rapidly use more turpentine; if it rolls under the dauber, use more soap, and so on.

Beeswax	1 lb.
Ceresin	1 lb.
Carnauba Wax	6 oz.
Turpentine	3 pt.
Yellow Soap	6 oz.
Oil-Soluble Black Anilin	enough to color
Water	sufficient

Shave the soap and dissolve in the smallest possible quantity of water by means of heat, melt the waxes together, add the turpentine and stir well, then add the anilin dye and stir in the soap solution, continuing to stir until cold.

Black Shoe Polish

Montan Wax	15
Paraffin Wax	10
Beeswax	4
Japan Wax	4
Nigrosine Base	3
Turpentine	64

Shoe Polish

Double Pressed Stearic Acid	2 parts
Linseed Oil	1 part
Turpentine	6 parts
Soap Flakes	1 part
Water	10 parts
Pigment to Color	optional

---

* Silver Polish

1. Infusorial Earth	48	lb.
2. Diglycol Stearate	7	lb.
3. Soda Ash	1	lb.
4. Trisodium Phosphate	1	lb.
5. Water	70	lb.
6. Clovel	½	lb.

Heat 2 and 5 to 150° F. and stir until homogeneous. Add the other ingredients and mix to a smooth paste.

Silver Polish

Castile Soap	10 parts
Water	50 parts
Tripoli Powder	10 parts
White Rouge	5 parts
French Chalk	15 parts
Petroleum	5 parts

Polish, Silver

Water	1	qt.
Soap Flakes	4	oz.
Whiting	8	oz.
Ammonia	½	oz.

---

* Silver Cleaner

Infusorial Earth	20%
Sod. Oleate	20%
Salt	5–15%
Water	balance

Liquid Stove Polish

Crude Montan Wax	2
Rosin	1
Carnauba Wax	2

Heat to 90° C. with stirring and to it add slowly

Caustic Potash	2
Water (Boiling)	86
Nigrosine	3

Keep on heat and agitate vigorously until uniform. Cool and work in

Graphite Flake	5
Lampblack	3

Mix thoroughly until uniform.

Suede Cleaner

Precipitated Chalk or Whiting	12	lb.
Quilaya Bark	20	lb.
Cream of Tartar Powder	60	lb.
Oil Birch Tar	1½	oz.

---

* Tile and Marble Polish

Sod. Silicate	1
Linseed Oil	1
Precipitated Chalk	1
Magnesium Chloride	0.2
Water	10
Gelatin	0.1

Tripoli Composition No. 2

Stearic Acid	55 lb.
Edible Tallow	2 lb.
Oleo Stearine	5 lb.
Rosin	9 lb.
Petrolatum	40 lb.
Japan Wax	1 lb.
Flint	315 lb.
Tripoli Flour, Double Ground	93 lb.
Ponolith	2 lb.

Tripoli Buffing Stick

Double Pressed Saponified Stearic Acid	30 parts
Edible Tallow	25 parts
Paraffin Wax	25 parts
Tripoli Flour	20 parts

(or as much as will be absorbed)

A buffing or polishing paste may be made using the above formulae with the addition of a small amount of turpentine and of water to bring to the consistency desired.

Grease Stick for Buffing and Polishing Purposes

Single Pressed Saponified Stearic Acid	25 parts
Edible Tallow	70 parts
Paraffine Wax	5 parts

Vienna Lime Composition

Double Pressed Saponified Stearic Acid	45 lb.
Edible Tallow	15 lb.
Vienna Lime	200 lb.
Ponolith	2½ lb.

Polishing Wax

Montan Wax	15
Carnauba Wax	5
Candelilla Wax	2
Paraffin Wax	3
Japan Wax	1
Turpentine	75

Liquid Polishing Wax

Beeswax	5
Ceresin	20

Melt together and cool to 65° C. Stir in slowly

Turpentine	85
Pine Oil	2.5

Window Cleanser

Castile Soap	2 parts
Water	5 parts
Chalk	4 parts
French Chalk	3 parts
Tripoli Powder	2 parts
Petroleum Spirits	5 parts

Wood Polish

Carnauba Wax	33 parts
Beeswax	66 parts
Dipentene	75 parts
Turpentine or White Spirit	225 parts
Soap	1 part
Water	10 parts

The soap is dissolved in water (hot) and the waxes are dissolved in the dipentene. When cool the solutions are mixed with vigorous shaking or stirring.

* Wood Preservative and Finish

Creosote Oil	4
Alcohol	1
Paste Wood Filler	4
Turpentine	2
Hydrochloric Acid (Conc.)	1

Furniture Polish

Pale Paraffin Oil	3 parts by vol.
Benzol	2 parts by vol.

This polish is being used by one of the largest furniture houses in America. The benzol softens the surface permitting the oil to leave a thin film on surface.

* Synthetic Spinel

A synthetic spinel having a permanent aquamarine color has an approx. compn. of alumina 92, magnesia 8, chromic oxide 0.12, cobaltic oxide 0.025 and titanic oxide 0.3%.

Jewelry Polish Powder

Marble Dust	90%
Jeweler's Rouge	10%

Non-Slippery Rubless Floor Polish

Carnauba Wax Nos. 1 or 2	500 lb.
Hydromalin	276 lb.

Heat with stirring for ½ hour to 120–140° C. Cool to 100° C. and add slowly with vigorous mixing

Water (Boiling)	3560 lb.

Stir until uniform; allow to stand overnight and add slowly while stirring

Sodium Silicate	80 lb.
Tescol	20 lb.

Sand Papers and Emery Papers

For this line of work the demand is primarily for glues of the higher viscosities, but a strong jelly strength is deemed important. The first treatment consists of sizing paper with a 10% glue solution. Paper is festooned until dried. Upper surface is then coated with a 35–40% glue solution, upon which the abrasive grain is sprinkled. The whole is again dried. The third treatment consists of applying a 10% solution of the same glue to bind the grains firmly together and to the paper. Again abrasive grains are sifted over surface, and then paper passes into drying chambers.

Abrasive Wheels

For polishing steel, iron, copper, etc., wheels composed of paper or felt disks are coated with hide glues at a proportion of 1 part glue 2 parts water which has been dissolved in the customary manner. Glue is applied to wheel at temperature of 140° F., and then wheel is promptly rolled into desired sized abrasive grain, and then allowed to dry for 24 to 48 hours, after which it is ready for use.

REPAIRING, RENOVATING, REMOVING STAINS

Press-Marks on Celanese-Garments

In order to remove such lustrous spots from dull finish Acetate rayon often a good result is obtained (in case of plain colored garments) by soaking the whole garment for 1 hour in pure Methanol with addition of a little Castor Oil. The amount of liquid should be just enough to perfectly penetrate the garment without any excess liquid. Thus bleeding of colors is avoided. The spots will disappear due to swelling action. Sometimes it is advisable to rub and slightly pull the parts having marks, to loosen the fibers, melted by the heat, from each other. Then the garment is dried on a hanger with a fan.

A Non-Inflammable Cleaning Liquid

The following can be used for a variety of purposes. It removes grease spots from delicate fabrics, fat and tarnish from jewelry, tableware, copperware and ironware. It will also kill moths and insects:

Kerosene	1 oz.
Carbon Tetrachloride	3 oz.
Oil of Citronella	2 drm.

Mix and filter if necessary. The carbon tetrachloride must be free from carbon bisulphide. If the latter is present, a fact which can easily be ascertained by the smell, the carbon tetrachloride must be shaken with charcoal and filtered.

Cleaning Colored Concrete

Colored concrete surfaces may be cleaned and made more impervious by washing with liquid soap. When this treatment is used the soap should be applied and allowed to stand overnight, being washed off thoroughly the next morning.

The application of ordinary floor wax once a month after the concrete is dry and clean will produce deep colors, improve the wearing surface and make it easy to keep clean. After the first two or three waxings, unless the surface is to be subjected to unusually severe wear, waxing twice a year will be sufficient.

Marble, Cleaning

A solution of potassium permanganate about ½ per cent strength is made, the permanganate being dissolved in a little hot water. This is a product which can be obtained from almost any chemist; this is then brushed into the marble until uniform penetration is obtained. Before it is allowed to dry, it is treated with a solution of ammonia and a little sodium hydrosulphite in warm water. When making up this solution it is essential to add the ammonia first as otherwise the hydrosulphite will be decomposed; this is then sponged on to the marble when the violet coloration of the permanganate will entirely disappear leaving a clean white product. This method can be applied efficiently on floors which become discolored through age, etc. If one application is not enough it can easily be repeated without harming the marble in any way whatsoever. If the floor is very greasy an initial washing with soda ash may be resorted to being well rinsed with clean water before applying the permanganate solution.

Stains, Blacking Removing

The following will probably be effective:

1 part Nitrobenzene (Oil or Mirbane)
7 parts Phenol (Carbolic Acid, U. S. P. 90% Solution)

After application, rinse well with alcohol.

Removing Stains

Stain	Treatment

Albumen.—Soak for a few hours in Pepsin 25, Hydrochloric Acid (25%) 50, Water 100 at 45° C.

Antimony Compounds.—Ammonium Sulfide solution.

Arsenic Compounds.—Ammonium Sulfide solution followed by ammonium hydroxide if necessary.

Asphalt, Gilsonite.—Soften by rubbing with warm petrolatum or mineral oil or tetralin and dissolve with following: Benzol 1, Carbontetrachloride 1, Trichlorethylene 1, Ethylene Dichloride 1.

Balsams.—Ether, Toluol or Chloroform.

Beer, Champagne.—Ammonium Chloride 2, Glycerin 2, Alcohol 2, Water 7 followed by water.

Blood.—Sodium Hydrosulfite or Trisodium Phosfate and Hydrogen Peroxide.

Burnt Sugar.—Glycerin 10, Water 10, Isopropyl Alcohol 20.

Cadmium Compounds.—Pot. Cyanide (poisonous) and thorough removal with water.

Chromic Compounds, Chromates.—Sod. Bisulfite or Sod. Hyposulfite and dilute sulfuric acid.

Cobalt.—Pot. Cyanide (poisonous) Solution followed by water.

Copper.—Warm 25–30% Pot. Iodide Solution.

Egg Yolk.—Soften with glycerin and treat with Alcoholic soap solution.

Grass.—Alcohol or Chloroform or Zinc Chloride 2% solution.

Henna.—Hydrogen Peroxide 10% 20, Am. Chloride 4, Water 20.

Iodine.—10% Pot. Iodide followed by 10% Sod. Thio Sulfate folowed by water.

Iron Salts.—Sod. Hydrosulfite 8% solution.

Lacquer.—Trichlorethylene 5, Paraffin Wax 1, Acetone 1, Benzol 1, Tetralin 1, Methanol 1.

Lead Compounds.—Stain with Tinc. Iodine; dry and dissolve with concentrated pot. iodide solution.

Manganese.—10% Am. Sulfate Solution followed by dilute Hydrochloric Acid then water.

Mercury.—5–10% Solution Pot. Cyanide (poisonous) followed by water.

Milk.—Ether or Ethylenedichloride followed by warm borax solution.

Mold.—3% Hydrogen Peroxide, Am. Chloride 4, Alcohol 10, Water 70.

Nickel.—10% Solution Pot. Cyanide (poisonous) then water.

''Nicotine.''—On skin—Sodium Sulfite 25, Water 100, Hydrochloric Acid 2 or 10% Hydrogen Peroxide 10, Am. Chloride 1, Alcohol 5.

Oil or Fat.—Glycol Oleate 1, Hexalin 2, Carbon Tetrachloride 1 followed by any dry cleaning solvent.

Perspiration.—10% Borax Solution or 10% Am. Carbonate Solution.

Picric Acid.—20% Solution Sod. Sulfate followed by soap and water.

Rust.—Pot. Binoxalate 1, Water 44, Glycerin 1, allow to remain for a few hours and wash.

Silver.—10% Solution Sod. Hydrosulfite (warm) for 15 minutes followed by soap and water.

Urine.—Citric Acid 10% followed by hot water.

Varnish.—Rosin Oil 1, Ethyl Acetate 1, Tetralin 1, Amyl Alcohol 1, Ammonium Hydroxide 1, Alcohol 1.

Vomit.—Ammonium Chloride 10% solution, followed by alcoholic soap and then water.

Water.—Rub with flannel wet with 5% White Mineral Oil and 95 Toluol.

Wine, Fruit.—Acetic or Tartaric Acid (10%) or Hydrogen Peroxide (10%) 5, Am. Chloride 20, Water 75.

Marble and Concrete Stain Removal

While practically every type of stain can be removed from concrete without appreciable injury to either the texture or color, the eradication of old stains which have been long neglected may require considerable patience. It is often a matter of repeating the treatment day after day until the desired results are attained. It is not always possible to determine what the staining matter is, and hence the treatment sometimes has to be a matter of experimentation. Usually the staining matter will be found to exist in a stable form, and its removal may require several applications of a solvent which does not appreciably affect the surface. A considerable variety of chemicals may be applied to concrete without appreciable injury, but acids or those chemicals which develop an acid condition should be carefully avoided. Even weak acids, such as oxalic and acetic, may show their effects on the surface if left on concrete for a considerable length of time.

Usually stains penetrate to such an extent that they cannot be readily removed by merely applying the proper chemical to the surface or by scrubbing the stained part and it is necessary to resort to a poultice or bandage. A poultice is made by mixing one or more chemicals with a fine inert powder to a pasty consistency. This is applied to the stain in a thick layer. The bandage treatment consists of a layer of cotton batting or a few layers of cloth soaked in a chemical solution and pasted over the stain. A stain may be eradicated, first by dissolving the staining matter and drawing it out by capillary suction or driving it back from the surface; and, second, by converting the coloring matter into a form which does not show as a stain. In removing an oil stain it is usually necessary to apply a solvent and draw the dissolved oil out. An iron stain is more satisfactorily treated by applying a reducing agent, although means must be taken to prevent the reoxidation of the iron and the reappearance of the stain. This is accomplished by an application of sodium citrate solution. Some chemicals used for removing stains are very unstable and decompose under certain conditions, producing stains of their own which may be more troublesome than the original. This is particularly true of the hydrosulphite ($Na_2S_2O_4$) used in removing iron stains, but unless the method of application described is rather closely followed a yellow stain will result. If the poultice is left on several hours, a black stain may develop, which is probably due to the formation of a sulphide of iron. Some staining matter is easily dissolved by a surface scrubbing and apparently removed, but as the area dries the stain may reappear. Tobacco stains scrubbed with a solution of washing soda may disappear in this way, but reappear stronger than before due to the solvent driving the staining matter into the surface in stronger concentrations. The chief function of a poultice is to draw dissolved staining matter out of the surface. In some cases a porous paper or blotter pasted to the stained surface after the proper solvent has been applied may be made to answer the purpose. When a stain has to be treated with a very volatile solvent, such as benzol, ether, acetone, etc., it is best to use a slab of stone or brick over the solvent. This prevents a rapid evaporation of such solvents, prolonging their action and affording a capillary action similar to a poultice. When so used, the stone or brick should be thoroughly dry.

In some cases it may not be possible to determine the type of stain. Many stains are yellow or brown, resembling iron rust. Oil stains when new resemble the oil itself, but after a considerable period of time they are apt to become yellow or dark brown. Copper and bronze stains are usually green, although, due to the iron or manganese content, or due to the alteration of fine particles of pyrites in the concrete, bronze sometimes causes a brown stain. In experiments on copper stains, made with a solution of copper sulphate, a brown stain was found on the surface after the copper stain had been removed. This yielded readily to the treatment for iron stains, indicating that it was caused by the alteration of some element in the surface, since the copper salt applied was "chemically pure."

Concrete in certain parts of buildings is apt to become stained from the perspiration or oil from the hands. Such discolorations sometimes become very prominent and resemble iron stains. This stain is not as difficult to remove as those caused by lubricating or linseed oils.

Under damp conditions, wood will rot and finally produce a chocolate-colored stain. When pine wood burns, pitch from the wood may penetrate the surface and produce a stain which is almost black. The eradication of such stains is a slow process, but in many cases it may be entirely practical.

1. Treatment of Iron Stains

Iron stains can usually be recognized by their resemblance to iron rust or by their position with respect to steel members of the structure.

Method No. 1.—Dissolve 1 part sodium citrate in 6 parts of water and mix this thoroughly with an equal volume of glycerin. Mix a part of this liquid with whiting to form a paste just stiff enough to adhere in a thick coating to the surface. Apply this to the stained area with a putty knife or trowel. This will become dry in a few days and it should then be replaced with a new layer or softened by the addition of more of the liquid. While this treatment has no injurious effects, its action may be too slow to be practical in cases of intense stains. Ammonium citrate may be used instead of sodium citrate to obtain somewhat quicker results, but, due to the development of an acid condition, it may injure a polished surface slightly.

Method No. 2.—For deep and intense iron stains it is more satisfactory to employ sodium hydrosulphite ($Na_2S_2O_4$). Before applying the hydrosulphite to the stain the surface should be soaked for a few minutes with a solution of sodium citrate made by dissolving 1 part of the citrate crystals in 6 parts of water. To apply the citrate solution, dip a white cloth or piece of cotton batting into the solution and paste it over the stain for 10 or 15 minutes. If the stain is on a horizontal face, sprinkle a thin layer of the hydrosulphite crystals over it, moisten with water, and cover with a stiff paste of whiting and water. If the stain is on a vertical face, place a layer of the whiting paste on a plasterer's trowel, sprinkle on a layer of the hydrosulphite, moisten slightly, and apply it to the stain. Remove after one hour. If the stain is not all removed, repeat the operation. Unless the stain is deep, one treatment will be sufficient. When the stain disappears, rinse the surface thoroughly with clear water and make another application of the citrate solution as at first. Although the polish is apt to be dimmed somewhat by this treatment, it is not a difficult matter to repolish th treated portion.

2. Copper or Bronze Stains

Such stains are found where the was from bronze, copper or brass runs ove concrete. The stain is nearly alway green, being due to the formation of th carbonate of copper, but bronze apparently causes a brown stain in som cases. The green stains may be eradicated in the following way:

Method No. 1.—Mix dry 1 part of ammonium chloride (sal ammoniac) and parts of powdered talc. Add ammoni water and stir into a paste. Place thi over the stain and leave until dry. A stain of this kind that has been collecting for several years may require several repetitions of this procedure t completely remove it. Sometimes aluminum chloride is employed instead o sal ammoniac.

Method No. 2.—Dissolve 8 ounces o potassium cyanide in 1 gallon of water Saturate a thick white cloth in the solution and place it over the stain. Whe the cloth has become dry, soak it agai in the cyanide solution and repeat th operation until the stain disappears Sometimes it may be advantageous t combine this and the method above that is, remove the greater part of th stain with the poultice and finish wit the cyanide solution. This solution i very poisonous if taken into the system

3. Ink Stains

Inks are of various compositions, an require different treatments.

Ordinary writing inks usually consis of gallotannate of iron, a blue dye, mineral acid, phenol and a gum or glycerin. Such an ink may etch the sur face of concrete due to the acid content To remove a stain of this type, make strong solution of sodium perborate i hot water. Mix this with whiting to thick paste, apply in a layer ¼-inc thick, and leave until dry. If some o the blue color is visible after this poul tice is removed, repeat the process. I only a brown stain remains, treat it b Method No. 1 for iron rust. Sodium per borate can be obtained from any drug gist. Repolish the surface if necessary

Synthetic Dye Inks.—Many of th red, green, violet, and other bright col ored inks are water solutions of syn thetic dyes. These contain no acid an do not etch concrete. Stains made b this type of ink can usually be remove by the sodium perborate poultice de scribed above. Often the stain fro such inks can be removed by applying ammonia water on a piece of cotton batting. Javelle water may also be effectively used in the same way as ammonia water or mixed to a paste with whiting and applied as a poultice. A mixture of equal parts of chlorinated lime and whiting reduced to a paste with water may also be used as a poulticing material.

Prussian Blue Inks.—Some blue inks contain Prussian blue, which is a ferrocyanide of iron. Stains from this type of ink cannot be removed by the perborate poultice, Javelle water, or chlorinated lime poultice. Such stains yield to a treatment of ammonia water applied on a layer of cotton batting. A strong soap solution applied in the same way may also be effective.

Indelible Ink.—This type of ink often consists entirely of synthetic dyes. Stains from dye inks may be treated as outlined above for that type. However, some indelible inks contain silver salts which cause a black stain. This may be removed with ammonia water applied on a layer of cotton batting. Usually several applications will be necessary.

4. Tobacco Stains

Method No. 1.—The grit scrubbing powders, commonly used on marble, terrazzo, and tile floors are usually satisfactory for application as a poulticing material on this type of stain. Stir the powder into a pail of hot water until a mortar consistency is obtained. Mix thoroughly for several minutes, then apply to the stained surface in a layer about one-half inch thick. Leave this on until dry. In most cases two or more applications of the poultice will be necessary.

Method No. 2.—If the scrubbing powders called for in Method No. 1 are not at hand, the following procedure may be used. Make up a soap solution by dissolving about 1 cubic inch of soap in a quart of hot water. In another vessel dissolve one large tablespoonful of soda ash or two tablespoonfuls of washing soda in one pint of water. Combine equal parts of these two solutions and apply a portion of it to the stained surface with a mop, or saturate a piece of cotton batting in the liquid and place it over the stain for a few minutes. Make up a poultice by mixing a portion of the soap and soda solution with powdered talc or whiting. Apply this to the stain and leave until dry. Scrape it off and repeat if necessary. Powdered talc is preferable to whiting, since it holds the moisture longer and thus prolongs the action of the active chemicals. It also has the advantage of being easier to remove from the surface after it has dried. Whiting is apt to cling so firmly that it has to be moistened before it can be scraped off. This is an undesirable feature, since the dried poultice contains the staining matter, and if it has to be soaked loose from the surface some of the staining matter is apt to be driven back into the concrete. If the paste is made of the proper consistency, it can be applied with a paint brush. A whiting paste has the desired brushing properties, but in order to make the talc poultice work well as a brushing coat it is necessary to add a teaspoonful of sugar to each pound of talc. Powdered talc in the raw state is of low cost, but is not always easily obtained. When only a small amount is required, one may employ the cheaper grades of talcum powders or purchase the unscented grades from automobile tire distributers.

Method No. 3.—The following formula will be found to be somewhat more efficacious than either of the foregoing: Dissolve 2 pounds of trisodium phosphate crystals in 1 gallon hot water. Mix the contents of a 12-ounce can of chlorinated lime to a paste in a shallow enameled pan by adding water slowly and mashing the lumps. Pour this and the trisodium phosphate solution into a stoneware jar and add water until approximately 2 gallons are obtained. Stir well, cover the jar, and allow the lime to settle. For use add some of the liquid to powdered talc until a thick paste is obtained, and apply as a poultice ¼-inch thick with a trowel. If it is desired to apply this with a brush, add about one teaspoonful of sugar to each pound of powdered talc. When dry scrape off with a wooden paddle or trowel. This mixture is a strong bleaching agent and is corrosive to metals, hence in using it care should be taken not to drop it on colored fabrics or metal fixtures.

This formula is also valuable for treating other stains and will be frequently referred to in the following methods. Trisodium phosphate may be purchased at most drug stores, at chemical supply houses, or laundry supply houses.

5. Urine Stains

Use Method No. 3 as outlined above for tobacco stains. Should some part of the stain prove stubborn, saturate a layer of cotton batting in the liquids and paste over that part of the surface. Resaturate the cotton if necessary.

If the polish has been injured, moisten a piece of felt cloth or chamois skin with water, dip it into some FF carborundum or emery flour and rub the surface until it appears smooth and glossy. Then polish with putty powder in the same manner until the desired finish is obtained. When applying the putty powder, use a new piece of felt or chamois skin.

6. Fire Stains

Concrete is often badly discolored from smoke or pitch from burning wood. Sometimes the original appearance may be restored by the following process: Scour with powdered pumice or a grit scrubbing powder to remove the surface deposit, then make a solution of trisodium phosphate and chlorinated lime as described in Method No. 3 for tobacco stains. Fold a white Canton flannel cloth to form three or four layers and saturate it in the liquid. Paste this over the stain and cover it with a piece of pane glass or a scrap slab of concrete, making sure the cloth is pressed firmly against the surface. Resaturate the cloth as often as necessary. Deep pitch stains are difficult to remove, and hence several treatments will be necessary. To restore the polish, use the method described above under method of treating urine stains.

7. Lubricating Oil Stains

Lubricating oil penetrates quite readily, and if accidentally dropped on the surface of concrete it should be mopped off immediately with a cloth and covered with fuller's earth or other dry powdered material, such as hydrated lime or whiting. In some cases a layer of dry portland cement will serve the purpose. The oil that has penetrated may usually be removed in this way if treated soon after the stain occurs. However, when the oil has remained on the surface for a considerable period of time and thoroughly oxidized, other methods will be necessary.

Method No. 1.—Place over the stain a piece of white Canton flannel somewhat larger than the stain and saturated in a mixture of equal parts of acetone and amyl acetate. Cover with a piece of pane glass, or preferably a small slab of concrete. If the stain is on a vertical surface it will be necessary to improvise a means of holding the cloth and its covering in place. When the cloth becomes dry, it should be again saturated and covered as at first. Old oil stains are difficult to remove and their treatment may require a great deal of patience. If the solvent tends to spread the stain, a larger cloth should be used. In covering the saturated cloth with a piece of glass the stain is driven into the concrete, while if a dry slab of concrete is used, some of the oil will be drawn into it.

Method No. 2.—A method frequently used consists in mixing a solvent, such as benzol or gasoline, with a dry powder such as hydrated lime, marble dust, or whiting, to form a paste which is plastered over the stain. While this method is said to be satisfactory for such oil stains as occur in construction, it acts slowly on old oil stains which have dried and oxidized.

Method No. 3.—Lubricating oil stains can be removed with more facility where the following method can be used. Place a layer of asbestos fiber about one-fourth inch thick over the stained portion, saturate it with amyl acetate, and cover with a scrap slab of concrete. Place on top of the auxiliary slab a hot iron of about the temperature used for pressing fabrics. Apply more of the amyl acetate as the asbestos becomes dry and reheat the iron as often as necessary. A few layers of Canton flannel may be used instead of asbestos fiber if care is taken not to scorch the cloth. Stains from scorched cloth may be removed by the same method recommended for fire stains.

8. Linseed-Oil Stains

This type of stain is usually found around plumbing fixtures where putty has been used. The linseed oil from the putty may spread for some distance through the concrete and produce a stain that is very difficult to remove. The oil in oxidizing forms a "resinous matter" which practically seals the pores and effectively prevents the penetration of any solvent which may be applied. The use of putty for filling around pipes where they pass through concrete is objectionable because of the stains that are apt to occur. Grafting wax is more desirable for this purpose as it does not stain the concrete and can be easily removed.

Experiments have been made on several treatments applied to the inside walls of openings through concrete to prevent the penetration of linseed oil from putty. The only application of this kind that was found effective consisted of sodium silicate. At least two applications of the sodium silicate should be made, the first consisting of the commercial silicate diluted with twice its volume of water, and the second consisting of the undiluted silicate. This should be applied with a brush, and ample time should be allowed for each application to dry.

Method No. 1 recommended for use on lubricating oil stains will slowly dissolve this "resinous matter" and reduce the stain, but it is not well adapted to use around plumbing fixtures. The coloring matter in such stains may be bleached as follows:

Method No. 1.—Cut a piece of thick white cloth or a layer of cotton batting to fit around the fixture. Saturate this with hydrogen peroxide and paste it over the stain. The bleaching action may be accelerated by moistening another cloth in ammonia water and placing this over the first. Repeat the operation as described until the discoloration is removed.

Method No. 2.—Mix dry one part trisodium phosphate, 1 part sodium perborate, and 3 parts powdered talc. Make a strong soap solution in hot water and add enough of this to the dry mixture to form a thick paste. Cover the stain with the paste and leave until dry. The same material can be used over again by reducing it to a paste with some more of the soap solution. In some cases it may be found desirable to alternate this treatment with Method No. 1 for lubricating oil stains.

Method No. 3.—Combine equal parts of wood alcohol and a 10 per cent solution of trisodium phosphate. Make a paste of this mixture and asbestos fiber sufficient to cover the stain with a layer one-fourth inch thick. Place a scrap slab of concrete over this and apply a hot iron as described in Method No. 3 for lubricating oil stains. A few repetitions of this process may be necessary in cases of very pronounced stains.

9. Rotten Wood Stains

Under damp conditions wood will rot and cause a chocolate-colored stain on concrete which is readily distinguished from most other stains by its dark color. The best treatment found for this type of stain is that recommended for fire stains. The action may be accelerated by first scrubbing the surface thoroughly with glycerin diluted with four times its volume of water.

10. Coffee Stains

Coffee stains can be removed by saturating a cloth in glycerin diluted with four times its volume of water and pasting it over the stained portion. Javelle water, or the solution used on fire stains, will also prove effective.

11. Iodine Stains

This stain will gradually disappear of its own accord within a few weeks time. It may be quickly removed by applying alcohol and covering with whiting or talcum powder. If the stain is on a vertical wall, mix the talcum to a paste with alcohol, apply some alcohol to the stain, and then cover it with the paste. One application will usually prove sufficient.

12. Barium Sulphide Stains

The yellow stain left by barium sulphide and other alkaline sulphides may be removed by applying a weak solution of potassium cyanide. Dissolve a teaspoonful of potassium cyanide in a glass of water, saturate a piece of cotton batting in the liquid, paste it over the stain, and leave until dry. One or two applications will usually suffice. The cyanide is very poisonous if taken into the system.

13. Perspiration Stains

Secretions from the hands or oil from the hair may produce stains on concrete. The stain is brown or yellow and may be mistaken for an iron stain. The best treatment found is that recommended for fire stains. Bad stains of this kind are rather stubborn and may require several treatments.

14. General Service Stains

The general cleaning and care of terrazzo floors is discussed in another data sheet. However, when certain areas become yellow while adjacent slabs remain free from discoloration, the trouble is probably due to the original finishing of the floor. Such discolorations are not usually hard to remove by poultice methods, or they may yield to a surface scrubbing with Javelle water. Javelle water can usually be purchased at drug stores or may be prepared as follows:

Dissolve 3 pounds of washing soda in 1 gallon of water. Mix the contents of a 12-ounce can of chlorinated lime to a paste in a shallow enameled pan by adding water slowly and mashing the lumps with a spatula or pointing trowel. Add the paste to the soda solution, make up to 2 gallons by adding water, and place in a covered stoneware jar to settle. Pour off the clear liquid when required for use and dilute with six times its volume of clear water. Use this as a soap or other scrubbing solution. In using this solution it is advisable to first rinse the surface with clear water. Javelle water is a strong bleaching material, hence it should not be allowed to drop on colored fabrics. It is not recommended for general cleaning purposes, but its occasional use on stained concrete is believed to be entirely safe.

Poulticing with commercial grit scrubbing powders, such as those commonly used for cleaning marble floors, will prove satisfactory for removing most stains of this class. In poulticing with these, the material is slowly stirred into a pail of hot water until a thick paste of mortar consistency is obtained. A small addition of whiting will add somewhat to the working qualities of the poultice. This is applied to the surface with a trowel in a layer ¼ inch thick or more and allowed to remain until dry, when it is scraped off with a wooden paddle.

Should it be deemed expedient to use a poultice that may be applied with a brush instead of a trowel, Method No. 3 for tobacco stains is well adapted to this purpose.

Stains, Removing

Argyrol stains can be removed by applying potassium iodide solution followed by hypo crystals.

Blood stains can be removed in water with ammonia.

Candle drippings are removed with lard and benzol.

Cod liver oil stains are removed with soap dissolved in amyl acetate.

Enamel stains are removed with amyl acetate and acetone.

Fruit stains are removed by pouring boiling water through the garment from a height of several feet. Use peroxide of hydrogen.

Grass stains are removed with ether or soap and alcohol.

Gum stains are removed with carbon tetrachloride, benzol.

To remove ink stains apply hydrogen peroxide and hold in steam issuing from a kettle until yellowish. Repeat. Then apply oxalic acid solution and wash with water. Repeat if needed.

To remove iodine stains use sodium thiosulphate.

Lacquer stains can be removed easily with amyl acetate (banana oil), lacquer thinner.

To remove mercurochrome stains, 1st, boil ¾ hour in soapy water, and, 2nd, apply benzaldehyde, then a 25% hydrochloric acid solution. Rinse thoroughly afterward.

Mildew is removed in one minute with Javelle water, but *not* from silk or wool.

Paint or varnish is removed with carbon tetrachloride, benzol, Stoddard's Solvent, amyl acetate; *not* for Rayon, which should be scrubbed with two parts carbon tetrachloride, two of alcohol, one part of oleic acid.

Perfume can be removed with alcohol.

Perspiration stains are removed with soapy water and hydrogen peroxide.

Scorched stains are removed with potassium permanganate followed by hydrogen peroxide.

Shoe polish stains are removed the same as candle drippings, or use benzol.

Developer Stains, Removal of

Treatment with I as follows is claimed to remove developer stains from fabrics. Soln. 1: KI 35 g.; I (crystals) 10 g.; water to 1 l. Soln. 2: $Na_2S_2O_3 \cdot 5H_2O$ 25 g.; water to 1 l. The stained material is treated in soln. 1 for a few min., then placed in soln. 2 for 15–20 min., and subsequently washed for 30–40 min. Both new and old stains are said to yield to the treatment.

Hectograph Stains from Skin, Removing

Sodium Hydrosulfite	5–10
Water	95–90

General Spot Remover (Egg, Blood, Candy, General Dirt)

2% Liquid Soap Solution

Wet the spot and place folded cloth underneath. Dip clean cloth in soap solution and gently rub spot until lather forms. Remove suds by rubbing with wet cloth. Repeat if necessary.

Grass, and Fruit Stain Remover

Immerse spot in 95% denatured alcohol and then follow with 2% soap solution.

Grease, Oil, Paint and Lacquer Spot Remover

10 lb. Alcohol
20 lb. Ethyl Acetate
20 lb. Butyl Acetate
20 lb. Toluol
30 lb. Carbon Tetrachloride

Mercurochrome Stains, Removing

It is stated that two treatments with benzaldehyde, followed with a 25 per cent hydrochloric acid applications and an alcohol rinse, with a final bath in water will remove fresh mercurochrome stains from silk. Glacial acetic acid followed by ether is also recommended as a remover of mercurochrome stains, as is phosphoric acid in rubbing alcohol.

Rust and Ink Remover

Immerse portion of fabric with rust or ink spot alternately in Solution A and B rinsing with water after each immersion.

Solution A

5% Ammonium Sulfide Solution
95% Water

Solution B

5% Oxalic Acid
95% Water

Scorch Remover

Slight scorch spots can be removed by immersing for about an hour or more in a 3% Hydrogen peroxide solution.

Wood Preservative Finish

Creosote, Oil	4
Alcohol	1
Turpentine	2
White Lead	3
Paste Wood Filler	4

Leather Soles, Impregnant for

Crepe Rubber	15
Rosin	30
Linseed Oil	35
Turpentine	17
Paraffin	3

Keep melted with occasional stirring until rubber has dissolved.

Leather "Nourisher"

For leggings, boots, base-ball gloves, etc.

Menhaden Oil	39
Tallow	60
Clovel	1

* Preservative, Leather

Oleyl or Cetyl Phthalate	50
Light Mineral Oil	35
Montan Wax	10
Ceresin	5

Protecting Leather during Manufacture

Shoes, bags, novelties, etc., made of leather are soiled readily while being handled in various "putting together" operations.

To avoid this they are dipped or sprayed with following and dried

Rubber Latex	20
Carnauba Wax Emulsion	10
Water	40

After articles are finished the deposited film is easily stripped off.

Cleaning Stained Limestone

1. Scrub surface with

Washing Soda 5–10% Solution

using a bristle brush according to the intensity of the stain. After half an hour use a steam jet, applying the treatment uniformly to remove the stain. After this treatment the stone usually appears clean and fresh, but if left to itself the stain tends to come back. To prevent this the surface should be scrubbed uniformly with the 10% formic acid solution.

2. A poultice method has been worked out which can be used advantageously under certain conditions for indurated stains, especially for localized or interior stains. The material for poultices can be conveniently prepared by shredding old newspapers or similar paper stock under a steam jet, sufficient fireclay being added to make the mass plastic. Washing soda is then added, according to the intensity of the stain, in amounts of from 5 to 10 per cent, and the whole is plastered over the stained surface with a trowel. The alkaline poultice is easily stripped off after 24 hours and a similar poultice containing 10% formic acid is applied in the same way and removed after another 24 hours. If the wall is dry at the start this treatment is usually successful if carried out by a workman experienced in its use.

Rust Stains

Rust stains are produced by corroding fire escapes, lamp brackets, and similar attachments of iron or steel in contact with limestone walls. These can be prevented by keeping the iron work protected from rusting, and can be removed by suitable treatment, although they sometimes become so thick and so hard that drastic methods are required. Scrubbing with hot concentrated oxalic acid will usually remove all rust stains, the wall being washed thoroughly after the treatment. Hydrofluoric acid put up in lead tubes under various trade names for dry cleaners may also be used, but the corrosive character of the acid demands caution.

Copper Stains

Copper stains are occasionally observed on limestone surfaces below copper roofs or gutters, adjoining copper down-spots, or around copper, bronze, or brass name plates, lamp standards, and the like. The following methods of removing copper stains have been developed in our laboratory. A potassium cyanide solution will wash off this stain very satisfactorily but must be used with caution because of its poisonous nature.

Cigarette Stain Removal

The following method removes cigarette stains from fingers.

A. Pot. Permanganate (2% Soln.)

B. Sod. Bisulfite 10
Orris Root, Powd. 10
Perfume to suit

Apply solution A with a swab and after a few minutes rub with B moistening with water if necessary. Wash well with soap and water.

Dry Cleaning Soap

	Parts
1. Oleic Acid-white	10
2. An alcohol solution of pot. Hydroxide (2 oz. by wt. of pot. Hydroxide in 10 oz. of denatured alcohol)	10
3. Carbon Tetrachloride	50

Mix 1 and 2 then add 3.

Use plain then rinse article with gasoline or better still with carbontetra chloride allow to dry.

Dry Cleaner

Use

Glycololeate	2 parts
Carbontetra Chloride	60 parts
Varnoline	20 parts
Benzine	18 parts

An excellent cleaner that will not injure the finest fabrics.

RESINS, GUMS, WAXES

Brewers' Pitch

A. Rosin	160
Pale Rosin Oil	30
B. Rosin	168
Paraffin Wax	22
Linseed Oil	10

---

* Synthetic Resins

Example 1

	Parts by Weight
Propylene Glycol (1–2 Propane Diol)	76
Phthalic Anhydride	148

This mixture, representing one mol. each of the glycol and phthalic anhydride, was heated together in a partially closed vessel to a maximum temperature of 290° C., over a period of approximately 2½ hours. The final product was a soft, pale, straw-colored resin having an acid number of 56.3. This product was freely soluble in n-butyl and amyl acetates, and in n-butyl propionate. This resinous material is not substantially soluble in toluene alone, but solutions of the resins in the previous solvents may be diluted with toluene. Accordingly this resin may be used to advantage in com-

All formulae preceded by an asterisk (*) are covered by patents.

positions containing the usual solvent mixtures in which a large proportion of hydrocarbon diluent is used. This resinous material is particularly valuable in view of its compatibility with nitrocellulose.

Example 2

	Parts by Weight
Trimethylene Glycol (1–3 Propane Diol)	76
Phthalic Anhydride	148

This mixture of equivalent combining proportions was heated as in Example 1, yielding a product having substantially the same characteristics as that obtained in Example 1. This material likewise is compatible with nitro-cellulose and is suitable for use in lacquer compositions.

Example 3

	Parts by Weight
2–3 Butylene Glycol	100
Phthalic Anhydride	148

This mixture was reacted as described in Example 1 and yielded a product of a softer nature than those prepared in accordance with Examples 1 and 2. The resinous material so prepared was found to be soluble in toluene as well as in such solvents as butyl acetate and the like. It displays excellent compatibility with nitro-cellulose.

* Resin, Synthetic

Dihydroxystearic Acid	45
Phthalic Anhydride	80
Glycerol	50

Heat for two hours at 242° C. in a kettle fitted with a short air condenser. The resin formed is hard, tough and light in color.

* Resin, Synthetic

A hard, inert resin may be made by causing resinification to occur by heating in the usual well understood manner between 24.3 parts of phthalic anhydride, 10 parts borneol and 5 parts of glycerine. This is a dark-red resin which quickly reaches the B-stage on heating at 150° C. more rapidly than without the addition of borneol.

A reddish-brown, tough, water resistant resin may be prepared by the interaction of 3 parts of phthalic anhydride, 2 parts of terpene hydrate and 1 part glycerine. The first two ingredients may be caused to react separately at 240° C. and the glycerine then may be added to cause a second reaction to take place. Upon continued heating a fusible soluble resin is formed which is convertible.

* Resin, Water Soluble Synthetic

Four hundred parts of formaldehyde of 30 per cent strength are mixed with 100 parts of acetaldehyde. Into this mixture 5 parts of barium hydroxide are slowly introduced while well stirring. The temperature is kept at about 40° to 50° C. and care is taken that the temperature does not exceed 50° C., if necessary by external cooling. If after about 5 hours of test shows that only a small quantity of formaldehyde is still present, the barium is precipitated in the form of carbonate by introducing carbon dioxide and the carbonate is removed by filtration. The filtrate is evaporated in a vacuum at about 60° to 65° C. Together with the water which is eliminated by distillation small quantities of unaltered aldehyde likewise pass over. The filtrate is then allowed to cool whereby a limpid, highly viscous and colorless syrup is obtained which is very easily soluble in water but insoluble in organic solvents. It does not alter its properties, even after the lapse of years.

* Resin, Synthetic

Cresol	100
Formaldehyde	100
Triethanolamine	7½

Heat under a reflux to 100° C. for an hour. Allow to settle and separate the supernatant solution. Drive off water by heating in a vacuum.

The above resin may be mixed with wood flour in a heavy heated mixer. It is then cooled; ground and heated in molds at 100°–140° C. under pressure.

Rosin Emulsion

1. Rosin	100
2. Naphtha	100
3. Am. Linoleate	3
4. Ammonium Hydroxide	2½
5. Water	200

Heat one to 150° C. and turn off flame; run two (which has been previously heated on a water-bath to 90°–100° C.) into it slowly and stirring until all rosin has dissolved; cool and add three, four and five mixed together slowly with vigorous stirring. This gives a thick brown transparent emulsion which may be diluted infinitely with water.

Shellac, Reconditioning Insoluble

Shellac which has become infusible and insol. in EtOH through prolonged storage, overheating, or other cause is added slowly to rosin at 270°. The product is completely soluble in C_6H_6 and PhMe and can be used as substitute for rosin in making varnishes, adhesives, etc.

---

* Wax, Carving
(For Statuettes and Models)

Stearic Acid	6
Ceraflux	24
Carnauba Wax	1
Terra Alba	75

Dance Floor Wax

Ceresin	44
Stearic Acid	12
Scale Wax	140
Carnauba Wax	4
Oil Soluble Color	to suit

Dental Impression Wax

Paraffin Wax	90
Ceresin	39
Beeswax	40
Venice Turpentine	30
Japan Wax	20

Wax, Dental Impression

Shellac	45 %
Talc	30 %
Glycerin	2½%
Coloring	sufficient
Tallow Fatty Acids (to make)	100%

Flexible Wax

Methyl Abietate	10
Gelowax	90

Heat together and stir until homogeneous. The finished product has a softening point of 58° C. and a melting point of 67° C.

Grafting Wax Solid

Lanolin	22
Rosin	44
Ceresin	13
Beeswax	8
Japan Wax	2
Rozolin	9
Pine Oil	1

Grafting Wax Sticky

Lanolin	40
Rosin	26
Rozolin	10
Turpentine	11

Modeling Wax

Venice Turpentine	90
Rosin	16
Beeswax	60
Tallow	14
Thin Mineral Oil	4
Color	to suit

Plastic Modeling Wax

Gum Mastic	3
Beeswax	3
Ozokerite	2
Paraffin Wax	4
Tallow	19

Melt together and keeping hot work in

Sulfur Flowers	22
Gypsum	12
Pipe Clay	33
Mineral Pigment	4

Modelling Wax

Beeswax	4
Venice Turpentine	9
Lard	4
China Clay	3.5

Wax Putty

Beeswax	4 lb.
Oleostearin	2 lb.
Turpentine	1 lb.
Venice Turpentine	6 lb.

---

* Synthetic Wax

In a flask equipped with a return condenser, 56.8 parts by weight of stearic acid and 18.6 parts by weight of aniline are heated to substantially from 170° to 200° C. for approximately one hour. Water is formed as a result of the reaction between the organic acid and the amine. In order to eliminate the water so formed, it is desirable to so arrange the condenser that the water may escape, but so that any aniline being volatilized will be returned to the flask. At the end of the heating period, and after some cooling, 19 parts by weight of furfural are added and the whole is heated to about 200° C. for approximately one-half hour. At this temperature the product is a thin liquid, which, upon cooling, solidifies to a waxy, dark brown solid at room temperature.

Wine or Liquor Barrel Wax

Tallow	24
Paraffin	50
Japan Wax	5
Beeswax	5
Venice Turpentine	4
Rosin Oil	1
Talc	10

Thread Wax

Beeswax	40
Japan Wax	10
Paraffin Wax	150

Beeswax Substitute

Glyceryl Stearate	20
Beeswax	8
Japan Wax	10

Pure Stearic Acid Candles

Use Triple Pressed Saponified Stearic Acid. After melting down the Stearic Acid should be stirred or agitated until "milky" in appearance to destroy the large crystals. It should then be poured in moulds which have been heated to approximately the same temperature and cooled. A better appearance will be noted on more rapid cooling.

Standard Candle Formula

60 lb. Paraffin Wax
35 lb. Double Pressed Stearic Acid
5 lb. Beeswax

The above are melted together and agitated to insure complete blending. When melted an oil soluble dye of the desired hue is added and then the combination is poured in moulds and cooled. Care in the selection of the dye should be exercised to eliminate "bleeding" or fading, but many good dyes are available. It may be desirable to make up known strength of dyes in blocks of paraffine by merely adding the dye to the melted wax and then pour in moulds, forming blocks of uniform size. This permits easy storing and somewhat facilitates the complete blending of the color when introduced to the melting kettle.

A better grade of candles are made by increasing the amount of Stearic Acid and decreasing the amount of paraffin, or vice versa.

Pure Beeswax Candles

Are made from the pure wax and range down to combinations as low as 40% Beeswax, 50% Paraffin and 10% Stearic Acid.

Virgil Lights

Eighty per cent Paraffin, 15% Double Pressed Stearic Acid and 5% Beeswax. This can be varied to as much as 95% Paraffin and 5% Stearic Acid.

Tapered Candles

These are usually a hand-dipped operation entirely. The combination of waxes and color is melted in the kettle and a constant temperature maintained at slightly above the melting point. Dipping proceeds from the bottom and progresses up the wick to the desired length in order to attain the desired taper.

---

* Non-Fading Colored Candles

Candles or other wax products colored with Rhodamine B or chinoline yellow are prevented from fading by the incorporation of a 0.025% Betanaphthol or 0.1% Sulfur.

Candle Wicks

The matter of the selection of the wick for various compositions of candles is one of careful consideration. For instance, the wick used in a pure stearic acid candle, usually a 48 to 51 ply—meaning three strands of 16 or 17 threads each, would be entirely unsuited for a candle containing very much paraffin, which would require a smaller wick. The wick should be treated with Boracic Acid, the object of which is to prevent the wick from continued glowing and smoking when blown out. One of the strands of the wick should be woven tighter than the other two in order to force the wick into separation while burning to dissipate the ash.

Birthday Candles

Are made entirely of paraffine and the proper oil soluble dye. The procedure, though, is entirely different than in the case of other candles. The thin threads, forming the wicks are formed into endless belts and placed over two drums. These drums are spaced a few feet apart and are set up to revolve slowly, allowing the "endless belt" wicks to run through a tank of the melted wax. This operation is continued until the series of wicks have picked up the desired amount of wax and have reached the required diameter. The "belts" are then cut and laid out on tables where the candles are cut to length. The head of the candle is then inserted into a revolving cutter or a revolving hot mould to properly shape the head.

Dewaxing Gum Damar

Ten pounds of damar gum are dissolved in 1 gallon of solvent mixture made up as follows:

24 oz. fl. ethyl acetate
24 oz. fl. acetone
112 oz. fl. toluol

When the solution is complete, 120 oz. (fluid) of methyl alcohol are added, when a white precipitate is formed which settles down to the bottom of the container in the form of a slimy mass. After standing for a few days, this mass becomes quite hard and may be removed. The resultant gum solution is perfectly clear and is miscible with nitro-cellulose solutions without the formation of a precipitate.

* Raising Melting Point of Rosin

The m.p. is raised from about 52° to about 66° by heating the rosin at 260°-300° for 1–8 hr. and then distilling *in vacuo* or with superheated steam until the original wt. of the rosin has been reduced by 10%–16%.

* Synthetic Thiourea Resins

Example 1

One part of barium hydroxide is dissolved in 160 parts of 36% formaldehyde solution. One hundred and twenty parts of thiourea are then added and the mixture stirred. The temperature of the solution at first drops, due to the absorption of heat by the thiourea going into solution. The temperature soon rises, however, due to the heat of the reaction of the thiourea and formaldehyde. If the original temperatures of all of the materials used is approximately 20° C., the maximum temperature reached during the reaction may be as high as 40° C. or even higher. It is preferable, but not at all essential, that the temperature of the reaction mixture be held below 40° C. by cooling the mixture during the reaction if necessary. In any case, no heat is used in bringing about this reaction other than that generated by the reaction itself.

The reaction is apparently complete in about two hours. The clear solution may be kept over long periods of time without harm. There is some tendency, however, to develop slight acidity, so that it is preferable to add about two parts of ammonia solution (specific gravity 0.9).

In order to hold the solution nearly neutral, carbon dioxide is passed in. This serves a double purpose in precipitating out the barium as barium carbonate, and forming ammonium carbonate which acts as a buffer to hold the solution substantially neutral during the evaporation of the water.

In some cases after following the above procedure the water is removed by boiling in an open evaporator. A thermometer inserted in the boiling solution showed a maximum boiling point 106° C., at which time substantially all of the water of solution and reaction had been removed.

Care must be taken if the water is removed by this process, since there is a considerable tendency to foam during the last stages. If overheating occurs where the heat is applied to the evaporation vessel, the resin will cure to its infusible form in a layer over the vessel. This is indicated by a drop in the temperature of the solution, due to the poor heat transference of the cured portion of the resin.

The resin, which is very viscous at 106° cools to an almost colorless brittle product which is potentially reactive.

The clear solution, preferably stabilized and preferably treated with carbon dioxide, will keep over long periods of time at ordinary room temperature and is, therefore, valuable as a varnish or coating material or may be used for the purpose of impregnation into various sheetlike bodies such as paper, cloth, asbestos, etc., the water being evaporated and the sheetlike material may be pressed into form of any desired shape. The clear varnish makes it possible to ship the product to the ultimate user, and when properly stabilized makes a material of considerable value to the fabricators of laminated material, etc.

Example 2

One part of barium hydroxide, 104 parts of 36% formaldehyde solution, 160 parts of thiourea.

As in Example 1, the barium hydroxide is dissolved in the formaldehyde and the thiourea added. The mixture should be stirred until all of the thiourea is in solution. The solution first cools and then warms up during the reaction. In about two hours the reaction is apparently complete, but it is usually convenient to allow the solution to stand overnight.

Twenty-five parts of hexamethylenetetramine is added to the solution and carbon dioxide passed in until the solution is neutral to litmus. Besides the barium carbonate, where commercial formaldehyde is used, there is usually a small amount of colored insoluble material present. Where a very light colored resin is desired, the solution should be filtered or centrifuged.

The water is then evaporated from the filtered solution by boiling in an open evaporator. When the temperature reaches about 100° C., the product has a tendency to turn milky, but this may be disregarded. Evaporation of the water is continued until a product of the desired viscosity is obtained. Since this resin cures at temperatures above about 110°, care must be used during the last stages of evaporation.

Example 3

Place in a suitable mixing device,

Thiourea	305 parts
Urea	120 parts
36% Formaldehyde Solution	835 parts
Ammonia Solution (sp. gr. 0.9)	8 parts
Calcium Hydrate ($Ca(OH)_2$)	1 part

Agitation should be started as soon as the calcium hydrate has been added. A reaction starts almost as soon as the calcium hydrate is added, the mixture warms up and both the urea and thiourea go into solution.

The addition of ammonia is desirable but not essential. Where ammonia is not used, the formaldehyde solution should be neutral, or slightly alkaline. In any case the solution should be sufficiently close to the neutral point that it becomes alkaline to litmus upon the addition of one part of calcium hydrate or of calcium oxide to the quantities of formaldehyde, urea and thiourea shown in this example. Rather than adjust the hydrogen ion concentration of the solution, it is more convenient to add ammonium hydroxide solution. A greater amount than 8 parts of ammonium hydroxide may be added, if desirable, without changing to any appreciable extent the nature of the product.

The initial reaction is usually complete in about two hours, but the solution should preferably stand eight hours or longer before the water is removed in order that additional polymerization may take place. Carbon dioxide may be passed in and the solution filtered or centrifuged where a very clear colorless product is desired.

The water may be rapidly and conveniently removed by distillation, preferably under reduced pressure. During the removal of the last portions of the water, foaming is likely to occur, due to the high viscosity of the solution. This tendency to foam can be very largely prevented by the addition of a very small amount of paraffin. Usually an amount of paraffin equivalent to less than 0.01% of the weight of the batch is ample to prevent excessive foaming. Other well known oily or water insoluble materials may be substituted for the paraffin for the purpose of reducing the tendency to foam.

Where a vacuum distillation is used to remove the water, the solution may be heated more rapidly without danger of curing the resin than is the case where no vacuum is used. In any case the solution should not be heated above 110° C. for any period of time, as there is danger of the resin going over to the insoluble infusible state above this temperature. Samples removed from time to time during the distillation of the water show a product of increasing hardness. Toward the end of the distillation the temperature rises more rapidly and the rate of distillation decreases. Where a hard grindable resin is desired, the temperature of the resin may be allowed to raise as high as 105° C. toward the end of the distillation in order to drive off substantially all of the water.

---

* Resin, White Synthetic

100 parts by weight of phenol, 25 parts urea and 160 parts of 40 per cent formaldehyde solution were boiled in an open flask in the presence of about 1 part of concentrated hydrochloric acid. After boiling for a short time a white mass separated and the boiling was continued for 15 minutes. When cold a white, rather brittle porcelainlike soluble resin was obtained. It was washed first with a 2 per cent solution of sodium carbonate and then with water. The yield of the resin was 178 parts. This resin was opaque and pure white in color. It was exposed to sunlight for a period of nearly two months and during that time there was no discoloration. The opacity of the exterior layers disappeared and a white glass-like coating resulted. This appears to be due to the removal of a small amount of moisture present in the mass.

RUBBER

* Latex, Artificial Rubber

In 750 grammes of benzene (or commercial "benzol"), dissolve 250 grammes of crude rubber (balata or gutta percha) and 25 grammes of oleic acid, with proper agitation until the oleic acid is diffused throughout the mass. Thoroughly mix 20 grammes of 26° aqua ammonia with 750 grammes of water. Then add and thoroughly mix the water with the rubber solution. The dispersed or diffused particles of oleic acid are saponified by the ammonia in situ, forming an ammonia soap which acts as a dispersing agent and stabilizes the final dispersion. As the ammoniated water is added to and stirred in the rubber-benzol solution it will be observed that at first the water forms the disperse phase of the dispersion, but as the total volume of water increases, there is a change of phase, and the water then constitutes the continuous phase. The final dispersion is a white milky mass which may be diluted practically to any reasonable or operative extent with water. It may be used as thus produced, but, if desired, the solvent may be removed by evaporation, but preferably in a vacuum still at a low temperature (say, not over 50° C.) for recovery of the solvent.

Coloring Latex Black

Colloidal Micronex is a dispersed carbon black suitable for use with rubber. It does not require grinding. It is merely stirred into the latex in amounts varying with the depth of color desired.

---

* Compounded Latex

Example 1: To latex preserved with ½% ammonia and having a concentration of about 35% is added 1% of lauric acid in the form of ammonium laurate, and ¾% of ammonium chloride, these latter figures being based on 100 parts of solids in the latex. The latex is then spray dried, and the resulting rubber has a quick breakdown and a high abrasion and flexing resistance when vulcanized.

Example 2: To a similar latex is added 1% of lauric acid as ammonium laurate and then ½% of phosphoric acid as secondary ammonium phosphate. The latex is then spray dried, and the resulting crude rubber has a quick breakdown, good calendering, and extruding properties; and the vulcanized rubber has a good abrasion and flexing resistance.

Example 3: To a similar latex 1% of lauric acid as ammonium laurate is added and then 1% of monochloracetic acid as the ammonium salt. The latex is spray dried, and the resulting crude rubber has excellent breakdown and milling properties and, when vulcanized, a good abrasion and flexing resistance.

Latex as received from the tree is treated with 0.2-part of formaldehyde and allowed to stand for about 24 hours, and then 0.5-part ammonia is added. The latex is spray dried, producing a rubber having its proteins tanned or reacted upon by formaldehyde and which rubber is less absorptive to water. If desired suitable compounding and curing agents may be added to the latex before drying.

---

* Softened Rubber

Softened rubber is now being produced from ordinary plantation crepe and sheet. The pieces of crepe or sheet are first soaked in tanks to soften them and then rapidly reduced to crumb by a machine consisting of a pair of rolls working in a hopper. The crumb is placed in trays in a heater which can treat 1,000 pounds in one charge, heated for about one hour in vacuum and then under controlled conditions for about 40 minutes. At the end of this time the mass looks like toasted cheese, and it is finally passed through sheeting rolls.

Greater plasticity than ordinary masticated rubber is claimed for the new product; it facilitates masticating, mixing, calendering, tubing, molding, and spreading, saving time, power, labor, and solvent and increasing output and efficiency. The danger of scorching is said to be reduced; calendering and tubing give smooth surfaces. Spreading doughs and solutions

All formulae preceded by an asterisk (*) are covered by patents.

having much lower viscosity than usual permit varied mixings and the addition of a much greater proportion of mineral fillers; while unvulcanized softened rubber dough holds its shape much better, a matter of importance in calendered, forced, or stamped goods and in molding ebonite.

Mechanical properties and aging are said to be unimpaired. But practically pure mixes, containing little filler, show a slight falling off in breaking strain as compared with ordinary rubber. More heavily compounded rubber, however, as tire treads, shows no difference in mechanical properties when compared with ordinary mixes. For comparatively pure mixes, therefore, blending softened rubber with ordinary rubber is advised; thus a 50/50 mixture is recommended for high grade inner tubes.

Protection of Rubber Belting in Storage

Shellac	1 qt.
Alcohol	1 pt.
Ammonia	1½ qt.
Water	3 qt.

Apply with a brush.

---

* Rubber Cleaner

The following composition will clean rubber and reduce swelling

Castor Oil	10
Paraldehyde	10

---

* Factice Emulsion

85 kilograms of Colza oil and 15 kilograms of elemental sulphur were heated together with stirring for five hours at about 150° C. The mixing was allowed to cool to 95° C. and the mass was then slowly poured into a homogenizing plant into which were simultaneously introduced 150 kilograms of an approximately 3 per cent. aqueous solution of neutral soap. The product was a viscous fluid, which was again passed through the plant. In this case also the viscosity slightly increased in a period of 48 hours after the preparation.

---

85 kilograms of Colza oil and 15 kilograms of elemental sulphur were heated together with stirring for five hours at about 150° C. The mixing was allowed to cool to 95° C. and the still fluid mass was then slowly poured into a homogenizing plant, into which were simultaneously introduced 100 kilograms of a 10 per cent. solution of casein in ammoniacal water. There resulted a viscous fluid which was passed a second time through the plant. The viscosity slightly increased during 48 hours after the preparation.

---

* Rubber Substitute (Factice)

A white rubber factice is made by mixing non-mineral oil, *e.g.*, rape-seed oil (100 pts.), a low-temp. vulcanising agent, *e.g.*, S_2Cl_2 (20 pts.), a stabilising agent, *e.g.*, MgO (5 pts.), and an NH_4 salt, *e.g.*, NH_4HCO_3 (10 pts.), and maintaining the temp. below that at which NH_3 is materially generated, until vulcanisation is complete.

---

* Latex Factice Compound

85 kilograms of Colza oil and 15 kilograms of elemental sulphur were heated together with stirring for five hours at about 150° C. The mixing was allowed to cool to 95° C. and the still fluid mass was then slowly poured into a homogenizing plant, into which were simultaneously introduced 40 kilograms of a 3 per cent. aqueous solution of saponin. There resulted a still fluid somewhat viscous substance which, on a second passage through the plant, effected at once, commenced to display an increase in viscosity. Left to stand for 48 hours, a product of a paste-like consistency was obtained.

A compounded final-dispersion was made up as follows:

Normal rubber latex (about 33 per cent. dry rubber)	10.00 kg.
Substitute-dispersion at 75 per cent (prepared according to Example 1)	2.00 kg.
Sulphur	60 gr.
Zinc Oxide	100 gr.
Ultra-accelerator	10 gr.
Calcium Sulphate	50 gr.

A stainless steel former for a finger stall, previously heated to 95° C., was immersed in the above compounded final-dispersion for 10 seconds. There was deposited upon the former a coating of a thickness of about 1 mm. which, after drying and vulcanization, presented great smoothness to the touch.

A compounded final-dispersion was made up as follows:

Concentrated rubber latex (about 50 per cent. dry rubber)	10.00 kg.

Substitute-dispersion at 75 per cent. (as in the previous example)	3.00 kg.
Calcium carbonate (in fine subdivision)	3.00 kg.
Sulphur	60 gr.
Zinc Oxide	100 gr.
Ultra-accelerator	10 gr.
Organic dyestuff	10 gr.
Calcium Sulphate	50 gr.

This compounded final-dispersion was proved by immersion of differently shaped heated formers to be suitable for the manufacture of articles of various kinds such, for example, as bathing caps, tobacco pouches and hand-grips, all of which proved in the finished state to be very smooth and of great softness.

---

* Latex, Powdering Rubber

6–12% of dextrin is added to the latex which is then sprayed into a heated chamber to give a rubber powder.

---

* Latex, Removing Ammonia Odor

Assuming the latex contains 0.75% ammonia it may be treated as follows:

Latex	100
Water	25
Boric Acid	2.75
Dextrose	8.25

Oil-Resisting Materials

Mention has previously been made of new products designed to resist practically all solvents, oils and fats, such as Ethanite, a reaction product of ethylene dichloride and calcium polysulfide, and Thiokol, a polymethylene polysulfide. Although different claims may be made for the individual products now on the market, in general these polysulfides may be vulcanized in a similar manner to rubber requiring no sulfur, but zinc oxide in proportions of one to twenty per cent. is necessary; the material in appearance is similar to rubber, being homogeneous and pliable, but the gravity is much higher, viz., 1.6. The suitable vulcanizing temperatures are similar to those with rubber mixings, such as one hour at forty pounds steam pressure. The addition of rubber is not necessary, although milling is facilitated thereby. In the case of Ethanite it is stated that the addition of five per cent of rubber gives a product which is as resistant to oil as Ethanite alone, but generally speaking, the oil resistance deteriorates according to the amount of rubber present. Carbon black may be added to increase tensile strength and decrease porosity, and a mix which is stated to be resistant to practically all oils and solvents is: Ethanite 20, pale crepe 1, zinc oxide 2, carbon black 5.

When cured these products show practically no dimensional increase when immersed in such solvents as benzol, toluol, and carbon tetrachloride, and acids, with the exception of strong nitric or chromic acids, are without action. A 20 per cent. caustic soda solution or concentrated ammonia attacks the material, but the latter does not appear to suffer from aging in the usual manner of rubber goods. The particular advantages obtained are offset to some extent by the objectionable characteristic odor which, besides rendering the use of the products impracticable in many instances, for example foodstuffs, renders the general atmosphere where it is in process, particularly in the region of the mill, decidedly unpleasant. Possibly means will be found of overcoming this, at any rate to a considerable extent.

Rubber Goods, Non-sticking

Sprinkling with talc prevents rubber goods and sheets from sticking.

---

* Resin, Rubber Compound

Dissolve 100 grams of Rosoap (60% dry matter) in 500 c.c. of water: add 10 grams of latex (containing 30% rubber and a trace of ammonia) with thorough stirring; add enough hydrochloric acid to neutralize the free alkali and to decompose the rosin soap; boil the mixture with formation of viscous layer of rosin and rubber disseminated therethrough; remove the rosin and rubber mixture and dry the same in an oven to drive off the moisture. The product is a clear dry solution in viscous form and has properties that are not found in either constituent alone. When cooled and set it is tough, hard, does not absorb water to the same extent as rosin, and does not deteriorate readily. It can be used with oil and turpentine to produce varnish. It is suitable also for electrical insulation.

---

* Resin, Synthetic

PhOH 100, tung oil 150 and H_3PO_4 1 part refluxed for 6 hrs., 100 parts of 40% CH_2O soln., 50 parts of colophony and 3 parts of aq. NH_3 are added and refluxing is continued for 5 hrs., the mixt. is then evapd. until anhyd. and is heated at 150° until a product is obtained which is clear and non-tacky at room temp.

Synthetic Resin, Fusible

Solid, permanently fusible resins are made by heating CH_2O (7 mols.) and commercial PhOH (13 mols.) in the presence of 25% aq. NH_3 (5 mols.) so that NH_3 escapes during the reaction, thereby evaporating the product. The PhOH can, in part, be substituted by urea etc.

---

* Resin, Synthetic (Alkyd)

Rosin	1340
Phthalic Anhydride	308
Glycerol	348

Heat with stirring to 290° C. When acid number has dropped to 10–20 cool quickly to 200° C. and then allowed to cool naturally.

This resin is soluble in benzol and lacquer thinners. It is light in color and hard.

---

* Resin, Synthetic (Sugar)

Glucose	80
Water	60
Rosin	60
Aniline	60

Reflux for 5 hours. Allow to settle; draw off and discard aqueous layer. The resinous reaction product upon melting and continued heating becomes infusible.

Rubber Goods

A single rubber product may be compounded with any number of mixtures, combining various grades of rubber, reinforcing agents, pigments and vulcanizing agents. For most items, a number of different compounds will serve with equal satisfaction. All of the possible combinations cannot be included here, but the following compounds are representative and can be readily adapted to commercial factory production by slight modifications to suit specific conditions. Adjustments as to curing conditions, temperature, or time of cure may be desirable depending on prevailing factory conditions. The curing data given for the various compounds is not intended to be specific and may be modified as desired.

Hospital Sheeting

Pale Crepe	100
Petrolatum	1.00
Zinc Oxide	10
Lithopone	75
Whiting	63
Color	as desired
Monex	0.50
Sulfur	2.00

Cure—In air—60 minutes, rise to 245° F. and hold 60 minutes.

Rubber Clothing

Pale Crepe	100
Plastogen	6.00
Stearic Acid	1.00
Zinc Oxide	5.00
Dixie Clay	40.00
Kalite—No. 1	40.00
Captax	1.00
Zimate	0.10
Sulfur	1.50

Cure—60 minutes rise to 260° F. and 30 to 60 minutes at 260° F.

White Tiling

Pale Crepe	15.00
Paraffin	0.3125
Whiting	50.00
Ti-Tone	25.00
Zinc Oxide	6.50
Magnesium Carbonate	1.50
10% Thionex Master Batch	0.625
Anti-Scorch-T	0.0625
Sulfur	1.00

Cure—11 to 12 minutes at 40 lb. steam.

Tire Cushion Stocks

Smoked Sheets	60.00
Amber Crepe	40.00
Cumar Resin	1.00
Mineral Rubber	2.00
Stearic Acid	0.50
Neozone A	1.00
Zinc Oxide	30.00
Accelerator 808	0.6875
Sulfur	3.25

Cure—45 minutes at 281° F.

White Tubing

Pale Crepe	100
Petrolatum	7.50
Agerite Gel	1.00
Zinc Oxide	15.00
Lithopone	130.00
Dixie Clay	40.00
Kalite No. 1	200.00
Altax	1.25
Sulfur	3.00

Cure—In talc 30 minutes at 20 lb.

Belt Friction

Smoked Sheets	9.4375
Thin Brown Crepe	10.00
Whole Tire Reclaim	59.00
Paraflux	5.00
Stearic Acid	0.50
Neozone D	0.5625

Litharge	0.0625
Whiting	10.3125
Zinc Oxide	2.25
10% Thionex Master Batch	0.6250
Sulfur	2.25

Cure—15 minutes at 274° F.

Transparent Rubber

Pale Crepe	100.00
Plastogen	5.00
Rodo No. 10	0.10
Stearic Acid	1.00
Zinc Carbonate	2.00
Zimate	0.25
Captax	0.50
Sulfur	1.50

Cure—Approximately 15 minutes at 15 lb.

High Grade Comb

Smoked Sheets	100.00
Cottonseed Oil	2.00
Beeswax	2.00
Accelerator 833	1.50
Sulfur	45.00

Cure—Approximately 6 hours in water at 274° F.

Tire Carcass

Pale Crepe	50.00
Smoked Sheets	50.00
Plastogen	4.00
Stearic Acid	2.00
Agerite Powder	1.00
Zinc Oxide	5.00
Tuads	.05
Captax	1.00
Sulfur	2.50

Cure—45 minutes at 274° F.

Black Footwear

Rubber	100.00
Plastogen	6.00
Agerite Powder	1.00
Zinc Oxide	5.00
Whiting	40.00
Kalite No. 1	20.00
Dixie Clay	25.00
Gas Black	2.00
Zimate	0.10
Altax	0.50
Captax	0.50
Sulfur	2.50

Cure—Dry heat. 60 minutes rise to 260° F. and one hour at 260° F. under 30 lb. air pressure.

Black Heel

Smoked Sheets	11.50
Whole Tire Reclaim	64.00
Refined Asphalt	3.00
Paraffin	0.25
Stearic Acid	0.375
Neozone A	0.50
Carbon Black	9.375
Whiting (Natural)	7.25
Zinc Oxide	1.00
Litharge	0.125
10% Thionex Master Batch	1.125
Sulfur	1.50

Cure—12 minutes at 40 lb. steam.

Bathing Cap

Rubber	100.00
Stearic Acid	1.00
Cycline Oil-softener	4.00
Zinc Oxide	5.00
Whiting	15.00
Lithopone	15.00
Barytes	15.00
Ureka C	1.25
D. P. G.	.25
Sulfur	2.00

Cure—8 minutes at 40 lb. steam.

Hard White Sole

Pale Crepe	28.75
Stearic Acid	0.25
Magnesium Carbonate	43.00
Lithopone	21.40
Zinc Oxide	1.50
Glue	2.88
Ultramarine Blue	0.09
Diphenylguanidine	0.28
10% Thionex Master Batch	0.35
Sulfur	1.50

Cure—8 to 10 minutes at 316° F.

High Grade Black Sole

Pale Crepe	50.00
Smoked Sheets	50.00
Agerite Gel	1.25
Zinc Oxide	60.00
Gas Black	10.00
Dixie Clay	40.00
Kalite No. 1	60.00
Captax	1.25
Tuads	.0125
Sulfur	2.50

Cure—60 minutes rise and 45 to 60 minutes at 255° F. under 30 lb. air pressure.

Soft Rubber Sponge

Rubber	100.00
Stearic Acid	1.00
Red Oil	1.00
Petrolatum	18.00
White Substitute	5.00
Zinc Oxide	2.50
Sodium Bicarbonate	15.00

Whiting	25.00
Ureka C	.625
Guantal	.375
Sulfur	4.00

Cure—¾ inch thick, 20 minutes at 70 lb. steam.

Packing

Smoked Sheets	35.125
Whole Tire Reclaim	10.00
Paraffin	1.00
Paraffin Oil	5.00
Stearic Acid	0.375
Clay	20.00
Whiting	20.00
Red Iron Oxide	6.00
Zinc Oxide	1.50
Beutene	0.75
Sulfur	0.75

Cure—12 minutes at 45 lb.

Tire Tread

Smoked Sheets	100.00
Pine Tar	4.00
Stearic Acid	2.00
Neozone A	1.25
Carbon Black	40.00
Zinc Oxide	10.00
Accelerator 808	0.875
Sulfur	3.25

Cure—60 minutes at 281° F.

White Sidewall

Pale Crepe	100.00
Plastogen	4.00
Stearic Acid	1.00
Zinc Oxide	5.00
Kalite No. 1	40.00
Dixie Clay	30.00
Titanium Dioxide	25.00
Captax	1.00
Sulfur	2.25

Cure—Press Cure—Approximately 45 minutes at 30 lb. steam.

Code Wire Compd.

Smoked Sheets	5.00
Blended Reclaim	48.00
Mineral Rubber	20.00
Stearic Acid	0.25
Paraffin	0.25
Neozone A	0.3125
Whiting	23.625
Zinc Oxide	1.00
Accelerator 808	0.3125
Sulfur	1.25

Cure—30 minutes rise to 275°F. plus 105 minutes at 275° in soapstone.

30% Wire

Smoked Sheets	32.00
Paraffin	1.00
Agerite Gel	0.60
Kalite No. 1	33.00
Zinc Oxide	32.00
Carbon Black—P-33	0.20
Captax	0.20
Sulfur	0.80

Cure—Steam Cure in talc. 30 minutes at 260° F.

Red Molded Tube

Smoked Sheets	97.75
Medium Process Oil	1.50
Stearic Acid	1.25
Blanc Fixe	40.00
Zinc Oxide	5.00
Du Pont Rubber Orange 2R	.75
10% Thionex Master Batch	2.50
Sulfur	1.75

Cure—5 minutes at 292° F.

Passenger Car Inner Tube

Pale Crepe	50.00
Smoked Sheets	50.00
Plastogen	4.00
Stearic Acid	.50
Agerite Powder	1.00
Kalite No. 1	50.00
Zinc Oxide	5.00
Tuads	.10
Altax	.50
Captax	.50
Sulfur	1.00

Cure—3 minutes at 55 lb.

High Grade Hose Tube

Smoked Sheets	14.00
Amber Crepe	10.00
Whole Tire Reclaim	20.00
Petrolatum	2.00
Paraffin	0.50
Stearic Acid	0.25
Neozone D	0.375
Whiting	20.00
Soft Clay	20.25
Carbon Black	7.25
Zinc Oxide	3.00
Litharge	0.125
10% Thionex Master Batch	1.000
Sulfur	1.250

Cure—15 minutes at 274° F.

Fire Hose

Pale Crepe	23
Smoked Sheets	23
Zinc Oxide	32
Whiting—Precipitated	10

Litharge	10
Sulfur	2.00

Cure—45 minutes at 274° F. in steam.

Hot Water Bottle

Pale Crepe	34.375
Medium Process Oil	0.50
Barytes	34.00
Whiting	25.25
Zinc Oxide	3.00
Du Pont Rubber Orange AD	0.75
10% Thionex Master Batch	1.4375
Sulfur	0.6875

Cure—7 minutes at 287° F.

Electricians Gloves

Pale Crepe	100.00
Mineral Rubber	4.50
Paraffin	0.75
Zinc Stearate	1.50
Agerite Gel	1.00
Zinc Oxide	15.50
Blanc Fixe	9.25
Tuads	3.00
Vandex	1.50

Cure—Press—15 minutes at 30 lb.

Bands and Thread

Pale Crepe	100.00
Agerite White	1.00
Zinc Oxide (fine particle size)	2.00
Color	to suit
Zimate	0.10
Altax	0.50
Captax	0.50
Sulfur	2.00

Cure—Open steam. 10 minutes rise to 260° F. and 30 minutes at 260°.

Wringer Roll Compd.

Smoked Sheets	38.00
Paraffin	0.50
Mineral Oil	1.25
Du Pont Antox	0.375
Zinc Oxide	2.00
Lithopone	35.00
Whiting	21.50
Accelerator 808	0.125
Sulfur	1.25

Cure—45 minutes at 292° F.

Black Combining Cement for Double Texture Pyroxylin Goods

Smoked Sheets	15 lb.
Boot and Shoe Reclaim	20 lb.
Soft Factice	10 lb.
Soft Mineral Rubber	8 lb.
Carbon Black	1 lb.
Lime	1 lb. 8 oz.
By Product Whiting	65 lb.

Dissolve in petroleum naphtha to spreader consistency.

Light Color Combining Cement for Double Texture Pyroxylin Goods

Smoked Sheets	15 lb.
White Reclaim	20 lb.
Soft Factice	10 lb.
Hard Mineral Rubber	8 lb.
Cliffstone Whiting	25 lb.
By Product Whiting	50 lb.
Lime	1 lb. 8 oz.
Raw Sienna	2 lb.

Dissolve in Petroleum Naphtha.

Black Combining Cement for Double Texture Rubber Goods

Smoked Sheets	15 lb.
Boot and Shoe Reclaim	25 lb.
Soft Factice	8 lb.
Litharge	8 lb.
Cliffstone Whiting	65 lb.
Rosin Oil	2 lb.
Sulfur	8 oz.

Dissolve in petroleum naphtha.

Light Colored Combining Cement for Double Texture Rubber Goods

Smoked Sheets	15 lb.
White Reclaim	30 lb.
Soft Factice	8 lb.
Litharge	2 lb.
Zinc Oxide	10 lb.
Magnesium Oxide	5 lb.
Raw Sienna	4 lb.
By Product Whiting	50 lb.
Sulfur	8 oz.
Rosin Oil	8 oz.

Dissolve in petroleum naphtha.

Solution for Application on Rubber Materials to Be Embossed to Prevent Sticking on Rolls

Glycerine	5 lb.
Denatured Alcohol	95 lb.

Anchor Rubber for Artificial Suede

Pale Crepe	40 lb.
White Reclaim	20 lb.
Tube Reclaim	15 lb.
Hard Factice (Brown)	8 lb.
Zinc Oxide	5 lb.
Lithopone	6 lb. 4 oz.
Cottonseed Oil	1 lb.
Stearic Acid	8 oz.
Sulfur	14 oz.
Captax or Ureka	14 oz.
Anti Oxidant	8 oz.

About 4 oz. per square yard of this compound is calendered onto a backing fabric. A cement of the same compound is then applied and closely followed with thorough dusting of finely divided cotton flock. The material is then festooned in an oven and cured ½ hour, rise to 250° F. and 1 hour at 250° F.

* Rubber, Artificial

750 grams of hydrated sodium sulfide ($Na_2S.9H_2O$) is dissolved in approximately a liter of water and the solution is boiled with 300 grams of sulfur to produce a solution of polysulfide believed to be largely Na_2S_4, although a certain amount of Na_2S_5 is doubtless formed. If larger amounts of sulfur are used in this example, still greater proportions of Na_2S_5 will be formed.

Water is added to make the specific gravity at 70° C. approximately that of ethylene dichloride producing about 1200 to 1300 c.c. of solution. About 300 c.c. of ethylene dichloride are added and the mixture gradually heated to about 70° C., preferably in a vessel having a reflux condenser. The reaction proceeds rapidly and is completed after digesting for an hour or more at such a temperature that active refluxing of the ethylene dichloride and steam occurs. The mixture is then cooled and the liquid portion is drawn off, leaving a yellow plastic. This is boiled with water to drive off occluded volatile compounds and to extract soluble salts, the boiling preferably being repeated several times, and the plastic being comminuted between boilings. The purified plastic is substantially free from halogen, is of high coherence, resiliency and pliability, and has elasticity somewhat similar to that of soft rubber. It is only slightly soluble in most ordinary organic solvents, although somewhat swollen by carbon disulfide. It can be worked, molded and rolled into sheets at temperatures around 130°–140° C.

* Rubber Belts, Noiseless

The surface of a rubber belt is covered with Zinc Stearate and it is heated at 280–300° F. to cause penetration. This treatment may be repeated a number of times.

Rubber Cement, Reducing Viscosity of

The addition of 2–3% alcohol reduces the viscosity of thick rubber cements.

Cheap Rubber Topping Formula

Smoked Sheets	7 lb.
Boot and Shoe Reclaim	57 lb.
Cliffstone Whiting	55 lb.
Sublimed Litharge	9 lb.
Hard Mineral Rubber	3 lb.
Palm Oil	2 lb.
Tar Oil	2 lb.
Paraffin	1 lb.
Sulfur	11 oz.
Carbon Black	1 lb. 8 oz.

Cure ½ hour. Rise to 250° F., one hour at 250°.

Rubber Pencil Eraser

Crepe Rubber	4
Starch	10
Petrolatum	4
Vulcanized Waste Rubber	2
Factice	1
Abrasive	2
Lithopone	3
Sulfur	0.1
Accelerator	0.05

* Rubber Flooring Composition

Pale Crepe Rubber	120
Ground Cork	260
Venetian Red	30
Zinc Oxide	30
Sulfur	6
Accelerator	1

* Imitation Rubber

Isocolloids are transformed to emulsions or emulsion-like compns. *E.g.*, 400 parts of linseed oil contg. $NaHSO_3$, 5.5 parts NaI, 15 parts KH_4 oleate, 3 parts gelatin, 800 parts water and 8 parts MgO_2 are mixed, coagulated and dried in a CO_2 atm. The rubber-like product is plastic, can be mixed with filling materials and can be vulcanized at 80°. Or vulcanization can be carried out at 120–160° or at lower temp. in presence of piperidine-piperidyldithocarbamide, heptaldehyde, aniline, etc. as ultraaccelerators; protective colloids may be added.

* Rubber Matrix

The composition employed can be poured cold into a mold or upon a backing sheet. When set, it is of somewhat wax-like character, more or less tough or tenacious, and unaffected by the temperature at which molten metal or alloy for producing printing plates is commonly poured.

The composition is made to the following formula:

Commercial Rubber Cement	3 lb.
Carbon Tetrachloride	2 lb.
Benzol	2 lb.
Chemically Pure Talcum Powder	4 lb.
Carbon Black	½ oz.

These ingredients are mixed in a suitable mill, and, while in a fluid state, the composition is flowed over a metal sheet, pulp board, etc., to which it adheres quite closely. Thus prepared, the coated sheets may be stored for use.

When a matrix or mold is to be produced, a section of the coated stock is impressed with the desired form, pattern, or design. It is then supported and encompassed by guards to receive the molten metal, which is poured upon the composition as in the usual way of pouring stereotype plates. In actual practice cast printing plates have been produced by this process in from 3 to 5 minutes, starting with the backed composition, and perfect impressions have been taken on the composition from surfaces in which the lines or markings showing the design are so slightly out of the common plane that reproduction would be deemed impossible.

Owing to the fact that the composition neither expands nor contracts during or after molding or application to the backing surface, the cast plate reproduces absolutely the original pattern and will fit with precision its place in a press or in a form of which it constitutes part. This is a feature of importance, in that where the plate is to be used as part of a general make-up, difficulty has been experienced in causing it to register or in positioning it to occupy the space intended.

---

* Heat Exchange Medium

Diphenyl Oxide	70
Diphenylene Oxide	30

This may be reheated and revaporized without decomposition.

---

* Plasticized Rubber

Milled Plantation Rubber	100
Phenol Sulfonic Acid	7½

Form in sheets and heat to 135–140° C. for 6 hrs. This product disperses in benzol to form an extremely liquid solution.

---

* Plasticizing and Activating Rubber

Agents that are both plasticizing and activating, *e.g.*, stearic, oleic and lauric acids and oxidized paraffins, are incorporated into sheet rubber prior to milling by dipping the latter into a molten bath of the agent and allowing to stand to permit penetration. In an example crepe rubber sheets are immersed in molten stearic acid at 225° F. (or lauric acid at 180° F. or oleic acid at 150° F.) for 2–3 min., acid being absorbed to about ½ wt. of the rubber. The bath is allowed to drain 4–6 min. and then stored in a chamber at 125° F. for 5 hrs.

---

* Porous Rubber

15–20% Urea is incorporated in the raw rubber mixture and vulcanization is effected at 122°.

---

* Rubber, Quick Blending

Creped sheet rubber contg. approx. 1–2% of moisture is dipped in a bath of molten stearic acid at 225° F. for a few min. The rubber is removed and subjected to a temp. of 125° F. for 5 hrs. Lauric acid, pine oil and similar oil substances employed as softeners and plasticizers may be incorporated in a similar manner. The time and labor expended to obtain uniformly blended material is greatly reduced.

Raincoat Rubber Compound

Hevea Rubber	48
Litharge	10
Zinc Oxide	20.5
Mineral Rubber	5
Sulfur	1.5
Whiting	15

---

* Sponge Rubber

Compn. comprising rubber 60%, S 25, hydrocarbon 6, calcined MgO 3, ceresin 1 and coloring matter 5 is placed in a mold which is inserted in an autoclave to which steam is gradually admitted at 8 lb., and the pressure of gas, *e.g.*, air or N, injected into the mold is 180 atm. After 400 min. the steam is shut off and the autoclave cooled. The material, now about 6 times its original bulk, is inserted into a larger mold and heated again with steam under 85 lb. pressure for 45 min. The resultant material weighs not more than 5 lb. per cubic foot.

---

* Rubber, Porous Sponge

A dough of the following composition is heated under pressure.

Rubber	55
Sulfur	3.5
Vulcanized Oil	9
Golden Antimony	13
Adheso Wax	2
Magnesium Carbonate	17.5

---

* Thermo Plastic Rubber

Crepe Rubber	8
Benzol	24

Bubble Chlorine through slowly while cooling. Stop when chlorine no longer combines and escapes. Pour in pans in thin layers and evaporate solvent. This gives a thermoplastic chlorinated rubber.

* Rubber, Thermoplastic

Crepe Rubber	100
Diethyl Sulfate	10–15

Heat while on mill to 125–140° C. for 8 hrs.

Transparent Rubber Goods

Jatex, a concentrate obtained by centrifuging latex which after evaporation to 40 per cent gives a film as clear as glass, is used as dipping fluid. The articles are dipped at 40° C. followed by vulcanization in a bath made by dissolving 100 grams or more of the finest sulphur in 1000 c.c. benzol. Part of the sulphur remains on the bottom of the vessel and maintains saturated solution when the temperature goes up, and as sulphur is taken up during the vulcanization process. To promote the reaction is used an addition of 20 grams Vulcafor ZDC (zinc di-ethylene carbamate).

* Rubber Wax Mixture

Rubber can be introduced into waxes or high boiling oils by heating the molten wax or oil to 120 to 130° C., stirring, and introducing rubber latex in a fine stream at a rate which allows the water in the latex to boil off. Heating and stirring is continued until all the water is out of the mix. Up to 4% by weight of rubber can thus be introduced into molten paraffin wax, yielding a very viscous mass. The rubber is disseminated in a fine condition throughout the oil or wax. In waxes, the rubber serves to give the product additional strength and cuts down brittleness. The rubber can be vulcanized by the addition of vulcanizers.

* Rubber-Scorching, Prevention of

To prevent scorching during milling of rubber 1–2% Glyceryl Phthatlate is used.

Shoemaker's Wax, Hard

Rosin	8
Ester Gum	2
Montan Wax Crude	30
Paraffin Wax	45
Stearin Pitch	10
Beeswax	5
Oil Soluble Color	to suit

Shoemaker's Wax, Soft

Rosin	5
Paraffin Wax	65
Japan Wax	5
Stearin Pitch	20
Beeswax	5
Oil Soluble Color	to suit

* Rubber Flooring Composition

The method of producing floor coverings which consists in mixing together dry raw rubber with not less than 15 per cent by weight of sulphur and with 50 per cent to 85 per cent of the whole mass of cork granules by kneading and rolling giving the whole mass a desired shape and then vulcanizing said mass under a pressure of from 425 to 850 pounds per square inch and a temperature of approximately 145° C. and finally cooling it while the pressure is sustained.

RUST PREVENTION, PICKLING

* Corrosion Proofing Aluminum, Zinc, Magnesium and Their Alloys

Sod. Phospho-Chromate	0.75
Sod. Sulfo-molybdate	0.75
Trisodium Phosfate	0.40
Soda Ash	1.80
Sod. Tartrate	1.80
Water	94.50

Dissolve salts in water and bring to a boil. The metal to be protected is immersed in this hot solution until a sufficiently thick protective coating is formed.

---

* Preventing Corrosion of Aluminum Tubes

To toothpastes or other mildly alkaline preparations packed in aluminum tubes, the addition of 0.07–0.4% sod. silicate prevents corrosion.

Battery Terminals, Prevention of Corrosion

Slaked Lime	7
Sod. Bicarbonate	2
Borax	1
Rezinel No. 2 sufficient to make a paste	

Rustproofing Small Iron Parts

The articles are immersed in an aq. soln. contg. $FeCl_2$ 2% together with 2% of a salt of a metal below Fe in the electrochemical series, such as $HgCl_2$ and are then withdrawn and dried in a warm atm. They are then heated to about 100° and subjected to a humidity of 80% and then immediately immersed in boiling water to fix the resulting Fe oxides adhering to the surfaces.

To Prevent Gray Iron Castings from Rusting

The following mixture should be applied to the castings.

Carbonate of Soda	1 lb.
Lard Oil	1 qt.
Soft Soap	1 qt.
Water sufficient to make	10–12 gal.

Boil the above for half an hour, preferably using a steam coil. If the smell is objectionable add 2 lb. unslaked lime.

Rust Remover

Orthophosphoric Acid	35%
Water	30%
Ethyl Methyl Ketone	10%
Monoethylether of Ethylene Glycol	25%

---

* Corrosion Inhibitor

Sod. Chromate	20
Mineral Oil	15
Sulfonated Red Oil	50
Diglycol Oleate	2
Water	9
Soap	1

---

* Tarnishing of Magnesium, Prevention of

Magnesium articles are subjected to the action of 10–30% Sulfuric acid solution and then washed thoroughly.

Magnesium and Its Alloys, Prevention of Corrosion by Water

1% Pot. Dichromate is dissolved in the water used.

Rust Prevention

To give temporary protection from rusting metal articles are coated with a 50% solution of lanolin in naphtha.

---

* Tin Cans, Corrosion Preventing Coating for

A coating of glue containing 0.5% paraldehyde prevents corrosion of cans containing oil.

Rust Remover

100 parts of stannic chloride are dissolved in 1,000 parts of water. This solution is added to one containing 2 parts of tartaric acid dissolved in 1,000 parts of water and 2,000 parts of water are

All formulae preceded by an asterisk (*) are covered by patents.

added. The solution is applied by means of a brush, after removing grease, and is allowed to remain on for a few moments when the article is rubbed clean, first with a moist cloth and then with a dry cloth, and, if necessary, repolished in the usual way.

* Steel Pickling, Inhibitor for

About 0.05% Dibenzyl formaldehyde mercaptal is used with the diluted sulfuric acid.

* Steel, Pickling

In bronzing iron or steel the grease is removed and the iron or steel pickled, cleaned and introduced into a bronzing bath of NaOH 60, trinitrotoluene 2, PbO_2 0.8 and HNO_3 2.95 parts.

* Steel, Cleaning (Prior to Galvanizing)

The iron or steel is passed through a cold bath containing 35–250 gm. H_2SO_4 per liter and is made the anode with a current density of 20 amps. per sq. dm.

* Iron and Steel, Phosphate Coating on

The article is made the cathode in a boiling solution containing $Zn(H_2PO_4)_2$ and 0.05–0.13% of free H_3PO_4, with or without NaH_2PO_4, until a dense black coating is produced.

Diminishing Corrosion of Aluminum

Aluminum or its alloys are protected against corrosion by chlorine or bromine water by the addition of 0.5 and 5% of sod. silicate respectively.

* Rustproofing Iron and Steel

Iron or metal parts are dipped in a water solution of ammonium linoleate, oleate or palmoleate. On exposure to air the water and ammonia evaporate leaving a protective fatty film.

SILK, RAYON, COTTON, FIBRE, ETC.

* Wrinkle or Crease Proof Fabrics

Example 1.—A piece of printed satin made from artificial silk viscose is passed on the jigger through a bath containing a zirconium salt in solution. The temperature of the bath is about 18° C.; the time of passage through the bath is about 5 minutes; the concentration of the bath is about 50 grams of zirconium acetate per liter. After wringing, the piece passes through a second bath which contains an aqueous shellac-containing borax solution. Thereafter the piece so treated is dried on a cylinder drying machine and the goods are then passed over a solid mixture of Japan wax to which, in order to reduce its softening point, paraffin is added. Sufficient paraffin is added to give a softening point of about 30° C. The shellac-containing borax solution may be produced by dissolving 12 kg. of shellac in a solution of 3 kg. of borax in 40 kg. of water. In place of this a solution of 30 parts of shellac in a solution of 6 parts of trisodium phosphate to 100 parts of water may be used. Either of these shellac solutions for the purpose of softening may contain about 0.1% of olive oil as an emulsion.

The printed goods as so prepared are provided with a finish which has not been obtained hitherto. This finish is, for example, exceedingly useful for the manufacture of umbrellas. However, goods finished in this manner are also exceptionally valuable for blouses and other wearing apparel.

Example 2.—Boiled cotton goods are shrunk in a well-known manner with mercerizing lye—caustic soda solution—of approximately 30° B°. with the addition of about 1% sodium peroxide. The goods remain in this liquor for about 30 seconds at about 15° C. Thereupon they are rinsed and dyed in a manner customary in the textile finishing industry. This is followed by the treatment in ac-

All formulae preceded by an asterisk (*) are covered by patents.

cordance with Example 1 without prior drying of the fabric.

Example 3.—A piece of artificial silk with cotton warp of a weight of about 12.5 kg. is treated on the jigger at 60° C. in a bath which contains per liter 200 grams of urea and 4 grams of aluminum acetate free from sulfuric acid and also free from aluminum sulfate. After letting the liquid act for 10 minutes, the piece is passed through a second bath for ½ minute containing 250 ccm. of 40% formaldehyde solution and 8 grams of aluminum acetate per liter at 60° C. After letting the material lie or hang in the air for half an hour there takes place a strong condensation between the urea and the formaldehyde. Thereupon the fabric is dried hot at about 80° C., without prior rinsing, in a suitable device such as a drying room, tentering frame or the like.

In order to remove the surplus of the condensation product, the fabric is now treated at 80° C. with a liquor which contains 10 parts of 40% caustic soda solution per liter of water. After a passage of 5 minutes, the piece, after removal of the surplus, shows the desired feel so that it is only necessary to rinse well and to dry. If one works with lesser quantities of the substances mentioned, there suffices in place of the lye treatment, a passage through a boiling 3% soap solution. After this also, as stated above, it is thoroughly washed and dried on the tentering frame. Finally, the fabric is calendered in the customary manner.

Example 4.—40 kg. of urea are dissolved in the cold in 20 liters of 20% formaldehyde solution. This solution is left to stand for 12 hours and is subsequently diluted with 4 times its quantity of water and thereupon heated to 80° C. A cotton fabric which has been subjected to a prior treatment for ¼ hour with a cold 0.2% aluminum acetate solution, is agitated for a short time in the above hot solution of this pre-condensate and thereupon pressed between a pair of rollers. The fabric while still wet, is left to lie for one hour and is thereupon dried at 80° C. Finally the fabric is calendered in the customary manner.

Example 5.—Artificial silk fabric is run into a solution heated to about 70° C. containing 200 grams of urea and 2 grams of aluminum acetate per liter. It is there treated for about ¼ hour. Thereupon the fabric is wrung and passed on a slop-padding machine through a cold 40% formaldehyde solution. After previous wringing, the fabric is hung up for 2½ hours at room temperature. This is followed by a drying at 80° C. and then by the customary finishing.

Example 6.—A liquor containing 200 grams of urea and 2 grams of tin chloride, is heated to 60° C. and a rayon yarn is agitated in same for ¼ hour. After wringing, the further treatment is continued with formaldehyde solution and completed as described in Example 5.

Example 7.—An artificial silk fabric is treated at 60°–70° C. for 10 minutes in a bath consisting of an aqueous solution with a content of 200 grams of urea and 2 grams of aluminum acetate per liter. This is followed by a wringing of the fabric and by a slop-padding with a cold 40% formaldehyde solution which also contains 2 grams of tin chloride per liter. The fabric while still moist, is rolled up and is left to itself while being turned continuously and slowly for 2 hours. Thereupon the fabric is dried in a drying room and is left exposed a short while longer to a temperature of 80° C., whereupon the customary finishing treatment can follow.

Example 8.—20 kg. of urea are dissolved in 50 kg. of 40% formaldehyde solution and to the clear solution ammonia is added until a slight alkalinity is shown. The solution is now permitted to stand for 3 hours at room temperature. Thereupon it is acidulated slightly with acetic acid and 175 gr. of aluminum acetate dissolved in 50 kg. of water are added. In a bath thus prepared, a cotton-artificial silk-mixed fabric is treated for 10 minutes at room temperature, then squeezed and left overnight. This is followed by a hot drying at about 80° C. finally by a calendering on a highly heated calender at about 120° C.

Example 9.—A viscose fabric is put into a bath in the jigger consisting of 200 grams of urea and 4 grams zinc acetate per liter and left therein at 60° C. during 10 minutes. The fabric is then squeezed and passed through a second bath for ½ minute containing 300 ccm. of 40% formaldehyde solution and 8 grams of aluminum acetate per liter at 60° C. The subsequent treatment is done as stated in Example 3.

The silk, cotton and mixtures thereof finished in accordance with the invention are much more flexible than the corresponding untreated materials. They have acquired properties of animal fibers such as silk and wool. It is possible to crush the fabrics much more firmly together without causing them to wrinkle.

The artificial silk finished in accordance with the invention is very much better adapted for hosiery purposes than such silk hitherto found on the market. It has above all the important property of greater mobility in the meshes and a far greater lack of sensitivity to moisture and street dirt. A special property of the artificial silk obtained in accordance with this process lies in the fact that when moistened with water no rings form on the fabric, whereas when ordinary artificial silk is moistened in this way spots immediately become noticeable which leave rings on drying. Accordingly fabrics and dress materials prepared from it are considerably more valuable than hitherto. The goods thus finished dye excellently almost invariably. It is well-known that dyeing usually entails difficulties in connection with textile goods which have been treated in accordance with other finishing processes.

In general the threads treated in accordance with the present process are not very much harder than the untreated goods. Artificial silk, however, which has been purposely given a hard finish, can be easily softened in a well-known manner, viz., either in a mechanical way by passing through a breaking machine or by a subsequent impregnation with one of the paraffine emulsions. Above all, however, the artificial silk fabrics treated in accordance with the new process are very similar to real silk in connection with its resistance to crushing. It is a well-known fact that neckties or ribbons made of rayon are crushed and wrinkled after having been tied two or three times, to such an extent that they cannot be used again without first ironing them. As compared with this, genuine silk goods, as is well-known, even after having been tied frequently, possess this defect to a very much lesser extent.

Metallic Printing on Textiles

A certain number of fabrics are adorned with metallic powders printed with the aid of hot solutions of glue or gelatine, containing powders of aluminum, copper, bronze or brass in suspension, which remain fixed on the material after cooling. Cylinders of copper, aluminum or brass are used for applying the paste and are hollow so that steam or hot air may be introduced. The color-feed rollers are also heated. The trough for the metallic paste has a double bottom and it, too, is heated. All the heating elements are maintained at about the same temperature.

The printing completed, the cotton fabrics are passed through a drying machine.

Use of Glue

It has been found by experience that the use of a glue or gelatine paste at a high temperature has the great advantage of causing the metallic powder to adhere more easily to the surface of the fabric. But, to increase the fixation still more, the cloth is submitted, immediately after drying, to a certain pressure by passing it through a pair of calender rolls, which at the same time give it a slightly glazed finish.

If the metallic powder used is sufficiently fixed, the designs are very smooth and glossy, and if they are geometrical shapes they form a collection of fine lines almost imperceptible to the eye, but giving more attraction to the cloth. It is the impression of the rollers which produces this effect.

(1) *Dress goods with metallic effects.*—Certain garments for daily use gain much from the discreet use of metallic fabrics, and as these give a rather exclusive air their use has developed of late. The printing of these fabrics must be done with greater care than of those destined for carnival wear. The fixation of the powders must be absolutely complete, to the point of being able to resist a soaping without risk of the powder bleeding, even partially.

The designs used are most frequently flowers or leaves on a background of accentuated lines, to which a very special finish is obtained by pressure. The cheapness of the powders permits their use for muslins, tulles and voiles. When these more common fabrics are manufactured with care there is not much to choose between them and the older and more expensive goods. Their appearance in light, after they have passed through the calender, is remarkable.

(2) *The Printing Pastes.*—The printing pastes employed for the manufacture of these goods are very varied, but the majority of them permit the ordinary use of the metallic powders just enumerated. These are finally fixed with albumen, casein, rubber, or even with resin, bakelite or cellulose acetate.

One can, in this case, obtain very good results by printing in the cold, followed by drying and steaming. The goods produced in this way have sufficient resistance to washing and rubbing.

Sometimes, in the preparation of the pastes blood albumen (*e.g.*, 10 parts of the commercial quality, inodorous as far as possible) is used. It is wetted with 15 parts of water and mixed with a wooden rod twelve hours later, until a uniform mass is formed. This is then filtered through a sieve and ¾ part of essence of terebenthine and 1 to 3 parts of bronze, brass, aluminum or other powder are added. This mixture is used for direct printing from engraved rollers.

The smell left by blood albumen in the fabric sometimes gives rise to complaints. It is avoided by mixing an egg albumen with the blood albumen, or by using the former exclusively. This leads to a marked economy, but the results are less certain and sales more difficult. One or other of these albumens is sometimes replaced by casein dissolved in a weak ammonia solution. In these various cases, the fixation of the powders is not so good. When it is wished to use rubber for the fixation, 150 to 200 parts of the powder are mixed with 1,000 parts of a solution of this substance in benzine; fixation takes place after the solvent has evaporated.

(3) *Production of Metallic Designs with the Aid of Acetyl Cellulose.*—Solutions of cellulose or of its esters give excellent results, when it is a question of producing fine designs. The cellulose is dissolved in ammoniacal copper oxide, the metallic powder is added in the desired proportion, and the paste used on a color printing machine. The copper oxide in the fabric is eliminated with the aid of acid. The objection to this procedure is its high cost.

Instead, one may use acetyl cellulose dissolved in an appropriate solvent. The paste is prepared by mixing 12 parts of the acetyl cellulose solution, 24 parts of resorcine, 16 parts of water (added later), and 48 parts of denatured alcohol. The mixture is agitated, allowed to stand until the constituents are entirely dissolved and 15 parts of fine metallic powder are then added. One hundred parts of this paste are used on the roller printing machine, together, if wished, with pastes containing basic colors or others.

If colors are being used, one proceeds as in the following example: 1 part of Rhoduline Blue 3GO, 2 parts of Rhoduline Yellow 6G, 3 parts of a good commercial acetic acid; to this mixture add 20 parts of iron-free water and, later, 10 parts of hydrolite dissolved in 10 parts of water. After mixing these substances well, 10 parts of aniline oil, 10 parts of alcohol and 12 parts of tannin powder are added. The paste is then ready for use.

When the designs have been printed on the cotton fabric, this is dried, steamed for four minutes, and passed through a tartar emetic bath, if the color must possess good fastness; finally, the fabric is rinsed in running water, dried and calendered.

It is simple to vary the effect by mixing color of various kinds with the powders, so as to shade or modify these. Interesting effects are also obtained by confining the powders to certain parts of the print, obtained with basic colors or others on cotton, and by limiting the print to points, circles and so on, with lines of gold or silver, applied on the bench and giving the appearance of original oriental goods.

Finishing Compound for Light Woolen Fabrics

Soyabean-Lecithin	5	lb.
Olive Oil	2	lb.
White Mineral Oil	2	lb.
Triple-Sulphonated Castor Oil	½	lb.
Butyl Cellosolve	2½	lb.

This compound forms a white stable emulsion with warm water. About 5 parts of about compound in 100 parts of water are used.

The Dyeing of Cotton

The preparation of the fiber for dyeing depends upon the form in which it comes into the dyehouse and differs in the handling as well as processing.

Skeins are boiled out under pressure of 2–3 pounds with 0.25–0.5% calc. soda and 1% of a sulphonated oil or suitable wetting-out agent for 3 hours. The boiling liquor should be at least 15–20 inches above the check-chain before the kier is closed.

Piece-goods must be thoroughly desized before dyeing to prevent "Landscapes" or cloud effects. An addition of 0.1–0.2% of Activin based upon the weight of the goods will aid in a rapid and more complete desizing of the material. Piece-goods which must be bleached are best boiled out with 3% caustic soda, 2% calc. soda, 1% of a wetting agent and 0.1–0.2% Activin for 4 hours under 3 pounds of pressure. It may be said here that the degree of desizing can be successfully tested with a solution of potassium iodide.

When piece-goods are to be dyed with vat colors it is well to note that the ends of the pieces when sewn together should lie over one another, somewhat in the manner of roof-shingles. Pieces sewn together side by side, *i.e.*, against each other, will show ''airstripes'' after dyeing, evidenced by a deeper shade.

Tubular knit goods (jersey) and delicate materials are, of course, not boiled out under pressure, but are boiled out on the reel with 1% of calc. soda and 1% of sulph. oil for one hour.

Raw cotton, slubbing, cops, bobbins, and warp on the beam are usually handled in mechanical apparatus and are boiled out with 1% calc. soda and 1% sulph. oil for one hour.

Preferred and often used is the cold-wetting-out method for raw cotton and slubbing, which has the advantage of preserving the spinning qualities of the fiber. During the packing of the material attention should be paid that no channels develop, as this will interfere not only with the proper boiling-out process but also will give unsatisfactory results in dyeing.

Bobbins and warps on beams can, of course, be dyed with vat colors in mechanical apparatus, however, certain irregularities must be overlooked, and the same is true when dyeing skeins in apparatus which employ so-called ''Hang-systems.''

Dyeing skeins with vat colors in the dye kettle offers, of course, also certain difficulties such as unevenness, and an aid to good results are levelling and protecting agents such as Tetracarnit, Glue, Sulfite-cellulose-waste liquors, Soap, Sulphonated oils, etc. It must, however, be remembered that Soap or Sulphonated oils can be used only to limited amounts in the dyebath, as they will induce the material to swim and thereby only hinder the dyeing process. An addition of Glucose to the dyebath will often aid in overcoming unevenness, however, the amount of caustic soda must be increased about 30%, as the Glucose will use up this amount. A further aid to level uneven dyeings is to remove the lot from the dye liquor, squeeze, and return to the dyebath under addition of more sodium hydrosulfite, and raising the dyeing temperature from 60–100° F. It must be mentioned, however, that most of the vat color types will lose their brilliancy and also give up part of their fastness qualities should the temperature be raised above their regular dyeing temperature. It is perhaps more advisable, providing the dyeing qualities of the dyestuffs are accurately known to the dyer, to begin dyeing at a lower temperature and gradually raise to the dyeing temperature, as in this manner no complications will have to be feared, provided the condition of the vat is constantly observed.

After dyeing the material is squeezed and hung on sticks to oxidize. Should oxidation be too sluggish the process can be hastened by passing the lot through a bath made up with 0.3–0.5 cc. per liter of 30% Hydrogen Peroxide, at a temperature of 80–100° F. Sodium Perborate (1–3% from the weight of the goods) can be used instead of Hydrogen Peroxide. After the material has been handled in such a bath for 10–15 minutes, the temperature can be raised to the boil and the subsequent soaping be carried out without fear of complications, as the perborate will give up its oxygen quickly at a temperature of 150° F. It may be pointed out that such a method is also more economical as it eliminates one extra handling of the material.

---

* Air-ship Fabric, Coating for

Cotton or silk is coated with

Polyglycerols	2.5
Gelatin	1

This gives a flexible, adherent, gas-tight finish.

Penetration and wetting out agents suitable for dyeing cotton and rayon goods in various forms (hosiery, package yarns, skein, etc.).

A.	Sulf. Castor or Red Oil	35 parts
	Steam Distilled Pine Oil	35 parts
	Water	30 parts

Heat the castor or red oil agitate while adding the pine oil until thoroughly blended, add water—then adding a 25% solution of NaOH solution with stirring until the solution becomes clear. Test 10 cc. in 50 or 100 cc. of cold water should dissolve instantly and no separation should occur.

B.	Water	50 parts
	KOH or NaOH	16 parts
	75% Sulf. Castor Oil or Red Oil	6– 8 parts
	Cresylic Acid	25–32 parts

Caustic to water then castor or red oil is added while being stirred until solution clears. Cresylic is best added before oil.

C. Water 50 parts
KOH or NaOH 5 parts
Sulf. Red Oil (75% Strength) 20 parts
Steam Distilled Pine Oil 15 parts
Cresylic Acid 12–18 parts
Sulf. Red Oil or Castor Oil (75%) 8–10 parts

These materials added in order named with constant stirring until solution clears. Then solution should be tested for stability and solubility in cold water as well as wetting out properties by some approved method.

Removing Cotton from Cotton Wool Mixture

Cotton can be removed from wool cloth by holding the cloth in hot vapors of hydrochloric acid at a temperature of about 100° C. for 3 hours. The treated material can then be soaked in water with beating, whereupon the cotton fibers disintegrate and become dislodged. The wool fibers will retain their shape and strength.

SCROOP

Cotton Hosiery

Wash after dyeing for one-half hour at 120° F. in a bath containing 7% soap based on weight of goods. Extract, but do not rinse. Then place in a cold bath of 10% acetic acid and run one-half hour and rinse. This imparts a scroop like silk.

Boil-off Liquor

For cotton yarn chain form 200 gallons of water, 2 quarts 75% Sulphonated Oil; 4 pounds of Soda Ash, powdered; 2 pounds Caustic Soda, flake. Run this at a boil. Second boil-off in dye bath before dyeing; 200 gallons of water; 1 quart of 75% Sulphonated Oil; 4 pounds of Soda Ash.

Finish on Sulfur-dyed Cotton Khaki

A. Corn Starch 45–50 lb.
Dextrin 34–36 lb.
50% Sulfonated Castor Oil (Turkey Red) 25 lb.
Water 100 gal.

B. Dextrin 45–50 lb.
50% Sulfonated Tallow 45–50 lb.
Water 100 gal.

A and B will give fair increase in weight on finished goods if "feel" is too harsh, increase proportion of starch and for softer feel use 5–20 lb. of emulsified Japan Wax. Chemical finishes for increasing weight are not recommended.

C. Sulfonated Castor or Sulfonated Tallow 20 lb.
Corn Starch 30– 35 lb.
Dextrine 70– 85 lb.
Epsom Salts 90–100 lb.
Glucose 8– 12 lb.
Formaldehyde 1½– 3 lb.

This formula C can be used on cheaper goods for large weight increases.

Dyeing Cotton Black (Chrome)

Dissolve 3.3 lb. of bichromate of potash in a small quantity of water, mix the solution with 100 gallons of logwood decoction at 3° Tw., and add 7.7 lb. hydrochloric acid, 34° Tw. The cotton is introduced into the cold solution, and the temperature is very gradually raised to boiling point. The cotton acquires at first a deep indigo-blue shade, which changes to a blue-black on washing with a calcareous water.

A slight modification of this process consists in working the cotton in a solution containing at first only the bichromate of potash and hydrochloric acid, and adding the decoction of logwood to the dye bath in small portions from time to time, gradually raising the temperature as before.

Anti-Seize Compound

Used in threads to prevent seizing.

Petrolatum 50%
Zinc Dust 50%

Scouring Cotton-Rayon Fabrics

Turkey Red Oil 5
Olive Oil Soap 5
Soda Ash 1
Water 100 gal.

Use at 200° F. for 1–2 hrs. If fabric contains celanese keep temperature below 175° F. and leave out soda ash.

FINISHING OF COTTON YARNS OR CLOTH

White Yarn and Cloth

Water 60 gal.
Potato Starch 20 lb.
Lupogum 4 lb.
Tallow 10 lb.

Japan Wax	4 oz.
Olive Oil Soap	4 oz.

Dry on the tenter frame, let the cloth or yarn pass over a 3 cylinder roller and mangle with pressure.

Flannels, Finish for

Water	15 gal.
Lupogum	14 oz.
Soap	1 lb.

Back-Filling Cotton Cloth, Linings, Etc.

Water	15 gal.
Wheat Starch	9½ lb.
Lupogum	10 oz.
China-Clay	62½ lb.
Chalk	12½ lb.

or

Water	1000 parts
Lupogum	10 parts
Wheat Starch	20 parts
China-Clay	20 parts
Japan Wax	1 part

Ticking, Finish for

Water	15 gal.
Lupogum	10 oz.
Potato Starch	3¾ lb.
White Dextrine	2½ lb.
Sulphate of Magnesia	2½ lb.
China-Clay	5 lb.
Helveteen	10 oz.

or

Rice Starch	11¼ lb.
Lupogum	10 oz.
China Clay	4 lb. 6 oz.
Salicylic Acid	2 oz.

Sizing of Rayon Hanks to be Used as Warps

1 lb. Lupogum is stirred thoroughly into 9 gal. cold water and dissolved;

1 lb. Glucose is dissolved in lukewarm water;

1¼ lb. Olive Oil Emulsion.

All three are mixed, brought to a boil and boiled for 1 minute. The whole mass will be about 12 gal. due to condensed steam. This mixture of 12 gal. is sufficient for 50 lb. rayon, *i.e.*, for a bath of 120 gal.

Scouring Cotton-Rayon Fabrics

Turkey Brown Oil	10 lb.
Olive Oil Soap	10 lb.
Soda Ash	1 lb.
Water	100 gal.

Treat for 1–2 hrs. at 200° F. If fabric contains Celanese leave out the soda ash and do not heat above 175° F.

Flax Waste, "Cottonizing"

Treat flax waste one hour at 40–90° C. with

Caustic Soda	10
Sod. Silicate	5
Water	85

Keep at 90° C. for 1 hr.

* Creaseproof Fabrics

This is achieved by impregnating the material while the fibers are in a swollen condition. The cloth is therefore treated with a mercerizing liquid, *e.g.*, caustic soda with or without tension, whereby the cellulose is swollen to the greatest possible extent The excess of caustic is removed by squeezing till the material contains an equal weight or a little more of water and then immediately mangled with the following resin:

Phenol	100
Formalin	100
Potassium Carbonate	4

which is boiled for 5 minutes and rapidly cooled. The fabric is then squeezed till it contains about an equal quantity of liquor (*i.e.*, its own weight of resin sol) and dried at a low temperature, finally being heated at 170° C. on drying tins for 2 minutes in order to complete the reaction. Lastly, the excess of resin is removed by boiling with soap as previously. Under these conditions the fabric retains just under 15 per cent of resin.

Olive Oil Emulsion

May be used for finishing blankets, hosiery, mercerized cottons, etc.

25% Tri-sodium-phosphate Solution	50 parts
Olive Oil	30 parts
50% Sulf. Tallow	10–15 parts

Add half of olive oil and mix thoroughly in TSP solution then boil and agitate until saponification takes place and add in the remaining half; then add in sulf. tallow and mix until a smooth blended emulsion is formed. Test—10 cc. in 100 cc. lukewarm water; should emulsify and not separate out in oily spots, etc. Should have consistency of soft lard or butter.

Crepe Dye Resist

Resist White

Precipitated Chalk	200 gm.
Potassium Sulphite 90° Tw.	50 gm.
Acetate of Soda	50 gm.
Water	265 gm.
Dark British Gum	325 gm.
	1000

Beat the whole into a smooth paste, heat until the gum is dissolved, and cool.

Resist White gives a better white under the black than zinc oxide. Zinc oxide, however, is to be preferred for colors, because it works better in printing and yields brighter shades. It is usually ground up with a little glycerin, and turpentine is added to minimize the tendency to froth.

Paste (For Colors) Standard

Zinc Oxide	200 gm.
Water	170 gm.
Glycerin	25 gm.

Beat into a paste, and add

Dark British Gum	200 gm.
Gum Senegal 50% Solution	150 gm.
Turpentine	30 gm.
	775

Heat to dissolve the gum, and then use warm or cold.

* Increasing Ironing Resistance of "Celanese"

The material is treated with a 5½% caustic soda solution and dried immediately and quickly.

Scouring and Dyeing Assistant

For use with Acetate yarns and materials (hosiery, etc.).

Good Grade Soluble Pine Oil	50 lb.
Trisodium Phosphate	10–20 lb.
Dichlorethylether	4– 8 lb.

Add the Tri Sodium Phosphate in a concentrated solution with constant stirring and warming until complete saponification takes place. Then add the solvent slowly with stirring. The pH should be kept in a 10% solution to 11. or below.

Test.—A complete dispersion in cold water when mixed.

This is an inexpensive scouring and dyeing assistant on hosiery, knit-wear, etc.

Finish for Fancy Woven Goods

1. Composition of the finish:

Dextrine	150 parts
Epsom Salt	80–90 parts
Monopole Soap	6– 7 parts

per 1000 parts paste or brought up to the required degree of Tw.

2. Thicker finish:

Dextrine	200 parts
Epsom Salt	110–130 parts
Glucose	50 parts
Monopole Soap	6– 7 parts

per 1000 parts paste or brought up to the required degree of Tw.

3. Cheap finish:

Potato Flour	50 parts
Epsom Salt	50 parts
Monopole Soap	5–6 parts

per 1000 parts paste.

Dissolve the different constituents separately in water and mix them together by good stirring. In cases where the products cannot be dissolved separately owing to want of accommodation dissolve the dextrine or potato flour together with the Epsom Salt and boil, then add the glucose and finally the Monopole Soap. The latter is dissolved with direct steam in a small quantity of water but before adding it to the finish, dilute the dissolved soap with as much water as possible in order that the fatty matter may be finely and uniformly divided and thus render same particularly stable. The dissolving of a little dextrin (4–5 oz. dextrine per 1 lb. of soap) together with the Monopole Soap will be found advantageous.

It is not necessary to boil the finish again after the addition of the soap, although a boiling is not detrimental. The temperature of the size ready for use should be 95–115° F.

Scouring Knit Goods

Scour at 160° F. for 20 minutes in

Trisodium Phosfate	1
Olive Oil Soap	2
Water	97

Rinse well in soft water.

Dyeing Knit Fabrics

Using direct colors. For light shades dissolve dyes separately and strain into bath. Dye goods for 10 minutes at 80° F. Add glauber salts (5% of weight of goods) and raise temperature to 120° F. Shade should be reached in 15 minutes.

For dark shades increase glauber salts o 15% and increase temperature to .60° F.

½ of 1% neutral olive oil soap may e used for improving feel of finished goods. Dry at 100° F.

SCROOP

Rayon Products

The fabric should be run first through lukewarm bath of turkey red oil. Then mmerse for 5 minutes in a 1% solution f glycerine or glucose to which has been dded ½% of acetic acid. After which emove the goods, extract, and dry at a ow temperature, but do not wash.

* Rayon, Delustering

The rayon (500 g.) is introduced into 1. of cold aq. soln. of $MgSiF_6 . 6H_2O$ 10 to 40%). After 10 min. the soln. s slowly heated to 70–90° and kept at his temp. for 10 min., whereby hydrolysis akes place and the SiO_2 deposits in the hread. Subsequently the material is washed out.

* Delustering Rayon

An acetate fabric is worked for an our at 75° C. in a bath containing 30% atex and 0.5% ammonium thiocyanate to ct as swelling agent on the silk fibers, r an acetate fabric may be treated for 5 minutes at 75° C. with 50 times its weight of

Aqueous Dispersion of Colloidal Graphite	=10 %
Ammonium Thiocyanate	= .1%
30% latex	= 5.5%

which produces a non-rubbing medium gray color, fast to light, and washing.

One per cent of zinc oxide with 2 per ent of latex (30 per cent) and a swelling agent gives excellent results as far s delustering is concerned.

Scouring and Dyeing Rayon Pile Fabrics

A continuous full width scouring or dyeing machine was used for the entire process. The machine consisted of seven boxes holding approximately 540 gallons ach at the working height. The first wo boxes containing 24 pounds Trisodium Phosphate and 16 pounds Olive Soap each. The major part of the soil nd dirt in the cloth came off in the first wo boxes. In order to avoid contamination of the next four boxes, nip rolls were placed between the first two boxes nd after the second. To prevent distorting or damaging the pile in the nip, a barrel spreader and a rotating bristle brush were placed before each set of squeeze rolls in the machine. The next two boxes contained 24 pounds of Trisodium Phosphate each. Most of the grease was emulsified in the first two boxes. The small amount remaining was easily removed by the fairly alkaline baths in boxes 3 and 4. Due to the quite heavy nip after box 2 little soapy liquor is carried over into box 3, while the percentage of soap in box 4 is negligible. The temperature in the first four boxes was maintained at 200° F. by means of closed steam coils, while the remaining three boxes were all cold. A nip roll is placed after box 4 to squeeze out as much of the alkaline liquor as possible. Box 5 contains 12 pounds of 28% acetic acid to neutralize any alkaline residue. .04 pounds of an acid violet (color index number 698), having practically no affinity for either rayon or cotton, was also placed in the box. This dyestuff was used so as to prevent any exhaustion of the color. The latter would necessitate feeding dyestuff into box 5 which might in turn result in uneven pieces from end to end ("tailing off"). A nip roll was placed between this box and the next. Box 6 contained .04 pounds of acid violet and no acid. This box and box 7, which contained water only, were intended to level out any slight unevenness in color which might result from the possible unevenness in the acidity of the cloth in box 5. Nip rolls were used before and after box 7. After passing through the last nip, the cloth was plaited on a flat truck and was then ready for finishing. The cloth travelled at a speed of 15 yards per minute and took about 20 seconds to pass through each box.

Boiling Off Silk

Raw silk consists chiefly of two substances, the true silk fiber, called fibroin," and an outer layer of material known as "sericin." It also contains a very small amount of wax, fat, coloring matter and ash. Most of the coloring matter is in the outer sericin layer.

Sericin is a substance resembling gelatine in its properties, and is soluble in water only by prolonged boiling.

Fibroin is a proteid and is not noticeably affected by prolonged boiling in water, but is somewhat readily attacked by caustic alkalies even in weak solutions, their action rendering it more brittle and rough and diminishing its gloss. Fibroin is also attacked by soap solutions if boiled for a long time, but it is not acted upon by weak acid solutions.

In preparation of silk for the dye bath it has been customary to "boil off."

This process consists in boiling in a bath of soap and water, sometimes with the addition of Carbonate of Soda, the purpose of such treatment being to remove the outer layer of sericin, whereby the silk becomes lighter in color and the luster is developed, and it becomes softer and more suitable for dyeing.

During the process of boiling off, the sericin first swells up, making the silk sticky. It then dissolves, leaving the lustrous and internal thread exposed.

In treating piece goods which are composed partly of cotton or wool, the boiling off process serves the further purpose of cleansing from the material whatever dust may be adhering to the silk.

It tends also to improve the quality of the cotton or wool mixture. It is customary to put the goods through a washing process after boiling off. The boiling off and washing processes consume much time and labor, and employ materials which, while not expensive in themselves or in small quantities, become expensive when used in large quantities, as they must be used in the customary practice of the art.

It is claimed by users of Sulphonated Castor Oil AA that silk left to soak in a bath made up to consist of:

One part of the Oil to 1000 parts of water, with the addition of sufficient soda ash, or about two parts, to make the bath slightly alkaline at a temperature of about 98° C. for one-half hour, the degumming process will become complete during the dyeing.

The solution is very mild in its action upon the fibroin, leaving it coated with a very thin layer of nitrogenous material which is repellent to water, though soluble on prolonged boiling. The protective layer is of extreme thinness, and is removed in whole or in part in the ordinary operations to which silk goods are subjected subsequent to boiling off. This layer also probably protects the fibroin from weakening not only during the time that it is in the bath, but during the subsequent operation of dyeing.

---

* Silk, Degumming

Silk is treated at 50° with a solution of papain with Sod. Sulphoxylate equal to 25% of papain used.

---

* Silk and Rayon, Delustering

Delustering of artificial silk is effected by treatment, at a temp. within about the range of 20–100°, with a soln. formed of approx. equal proportions of alum and $BaCl_2$ (the total quantity of which may be from less than 1% to about 5% the wt. of the artificial silk treated).

---

* Delustering Cellulose Acetate

The material is steeped at 80–100° for a short time in a 5% pine oil emulsion.

Dyeing Silk Black (Lyons)

About 10 to 20 per cent yellow prussiate of potash is used in proportion to the weighting with oxide of iron which the silk has received previously. In addition, a quantity of hydrochloric acid equal to the prussiate, is required. Prepare the bath with the prussiate and half the hydrochloric acid. Enter at 30° C. turn the silk about ten times, heat to 45° turn a few times, add the other half of the acid and heat to 50 to 55° C., turn again a few times, wring out and wash well in water.

A weighting of 16 to 24 per cent is obtained; or by a threefold treatment with nitrate, etc., the loss sustained by the discharging is recovered, and the silk brought to "pari." A further weighting of 4 per cent may be added by one more treatment with "nitrate of iron" after the blue dyeing, and subsequent rinsing with water to precipitate the ferric hydroxide (hot soaping would affect the Prussian blue). Work the silk after these treatments one hour in an old bath of catechu (gambier) standing at 4 to 7½° Tw., the temperature of which should not exceed 50° C., so that the Prussian blue may not be decomposed and the shade become too dark; rinse and hydro-extract. The silk acquires in the catechu bath an over-charge (over pari) of 15 per cent and becomes more greenish.

---

* Silk Weighting and Waterproofing

240 grams of nickel sulphate are dissolved in 9320 cubic centimeters of a 14–15% solution of ammonia, and 680 cubic centimeters of an aluminate solution containing 68 grams of sodium hydroxide and 2.5 grams of aluminum are added, under agitation. A Bordeaux-red liquid is obtained which may be used directly, if pure reagents have been employed, or after filtration, if the reagents employed

were such as to render filtration necessary.

Boil Off, Celanese Velvet

Here the boil off bath is adjusted to a pH of 7.6 after adding 3 lb. of sodium sulphide per 1,000 gallons of water and approximately the same amount of 84% commercial acetic acid. Then 3 lb. of seritex (probably the enzyme papaine) per 100 gallons is added and the bath heated to 165° F. The velvet, which has previously been soaked for 30 minutes in a weak olive soap solution, is immersed in this bath for 2½ to 3 hours. The goods, generally hooked on a vertical star frame, are kept slowly moving all this time. After this treatment, the velvet is immersed in a 0.5% olive soap solution at 170° to 175° F. for 30 minutes. Then it is rinsed in soft warm water which is gradually cooled by a steady influx of cold water. After this rinse the goods are ready for dyeing.

Viscose Manufacture
For Rayon and Cellophane

Steep 2 lb. cotton or pure wood pulp fiber in 18% NaOH solution at 20° C. for 1 hour.

Press excess caustic out till pulp weighs 6.5 lb.

Keep in a closed container for 70 hours at 20° C.

Place in large mason jars, first breaking pulp up. Add ¾ lb. Carbon Bisulfide; close jar and shake for 2 hours till orange color appears.

Dissolve this xanthate in a 3½% NaOH so as to finally have 7% cellulose in solution, approximately use 16 lb. to 18 lb. of 3½% NaOH solution.

Keep this viscose for 3 days at 18° C.

For coagulation use a spin bath of following specifications:

H_2SO_4	9%
Na_2SO_4	18%
$Zn\ SO_4$	1%
Glucose	5%
Temp.	45° C.

Then rinse acid out of thread.

For transparent films spread very thin on a plate of glass. Place glass in a solution of 30% $(NH_4)_2SO_4$. Then place in saturated salt solution. Then place in 3% H_2SO_4 solution till film is clean. Wash acid free and dry.

* Viscose Sponge

Viscose solution containing $<$ 6% of NaOH (*e.g.*, 3.7% of NaOH and 7–7.5% of cellulose) and ripened to $<$ 5° is diluted, *e.g.*, with an equal vol. of H_2O, mixed with a foaming agent, *e.g.*, 0.5% of oleic acid, and worked into a foam. This is run into moulds and allowed to coagulate spontaneously, coagulation being accelerated, if desired, by heat or the addition of salts, *e.g.*, NaCl, to the foam. After washing and drying, a light (apparent d 0.02–0.1), porous, elastic product is obtained particularly suitable for use as a heat-insulating material. Fillers, softeners, colouring agents, or fire- or H_2O-proofing agents may be incorporated.

Viscose Skeins, Weighting

Light Scour (Based on Weight of Material)

Neutral Olive Oil Soap	5–6 %
Soda Ash or Trisodium Phosphate	1–1½%

Rinse thoroughly in warm water and hydro-extract; place these skeins (not dried) in solution of 7–8° Bé. Aluminum sulfate. Keep at room temperature (70°–75° F.) for one hour. Place skein on rack and drain; and turn occasionally. Rinse in 4–6 parts cold water washes. Place in 5–6° Bé. Silicate of Soda (iron free) solution warm from room temperature (70° F.) to 100° F. Allow to stand for one hour. Wash thoroughly in warm water at 130°–150° F. Direct or basic dyestuffs may be used after this.

Stripping of Textiles or
Discharging of Colors

Heat a solution containing one gallon of stripper T. S. (Arkansas Co.) per 100 gallons of H_2O to 180° F. Rayons, cottons, silks will strip and certain amounts of celanese colors.

Sodium hydro sulphite can be used at 3% to 4% strength and same temperature. Three per cent Formaldehyde-Sulfoxylates solutions containing 1% NH_4OH will strip the majority of colors at high temperatures, especially the acid colors.

To dye materials so that a white design will be left, the method is to use a mixture of 10% zinc acetate and 10% Hydrosulfite in paste form at the design. Then dye with a vat color. The metal resists dye and hydrosulphite discharges color at that design.

Textile Materials, Identifying

	Vegetable Fibres					Artificial Fibres			Animal Fibres	
	Cotton	Linen	Jute	Hemp	Ramie	Viscose	Chardonnet	Acetate Silk	Wool	Silk
Burning	Burn rapidly with pungent smell					Burn rapidly with pungent smell		Forms beads	Burn slowly with characteristic smell	
Caustic soda, 76° Tw.	Insoluble	Insoluble	Brown. Insoluble	Yellow. Insoluble	Insoluble	Unchanged	Disintegrated and partly dissolves	Fibre swells	Soluble cold	Soluble hot
Alkaline lead	....	....	....	....	....	....	....	....	Black	
Sulphuric acid, 168° Tw.	Dissolves rapidly	Dissolves slowly	Dissolves slowly	Dissolves slowly	Dissolves slowly	Rapidly dissolve			Insoluble	Dissolves
Nitric acid	Insoluble	Insoluble	Brown. Insoluble	Yellow. Insoluble	Insoluble	Dissolve rapidly with yellow coloration			Yellow. Insoluble	Yellow. Dissolves
Ammoniacal copper solution	Soluble	Soluble	Insoluble	Insoluble	Insoluble	Swells, disintegrates and is partly dissolved		Unchanged	Insoluble cold	Soluble cold
Aniline sulphate	....	....	Yellow	Yellow						
Acetone	....	....	....	....	....	Unchanged	Unchanged	Dissolves rapidly		
Iodine and sulphuric acid	Blue	Blue	Yellow	Yellow	Blue					
Diaphenylamine and sulphuric acid	....	....	....	....	....	....	Blue			

* Horse-hair Substitute

Hard vegetable fibers, such as coconut fibers are heated with dil. lyes, *e.g.*, 7–8% NaOH in a closed vessel to 120°–135°, and then treating with oily or hygroscopic substances.

Wool, Silk and Cotton, Determining in Textiles

Use Dreaper's reagent which is made by adding 2 grms. of sodium hydroxide dissolved in 30 c.c. of water to 2 grms. of lead acetate dissolved in 50 c.c. of water. The mixture is boiled until it becomes clear, cooled to about 60° C., and 0.3 grm. of magenta dissolved in 5 c.c. of alcohol added. The solution is made up to 100 c.c and filtered if necessary. A piece of the fabric to be tested is heated in this solution nearly to the boiling-point for 2 minutes, washed with water, then with dilute acetate acid, and dried. Silk will be colored red and wool black, while vegetable fibers remain white. The magenta may be replaced by picric acid.

* Artificial Wool

Cotton thread or cloth is given a wool-like appearance by treating it with NaOH soln. (35°–40° Bé.), with aq. soln. of $ClCH_2CO_2H$ (5°–10° Bé.) and then with NH_4OH (2%–5%).

Carbonizing Wool in Cotton Mixture

Some kinds of burnt out embroideries which consist partly of pure cotton and partly also of artificial silk and cotton, are prepared on a ground of wool or cotton. The ground is then usually carbonized before the dyeing, that is to say, removed so that the actual embroidery alone remains standing out.

For cotton embroidery, a wool ground is usually used, and is carbonized by a hot treatment or by boiling for 20 to 30 minutes with caustic soda lye of 3°–5° Tw. The embroidery is then rinsed thoroughly, soured off and dried, the destroyed wool then being removed by heating.

Bleaching Wool and Silk

Treat cold for 30 min. the well-degreased wool with 20 parts of a soln. contg. 3 g.

$KMnO_4$ and 3.5 g. $MgSO_4$ per l., expose for 3–4 hrs. to the sun, treat in a bath contg. 40 c.c. $NaHSO_3$ 35° Bé. and 4 c.c. H_2SO_4 66° Bé. and rinse. Add more $NaHSO_3$ if the goods are still colored. For silk the first bath contains per l. of water 1.5 g. $KMnO_4$, 2 c.c. H_2SO_4 66° Bé.; the second bath 20 g. $NaHSO_3$ 35° Bé. and 2 g. H_2SO_4.

Woolens, Finish for

Water	15 gal.
Lupogum	6 oz.

dry on felt covered rollers.

Worsteds and Cheviots, Finish

Water	15	gal.
Potato Starch	2½–2¾	lb.
Lupogum	6	oz.

Heavy Woolen Cloth, Finish

Water	15	gal.
Potato Starch	3¾	lb.
Lupogum	6	oz.
Glauber's Salt	3 lb. 2	oz.
Sulphate of Magnesia	2½	lb.
Glycerine	10	oz.

* Wool, Oil Treatment for

Wool fiber is treated with a saponaceous aq. emulsion contg. soap 5, olive oil 10 and water 1000 parts, and is then treated with a soln. of Al formate, and dried at 60–80°.

Removing Oil and Grease Spots

Immerse the goods for one hour in a warm saturated solution of sodium aluminate, diluted to about ½ strength. Then rinse in warm water, extract and dry. Much better results are obtained when the solution is lukewarm, although it can be used cold.

Solutions made by this same formula may also be bottled and used for removing small spots, as it leaves no fringe or ring. Put a piece of blotting paper under the spot and apply solution with a cloth.

* Wetting (Penetration) Agent

A penetrating or wetting agent useful in mercerizing textiles consists of

Cresols	90
Pine Oil	6
Red Oil	4

Cotton, Coloring

Cotton and cotton materials are generally dyed with *direct* dyes, sometimes called substantive dyes. They do not need any chemical to develop or lock the dye into the fiber. Common salt, however, is used as an auxiliary to aid dyeing.

Dyeing instructions: Prepare dye bath using about four gallons of water to each pound of material.

Add five pounds of salt for each pound of dye used.

Bring temperature up to 140° F. Introduce the material. Bring temperature up to a boil and keep at boiling point three-quarters of an hour. Rinse and dry.

Average Yellow requires
1 lb. of dye to 100 lb. material

Average Red requires
2 lb. of dye to 100 lb. material

Average Blue requires
2 lb. of dye to 100 lb. material

Average Green requires
2 lb. of dye to 100 lb. material

Average Black requires
5 lb. of dye to 100 lb. material

Representative dyes are:

Direct Fast Yellow NN
Chrysophinine (Yellow)
Direct Blue 2B
Direct Sky Blue 5B
Direct Orange 2R
Direct Green
Congo Red
Direct Black E
Direct Pink E
Direct Violet N
Direct Brown

Wool, Coloring

Wool and woolen materials, for the most part, are dyed with acid dyes; the acid used is Sulphuric. In some cases acetic acid is used. Glauber salts are added as an auxiliary in dyeing.

Dyeing instructions:

For each 100 lb. of material
use 4 gallons of water.
add 3 lb. of Sulphuric Acid.
add 10 lb. of Glauber Salts.
add 1 to 5 lb. of color depending on shade and color strength.

Yellow generally requires 1 lb.
Red, blue, green generally require 2 lb.
Black generally requires 5 lb.

Bring temperature of dye bath to 140° F. Immerse material, bring to boil and boil three-quarters of an hour and rinse.

Representative dyes are:

Yellow—Tartrazine
Lemon Yellow—Erio Flavine
Orange—Orange II
Red—Ponceau 2R
Red—Crocein Scarlet
Magenta—Acid Magenta B
Violet—Acid Violet 6 BN
Green—Patent Blue A
Black—Acid Black J
Black—Acid Black 10 BX

Silk, Coloring

Silk may be colored with Direct, Acid, or Basic colors. The Direct colors are dyed in a neutral bath. Some direct colors require the addition of Acetic Acid to the dye bath toward the end of the operation. Temperature 180 to 200° F. Time about 30 minutes.

Acid Colors.—Dyed in bath acidulated with Sulphuric Acid. Temperature 180 to 200° F. Time about 30 minutes.

Basic Colors.—Dyed in bath acidulated with Acetic Acid. Temperature start at 100° F., go to 140 to 175° F. slowly. For Auramine, temperature must not exceed 140° F.

Direct dyes (see dyes for cotton).

Acid dyes (see dyes for wool).

Basic dyes:

Yellow—Auramine
2 lb. per 100 lb. material
Orange—Chrysoidine Y
2 lb. per 100 lb. material
Brown—Bismark Brown
2 lb. per 100 lb. material
Pink—Rhodamine B
2 lb. per 100 lb. material
Blue—Methylene Blue 2B
2 lb. per 100 lb. material
Violet—Methyl Violet
2 lb. per 100 lb. material
Green—Malachite Green X
2 lb. per 100 lb. material
Black—Basic Black
2 lb. per 100 lb. material

Dyeing Tussah Pile Fabric

Goods are entered into the dyebath at 120° to 125° F. After running for 30 minutes to thoroughly wet the cloth, 37.5% Fustic Extract and 5% bluestone are added in the order named, but a few minutes apart. The cloth is run in this liquor for 15 minutes when 7.5% copperas and 3% oxalic acid (previously dissolved and mixed together) are added. The temperature of the bath is raised to 175° F. in the next 45 minutes after which 75% Hemastine Extract is added and the temperature raised to a boil. The dyebath is kept at a boil for an additional 1.5 hours. The goods are then rinsed twice. After hydroextracting the cloth is ready for finishing.

During the dyeing process the dyebath must be kept a clear amber color. Any darkening would indicate insoluble lake which is rectified by the addition of more oxalic acid. Care must be taken not to add too large an excess of acid as this would tend to redden the shade. If a bluer shade of black is desired, this may be obtained by cooling the bath to 180° F. after it has boiled for 75 minutes and then adding one per cent soda ash. The bath is then raised to a boil again for an additional 15 minutes. The shade of the black is regulated by the amount of Fustic Extract used.

---

* Protecting Wool in Vat Dyeing

Wood fabric is first printed with the following paste:

Indigo Pure 20% Paste	15.0 parts
Glycerine	5.0 parts
50% Gum Thickening	10.0 parts
Potassium Carbonate	7.5 parts
Formosul	10.0 parts
Sodium Aminoacetate 50%	5.0 parts
Water	7.5 parts

The fabric is then partially dried and steamed; afterwards it is oxidized in an acidified hydrogen peroxide or perborate bath, soaped, and dried. It is found that the wool material printed by this method suffers no loss of strength and does not acquire the harsh handle which it otherwise would.

Direct Wool Printing

For direct printing on wool, the following formula is recommended for the Chrome Fast Dyes:

Dyestuff	20 gm.
Glycerine	50 gm.
Water	408 gm.
Neutral Starch Tragacanth Thickening	500 gm.
Ammonium Oxalate	12 gm.
Neutral Ammonium Chromate	10 gm.
Total	1000

Steam one hour and wash.

For heavier shades the quantity of dyestuff is proportionately increased. With Erio Chrome Print Black a full bloomy shade can be produced with 60–80 grams dyestuff per 1000.

Blue Linen Finishing

Cheap Finish

Water	100	parts
Potato Flour	6	parts
Gluten	6	parts
Monopole Soap	0.6	part

Cheap Finish with a Heavy Weighting

Water	100	parts
Potato Flour	10	parts
Epsom Salt	6	parts
or		
Chloride of Magnesium	4–5	parts
Sirup (Treacle)	2–3	parts
Monopole Soap	0.8	part

Superior Finish

Water	100	parts
Dextrine	14	parts
Epsom Salt	6–7	parts
Monopole Soap	0.6–0.7	part

Finish with a Very Heavy Weighting

Water	100	parts
Wheat Starch	5	parts
Potato Starch	7	parts
China Clay	10	parts
Chloride of Magnesium	3	parts
Monopole Soap	0.8	part

It is advisable to color the finish with a little substantive Blue and basic Violet, say with ½ gm. Benzo Blue RW and ⅜ gm. Methyl Violet B p. lb. paste.

To prepare the finish proceed as follows:

Dissolve the different constituents separately in water and pour them together while stirring well. In cases where the constituents cannot be dissolved separately owing to want of accommodation, dissolve the dextrine or potato flour together with the Epsom salt and boil; finally add the Monopole Soap. The latter is dissolved with direct steam in a small quantity of water, but before adding it to the finish, dilute the dissolved soap with as much water as possible in order that this weak soap solution may finely and uniformly divide the fatty matter and thus render the size particularly stable. The dissolving of a little dextrine (4–5 oz. dextrine per 1 lb. of. Soap) together with the Monopole Soap will be found advantageous.

It is not necessary to boil the finish again after the addition of the soap, although a boiling is not detrimental. The temperature of the size ready for use should be 95–115° F.

---

* Delustered Cellulose Acetate Yarn

Cellulose acetate is dissolved in acetone contg. approx. 2.5% water. Before this process of soln. is completed there is added Halowax (chlorinated naphthalene), to the amt. of about 12% of the cellulose acetate, dissolved in about 3 times its own wt. of acetone. The two solns. are thoroughly mixed giving a spinning soln. A delustered cellulose acetate yarn is produced by spinning.

SIZING AND STIFFENINGS, SOFTENERS

Backing for Sheet Plastics

Pigment	7 lb.
Ethyl Lactate	25 lb.
Methanol	50 lb.
Ethyl Acetate	25 lb.
Cellulose Acetate	7 lb.

Fine Cotton Size

Potato Starch	75
Tallow	7½
Pine Oil or Turkey Red Oil	1
Water	830

Size, Alkali

Dextrin (Potato)	30 lb.
Castor Oil	1 lb.
Caustic Soda	30 lb.
Pot. Carbonate	30 lb.
Water	65 lb.

Dissolve dextrin in part of water and emulsify oil in this.

Dissolve alkalies in balance of water and stir in.

Concentrated Warp Sizing
(For Cotton Warps)

36–42 lbs. Sul. Tallow (75%, if 50% used increase proportion)
18–24 lb. Raw Beef Tallow—good quality preferred, otherwise size may be discolored slightly.
14–20 oz. of Dry Gum Tragacanth
38–45 lb. of Water.

The gum tragacanth should be placed in separate vessel and heated up to boil and allowed to stand until complete jell has been reached, then it is ready to add to mix.

Mix the sulfonated tallow and raw tallow in kettle and heat while mixing until thoroughly blended and syrupy.

Add the gum trag jell and mix until blended.

Add the necessary amount of preservative and place in closed barrels until ready for use.

Concentrated Warp Size Lubricant

10–14 lb. Sulf Tallow (75%)
18–22 lb. Mineral Oil Softener
18–24 oz. Dry Gum Tragacanth
14–16 lb. Raw Tallow
4–45 lb. Water.

Prepare gum tragacanth in separate vessel as noted above.

Place the two tallows in kettle, agitate and heat until blended, then add the mineral oil softener continue agitation and heat until blended.

Add Gum trag jell and additional heat may be necessary for a thorough blend.

Concentrated Finishing Compound
(For Cotton Piece Goods)

22–26 lb. Sulf. Tallow (75%)
12–15 lb. Japan Wax
20–24 lb. 25% Tri-Sodium-Phosphate Solution
50–60 lb. Water.

The Japan wax should be emulsified in a separate vessel.

Mix the tallow, ⅓ of the Japan wax (emulsified) and required amount of T. S. P. solution until thoroughly blended.

Add the remainder of the Japan wax emulsion, agitate and heat; it is best not to boil.

Stir until a creamy mix is secured.

Sizing Compound for Cotton Warp Yarns (To be Used With Starches)

40–50 parts good quality Beef Tallow
8–12 parts good quality Sulfonated Tallow (50% commercial grade)
1½–2½ parts solvent and emulsifier Di-Ethylene Glycol for example)
½–2 parts Locust Bean Gum (Gum Trag) made up into a 8:100 Water Gel and in thorough solution before adding.
1½ parts Steam Distilled Pine Oil

All formulae preceded by an asterisk (*) are covered by patents.

2 parts Japan Wax (made into thorough emulsion before adding
1–1½ parts cresylic acid
40–50 parts of Water added with thorough agitation and sufficient heating.

Sizing Compound for Cotton Warps
(To be Used in Combination With Type of Starch Needed)

30 parts good quality White Beef
30 parts good quality 50% Sulfonated Tallow
6– 8 parts Japan Wax Emulsion
32–34 parts Water
1– 2 parts disinfectant or deodorant should be used.

Melt Japan wax and sulf. tallow while agitating, when thoroughly melted add beef Tallow and stir until thoroughly mixed. Then add water gradually and agitate until a full white creamy mix is secured.

Cotton Warp Sizing

14–20 lb. Tapioca Flour
1⅓–3 lb. Animal Glue (ground)
3–5 lb. 50% Sulfonated Tallow. (May substitute Tallow emulsion.)
3–5 lb. Paraffin Wax
90–150 gal. Water

Warp Sizing for Durene
(Mercerized Cotton)

100 lb. Corn Starch
12–15 lb. Raw Beef Tallow (Tallow Emulsion can be used)
200–250 gal. Water.

Sizing for Polishing
(Cotton Cordage)

2 oz. Tri-Sodium-Phosphate
4 lb. Irish Potato Starch
1¾ lb. Japan Wax
1¾ lb. Paraffin Wax (127° M. P.)
6 oz. Narobin
14 oz. Mineral Oil Softener

Mix thoroughly and make up to 10 gallons with necessary amount water. Use sufficient amount of water to dissolve starch and heat with constant stirring until all products are thoroughly mixed—then allow to cool and use cold.

The Tri-Sodium Phosphate is used primarily because of ''hard water.''

Cotton Size

Wheat Starch	4 lb.
Narobin	1 lb.
Water	25 gal.

---

* Linseed Oil Size

Linseed Oil	100
Trichlorethylene	100
Am. Linoleate	16
Water	100–200

Size, Newspaper

The pulp is sized with a mixt. of 1.2% Na_2SO_3 in 4 pt. of water, 0.5% NaOH (5% soln.) and 3.2% $Al_2(SO_4)_3$ (6–8° Bé.) (all wts. are based on the wt. of fiber). The method produces better results, and a considerable economy than the use of rosin.

Sizing for Rayons

75 lb. Coconut Oil
11 lb. Tri-Ethanolamine
20 lb. Red Oil (Oleic Acid)
2–3 lb. Preservative (Sodium Benzoate, etc.)
50 lb. good grade Gelatine
make up to 100 gallons sizing.

The tri-ethanolamine and red oil are mixed first—then added to the melted coconut oil with stirring.

The gelatine is dissolved and added to the above mixture with stirring on reaching a well blended size it is diluted to 100 gallons and stirred further. When used, water is added two to one to secure proper take up in sizing in slashing machine.

Skein Sizing Rayon Yarns

Take 60 lb. Gelatin and soften it by allowing it to soak for 2 hours in water. Boil for 20 minutes in 200 gal. water. Dip yarn in this, centrifuge and dry.

Size, Concentrated Rosin

Rosin	70
Soda Ash	7
Beeswax	2
Water	21

Boil together until a sample solidifies on cooling. This may be shipped solid and is dispersed in hot water when needed.

Soap, Rosin Size

Into a suitable boiler or heater an amount for instance 100 kilogrs. of resin is placed and as much water, then a mixture of carbonated and bicarbonated alkalis is added in a quantity necessary for saturating say 88% of the resin put in operation. If the bicarbonate is employed in about the proportion of half the carbonate, then approximately 11 kilogrs. of carbonate of soda and 5 kilogrs. of bicarbonate of soda will be required.

The boiler is heated by steam for example and when cooking is considered sufficient, water and a volatile alkali (ammonia) are added, the amount of alkali being sufficient to saturate the 12 kilogrs. of resin which have not been affected by the carbonated alkali. For this second phase of saponification by means of ammonia liquid it is necessary to employ about 4 kilograms of aqueous ammonia solution having a density of 0.930 (which would contain about 18% of pure ammonia) when the quantity of hydrated resin to saponify is 12 kilogr. that is to say, the proportion of ammonia liquid is ⅓ to ⅔ hydrated resin. The heating by steam is continued so as to bring the mixture up to boiling point for some minutes, at the end of which time the product is finished.

---

* Sizing, Textile

Rosin	24–60
Linseed Oil	24–60
Borax Casein Solution	8–10
Sod. Silicate	5–10
Water	500

Size, Textile

Corn Starch	85 gm.
Sulfuric Acid (66° Bé)	0.4 gm.
Glycerol	10 gm.
Water	1000 c.c.
Caustic Soda	to make neutral

Woolen Yarn Size

Potato Starch	12 lb.
Narobin	2 lb.
Water	25 gal.

Jute Size

Potato Starch	12 gal.
Narobin	1 gal.
Water	25 gal.

Size for Mercerized or Dyed Yarn

Narobin	3 lb.
Water	25–50 gal.

Warp Sizing

1. Potato Starch	40 lb.
2. Narobin	10 lb.
3. Water	125 gal.

Boil two in 100 gal. of three for ½ hr. Stir one in 25 gal. three mix both solutions, stir and boil until uniform.

---

* Wax Size

The following is used for treating paper-cloth.

Japan Wax	100
Soap	10
Water	40

Boil and stir until homogeneous. This is diluted with boiling water and stirred before use.

Sizing, Window Shade and Automobile Top

Tung Oil	2 gal.
Casein	10 lb.
Borax	8 lb.
Paraformaldehyde	1 oz.
Animal Glue	26 lb.

Solubilizing Starch

The starch is mixed with required amount of water and 1% Aktivin S on amount of starch used.

A wooden vat with mechanical agitator preferred, copper can be used but wood keeps solution hot the longest. Direct steam may be used in boiling up starch. A thick paste is made first, this becomes thinner and after boiling 20 minutes or longer the starch becomes thin flowing. Do not fail to actually boil starch and covering to prevent splashing.

100	lb.	Starch
150	gal.	Water
1	lb.	Aktivin S

Stirring and boiling is discontinued when desired thinness is reached.

Textile Size (Soluble Starch)

Method No. 1

200 gal. Water 200 lb. Tapioca Starch 1 lb. Polyzime	agitating constantly while mixing these materials

Warm to 75° C. (167° F.) over a period of 15 minutes and cool to 55° C. (131° F.) and then add 1 lb. polyzime and keep it at this temp. until liquid has reached suitable consistency (15 to 30 minutes is usually sufficient). Then increase temp. to 80° C. (178° F.) and keep it at this temp. for 15 minutes to stop enzymatic action. Cool down and if desired to preserve add a small amount of salicylic acid or zinc chloride.

Method No. 2

200 gal. Water 200 lb. Tapioca Starch 2 lb. Polyzime	mix thoroly and agitate

Warm to 72° C. (162° F.) for 20–30 minutes, then keep at this temp. for 15 to 30 minutes when starch will be dextrinized to desired degree. Now raise temp. to 80° C. (178° F.) and retain this temp. for 15 mins. to stop enzymatic action.

Note: If potato starch used, add 50% more polyzime; if corn starch used, add 100% more polyzime

If flour containing gluten is used, polyzime is supposed to possess a high degree of proteoclastic properties and will naturally bring about a conversion as above.

Precautions: Starch liquid should be neutral or faintly acid.

Enzymatic action will be destroyed at 80° C. in ten minutes but cannot be destroyed at 75° C. even if heated 1 hour.

Water quantity can be changed to any ratio with starch and a good paste be made at high concentration of 1 part starch to 2 parts of water. Polyzime must always be added in ratio to starch used and not to water.

Cotton Good Softeners

The saponified cocoanut oil softeners are easily made by heating the melted oil with the required amount of a concentrated caustic soda solution until saponification is complete, following which the mixture is diluted to approximately 20 per cent fat content.

Coconut Oil Softener

Cocoanut Oil	2060 lb.
Soda Ash	135 lb.
Caustic 39° Bé	1090 lb.
Dilute to produce	9000 lb.

These products are finished off alkaline or neutral as desired and are exceptionally well suited for use in hard water or in mixes containing excessive amounts of salts, such as Epsom and others. Their excellent solubility, moreover, permits of easy removal on washing when this is necessary. Cocoanut oil soaps almost invariably become rancid with age, although this can be retarded by complete saponification. Softeners made from the completely neutralized fatty acids are less liable to this fault than those made from the oil itself. The great fluidity of the soap with its capacity for holding water enhances the value of this material as a softener, as well as for the lustrous sheen imparted on calendering. A shirting formula containing this oil is given here:

Shirtings

1 lb. 10 oz. Wheat Starch
15 lb. Potato Starch
60 lb. Talc
2 lb. 8 oz. Stearic Acid Softener
13 oz. Cocoanut Oil Softener
40 gal. Mix

Softener, Textile

150 lb. Water, add
180 lb. Castor Soap Oil, add
½ gal. Caustic Soda, 25° Bé., and add
80 lb. Stearic Acid, and heat up and cook slowly until the Stearic Acid is melted, mixing the contents meanwhile.

You have in this compound the added softening properties of the Castor Soap Oil, resulting in a more efficient softener than can be produced when Stearic Acid is used alone. It is neutral.

Textile Softener

65 lb. Double Pressed Stearic Acid
10 lb. Ammonia
1 lb. Formaldehyde
450 lb. Water.

---

* Textile Size

A substantially non-acid strengthening adhesive size for textile fibers which is soluble in the alkaline solvent used for removing size from textiles, comprising a boiled mixture of cobalt drier and linseed oil in substantially the proportions of from 200 to 500 grams of linseed oil and substantially 25 grams of cobalt drier and 100 kilograms of boiled linseed oil free from driers.

Sizing of Wooden Containers

Barrels and Casks that are to be used as containers for anhydrous and certain organic liquids are sized with a solution of either hide or bone glue before use, as otherwise the liquid would penetrate the wood and be lost, besides resulting in a decay of the wood. A first treatment is given to fill all of the cracks and imperfections, and a second to size the whole inner surface. A few quarts of the glue solution are introduced into each barrel and steam applied under a low pressure to force the solution well into the pores of the wood. The barrels are rotated and finally drained while still hot.

Glue as a Size in Paints and Calsomine

In the painter's trade glue is employed both as a size for the treatment of walls prior to the application of paint, merely to fill up the pores of the wall, for which bone glue is satisfactory; or it may be mixed with a little paint, an insoluble base, and water, in the preparation of a calsomine. In the higher grades of these calsomines which must be used with hot water, the better grades of hide glue are used.

WATERPROOFING

Waterproofing Composition

To thirty parts of commercial petrolatum fifteen parts, by weight, of aluminum palmitate are added and the mixture kneaded into a smooth paste free from lumps. Or the petrolatum may be heated to about 130° F., whereupon the consistency of the petrolatum is such that a smooth mixture is produced by introducing the palmitate and stirring. To this mixture is added fifty parts of commercial yellow beeswax and one hundred five parts of soft paraffin wax, such as white scale wax, and the resulting mixture agitated in a steam heated container. The temperature is brought up to 250° to 270° F. and the agitation continued until a smooth, homogeneous mass is obtained. The mixture is then allowed to cool to about 220° F. and about eight hundred parts by weight of a petroleum thinner having a boiling range in this instance of 275° to 450° F. added. It will be found that the resulting product is stable and homogeneous, of proper viscosity for application by hand or machine, and extremely suitable as a saturant for waterproofing fabrics. It acts as a preservative to fabrics to which it is applied and forms a water-repellent and impervious coating on each of the fibers making up the material.

* Waterproofing Composition

Celluloid (16 oz.) is dissolved in 35 oz. of acetone and 40 oz. of alc., and 5 oz. of castor oil is added. A second soln. is formed by dissolving 6 oz. of gum sandrac and 6 oz. of gum mastic in 15 oz. of amyl acetate, 15 oz. of butyl acetate and 15 oz. of butyl alc. This soln. is strained and mixed with the first soln. for about 1 hr. Benzene (35 oz.) is slowly added to the compn. and thoroughly mixed for 30 min.

Waterproofing

Gelowax	17
Carbon Tetrachloride	10
Ethylene Dichloride	10
Benzol or Naphtha	60

Digest until dissolved.

* Waterproofing

A composition for application to textiles, paper, etc., consists of

Latex	65
Caustic Soda	1
Water	2
Precipitated Chalk	20
Castor Oil	5
Phenol	0.5
Rosin	4.5
Rapeseed Oil	4

Waterproofing Liquid (Cloth or Wood)

Paraffin	2/5 oz.
Gum Damar	1 1/5 oz.
Pure Rubber	1/8 oz.
Benzol	13 oz.

Carbon tetrachloride q. s. 1 gallon. Dissolve rubber in benzol; add other ingredients and allow to dissolve. (Inflammable).

Waterproofing Liquid

This may be used on fabrics, paper and other fibrous bases. It penetrates quickly and leaves a flexible, odorless product which is highly water repellent.

Example 1.—Use of high melting paraffin wax and plasticizer for the cellulose nitrate.

	Per cent
Nitrocotton (15–20 seconds)	1.0
High Melting Paraffin Wax	4.0
Naphthene Base Mineral Oil	6.0
Butyl Stearate	2.0
Butyl Acetate	4.0
Ethyl Acetate	25.0
Gasoline	13.0
Toluol	40.0
Ethanol (Denatured)	5.0
	100.0

Example 2.—Use of Japan wax and no plasticizer for the cellulose nitrate.

	Per cent
Nitrocotton (15–20 seconds)	1.0
Japan Wax	3.0
Naphthene Base Mineral Oil	3.0
Toluol	30.0
Ethyl Acetate	33.0
Butyl Acetate	30.0
	100.0

The compositions of the above examples are prepared by a simple mixing operation. Preferably the wax is added to the toluol in a mixer and agitated until dissolved, and the cellulose nitrate is separately dissolved in the ester solvents and alcohol, the other materials then being added to the nitrocellulose solution, which is then combined with the wax solution.

The compositions may be applied to fabrics by a number of known methods but it is preferred to apply these compositions simply by immersing the fabric, or paper, or material to be treated until it is thoroughly saturated and then wringing out the excess coating material by squeeze rolls or centrifuging. This process is conducted at room temperature generally, although in using the composition in Example 1, it is preferred to carry out the process at a temperature not lower than 73° F., since there is some tendency for the high melting paraffin wax to precipitate out if the operating temperature is below 73° F. In the case of the composition in Example 2, it is not necessary to observe this temperature requirement since the Japan wax does not show any tendency to precipitate out. After the excess coating material has been removed the volatile solvents of the composition are then removed by drying the fabric, or paper, at ordinary or slightly elevated temperatures.

Canvas Waterproofing

Raw Linseed Oil	1 gal.
Beeswax Crude	13 oz.
White Lead	1 lb.
Rosin	12 oz.

Boil the above and apply warm to upper side of canvas, wetting the canvas with a sponge on the underside before applying.

Waterproofing Canvas

Gilsonite	80 lb.
Stearine Pitch	62 lb.
Scale Wax	34 lb.
Mineral Oil	10 lb.
Creosote Oil	10 lb.
Copper Linoleate	9 lb.

Melt together.

Apply at a temperature of 300° F. Scrape off excess while hot.

Waterproofing Canvas

Beeswax	25 lb.
Glyceryl Stearate	5 lb.
Stearine Pitch	102 lb.
Copper Oleate	15 lb.
Castor Oil	48 lb.
Naphtha	50 lb.

Waterproofing Canvas

For canvas paulins or large portable covers:

Formula 1

Petrolatum (Vaseline), Dark or Amber	8½ lb.
Beeswax, Yellow Refined	1½ lb.
Earth Pigment, Dry (Ochre, Sienna, or Umber)	5 lb.
Volatile Mineral Spirits (Painters' Naphtha)	5 gal.

Formula 2

Petroleum Asphalt, Medium Hard	7½ lb.
Petrolatum, Dark or Amber	2½ lb.
Lampblack, Dry	1 lb.
Volatile Mineral Spirits (Painters' Naphtha)	5 gal.

The quantities specified are sufficient to treat about 40 square yards of canvas on one side.

A mixture of 3 gallons of gasoline and 2 gallons of kerosene can be substituted for the volatile mineral spirits, but will evaporate more slowly. Canvas treated according to the first formula will be colored buff by ochre, khaki by raw sienna, drab by raw umber, and brown by burnt umber. If a white treatment is preferred, use dry zinc oxide in place of earth pigment. For some purposes, Formula 1 with a light-colored pigment will be preferable to Formula 2, because canvas treated with the latter will absorb more heat from sunlight, owing to its black color.

For permanently fixed canvas covers:

Formula 3

Boiled Linseed Oil	1 gal.
Lampblack, Ground in Linseed Oil	2 lb.
Japan Drier	1 pt.

Formula 4

Boiled Linseed Oil	1 gal.
Aluminum Bronzing Powder	1 lb.
Japan Drier	½ pt.

For lightweight fabrics not continuously or frequently exposed to sunlight:

Formula 5

Beeswax, Yellow Refined	½ lb.
Spirits of Turpentine	1 gal.

Mixing the Materials

In the preparation of waterproofing solutions according to Formulas 1, 2, and 5, place the specified weights of waterproofing materials in a suitable metal container and melt slowly and carefully at as low a temperature as possible, with constant stirring. Then remove to a place where there is good ventilation and no fire or open flame and pour the melted material into the solvent while stirring. When a pigment is used, thin the pigment in a separate container by mixing with it small additions of the liquid, and when the pigment mixture is sufficiently thinned strain it through fine-mesh wire screen or several thicknesses of cheesecloth into the waterproofing liquid. In Formulas 3 and 4 the pigments should be thinned in a similar manner with linseed oil before they are added to the bulk of the oil.

When the waterproofing material settles to the bottom of the container or thickens, it will be necessary to warm the mixture just before applying it to the canvas. This must be done in the open air by placing the container in a tub or can of hot water. Be sure that the container is open, and *never place it over or near a flame.*

Application

The mixture must be thoroughly stirred before and during application, in order to keep the undissolved material in suspension. These preparations may be applied to the canvas by means of a paint brush or by spraying. Wagon covers, shock covers, etc., may be treated best by stretching the canvas against the side of a barn or attaching it to a frame and applying the material with a brush. Once the canvas is fixed in position, no more time is required to treat it than is necessary to apply a first coat of paint to a rough board siding having the same area. Much time may be saved in treating large paulins and standing tents by applying the material with a spray pump, with which a pressure of at least 50 pounds is developed. Some loss of material, however, results from this method.

The experience has been that one coat applied to one side of the canvas usually is sufficient. With one coat applied to one side, using the strength of solution as given in the formulas, there will be an increase in weight of approximately 40 to 50 per cent when Formula 1 or 2 is used. When Formula 3 or 4 is used the fabric will gain about 75 per cent in weight. When Formula 5 is used the gain in weight will be around 10 per cent.

When canvas is treated with linseed-oil preparations it should be allowed to dry thoroughly (for two or three weeks) while freely exposed to the air. If folded and stored in a warm place before drying is complete the accumulated heat from continued oxidation may result in spontaneous combustion.

---

*Waterproofing Cement Walls

Cement walls are waterproofed and freshened by painting or spraying with following:

Soda Ash	9
Alum. Sulfate	1
Pot. Permanganate	0.03
Water	20

Cement enough to still keep fluid.

Integral Waterproofing for Concrete

Al or Ca Stearate

About ¼ to ½ lb. to the bag of cement.

Cement Waterproofing (Integral)

Dissolve in gauging water about ½ gal. Ammonium Stearate 28% to every bag of cement.

Dampproofing (Concrete, etc.)

1 lb. Paraffin Wax
¼ gal. China Wood Oil
½ gal. Bodied Linseed Oil (3 Hour heat)
¼ gal. Varnolene
1 gal. Benzol

2⅛ gal. Yield

Heat slightly to dissolve wax.

---

* Waterproofing for Cordage

Montan wax emulsions in H_2O, prepd. with rosin and Na_2CO_3, are used to impregnate ropes, nets, etc. Example: Eight kg. of montan wax, 2 kg. of rosin and 1.3 kg. of calcined Na_2CO_3 are ground together and 2 kg. of train or linseed oil are added. The mixt. is dispersed in 50 l. of b. H_2O. This basic emulsion may be dild. with 10 times its vol. of H_2O.

Waterproofing for Cloth

Naphtha	100
Rubber Cement	45
Ester Gum	20
Cumar	4
Paraffin Wax (128°)	32

---

* Waterproofing Cellulose Articles

Cellulose fibre articles are impregnated at 150–232° with a mixture of blown petroleum asphalt (80–90%), rubber (5–15%), and wax (about 5%).

---

* Waterproofing Cloth

Glycerol	31
Phthalic Anhydride	74

heated together at 185° C. till the product has an acid value of 126. The resin is then cooled by pouring into trays and ground. A solution is then made by stirring together at 70° C.

Resin	25 parts
Gaseous Ammonia	1.4 parts
Water	100 parts

This gives a viscous and practically water-white solution which, when applied to glass and dried for one hour at 100°, gives a clear, hard, adherent film.

The resin thus obtained is used in conjunction with latex. For example:

10 parts of the Resin Syrup obtained as above is mixed with
20 parts Natural Rubber Latex,

and the resulting stable compound is used for producing a flexible non-tacky waterproof finish on cloth.

The usual rubber compounding materials, such as plasticizers, vulcanizing agents, anti-oxidants and fillers may, of course, be added if required, e.g., a mixture of:

	Parts Dry Weight
Latex to give	100
Zinc Oxide	5
Colloid Sulphur	2
Tetramethylthiuram Disulphide	2
Resin Solution	75

may be used to coat the backs of carpets and the like and dry-cured at 120° C. for 30 minutes.

Cloth Waterproofing

Aluminum acetate is used for waterproofing cloth, the usual procedure being to immerse the well cleaned material in a solution of aluminum acetate of 4 to 5 degrees Baumé strength. The material is soaked for a period of about twelve hours and then dried in a warm room. The cloth is then introduced into a soap solution made up of about five pounds of soap in 13 gallons of water, the excess liquid wrung out and the cloth then given a bath in a 2% alum solution, followed by drying. This latter process precipitates aluminum stearate into the fibers of the cloth.

Another process, somewhat similar to the one above, consists in first immersing the cloth in a solution of:

White Soap Chips	10 lb.
Dextrine	20 lb.
Water	16 gal.

To cause thorough solution, the above is heated. After passing the cloth into this first solution, it is hung to drain and while still wet immersed in:

Zinc Sulphate (White Vitriol)	6 lb.
Dissolved in Water	9 gal.

The material is then removed after thorough penetration by the second solution, and dried, any coarse precipitated particles being brushed out.

Another method uses the following formula:

Lead Acetate (Sugar of Lead)	1 lb.
Tannic Acid	2 oz.
Sodium Sulphate (Glauber's Salts)	1 oz.
Alum	10 oz.
Water	1 gal.

Waterproofing Duck

Boiled Linseed Oil	100 lb.
Carbon Black	18 lb.
Turkey Brown Oil	20 lb.
Naphtha	46 lb.
Water	10 lb.
Ammonium Hydroxide	2 lb.

Agitate with a high-speed stirrer until completely emulsified. Apply two coats to each side of the material.

* Waterproofing for Fabrics

Dissolve 34 ounces sliced pale crêpe rubber in 1½ gallons linseed oil by boiling and add 4 ounces liquid drier.

* Leather, Waterproofing

Unfilled leather is impregnated with following:

Rubber Latex	100
Gasoline	100
Paraffin	25
Mineral Oil	10
10% Soap Solution	50

* Waterproofing Leather

Rubber Latex	100 cc.
Gasoline	100 cc.
Paraffin Wax	25 gm.
Paraffin Oil	10 gm.
10% Soap Solution	50 cc.

The wax is dissolved in the gasoline and paraffin oil, and the soap solution is added to the latex after which the mixture of gasoline, wax and paraffin oil is introduced gradually into the combined latex and soap solution with vigorous stirrings.

This gives a composition of substantially the proper consistency for ordinary waterproofing purposes and having high penetration characteristics. The rubber in this form freely permeates leather and like materials without being filtered out and left on the surface as is the case with the ordinary solutions of crude rubber in solvents and on account of this penetrating capability of the rubber in this form of composition, a highly effective waterproofing occurs. The residue of the composition which remains in the leather after the solvent has evaporated in sufficiently plastic to preserve the softness or pliability of the leather and its plasticity is not materially affected by usual changes in temperature and it therefore does not become stiff when subjected to cold or too soft when subjected to heat. It is not affected by atmospheric oxidizing agents and its adhering properties are such that it is not washed out by wetting and drying of the impregnated material, in service, as are the waterproofing compositions commonly used.

* Masonry, Waterproofing

Cement, concrete, etc., is painted with a solution of

Aluminum Stearate	3
Naphtha	100
Acetic Acid	1½

* Paper, Waterproofing

Previously blown petroleum asphalt 80–90, rubber 5–15 and waxy material such as beeswax about 5 parts are heated together.

Waterproofing Shoes

Natural Wool Grease	8 oz.
Dark Petrolatum	4 oz.
Paraffin Wax	4 oz.

Melt the ingredients together by warming them carefully and stirring thoroughly. Apply grease when it is warm but never hotter than the hand can bear.

* Shotgun-Shells, Waterproofing for

M-Styrene	18
Tricresyl Phosphate	3.6
Ethyl Acetate	30
Butyl Acetate	20
Toluene	25
Xylene	25

Straw Hats, Waterproofing for

Bleached Shellac	75 parts
White Rosin	15 parts
Venice Turpentine	15 parts
Castor Oil	2 parts
Alcohol (Denatured)	250 parts

Gum Sandarac	135 gm.
Gum Elemi	45 gm.
Castor Oil	11 gm.
Rosin, Bleached	45 gm.
Alcohol (Denatured)	1,000 cc.

White Shellac	4 oz.
Gum Sandarac	1 oz.
Gum Thus	1 oz.
Alcohol (Denatured)	1 pt.

---

* Waterproofing for Textiles

Casein	4
Water	6
Am. Hydroxide	0.45
Rubber Latex	24

Waterproofing for Textiles

Rubber Cement	46
Ester Gum	22
Cumar	2
Paraffin	31
Naphtha	100

Waterproofing Cloth

The process is carried out in two padding machines.

The first padder contains a soap emulsion made up as follows:

Twenty-five pounds Soap (stearic acid type) is dissolved in 100 gallons boiling water. Twelve pounds Japan wax is added a little at a time with stirring so that an emulsion is obtained.

The second padder contains the following solution:

Fifty pounds Lead Acetate and 40 lb. Aluminum Acetate are dissolved in 100 gallons water. The clear solution is siphoned off the lead sulphate which is formed in the reaction and is run into the second padder.

The cloth is entered into the first bath at the rate of about 15 yards per minute so that it is in contact with the emulsion for about 12 seconds. This rate has to be varied with the type of cloth treated. The cloth is squeezed between rollers and without rinsing is passed into the second bath. It is squeezed between rollers again and dried.

Waterproofing Textiles

Fabrics may be rendered waterproof with glue and tannin. Both should penetrate the fabric. If fabric is dipped in strong solution of glue and then in tannin, the glue only will become insoluble on the outside, and that which has penetrated deeper in fibre will be unchanged. Treatment is thus commenced with a very weak solution composed of 5 parts of glue in 100 parts of water and fabric immersed 10 to 15 minutes.

Fabric wrung out and when nearly dry passed into tannin solution. This solution can be strong as only so much of it is taken up as corresponds to glue present. Tannin reacts quickly with glue so that only a short period of immersion is necessary. The fabric again hung to dry and then washed in water to remove excess tannin. Process is twice repeated. Fabric is now passed through a stronger glue solution, 5%, and then again tannin. By repeating the process as many times as desired the coating can be made as thick as desired.

Another Method: Potash alum 100 lb. dissolved in 10 gallons of boiling water in one pot; in another pot 100 lb. glue, 200 lb. water. Solution is affected when glue is hot, add 5 lb. tannin and 2 lb. sodium silicate. Two solutions are boiled together with constant stirring. When mixture is complete, allow to jell. To waterproof: 1 lb. jelly to 1 lb. water is boiled, bath cooled to 176° F. and fabric soaked ½ hour and then stretched out horizontally for 6 hours to drain. If drying room is used keep temperature below 122° F.

Another Method: Dissolve 10 lb. gelatine, 10 lb. tallow soap in 30 gal. boiling water and mix solution in 4 gal. water in which 15 lb. alum has been dissolved. The whole is boiled for ½ hour and cooled to 104° F. At that temperature fabric is soaked in it, dried, rinsed, dried, and finally calendered. In this process the alum partially decomposed the soap, forming either free fatty acid or an acid alumina soap. The gelatine forms an insoluble compound with the alum. The free fatty acid or acid soap is mostly carried down on the fibre by the precipitate formed by the alum and gelatine.

TABLES

Conversion Factors

1. Grams per litre (g./l.) multiplied by 0.134=avoirdupois ounces per gallon (oz./gal.).

2. Avoirdupois ounces per gallon (oz./gal.) multiplied by 7.5=grams per litre (g./l.).

3. Grams per litre (g./l.) multiplied by 0.122=troy ounces per gallon (troy oz./gal.).

4. Troy ounces per gallon (troy oz./gal.) multiplied by 8.2=grams per litre (g./l.).

5. Grams per litre (g./l.) multiplied by 2.44=pennyweights per gallon (dwt./gal.)

6. Pennyweights per gallon (dwt./gal.) multiplied by 0.41=grams per litre (g./l.).

7. Amperes per square decimeter (amp./dm.^{2}) multiplied by 9.29=amperes per square foot (amp./sq. ft.).

8. Amperes per square foot (amp./sq. ft.) multiplied by 0.108=amperes per square decimeter (amp./dm.^{2}).

Thermometer Readings:

Degrees Centigrade $\times$ 1.8 + 32 = deg. Fahr.

$$\text{Degrees } \frac{\text{Fahrenheit} - 32}{1.8} = \text{deg. Cent.}$$

$$\text{Degrees } \frac{\text{Reamur} \times 9}{4} + 32 = \text{deg. Fahr.}$$

$$\text{Degrees } \frac{(\text{Fahrenheit} - 32)4}{9} = \text{deg. Reaumur.}$$

$$\text{Degrees } \frac{\text{Reaumur} \times 5}{4} = \text{deg. Cent.}$$

$$\text{Degrees } \frac{\text{Centigrade} \times 4}{5} = \text{deg. Reaumur.}$$

SPECIFIC GRAVITY
WEIGHT REQUIRED TO MAKE A GALLON

	Specific Gravity	Pounds to Gallon
Litharge	9.3	77.5
Red-Lead	8.7 to 8.8	72.5
Orange Mineral (orange lead)	8.6 to 8.7	73.0
White-Lead	6.7	55.8
Basic Lead Sulphate	6.4	53.3
Chrome Yellow (medium)	6.0	50.0
Zinc Oxide (white zinc)	5.6	46.6
Basic Lead Chromate	6.8	56.6
English (mercury) Vermillion	8.2	68.3
Bright Red Oxide of Iron	4.9 to 5.26	42.0
Indian Red Oxide of Iron	5.26	43.8
Brown Oxide of Iron (Prince's)	3.2	26.6
Ultramarine	2.4	20.0
Prussian Blue	1.85	15.4
Chrome Green (blue tone)	4.44	37.0
Chrome Green (yellow tone)	4.0	33.0
Lithopone	4.25	35.4
Ochre	2.94	24.5
Barytes	4.35 to 4.46	35. to 37.0
Blanc Fixe	4.25	35.4
Gypsum (terra alba)	2.3	19.0
Asbestine (magnesium silicate)	2.75	23.0
China Clay (aluminum silicate)	2.6 to 2.7	22.5
Whiting	2.65	22.0
Silica	2.65	22.0
Natural Graphite	2.1 to 2.4	18.0
Acheson's Graphite	2.2	18.3
Lampblack	1.85	15.4
Carbon Black	1.85	15.4
Keystone Filler (ground slate)	2.66	22.0
Titanox	4.3	35.8
Titanium Oxide	3.9 to 4.0	33.3
Drop Black	2.5	20.8

To this table the following data may be added: The weight of one gallon of paste made with

	Pounds
Red-Lead	44.8
White-Lead (heavy paste)	34.0
White-Lead (soft paste)	30.8
White Zinc	25.0
Chrome Yellow (medium)	24.0
Chrome Green	24.0
Venetian Red	19.0
French Ochre	15.0
Prussian Blue	10.0
Lampblack	9.1
Drop Black	11.7

WEIGHTS AND MEASURES
ENGLISH SYSTEM

Avoirdupois and Commercial Weights

16 drams, or 437.5 grains	=1 ounce, oz.
16 ounces, or 7000 grains	=1 pound, lb.

WEIGHTS AND MEASURES, ENGLISH SYSTEM—*Continued*

28 pounds	=1 quarter, qr.
4 quarters (English)	=1 hundredweight, cwt.—112 lbs.
20 hundredweight	=1 ton of 2240 lbs., gross or long ton
2000 pounds	=1 net, or short, ton
2204.6 pounds	=1 metric ton=1000 kilos

1 stone=14 pounds; 1 quintal=100 pounds

Troy Weights

24 grains	= 1 pennyweight, dwt.
20 pennyweights	= 1 ounce, oz. = 480 grains
12 ounces	= 1 pound, lb. = 5760 grains
1 carat	= 3.168 grains = 0.205 gram

Troy weight is used for weighing gold and silver. The grain is the same in Avoirdupois, Troy and Apothecaries' weights.

Apothecaries' Weights

20 grains	=1 scruple
2 scruples	=1 drachm, ʒ=60 grains
8 drachms	=1 ounce, ℥=480 grains
12 ounces	=1 pound, lb.=5760 grains

Apothecaries' Measures

60 minims (min.)	=1 fluid drachm (fl. dr.)
8 fluid drachms	=1 fluid ounce (fl. oz.)
20 fluid ounces	=1 pint (O) +
8 pints	=1 gallon (C) +

Relations of Apothecaries' Measures to Weights

(All liquids to be measured at 62° Fahr.)

1 minim is the measure of	0.0115	grains of distilled water
1 fluid drachm " "	54.687	" " " "
1 fluid ounce " "	437.5	" " " "
1 pint " "	8750	" " " "
1 gallon " "	70000	" " " "

Linear Measure

12 inches	=1 foot
3 feet	=1 yard
6 feet	=1 fathom
5½ yards	=1 rod pole, or perch
4 poles	=1 chain
40 poles	=1 furlong
8 furlongs	=1 mile=1760 yards

Square Measure

144 square inches=1 square foot
9 square feet =1 square yard
30.25 square yards or 272.5 sq. feet=1 square rod
160 square rods or 4840 sq. yards or 43560 sq. feet=1 acre
640 acres=1 square mile

An acre equals a square whose side is 208.7 feet

Cubic Measure

1728 cubic inches =1 cubic foot
27 cubic feet =1 cubic yard
1 cord of wood=a pile 4×4×8 feet=128 cubic feet
1 perch of masonry=16.5×1.5×1 foot=24.75 cubic feet

1 cubic inch of water at 62° Fahr. weighs	252.286	grains
" " " " " " "	0.57665	oz. (av.)
" " " " " " "	0.036041	lb.
1 cubic foot " " " " " "	996.458	oz. (av.)
" " " " " " "	62.2786	lb.
1 cubic yard " " " " " "	0.75068	tons

CAPACITY MEASURE

Liquid

4 gills =1 pint
2 pints =1 quart
4 quarts=1 gallon

CONVERSION OF THERMOMETER READINGS

F°	C°	F°	C°	F°	C°	F°	C°	F°	C°	F°	C°
-40	-40.00	30	-1.11	80	26.67	250	121.11	500	260.00	900	482.22
-38	-38.89	31	-0.56	81	27.22	255	123.89	505	262.78	910	487.78
-36	-37.78	32	0.00	82	27.78	260	126.67	510	265.56	920	493.33
-34	-36.67	33	0.56	83	28.33	265	129.44	515	268.33	930	498.89
-32	-35.56	34	1.11	84	28.89	270	132.22	520	271.11	940	504.44
-30	-34.44	35	1.67	85	29.44	275	135.00	525	273.89	950	510.00
-28	-33.33	36	2.22	86	30.00	280	137.78	530	276.67	960	515.56
-26	-32.22	37	2.78	87	30.56	285	140.55	535	279.44	970	521.11
-24	-31.11	38	3.33	88	31.11	290	143.33	540	282.22	980	526.67
-22	-30.00	39	3.89	89	31.67	295	146.11	545	285.00	990	532.22
-20	-28.89	40	4.44	90	32.22	300	148.89	550	287.78	1000	537.78
-18	-27.78	41	5.00	91	32.78	305	151.67	555	290.55	1050	565.56
-16	-26.67	42	5.56	92	33.33	310	154.44	560	293.33	1100	593.33
-14	-25.56	43	6.11	93	33.89	315	157.22	565	296.11	1150	621.11
-12	-24.44	44	6.67	94	39.44	320	160.00	570	298.89	1200	648.89
-10	-23.33	45	7.22	95	35.00	325	162.78	575	301.67	1250	676.67
-8	-22.22	46	7.78	96	35.56	330	165.56	580	304.44	1300	704.44
-6	-21.11	47	8.33	97	36.11	335	168.33	585	307.22	1350	732.22
-4	-20.00	48	8.89	98	36.67	340	171.11	590	310.00	1400	760.00
-2	-18.89	49	9.44	99	37.22	345	173.89	595	312.78	1450	787.78
0	-17.78	50	10.00	100	37.78	350	176.67	600	315.56	1500	815.56
1	-17.22	51	10.56	105	40.55	355	179.44	610	321.11	1550	843.33
2	-16.67	52	11.11	110	43.33	360	182.22	620	326.67	1600	871.11
3	-16.11	53	11.67	115	46.11	365	185.00	630	332.22	1650	898.89
4	-15.56	54	12.22	120	48.89	370	187.78	640	337.78	1700	926.67
5	-15.00	55	12.78	125	51.67	375	190.55	650	343.33	1750	954.44
6	-14.44	56	13.33	130	54.44	380	193.33	660	348.89	1800	982.22
7	-13.89	57	13.89	135	57.22	385	196.11	670	354.44	1850	1010.00
8	-13.33	58	14.44	140	60.00	390	198.89	680	360.00	1900	1037.78
9	-12.78	59	15.00	145	62.78	395	201.67	690	365.56	1950	1065.56
10	-12.22	60	15.56	150	65.56	400	204.44	700	371.11	2000	1093.33
11	-11.67	61	16.11	155	68.33	405	207.22	710	376.67	2050	1121.11
12	-11.11	62	16.67	160	71.11	410	210.00	720	382.22	2100	1148.89
13	-10.56	63	17.22	165	73.89	415	212.78	730	387.78	2150	1176.67
14	-10.00	64	17.78	170	76.67	420	215.56	740	393.33	2200	1204.44
15	-9.44	65	18.33	175	79.44	425	218.33	750	398.89	2250	1232.22
16	-8.89	66	18.89	180	82.22	430	221.11	760	404.44	2300	1260.00
17	-8.33	67	19.44	185	85.00	435	223.89	770	410.00	2350	1287.78
18	-7.78	68	20.00	190	87.78	440	226.67	780	415.56	2400	1315.56
19	-7.22	69	20.56	195	90.55	445	229.44	790	421.11	2450	1343.33
20	-6.67	70	21.11	200	93.33	450	232.22	800	426.67	2500	1371.11
21	-6.11	71	21.67	205	96.11	455	235.00	810	432.22	2550	1398.89
22	-5.56	72	22.22	210	98.89	460	237.78	820	437.78	2600	1426.67
23	-5.00	73	22.78	215	101.67	465	240.55	830	443.33	2650	1454.44
24	-4.44	74	23.33	220	104.44	470	243.33	840	448.89	2700	1482.22
25	-3.89	75	23.89	225	107.22	475	246.11	850	454.44	2750	1510.00
26	-3.33	76	24.44	230	110.00	480	248.89	860	460.00	2800	1537.78
27	-2.78	77	25.00	235	112.78	485	251.67	870	465.56	2850	1565.56
28	-2.22	78	25.56	240	115.56	490	254.44	880	471.11	2900	1593.33
29	-1.67	79	26.11	245	118.33	495	257.22	890	476.67	2950	1621.11

EQUIVALENTS OF TWADDLE, BAUMÉ AND SPECIFIC GRAVITY SCALES

Twaddle	Baumé	Specific Gravity	Twaddle	Baumé	Specific Gravity	Twaddle	Baumé	Specific Gravity	Twaddle	Baumé	Specific Gravity
0	0	1.000	44	26.0	1.220	88	44.1	1.440	131	57.1	1.655
1	0.7	1.005	45	26.4	1.225	89	44.4	1.445	132	57.4	1.660
2	1.4	1.010	46	26.9	1.230	90	44.8	1.450	133	57.7	1.665
3	2.1	1.015	47	27.4	1.235	91	45.1	1.455	134	57.9	1.670
4	2.7	1.020	48	27.9	1.240	92	45.4	1.460	135	58.2	1.675
5	3.4	1.025	49	28.4	1.245	93	45.8	1.465	136	58.4	1.680
6	4.1	1.030	50	28.8	1.250	94	46.1	1.470	137	58.7	1.685
7	4.7	1.035	51	29.3	1.255	95	46.4	1.475	138	58.9	1.690
8	5.4	1.040	52	29.7	1.260	96	46.8	1.480	139	59.2	1.695
9	6.0	1.045	53	30.2	1.265	97	47.1	1.485	140	59.5	1.700
10	6.7	1.050	54	30.6	1.270	98	47.4	1.490	141	59.7	1.705
11	7.4	1.055	55	31.1	1.275	99	47.8	1.495	142	60.0	1.710
12	8.0	1.060	56	31.5	1.280	100	48.1	1.500	143	60.2	1.715
13	8.7	1.065	57	32.0	1.285	101	48.4	1.505	144	60.4	1.720
14	9.4	1.070	58	32.4	1.290	102	48.7	1.510	145	60.6	1.725
15	10.0	1.075	59	32.8	1.295	103	49.0	1.515	146	60.9	1.730
16	10.6	1.080	60	33.3	1.300	104	49.4	1.520	147	61.1	1.735
17	11.2	1.085	61	33.7	1.305	105	49.7	1.525	148	61.4	1.740
18	11.9	1.090	62	34.2	1.310	106	50.0	1.530	149	61.6	1.745
19	12.4	1.095	63	34.6	1.315	107	50.3	1.535	150	61.8	1.750
20	13.0	1.100	64	35.0	1.320	108	50.6	1.540	151	62.1	1.755
21	13.6	1.105	65	35.4	1.325	109	50.9	1.545	152	62.3	1.760
22	14.2	1.110	66	35.8	1.330	110	51.2	1.550	153	62.5	1.765
23	14.9	1.115	67	36.2	1.335	111	51.5	1.555	154	62.8	1.770
24	15.4	1.120	68	36.6	1.340	112	51.8	1.560	155	63.0	1.775
25	16.0	1.125	69	37.0	1.345	113	52.1	1.565	156	63.2	1.780
26	16.5	1.130	70	37.4	1.350	114	52.4	1.570	157	63.5	1.785
27	17.1	1.135	71	37.8	1.355	115	52.7	1.575	158	63.7	1.790
28	17.7	1.140	72	38.2	1.360	116	53.0	1.580	159	64.0	1.795
29	18.3	1.145	73	38.6	1.365	117	53.3	1.585	160	64.2	1.800
30	18.8	1.150	74	39.0	1.370	118	53.6	1.590	161	64.4	1.805
31	19.3	1.155	75	39.4	1.375	119	53.9	1.595	162	64.6	1.810
32	19.8	1.160	76	39.8	1.380	120	54.1	1.600	163	64.8	1.815
33	20.3	1.165	77	40.1	1.385	121	54.4	1.605	164	65.0	1.820
34	20.9	1.170	78	40.5	1.390	122	54.7	1.610	165	65.2	1.825
35	21.4	1.175	79	40.8	1.395	123	55.0	1.615	166	65.5	1.830
36	22.0	1.180	80	41.2	1.400	124	55.2	1.620	167	65.7	1.835
37	22.5	1.185	81	41.6	1.405	125	55.5	1.625	168	65.9	1.840
38	23.0	1.190	82	42.0	1.410	126	55.8	1.630	169	66.1	1.845
39	23.5	1.195	83	42.3	1.415	127	56.0	1.635	170	66.3	1.850
40	24.0	1.200	84	42.7	1.420	128	56.3	1.640	171	66.5	1.855
41	24.5	1.205	85	43.1	1.425	129	56.6	1.645	172	66.7	1.860
42	25.0	1.210	86	43.4	1.430	130	56.9	1.650	173	67.0	1.865
43	25.5	1.215	87	43.8	1.435						

Relation of Capacity, Volume and Weight

1 pint =28.875 cubic inches
1 quart =57.75 cubic inches
1 gallon (U. S.) =231 cubic inches
1 gallon (English) =277.274 cubic inches
7.4805 gallons =1 cubic foot
1 gallon water at 62° Fahr. weighs 8.3356 lbs.

Dry

2 pints =1 quart
8 quarts=1 peck
4 pecks =1 bushel
1 U. S. standard bushel (struck)=2150.42 cubic inches.
0.80356 U. S. bushels (struck) =1 cubic foot

METRIC EQUIVALENTS

Linear Measure

1 centimeter=0.3937 in.
1 decimeter=3.937 in.=0.328 ft.
1 meter=39.37 in.=1.0936 yds.
1 decameter=1.9884 rods
1 kilometer=0.62137 miles
1 inch=2.54 centimeters
1 foot=3.048 decimeters
1 yard=0.9144 meters
1 rod=0.5029 decameters
1 mile=1.6093 kilometers

(The meter, as used in Europe, is 39.370432 inches.)

Square Measure

1 sq. centimeter=0.1550 sq. inches
1 sq. decimeter=0.1076 sq. feet
1 sq. meter=1.196 sq. yards
1 are=3.954 sq. rods
1 hectare=2.47 acres
1 sq. kilometer=0.386 sq. miles
1 sq. inch=6.452 sq. centimeters
1 sq. foot=9.2903 sq. decimeters
1 sq. yard=0.8361 sq. meters
1 sq. rod=0.2529 ares
1 acre=0.4047 hectares
1 sq. mile=.259 sq. kilometers

Weights

1 decigram=0.003527 oz.=1.5432 grains
1 gram=0.03527 oz. avoir., or about 15½ troy grains
1 kilogram=2.2046 lbs. avoir.
1 metric ton=1.1023 English short tons
1 ounce avoir.=28.35 grams
1 pound avoir.=0.4536 kilograms
1 English short ton=0.9072 metric tons

Approximate Metric Equivalents

1 decimeter=4 inches
1 meter=1.1 yards
1 kilometer=⅝ of a mile
1 hectare=2½ acres
1 stere, or cu. meter=¼ of a cord
1 liter=1.06 qt. liquid, 0.9 qt. dry
1 hectoliter=2⅚ bushels
1 kilogram=2⅕ lbs.
1 metric ton=2200 lbs.

Comparison of Avoirdupois and Metric Weights

Grains	Drams	Oz. Av.	Lbs. Av.	Deniers	Grams
1.000	.....	......	.....	1.296	0.065
27.340	**1.000**	......	.....	35.437	1.772
437.500	16.000	**1.000**	.....	566.990	28.350
7000.000	256.000	16.000	**1.000**	9071.840	453.592
0.772	.....	......	.....	**1.000**	0.050
15.432	.....	0.03527	.....	20.000	**1.000**

pH Values of Chemicals

Solution Strength	Reagent	pH
1%	Commercial Olive Oil Soap (Neutral)	10.1 –10.3
1%	Commercial Olive Oil Soap (Neutral)	10.1 –10.3
1%	Commercial Olive Oil or Tallow Soap Containing 20% Soda Ash	10.75–10.88
1%	Commercial Olive Oil or Tallow Soap Containing 5% Caustic	12.0–12.2
½%	Commercial Olive Oil or Tallow Soap	10.0 –10.2
¼%	Commercial Olive Oil or Tallow Soap	9.9 –10.1
1%	Sulphonated Oils (Neutral)	6.0 –7.0
1%	Sulphonated Oils Containing Free Acid	Below 6.0
1%	Sulphonated Oils Containing Soap or Alkalies	Above 7.0
¼%	Trisodium Phosphate	12.3
¼%	Sodium Silicate	12.2
¼%	Sodium Carbonate	11.3
¼%	Sodium Sulphite	9.7
¼%	Disodium Phosphate	8.9
¼%	Borax	8.8
¼%	Monosodium Phosphate	5.0

pH Ranges of Common Indicators

	Useful pH Range
Thymol Blue	1.2–2.8
Bromphenol Green	2.8–4.6
Methyl Orange	3.1–4.4
Bromcresol Green	4.0–5.6
Methyl Red	4.4–6.0
Propyl Red	4.8–6.4
Brom Cresol Purple	5.2–6.8
Brom Thymol Blue	6.0–7.6
Phenol Red	6.8–8.4
Litmus	7.2–8.8
Cresol Red	7.2–8.8
Cresolphthalein	8.2–9.8
Phenolphthalein	8.6–10.2
Nitro Yellow	10.0–11.6
Alizarin Yellow R	10.1–12.1
Sulfo Orange	11.2–12.6

Melting Points of Resins, Etc.

Material	Melting Point ° C.
Amber	250–325
Benzoin	75–100
Copal (Zanzibar)	280
Copal (Congo)	220
Copal (Kauri)	165
Copal (Manila)	120
Cumarone	127–142
Dammar (Batavia)	100
Dammar (Singapore)	95
Dragon's Blood	120
Elemi	75–120
Ester Gum	120–140
Gilsonite	123° C.
Guiac	85–90
Indene	127–142
Mastic	105–120
Pontianak	135
Rosin (Colophony)	100–140
Sandarac	135–150
Shellac	120

* Melting Points of Common Waxes

Wax	Melting Point ° C.
Bayberry Wax	40–44
Beeswax White	67.2
Beeswax Yellow	61
Candelilla Wax	64–67
Carnauba Wax	85
Ceresine	74–80
Chinese Insect Wax	92.2
Cocoa Butter	21.5–27.3
Japan Wax	54.5–59.6
Montan Wax Refined	95–96
Myrtle Wax	47–48
Ozokerite	65–110
Paraffin	55–65° C.
Spermaceti	44–47.5
Tallow (Beef)	42.5–44

* Very often there is considerable difference between the melting and solidifying point. Natural and commercially adulterated articles will also show variations.

REFERENCES CONSULTED

Aircraft Engineering
Allgem. Photo-Zeitung
American Electroplaters Society Review
American Gas Assoc. Proc.
American Machinist
American Paint & Varnish Mfrs. Assn.
American Perfumer
American Society Testing Materials
Atelier Photography
Austrian Patent Office

Belgian Patent Office
Berichte Ges. Kohlentech.
Bied. Zentralblatt
Brass World
Brewery Age
British Industrial Finishing
British Journal of Photography
British Patent Office
British Plastics
British Soap Mfr.
Bureau of Standards Publications

Canadian Patent Office
Chemical & Metallurgical Engineering
Chemiker-Zeitung
Chemist Analyst
Chimie Industrie
Cotton

Der Chemisch Technische Fabrikant
Der Parfumer
Deutscher Zuckerind.
Drug Trade News
Dutch Patent Office
Dyestuffs

Food Products Journal
French Patent Office

Gas Journal
German Patent Office

Hungarian Patent Office

Idaho Agricultural Experiment Station
Industrial Chemist
Industrial Finishing
Industrial Woodworking

Japanese Patent Office
Journal American Ceramic Society
Journal Appl. Chem. Russ.
Journal Chemical Industry
Journal Council Sci. Industrial Research
Journal Dept. Agriculture Ireland
Journal Econ. Entomology
Journal Institute of Metals
Journal of Industrial & Eng. Chemistry
Journal of Society of Chemical Industry
Journal Society Chemical Industry (Japan)

Khimstroi
Korrosion

Lancet
Laundry Owner's National Association

Manufacturing Chemist
Melliand Textile Monthly
Metal Industry
Metals & Alloys
Minn. Agricultural Experiment Station
Monats-Bull. Schweiz. Ver. Gas Wass.
Museum Technique

New York Agricultural Experiment Station

Oil & Colour Trade Journal
Oils, Drugs & Paints

Paint Mfgr.
Paint & Varnish Production Mgr.
Paper Maker
Perf. & Ess. Oil Record
Pharm. Journal

Phot. Chonik
Photofreund
Plater's Guide Book
Portland Cement Association
Practical Druggist
Purdue Agricultural Experiment Station

Quart-Journal Pharm. Pharmacologie

Revue Applied Mycology
Russian Patent Office

Science
Seifen Sieder Zeitung
Soap
Soap Gazette
Swedish Patent Office
Synthetic & Applied Finishes

Tex. Agricultural Exp. Station

U. S. Dept. of Agric.
United States Patent Office

Welsh Agricultural Journal

Zeit. Untersuch. Lebensm.

INDEX

For Chemical Advisors, Special Raw Materials, Equipment, Containers, etc., consult Supply Section at end of book.

All formulae preceded by an asterisk (*) are covered by patents.

B

For Chemical Advisors, Special Raw Materials, Equipment, Containers, etc., consult Supply Section at end of book.

All formulae preceded by an asterisk (*) are covered by patents.

C

For Chemical Advisors, Special Raw Materials, Equipment, Containers, etc., consult Supply Section at end of book.

All formulae preceded by an asterisk (*) are covered by patents.

For Chemical Advisors, Special Raw Materials, Equipment, Containers, etc., consult Supply Section at end of book.

All formulae preceded by an asterisk (*) are covered by patents.

For Chemical Advisors, Special Raw Materials, Equipment, Containers, etc., consult Supply Section at end of book.

E

F

All formulae preceded by an asterisk (*) are covered by patents.

All formulae preceded by an asterisk (*) are covered by patents.

For Chemical Advisors, Special Raw Materials, Equipment, Containers, etc., consult Supply Section at end of book.

H

All formulae preceded by an asterisk (*) are covered by patents.

I

All formulae preceded by an asterisk (*) are covered by patents.

J

K

L

For Chemical Advisors, Special Raw Materials, Equipment, Containers, etc., consult Supply Section at end of book.

All formulae preceded by an asterisk (*) are covered by patents.

For Chemical Advisors, Special Raw Materials, Equipment, Containers, etc., consult Supply Section at end of book.

All formulae preceded by an asterisk (*) are covered by patents.

P

All formulae preceded by an asterisk (*) are covered by patents.

For Chemical Advisors, Special Raw Materials, Equipment, Containers, etc., consult Supply Section at end of book.

All formulae preceded by an asterisk (*) are covered by patents.

For Chemical Advisors, Special Raw Materials, Equipment, Containers, etc., consult Supply Section at end of book.

All formulae preceded by an asterisk (*) are covered by patents.

Q

R

For Chemical Advisors, Special Raw Materials, Equipment, Containers, etc., consult Supply Section at end of book.

All formulae preceded by an asterisk (*) are covered by patents.

For Chemical Advisors, Special Raw Materials, Equipment, Containers, etc., consult Supply Section at end of book.

All formulae preceded by an asterisk (*) are covered by patents.

For Chemical Advisors, Special Raw Materials, Equipment, Containers, etc., consult Supply Section at end of book.

All formulae preceded by an asterisk (*) are covered by patents.

For Chemical Advisors, Special Raw Materials, Equipment, Containers, etc., consult Supply Section at end of book.

For Chemical Advisors, Special Raw Materials, Equipment, Containers, etc., consult Supply Section at end of book.

All formulae preceded by an asterisk (*) are covered by patents.

For Chemical Advisors, Special Raw Materials, Equipment, Containers, etc., consult Supply Section at end of book.

PRESS OF
BRAUNWORTH & CO., INC.
BOOK MANUFACTURERS
BROOKLYN, NEW YORK

ADDENDA

ALCOHOLIC LIQUORS

The most important constituent of alcoholic beverages is the alcohol. Its strength depends upon the character of the beverage. If the alcohol is inferior in quality or has an oily taste and odor, the finished product will be unsatisfactory. Be sure to use good alcohol. Sugar is used to sweeten the liqueurs and, in many cases thickens the liqueurs as well, which is desirable.

The colors used should be certified, pure food colors. For brown coloring the most predominant color is burnt sugar color or caramel. Sometimes its taste helps to mellow or round out the taste of liqueurs. Wines and fruit juices also may be used sometimes to bring out the fuller taste.

The quantities of essences or flavoring oils called for in each formula should be carefully measured. It is the essence or oils that gives the alcohol in the finished beverage its characteristic taste and aroma. The skill employed in making these beverages usually decides success or failure. As with all formulas, carelessness, inaccuracy and haste will only result in failure. A formula that imparts good taste and aroma is one always sought for. Good recipes never grow old. They do not change as the science of Chemistry does. And so an old formula when tried and found to be true never grows old.

Some of the liquor formulas in this book may call for substances other than simple oils or simple ingredients. By referring to the first section of this book in the chapter of non-alcoholic flavors beginning on page 30, you will find formulas for making these products. When difficulty arises or should you desire to become more expert in mixing, blending and compounding, call in a reliable, reputable chemist. He will be able to assist you and render valuable service.

Even a freshly prepared mixture of aromatic substances lacks homogeneousness and only after some period of time are the ingredients well mixed and blended. However, storage is necessary in every case to round out taste, flavor and brilliancy—to produce an equilibrium of the reactants present, to give the proper bouquet which characterizes a good product.

When beverages are stored in barrels, the tannin of the wood appears to possess the power of hastening, ageing and improving the taste. Oak barrels are best to use to clear or make liqueur brilliant. Storage is usually sufficient but the clearness can be hastened by the addition of 1 pint of skimmed milk. The clear liquid is then siphoned off later. Where rapid clearing is desired filtration must be resorted to.

Essence Aromatic

No. 1

Cardamom	83 gm.
Clove	166 gm.
Mace	166 gm.
Cinnamon	580 gm.
95% Alcohol	10 kilos

No. 2

Curacao Peels	460 gm.
Cloves	83 gm.
Mace	83 gm.
95% Alcohol	10 kilos

No. 3

Angelica Root	120 gm.
Galgant Root	120 gm.
Ginger Root	10 gm.
Calamus Root	120 gm.
Chamomile	100 gm.
Laurel Leaves	120 gm.
Mace	20 gm.
Cloves	60 gm.
Orange Peels	80 gm.
Peppermint	160 gm.
Cinnamon	100 gm.
Zedoary Plant	200 gm.
95% Alcohol	10 kilos

No. 4

Orange Peels	450 gm.
Cloves	90 gm.
Mace	90 gm.
95% Alcohol	10 kilos

No. 5

Angelica Root	100 gm.
Ginger Root	50 gm.
Calamus Root	100 gm.
Cardamom	100 gm.
Lavender	200 gm.
Mace	15 gm.
Nutmeg	25 gm.
Orange Peels	300 gm.
Peppermint	200 gm.
Cinnamon	50 gm.
Zedoary Plant	100 gm.
95% Alcohol	10 kilos

Absinthe Essence a la Turine

No. 1

Oil Angelica	3 gm.
Oil Anise	5 gm.
Oil Fennel	5 gm.
Oil Cardamom	1 gm.
Oil Coriander	5 gm.
Oil Marjoram	3 gm.
Oil Star Anise	6 gm.
Oil Wormwood	3 gm.
95% Alcohol	10 kilos

For INDEX to Addenda see page 587.

For Chemical Advisors, Special Raw Materials, Equipment, Containers, etc., consult Supply Section at end of book.

No. 2

Anise Seed	160	gm.
Bitter Almond	70	gm.
Fennel	100	gm.
Calamus	20	gm.
Coriander	50	gm.
Peppermint	10	gm.
Sassafras Wood	100	gm.
Wormwood Herb	20	gm.
Sugar	700	gm.
95% Alcohol	10	kilos

Vienna Absinthe Essence

No. 1

Oil Angelica	1½	gm.
Oil Anise	2	gm.
Oil Fennel	1½	gm.
Oil Ginger	1	gm.
Oil Coriander	1½	gm.
Oil Marjoram	1½	gm.
Oil Star Anise	2	gm.
Oil Wormwood	3½	gm.
95% Alcohol	10	kilos

No. 2

Angelica Root	100	gm.
Anise Seed	200	gm.
Calamus	120	gm.
Marjoram	50	gm.
Peppermint	30	gm.
Star Anise Seed	50	gm.
Wormwood	200	gm.
Sugar	2	kilos
95% Alcohol	10	kilos

Swiss Absinthe

No. 1

Oil Angelica	5	gm.
Oil Anise	10	gm.
Oil Fennel	10	gm.
Oil Cardamom	3	gm.
Oil Coriander	10	gm.
Oil Marjoram	10	gm.
Oil Star Anise	12	gm.
Oil Wormwood	15	gm.
95% Alcohol	10	kilos

No. 2

Oil Angelica	8	gm.
Oil Anise	15	gm.
Oil Tincture Arrac No. 5	100	gm.
Oil Fennel	15	gm.
Oil Marjoram	15	gm.
Oil Orange	20	gm.
Oil Wormwood	20	gm.
Oil Lemon	10	gm.
95% Alcohol	10	kilos

Swiss Absinthe Essence

No. 1

Oil Angelica	1	gm.
Oil Anise	1	gm.
Oil Marjoram	1	gm.
Oil Orange	1½	gm.
Oil Ether Oenanthic	1/10	gm.
Oil Star Anise	1	gm.
Oil Wormwood	3	gm.
Oil Lemon	1	gm.
95% Alcohol	10	kilos

No. 2

Oil Angelica	2	gm.
Oil Anise Russian	5	gm.
Oil Fennel	3	gm.
Oil Calamus	20	gm.
Oil Caraway	3	gm.
Oil Marjoram	5	gm.
Oil Mace	2	gm.
Oil Clove	1	gm.
Oil Orange	20	gm.
Oil Pimento	½	gm.
Oil Juniper Berry	2	gm.
Oil Wormwood	25	gm.
Oil Lemon	3	gm.
95% Alcohol	10	kilos

Alant Essence

Alant Root	5	gm.
Cinnamon	½	gm.
95% Alcohol	10	kilos

Color: Red.

Angelica Essence

Angelica Root	1	kilo
Coriander	100	gm.
Caraway Seed	200	gm.
95% Alcohol	10	kilos

Anise Essence

Anise Seed	4	gm.
Oil Star Anise	1	gm.
95% Alcohol	10	kilos

Color: Green.

Barbado Essence

No. 1

Mace	3	gm.
Cloves	5	gm.
Orange Peel Fresh	100	gm.
Cinnamon	16	gm.
Lemon Peel Fresh	100	gm.
95% Alcohol	10	kilos

Color: Brown.

No. 2

Oil Bergamot	4	gm.
Oil Cloves	1	gm.
Oil Nutmeg	1	gm.
Oil Cinnamon	1	gm.
Oil Lemon	4	gm.
95% Alcohol	10	kilos

Angostura Bitter Essence

Angostura Bark	1000	gm.
Cardamom	200	gm.
Clove	50	gm.

Cinnamon Buds 500 gm.
Water 5 litres
Alcohol 5 litres
Color: Dark Brown.

To get the correct and agreeable aroma it has to be cut down 4 to 5 times with 50% Alcohol.

BRANDIES

Anise Brandy

Alcohol 90% by Volume 36 lit.
Anise Oil Essence 30 gm.*
Sugar Syrup 65% 4 lit.
Water 60 lit.

Lemon Brandy

Alcohol 90% by Volume 36 lit.
Lemon Essence 50 gm.*
Sugar Syrup 65% 4 lit.
Water 60 lit.
Color Yellow to suit.

Raspberry Brandy

Alcohol 90% by Volume 17 lit.
Cherry Whiskey 3 lit.*
Raspberry Juice 27 lit.
Sugar Syrup 65% 7 lit.
Water 46 lit.

Kummel Brandy

Alcohol 90% by Volume 36 lit.
Coriander Essence ½ lit.*
Sugar Syrup 65% 4 lit.
Water 60 lit.

Cherry Brandy

Alcohol 90% by Volume 16 lit.
Bitter Almond Oil Essence 10 gm.*
Cinnamon Oil Essence 20 gm.*
Clove Oil Essence 10 gm.*
Sugar Syrup 65% 3½ lit.
Water 32½ lit.
Cherry Juice 48 lit.

Clove Brandy

Alcohol 90% by Volume 36 lit.
Clove Oil Essence 100 gm.*
Cinnamon Oil Essence 50 gm.*
Sugar Syrup 65% 4 lit.
Water 57½ lit.
Cherry Juice 2½ lit.
Color: Brown.

Corn Brandy (30% Alcohol)

Alcohol 90% by Volume 33¼ lit.
Coriander Oil Essence 85 gm.*

* In this formula and the others that follow where an essence is used dissolve latter in alcohol first, then add balance of ingredients and then filter.

Rum Essence ¼ lit.
Water 66½ lit.

Peppermint Brandy

Alcohol 90% by Volume 36 lit.
Peppermint Oil Essence 150 gm.*
Sugar Syrup 65% 4 lit.
Water 60 lit.
Filter and clarify with 10 grams Alum. Color green or leave white.

Orange Brandy, White

Alcohol 90% by Volume 36 lit.
Bitter Orange Oil Essence ½ lit.*
Sugar Syrup 65% 4 lit.
Water 59½ lit.
For brown, color with caramel color.

Absinthe Brandy

Alcohol 90% by Volume 36 lit.
Absinthe Essence ½ lit.*
Sugar Syrup 65% 2½ lit.
Water 61 lit.
Color: Green.

Juniper Brandy

Alcohol 90% by Volume 40 lit.
Juniper Berry Essence ½ lit.*
Sugar Syrup 65% 3 lit.
Water 56½ lit.

Color is white. For brown use caramel color.

Calamus Brandy

Alcohol 90% by Volume 36 lit.
Calamus Essence ½ lit.*
Sugar Syrup 65% 4 lit.
Water 59½ lit.
Color: Brown.

Bergamot Brandy

Alcohol 90% by Volume 38 lit.
Bergamot Oil Essence 25 gm.*
Sugar Syrup 65% 6 lit.
Water 56 lit.

Anise Liqueur

Alcohol 90% by Volume 50 lit.
Anise Essence 60 gm.*
Fennel Essence 20 gm.*
Cinnamon Essence 5 gm.*
Sugar Syrup 65% 25 lit.
Water 25 lit.

Anisette

Oil Anise Russian, Rectified 465 mils
Oil Sweet Fennel, Rectified 20 mils
Oil Coriander, Pure 10 mils

Oil Star Anise, Leadfree 465 mils
Oil Angelica Root 30 mils
Oil Bitter Almonds, F.F.P.A. 8 mils
Oil Rose, Artificial 2 mils

Dissolve ½ oz. of above mixture in 22 gallons alcohol. Then add 28 gallons water in which has been dissolved 112 lb. sugar.

Peppermint Liqueur

Alcohol 90% by Volume 50 lit.
Peppermint Essence 400 gm.*
Sugar Syrup 65% 30 lit.
Water 20 lit.

Creme de Menthe

Oil Peppermint, Twice Rectified 2 oz.
Menthol 2 dr.
Alcohol 35 oz. 4 dr.

Green Coloring.

Dissolve 1 oz. of this mixture in 1½ gallons alcohol. Then add 1½ gallons water in which has been dissolved 5½ lb. sugar.

Ginger Liqueur

Alcohol 90% by Volume 30 lit.
Ginger Extract 20 lit.*
Sugar Syrup 65% 40 lit.
Water 10 lit.

Color: Brown.

Chartreuse

Alcohol 90% by Volume 22½ lit.
Chartreuse Essence 1650 gm.
Sugar Syrup 10 lit.
Water 17½ kilos

Yellowish Color.

Chartreuse

Oil Peppermint, Rectified 1½ dr.
Oil Lemon, Handpressed 2 dr.
Oil Cassia, Leadfree 1 dr.
Oil Cloves Pure 1 dr.
Oil Mace Distilled 1½ dr.
Oil Anise Seed, Russian, Rectified 1 dr.
Oil Angelica Root 40 dr.
Oil Bitter Almonds, F.F.P.A. ½ dr
Oil Wormwood, American 20 dr.
Oil Neroli Bigrade, Petale, Extra 1 dr.
Oil Cognac, Genuine, White 15 dr.
Alcohol 20 oz.

Dissolve 1 oz. of this mixture in 7 gallons alcohol. Then add 9 gallons water in which has been dissolved 38 lb. sugar.

Lemon Brandy

Alcohol 90% by Volume 21½ lit.
Lemon Essence 600 gm.
Sugar Syrup 5½ lit.
Water 23 lit.

Color: Yellow.

Cognac

Alcohol 90% by Volume 22 lit.
Cognac Essence 500 gm.
Citric Acid 12½ gm.
Rock Candy 1 kilo
Water 28 lit.

Dissolve the Citric Acid in ¼ liter of water. Dissolve the Rock Candy in 1 liter of water. Mix the ingredients thoroughly and allow to remain in the vessel for several weeks.

Cognac Brandy

Essence Brandy 20 oz.
Extract Vanilla 4 oz.
Tinct. Orrisroot, Florentine (2 lb. to 1 gal.) 2 oz.
Oil Cognac, Genuine 1 oz.
Oil Bitter Almonds, Free from Prussic Acid 2 dr.
Essence Rum, New England 6 dr.
Acetic Ether, Absolute 2 oz. 2 dr.
Nitrous Ether, Absolute 2 oz.
Alcohol 10 oz.

Dissolve 1 oz. of above mixture in 10 gallons alcohol. Then add 10 gallons water. Mix. Filter through magnesium carbonate. Color with caramel.

Cognac

Oil Bitter Almond 20 dr.
Oil Cognac 50 gm.
Violet Flower Essence 25 gm.
Woodruff Essence 50 gm.
Oenanthic Ether 15 gm.
Acetic Ether 120 gm.

Dissolve 1 oz. of above mixture in 30 gallons alcohol. Then add 30 gallons water. Mix. Filter and color with caramel.

Geneva Gin

Alcohol 90% by Volume 22½ lit.
Geneva Essence 150 gm.
Water 27½ lit.

Mix well and store for several weeks.

Goldwasser

Alcohol 90% by Volume 23¼ lit.
Goldwasser Essence 750 gm.
Rose Water 1¼ lit.

Orange Blossom Water	750	gm.
Sugar Solution	5	lit.
Water	20½	lit.

After the mixture has been stored for some time there is added to it a small quantity of genuine Gold Leaf.

Hamburger Bitters

Alcohol 90% by Volume	21½	lit.
Hamburger Bitter Essence	550	gm.
Sugar Solution	4½	lit.
Water	24	lit.

Color Brown with Caramel.

Absinthe Brandy (Swiss)

Alcohol 90% by Volume	25	lit.
Absinthe Essence	365	gm.
Water	25	lit.

Color Green to suit.

Absinthe Brandy (French)

Alcohol 90% by Volume	21¾	lit.
Swiss Absinthe Essence	375	gm.
Sugar Syrup	3¼	lit.
Water	25	lit.

Color Green to suit.

Absinthe (French)

Oil Wormwood, American	10	oz.
Oil Star Anise, Leadfree	16	oz.
Oil Anise Russian, Rectified	12	oz.
Oil Fennel, Rectified	6	oz.
Oil Neroli, Artificial	½	dr.
Alcohol	3	oz.
Tinct. Gum Benzoin, Siam 2 lb. to 1 gal.	3	oz.

Dissolve ½ oz. of above mixture in 26 gallons alcohol. Then add 24 gallons water. Mix. Filter through magnesium carbonate. Color to suit.

Pineapple Brandy

Alcohol 90% by Volume	21¾	lit.
Pineapple Ester (Conc.)	265	gm.
Pineapple Essence from Fresh Fruit	145	gm.
Sugar Solution	3¼	lit.
Water	25	lit.

Italian Orange Brandy

Alcohol 90% by Volume	21½	lit.
Orange Essence	500	gm.
Sugar Solution	8½	lit.
Water	20	lit.

Color Yellow with Tincture of Saffron.

Aromatique

Alcohol 90% by Volume	21¼	lit.
Aromatique Essence	750	gm.
Sugar Solution	7¾	lit.
Water	21	lit.

Colored Brown with Caramel.

Calamus

Alcohol 90% by Volume	21½	lit.
Calamus Essence	500	gm.
Sugar Syrup	4½	lit.
Water	24	lit.

Color Light Brown with Caramel.

Cardinal

Rhine or Moselle Wine	75	lit.
Cardinal Essence	400	gm.
Sugar	10	kilos
Water	10	lit.

Dissolve Sugar in the water and the essence in the Wine and mix the two solutions.

Benedictine

Oil Sweet Orange, Hand-pressed	72	oz.
Oil Angelica Root	6	oz.
Oil Calamus	3	oz.
Oil Cinnamon, Ceylon	3	oz.
Oil Mace, Distilled	3	oz.
Oil Celery	3	oz.
Alcohol	12	oz.

Dissolve 1 oz. of above mixture in 5 gallons alcohol. Then add 5 gallons water to which has been added 24 lb. sugar.

Slivovitz

Oil Bitter Almonds, F.F.P.A.	2	mils
Oil Neroli, Artificial	1	mil
Oil Cognac, Genuine, Green	2	mils
Vanillin	5	gm.
Essence Raspberry Aroma	300	mils
Essence Plum	300	mils
Essence Jamaica Rum	25	mils
Essence Raisin Wine	50	mils
Prune Spirit	100	mils
Alcohol	100	mils

Dissolve 1 oz. of above mixture in 8 gallons alcohol. Then add 8 gallons water. Mix. Filter through magnesium carbonate.

Jamaica Rum

Oil of Cassia	1	dr.
Oil of Birch Tar	25	dr.
Oil of Ylang Ylang Natural	3	dr.
Oil of Orange Flower Natural	20	dr.

Oil of Ceylon Cinnamon	15 dr.
Rum Ether Pure	3 pt.
Acetic Ether	2½ oz.
Butyric Ether	1 oz. 1 dr.
Tincture of Saffron 1 lb. to a gal.	4 oz.
Extract of Vanilla Pure	3 oz.
Balsam Peru	2 dr.
Tincture Styrax U.S.P.	2 dr.
Coumarin	5 dr.

Dissolve 1 oz. of above mixture in 4½ gallons of alcohol. Then add 5½ gallons water. Mix. Filter through magnesium carbonate. Allow to age in barrel.

Whiskey "Scotch"

Guaiacol, Pure	4 dr.
Oil Cade, Pure	1 oz.
Butyric Ether, Pure	4 oz.
Essence Rye Whiskey	2 gal.

Dissolve 1 oz. of above mixture in 2¼ gallons of alcohol. Then add 2¾ gallons water. Mix. Filter through magnesium carbonate. Color with caramel.

Scotch

Oil Corn Fusel	6 oz.
Oil Bitter Almonds	4 dr.
Oil Coriander	4 dr.
Oil Cade	1 oz.
Guaiacol	2 dr.
Butyric Ether	4 oz.
Alcohol	4 oz.

Dissolve 1 oz. of above mixture in 14 gallons alcohol. Then add 16 gallons water. Mix. Filter through magnesium carbonate. Color with caramel.

Scotch Whisky Mix

Oil Fusel	6 oz.
Oil Bitter Almond	4 dr.
Oil Coriander	4 dr.
Oil Cade Pure	1 oz.
Guaiacol Pure	2 dr.
Butyric Ether	4 oz.

1 oz. to 60 gal. (50% alcohol).

Gin, Old Tom

Oil Coriander, Pure	3 oz. 4 dr.
Oil Angelica Root	3 dr.
Oil Anise, Russian, Rectified	1 oz.
Oil Caraway, Dutch	4 dr.
Oil Juniper Berries, Rectified	7 oz. 4 dr.
Alcohol	1 pt. 8 oz.

Dissolve 1 oz. of above oil in 4½ gallons alcohol. Then add 5½ gallons water. Mix. Filter through magnesium carbonate.

Gin, Old Tom

Essence Gin, Holland	1 gal.
Alcohol	1 pt.
Oil Coriander, Pure	1 oz.
Oil Calamus	1 oz.

Dissolve 1 oz. of above oil in 5½ gallons alcohol. Then add 6½ gallons water. Mix. Filter through magnesium carbonate.

Gin, London Dock

Oil Gin, Old Tom	6 oz.
Oil Gin, Holland	18 oz.
Oil Cassia, Rectified	4 dr.
Alcohol	64 oz.

Dissolve 1 oz. of above oil in 3 gallons alcohol. Then add 4 gallons water. Mix. Filter through magnesium carbonate.

Gordon Gin

Oil Juniper Berries	16 oz.
Oil Angelica Root	20 cc.
Oil Angelica Seed	20 cc.
Oil Coriander	40 cc.
Oil Lemon	60 cc.
Sweet Orange	20 cc.
Neroli	5 cc.
Geranium Rose	5 cc.
Alcohol to make 1 gal.	

4 oz. of above is used to 50 gal. 50% alcohol.

Oil Gin Holland

Oil Lemon	1 dr.
Oil Anise	1 dr.
Oil Angelica Root	6 dr.
Oil Fusel	4 dr.
Oil Juniper Berries	20 oz.
Oil Rosemary Flavor	6 dr.
Oil Coriander	4 dr.
Alcohol	10 oz.

Dissolve 1 oz. of above oil in 7 gallons alcohol. Then add 8 gallons water. Mix. Filter through magnesium carbonate.

Holland Gin

Oil Gin	1000 mils
Glycerine C.P.	200 mils
Alcohol	216 oz.

Dissolve 5 oz. of above in 2¼ gallons alcohol. Then add 2¾ gallons water. Mix. Filter through magnesium carbonate.

Whiskey "Rye"

Oil Fusel Potato	2 pt.
Oil Fusel Rye	18 pt.
Rum Ether, Pure	20 pt.

Oil Coriander, Pure	5 oz.
Oil Bitter Almonds, F.F.P.A.	2 oz. 4 dr.
Alcohol	50 pt.
Tinct. Catechu	1 pt.
Vanillin	2 dr.
Heliotropin	4 dr.
Tinct. Balsam, Peru, True	1 dr.

Dissolve 1 oz. of above in 7¼ gallons alcohol. Then add to it 7¾ gallons water. Mix; filter; and color with caramel.

Bourbon

Oil Bourbon	6 oz.
Alcohol	32 oz.
Sugar Color	20 oz.
Citric Acid Solution	8 oz.
Tannic Acid Solution	1 oz.
	67 oz.
Water	61 oz.
	128 oz.

Filter. Then dissolve 1 oz. of above in ½ gal. alcohol and then add ½ gal. water.

Super Aroma Bourbon

Oil Fusel Rectified	240 oz.
Ess. Pineapple	½ oz.
Ess. Peach Blossom	½ oz.
Citric Acid Solution 50%	240 oz.
Solution Saccharin Saturated	¼ oz.
Oil Jam. Rum	13 oz.
Alcohol	133 oz.
Tannic Acid Solution	1 oz.
	626 oz.

Filter. Then 1 oz. of this will flavor 5 gallons of 50% alcohol.

Bourbon

Oil Bourbon	40 oz.
Oil Combindlion	20 oz.
Alcohol	10 oz.
Tannic Acid Solution 1 lb. C.P. Tannic Acid Dissolved in 1 gal. Hot Water	10 oz.
Saccharin Solution 1 lb. Soluble Water Saccharin 5 gal. Boiling Water	½ oz.
Citric Acid Solution	10 oz.
Sugar Color 100%	200 oz.
Vanilla Ext. Imitation	2 oz.

Imit. Vanilla Ext. 1 oz. Vanillin. Dissolve in ½ gal. Alcohol; ½ gal. Water.

Whiskey Bourbon

Fusel Oil	1 gal.
Oil Bitter Almond	1½ oz.
Oil Rose Art.	48 min.
Vanilla Extract	32 oz.
Ess. Jamaica Rum	40 oz.
Pineapple Aroma	40 oz.
Acetic Ether	12 oz.

Dissolve 1 oz. of above in 12 gallons alcohol. Then add 13 gallons water. Mix. Filter through magnesium carbonate. Store in charred barrel until color becomes caramel.

Cherry Brandy Liqueur

Genuine Cherry Brandy	1 pt.
Cherry Fruit Juice	1½ pt.
Alcohol	2 pt.
Sugar Syrup 65%	2 pt.
Water	2 pt.

Essence for Artificial Cherry Brandy

(1 oz. per gallon)

Oil of Neroli	2 drops
Oil of Cloves	¼ dram
Oil of Cinnamon	¼ dram
Oil of Bitter Almonds	2 oz.
Rum Ether	14 oz.
Wine Brandy	16 oz.
Colorless Cherry Flavor	3 lb.
Genuine Bitter Almond Water	5 lb.

Cherry Liqueur Essence

(2 oz. per gallon)

Vanillin	1½ dram
Oil of Cloves	2 oz.
Oil of Cinnamon	3 oz.
Benzaldehyde	5 oz.
Rum Essence	14 oz.
Alcohol	16 oz.
Cherry Juice	2½ lb.
Cherry Flavor	5 lb.

Essence for Artificial Slivovitz

(1 oz. per gallon)

Oil of Cognac	2 oz.
Benzaldehyde	4 oz.
Rum Essence Ethyl Acetate	6 oz.
Orris Root Tincture	12 oz.
Wine Brandy	1 lb.
Pineapple Essence	1 lb.
Carob Tincture (1 to 5)	2 lb.
Alcohol	2 lb.
Distilled Water	2 lb.

Ginger Liqueur

Alcohol 90% by Volume	31 lit.
Ginger Extract	17 lit.*
Sugar Syrup 65%	27 lit.
Water	25 lit.

Color: Brown.

Kummel Liqueur

Alcohol 90% by Volume	45	lit.
Kummel Essence	1	lit.*
Orange Peel Essence	¼	lit.*
Sugar Syrup 65%	38¾	lit.
Water	15	lit.

Turko-Liqueur

Alcohol 90% by Volume	31	lit.
Hamburger Bitter Extract	½	lit.*
Sugar Syrup 65%	19	lit.
Water	35	lit.
Ginger Extract	9	lit.*
Caramel Color	2	lit.
Swiss Absinthe	4	lit.*

Maraschino Liqueur

Alcohol 90% by Volume	36	lit.
Bitter Almond Oil Essence	115	gm.*
Steroli Oil Essence	200	gm.*
Rose Oil	30	drops
Sugar Syrup 65%	64	lit.

Vanilla Liqueur

Alcohol 90% by Volume	20	lit.
Vanilla Extract	10	lit.*
Raspberry Juice	1	lit.
Sugar Syrup 65%	40	lit.
Water	29	lit.
Caramel Color	2	oz.

Lemon Liqueur

Alcohol 90% by Volume	34	lit.
Lemon Essence	7	lit.*
Sugar Syrup 65%	26	lit.
Corn Syrup	13	lit.
Water	20	lit.

Color: Yellow.

Spanish Bitter Liqueur

Alcohol 90% by Volume	43	lit.
Spanish Bitter Oil Essence	¾	lit.*
Sugar Syrup 65%	28	lit.
Water	28	lit.
Color (Brown)	8	oz.

Rose Liqueur

Alcohol 90% by Volume	40	lit.
Rose Oil Essence	80	gm.*
Sugar Syrup 65%	32	lit.
Corn Syrup	10	lit.
Water	18	lit.

Color: Red.

Sherry Cordial

Alcohol 90% by Volume	35	lit.
Bitter Almond Oil Essence	56	gm.*
Ethyl Acetate	65	gm.*
Sugar Syrup 65%	45	lit.
Water	20	lit.

French Liqueur (Cremes) as below:

Fleur d'Amour (Flower of Love)

Alcohol 90% by Volume	34	lit.
Lemon Oil Essence	½	lit.*
Clove Oil Essence	150	gm.*
Nutmeg Oil Essence	150	gm.*
Sugar Syrup 65%	45	lit.
Water	20½	lit.

Color: Bluish Red.

Anisette d'Hollande

Alcohol 90% by Volume	40	lit.
Anise Oil Essence	700	gm.*
Fennel Oil Essence	300	gm.*
Cinnamon Oil Essence	10	gm.*
Sugar Syrup 65%	40	lit.
Water	18½	lit.

Creme de Rose

Alcohol 90% by Volume	4⅛	lit.
Genuine Turkish Rose Oil Essence	12½	gm.*
Sugar Syrup 65%	10	lit.
Water	⅞	lit

Color Red with Aniline.

Creme de Chocolat

Alcohol 95% by Volume	4¾	lit.
Cocoa Powder	375	gm.
Bitter Chocolate	250	gm.
Cinnamon Essence	A few drops	
Vanilla Extract	A few drops	
Sugar Syrup 65%	12	lit.
Water	1¼	lit.

Cook together the cocoa and chocolate with the water. When cold add the alcohol with stirring. After one half hour filter. Then add to the filtrate the sugar syrup and essence.

Creme de Noix

Alcohol 95% by Volume	4	lit.
Nut Essence	100	gm.*
Sugar Syrup	11	lit.
Nut Extract	2	lit.*
Water	1	lit.

Color faint brown.

Schiedamer Geneva Holland Gin

Alcohol by Volume 78%	20¼	lit.
Oil of Juniper	3	gm.
Lemon Balm Oil	3	gm.
Genuine Cognac	¾	lit.
Sugar Syrup	¾	lit.
Water	8¼	lit.

Extract d'Absinthe

Alcohol by Volume 90%	80	lit.
Vermouth Essence	710	gm.
Anise Essence	1250	gm.
Fennel Essence	65	gm.
Coriander Essence	65	gm.
Ethyl Acetate	210	gm.
Water	20	lit.

Color: Green.

Goldwasser Whiskey

Alcohol by Volume 90%	7	lit.
Goldwasser Essence	130	gm.
Sugar Syrup 65%	7	lit.
Water	3	lit.

Arrack

Ethyl Acetate	100	gm.
Black Balsam Peru	130	gm.
Vanilla	16	gm.
Oil of Neroli	5	gm.
Oil of Birch	1	gm.
Ground Horseradish	500	gm.
Onions	125	gm.
Iron Filings	2	kg.
Cocoa	25	gm.
Raisin Stems	1	kg.
Alcohol by Volume 90%	41	lit.
Water	27½	lit.

The above are mixed together and then filtered.

Arrack—No. 1

Alcohol by Volume 90%	6	lit.
Arrack	21	lit.
Vanilla Spirit	1/10	lit.
Oil Bitter Almonds	2	drops
Water	3	lit.

No. 2

Alcohol by Volume 90%	12	lit.
Arrack	16½	lit.
Vanilla Spirits	125	gm.
Oil Bitter Almonds	2	drops
Water	11½	lit.

Cognac

Alcohol by Volume 90%	31	lit.
Cognac	16	lit.
Cognac Essence	1/16	lit.
Oil of Grapeseed	16	gr.
Sugar Syrup 65%	¼	lit.
Water	22¾	lit.

Cognac

Alcohol by Volume 90%	11	lit.
Acetic Acid	16	gm.
Ethyl Acetate	8	gm.
Brown Sugar to be Dissolved in ¼ liter Water	120	gm.
Water	5½	lit.

Color: Yellow.

Cognac

Alcohol by Volume 90%	5	lit.
Ethyl Acetate	20	gm.
Pyroligneous Acid	20	gm.
Water	5	lit.

Color: Yellow and age 5–6 weeks.

Rum Essence

To 103 litres Rum 60% by Volume add:

Butyric Ether	187	gm.
Formic Ether	312	gm.
Birch Oil	1	gm.
Vanilla Essence	¼	lit.
Alcohol by Volume 90%	144	lit.
Balsam Peru	65	gm.
Ethyl Ether	165	gm.
Raisin Stems	1	kg.
Cedar Wood Shavings	250	gm.

The above is then added with 102 litres of brandy or alcohol 60% by volume, mixed and colored with caramel.

Rum

Alcohol by Volume 90%	4	lit.
Jamaica Rum	1	lit.
Spirit of Birch Oil	12	drops
Tincture of Lamp Black	12	drops
Ethyl Acetate	120	drops
Vanilla Extract	90	drops
Sugar dissolved in a little water	40	gm.

Mix the above with 3 litres of distilled water, filter and allow to remain in storage for awhile.

Rum New England

Oil Cinnamon, Ceylon	2	dr.
Oil Cloves, Pure	2	dr.
Oil Chamomile, Roman	4	dr.
Rum Ether, Pure	4	pt.
Butyric Ether, Absolute	3	oz.
Extract Vanilla	4	dr.
Acetic Ether, Absolute	3	oz.
Alcohol	8	oz.

Dissolve 1 oz. of above mixture in 4½ gallons alcohol. Then add 5½ gallons water. Mix. Filter through magnesium carbonate. Color with caramel.

New England Rum

Nitrous Ether	250	gr.
Butyric Ether	250	gr.
Acetic Ether	250	gr.
Oil Lemon	3	gr.
Oil Cinnamon	3	gr.
Oil Neroli	1	gr.
Balsam of Peru	2	gr.
Rum Ess. No. 10	500	gr.

Dissolve 1 oz. of above in 2¾ gallons

alcohol. Then add 3¼ gallons water. Mix. Filter through magnesium carbonate. Color with caramel.

Rum—No. 1

Alcohol by Volume 90%	60 lit.
Vanilla Spirit	1 lit.
Sugar Syrup 65%	1 lit.
Jamaica Rum	18½ lit.
Rum Essence	½ lit.
Water	19 lit.

Color with Caramel.

Rum Punch Extract—No. 1

Alcohol by Volume 90%	21 lit.
Lemon Oil	15 gm.
Oil of Rose	15 drops
Jamaica Rum	42 lit.
Sugar	52 kg.
Citric Acid	390 gm.
Water	24 lit.

Color to suit.

No. 2

Alcohol by Volume 90%	34½ lit.
Lemon Oil	15 gm.
Oil Rose	15 drops
Jamaica Rum	11¼ lit.
Coarse Sugar	52 kg.
Citric Acid	390 gm.
Water	30¾ lit.

Color to suit.

Arrack Punch Extract—No. 1

Alcohol 90%	14 lit.
Lemon Oil	10 gm.
Rose Oil	10 drops
Arrack de Goa	28 lit.
Sugar	35 kg.
Citric Acid	260 gm.
Water	16 lit.

No. 2

Alcohol 90%	23 lit.
Lemon Oil	10 gm.
Rose Oil	10 drops
Arrack de Goa	4½ lit.
Sugar	35 kg.
Citric Acid	260 gm.
Water	20½ lit.

Victoria Punch Extract—No. 1

Alcohol 90%	21¾ lit.
Lemon Essence	85 gm.
Pineapple Ether	2 gm.
Arrack de Goa	28½ lit.
Sugar	47½ kg.
Cherry Juice	8 lit.
Raspberry Juice	2¼ lit.
Water	9 lit.
Tartaric Acid dissolved in 1 litre of water	¾ lb.

Rum Grog Extract

Alcohol 90%	15 lit.
Jamaica Rum	75 lit.
Water	27 lit.
Sugar	45 kg.

Arrack Grog Extract

Alcohol 90%	6 lit.
Arrack de Goa	30 lit.
Water	18 lit.
Sugar	30 kg.

Arrack Punch Extract Ordinary Type

Alcohol 90%	30½ lit.
Lemon Oil	55 gm.
Arrack de Goa	1½ lit.
Sugar Syrup 65%	24 lit.
Corn Syrup	4 lit.
Water	7 lit.
Vanilla Spirit	1 lit.
Pineapple Ether	45 gm.
Tartaric Acid dissolved in ½ litre water	150 gm.

Rum Punch Extract Ordinary Type

Alcohol 90%	83½ lit.
Rum Essence	1 lit.
Lemon Oil	280 gm.
Sugar Syrup 65%	41 lit.
Vanilla Spirit	2 lit.
Tartaric Acid dissolved in 1½ litres water	300 gm.
Water	81 lit.

Angostura Bitter

Angelica Root	25 gm.
Angostura Bark	500 gm.
Cinnamon Ceylon	60 gm.
Gentian	40 gm.
Galgant	150 gm.
Hops	40 gm.
Ginger	10 gm.
Cardamom	60 gm.
Clove	10 gm.
Pimento	70 gm.
Orange Peel Fresh	250 gm.
Raisin	2000 gm.
or	
Honey	250 gm.
Rum	1760 gm.
Woodruff	150 gm.
Cinnamon Buds	150 gm.
Alcohol	17 lit.

Angostura Bitter American

Angostura Bark	18½ gm.
Gentian	7½ gm.
Galgant	17½ gm.
Hazel Root	7½ gm.
Honey	250 gm.

Cardamom	18½	gm.
Catechu	7.6	gm.
Coriander	7½	gm.
Caraway	7½	gm.
Curcuma	100	gm.
Dandelion Root	7½	gm.
Mace Buds	3½	gm.
Nutmeg	7½	gm.
Cloves	1	gm.
Pimento	22	gm.
Orange Peel	30	gm.
Sandalwood Red	30	gm.
Snake Root	7½	gm.
Licorice	7½	gm.
Wormwood	7½	gm.
Cinnamon	7½	gm.
Alcohol 65%	7.2	lit.

Angostura Bitter a la Siegert

Angelica Root	3	gm.
Gentian Root	15	gm.
Galgant Root	15	gm.
Ginger Root	3	gm.
Cardamom Small	20	gm.
Cinnamon	20	gm.
Cloves	3	gm.
Orange Peel Bitter	25	gm.
Sandalwood Red	80	gm.
Tonka Beans	80	gm.
Zedoary Plant	15	gm.

Everything roughly cut and put into 5000 grams of 60% Alcohol. This mixture has to stand 15 days, then filtered. After this add 200 grams Sugar Color, 500 grams Malaga Wine. Let it stand for an additional few days and filter it again.

Angostura Bitter

Angostura Bark Genuine	90	gm.
Chamomile	24	gm.
Cardamom	8	gm.
Cinnamon Ceylon	7	gm.
Orange Peel	24	gm.
Raisins	300	gm.
Water	5	kilos
Alcohol	5	kilos

Bitter Essence Simple

Curacao Peels	50	gm.
Calamus Root	50	gm.
Lesser Centaury	50	gm.
Alcohol	10	kilos

Colored: Dark Brown.

Bitter Essence Double—No. 1

Buck Bean	100	gm.
Orange Peel Dry	50	gm.
Gentian Root	20	gm.
Wormwood	50	gm.
Cinnamon	20	gm.
Alcohol	10	kilos

Colored: Dark Brown.

No. 2

Holy Thistle	400	gm.
Gentian Root	400	gm.
Lesser Centaury	400	gm.
Vermouth	400	gm.
Alcohol	10	kilos

Colored: Dark Brown.

Stomach Bitter Essence—No. 1

Angelica Root	100	gm.
Gentian Root	100	gm.
Holy Thistle	20	gm.
Buck Bean	80	gm.
Wormwood	80	gm.
Bitter Orange Peel	80	gm.
Lemon Peel	50	gm.
Alcohol	10	kilos

No. 2

Angelica Root	30	gm.
Gentian Root	140	gm.
Holy Thistle	40	gm.
Buckbean	40	gm.
Bitter Orange Peel	200	gm.
Alcohol	10	kilos

Both Bitters Colored Brown Green.

Bitter Essence English—No. 1

Holy Thistle	50	gm.
Gentian	30	gm.
Lesser Centaury	50	gm.
Wormwood Herb	50	gm.
Orange Peel	30	gm.
Orris Root	30	gm.
Grains of Paradise	60	gm.
Alcohol	10	kilos

Colored: Dark Brown.

No. 2

Curacao Peel	100	gm.
Gentian Root	40	gm.
Lesser Centaury	30	gm.
Orris Root	80	gm.
Holy Thistle	10	gm.
Wormwood	40	gm.
Alcohol	10	kilos

Colored: Red Brown.

No. 3

Benedictine Herb	8	gm.
Cardamom	4	gm.
Gentian	16	gm.
Orange Peel	40	gm.
Grains of Paradise	10	gm.
Lesser Centaury	20	gm.
Orris Root	20	gm.
Wormwood	5	gm.
95% Alcohol	10	kilos

Bitter Essence Spanish—No. 1

Horse Heel	80 gm.
Angelica Root	40 gm.
Holy Thistle	80 gm.
Calamus Root	250 gm.
Gentian	40 gm.
Polypodium	10 gm.
Galgant Root	80 gm.
Masterwort	40 gm.
Burnt Saxifraga	40 gm.
Lesser Centaury	150 gm.
Wormwood	40 gm.
Alcohol	10 kilos

Colored: Brown.

No. 2

Horse Heel	40 gm.
Galgant Root	30 gm.
Spearmint	100 gm.
Melissa	40 gm.
Curacao Peel	100 gm.
Wormwood	10 gm.
Alcohol	10 kilos

Colored: Brown.

No. 3

Horse Heel	30 gm.
Angelica Root	60 gm.
Benedictine Herb	30 gm.
Calamus Root	120 gm.
Gentian Root	30 gm.
Galgant Root	30 gm.
Burnt Saxifraga	15 gm.
Lesser Centaury	60 gm.
Tormentilla Root	15 gm.
Orris Root	50 gm.
Wormwood	15 gm.
Alcohol	10 kilos

Colored: Brown.

Flower Essence

Vanilla Tincture No. 46	200 gm.
Oil Rose	5 gm.
Jasmine Spirit	10 kg.

Colored: Rose Red or Violet.

Curacao Peels Essence

Curacao Peels	1 kg.
Orange Flower Water	1 kg.
Alcohol 95%	10 kg.

Colored: Golden Brown.

Essence Elixer de Suede

Inula (Horse Heel)	8 gm.
Gentian Root	8 gm.
Saffron	5 gm.
Cinnamon	5 gm.
Zedoary Plant	10 gm.
Alcohol 95%	10 kg.

Colored: Green.

Raspberry Essence

Raspberry Squashed	10 kg.
Orris Root	200 gm.
Alcohol 90%	10 kg.

Colored: Red.

Grunewald Essence—No. 1

Buck Bean	40 gm.
Calamus Root	5 gm.
Holy Thistle	8 gm.
Gentian Root	40 gm.
Galgant Root	40 gm.
Orange Peels	40 gm.
Wormwood Root	8 gm.
Alcohol 95%	10 kg.

Colored: Green.

No. 2

Oranges Unripe Green	500 gm.
Gentian Root	50 gm.
Galgant Root	40 gm.
Cassia	40 gm.
Ginger	40 gm.
Nutmeg	10 gm.
Cloves	30 gm.
Alcohol 95%	10 kg.

Colored: Green.

Harts Content Essence

Angelica Root	60 gm.
Calamus Root	120 gm.
Catachou	20 gm.
Gentian Root	120 gm.
Ginger	10 gm.
Cloves	10 gm.
Melissa	50 gm.
Orange Peels	60 gm.
Juniper Berries	10 gm.
Wormwood	10 gm.
Alcohol 95%	10 kg.

Colored: Dark Brown.

Strawberry Essence

Strawberry Squashed	10 kg.
Orris Root	200 gm.
Alcohol 90%	10 kg.

Colored: Red.

Virgin Essence

Vanilla Tincture No. 46	100 gm.
Oil Anise	20 gm.
Jasmine Water	100 gm.
Oil Neroli	5 gm.
Rose Oil	1 gm.
Alcohol 95%	10 kg.

No Color.

Coffee Essence—No. 1

Coffee Burned and Ground	1 kg.
Vanilla Tincture No. 46	50 gm.

Mace Tincture No. 28	5	gm.
Cinnamon Tincture No. 2	10	gm.
Alcohol 95%	10	kg.

No Color.

No. 2

Clove Tincture No. 23	20	gm.
Mace Tincture No. 28	20	gm.
Cinnamon Tincture No. 52	20	gm.
Coffee Burned and Ground	600	gm.
Alcohol 95%	10	kg.

No Color.

No. 3

Coffee Burned and Ground	400	gm.
Cinnamon	5	gm.
Vanilla	2	gm.
Alcohol 95%	10	kg.

No Color.

Coffee Triple Essence

Coffee Burned and Ground	5	kg.
Vanilla	10	gm.
Alcohol 95%	10	kg.

No Color.

Calamus Essence—No. 1

Angelica Root	40	gm.
Calamus Root	600	gm.
Alcohol 95%	10	kg.

Colored: Brown.

No. 2

Calamus Root	300	gm.
Ginger Root	20	gm.
Fresh Orange Peels	50	gm.
Alcohol 95%	10	kg.

Colored: Brown.

Cardinal Essence

Orange Peels Dry	1000	gm.
Oranges Green Unripe	600	gm.
Lemon Peels	50	gm.
Alcohol 95%	10	kg.

Colored: Red-Yellow.

Carmelite Essence

Lemon Peels	500	gm.
Coriander	100	gm.
Nutmegs	50	gm.
Pimento	10	gm.
Orange Peels	500	gm.
Alcohol 95%	10	kg.

Colored: Green.

Contuszawka Essence

Ethyl Butyrate	150	gm.
Anise Oil	80	gm.
Lemon Oil	40	gm.
Oil Coriander	60	gm.
Oil Fennel	50	gm.
Oil Caraway	60	gm.
Oil Orange Flower	5	gm.
Oil Peppermint	30	gm.
Oil Rose	5	gm.
Oil Star Anise	150	gm.
Oil Juniper Berry	20	gm.
Oil Wormwood	20	gm.
Oil Cinnamon	5	gm.
Alcohol 95%	10	kg.

No Color.

Herb Essence

Angelica Root	2½	gm.
Anise Seed	10	gm.
Calamus Root	20	gm.
Lemon Peels	25	gm.
Coriander Seed	2½	gm.
Galgant Root	3	gm.
Ginger Root	2½	gm.
Marjoram Herb	3	gm.
Orange Peels	25	gm.
Rosemary Herb	3	gm.
Orris Root	2½	gm.
Juniper Berries	2½	gm.
Alcohol 95%	10	kg.

Colored: Grass Green.

Spearmint Essence

Spearmint	4	kg.
Peppermint	500	gm.
Melissa	200	gm.
Alcohol 95%	10	kg.

Colored: Dark Green.

Caraway Essence

Caraway Seed Squashed	500	gm.
Anise Squashed	30	gm.
Coriander Squashed	30	gm.
Fennel	30	gm.
Orris Root	50	gm.
Cinnamon	20	gm.
Alcohol 95%	10	kg.

No Color.

Life Essence—No. 1

Angelica Root	120	gm.
Calamus Root	20	gm.
Cardamom	20	gm.
Gentian Root	120	gm.
Zedoary Plant	120	gm.
Alcohol 95%	10	kg.

No Color.

No. 2

Buck Bean	250	gm.
Calamus Root	20	gm.
Orange Peels Fresh	60	gm.
Coriander	30	gm.
Ginger	10	gm.
Oranges Unripe	60	gm.
Juniper Berries	30	gm.

Wormwood	250 gm.
Cinnamon	30 gm.
Alcohol 95%	10 kg.

No Color.

Flower of Love Essence

Oil Cloves	10 gm.
Oil Nutmeg	10 gm.
Oil Cinnamon	3 gm.
Alcohol 95%	10 kg.

Colored: Light Red.

Stomach Bitter Essence

Angelica	150 gm.
Anise	100 gm.
Calamus	300 gm.
Peppermint	50 gm.
Orange Bitter	300 gm.
Cinnamon	50 gm.
Alcohol 95%	10 kg.

Stomach Bitter Essence French

Anise	20 gm.
Cardamom	25 gm.
Lemon Peels	45 gm.
Fennel	40 gm.
Galgant Root	10 gm.
Ginger	20 gm.
Mace	5 gm.
Nutmeg	5 gm.
Cloves	10 gm.
Orris Root	15 gm.
Woodruff Herb Dry	100 gm.
Cinnamon	10 gm.
Alcohol 95%	10 kg.

Stomach Bitter Essence Breslau—No. 1

Anise	25 gm.
Basilicum Herb	25 gm.
Calamus Root	5 gm.
Chamomile	25 gm.
Cardamom	3 gm.
Lemon Peels	50 gm.
Coriander	15 gm.
Galgant Root	5 gm.
Mace	3 gm.
Nutmeg	3 gm.
Orange Peels	50 gm.
Rosemary Herb	25 gm.
Orris Root	5 gm.
Alcohol 95%	10 kg.

No Color.

No. 2

Inula (Horse Heel)	5 gm.
Angelica Root	1.5 gm.
Basilicum Herb	20 gm.
Calamus Root	5 gm.
Lemon Peels	35 gm.
Galgant Root	1.5 gm.
Coriander Seeds	3 gm.
Ginger Root	1.5 gm.
Caraway	3 gm.
Spearmint	20 gm.
Pimento	3 gm.
Orange Peels	35 gm.
Juniper Berries	3 gm.
Alcohol 95%	10 kg.

No. 3

Anise	20 gm.
Basilicum Herb	20 gm.
Lemon Peels	50 gm.
Calamus	20 gm.
Chamomile	20 gm.
Cardamom	5 gm.
Coriander	20 gm.
Galgant	15 gm.
Lavender Herb	5 gm.
Mace	5 gm.
Nutmeg	5 gm.
Orange Peels	50 gm.
Rosemary	20 gm.
Orris Root	15 gm.
Cinnamon	5 gm.
Alcohol 95%	10 kg.

No. 4

Inula (Horse Heel)	15 gm.
Angelica Root	10 gm.
Anise	30 gm.
Basilicum Herb	10 gm.
Calamus	25 gm.
Lemon Peels	45 gm.
Coriander	25 gm.
Galgant	20 gm.
English Spice	15 gm.
Ginger	10 gm.
Spearmint	10 gm.
Caraway	15 gm.
Lavender Herb	10 gm.
Grains of Paradise	10 gm.
Orange Peels	45 gm.
Juniper Berries	10 gm.
Alcohol	10 kg.

Stomach Bitter Danzig—No. 1

Inula (Horse Heel)	25 gm.
Anise	70 gm.
Calamus	12 gm.
Chamomile	5 gm.
Lemon Peels	45 gm.
Dill Seed	12 gm.
Caraway	15 gm.
Nutmeg	12 gm.
Pimento	6 gm.
Orange Peels	45 gm.
Oil Rose	1 gm.
Orris Root	15 gm.
Cinnamon	15 gm.
Zedoary Plant	15 gm.
Alcohol	10 kg.

No. 2

Angelica Root	50 gm.
Anise	20 gm.

Coriander	25	gm.
Lemon Peels	70	gm.
Fennel	50	gm.
Galgant Root	10	gm.
Mace	50	gm.
Nutmeg	20	gm.
Pimento	20	gm.
Orange Peels	60	gm.
Rose Oil	½	gm.
Cinnamon	60	gm.
Alcohol 95%	10	kg.

Stomach Essence Vienna

Inula (Horse Heel)	25	gm.
Anise	35	gm.
Calamus	30	gm.
Coriander	15	gm.
Dill Seed	10	gm.
Fennel	30	gm.
Galgant Root	15	gm.
Caraway	20	gm.
Mace	15	gm.
Nutmeg	15	gm.
Cloves	20	gm.
Pimpinele	10	gm.
Orris Root	15	gm.
Cinnamon	45	gm.
Zedoary Plant	15	gm.
Alcohol 95%	10	kg.

Color: Brown or Greenish Brown to all Stomach Essences.

Alp-Herbs Stomach Essence

Angelica Root	20	gm.
Benedictine Herb	20	gm.
Calamus Root	30	gm.
Lemon Peels	70	gm.
Coriander	20	gm.
Cardamom	2	gm.
Galgant Root	20	gm.
Ginger Root	20	gm.
Marjoram	20	gm.
Orange Peels	70	gm.
Rosemary	20	gm.
Thyme	20	gm.
Tonka Beans	50	gm.
Orris Root	20	gm.
Juniper Berries	20	gm.
Alcohol 95%	10	kg.

Color: Brownish-Green.

Stomach Elixir Essence

Cardamom	10	gm.
Calamus Root	120	gm.
Calumba Root	60	gm.
Gentian	60	gm.
Galgant	60	gm.
Ginger	10	gm.
Pimpinele	120	gm.
Tormentilla	120	gm.
Wormwood	20	gm.
Orange Peels	60	gm.
Zedoary Plant	60	gm.
Cinnamon	10	gm.
Alcohol 95%	10	kg.

Color: Brown.

Stomach Essence—No. 1

Calamus Root	250	gm.
Coriander	30	gm.
Gentian Root	200	gm.
Galgant Root	200	gm.
Lesser Centaury	60	gm.
Orris Root	60	gm.
Zedoary Plant	120	gm.
Alcohol	10	kg.

Color: Brown.

No. 2

Angelica Root	60	gm.
Benedictine Herb	120	gm.
Buck Bean	200	gm.
Cardamom	15	gm.
Gentian Root	200	gm.
Ginger	30	gm.
Orange Peels Fresh	60	gm.
Oranges Unripe	60	gm.
Lesser Centaury	200	gm.
Wormwood	120	gm.
Alcohol	10	kg.

Color: Brown.

No. 3

Inula (Horse Heel)	30	gm.
Angelica Root	20	gm.
Calamus Root	250	gm.
Galgant Root	40	gm.
Juniper Berries	60	gm.
Alcohol 95%	10	kg.

Color: Brown.

Musk Essence—No. 1

Musk	40	gm.
Vanilla	40	gm.
Amber	15	gm.
Alcohol 95%	1	kg.

No. 2

Musk	20	gm.
Ambra	10	gm.
Alcohol 95%	1	kg.

Clove Essence

Cloves	200	gm.
Cinnamon	50	gm.
Alcohol 95%	1	kg.

Color: Red-Brown.

Persico Essence—No. 1

Bitter Almonds	400	gm.
Water	4	kg.
Alcohol 95%	10	kg.

Chopped bitter almonds must stand in water one day in a warm place.

No. 2

Apricot Pits Crushed	2 kg.
Cherry Pits	200 gm.
Cloves	5 gm.
Mace	5 gm.
Alcohol	10 kg.

The apricot pits may be replaced by cherry pits because the latter have a finer taste. All Persico Essences stay uncolored and are not to be taken alone, having a certain content of persico acid which has a bad effect on the health and are only harmless when considerably thinned down.

No. 3

Sweet Almonds	1 kg.
Bitter Almonds	2 kg.
Lemon Peels	500 gm.
Alcohol 95%	10 kg.

Sweet Almonds are to be roasted until they have a light brown color inside.

Peru Essence

Orris Root Tincture	2 kg.
Peru Balsam Tincture	1 kg.
Alcohol 95%	10 kg.

Color: Red-Brown.

Rose Essence

Rose Leaves Salted	150 gm.
Orange Flowers	15 gm.
Cloves	2 gm.
Vanilla	2 gm.
Alcohol 95%	10 kg.

Color: Red.

Red Carnation Essence

Red Carnations	2 kg.
Cloves	100 gm.
Alcohol 95%	10 kg.

No Color.

Chocolate Essence—No. 1

Cocoa Beans Roasted and Ground	2 kg.
Cinnamon	25 gm.
Cloves	20 gm.
Vanilla Tincture	50 gm.
Alcohol 95%	10 kg.

No. 2

Vanilla Tincture No. 46	100 gm.
Cocoa Beans Roasted and Ground	2 kg.
Alcohol 95%	10 kg.

No. 3

Peru Balsam Tincture No. 35	50 gm.
Cocoa Beans Roasted and Ground	2 kg.
Alcohol 95%	10 kg.

No. 4

Aromatic Essences as Before	100 gm.
Cocoa Beans Roasted and Ground	2 kilos
Alcohol 95%	10 kilos

Spanish Bitter Essence

Oil Angelica Root	50 gm.
Oil Anise	30 gm.
Oil Orange Bitter	300 gm.
Oil Calamus	30 gm.
Oil Cassia	30 gm.
Ethyl Acetate	100 gm.
Oil Caraway (Roman)	30 gm.
Oil Peppermint	30 gm.
Oil Wormwood	100 gm.
Alcohol 95%	10 kg.

Color: Dark Green.

Sultan Essence

Benzoin Tincture	1 kg.
Musk Tincture	10 gm.
Amber Tincture	20 gm.
Oil Rose	1 gm.

Color: Green.

Venus Essence

Vanilla Tincture	1 kg.
Oil Rose	2 gm.
Oil Cinnamon	5 gm.

Color: Red.

Violet Flower Essence

This essence can be produced by extracting the fresh violet flowers with fat and later on extracted over with full proof alcohol.

Woodruff Essence

Fresh Woodruff	4 kg.
Tonka Beans	100 gm.
Alcohol 95%	10 kg.

Color: Grass-Green.

Vermouth di Torino Essence

Angelica Root	30 gm.
Valerian Root	15 gm.
Benedictine Herbs	200 gm.
Cardamom	10 gm.
Guaiac Wood	30 gm.
Orange Peels	60 gm.
Peppermint Herbs	100 gm.
Lesser Centaury	100 gm.
Wormwood	120 gm.
Alcohol 95%	10 kg.

Color: Dark Brown.

Wormwood Essence

Angelica Root	60 gm.
Anise	20 gm.
Benedictine Herb	60 gm.
Calamus Roôt	30 gm.
Coriander	20 gm.
Gentian Root	30 gm.
Marjoram	50 gm.
Orange Peels	50 gm.
Peppermint Herbs	50 gm.
Lesser Centaury	60 gm.
Wormwood	100 gm.
Cinnamon	30 gm.
Alcohol 95%	10 kg.

Color: Dark Brown.

Civet Essence

Civet	30 gm.
Rose Oil	1 gm.
Alcohol 95%	1 kg.

Cinnamon Essence

Cinnamon	1 kg.
Orange Flowers	100 gm.
Alcohol 95%	10 kg.

Color: Cinnamon-Brown.

Allash Caraway Essence—No. 1

Oil Anise	10 gm.
Oil Angelica	5 gm.
Oil Coriander	5 gm.
Oil Caraway	100 gm.
Vanilla Tincture	20 gm.
Alcohol	10 kg.

No. 2

Oil Anise	8 gm.
Oil Angelica	2 gm.
Oil Coriander	2 gm.
Oil Caraway	80 gm.
Vanilla Tincture	10 gm.
Alcohol	10 kg.

Bishop Essence

Oil Orange Peels	50 gm.
Oil Bitter Orange Peels	20 gm.
Alcohol	10 kg.

Essence Spice

Oil Cardamom	10 gm.
Oil Cloves	15 gm.
Oil Mace	10 gm.
Oil Cinnamon	30 gm.
Alcohol 95%	10 kg.

Gold Water Essence

Oil Calamus	3 gm.
Oil Lemon	5 gm.
Oil Lavender	2 gm.
Oil Cloves	1 gm.
Oil Nutmeg	5 gm.
Oil Orange	10 gm.
Oil Rose	5 gm.
Oil Juniper Berries	3 gm.
Oil Cinnamon	5 gm.
Alcohol 95%	10 kg.

Corn Essence

Ethyl Acetate	500 gm.
Ethyl Oenanthic	10 gm.
Oil Juniper Berry	50 gm.
Alcohol	10 kg.

Spearmint Essence

Oil Spearmint	4 gm.
Oil Peppermint	2 gm.
Alcohol 95%	10 kg.

Caraway Essence

Oil Anise	1 gm.
Oil Coriander	1 gm.
Oil Caraway	7 gm.
Orris Root Tincture	10 gm.

Essence Parfait d'Amour

Oil Anise	40 gm.
Oil Cardamom	40 gm.
Oil Chamomile	5 gm.
Oil Lemon	5 gm.
Oil Lavender	5 gm.
Oil Cloves	5 gm.
Oil Orange	5 gm.
Oil Rosemary	40 gm.
Oil Cinnamon	80 gm.
Alcohol 95%	10 kg.

Rum Essence

Ethyl Butyrate	80 gm.
Ethyl Acetate	15 gm.
Vanilla Tincture	5 gm.
Orris Root Tincture	15 gm.
Alcohol 95%	10 kg.

Liqueur Body for Cremes and Huiles

No. 1

Sugar Sol. = 437 Grams Sugar in 1 Litre Water.

Sugar Sol. above	57.20 lit.
Alcohol	45.76 lit.
Water	11.40 lit.

No. 2

Sugar. Sol. = 393.3 Grams Sugar in 1 Litre Water.

Sugar Sol. above	51.48 lit.
Alcohol	45.76 lit.
Water	28.60 lit.

No. 3

Sugar Sol. = 349.6 Grams Sugar in 1 Litre Water.

Sugar Sol. above	45.76 lit.
Alcohol	48.05 lit.
Water	22.88 lit.

Liqueur Body for Fine Liqueurs

No. 4

Sugar Sol. = 327.7 Grams Sugar in 1 Litre Water.

Sugar Sol. above	42.90 lit.
Alcohol	50.91 lit.
Water	20.59 lit.

No. 5

Sugar Sol. = 305.9 Grams Sugar in 1 Litre Water.

Sugar Sol. above	40.08 lit.
Alcohol	50.33 lit.
Water	24.02 lit.

No. 6

Sugar Sol. = 262.2 Grams Sugar in 1 Litre Water.

Sugar Sol. above	34.32 lit.
Alcohol	50.33 lit.
Water	27.25 lit.

Liqueur Body for Ordinary Liqueur

No. 7

Sugar Sol. = 218.5 Grams Sugar in 1 Litre Water.

Sugar Sol. above	28.60 lit.
Alcohol	53.77 lit.
Water	32.03 lit.

No. 8

Sugar Sol. = 174.8 Grams Sugar in 1 Litre Water.

Sugar Sol. above	22.88 lit.
Alcohol	50.08 lit.
Water	35.46 lit.

For Double Spirits or Whiskey

No. 9

Sugar Sol. = 131 Grams Sugar in 1 Litre Water.

Sugar Sol. above	17.16 lit.
Alcohol	57.20 lit.
Water	40.04 lit.

No. 10

Sugar Sol. = 109.25 Grams Sugar in 1 Litre Water.

Sugar Sol. above	14.30 lit.
Alcohol	58.31 lit.
Water	41.18 lit.

No. 11

Sugar Sol. = 87.4 Grams Sugar in 1 Litre Water.

Sugar Sol. above	11.44 lit.
Alcohol	59.48 lit.
Water	43.42 lit.

For Ordinary Spirits or Whiskey

No. 12

Sugar Sol. = 65.55 Grams Sugar in 1 Litre Water.

Sugar Sol. above	6.86 lit.
Alcohol	60.62 lit.
Water	46.90 lit.

No. 13

Sugar Sol. = 43.7 Grams Sugar in 1 Litre Water.

Sugar Sol. above	5.72 lit.
Alcohol	61.77 lit.
Water	46.90 lit.

Creme de Angelica

Oil Angelica	2.5 gm.
Oil Lemon	0.5 gm.
Oil Coriander	0.5 gm.
Oil Mace	0.2 gm.
Oil Nutmeg	0.2 gm.
Oil Cinnamon	0.5 gm.
Liqueur-body	11.5 lit.

Color: Yellow.

Angelica Liqueur

Oil Angelica	1 gm.
Oil Lemon	1 gm.
Oil Cardamom	0.5 gm.
Oil Calamus	0.5 gm.
Oil Mace	0.5 gm.
Oil Melissa	0.5 gm.
Oil Wormwood	0.5 gm.
Liqueur Body	11.5 lit.

Color: Green.

Huile d'Angelica

Oil Angelica	3 gm.
Oil Lemon	0.5 gm.
Oil Cloves	0.1 gm.
Oil Orange	0.5 gm.
Oil Peppermint	0.1 gm.

Color: Grass-Green.

Anise Liqueur

Oil Anise	4 gm.
Oil Star Anise	4 gm.

Dissolved in 0.25 lit. Alcohol 95%.

Liqueur Body	11.5 lit.

No Color.

Anisette Double

Oil Anise	2 gm.
Oil Star Anise	3 gm.
Liqueur Body	11.5 lit.

Color: Yellow.

Anisette de Martinique

Oil Anise	2.6 gm.
Oil Fennel	0.4 gm.
Oil Cinnamon	0.4 gm.
Liqueur Body	11.5 lit.

No Color.

Creme d'Anisette Melee

Oil Anise	16 gm.
Oil Fennel	4 gm.
Liqueur Body	11.5 lit.

No Color.

Anisette de Bordeaux

Oil Anise	5 gm.
Oil Star Anise	1 gm.
Alcohol 95%	3 kg.
Water	3.5 kg.
Sugar	1.5 kg.

Color: Yellow.

Anisette de Bordeaux Francais

Oil Anise	16 gm.
Oil Coriander	4 gm.
Liqueur Body	11.5 lit.

No Color.

Adieu de Bertrand

Oil Calamus	2.5 gm.
Oil Wormwood	4 gm.
Liqueur Body	11.5 lit.

No Color or Violet.

Amourette

Oil Lemon Italian	2.5 gm.
Oil Orange Italian	2.5 gm.
Oil Star Anise	0.5 gm.
Oil Peppermint U.S.P.	0.5 gm.
Liqueur Body	11.5 lit.

Color: Dark Red.

A Propos

Oil Lemon	2 gm.
Oil Fennel	0.5 gm.
Oil Mace	0.2 gm.
Oil Cloves	0.1 gm.
Oil Orange	3 gm.
Oil Cinnamon	0.2 gm.
Liqueur Body	11.5 lit.

Agua Bianca

Amber Tincture	2 gm.
Oil Bergamot	1 gm.
Oil Lemon	2 gm.
Oil Peppermint	2 gm.
Liqueur Body	11.5 lit.

With six Silver-leaves (ground) and mixed.

Absinthe Creme

Oil Anise	0.5 gm.
Oil Lemon	0.5 gm.
Cognac Essence	1 gm.
Oil Coriander	0.5 gm.
Oil Mace	0.2 gm.
Oil Melissa	0.2 gm.
Oil Orange Peels	1 gm.
Oil Star Anise	1 gm.
Oil Wormwood	1 gm.
Oil Cinnamon	0.5 gm.
Liqueur Body	11.5 lit.

Color: Green.

Swiss Double Absinthe

Oil Anise	16 gm.
Oil Coriander	1 gm.
Oil Fennel	1 gm.
Oil Wormwood	16 gm.
Alcohol 90%	1.25 lit.
Sugar Dissolved in 2.5 Litres Water	250 gm.

Color: Green.

Benevento Liqueur

Sugar Solution	45 lit.
Alcohol 90%	35 lit.
Water	20 lit.
Benevento-liqueur Oil	50 gm.

Color: Green.

Creme de Bergamot

Oil Bergamot	3 gm.
Jasmine Water	5 gm.
Rose Water	5 gm.
Vanilla Tincture	5 gm.
Liqueur Body	11.5 lit.

Color: Yellow.

Bergamot Liqueur

Oil Bergamot	5 gm.
Oil Neroli	1 gm.
Oil Rose	0.5 gm.
Vanilla Tincture	10 gm.
Liqueur Body	11.5 lit.

Color: Yellow.

Berliner Bitter

Oil Angelica	0.5 gm.
Oil Coriander	0.5 gm.
Oil Ginger	0.5 gm.
Oil Mace	0.5 gm.
Oil Star Anise	1 gm.
Oil Juniper Berry	0.5 gm.
Oil Wormwood	1 gm.
Liqueur Body	11.5 lit.

Color: Brown.

Boonekamp (Stomach Bitter)

Oil Angelica	0.5 gm.
Oil Orange Bitter	0.5 gm.
Oil Lemon	0.5 gm.
Oil Coriander	0.5 gm.
Oil Galgant	0.2 gm.

Oil Ginger	0.4	gm.
Oil Mace	0.4	gm.
Oil Marjoram	0.4	gm.
Oil Peppermint	0.4	gm.
Oil Star Anise	0.5	gm.
Oil Juniper Berry	0.5	gm.
Oil Wormwood	0.6	gm.
Liqueur Body	11.5	lit.

Color: Yellow.

Boonekamp Dutch (Stomach Bitter)

Oil Angelica	1	gm.
Oil Orange Bitter	1	gm.
Oil Calamus	0.5	gm.
Oil Coriander	0.5	gm.
Oil Ginger	1	gm.
Oil Mace	0.5	gm.
Oil Nutmeg	0.5	gm.
Oil Juniper Berry	1	gm.
Oil Wormwood	1.5	gm.
Oil Cinnamon	0.2	gm.
Liqueur Body	11.5	lit.

Color: Amber-Yellow.

Bouquet de Dames

Oil Mace	0.5	gm.
Oil Cloves	0.5	gm.
Oil Rose	1	gm.
Oil Cinnamon	0.5	gm.
Liqueur Body	11.5	lit.

No Color.

Water Cress Liqueur

Sugar Solution	15	lit.
Alcohol 90%	44.5	lit.
Water	40.5	lit.
Oil Water Cress	50	gm.

Color: Green.

Creme de Canelle

Oil Neroli	0.5	gm.
Oil Cinnamon	3	gm.
Liqueur Body	11.5	lit.

Color: Cinnamon-Brown.

Liqueur de Canelle

Oil Mace	0.5	gm.
Vanilla Tincture	5	gm.
Oil Cinnamon	2	gm.
Liqueur Body	11.5	lit.

Color: As Above.

Creme de Cassia

Oil Cassia	3	gm.
Rosewater	100	gm.
Liqueur Body	11.5	lit.

Christopher

Oil Lemon	5	gm.
Oil Mace	1	gm.
Oil Melissa	1	gm.
Oil Cloves	1	gm.
Oil Cinnamon	1	gm.
Liqueur Body	11.5	lit.

Curacao Simple

Oil Orange	4	gm.
Oil Mace	1	gm.
Oil Cloves	0.5	gm.
Liqueur Body	11.5	lit.

Color: Light Brown.

Curacao de Hollande

Oil Orange	18	gm.
Oil Neroli	0.5	gm.
Oil Cinnamon	0.25	gm.
Liqueur Body	11.5	lit.

Color: Light Brown.

Creme de Curacao Dutch

Oil Pear	1	gm.
Oil Bitter Orange	1.5	gm.
Raspberry Ether	2	gm.
Oil Neroli	0.4	gm.
Oil Mace	0.4	gm.
Oil Orange	1.5	gm.
Vanilla Tincture	5	gm.
Oil Cinnamon	0.5	gm.
Liqueur Body	11.5	lit.

Color: Yellow.

Curacao de Marseille

Raspberry Ether	10	gm.
Oil Mace	1	gm.
Oil Orange	4	gm.
Vanilla Tincture	10	gm.
Oil Cinnamon	1	gm.
Liqueur Body	11.5	lit.

Color: Light Brown.

Curacao Imperial

Oil Bitter Orange	4	gm.
Oil Lemon	2	gm.
Raspberry Ether	4	gm.
Oil Nutmeg	0.5	gm.
Oil Neroli	1	gm.
Oil Orange	2	gm.
Tonka Bean Tincture	10	gm.
Vanilla Tincture	10	gm.
Oil Cinnamon	0.5	gm.
Liqueur Body	11.5	lit.

Color: Dark Brown.

Creme de Dames

Oil Anise	0.5	gm.
Oil Cardamom	0.5	gm.
Oil Lemon	0.5	gm.
Raspberry Ether	5	gm.
Oil Mace	0.5	gm.

Oil Cloves	0.5 gm.
Oil Neroli	0.5 gm.
Vanilla Tincture	5 gm.
Oil Cinnamon	0.5 gm.
Liqueur Body	11.5 lit.

Color: Yellow.

Gold Water a la Danzig

Sugar Solution	25 lit.
Alcohol 90%	40 lit.
Water	28 lit.
Cherry Water	4 lit.
Cognac	3 lit.
Oil Danzig Gold Water	50 gm.

Ground Gold Leaves genuine to be dissolved.

Eau d'Argent

Oil Bitter Almond	8 gm.
Oil Lemon	½ gm.
Oil Mace	½ gm.
Liqueur Body	11.5 lit.

Five genuine silver leaves to be ground and mixed in to alcohol in which the oils have to be dissolved.

Eau d'Amour

Oil Bitter Almond	1 gm.
Oil Lemon	2 gm.
Oil Coriander	½ gm.
Oil Lavender	½ gm.
Oil Mace	½ gm.
Oil Cinnamon	½ gm.
Liqueur Body	11.5 lit.

Color: Red.

Eau d'Ardelle

Oil Lemon	2 gm.
Oil Mace	1 gm.
Oil Clove	1 gm.
Oil Orris Root Tincture	50 gm.
Liqueur Body	11.5 lit.

Color: Violet.

Eau de Diane

Oil Bitter Almond	0.5 gm.
Oil Rose	1 gm.
Oil Neroli	0.5 gm.
Liqueur Body	11.5 lit.

No Color.

Eau de Milles Fleurs

Oil Bergamot	0.5 gm.
Oil Lemon	1 gm.
Oil Lavender	1 gm.
Oil Mace	0.5 gm.
Tincture Musk Tonquin	0.5 gm.
Oil Coves	0.5 gm.
Balsam Pine	1 gm.
Rosemary Oil	1 gm.
Oil Cinnamon	1.5 gm.
Liqueur Body	11.5 lit.

Color: Rose Red.

Eau d'Or—A

Oil Lemon	8 gm.
Oil Rose	0.5 gm.
Oil Cinnamon	4 gm.
Liqueur Body	11.5 lit.

Color: Yellow.

Five genuine gold leaves ground and added to liqueurs.

B

Oil Calamus	2 gm.
Oil Cardamom	1 gm.
Oil Lemon	4 gm.
Oil Lavender	2 gm.
Oil Mace	4 gm.
Oil Cloves	1 gm.
Oil Orange Peels	12 gm.
Oil Rose	4 gm.
Oil Rosemary	2 gm.
Oil Juniper Berry	2 gm.
Oil Cinnamon	4 gm.

Distilled in 1 Litre of Alcohol.

Liqueur Body	11.5 lit.

Gold leaves as before.

Eau de Capucine

Oil Anise	1 gm.
Oil Fennel	1 gm.
Oil Mace	1 gm.
Oil Neroli	0.5 gm.
Oil Peppermint	0.5 gm.
Oil Cinnamon	1 gm.
Liqueur Body	11.5 lit.

Color: Dark Brown.

Eau des Chasseurs—A

Oil Mace	1 gm.
Oil Peppermint	2 gm.
Liqueur Body	11.5 lit.

Color: Green or no color.

B

Oil Cloves	1 gm.
Oil Mace	2 gm.
Oil Peppermint	5 gm.
Liqueur Body	11.5 lit.

Color: Dark Brown.

Eau Celeste

Oil Anise	1.5 gm.
Oil Cardamom	0.5 gm.
Oil Coriander	0.5 gm.
Oil Cloves	0.5 gm.
Oil Neroli	1 gm.
Oil Cinnamon	1 gm.
Liqueur Body	11.5 lit.

Eau Cordiale

Anise	50	gm.
Lemon Peels	400	gm.
Coriander Seed	50	gm.
Melissa	100	gm.
Nutmegs	20	gm.
Cinnamon	50	gm.
Liqueur Body	11.5	lit.

Color: Blue.

Eau de la Cote

Oil Bergamot	1	gm.
Oil Lemon	2	gm.
Oil Peppermint	0.5	gm.
Oil Cinnamon	1.5	gm.
Liqueur Body	11.5	lit.

Color: Yellow.

Agua Turca

Amber Tincture	10	gm.
Angelica Tincture	10	gm.
Musk Tincture	2	gm.
Tea Chinese	100	gm.
Vanilla Tincture	10	gm.
Alcohol 95%	4.6	lit.
Water	2.3	lit.
Sugar	2	kg.

No Color. The tea has to be extracted in a cold process for 8 days in the alcohol.

English Bitter

Benedictine Herb	10	gm.
Gentian Root	20	gm.
Orange Peels	100	gm.
Calamus Root	40	gm.
Lesser Centaury	50	gm.
Orris Root	50	gm.
Wormwood	20	gm.
Cinnamon	10	gm.
Alcohol 95%	0.6	lit.
Liqueur Body	11.5	lit.

Color: Brown.

Fine Bitter

Orange Peels	500	gm.
Oranges Unripe	100	gm.
Calamus Root	40	gm.
Cinnamon	20	gm.
Zedoary Plant	20	gm.
Alcohol 95%	1.15	lit.
Liqueur Body	11.5	lit.

Color: Dark Red Brown.

Gold Water

Anise	50	gm.
Lemon Peels	100	gm.
Coriander	50	gm.
Mace	30	gm.
Cloves	20	gm.
Cinnamon	20	gm.
Alcohol 95%	0.6	lit.
Liqueur Body	11.5	lit.

Five ground gold leaves.

Prinzess Water

Amber Tincture to be added to the finished liqueur	10	gm.
Chamomile	50	gm.
Lemon Peels	80	gm.
Coriander	40	gm.
Figs	100	gm.
Almonds Bitter	40	gm.
Melissa	60	gm.
Cloves	20	gm.
Rosemary	100	gm.
Cinnamon	20	gm.
Alcohol 95%	0.6	lit.
Liqueur Body	11.5	lit.

Six ground silver leaves.

Silver Water

Angelica Root	20	gm.
Lemon Peels	200	gm.
Cloves	20	gm.
Star Anise	20	gm.
Orris Root	50	gm.
Cinnamon	20	gm.
Alcohol 95%	0.6	lit.
Liqueur Body	11.5	lit.

10 Silver Leaves ground.

Greek Water

Angelica Root	20	gm.
Calamus Root	40	gm.
Cardamom	20	gm.
Cloves	20	gm.
Mace	20	gm.
Bitter Almonds	80	gm.
Wormwood	20	gm.
Cinnamon	20	gm.
Alcohol 95%	0.6	lit.
Liqueur Body	11.5	lit.

Color: Red Violet.

Eau de Sante

Angelica Root	40	gm.
Lemon Peels	100	gm.
Cardamom	40	gm.
Jasmine	100	gm.
Lavender	80	gm.
Marjoram	60	gm.
Grains of Paradise	40	gm.
Peppermint	80	gm.
Rosemary	100	gm.
Alcohol 95%	0.6	lit.
Liqueur Body	11.5	lit.

Color: Green.

Fleurs de l'Orient

Calamus	30	gm.
Lemon Peels	200	gm.
Dates	200	gm.
Fennel	100	gm.
Cloves	30	gm.
Orange Peels	300	gm.
Cinnamon	30	gm.
Alcohol 95%	0.6	lit.
Liqueur Body	11.5	lit.

Creme Aux Macarons

Cardamom	15	gm.
Bitter Almond	150	gm.
Cloves	10	gm.
Cinnamon	10	gm.

Digest with 150 gm. Orange Flower Water.

Rosewater	100	gm.
Liqueur Body	11.5	lit.

No Color.

Maraschinodella Boche de Cattaro

Bitter Almond Water	1.15	lit.
Raspberry Water	3.45	lit.
Orange Flavor Water	2.25	lit.
Alcohol 90%	4.60	lit.
Sugar dissolved in the Raspberry Water	4	kg.

No Color.

Persico Adriatico

Bitter Almond Water	1.15	lit.
Sugar Solution	4.60	lit.
Alcohol 95%	5.20	lit.

Lemon Absinthe

Lemon Peels	200	gm.
Peppermint Herb	100	gm.
Wormwood	50	gm.
Alcohol 95%	0.6	lit.
Liqueur Body	11.5	lit.

Color: Green.

Lemon Liqueur

Lemon Peels Fresh	400	gm.
Alcohol 95%	0.6	lit.
Liqueur Body	11.5	lit.

Lemon Peels to be extracted for 8 days with the alcohol. Color: Yellow.

China Liqueur

Angelica	50	gm.
Anise	50	gm.
Mace	20	gm.
Bitter Almonds	200	gm.
Cinnamon	20	gm.

Digested with 1.1 lit. Water.

Alcohol 95%	0.6	lit.
Calamus Tincture	5	gm.
Oil Neroli	0.5	gm.
Liqueur Body	11.5	lit.

Color: Brown.

Pineapple Fruit Liqueur

Pineapples	2	
Alcohol 95%	4.6	lit.
Water	3.4	lit.
Sugar	4.5	kg.
Vanilla Tincture	50	gm.
Pear Ether	5	gm.

Color: Yellow.

Apricot Fruit Liqueur

Apricots	6	kg.
Sugar	4	kg.
Alcohol 95%	3.3	lit.
Water	1.1	lit.
Cinnamon Tincture	50	gm.

Color: Rose Red.

Bergamot Fruit Liqueur

Ripe Bergamots	5	kg.
Alcohol 95%	4.6	lit.
Water	2.3	lit.
Sugar	5	kg.
Apple Ether	20	gm.

Color: Golden Yellow.

Blackberry Fruit Liqueur

Blackberries	3	kg.
Alcohol 95%	1.5	lit.
Sugar	0.8	kg.

or

Blackberry Juice	2	lit.
Sugar	0.8	kg.
Alcohol 95%	1.5	lit.

Pineapple Ratafia

Pineapple Ether	20	gm.
Cognac Essence	10	gm.
Ethyl Oenanthic Solution (1 gr. to 1 lit. Alcohol 95%)	10	gm.
Rose Water	40	gm.
Tartaric Acid	40	gm.
Liqueur Body	11.5	lit.

Color: Yellow.

Ratafia Aux Bergamottes

Oil Bergamot	3	gm.
Rose Water	10	gm.
Vanilla Tincture	6	gm.
Cinnamon Tincture	5	gm.
Tartaric Acid	20	gm.
Liqueur Body	11.5	lit.

Color: Yellow.

Ratafia Aux Chocolate

Cocoa Burned	1	kg.
Vanilla Tincture	10	gm.
Cinnamon Tincture	5	gm.
Liqueur Body	11.5	lit.

Color: Dark Brown.

Ratafia de Curacao

Oil Bitter Almond	2	gm.
Oil Clove	0.5	gm.
Oil Neroli	0.4	gm.
Oil Orange	2	gm.
Vanilla Essence	4	gm.
Oil Cinnamon	0.5	gm.
Liqueur Body	11.5	lit.

Ratafia Aux Citrons

Lemon Peels	4	gm.
Orange Flower Water	10	gm.
Tartaric Acid	60	gm.
Liqueur Body	11.5	lit.

Color: Yellow.

Tonka Ratafia

Tonka Beans	50	gm.
Vanilla Tincture	10	gm.
Cinnamon Tincture	10	gm.
Liqueur Body	11.5	lit.

Color: Green.

Raspberry Ratafia

Raspberry Ether	100	gm.
Orange Flower Water	10	gm.
Tartaric Acid	10	gm.
Cinnamon Tincture	5	gm.
Liqueur Body	11.5	lit.

Color: Raspberry Red.

Bishop

Cherry Juice	4.6	lit.
Curacao Peels	50	gm.
Cloves	10	gm.
Oranges	10	pieces
Cinnamon	10	gm.
Alcohol 95%	4.6	lit.
Sugar	2	kg.

Color: Yellow.

Calamus Ratafia

Oil Calamus	0.5	gm.
Oil Cardamom	3	gm.
Vanilla Tincture	4	gm.
Cinnamon Tincture	4	gm.
Liqueur Body	11.5	lit.

Color: Yellow.

Bitter Ratafia

Oil Angelica	1	gm.
Oil Cardamom	1	gm.
Cognac Essence	2	gm.
Oil Marjoram	0.5	gm.
Oil Melissa	0.5	gm.
Oil Wormwood	0.5	gm.
Raspberry Ether	5	gm.
Liqueur Body	11.5	lit.

Color: Brown.

Ginger Ratafia

Ginger	100	gm.
Mace	10	gm.
Vanilla	5	gm.
Cinnamon	10	gm.
Liqueur Body	11.5	lit.

Color: Yellow.

Grunewald

Grunewald Essence	1.1–2.25	lit.
Liquer Body	11.5	lit.

Color: Green.

Sailors Hearts-Content

Hearts Content Essence	2.55–3	lit.
Liqueur Body	11.5	lit.

Maraschino

Oil Bitter Almond	3	gm.
Cognac Essence	2	gm.
Raspberry Ether	2	gm.
Oil Neroli	1	gm.
Vanilla Tincture	5	gm.
Liqueur Body	11.5	lit.

No Color.

Stomach Creme

French Stomach Essence	1.1	lit.
Liqueur Body	11.5	lit.

Color: Light Brown.

Creme de Mocca

Coffee Essence	2.25	lit.
Liqueur Body	11.5	lit.

Color: Brown.

Double Carnation Liqueur

Cloves Tincture	0.52	lit.
Liqueur Body	11.5	lit.

Color: Brown.

Creme de Sultan

Sultan Essence	0.52	lit.
Liqueur Body	11.5	lit.

Color: Dark Red.

Vanilla Creme

Vanilla Essence	50	gm.
Balsam Peru Essence	100	gm.
Liqueur Body	11.5	lit.

Color: Red.

Venus Creme

Venus Essence	1.1	gm.
Liqueur Body	11.5	lit.

No Color.

Amber Liqueur

Amber Essence	5	gm.
Musk Essence	0.5	gm.
Civet Essence	0.5	gm.
Liqueur Body	11.5	lit.

Color: Light Brown.

Friends Drink

Oil Bergamot	2	gm.
Oil Lemon	1	gm.
Cognac Essence	5	gm.
Vanilla Tincture	10	gm.
Orris Root Tincture	10	gm.
Liqueur Body	11.5	lit.

Color: Brown.

English Bitter

Essence Bitter Orange	40	gm.
Essence Holy Thistle	60	gm.
Essence China Bark	40	gm.
Essence Gentian Root	40	gm.
Essence Lesser Centaury	60	gm.
Essence Orris Root	40	gm.
Essence Wormwood	60	gm.
Liqueur Body	11.5	lit.

Color: Brown.

Spanish Bitter Creme

Spanish Bitter Essence	1.1–1.6	lit.
Liqueur Body	11.5	lit.

Color: Brown.

Spanish Chocolate Creme

Chocolate Essence	2.25	lit.
Liqueur Body	11.5	lit.

Color: Brown.

Curacao of Java

Curacao	1.1	lit.
Oil Neroli	5	gm.

Dissolved in 50 gm. Alcohol 95%.

Liqueur Body	11.5	lit.

No Color.

Creme de Peru

Peru Essence	1.1	lit.
Liqueur Body	11.5	lit.

Color: Brown.

Chocolate Liqueur

Cocoa Beans Burned	200	gm.
Clove Tincture	5	gm.
Vanilla Tincture	16	gm.
Cinnamon Tincture	5	gm.
Liqueur Body	11.5	lit.

Color: Dark Red.

Creme de Vanilla Pure

Vanilla Tincture	150	gm.
Liqueur Body	11.5	lit.

Color: Red.

Creme de Vanilla Double

Vanilla Tincture	1.15	lit.
Liqueur Body	11.5	lit.

Color: Red.

Liqueur de Vanilla

Balsam Peru Tincture	200	gm.
Vanilla Tincture	200	gm.
Liqueur Body	11.5	lit.

Color: Red.

Huile de Vanilla Surfine

Benzoin Tincture	10	gm.
Rosewater	60	gm.
Vanilla Tincture	20	gm.
Liqueur Body	11.5	lit.

Color: Red.

Cinnamon Liqueur Simple

Cinnamon Tincture	0.55	lit.
Liqueur Body	11.5	lit.

Color: Cinnamon Brown.

Cinnamon Liqueur Double

Balsam Peru Tincture	0.1	lit.
Cinnamon Tincture	0.9	lit.
Liqueur Body	11.5	lit.

Color: Cinnamon Brown.

Cream of Lemon Fruits

Lemon Peels absolutely fresh extracted for 6 days in 0.6 lit. Alcohol 95%	10	pieces
Liqueur Body	11.5	lit.

Color: Yellow.

Cherry Liqueur

Oil Bitter Almond	1	gm.
Vanilla Tincture	2	gm.

For Chemical Advisors, Special Raw Materials, Equipment, Containers, etc., consult Supply Section at end of book.

Orris Root Tincture	5	gm.
Cinnamon Tincture	0.5	gm.
Liqueur Body	11.5	lit.

Color: Cherry Red.

Fleur de Montpelier

Angelica Tincture	10	gm.
Oil Bergamot	1	gm.
Oil Lemon	1	gm.
Oil Cloves	0.5	gm.
Oil Neroli	1	gm.
Oil Rose	0.2	gm.
Vanilla Tincture	5	gm.
Cinnamon Tincture	5	gm.
Liqueur Body	11.5	lit.

Color: Blue.

Creme de Girofles

Clove Tincture	600	gm.
Cinnamon Tincture	50	gm.
Liqueur Body	11.5	lit.

Color: Brown.

Nut Creme

Nut Tincture	1.15	lit.
Liqueur Body	11.5	lit.

Color: Green.

Creme de Peru

Balsam Peru Tincture	120	gm.
Orris Root Tincture	50	gm.
Liqueur Body	11.5	lit.

Color: Brown.

Creme de Chocolate

Aromatic Tincture	30	gm.
Cocoa Tincture	1.5	gm.
Balsam Peru Tincture	20	gm.
Liqueur Body	11.5	lit.

Color: Brown.

Cream of Raspberry

Raspberry Tincture	10	gm.
Vanilla Tincture	2	gm.
Cinnamon Tincture	2	gm.
Liqueur Body	11.5	lit.

Color: Red.

Indian Ginger

Amber Tincture	1	gm.
Musk Tincture	0.5	gm.
Oil Ginger	2	gm.
Liqueur Body	11.5	lit.

Color: Brown.

Cream of Virgins

Oil Anise	20	gm.
Oil Neroli	2	gm.

Dissolved in 100 gm. Alcohol 95%.

Vanilla Tincture	8	gm.
Liqueur Body	11.5	lit.

Colorless.

Creme de Coffee

Clove Tincture	10	gm.
Mace Tincture	10	gm.
Cinnamon Tincture	15	gm.
Coffee Tincture	1.15	lit.
Liqueur Body	11.5	lit.

Color: Dark Brown.

Coffee Liqueur

Coffee Burned	200	gm.
Tincture Cloves	5	gm.
Tincture Mace	5	gm.
Tincture Vanilla	15	gm.
Tincture Cinnamon	5	gm.
Liqueur Body	11.5	lit.

Color: Dark Brown.

Creme de Mocca

Vanilla Tincture	10	gm.
Orris Root Tincture	50	gm.
Cinnamon Tincture	15	gm.
Coffee Tincture	1.15	lit.
Liqueur Body	11.5	lit.

Color: Dark Brown.

Creme de Cassia

Cinnamon Tincture	500	gm.
Liqueur Body	11.5	lit.

Color: Brown.

Creme of China

Cassia Tincture	800	gm.
Liqueur Body	11.5	lit.

Color: Brown.

Creme de Cocoa

Vanilla Tincture	25	gm.
Cocoa Tincture	1.15	gm.
Liqueur Body	11.5	lit.

Color: Brown.

Cream of Flower

Jasmine Tincture	10	gm.
Vanilla Tincture	10	gm.
Oil Rose	1	gm.

Dissolved in 0.15 lit. Alcohol 95%.

Liqueur Body	11.5	lit.

No Color.

Cream of Sulton

Amber Tincture	1.5	gm.
Benzoin Tincture	1.5	gm.

Musk Tincture	0.5 gm.
Oil Rose	0.5 gm.

Dissolved in 50 gm. Alcohol 95%.

Liqueur Body	11.5 lit.

Color: Dark Red.

Creme of Lemon a la Malta

Oil Lemon	1.5 gm.
Oil Coriander	0.5 gm.
Oil Neroli	0.2 gm.
Oil Orange	1.5 gm.
Vanilla Tincture	5 gm.
Cinnamon Tincture	5 gm.
Liqueur Body	11.5 lit.

Color: Yellow.

Extract of Lemon Double

Oil Lemon	4 gm.
Oil Coriander	0.5 gm.
Oil Neroli	1 gm.
Oil Orange	2 gm.
Oil Star Anise	0.5 gm.
Tonka Bean Tincture	10 gm.
Vanilla Tincture	10 gm.
Liqueur Body	11.5 lit.

Color: Lemon Yellow.

Pineapple Liqueur

Pineapple Ether	15 gm.
Cognac Essence	10 gm.
Oil Rose	1 gm.
Liqueur Body	11.5 lit.

Color: Yellow.

Creme of Pineapple

Pineapple Ether	20 gm.
Pear Ether	5 gm.
Acetic Ether	10 gm.
Raspberry Ether	20 gm.
Liqueur Body	11.5 lit.

Color: Yellow.

Creme de Barbados

Oil Bergamot	1 gm.
Oil Lemon	1 gm.
Oil Mace	0.2 gm.
Oil Cloves	0.5 gm.
Oil Neroli	0.2 gm.
Balsam Peru Tincture	5 gm.
Orris Root Tincture	5 gm.
Oil Cinnamon	1 gm.
Liqueur Body	11.5 lit.

No Color or Brown.

Creme of Cinnamon

Cinnamon Tincture	100 gm.
Liqueur Body	11.5 lit.

Color: Brown.

Absinthe Fine

Oil Calamus	1 gm.
Oil Coriander	1.5 gm.
Oil Ginger	1 gm.
Oil Wormwood	1 gm.
Liqueur Body	11.5 lit.

Color: Green.

Cinnamon Liqueur

Oil Cinnamon	4 gm.

Dissolved in 0.1 lit. Alcohol 95%.

Liqueur Body	11.5 lit.

Color: Light Brown.

Dutch Cinnamon Liqueur

Oil Rosewood	1.5 gm.
Oil Cinnamon	2.5 gm.
Liqueur Body	11.5 lit.

Color: Light Brown.

Creme of Cinnamon Extra Fine

Genuine Oil Rose	0.5 gm.
Oil Cinnamon	3 gm.
Liqueur Body	11.5 lit.

Color: Brown or Red.

Cinnamon Liqueur Super Fine

Oil Mace	0.5 gm.
Oil Cloves	0.5 gm.
Oil Cinnamon	2 gm.
Liqueur Body	11.5 lit.

Color: Cinnamon Brown.

Lemon Liqueur

Oil Lemon	8 gm.

Dissolved in 0.1 lit. Alcohol 95%.

Liqueur Body	11.5 lit.

Color: Yellow.

Creme de Citron

Oil Lemon	14 gm.
Oil Neroli	0.4 gm.
Liqueur Body	11.5 lit.

Color: Light Yellow.

Huile de Citron

Oil Lemon	15 gm.
Liqueur Body	11.5 lit.

Color: Light Yellow.

Usquebaugh—No. 2

Oil Anise	1 gm.
Oil Calamus	0.5 gm.
Oil Cardamom	1 gm.
Oil Lemon	1 gm.
Oil Mace	0.5 gm.
Oil Nutmeg	0.5 gm.

Oil Cloves	0.5	gm.
Oil Cinnamon	1	gm.
Liqueur Body	11.5	lit.

Color: Yellow.

Venus Creme

Oil Cloves	2	gm.
Oil Cinnamon	2	gm.
Liqueur Body	11.5	lit.

Color: Rose Red.

Creme de Juniper Berry

Oil Lemon	4	gm.
Oil Orange	4	gm.
Oil Juniper Berry	10	gm.
Liqueur Body	11.5	lit.

Colorless or Faint Green.

Juniper Berry Liqueur—A

Oil Coriander	1	gm.
Cognac Essence	4	gm.
Oil Juniper Berry	3	gm.
Liqueur Body	11.5	lit.

Color: Green.

B

Oil Calamus	0.5	gm.
Oil Cardamom	0.5	gm.
Cognac Essence	2	gm.
Oil Coriander	0.5	gm.
Oil Juniper Berry	2	gm.
Oil Ginger	0.5	gm.
Liqueur Body	11.5	lit.

Color: Green.

English Absinthe

Oil Anise	8	gm.
Oil Wormwood	8	gm.
Liqueur Body	11.5	lit.

Color: Green.

Rostopschin

Oil Anise	1	gm.
Oil Cardamom	1	gm.
Oil Lemon	1	gm.
Oil Coriander	1	gm.
Oil Mace	1	gm.
Oil Cinnamon	0.5	gm.
Liqueur Body	11.5	lit.

No Color.

Creme de Celery

Oil Anise	0.5	gm.
Oil Bitter Almond	0.5	gm.
Oil Coriander	0.5	gm.
Oil Caraway	0.5	gm.
Oil Celery Seed	1	gm.
Liqueur Body	11.5	lit.

No Color.

Creme of Seven Fruits

Oil Anise Russian	2	gm.
Oil Lemon	2	gm.
Oil Coriander	2	gm.
Oil Caraway	5	gm.
Oil Muscat	1	gm.
Oil Cloves	1	gm.
Oil Cinnamon	1	gm.
Liqueur Body	11.5	lit.

Swiss Creme

Oil Angelica	0.5	gm.
Oil Bitter Almond	1	gm.
Oil Calamus	0.5	gm.
Oil Cardamom	0.5	gm.
Oil Cloves	0.2	gm.
Oil Peppermint	0.2	gm.
Oil Rosemary	0.2	gm.
Oil Thyme	0.5	gm.
Oil Juniper Berry	0.5	gm.
Oil Wormwood	0.5	gm.
Vanilla Tincture	6	gm.
Oil Cinnamon	0.5	gm.
Liqueur Body	11.5	lit.

Color: Green.

Usquebaugh—No. 1

Oil Anise	1	gm.
Oil Cardamom	0.5	gm.
Oil Lemon	0.5	gm.
Oil Coriander	0.5	gm.
Oil Mace	0.5	gm.
Oil Cloves	0.5	gm.
Oil Cinnamon	0.5	gm.
Liqueur Body	11.5	lit.

Color: Yellow.

Liqueur d'Oranges

Oil Lemon	2	gm.
Oil Orange	3	gm.
Liqueur Body	11.5	lit.

No Color.

Creme d'Oranges

Oil Neroli	0.5	gm.
Oil Orange	2.8	gm.
Liqueur Body	11.5	lit.

No Color.

Creme de Roses

Oil Geranium	0.5	gm.
Oil Rose	0.5	gm.
Liqueur Body	11.5	lit.

Color: Rose Red.

Huile de Roses

Oil Rose	1.5	gm.
Liqueur Body	11.5	lit.

Color: Pale Rose Red.

Rosa Bianca

Oil Cloves	0.5	gm.
Oil Rose	1	gm.
Liqueur Body	11.5	lit.

Colorless.

Creme de Roses de Bassora

Oil Neroli	0.5	gm.
Oil Rose	2	gm.
Liqueur Body	11.5	lit.

Color: Rose Red.

Creme de la Rose Mousseuse

Oil Neroli	0.2	gm.
Oil Rose	0.4	gm.
Vanilla Tincture	2	gm.
Liqueur Body	11.5	lit.

Color: Rose Red.

Rosemary Liqueur

Oil Lemon	1	gm.
Oil Coriander	1	gm.
Oil Rosemary	3	gm.
Liqueur Body	11.5	lit.

Color: Green.

Persico de Cattaro

Oil Bitter Almond	4.5	gm.
Liqueur Body	11.5	lit.

No Color.

Creme de Persico

Oil Bitter Almond	2	gm.
Oil Mace	0.4	gm.
Oil Cloves	0.2	gm.
Oil Neroli	0.2	gm.
Vanilla Tincture	4	gm.
Orris Root Tincture	6	gm.
Cinnamon Tincture	4	gm.
Liqueur Body	11.5	lit.

No Color.

Huile de Menthe

Oil Peppermint	4	gm.
Liqueur Body	11.5	lit.

Color: Green.

Mentha Bianca

Oil Spearmint	2	gm.
Oil Cloves	1	gm.
Oil Peppermint	4	gm.
Liqueur Body	11.5	lit.

No Color.

Creme de Menthe Anglaise

Oil Cloves	0.5	gm.
Oil Peppermint	4	gm.
Liqueur Body	11.5	lit.

Colorless or Green.

Liqueur de Menthe

Oil Peppermint	9	gm.

Dissolved in 0.1 lit. Alcohol 95%.

Liqueur Body	11.5	lit.

No Color.

Superior English Peppermint

Oil Spearmint	1	gm.
Oil Peppermint	4	gm.
Liqueur Body	11.5	lit.

Color: Grass Green.

Creme de Fleurs d'Oranges
(Creme of Orange Flower)

Oil Neroli	4.5	gm.
Oil Rose	0.5	gm.
Liqueur Body	11.5	lit.

No Color.

Fleurs d'Oranges

Oil Neroli	2	gm.
Oil Orange	3	gm.
Liqueur Body	11.5	lit.

No Color.

Huille de Fleurs d'Oranges

Oil Neroli	4.5	gm.
Oil Orange	2.5	gm.
Liqueur Body	11.5	lit.

No Color.

Creme de Fleurs d'Oranges

Pear Ether	1	gm.
Oil Bitter Almond	1	gm.
Raspberry Ether	2	gm.
Oil Neroli	1	gm.
Oil Cinnamon	1	gm.
Liqueur Body	11.5	lit.

Color: Yellow.

Fleur d'Oranges de la Riviere

Jasmine Water	50	gm.
Oil Neroli	5	gm.
Oil Rose	1	gm.
Liqueur Body	11.5	lit.

No Color.

Parfait Amour

Oil Lemon	8	gm.
Oil Mace	1	gm.

Oil Cloves	4	gm.
Liqueur Body	11.5	lit.

Color: Light Red.

Parfait Amour Liqueur

Oil Anise	4	gm.
Oil Chamomile	2	gm.
Oil Cardamom	4	gm.
Oil Lemon	2	gm.
Oil Lavender	2	gm.
Oil Cloves	2	gm.
Oil Orange	2	gm.
Oil Rosemary	4	gm.
Oil Cinnamon	20	gm.

Dissolve in 1.15 lit. Alcohol 95%.

Liqueur Body	11.5	lit.

Color: Rose Red.

Melisse Romaine

Oil Lemon	1	gm.
Oil Coriander	0.5	gm.
Oil Melissa	3	gm.
Vanilla Tincture	5	gm.
Oil Cinnamon	0.5	gm.
Liqueur Body	11.5	lit.

Color: Green.

Milk Liqueur

Oil Anise	0.5	gm.
Oil Cloves	0.5	gm.
Oil Orange	2	gm.
Oil Rose	0.5	gm.
Oil Cinnamon	1	gm.
Milk	1.15	lit.
Liqueur Body	11.5	lit.

Nordhauser Corn Liqueur

Acetic Ether	10	gm.
Oil Juniper Berry	0.5	gm.
Oil Cinnamon	0.5	gm.
Liqueur Body	11.5	lit.

Nordhauser Corn Liqueur Double

Acetic Ether	15	gm.
Raspberry Ether	10	gm.
Oil Mace	0.5	gm.
Oil Cloves	0.2	gm.
Oil Cinnamon	0.5	gm.
Liqueur Body	11.5	lit.

Nut Creme

Oil Cloves	10	gm.
Green Nuts	250	gm.
Mace	40	gm.
Orris Root	10	gm.
Oil Cinnamon	10	gm.

To be dissolved in 1 lit. Alcohol 95% and extracted in cold process or 14 days, then add 11.5 lit. Liqueur Body.

Color: Green.

Creme de Fleurs d'Oranges

Orange Flower Water	1	lit.
Liqueur Body	11.5	lit.

No Color.

Creme de Muscat—A

Oil Mace	2	gm.
Oil Nutmeg	1	gm.
Vanilla Tincture	5	gm.
Liqueur Body	11.5	lit.

Color: Red-Brown.

B

Cognac Essence	5	gm.
Oil Mace	1	gm.
Oil Nutmeg	1	gm.
Oil Neroli	0.5	gm.
Oil Cinnamon	0.5	gm.
Liqueur Body	11.5	lit.

Carnation Creme Liqueur

Oil Cloves	1.6	gm.
Liqueur Body	11.5	lit.

No Color or Light Yellow.

Liqueur Aux Fleurs d'Oeillets

Oil Cloves	5	gm.

Dissolved in 0.1 lit. Alcohol 95%.

Liqueur Body	11.5	lit.

Color: Brown.

Huile d'Oeillets

Oil Cloves	1.5	gm.
Oil Cinnamon	0.25	gm.
Liqueur Body	11.5	lit.

No Color or Light Yellow.

Creme de Clous de Girofle

Oil Bitter Almond	0.5	gm.
Oil Mace	0.2	gm.
Oil Cloves	1.5	gm.
Oil Cinnamon	0.5	gm.
Liqueur Body	11.5	lit.

Color: Brown.

Non Pareille

Oil Mace	1	gm.
Oil Cloves	1	gm.
Oil Rose	0.5	gm.
Liqueur Body	11.5	lit.

Color: Dark Cherry Red.

Almond Creme

Oil Bitter Almond	1.5	gm.
Oil Mace	0.5	gm.
Oil Peppermint	0.5	gm.
Tincture Balsam Peru	5	gm.
Tincture Cinnamon	5	gm.
Liqueur Body	11.5	lit.

No Color.

Maraschino Dalmatico

Oil Bitter Almond	1.5	gm.
Oil Neroli	0.5	gm.
Oil Rose	0.5	gm.
Oil Cinnamon	0.8	gm.
Liqueur Body	11.5	lit.

No Color.

Maraschino di Zara

Oil Bitter Almond	3	gm.
Oil Neroli	0.8	gm.
Jasmine Water	40	gm.
Rose Water	30	gm.
Liqueur Body	11.5	lit.

No Color.

Liqueur de Melisse

Oil Lemon	0.5	gm.
Oil Melissa	3	gm.
Oil Nutmeg	0.5	gm.
Liqueur Body	11.5	lit.

Color: Green.

Creme de Melisse

Oil Cardamom	0.5	gm.
Oil Lemon	0.5	gm.
Oil Coriander	0.5	gm.
Oil Mace	0.5	gm.
Oil Melissa	2	gm.
Oil Cinnamon	0.5	gm.
Liqueur Body	11.5	lit.

Color: Green.

Muscat Liqueur Simple

Tincture Benzoin	5	gm.
Oil Coriander	0.5	gm.
Oil Mace	1	gm.
Oil Nutmeg	1	gm.
Liqueur Body	11.5	lit.

Color: Brown.

Creme de Caraway Simple

Oil Caraway	16	gm.
Liqueur Body	11.5	lit.

No Color.

Creme de Caraway Double

Oil Anise	2	gm.
Oil Caraway	16	gm.
Liqueur Body	11.5	lit.

No Color.

Double Caraway a la Danzig

Oil Cardamom	0.5	gm.
Oil Coriander	0.5	gm.
Oil Fennel	0.5	gm.
Oil Caraway	2	gm.
Oil Orange	1	gm.
Liqueur Body	11.5	lit.

No Color.

Fine Caraway Liqueur

Oil Fennel	1	gm.
Oil Caraway	4	gm.
Vanilla Tincture	10	gm.
Oil Cinnamon	1	gm.
Liqueur Body	11.5	lit.

No Color.

Triple Caraway Essence

Oil Anise	2	gm.
Oil Lemon	5	gm.
Oil Coriander	3	gm.
Oil Caraway	150	gm.
Oil Mace	1	gm.

48 gm. of this mixture are to be mixed with 25 lit. of 60% Alcohol, 25 lit. Water, 1 lit. Bourbon Whiskey and 4 kilos Sugar.

Mixed Caraway Essence

Cognac Essence	5	gm.
Oil Coriander	1	gm.
Oil Fennel	1	gm.
Oil Caraway	3	gm.
Oil Neroli	0.5	gm.
Oil Orange	1	gm.
Oil Cinnamon	1	gm.
Liqueur Body	11.5	lit.

No Color.

Spearmint Buds Liqueur

Oil Spearmint	3	gm.
Oil Lavender	1	gm.
Oil Melissa	0.5	gm.
Oil Peppermint	0.5	gm.
Oil Orange	1	gm.
Oil Cinnamon	0.5	gm.
Liqueur Body	11.5	lit.

Color: Green.

Crambambuli—A

Oil Cardamom	1	gm.
Oil Lemon	1	gm.
Oil Mace	1	gm.
Oil Cloves	0.5	gm.
Oil Orange	1	gm.
Oil Cinnamon	0.5	gm.
Liqueur Body	11.5	lit.

Color: Dark Red.

B

Oil Calamus	0.5	gm.
Oil Cardamom	1	gm.
Oil Lemon	1	gm.
Oil Fennel	0.5	gm.
Oil Nutmeg	0.5	gm.

For Chemical Advisors, Special Raw Materials, Equipment, Containers, etc., consult Supply Section at end of book.

Oil Orange	1 gm.
Oil Peppermint	2 gm.
Oil Star Anise	0.5 gm.
Liqueur Body	11.5 lit.

Color: Dark Red.

Liqueur Polonaise (Kontuszowka)

Oil Lemon	0.4 gm.
Oil Cubeb	0.5 gm.
Oil Ginger	0.5 gm.
Oil Lavender	1 gm.
Oil Mace	0.5 gm.
Oil Marjoram	0.5 gm.
Oil Juniper Berries	0.5 gm.
Oil Wormwood	0.5 gm.
Liqueur Body	11.5 lit.

Color: Brown.

Caraway Liqueur

Oil Caraway 8 gm.
Dissolved in 0.1 lit. Alcohol 95%.
Liqueur Body 11.5 lit.

Liqueur des Carmelites Romains

Oil Lemon	2 gm.
Oil Coriander	1 gm.
Oil Fennel	1 gm.
Oil Mace	0.5 gm.
Oil Melissa	0.5 gm.
Oil Peppermint	0.5 gm.
Oil Orange	2 gm.
Zedoary Plant	1 gm.
Liqueur Body	11.5 lit.

Color: Yellow-Green.

Creme Carminative

Oil Anise	1 gm.
Oil Lemon	2 gm.
Oil Coriander	1 gm.
Oil Fennel	1 gm.
Oil Caraway	1 gm.
Oil Neroli	0.5 gm.
Oil Orange	1 gm.
Oil Cinnamon	0.5 gm.
Liqueur Body	11.5 lit.

Coriander Liqueur

Oil Lemon	1.5 gm.
Oil Coriander	4 gm.
Liqueur Body	11.5 lit.

Color: Yellow.

Creme de Coriander Double

Oil Lemon	1 gm.
Oil Coriander	4 gm.
Oil Cloves	0.5 gm.
Oil Cinnamon	0.5 gm.
Liqueur Body	11.5 lit.

Color: Yellow.

Spearmint Liqueur

Oil Spearmint 8 gm.
Dissolves in 0.1 lit. Alcohol 95%.
Liqueur Body 11.5 lit.

Color: Green.

Cardamom Liqueur

Oil Anise	2 gm.
Oil Cardamom	3 gm.
Vanilla Tincture	10 gm.
Liqueur Body	11.5 lit.

No Color.

Cardamom Liqueur

Oil Cardamom	3 gm.
Oil Lemon	1 gm.
Oil Coriander	1 gm.
Liqueur Body	11.5 lit.

Color: Yellow.

Cardinal Liqueur

Amber Tincture	10 gm.
Oil Lemon	2 gm.
Oil Nutmeg	1 gm.
Oil Cloves	1 gm.
Oil Cinnamon	1 gm.
Liqueur Body	11.5 lit.

No Color.

Cardinal Essence

Amber Tincture	10 gm.
Oil Lemon	4 gm.
Oil Mace	1 gm.
Oil Cloves	0.5 gm.
Oil Peppermint	0.5 gm.
Vanilla Tincture	25 gm.
Oil Cinnamon	1.5 gm.
Liqueur Body	11.5 lit.

Color: Red-Yellow.

Liqueur des Carmelites

Oil Lemon	1 gm.
Oil Coriander	1 gm.
Oil Mace	0.4 gm.
Oil Melissa	0.4 gm.
Oil Orange	1 gm.
Liqueur Body	11.5 lit.

No Color.

Honey Liqueur

Sugar Solution	30 lit.
Alcohol 90%	30 lit.
Water	40 lit.
Honey Aroma according to strength	50–500 gm.

Color: Honey Yellow.

Jasmine de la Province

Oil Jasmine	2	gm.
Liqueur Body	11.5	lit.

Color: Brownish.

Calamus Liqueur Simple

Oil Calamus	5	gm.

Dissolved in 0.1 lit. Alcohol 95%.

Liqueur Body	11.5	lit.

Color: Yellow.

Calamus Creme

Oil Calamus	3	gm.
Oil Cinnamon	2	gm.
Liqueur Body	11.5	lit.

Color: Yellow.

Calamus Liqueur

Oil Calamus	3	gm.
Oil Cardamom	2	gm.
Oil Orange	1	gm.
Liqueur Body	11.5	lit.

Color: Red-Yellow.

Eau de Calame

Oil Calamus	4	gm.
Oil Coriander	0.5	gm.
Oil Orange	1.5	gm.
Oil Star Anise	0.5	gm.
Oil Cinnamon	1	gm.
Liqueur Body	11.5	lit.

Color: Brown-Red.

Calamus Liqueur Composed

Oil Angelica	2	gm.
Oil Calamus	4	gm.
Oil Cloves	0.5	gm.
Oil Cinnamon	0.5	gm.
Liqueur Body	11.5	lit.

Color: Light Brown.

Genevre de Hollande

Oil Cardamom	0.5	gm.
Cognac Essence	4	gm.
Oil Juniper Berries	2	gm.
Oil Wormwood	0.5	gm.
Liqueur Body	11.5	lit.

Color: Yellow.

Double Genevre

Oil Cardamom	1	gm.
Oenanthic Ether	0.5	gm.
Oil Juniper Berries	4	gm.
Oil Wormwood	1	gm.
Oil Cinnamon	0.5	gm.
Liqueur Body	11.5	lit.

Color: Dark Yellow.

Spice—Creme Liqueur

Oil Anise	0.2	gm.
Oil Cardamom	0.2	gm.
Oil Lemon	0.5	gm.
Oil Ginger	1	gm.
Oil Mace	0.5	gm.
Oil Nutmeg	0.4	gm.
Oil Neroli	0.4	gm.
Oil Cloves	0.5	gm.
Oil Wormwood	0.5	gm.
Liqueur Body	11.5	lit.

Color: Brown.

Creme de Girofle

Oil Cloves	2	gm.
Oil Mace	1	gm.
Oil Cinnamon	1	gm.
Liqueur Body	11.5	lit.

Color: Brown.

Liqueur Allemande (Grunewald)

Oil Angelica	0.5	gm.
Oil Lemon	1	gm.
Oil Cardamom	0.5	gm.
Cognac Essence	5	gm.
Oil Galgant	1	gm.
Oil Orange	1	gm.
Oil Cinnamon	0.5	gm.
Oil Wormwood	1	gm.
Liqueur Body	11.5	lit.

Color: Brownish-Green.

Eau de Pucelle

Oil Angelica	0.5	gm.
Oil Fennel	1	gm.
Oil Cloves	0.5	gm.
Oil Orange	1	gm.
Tincture Orris Root	10	gm.
Oil Juniper Berry	1	gm.
Oil Cinnamon	1.5	gm.
Liqueur Body	11.5	lit.

Color: Yellow.

Eau Royale

Essence Amber	4	gm.
Oil Lemon	1	gm.
Oil Mace	0.5	gm.
Oil Cloves	0.5	gm.
Oil Orange	2	gm.
Oil Cinnamon	1	gm.
Liqueur Body	11.5	lit.

No Color.

Eau de Valeriane

Oil Angelica	1	gm.
Oil Valerian	3	gm.
Oil Calamus	2	gm.
Oil Lemon	1	gm.
Liqueur Body	11.5	lit.

Color: Yellow.

Double Fennel Liqueur

Oil Anise	0.5 gm.
Oil Coriander	0.5 gm.
Oil Fennel	3 gm.
Oil Star Anise	0.5 gm.
Liqueur Body	11.5 lit.

Color: Yellow.

Springflower Liqueur

Oil Cardamom	0.5 gm.
Cognac Essence	5 gm.
Oil Lavender	1 gm.
Oil Mace	1 gm.
Oil Melissa	0.5 gm.
Oil Peppermint	1 gm.
Oil Orange	1 gm.
Tincture Vanilla	10 gm.
Oil Cinnamon	1 gm.
Liqueur Body	11.5 lit.

Color: Green.

Bishop Drink

Anise	20 gm.
Lemon Peels	100 gm.
Fennel	10 gm.
Peppermint Herb	60 gm.
Orange Peels	60 gm.
Savin Herb	50 gm.
Liqueur Body	11.5 lit.

Color: Light Red.

Amourette

Lemon Oil	2.5 gm.
Orange Oil	2.5 gm.
Peppermint Oil	.05 gm.
Star Anise Oil	.05 gm.
Liqueur Body	11.5 lit.

Color: Dark Red.

Flower Creme Liqueur

Jasmine Tincture	10 gm.
Vanilla Tincture	10 gm.
Oil Rose	1 gm.

Dissolve in 0.15 lit. Alcohol 95%.

Liqueur Body	11.5 lit.

No Color.

Oriental Flower Creme

Calamus	30 gm.
Lemon Peels	200 gm.
Dates	200 gm.
Fennel	100 gm.
Cloves	30 gm.
Orange Peel	300 gm.
Cinnamon	30 gm.
Alcohol 95%	0.6 kilo
Liqueur Body	11.5 lit.

No Color.

Flower Essence

Vanilla Tincture	200 gm.
Rose Oil	5 gm.
Alcohol 70%	10 kilos

Color: Rose Red.

Bouquet des Dames

Oil Cloves	0.5 gm.
Oil Mace	0.5 gm.
Oil Rose	1 gm.
Oil Cinnamon	0.5 gm.
Liqueur Body	10 lit.

Polish Brandy—"A"

Raisins	280 gm.
Licorice	35 gm.
Cinnamon	25 gm.
Cardamom	25 gm.
Cloves	8 gm.
Galgant	8 gm.
Ammonia Rubber	8 gm.
Anise Seed	8 gm.
Coriander	8 gm.
Alcohol 60%	3 lit.

Extracted for few days, pressed, filtered and mixed with sugar, the last to be dissolved in rose water.

"B"

Rosemary	70 gm.
Calamus	8 gm.
Anise	8 gm.
Raisins	280 gm.
Pepper	50 gm.
Caraway	110 gm.
Ginger	110 gm.
Corn Brandy	4 lit.

Manufacture as in "A."

Trester Brandy

Oil Cognac, Genuine	4 oz.
Oil Corn Fusel	5 oz.
Methyl Salicylate	3 oz.
Acetic Ether, Absolute	2 lb. 8 oz.
Alcohol	24 pt.
Water	3 pt. 12 oz.

Filter through magnesium carbonate.

Blackberry Essence

Nutmeg Essence	10 gm.
Coriander Essence	10 gm.
Cinnamon Essence	10 gm.
Blackberry Ether	100 gm.
Blackberry Juice	170 gm.
Alcohol 60%	500 cm.

Breadwater Liqueur

Lemon Peel	1750 gm.
Cloves	100 gm.
Cinnamon Ceylon	100 gm.

Mace	50 gm.
Coriander	50 gm.
Anise	25 gm.
Alcohol 90%	10 lit.

This mixture has to be extracted for 36 hours, then add 5 lit. water, distill off 12 lit., then add tincture manufactured as follows: 7.5 kilos toasted pumpernickel, 12.5 lit. 75% alcohol—then add 18 lit. alcohol 90% and 25 kilos sugar and enough water to make it 100 kilos. Color with Caramel.

Train Liqueur Oil

Oil Peppermint	15 gm.
Oil Star Anise	15 gm.
Oil Cloves	10 gm.
Oil Calamus	3 gm.
Oil Juniper Berry	3 gm.
Oil Coriander Seed	3 gm.
Oil Bitter Orange	2 gm.
Raisin Ether	5 gm.
Acetic Ether	5 gm.
Violet Flower Essence	3 gm.

To be distilled over burnt magnesia.

Train Liqueur

Train Liqueur Oil	35 gm.
Pineapple Ether	150 gm.
Raisin Essence	150 gm.
Vanilla Essence	50 gm.
Violet Flower Essence	30 gm.
Alcohol 95%	20 lit.

This mixture has to stay 48 hours.

Add:

Water	20 lit.
Alcohol 95%	5 lit.
Sugar dissolved in 5 lit. Water	8 kilos

Color with Caramel.

Fig Fruit Liqueur

Fresh Figs	5 kilos
Water	2–3 lit.
Alcohol 95%	4–6 lit.
Sugar	2 kilos

Spring Flower Liqueur

Oil Cardamom	0.5 gm.
Cognac Essence	5 gm.
Oil Lavender	1 gm.
Oil Mace	1 gm.
Oil Melissa	0.5 gm.
Oil Peppermint	1 gm.
Oil Orange	1 gm.
Vanilla Tincture	10 gm.
Oil Cinnamon	1 gm.
Liqueur Body	11.5 lit.

Greek Water

Angelica Root	20 gm.
Calamus Root	40 gm.
Cardamom	20 gm.
Mace	20 gm.
Cloves	20 gm.
Bitter Almond	80 gm.
Wormwood	20 gm.
Cinnamon	20 gm.
Alcohol 95%	0.6 lit.

Color: Red Violet.

Greek Bitter Essence

Cinnamon Flowers	50 gm.
Caraway Seed	200 gm.
Peppermint Herb	200 gm.
Orange Peels	150 gm.
Angelica Root	150 gm.
Gentian	200 gm.
Alcohol 70%	4 lit.

Sky Water

Oil Anise	1.5 gm.
Cardamom Oil	0.5 gm.
Oil Coriander	0.5 gm.
Oil Cloves	0.5 gm.
Oil Neroli	1 gm.
Oil Cinnamon	1 gm.
Liqueur Body	11.5 lit.

Hunting Liqueur—"A"

Cassia	450 gm.
Ginger	125 gm.
Galgant Root	125 gm.
Cardamom	75 gm.
Cloves	75 gm.

Extract with 10 lit. Alcohol 60%.

In 5 lit. Alcohol you dissolve

Oil Cinnamon Ceylon	5 gm.
Oil Cubeb	5 gm.
Oil Mace	5 gm.
Oil Coriander	3 gm.

Add this to above mixture after it is filtered. Then add 25 lit. 95% alcohol and 12.5 kilos sugar dissolved in boiling water and complete with water sufficient to make 100 lit.

Color: Light Green.

"B"

Oil Caraway	12 gm.
Oil Anise Russian	15 gm.
Oil Star Anise	15 gm.
Oil Wormwood	5 gm.
Oil Ginger	2.5 gm.
Oil Coriander	2.5 gm.
Oil Peppermint	0.5 gm.
Oil Rum Essence	0.5 gm.
Liqueur Body	10 lit.

For Chemical Advisors, Special Raw Materials, Equipment, Containers, etc., consult Supply Section at end of book.

Hunters Water—"A"

Oil Mace	1	gm.
Oil Peppermint	2	gm.
Liqueur Body	11.5	lit.

No Color or Green.

Cherry Water Black Forest

Alcohol 94%	74	lit.
Nut Essence	20	gm.

(12 parts oil bitter almond in 88 parts Alcohol 95%.)

Orange Flower Water	2	lit.
Water	124	lit.

Coast Water

Oil Bergamot	1	gm.
Oil Lemon	2	gm.
Oil Peppermint	.05	gm.
Oil Cinnamon	.05	gm.
Liqueur Body	11.5	lit.

Color: Yellow.

Mogador Essence—No. 1

Wormwood	500	gm.
Peppermint	250	gm.
Cassia	65	gm.
Cubeb	18	gm.
Pimento	65	gm.
Cloves	65	gm.
Galgant	65	gm.
Oranges	65	gm.
White Cinnamon	65	gm.
Orange Peels	135	gm.
Chamomile	135	gm.
Alcohol 90%	8.5	lit.
Water	3.5	lit.

No. 2

Oil Wormwood	10	gm.
Oil Calamus	25	gm.
Oil Cloves	25	gm.
Oil Mace	35	gm.
Oil Orange	70	gm.
Oil Lemon	50	gm.
Oil Savin	12	gm.
Oil Cinnamon Genuine	8	gm.
Oil Thyme	25	gm.
Oil Lavender	12	gm.
Wine Spirit Essence	50	gm.
Orange Ether	50	gm.
Acetic Ether	50	gm.
Alcohol	4.6	lit.

Color: Golden Yellow.

Non Pareille

Oil Mace	1	gm.
Oil Cloves	1	gm.
Oil Rose	0.5	gm.
Liqueur Body	11.5	lit.

Nordhauser Corn Liqueur

Acetic Ether	10	gm.
Oil Juniper Berry	0.5	gm.
Oil Cinnamon	0.5	gm.
Liqueur Body	11.5	lit.

Double Nordhauser Corn Liqueur

Acetic Ether	15	gm.
Raspberry Ether	10	gm.
Oil Mace	0.5	gm.
Oil Cloves	0.2	gm.
Oil Cinnamon	0.5	gm.
Liqueur Body	11.5	lit.

No Color.

Polka Liqueur Oil

Oil Anise Russian	150	gm.
Oil Fennel	150	gm.
Oil Cloves	150	gm.
Oil Cinnamon	150	gm.
Oil Rosemary	75	gm.
Oil Chamomile Roman	75	gm.
Oil Angelica	50	gm.
Oil Spearmint	50	gm.

Creme de Flauve d'Orange (Orange Liqueur)

Alcohol by Volume 95%	4¼	lit.
Neroli Oil Essence	132	gm.
Bitter Oil Almond Ess.	175	gm.
Sugar Syrup 65%	11¼	lit.
Water	1½	lit.

Creme de Framboise (Strawberry)

Alcohol by Volume 95%	2	lit.
Raspberry Juice	7	lit.
Sugar Syrup 65%	8½	lit.

Creme de Vanille (Vanilla)

Alcohol by Volume 95%	3½	lit.
Vanilla Extract	4	lit.
Raspberry Juice	½	pt.
Sugar Syrup 65%	11½	lit.
Water	3	lit.
Caramel Color	65	gm.

Creme de Canelle

Alcohol by Volume 95%	5	lit.
Cinnamon Oil Essence	45	gm.
Water	12	lit.

Color with Caramel.

Creme d'Ananas (Pineapple)

Alcohol by Volume 95%	5	lit.
Pineapple Essence	100	gm.
Sugar Syrup 65%	11½	lit.
Water	½	lit.

Chartreuse Liqueur

Alcohol by Volume 90%	12½	lit.
Chartreuse Essence	5	gm.
Sugar Syrup 65%	4½	lit.
Water	12½	lit.

Wherever the word essence appears you take one part of the essential oil and mix thoroughly with 7 parts of 95% alcohol and these mixtures or solutions constitute the essences as given in the formulas.

Details for preparation of extracts and essences for some of the above formulas.

Lemon Essence

Alcohol by Volume 95%	2⅓	lit.
Lemon Juice	2⅓	lit.

Mix the alcohol and lemon juice and then filter.

Spanish Bitter Oil Essence

Angelica Oil	160	gm.
Bitter Almond Oil	10	gm.
Lemon Oil	80	gm.
Calamus	160	gm.
Spearmint Oil	160	gm.
Coriander Oil	20	gm.
Clove Oil	40	gm.
Oil Bitter Orange	320	gm.
Wormwood Oil	160	gm.
Juniper Oil	160	gm.
Cinnamon Oil	40	gm.

Curacao Essence

Bitter Orange Oil	640	gm.
Neroli Oil	27	gm.
Orange Peel Sweet Oil	27	gm.
Cinnamon Oil	13½	gm.

Rhine Wine Extract

Alcohol by Volume 90%	3⅓	lit.
Strawberry Oil	75	gm.
Orange Peel Oil	50	gm.
Pineapple Essence	20½	lit.
Woodruff Extract	100	gm.
Neroli Oil	48	drops

Color: Yellow.

Polish Water

Currants	185	gm.
Anise	30	gm.
Cinnamon	30	gm.
Cloves	30	gm.
Fennel	30	gm.
Peppermint	30	gm.
Galgant	20	gm.

These drugs have to be crumbled and extracted for 14 days with 18 lit. alcohol 95% and 4 lit. rosewater, then press out the liquid. Add 10 lit. water and 10 lit. sugar syrup, then filter.

Rosemary Liqueur Essence

Oil Rosemary	80	gm.
Oil Lemon	30	gm.
Oil Neroli	10	gm.
Oil Mace	0.5	gm.
Acetic Ether	50	gm.
Raisin Essence	100	gm.
Violet Flower Essence	100	gm.
Alcohol 95%	2	lit.

Celery Liqueur

Oil Anise	0.5	gm.
Oil Bitter Almond	0.5	gm.
Oil Coriander	0.5	gm.
Oil Caraway	0.5	gm.
Oil Celery Seed	1	gm.
Liqueur Body	11.5	lit.

Swiss Creme

Oil Angelica	0.5	gm.
Oil Bitter Almond	1	gm.
Oil Calamus	0.5	gm.
Oil Cardamom	0.5	gm.
Oil Cloves	0.2	gm.
Oil Peppermint	0.2	gm.
Oil Rosemary	0.2	gm.
Oil Thyme	0.5	gm.
Oil Juniper Berry	0.5	gm.
Oil Wormwood	0.5	gm.
Vanilla Tincture	6	gm.
Oil Cinnamon	0.5	gm.
Liqueur Body	11.5	lit.

Color: Green.

Date Fruit Liqueur

Dates (Squashed)	4	kg.
Water	4.6	lit.
Alcohol 95%	4.6	lit.
Sugar	4	kg.

Strawberry Fruit Liqueur—No. 1

Strawberries	6.8	kg.
Sugar	4.5	kg.
Alcohol 95%	4.6	lit.
Water	3.4	lit.

According to taste 50 gm. Mace or Cinnamon Tincture.

No. 2

Strawberries	5	kg.
Alcohol 95%	5	lit.
Sugar in 3 lit. Water	2.5	kg.

According to taste a little Vanilla.

Fig Fruit Liqueur

Figs (Fresh)	5	kg.
Water	2.28	lit.

For Chemical Advisors, Special Raw Materials, Equipment, Containers, etc., consult Supply Section at end of book.

Alcohol	4.6 lit.
Sugar	2 kg.

Raspberry Fruit Liqueur

6 kg. Raspberries (squashed) are to be extracted with 2 lit. Water and 5 lit. Alcohol 95%. Shake daily for 14 days.

Raspberry Fruit Liqueur

Fresh Pressed Raspberry Juice	10 lit.
Alcohol 95%	8 lit.
Sugar	6 kg.

Cherry Fruit Liqueur

Cherries	5.5 kg.
Sugar	3 kg.
Alcohol 95%	4.6 lit.
Water	1.1 lit.
Bitter Almond Tincture	50 gm.

Very sweet dark cherries very ripe, have to be squashed including the pits, in a stone mortar. The mash has to stand a few days in a cool place then press it out and add the sugar and water and heat until it boils. After it cools off add Bitter Almond Tincture and alcohol.

Color: Dark Red.

Peach Fruit Liqueur

Peaches	6 kg.
Sugar	4.5 kg.
Alcohol 95%	4.6 lit.
Water	1.7 lit.
Bitter Almond Tincture	16 gm.

The fruits skinned and pits removed. Then to be squashed and ex-pressed. To the residue 1.7 lit. water added together with the stamped pits. This mash remains for 2 days then press it. Dissolve sugar in those liquids, add Bitter Almond Tincture. No heating.

Color: Pale Red.

Orange Fruit Liqueur

Fresh Orange Juice	1.1 lit.
Alcohol 95%	4.6 lit.
Water	3.4 lit.
Sugar	4.5 kg.
Curacao Tincture	100–200 gm.

The fruit juice stays with alcohol for 8 days, then filter. The clear liquid has to be mixed with a Sugar Solution then add carefully the Curacao Tincture to avoid bitter taste.

Color: Golden Yellow.

Apple Fruit Ether

Ethyl Acetate	10 gm.
Amyl Valeriate	100 gm.
Cold Saturated Malic Acid Solution	10 gm.
Alcohol 95%	1000 gm.

Pineapple Fruit Ether

Ethyl Butyrate	50 gm.
Amyl Butyrate	100 gm.
Alcohol 95%	1000 gm.

Apricot Fruit Ether

Ethyl Butyrate	100 gm.
Ethyl Valeriate	50 gm.
Oil Bitter Almond	10 gm.
Alcohol 95%	1000 gm.

Pear Fruit Ether

Ethyl Acetate	50 gm.
Amyl Acetate	100 gm.
Alcohol 95%	1000 gm.

Strawberry Fruit Ether

Ethyl Acetate	50 gm.
Ethyl Formate	10 gm.
Ethyl Butyrate	50 gm.
Ethyl Salicylate	10 gm.
Amyl Acetate	30 gm.
Amyl Butyrate	20 gm.
Alcohol 95%	1000 gm.

Current Fruit Ether

Ethyl Acetate	50 gm.
Ethyl Formate	10 gm.
Ethyl Butyrate	10 gm.
Ethyl Benzoate	10 gm.
Ethyl Oenanthate	10 gm.
Ethyl Salicylate	10 gm.
Ethyl Sebaciate	10 gm.
Amyl Butyrate	10 gm.
Cold Saturated Solution of Tartaric Acid	50 gm.
Alcohol 95%	1000 gm.

Cherry Fruit Ether

Ethyl Acetate	50 gm.
Ethyl Benzoate	50 gm.
Oil Bitter Almond	10 gm.
Cold Saturated Solution Benzoic Acid in Alcohol 95%	10 gm.
Alcohol 95%	1000 gm.

Melon Fruit Ether

Ethyl Formate	20 gm.
Ethyl Butyrate	40 gm.
Ethyl Valeriate	50 gm.
Ethyl Sebaciate	100 gm.
Alcohol 95%	1000 gm.

Orange Fruit Ether

Ethyl Acetate	50 gm.
Ethyl Formate	10 gm.
Ethyl Butyrate	10 gm.
Ethyl Benzoate	10 gm.
Methyl Salicylate	10 gm.
Amyl Acetate	10 gm.
Orange Flower Oil	100 gm.
Cold Saturated Solution of Tartaric Acid in Alcohol	10 gm.
Alcohol 95%	1000 gm.

Peach Fruit Ether

Ethyl Acetate	50 gm.
Ethyl Formate	50 gm.
Ethyl Butyrate	50 gm.
Ethyl Valeriate	50 gm.
Ethyl Sebaciate	10 gm.
Oil Bitter Almond	50 gm.
Alcohol 95%	1000 gm.

Grape Fruit Ether

Ethyl Formate	20 gm.
Ethyl Oenanthate	100 gm.
Methyl Salicylate	10 gm.
Cold Saturated Solution of Tartaric Acid in Alcohol	50 gm.
Succinic Acid	30 gm.
Alcohol 95%	1000 gm.

Goldwasser Essence

Angelica Oil	4 gm.
Anise Oil	32 gm.
Lemon Oil	290 gm.
Spearmint Oil	32 gm.
Laurel Oil	32 gm.
Lavender Oil	64 gm.
Nutmeg Oil	16 gm.
Balm Oil	20 gm.
Clove Oil	64 gm.
Orange Oil	16 gm.
Rose Oil	16 gm.
Rosemary Oil	32 gm.
Juniper Oil	32 gm.

Curacao Essence

Bitter Orange Oil	640 gm.
Neroli Oil	27 gm.
Sweet Orange Oil	27 gm.
Cinnamon Oil	13½ gm.

Rhine Wine Extract

Mix together:

Alcohol 90%	3½ lit.
Strawberry Oil	75 gm.
Orange Oil	50 gm.
Pineapple Essence	20½ lit.
Woodruff Essence	100 gm.
Neroli Oil	48 drops

Color: Slightly Yellow.

Birch Oil Spirit

Alcohol 90%	¼ lit.
Oil Birch	5 gm.

Lamp Black Tincture

Lamp Black	17 gm.
Alcohol 90%	¼ lit.

Vanilla Extract

Chopped Vanilla Bean	8 gm.
Alcohol 90%	¼ lit.

Weichxel Fruit Ether

Ethyl Acetate	100 gm.
Ethyl Benzoate	50 gm.
Oil Bitter Almond	20 gm.
Cold Saturated Solution Malic Acid in Alcohol	10 gm.
Benzoic Acid	30 gm.
Alcohol 95%	1000 gm.

Lemon Fruit Ether

Ethyl Acetate	100 gm.
Oil Lemon	100 gm.
Cold Saturated Citric Acid Solution	100 gm.
Alcohol 95%	1000 gm.

Mulled Wine Extract

Sugar	47½ kg.
Water	14 lit.
Cherry Juice	8 lit.
Raspberry Juice	2¼ lit.

Cook the above together and then add:

Alcohol 90%	37½ lit.
Clove Essence	⅙ lit.
Cinnamon Essence	⅙ lit.
Moselle Wine	11½ lit.

Color: Dark Cherry.

Orange Lemonade

Sugar Syrup 65%	45 lit.
Alcohol 90%	4 lit.
Citric Acid dissolved in 1 lit. water	750 gm.
Orange Oil Essence	1½ lit.

Lemon Lemonade

Sugar Syrup 65%	45 lit.
Alcohol 90%	4 lit.
Lemon Oil Essence	1½ lit.
Citric Acid dissolved in 1 lit. water	750 gm.

Strawberry Lemonade

Sugar Syrup 65%	30 lit.
Alcohol 90%	3 lit.

Strawberry Ether	25 gm.
Citric Acid dissolved in 1 lit. water	750 gm.

Color: Strawberry.

Raspberry Lemonade

Sugar with	75 kg.
Raspberry Juice	31 lit.
Cherry Juice	10 lit.
Water	7½ lit.
Tartaric Acid	1½ kg.

Heat together juices and sugar; then dissolve acid in water and then mix all together.

Champagne

Rhine Wine	32 lit.
Whole Lemons and peels cut up	4
Raisins	2 kg.
Orange Oil Essence	30 gm.
Oil of Neroli	10 drops
Sugar	8 kg.
Water	2 lit.

Cherry Lemonade

Cherry Juice cooked with	17 lit.
Sugar and	12½ kg.
Tartaric Acid dissolved in ¼ lit. water	125 gm.

Cardinal Wine

Moselle Wine	52 lit.
Alcohol 90%	7 lit.
Sugar Syrup 65%	11 lit.

Flavor with Cardinal Extract and make acid with Tartaric.

Bischof Wine

Red Wine	54 lit.
Alcohol 90%	6 lit.
Sugar Syrup 65%	10 lit.

Flavor with Bischof Extract and make acid with Tartaric.

Cardinal Extract

Alcohol 95%	8 lit.
Orange Blossom Oil	416 gm.
Sweet Orange Peel	266 gm.
Water	1⅛ lit.
Caramel Color	⅛ lit.

Bischof Extract

Cardinal Extract	3 lit.
Orange Peels	100 gm.
Bitter Orange Oil	100 gm.

Vanilla Spirit

500 gm. vanilla bean percolate with 42 lit. 90% alcohol, and 5 lit. water and distill over 34 lit.

Lemon Essence

Lemon Juice	2⅓ lit.
Alcohol 90%	2⅓ lit.

Mix the above and filter.

Maraschino Liqueur

Alcohol 90%	20 lit.
Maraschino Essence	625 gm.
Concentrated Raspberry Ether	100 gm.
Sugar Syrup 65%	15 lit.
Water	15 lit.

Orange Liqueur

Alcohol 90%	20 lit.
Orange Essence	875 gm.
Sugar Syrup 65%	7 lit.
Water	23 lit.

Clove Bitters

Alcohol 90%	20 lit.
Clove Bitter Essence	1½ lit.
Sugar Syrup 65%	4 lit.
Water	24½ lit.

Rose Liqueur

Alcohol 90%	20 lit.
Rose Essence	350 gm.
Sugar Syrup 65%	12½ lit.
Water	17½ lit.

Rum

Alcohol 90%	25 lit.
Jamaica Rum Essence	600 gm.
Butyric Ether	15 gm.
Water	25 lit.

Mix well, color well and let stand.

Maraschino

Alcohol 90%	20 lit.
Oil of Bitter Almonds	35 gm.
Sugar Syrup 65%	15 lit.
Water	15 lit.

Cloves

Alcohol 90%	20 lit.
Oil of Clove	25 gm.
Sugar Syrup 65%	5 lit.
Water	25 lit.

Color: Light Brown.

Hamburger Bitter Extract

Galgant Root	3000	gm.
Oak Bark	125	gm.
Ginger Root	150	gm.
Orris Root	375	gm.
Gentian	1000	gm.
Alcohol by Volume 90%	8½	lit.

Digest the above in alcohol for 5 days after which add 8½ lit. of water and let stand for 8 days. Then draw off the clear liquid and add to this the following solution:

Alcohol by Volume 90%	3	lit.
Oil of Bay	75	gm.
Oil Cinnamon	600	gm.
Oil Nutmeg	25	gm.
Oil Cloves	6	gm.
Ethyl Acetate	75	gm.
Oil Calamus	730	gm.

Hamburger Drops

Alcohol by Volume 90%	21½	lit.
Hamburger Bitter Essence	550	gm.
Sugar Solution	10½	lit.
Water	23	lit.

Color: Brown with Caramel Color.

Raspberry

Cook together the following:

Sugar	7½	kg.
Clear Raspberry Juice	7½	lit.
Cherry Juice	2½	lit.

Cool this and add to it the following mixture:

Raspberry Syrup	12½	lit.
Alcohol by Volume 90%	17½	lit.
Aromatic Essence	400	gm.
Water	20	lit.

Raspberry-Lemonade

Sugar Syrup	30	lit.
Raspberry-Lemonade Essence	1½	kg.
Citric Acid	250	gm.

Mix well and color Red.

TINCTURES

Amber Tincture

Amber Grease Gray	40	gm.
Alcohol 95%	600	gm.

Pineapple Tincture

Pineapple Ether	160	gm.
Alcohol 95%	10	kg.

Angelica Tincture

Angelica Root	1500	gm.
Alcohol 95%	10	kg.

Angostura Tincture

Angostura Bark	1000	gm.
Alcohol 95%	10	kg.

Anise Tincture

Anise Squashed	1600	gm.
Alcohol 95%	10	kg.

Arrac Tincture

Cognac Ether	1800	gm.
Alcohol 95%	10	kg.

Valerian Tincture

Valerian Root	2500	gm.
Alcohol 95%	10	kg.

Basil Tincture

Basil Leaves	6	kg.
Alcohol 95%	10	kg.

Benzoin Tincture

Benzoin Tears	600	gm.
Alcohol 95%	10	kg.

Bergamot Tincture

Bergamot Peels	3	kg.
Alcohol 95%	10	kg.

Castoreum Tincture

Castoreum	50	gm.
Alcohol 95%	500	gm.

Curacao Tincture

Curacao Peels	3	kg.
Alcohol 95%	10	kg.
or		
Orange Peels Fresh Green	3	kg.
Alcohol 95%	10	kg.

Fennel Tincture

Fennel Squashed	1600	gm.
Alcohol 95%	10	kg.

Orris Tincture

Orris Root Florentine	500	gm.
Alcohol 95%	10	kg.

Jasmine Tincture

Jasmine Flowers	5 kg.
Alcohol 95%	10 kg.

Coffee Tincture

Coffee Fresh, Burnt, Ground	4 kg.
Alcohol 95%	10 kg.

Calamus Tincture

Calamus Root	2500 gm.
Alcohol 95%	10 kg.

Cardamom Tincture

Cardamom	600 gm.
Alcohol 95%	10 kg.

Cassia Tincture

Cassia Cinnamon	1 kg.
Alcohol 95%	10 kg.

Catechu Tincture

Catechu	1500 gm.
Alcohol 95%	10 kg.

Spearmint Tincture

Spearmint Dry	1 kg.
Alcohol 95%	10 kg.

Caraway Tincture

Caraway Seed Squashed	1 kg.
Alcohol 95%	10 kg.

Lavender Tincture

Lavender Dry	1 kg.
Alcohol 95%	10 kg.

Mace Tincture

Mace	800 gm.
Alcohol 95%	10 kg.

Marjoram Tincture

Marjoram Dry	1 kg.
Alcohol 95%	10 kg.

Melissa Tincture

Melissa	2500 gm.
Alcohol 95%	10 kg.

Musk Tincture

Musk Tonquin	1 gm.
Alcohol 95%	1 lit.

Rosemary Tincture

Rosemary	1500 gm.
Alcohol 95%	10 kg.

Sage Tincture

Sage Dry	2500 gm.
Alcohol 95%	10 kg.

Celery Tincture

Celery Seeds	200 gm.
Alcohol 95%	10 kg.

Cocoa Tincture—No. 1

Cocoa	2 kg.
Alcohol 95%	10 kg.

No. 2

Cocoa Deoiled	1500 gm.
Alcohol 95%	10 kg.

Star Anise Tincture

Star Anise Crushed	1600 gm.
Alcohol 95%	10 kg.

Nutmeg Tincture

Nutmegs Pulverized	1 kg.
Alcohol 95%	10 kg.

Clove Tincture

Cloves	1500 gm.
Alcohol 95%	10 kg.

Nut Tincture

Nuts (Green-Soft)	1 kg.
Alcohol 95%	10 kg.

Peru Balsam Tincture

Balsam Peru	70 gm.
Alcohol 95%	1 kg.

Peppermint Tincture

Peppermint	1 kg.
Alcohol 95%	10 kg.

Orange Tincture

Orange Peels	4 kg.
Alcohol 95%	10 kg.

Rose Tincture

Rose Leaves Salted	1500 gm.
Alcohol 95%	10 kg.

Thyme Tincture

Thyme	2500 gm.
Alcohol 95%	10 kg.

Tonka Bean Tincture

Tonka Beans Crushed	1 kg.
Alcohol 95%	10 kg.

Vanilla Tincture

Vanilla Crushed	75 gm.
Alcohol 95%	10 kg.

Orris Root Tincture

Orris Root Crushed	1 kg.
Alcohol 95%	10 kg.

Juniper Berry Tincture

Juniper Berries	2500 gm.
Alcohol 95%	10 kg.

Woodruff Tincture—No. 1

Woodruff Fresh	2500 gm.
Alcohol 95%	10 kg.

No. 2

Woodruff Dry	1800 gm.
Alcohol 95%	10 kg.

Wormwood Tincture

Wormwood Dry	1 kg.
Alcohol 95%	10 kg.

Civet Tincture

Civet	40 gm.
Alcohol 95%	600 gm.

Cinnamon Tincture

Cinnamon (Fine Pulverized)	1 kg.
Alcohol 95%	10 kg.

Lemon Tincture

Lemon Peels of 80–100 fresh lemons to 10 kg. Alcohol 95%.

OIL TINCTURES

Angelica Oil Tincture

Oil Angelica	40 gm.
Alcohol 95%	10 kg.

Anise Oil Tincture

Oil Anise	30–40 gm.
Alcohol 95%	10 kg.

Bergamot Oil Tincture

Oil Bergamot	40 gm.
Alcohol 95%	10 kg.

Bitter Almond Oil Tincture

Oil Bitter Almond	100–150 gm.
Alcohol 95%	10 kg.

Fennel Oil Tincture

Oil Fennel	70–80 gm.
Alcohol 95%	10 kg.

Raspberry Ether Tincture

Raspberry Ether	1 kg.
Alcohol 95%	10 kg.

Calamus Oil Tincture

Oil Calamus	50–70 gm.
Alcohol 95%	10 kg.

Cassia Oil Tincture

Oil Cassia	100–150 gm.
Alcohol 95%	10 kg.

Coriander Oil Tincture

Oil Coriander	70–100 gm.
Alcohol 95%	10 kg.

Spearmint Oil Tincture

Oil Spearmint	50–60 gm.
Alcohol 95%	10 kg.

Caraway Oil Tincture

Oil Caraway	50–60 gm.
Alcohol 95%	10 kg.

Lavender Oil Tincture

Oil Lavender	120–150 gm.
Alcohol 95%	10 kg.

Mace Oil Tincture

Oil Mace	40–70 gm.
Alcohol 95%	10 kg.

Marjoram Oil Tincture

Oil Marjoram	140–160 gm.
Alcohol 95%	10 kg.

Melissa Oil Tincture

Oil Melissa	40–60 gm.
Alcohol 95%	10 kg.

Nutmeg Oil Tincture

Oil Nutmeg	40–60 gm.
Alcohol 95%	10 kg.

Clove Oil Tincture

Oil Cloves	50–60 gm.
Alcohol 95%	10 kg.

Neroli Oil Tincture

Oil Neroli	60 gm.
Alcohol 95%	10 kg.

Oenanthic Tincture

Ethyl Oenanthate	20 gm.
Alcohol 95%	10 kg.

Peppermint Oil Tincture

Oil Peppermint	50–60 gm.
Alcohol 95%	10 kg.

Rose Oil Tincture

Oil Rose	50 gm.
Alcohol 95%	10 kg.

Rosemary Oil Tincture

Oil Rosemary	160–200 gm.
Alcohol 95%	10 kg.

Sage Oil Tincture

Oil Sage	50–60 gm.
Alcohol 95%	10 kg.

Celery Oil Tincture

Oil Celery	50–60 gm.
Alcohol 95%	10 kg.

Star Anise Tincture

Oil Star Anise	50–60 gm.
Alcohol 95%	10 kg.

Lemon Oil Tincture

Oil Lemon	60–80 gm.
Alcohol 95%	10 kg.

Simple Tinctures

Anise:	750 gm.	Aniseseed	4½ lit. Alcohol
Angelica:	750 gm.	Angelica Root	7 lit. Alcohol
Lemon:	1 kg.	Lemon Peel	4½ lit. Alcohol
Calamus:	1 kg.	Calamus Root	7 lit. Alcohol
Strawberry:	58 lit.	Ripe Berries	14 lit. Alcohol
Raspberry:	58 lit.	Raspberries	14 lit. Alcohol
Coffee:	750 gm.	Roasted Coffee	9 lit. Alcohol
Cherry:	58 lit.	Sour Ripe Cherries	14 lit. Alcohol
Kummel:	750 gm.	Caraway Seeds	4½ lit. Alcohol
Balm Mint:	750 gm.	Balm Mint	9 lit. Alcohol
Nutmeg:	875 gm.	Nutmeg	4½ lit. Alcohol
Cloves:	750 gm.	Cloves	4½ lit. Alcohol
Peppermint:	3 kg.	Peppermint Leaves	14 lit. Alcohol
Orange:	1 kg.	Orange Peel	4½ lit. Alcohol
Quassia:	375 gm.	Quassia	4½ lit. Alcohol
Juniper:	750 gm.	Juniper Berries	9 lit. Alcohol

Mixed Tinctures

Cardamom:	750 gm. Cardamom Seeds, 400 gr. Anise Seed	7 lit. Spirit
Nutmeg:	200 gm. Nutmeg, 25 gm. Nutmeg Leaves, 50 gr. Cinnamon	4½ lit. Spirit

Absinthe Extract

Wormwood	500 gm.
Green Anise Seed	500 gm.
Star Anise	125 gm.
Fennel Seed	35 gm.
Coriander Seed	35 gm.
Nutmeg Leaves	20 gm.
Cinnamon	5 gm.
Alcohol	7 lit.
Water	3½ lit.

Allow to soak for 8 days. It is then distilled over.

ARTIFICIAL WINE FLAVORS

Claret Essence

Ambergris Tincture	¼ dr.
Ethyl Acetate	3¾ dr.
Carob Tincture	8½ oz.
Cherry Juice	7¼ oz.
Krameria Tincture	4 lb.
Wine Distillate	5 lb.

White Wine Essence

Cognac Oil	10 dr.
Ethyl Nitrite	22 dr.
Ethyl Acetate	1½ oz.

St. Johns Bread Tincture	12½ oz.
Wine Distillate	4½ lb.
Water	4½ lb.

Port Wine Essence

Ambergris Tincture	¼ dr.
Ethyl Acetate	7¾ dr.
Krameria Tincture	1½ oz.
Elder Flower Tincture	2 oz.
St. Johns Bread Tincture	3 oz.
Carob Tincture	3 oz.
Cacao Essence	3 oz.
Wine Distillate	3 oz.

Claret Lemonade

Clove Tincture	3 dr.
Cinnamon Tincture	5 dr.
Claret Essence	2 oz.
Cherry Juice	5½ oz.
Red Wine	8 oz.

Muscatel Lemonade

Honey Lemonade Essence	½ oz.
Claret Essence	2 oz.
Port Wine Essence	3 oz.
Grape Essence	10½ oz.

Nectar Lemonade

Honey Lemonade Essence	¼ oz.
Rum Essence	1¾ oz.
Port Wine Essence	3 oz.
Currant Essence	3 oz.
Apple Essence	8 oz.

WINE FONDANT FLAVORS

Burgundy Fondant

Ambergris Tincture	½ oz.
Rhatany Tincture	3½ oz.
Cherry Juice	4 oz.
Raspberry Essence	8 oz.
Black Currant Essence	1 lb.
Grape Essence	8 lb.

Claret Fondant

Civet Tincture	½ oz.
Ambergris Tincture	½ oz.
Rhatany Tincture	7 oz.
Black Currant Essence	1 lb.
Cherry Juice	2 lb.
Grape Essence (from dried grapes)	6½ lb.

Madeira Fondant

Pineapple Essence	4 oz.
Brown Cacao Essence	8 oz.
Elder Flower Tincture	8 oz.
Black Currant Essence	12 oz.
Grape Essence	8 lb.

Malaga Fondant

Civet Tincture	5 gm.
Ambergris Tincture	5 gm.
Vanillin	5 gm.
Cherry Water Genuine	9 gm.
Rhatany Tincture	6½ oz.
Black Currant Essence	1 lb.
Carob Tincture	1 lb.
Grape Essence	7½ lb.

Muscatel Essence

Coumarin	¼ oz.
Mace Tincture	1¾ oz.
Elder Flower Essence	6 oz.
Apple Essence	1½ lb.
Grape Essence	8 lb.

Port Fondant

Vanillin	¼ oz.
Ambergris Tincture	¼ oz.
Brown Cacao Essence	7½ oz.
Rhatany Tincture	8 oz.
Grape Essence	9 lb.

Rhine Wine Fondant

White Cognac Oil	2½ oz.
Heliotropin	3½ oz.
Ethyl Acetate	10 gm.
Apple Essence	1 lb. 7 oz.
Grape Essence	8½ lb.

Sherry Fondant

Civet Tincture	¼ oz.
Elder Flower Tincture	2¾ oz.
Black Currant Essence	13 oz.
Pineapple Essence	1 lb.
Grape Essence	8 lb.

Tokay Fondant

Civet Tincture	2½ dr.
Pineapple Essence	½ lb.
Raspberry Essence	½ lb.
Carob Tincture	1 lb.
Grape Essence	8 lb.

Pear Essence
(1 oz. per gallon)

Vanillin	½ dr.
Amyl Acetate	1½ dr.
Raspberry Distillate	5 oz.
Bergamot Essence	11 oz.
Orange Flower Water	1 lb.
Wine Brandy	1 lb.
Distilled Water	2½ lb.
Alcohol	4½ lb.

Burgundy Wine Punch Extract

Vanilla Essence	1	oz.
Lemon Juice	¼	gal.
Rum	¼	gal.
Arrac	¼	gal.
Water	2¼	gal.
Genuine Burgundy Wine	3	gal.
Sugar Syrup	4	gal.

Claret Punch Extract

Cardamom Tincture	1	oz.
Cinnamon Tincture	3	oz.
Clove Tincture	3	oz.
Lemon Juice	1	lb.
Genuine Rum	¼	gal.
Sugar Syrup	4	gal.
Dark Claret Wine	4¾	gal.

Glowing Wine Punch Extract

Cardamom Tincture	2	oz.
Pineapple Essence	3	oz.
Cinnamon Tincture	5	oz.
Clove Tincture	5	oz.
Genuine Arrac	¼	gal.
Alcohol	1½	gal.
Cherry Fruit Syrup	4	gal.
Claret Wine	4½	gal.

White Wine Punch Extract from Moselle, Rhine or Chablis Wine

Sweet Orange Juice	½	gal.
Genuine Arrac	1	gal.
Sugar Syrup 65%	3¾	gal.
Moselle, Rhine or Chablis	5	gal.

INDEX TO ADDENDA

For Chemical Advisors, Special Raw Materials, Equipment, Containers, etc., consult Supply Section at end of book.

For Chemical Advisors, Special Raw Materials, Equipment, Containers, etc., consult Supply Section at end of book.

D

E

For Chemical Advisors, Special Raw Materials, Equipment, Containers, etc., consult Supply Section at end of book.

F

For Chemical Advisors, Special Raw Materials, Equipment, Containers, etc., consult Supply Section at end of book.

For Chemical Advisors, Special Raw Materials, Equipment, Containers, etc., consult Supply Section at end of book.

T

For Chemical Advisors, Special Raw Materials, Equipment, Containers, etc., consult Supply Section at end of book.

For Chemical Advisors, Special Raw Materials, Equipment, Containers, etc., consult Supply Section at end of book.